The NE Atlantic Region: A Reappraisal of Crustal Structure, Tectonostratigraphy and Magmatic Evolution

Geological Society books refereeing procedures

The Society makes every effort to ensure that the scientific and production quality of its books matches that of its journals. Since 1997, all book proposals have been refereed by specialist reviewers as well as by the Society's Books Editorial Committee. If the referees identify weaknesses in the proposal, these must be addressed before the proposal is accepted.

Once the book is accepted, the Society Book Editors ensure that the volume editors follow strict guidelines on refereeing and quality control. We insist that individual papers can only be accepted after satisfactory review by two independent referees. The questions on the review forms are similar to those for *Journal of the Geological Society*. The referees' forms and comments must be available to the Society's Book Editors on request.

Although many of the books result from meetings, the editors are expected to commission papers that were not presented at the meeting to ensure that the book provides a balanced coverage of the subject. Being accepted for presentation at the meeting does not guarantee inclusion in the book.

More information about submitting a proposal and producing a book for the Society can be found on its website: www.geolsoc.org.uk.

It is recommended that reference to all or part of this book should be made in one of the following ways:

Péron-pinvidic, G., Hopper, J. R., Stoker, M. S., Gaina, C., Doornenbal, J. C., Funck, T. & Árting, U. E. (eds) 2017. *The NE Atlantic Region: A Reappraisal of Crustal Structure, Tectonostratigraphy and Magmetic Evolution.* Geological Society, London, Special Publications, **447**.

Gerlings, J., Hopper, J. R., Fyhn, M. B. W. & Frandsen, N. 2017. Mesozoic and older rift basins on the SE Greenland Shelf offshore Ammassalik. *In*: Péron-pinvidic, G., Hopper, J. R., Stoker, M. S., Gaina, C., Doornenbal, J. C., Funck, T. & Árting, U. E. (eds). *The NE Atlantic Region: A Reappraisal of Crustal Structure, Tectonostratigraphy and Magmetic Evolution.* Geological Society, London, Special Publications, **447**, 375–392. First published online April 13, 2017, https://doi.org/10.1144/SP447.15

GEOLOGICAL SOCIETY SPECIAL PUBLICATION NO. 447

The NE Atlantic Region: A Reappraisal of Crustal Structure, Tectonostratigraphy and Magmatic Evolution

EDITED BY

G. PÉRON-PINVIDIC
Geological Survey of Norway, Norway

J. R. HOPPER and T. FUNCK
Geological Survey of Denmark and Greenland, Denmark

M. S. STOKER
British Geological Survey, UK

C. GAINA
University of Oslo, Norway

J. C. DOORNENBAL
Geological Survey of The Netherlands, The Netherlands

and

U. E. ÁRTING
Faroese Geological Survey, Faroe Islands

2017
Published by
The Geological Society
London

THE GEOLOGICAL SOCIETY

The Geological Society of London (GSL) was founded in 1807. It is the oldest national geological society in the world and the largest in Europe. It was incorporated under Royal Charter in 1825 and is Registered Charity 210161.

The Society is the UK national learned and professional society for geology with a worldwide Fellowship (FGS) of over 10 000. The Society has the power to confer Chartered status on suitably qualified Fellows, and about 2000 of the Fellowship carry the title (CGeol). Chartered Geologists may also obtain the equivalent European title, European Geologist (EurGeol). One fifth of the Society's fellowship resides outside the UK. To find out more about the Society, log on to www.geolsoc.org.uk.

The Geological Society Publishing House (Bath, UK) produces the Society's international journals and books, and acts as European distributor for selected publications of the American Association of Petroleum Geologists (AAPG), the Indonesian Petroleum Association (IPA), the Geological Society of America (GSA), the Society for Sedimentary Geology (SEPM) and the Geologists' Association (GA). Joint marketing agreements ensure that GSL Fellows may purchase these societies' publications at a discount. The Society's online bookshop (accessible from www.geolsoc.org.uk) offers secure book purchasing with your credit or debit card.

To find out about joining the Society and benefiting from substantial discounts on publications of GSL and other societies worldwide, consult www.geolsoc.org.uk, or contact the Fellowship Department at: The Geological Society, Burlington House, Piccadilly, London W1J 0BG: Tel. +44 (0)20 7434 9944; Fax +44 (0)20 7439 8975; E-mail: enquiries@geolsoc.org.uk.

For information about the Society's meetings, consult *Events* on www.geolsoc.org.uk. To find out more about the Society's Corporate Affiliates Scheme, write to enquiries@geolsoc.org.uk.

Published by The Geological Society from:
The Geological Society Publishing House, Unit 7, Brassmill Enterprise Centre, Brassmill Lane, Bath BA1 3JN, UK

The Lyell Collection: www.lyellcollection.org
Online bookshop: www.geolsoc.org.uk/bookshop
Orders: Tel. +44 (0)1225 445046, Fax +44 (0)1225 442836

British Library Cataloguing in Publication Data

A catalogue record for this book is available from the British Library.
ISBN 978-1-78620-278-9
ISSN 0305-8719

Distributors
For details of international agents and distributors see:
www.geolsoc.org.uk/agentsdistributors

Typeset by Nova Techset Private Limited, Bengaluru & Chennai, India
Printed and bound by CPI Group (UK) Ltd, Croydon CR0 4YY

Contents

The NE Atlantic region: a reappraisal of crustal structure, tectonostratigraphy and magmatic evolution – an introduction to the NAG-TEC project

GWENN PÉRON-PINVIDIC[1]*, JOHN R. HOPPER[2], MARTYN STOKER[3], CARMEN GAINA[4], THOMAS FUNCK[2], UNI E. ÁRTING[5] & JOHANNES CORNELIS DOORNENBAL[6]

[1]*Geological Survey of Norway, Leiv Eirikssons vei 39, 7040 Trondheim, Norway*

[2]*Geological Survey of Denmark and Greenland, Øster Voldgade 10, 1350 Copenhagen K, Denmark*

[3]*British Geological Survey, The Lyell Centre, Research Avenue South, Edinburgh EH14 4AP, UK*

[4]*Centre for Earth Evolution and Dynamics, University of Oslo, PO Box 1028, Blindern, 0315 Oslo, Norway*

[5]*Jarðfeingi (Faroese Geological Survey) Brekkutún 1, PO Box 3059, FO-110 Tórshavn, Faroe Islands*

[6]*TNO – Geological Survey of The Netherlands, Princetonlaan 6, 3584 CB Utrecht, The Netherlands*

**Correspondence: gwenn@ngu.no*

The NE Atlantic region and its continental margins (Fig. 1) hold unique information for understanding many aspects of Earth science, from global geodynamics to palaeoceanography and global environmental change. It also holds some of the world's most important hydrocarbon reserves from the North Sea, along the Atlantic margins of Ireland, Britain and Norway, and into the Arctic in the Barents Sea. Historically, studies in the NE Atlantic were important for establishing many of the key ideas during the early part of the plate tectonic revolution. Linear magnetic anomalies along the Reykjanes Ridge were identified as early as in the 1960s (Heirtzler *et al.* 1966) and provided strong evidence for the seafloor spreading hypothesis (Dietz 1961), which by then had been established as a new and holistic theory (Ewing & Heezen 1956). At the same time, Iceland was already recognized as an intriguing anomalous entity (Böðvarsson & Walker 1964) and contributed to knowledge about how Earth's magnetic field reversed its polarity through time. The fact that rifting occurs in close association with old sutures and orogenic belts led Wilson to propose that the Atlantic Ocean closed and opened again, establishing the concept of the 'Wilson tectonic cycle' (Wilson 1966; Dewey 1969). The North Atlantic continental margins have long been considered as archetypal, and divergent margins world-wide are commonly described as 'Atlantic-type passive margins'. However, it is now accepted that these so-called 'passive' margins remain dynamic long after break-up, including post-rift vertical movements of up to kilometre scale. The type examples for such epeirogenic movements being, once again, the North Atlantic margins (Praeg *et al.* 2005).

Today, the NE Atlantic region remains at the centre of ongoing fundamental controversies regarding the scales of mantle convection, the origin and extent of mantle plumes and their roles in plate tectonics, and the process of plate break-up (e.g. White & McKenzie 1989; Foulger & Anderson 2005; Lundin & Doré 2005). Regardless of diverging views on how various processes operate to shape the region, understanding the geological and tectonic history of the NE Atlantic has had and will continue to have major implications for understanding global geodynamics.

From: PÉRON-PINVIDIC, G., HOPPER, J. R., STOKER, M. S., GAINA, C., DOORNENBAL, J. C., FUNCK, T. & ÁRTING, U. E. (eds) 2017. *The NE Atlantic Region: A Reappraisal of Crustal Structure, Tectonostratigraphy and Magmatic Evolution*. Geological Society, London, Special Publications, **447**, 1–9.
First published online July 12, 2017, updated August 18, 2017, https://doi.org/10.1144/SP447.17

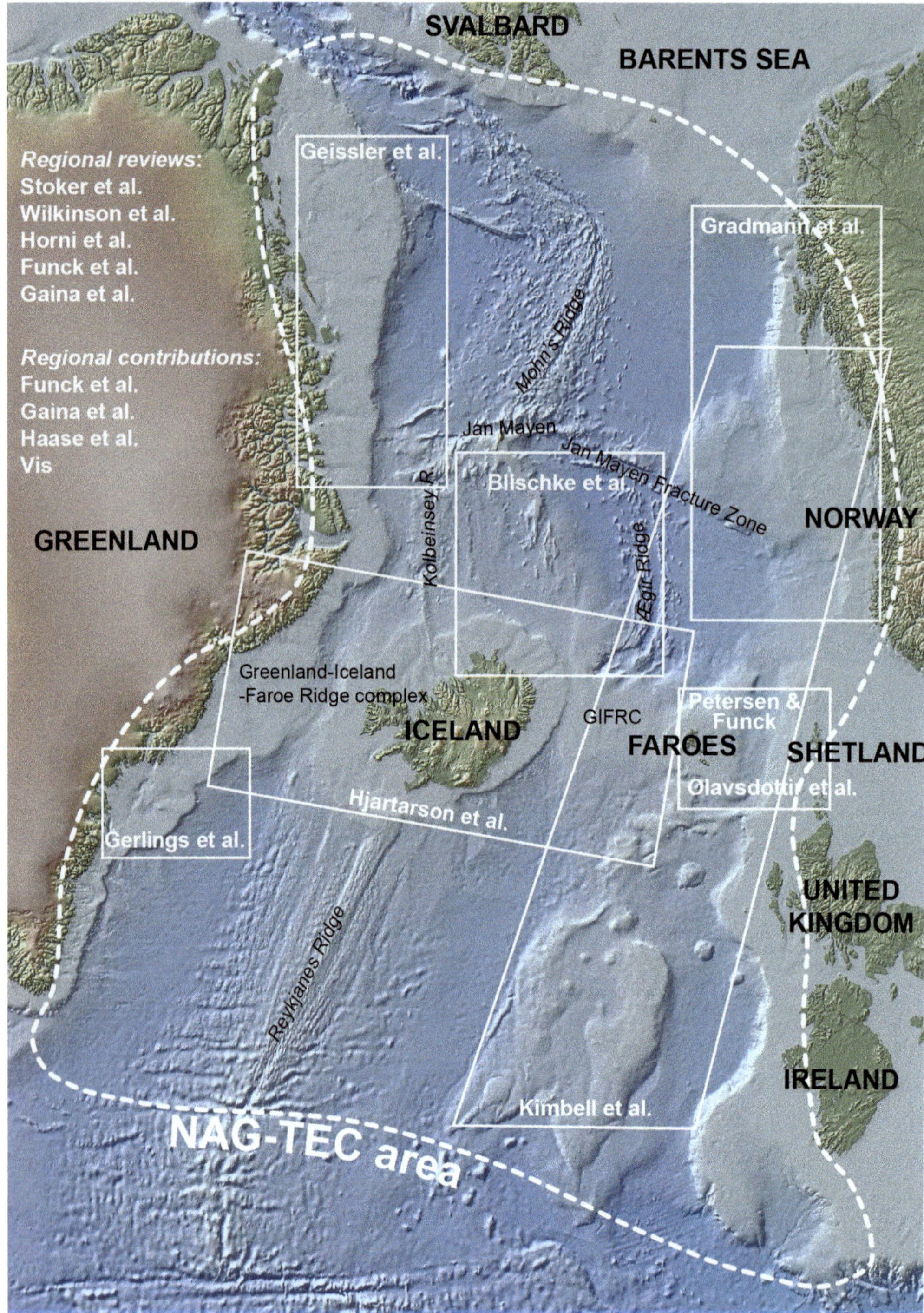

Fig. 1. Bathymetric map of the NE Atlantic. The top left-hand column lists the contributions of this Special Publication that deal with regional studies. The white squares on the map locate the local studies.

The NAG initiative and NAG-TEC project background

The Northeast Atlantic Geoscience (NAG) group comprises ten geological surveys from northern Europe under a cooperative framework that was initiated in September 2008. These include the BGR (Germany), BGS (Britain), GEUS (Denmark and Greenland), GSI (Ireland), GSNI (Northern Ireland), ÍSOR (Iceland), Jarðfeingi (the Faroe Islands), NGU (Norway), SGU (Sweden) and TNO (The Netherlands). The primary aim is the sharing of scientific knowledge and resources on major scientific and societal issues that are of common interest to all of the surveys.

Among the projects that have since initiated, an effort to understand the tectonic development of the NE Atlantic region resulted in the formation of the NAG-TEC Group, consisting of the BGS, GEUS, GSI, GSNI, ÍSOR, Jarðfeingi, NGU and TNO. In the spring of 2009, representatives of the surveys met to discuss the initiative with a particular emphasis on continental margin evolution and understanding deep-water basins. A series of further workshops and meetings were held during 2009 and 2010, accompanied by close discussions with, and recommendations by, representatives of the oil and gas industry, as well as ongoing dialogue with the survey directors. From all of these discussions, it quickly became apparent that in order to define fundamental questions that would motivate future projects, it was necessary to compile a comprehensive database of geoscientific information encompassing the entire region. The ultimate goal of the NAG-TEC project was to create a unified compilation of key geological and geophysical data relevant to the understanding of the tectonic, stratigraphic and magmatic development of the region, with a particular focus on the margins and sedimentary basins. This would provide comprehensive constraints to facilitate detailed conjugate margin comparisons and regional correlations, notably to de-risk scientific and industry exploration in poorly known areas.

After intense marketing with oil companies in the spring of 2011, the NAG-TEC project was officially launched during a workshop in Copenhagen on 1 June 2011. The project was led by the GEUS, partnered by the BGS, GSI, GSNI, ÍSOR, Jarðfeingi, NGU and TNO. The BGR secured funds to help sponsor the project, which, together with nine oil companies who signed up early, allowed the project to start. By the end of project, an additional five companies had joined the group, providing approximately half the funds needed to complete the Atlas. Another four companies purchased the database and Atlas after completion, showing the ongoing value of comprehensive regional studies such as this. Partners and collaborators from universities, other research institutes and government agencies (see the Acknowledgements) helped at various stages during the course of the project to ensure the completeness and quality of the compilation protocols and products.

The tectonostratigraphic Atlas and ArcGIS database

The original focus of the project was to collect and assess as much geophysical and geological information as possible, compiling all published and publically available information and datasets. In addition, a substantial amount of unpublished data owned by the geological surveys or other contributors were added to the database. This process created an overview of the state of knowledge of the entire region by listing the available data, revealing where major data gaps still exist. The final ArcGIS database provides a foundation from which ongoing research and exploration into the frontier regions of the Atlantic and into the Arctic can build.

The process of data gathering underlined the virtual non-existence of systematic compilations of tectonostratigraphic information over the entire NE Atlantic region, and that differences between local nomenclatures hampered regional correlations. These correlations were central to one of the main goals of the project, which was to provide an understanding of a full-rift system and conjugate margin evolution. The project provides quantitative constraints on key basin parameters, such as crustal thickness, burial and exhumation history, an understanding of key unconformities and their interpretation, information regarding the interaction between volcanic events and sediments, and the importance of the pre-existing structure. The key outcomes from the NAG-TEC project are the provision of a better understanding of potential links between surface processes and deeper crustal and mantle processes, constraints on continent–ocean transition zones, deep-water basin evolution, a unified stratigraphic and structural framework, and regional correlations and comparisons.

The compilation of the *NAG-TEC Tectonostratigraphic Atlas* (Hopper *et al.* 2014) and the GIS model behind it was completed on 1 June 2014. The Atlas, GIS database, and the accompanying geophysical datasets and database of references are today in wide use amongst the surveys, as well as the company sponsors. The Atlas offers thorough descriptions of the various datasets, related compilations, uncertainties and modelling results, and includes chapters dealing with the major geophysical and geological disciplines such as stratigraphy, structural elements, bathymetry, gravity, magnetic, seismic, heat flow and tomography. It also presents chapters and sub-chapters documenting the regional plate kinematics, the identification and characterization of basement types, regional structural elements maps, Devonian–Cenozoic stratigraphic distribution maps, stratigraphic correlation charts of the post-Caledonian succession, and, finally, regional volcanism, including geochronology, geochemistry and volcanic facies characterization. Following a 2 year confidentiality period, the Atlas was made publically available in June 2016 and can be purchased via the NAG-TEC website (https://www.nag-tec.org). The underlying GIS project will be released in June 2019, and is in active use by the surveys and companies for defining future work and exploration.

This Special Publication: towards an understanding of the NE Atlantic region

As noted above, the main goal of the NAG-TEC Atlas and GIS database project was to focus on the empirical data; however, one final objective of the project was also to publish aspects of the scientific questions, which is a crucial remit of all of the geological surveys. Following the release of the Atlas in 2016, this Geological Society of London Special Publication offers the first published scientific communications based on the NAG-TEC project or using results from the NAG-TEC compilation. The contributions presented in this Special Publication span a wide range of geoscience methods and disciplines, which offer scientific analyses, descriptions, modelling and interpretations that we consider as marking today's status of knowledge at a regional scale. As can be appreciated, the NAG-TEC project (Atlas and GIS database) involved much more compilation and research than is evident from the papers presented here. In addition to summarizing the main contributions of the volume, a short overview of additional important results is presented at the end. For further information, the reader is referred to the NAG-TEC Atlas (Hopper *et al.* 2014).

The 17 papers presented in this volume encompass a broad range of issues related to crustal structure, tectonostratigraphy and magmatic evolution of the NE Atlantic region. The mix of papers falls into two main categories: (1) regional reviews and contributions; and (2) more process-orientated and/or area-specific papers. The categorization and geographical spread of the papers is shown in Figure 1.

Stratigraphy

A major overview of the Late Palaeozoic–Mesozoic stratigraphy of the NE Atlantic region is presented by **Stoker *et al.* (2016)**, who describe the distribution, structural setting, stratigraphy, depositional environment and correlation of these rocks across the conjugate margins and along the axis of the proto-NE Atlantic rift system. The establishment of a regional stratigraphic framework, based on a strong observational record, has provided important new constraints on the timing and nature of sedimentary basin development in the NE Atlantic by challenging the 'classic' view of palinspastic reconstruction in the central part of the proto-NE Atlantic rift system, with implications for the break-up of the Pangaean supercontinent. In particular, there is currently a lack of evidence for a substantive and continuous rift system along the proto-NE Atlantic until the Late Cretaceous.

Geochronology and volcanism

Four papers detail various aspects of the regional volcanic history.

Wilkinson *et al.* (2016) present a thorough geochronological review of all available published ages related to magmatic activity in the NE Atlantic. They undertook an evaluation of the quality of each published age, and produced a filtered dataset and optimized age model for the North Atlantic Igneous Province (NAIP). In particular, $^{40}Ar/^{39}Ar$ and K–Ar age data have been recalculated to a common reference system to allow across province comparisons, and all data have been recalibrated to the Geological Time Scale 2012 (Gradstein *et al.* 2012).

A compilation of the NAIP volcanic activity is described by **á Horni *et al.* (2017)**, who undertook a regional summary of the volcanostratigraphy based on an established methodology of seismic facies analysis of the volcanic units (e.g. Symonds *et al.* 1998; Planke *et al.* 2000). This formed the basis for reassessing the eruption volumes and production rates of the province, along with an investigation of the variation of the spatial patterns of volcanism along the margins. The Greenland–Iceland–Faroe Ridge complex (GIFRC) was not included in the volcanostratigraphy but, instead, forms a separate contribution by **Hjartarson *et al.* (2017)**, who reveal several new potential abandoned rift centres, mapped as syncline and anticline structures. These formed by extensional processes through rifting and rift jumps, causing crustal variations through time, apparently more common within the ridge complex than in the adjacent oceanic floor. Rift-jump events and crustal volume variations across the GIFRC can be linked to mid-oceanic ridge opening processes since 55 Ma, which challenges the long-held view that it formed as a hotspot track (Bjarnason 2008).

Geissler *et al.* (2016) describe the NE Greenland margin, where the break-up-related volcanic complexes are comparable with the better-known Mid-Norwegian–SW Barents Sea conjugate. Seismic facies types match between the conjugate pair and show strong lateral variations. By focusing on the spatial and temporal evolution of the magmatic activity at late-rift and early oceanic drift stages, they are able to confirm that the initial oceanic rifting occurred subaerially in the area immediately north of the Jan Mayen Fracture Zone until about 49 Ma, when the spreading axis submerged below the sea level and normal oceanic crust was accreted.

Crustal structure

The crustal structure of the overall NE Atlantic has been one of the main targets of the NAG-TEC project and, together with the stratigraphic compilation, one of the major topics requested by our industry sponsors.

Seismic refraction profiles represent a unique dataset to constrain the depth to the basement. Unfortunately, the coverage is very inhomogeneous, with relatively good coverage along the NW European margins, but only a few profiles along the East Greenland margins and in the central parts of the ocean. **Funck *et al.* (2016*a*)** undertook a review and compilation of all existing refraction profiles in the NE Atlantic. The resulting seismic refraction database, combined with results from receiver functions and gravity information, was then used to compile various maps, such as the top-basement and Moho depths (**Funck *et al.* 2016*b***), from which a crustal thickness map was derived, providing a consistent comparison of conjugate and along-strike variations in structure.

The paper by **Haase *et al.* (2016)** describes a 3D regional crustal model for the NE Atlantic, based on the integration of seismic constraints and gravity data. For the NAG-TEC Atlas, the 'DTU10' gravity dataset provided by Andersen (2010) had a homogeneous coverage and sufficient good resolution for the purpose of the project. Haase & Ebbing (2014) used it as their main input to compute Bouguer, free air, isostatic and tilt derivative versions of the grid to the Atlas and GIS database. Based on this first-order compilation, and on additional information from the seismic refraction compilation (see Funck *et al.* 2014), the model presented by Haase *et al.* (2016) offers a 3D perspective of the basement and Moho geometries in the NE Atlantic that permits an improved resolution of some areas, such as basins along the NE Greenland margin, as well as the GIFRC. The maps generated are particularly useful in areas with poor seismic data coverage, in particular where the Moho and basement maps obtained from the seismic refraction database are very coarse and lack detailed structure (Funck *et al.* 2016*a*).

Focused studies

The following six papers present in more detail specific regions within the NE Atlantic or discuss topics related to the evolution of the oceanic basins and their continental margins.

The controversial distribution and causes for Cenozoic compressional structures along the NW European margin is discussed in **Kimbell *et al.* (2016)**. They compare the locations of the post-break-up anticlines with the crustal thickness maps derived from the NAG-TEC database. A correlation between the compressional structures and the distribution of high-velocity lower crust and/ or partially serpentinized upper mantle is noted, and they suggest that the lithospheric weakening resulting from crustal hyperextension and partial serpentinization of the upper mantle, commonly viewed as a probable factor in the location of the compressional deformation, cannot account for all the post-break-up reactivation mapped in the region. Based on first-order rheological considerations, these authors suggest that an alternative weakening mechanism could include ductile lower crust and lithospheric decoupling.

On a similar theme, **Gradmann *et al.* (2017)** use simple isostatic calculations to compare to various geophysical datasets in order to identify areas of inconsistencies. Along the Mid-Norwegian margin, they use the density model built for the Haase *et al.* (2016) gravity inversion to forward-calculate bathymetry. They propose that the misfits with the observed bathymetry point to poorly understood regions that require further investigations.

The Jan Mayen microcontinent (JMMC) is one of the most striking structural features of the NE Atlantic, although its origin, evolution and boundaries are still highly debated. **Blischke *et al.* (2016)** present a revision of the architecture and structural development of the JMMC, with a specific emphasis on its Cenozoic evolution and relationship to the rift–drift transition during the emplacement of the Kolbeinsey Ridge. The authors update the structural interpretation, including all available seismic datasets, combined with other available geophysical data, borehole records and plate tectonic reconstructions (Gaina 2014). On this basis, they are able to highlight similarities in structural trends between the microcontinent and the Jameson Land (East Greenland) and Faroe–Shetland basins, suggesting that the three areas formed an en echelon set of rift basins that developed prior to the final opening of the NE Atlantic Ocean.

The Faroe–Shetland region is of particular interest to the oil and gas industry; however, much of the Faroese continental margin remains poorly understood due to the widespread cover of volcanic rocks associated with the NAIP. In order to establish what is and what is not known, **Ólavsdóttir *et al.* (2016)** present a review of the stratigraphy and structure of the Faroese margin, with an emphasis on providing both context and constraints for ongoing discussions of the processes and events that have helped to shape this complex tectonic region. Seismic refraction data provide some indications of the pre-volcanic geology, which probably includes Archaean rocks overlain by a Mesozoic (Cretaceous?) cover. As a complement to this paper, **Petersen & Funck (2016)** provide a review of the seismic refraction models available in the Faroe–Shetland area and how the sub-basalt rocks were constrained in the individual studies. This allows for a better assessment of the pre-volcanic geology. Despite the seismic imaging problems due to the presence of the volcanic rocks, the review concludes that the Faroe–Shetland Basin is floored by thinned continental crust, overlain by sub-basalt sedimentary rocks.

Further west, **Gerlings *et al.* (2017)** describe the Ammassalik Basin of SE Greenland. This somewhat isolated basin is inferred to be conjugate to the basins in the northern Hebrides–Rockall region, and might contain at least 4 km of sediment. Proven sediments in the basin are of Cretaceous (Albian) age, but rocks as old as Permian–Triassic (as proven in the Hebridean basins) cannot be discounted. Sampling this basin would provide important information to better constrain the Upper Palaeozoic–Mesozoic history of this enigmatic southern segment of the NE Atlantic Ocean.

The oceanic domain and regional kinematics

Two papers by **Gaina *et al.* (2016, 2017)** gather information about the oceanic sub-basins in the NE Atlantic and propose a classification of oceanic crust domains linked to regional kinematics. Gaina *et al.* (2017) assess the break-up and oceanic spreading history of the NE Atlantic based on detailed mapping of magnetic anomalies, fracture zones and derived isochrons. A new oceanic lithosphere age grid and associated grids of seafloor spreading rates and asymmetry in crustal accretion were used to identify possible connections between tectonic and magmatic events and the formation and distribution of seamounts and volcanic edifices in the NE Atlantic oceanic domain (Gaina *et al.* 2016). They conclude that seamount formation and oceanic basement morphology are intimately related to the complex interplay between variations in relative plate motion and episodic plume activity.

The oil perspective

The paper by **Vis (2017)** offers an applied aspect to this volume, by combining the information of the NAG-TEC Atlas and GIS database with clustered oil-slick data from the Global Offshore Seepage Database, with the aim of identifying active oil seepage. In order to illustrate the strength of the NAG-TEC database to support studies in underexplored areas in the NE Atlantic region, the author focused on three regions with oil-slick observations but with no current hydrocarbon activities: the Western Barents Sea Margin, the Irish Atlantic Margin, and the East Greenland and Jan Mayen regions. The potential value of integrating such diverse datasets in a frontier exploration area is discussed.

Additional key Atlas results

The above contributions span a wide range of geoscience methods and disciplines to offer scientific analyses, descriptions, modelling and interpretations that we consider as marking today's state of knowledge of the NE Atlantic at a regional scale. It is worth mentioning that the NAG-TEC project involved more compilation and research than published in this Special Publication. Several ongoing investigations could not be completed in time to make this volume, but will hopefully be available in other journals in the near future. The reader is referred to the original NAG-TEC Atlas (Hopper *et al.* 2014) for additional information, and some of the key highlights are described below.

The refraction compilation (Funck *et al.* 2014) further enabled the construction of basement-type maps (oceanic, continental, volcanic and non-volcanic transitional) and an assessment of areas with high-velocity lower crust, exhumed mantle or serpentinized mantle. The basement-type map permitted the NAG-TEC group to also work on a revised identification of the continent–ocean boundary (COB) (Hopper *et al.* 2014). Published COB locations usually vary considerably as they depend in large part on the dataset used, including both processing and display parameters. The strength of the NAG-TEC approach resides in the regional uniformity of the compilation that allows consistent mapping. Based on the basement-type map of Funck *et al.* (2014), Hopper *et al.* (2014) constructed a regionally consistent line that connected the landwards-most points of seismically defined oceanic crust, which was then compared to the potential field grids, to the structural map and to plate kinematic reconstructions. Even with this regionally consistent approach, some problem areas remain, especially in areas with sparse seismic data.

The NAG-TEC project also included the compilation of the structural elements maps available over the entire NE Atlantic. G. Kimbell led the compilation shown in the Atlas. The resulting map includes GIS elements produced by the Norwegian Petroleum Directorate (Gabrielsen *et al.* 1990; Blystad *et al.* 1995), by the Petroleum Affairs Division (Naylor *et al.* 1999), and elements from the

Faroe–Shetland area based on the map published by the BGS and Jarðfeingi (Ritchie *et al.* 2011), as well as elements from the north Rockall area based on the map published by the BGS (Hitchen *et al.* 2013). For East Greenland, the structural elements are derived primarily from Hamann *et al.* (2005) and Tsikalas *et al.* (2005), with revisions based on re-analysis of available seismic, gravity and magnetic data compiled for the NAG-TEC project.

The contribution by Árting (2014) summarizes the compilation of geochemistry information related to the regional North Atlantic Igneous Province volcanic activity; the authors compiled, quality-controlled and cross-checked mapped occurrences of volcanic activity over the NE Atlantic.

Finally, the structure of the entire mantle from the Moho to the core–mantle boundary has been investigated by Spakman (2014) using seismic tomography. He provides an overview of the three-dimensional structure and shows a comparison between seven published models from different research groups, showing a miscorrelation between the short-wavelength details (<500–1000 km). Based on this, Spakman (2014) recommends caution when interpreting tomographical models of the North Atlantic mantle structure.

Our thanks go to every colleague who participated in the NAG-TEC project. This includes the more than 80 staff of the geological surveys who directly participated in the various tasks, including data compilation, quality control, uncertainty assessment, modelling, interpretation, graphic design and construction of the GIS database. We are also grateful to all of our colleagues from collaborating universities and other research institutes, including: the Alfred Wegener Institute (Germany); the Centre for Earth Evolution and Dynamics, University of Oslo (Norway); the University of Bergen (Norway); University College Dublin (Ireland); the University of Iceland (Iceland); and Utrecht University (The Netherlands).

The project also greatly benefited from the discussions and advice of 18 sponsoring companies that funded half of the budget necessary to accomplish this project, and we hereby acknowledge them (alphabetical order): Bayerngas Norge AS; BP Exploration Operating Company Ltd; Bundesanstalt für Geowissenschaften und Rohstoffe (BGR); Capricorn Norge AS; Chevron East Greenland Exploration A/S; ConocoPhillips Skandinavia AS; DEA Norge AS; Det norske oljeselskap ASA; DONG E&P A/S; E.ON Norge AS; ExxonMobil Exploration and Production Norway AS; Japan Oil, Gas, and Metals National Corporation (JOGMEC); Maersk Oil; Nalcor Energy – Oil and Gas Inc.; Nexen Energy ULC; Norwegian Energy Company ASA (Noreco); Repsol Exploration Norge AS; Statoil (UK) Ltd; and Wintershall Holding GmbH.

Correction notice: The original version was incorrect. The list of NAG partners was missing NGU (Norway). This has now been added.

References

Andersen, O.B. 2010. The DTU10 gravity field and mean sea surface. *Paper presented at the Gravity Field of the Earth: IGFS2 – 2010. Second International Symposium of the International Gravity Field Service*, 20–22 September 2010, University of Alaska Fairbanks, Alaska, USA.

Árting, U. 2014. Regional volcanism. *In*: Hopper, J.R., Funck, T., Stoker, M., Árting, U., Péron-Pinvidic, G., Doornenbal, J.C. & Gaina, C. (eds) *Tectonostratigraphic Atlas of the North-East Atlantic Region*. Geological Survey of Denmark and Greenland (GEUS), Copenhagen, Denmark, 223–292.

Bjarnason, I.Þ. 2008. An Iceland hotspot saga. Jökull, **58**, 3–16.

Blischke, A., Gaina, C. *et al.* 2016. The Jan Mayen microcontinent: an update of its architecture, structural development and role during the transition from the Ægir Ridge to the mid-oceanic Kolbeinsey Ridge. *In*: Péron-Pinvidic, G., Hopper, J.R., Stoker, M.S., Gaina, C., Doornenbal, J.C., Funck, T. & Árting, U.E. (eds) *The NE Atlantic Region: A Reappraisal of Crustal Structure, Tectonostratigraphy and Magmatic Evolution*. Geological Society, London, Special Publications, **447**. First published online September 8, 2016, https://doi.org/10.1144/SP447.5

Blystad, P., Brekke, H., Færseth, R.B., Larsen, B.T., Skogseid, J. & Tørudbakken, B. 1995. *Structural Elements of the Norwegian Continental Shelf, Part II. The Norwegian Sea Region*. Norwegian Petroleum Directorate Bulletin, **8**.

Böðvarsson, G. & Walker, G.P.L. 1964. Crustal drift in Iceland. *Geophysical Journal of the Royal Astronomical Society*, **8**, 285–300, https://doi.org/10.1111/j.1365-246X.1964.tb06295.x

Dewey, J.F. 1969. Continental margins: a model for conversion of Atlantic type to Andean type. *Earth and Planetary Science Letters*, **6**, 189–197.

Dietz, R.S. 1961. Continent and ocean basin evolution by spreading. *Nature*, **190**, 854–857, https://doi.org/10.1038/190854a0

Ewing, M. & Heezen, B.C. 1956. Some problems of Antarctic submarine geology. *In*: Crary, A.P., Gould, L.M., Hulburt, E.O., Odishaw, H. & Smith, W.E. (eds) *Antarctica in the International Geophysical Year: Based on a Symposium on the Antarctic*. American Geophysical Union, Geophysical Monograph Series, **1**, 75–81.

Foulger, G.R. & Anderson, D.L. 2005. A cool model for the Iceland hotspot. *Journal of Volcanology and Geothermal Research*, **141**, 1–22.

Funck, T., Hopper, J.R. *et al.* 2014. Crustal structure. *In*: Hopper, J.R., Funck, T., Stoker, M., Árting, U., Péron-Pinvidic, G., Doornenbal, J.C. & Gaina, C. (eds) *Tectonostratigraphic Atlas of the North-East Atlantic Region*. Geological Survey of Denmark and Greenland (GEUS), Copenhagen, Denmark, 69–126.

Funck, T., Erlendsson, Ö., Geissler, W.H., Gradmann, S., Kimbell, G.S., McDermott, K. & Petersen, U.K. 2016*a*. A review of the NE Atlantic conjugate margins based on seismic refraction data. *In*: Péron-Pinvidic, G., Hopper, J.R., Stoker, M.S., Gaina, C., Doornenbal, J.C., Funck, T. & Árting,

U.E. (eds) *The NE Atlantic Region: A Reappraisal of Crustal Structure, Tectonostratigraphy and Magmatic Evolution*. Geological Society, London, Special Publications, **447**. First published online October 12, 2016, updated October 19, 2016, https://doi.org/10.1144/SP447.9

Funck, T., Geissler, W.H., Kimbell, G.S., Gradmann, S., Erlendsson, Ö., McDermott, K. & Petersen, U.K. 2016*b*. Moho and basement depth in the NE Atlantic Ocean based on seismic refraction data and receiver functions. *In*: Péron-Pinvidic, G., Hopper, J.R., Stoker, M.S., Gaina, C., Doornenbal, J.C., Funck, T. & Árting, U.E. (eds) *The NE Atlantic Region: A Reappraisal of Crustal Structure, Tectonostratigraphy and Magmatic Evolution*. Geological Society, London, Special Publications, **447**. First published online July 13, 2016, https://doi.org/10.1144/SP447.1

Gabrielsen, R.H., Færseth, R.B., Jensen, L.N., Kalheim, J.E. & Riis, F. 1990. *Structural Elements of the Norwegian Continental Shelf. Part I: The Barents Sea Region*. Norwegian Petroleum Directorate Bulletin, **6**.

Gaina, C. 2014. Plate reconstructions and regional kinematics. *In*: Hopper, J.R., Funck, T., Stoker, M., Árting, U., Péron-Pinvidic, G., Doornenbal, J.C. & Gaina, C. (eds) *Tectonostratigraphic Atlas of the North-East Atlantic Region*. Geological Survey of Denmark and Greenland (GEUS), Copenhagen, Denmark, 53–66.

Gaina, C., Blischke, A., Geissler, W.H., Kimbell, G.S. & Erlendsson, Ö. 2016. Seamounts and oceanic igneous features in the NE Atlantic: a link between plate motions and mantle dynamics. *In*: Péron-Pinvidic, G., Hopper, J.R., Stoker, M.S., Gaina, C., Doornenbal, J.C., Funck, T. & Árting, U.E. (eds) *The NE Atlantic Region: A Reappraisal of Crustal Structure, Tectonostratigraphy and Magmatic Evolution*. Geological Society, London, Special Publications, **447**. First published online September 8, 2016, https://doi.org/10.1144/SP447.6

Gaina, C., Nasuti, A., Kimbell, G.S. & Blischke, A. 2017. Break-up and seafloor spreading domains in the NE Atlantic. *In*: Péron-Pinvidic, G., Hopper, J.R., Stoker, M.S., Gaina, C., Doornenbal, J.C., Funck, T. & Árting, U.E. (eds) *The NE Atlantic Region: A Reappraisal of Crustal Structure, Tectonostratigraphy and Magmatic Evolution*. Geological Society, London, Special Publications, **447**. First published online February 3, 2017, https://doi.org/10.1144/SP447.12

Geissler, W., Hopper, J. *et al*. 2016. Seismic volcanostratigraphy of the East Greenland continental margin revisited. *In*: Péron-Pinvidic, G., Hopper, J.R., Stoker, M.S., Gaina, C., Doornenbal, J.C., Funck, T. & Árting, U.E. (eds) *The NE Atlantic Region: A Reappraisal of Crustal Structure, Tectonostratigraphy and Magmatic Evolution*. Geological Society, London, Special Publications, **447**. First published online December 14, 2016, https://doi.org/10.1144/SP447.11

Gerlings, J., Hopper, J.R., Fyhn, M.B.W. & Frandsen, N. 2017. Mesozoic and older rift basins on the SE Greenland Shelf near Ammassalik. *In*: Péron-Pinvidic, G., Hopper, J.R., Stoker, M.S., Gaina, C., Doornenbal, J.C., Funck, T. & Árting, U.E. (eds) *The NE Atlantic Region: A Reappraisal of Crustal Structure, Tectonostratigraphy and Magmatic Evolution*. Geological Society, London, Special Publications, **447**. First published online April 13, 2017, https://doi.org/10.1144/SP447.15

Gradmann, S., Haase, C. & Ebbing, J. 2017. Isostasy as a tool to validate interpretations of regional geophysical datasets – applications to the mid-Norwegian continental margin. *In*: Péron-Pinvidic, G., Hopper, J.R., Stoker, M.S., Gaina, C., Doornenbal, J.C., Funck, T. & Árting, U.E. (eds) *The NE Atlantic Region: A Reappraisal of Crustal Structure, Tectonostratigraphy and Magmatic Evolution*. Geological Society, London, Special Publications, **447**. First published online February 23, 2017, https://doi.org/10.1144/SP447.13

Gradstein, F.M., Ogg, J.G., Schmitz, M.D. & Ogg, G. 2012. *The Geologic Time Scale 2012*. Elsevier Science, Amsterdam, The Netherlands.

Haase, C. & Ebbing, J. 2014. Gravity data. *In*: Hopper, J.R., Funck, T., Stoker, M., Árting, U., Péron-Pinvidic, G., Doornenbal, J.C. & Gaina, C. (eds) *Tectonostratigraphic Atlas of the North-East Atlantic Region*. Geological Survey of Denmark and Greenland (GEUS), Copenhagen, Denmark, 29–39.

Haase, C., Ebbing, J. & Funck, T. 2016. A 3D regional crustal model of the NE Atlantic based on seismic and gravity data. *In*: Péron-Pinvidic, G., Hopper, J.R., Stoker, M.S., Gaina, C., Doornenbal, J.C., Funck, T. & Árting, U.E. (eds) *The NE Atlantic Region: A Reappraisal of Crustal Structure, Tectonostratigraphy and Magmatic Evolution*. Geological Society, London, Special Publications, **447**. First published online October 12, 2016, updated October 19, 2016, https://doi.org/10.1144/SP447.8

Hamann, N.E., Whittaker, R.C. & Stemmerik, L. 2005. Geological development of the Northeast Greenland Shelf. *In*: Doré, A.G. & Vinning, B.A. (eds) *Petroleum Geology: North-West Europe and Global Perspectives – Proceedings of the 6th Petroleum Geology Conference*. Geological Society, London, Petroleum Geology Conference Series, **6**, 887–902, https://doi.org/10.1144/0060887

Heirtzler, J.R., Le Pichon, X. & Baron, J.G. 1966. Magnetic anomalies over the Reykjanes Ridge. *Deep Sea Research and Oceanographic Abstracts*, **13**, 427–443.

Hitchen, K., Johnson, H. & Gatliff, R.W. 2013. *Geology of the Rockall Basin and Adjacent Areas*. British Geological Survey, Keyworth, Nottingham, UK.

Hjartarson, A., Erlendsson, O. & Blischke, A. 2017. The Greenland–Iceland–Faroe Ridge Complex. *In*: Péron-Pinvidic, G., Hopper, J.R., Stoker, M.S., Gaina, C., Doornenbal, J.C., Funck, T. & Árting, U.E. (eds) *The NE Atlantic Region: A Reappraisal of Crustal Structure, Tectonostratigraphy and Magmatic Evolution*. Geological Society, London, Special Publications, **447**. First published online April 19, 2017, https://doi.org/10.1144/SP447.14

Hopper, J.R., Funck, T., Stoker, M., Árting, U., Péron-Pinvidic, G., Doornenbal, J.C. & Gaina, C. (eds) 2014. *Tectonostratigraphic Atlas of the North-East*

Atlantic Region. Geological Survey of Denmark and Greenland (GEUS), Copenhagen, Denmark.

Á Horni, J., Hopper, J. et al. 2017. Regional distribution of volcanism within the North Atlantic Igneous Province. *In*: Péron-Pinvidic, G., Hopper, J.R., Stoker, M.S., Gaina, C., Doornenbal, J.C., Funck, T. & Árting, U.E. (eds) *The NE Atlantic Region: A Reappraisal of Crustal Structure, Tectonostratigraphy and Magmatic Evolution*. Geological Society, London, Special Publications, **447**. First published online July 11, 2017, https://doi.org/10.1144/SP447.18

Kimbell, G.S., Stewart, M.A. et al. 2016. Controls on the location of compressional deformation on the NW European margin. *In*: Péron-Pinvidic, G., Hopper, J.R., Stoker, M.S., Gaina, C., Doornenbal, J.C., Funck, T. & Árting, U.E. (eds) *The NE Atlantic Region: A Reappraisal of Crustal Structure, Tectonostratigraphy and Magmatic Evolution*. Geological Society, London, Special Publications, **447**. First published online August 12, 2016, https://doi.org/10.1144/SP447.3

Lundin, E. & Doré, A. 2005. NE Atlantic break-up: a re-examination of the Iceland mantle plume model and the Atlantic–Arctic linkage. *In*: Doré, A. & Vining, B.A. (eds) *Petroleum Geology: North-West Europe and Global Perspectives – Proceedings of the 6th Petroleum Geology Conference*. Geological Society, London, Petroleum Geology Conference Series, **6**, 739–754, https://doi.org/10.1144/0060739

Naylor, D., Shannon, P. & Murphy, N. 1999. *Irish Rockall Basin Region: A Standard Structural Nomenclature System*. Petroleum Affairs Division, Special Publications, **1/99**.

Ólavsdóttir, J., Eidesgaard, Ó.R. & Stoker, M.S. 2016. The stratigraphy and structure of the Faroese continental margin. *In*: Péron-Pinvidic, G., Hopper, J.R., Stoker, M.S., Gaina, C., Doornenbal, J.C., Funck, T. & Árting, U.E. (eds) *The NE Atlantic Region: A Reappraisal of Crustal Structure, Tectonostratigraphy and Magmatic Evolution*. Geological Society, London, Special Publications, **447**. First published online July 22, 2016, https://doi.org/10.1144/SP447.4

Petersen, U.K. & Funck, T. 2016. Review of velocity models in the Faroe–Shetland Channel. *In*: Péron-Pinvidic, G., Hopper, J.R., Stoker, M.S., Gaina, C., Doornenbal, J.C., Funck, T. & Árting, U.E. (eds) *The NE Atlantic Region: A Reappraisal of Crustal Structure, Tectonostratigraphy and Magmatic Evolution*. Geological Society, London, Special Publications, **447**. First published online September 9, 2016, https://doi.org/10.1144/SP447.7

Planke, S., Symonds, P.A., Alvestad, E. & Skogseid, J. 2000. Seismic volcanostratigraphy of large-volume basaltic extrusive complexes on rifted margins. *Journal of Geophysical Research: Solid Earth*, **105**, 19 335–19 351.

Praeg, D., Stoker, M.S., Shannon, P.M., Ceramicola, S., Hjelstuen, B., Lasberg, J.S. & Mathiesen, A. 2005. Episodic Cenozoic tectonism and the development of the NW European 'passive' continental margin. *Marine and Petroleum Geology*, **22**, 1007–1030.

Ritchie, J.D., Ziska, H., Kimbell, G., Quinn, M. & Chadwick, A. 2011. Structure. *In*: Ritchie, J.D., Ziska, H., Johnson, H. & Evans, D. (eds) *Geology of the Faroe–Shetland Basin and Adjacent Areas*. British Geological Survey and Jarðfeingi Research Report **RR/11/01**. British Geological Survey, Keyworth, Nottingham, UK, 9–70.

Spakman, W. 2014. Earth's mantle structure under the North Atlantic. *In*: Hopper, J.R., Funck, T., Stoker, M., Árting, U., Péron-Pinvidic, G., Doornenbal, J.C. & Gaina, C. (eds) *Tectonostratigraphic Atlas of the North-East Atlantic Region*. Geological Survey of Denmark and Greenland (GEUS), Copenhagen, Denmark, 293–300.

Stoker, M.S., Stewart, M.A. et al. 2016. An overview of the Upper Paleozoic–Mesozoic stratigraphy of the NE Atlantic region. *In*: Péron-Pinvidic, G., Hopper, J.R., Stoker, M.S., Gaina, C., Doornenbal, J.C., Funck, T. & Árting, U.E. (eds) *The NE Atlantic Region: A Reappraisal of Crustal Structure, Tectonostratigraphy and Magmatic Evolution*. Geological Society, London, Special Publications, **447**. First published online August 11, 2016, updated August 12 2016, https://doi.org/10.1144/SP447.2

Symonds, P.A., Planke, S., Frey, Ø. & Skogseid, J. 1998. Volcanic evolution of the western Australian continental margin and its implications for basin development. *In*: Purcell, R.R. & Purcell, P.G. (eds) *The Sedimentary Basins of Western Australia 2: Proceedings of the PESA Symposium*. Petroleum Exploration Society of Australia (PESA), Perth, Australia, 33–54.

Tsikalas, F., Faleide, J.I., Eldholm, O. & Wilson, J. 2005. Late Mesozoic–Cenozoic structural and stratigraphic correlations between the conjugate mid-Norway and NE Greenland continental margins. *In*: Doré, A.G. & Vining, B.A. (eds) *Petroleum Geology: North-West Europe and Global Perspectives – Proceedings of the 6th Petroleum Geology Conference*. Geological Society, London, Petroleum Geology Conference Series, **6**, 785–801, https://doi.org/10.1144/0060785

Vis, G.-J. 2017. Geology and seepage in the NE Atlantic region. *In*: Péron-Pinvidic, G., Hopper, J.R., Stoker, M.S., Gaina, C., Doornenbal, J.C., Funck, T. & Árting, U.E. (eds) *The NE Atlantic Region: A Reappraisal of Crustal Structure, Tectonostratigraphy and Magmatic Evolution*. Geological Society, London, Special Publications, **447**. First published online April 7, 2017, https://doi.org/10.1144/SP447.16

White, R.S. & McKenzie, D. 1989. Magmatism at rift zones: the generation of volcanic continental margins and flood basalts. *Journal of Geophysical Research*, **94**, 7685–7729.

Wilkinson, C.M., Ganerød, M., Hendriks, B.W.H. & Eide, E.A. 2016. Compilation and appraisal of geochronological data from the North Atlantic Igneous Province (NAIP). *In*: Péron-Pinvidic, G., Hopper, J.R., Stoker, M.S., Gaina, C., Doornenbal, J.C., Funck, T. & Árting, U.E. (eds) *The NE Atlantic Region: A Reappraisal of Crustal Structure, Tectonostratigraphy and Magmatic Evolution*. Geological Society, London, Special Publications, **447**. First published online November 8, 2016, https://doi.org/10.1144/SP447.10

Wilson, J.T. 1966. Did the Atlantic close and then re-open? *Nature*, **211**, 676–681, https://doi.org/10.1038/211676a0

An overview of the Upper Palaeozoic–Mesozoic stratigraphy of the NE Atlantic region

M. S. STOKER[1*], M. A. STEWART[1], P. M. SHANNON[2], M. BJERAGER[3], T. NIELSEN[3], A. BLISCHKE[4], B. O. HJELSTUEN[5], C. GAINA[6], K. McDERMOTT[2,7] & J. ÓLAVSDÓTTIR[8]

[1]*British Geological Survey, The Lyell Centre, Research Avenue South, Riccarton, Edinburgh EH14 4AP, UK*

[2]*School of Geological Sciences, University College Dublin, Belfield, Dublin 4, Ireland*

[3]*Geological Survey of Denmark and Greenland, Øster Vøldgade 10, DK-1350 Copenhagen, Denmark*

[4]*ÍSOR (Iceland Geosurvey), Deptartment of Energy Technology, Rangarvellir, PO Box 30, 602 Akureyri, Iceland*

[5]*Department of Earth Sciences, University of Bergen, Allegaten 41, N-5007 Bergen, Norway*

[6]*Centre for Earth Evolution and Dynamics, University of Oslo, Postbox 1028, Blindern, N-0135 Oslo, Norway*

[7]*Present address: ION Geophysical, 1st Floor, Integra House, Vicarage Road, Egham TW20 9JZ, UK*

[8]*Jarðfeingi, Brekkutín 1, PO Box 3059, FO-110 Tórshavn, Faroe Islands*

**Correspondence: mss@bgs.ac.uk*

Abstract: This study describes the distribution and stratigraphic range of the Upper Palaeozoic–Mesozoic succession in the NE Atlantic region, and is correlated between conjugate margins and along the axis of the NE Atlantic rift system. The stratigraphic framework has yielded important new constraints on the timing and nature of sedimentary basin development in the NE Atlantic, with implications for rifting and the break-up of the Pangaean supercontinent. From a regional perspective, the Permian–Triassic succession records a northwards transition from an arid interior to a passively subsiding, mixed carbonate–siliciclastic shelf margin. A Late Permian–earliest Triassic rift pulse has regional expression in the stratigraphic record. A fragmentary paralic to shallow-marine Lower Jurassic succession reflects Early Jurassic thermal subsidence and mild extensional tectonism; this was interrupted by widespread Mid-Jurassic uplift and erosion, and followed by an intense phase of Late Jurassic rifting in some (but not all) parts of the NE Atlantic region. The Cretaceous succession is dominated by thick basinal-marine deposits, which accumulated within and along a broad zone of extension and subsidence between Rockall and NE Greenland. There is no evidence for a substantive and continuous rift system along the proto-NE Atlantic until the Late Cretaceous.

The NE Atlantic region incorporates the conjugate continental margins of NW Europe (between SW Ireland and northern Norway) and East Greenland, as well as the intervening NE Atlantic Ocean. The latter is an area of complex bathymetry and structure that includes both active and extinct spreading ridges, Iceland and its insular margin, the Greenland–Iceland–Faroe Ridge Complex, and the Jan Mayen microcontinent (JMMC) (Fig. 1; Table 1). To the north, the NE Atlantic Ocean is connected to the Arctic Ocean via the Fram Strait, which developed as a result of transtensional shear between North Greenland and the western Barents Sea; to the south, the NE Atlantic Ocean is separated from the North Atlantic Ocean by the Charlie-Gibbs Fracture Zone.

For much of the Phanerozoic eon, the NE Atlantic region experienced a shared geological development since its amalgamation – the collision of Baltica and Laurentia to form Laurussia – during the

From: PÉRON-PINVIDIC, G., HOPPER, J. R., STOKER, M. S., GAINA, C., DOORNENBAL, J. C., FUNCK, T. & ÁRTING, U. E. (eds) 2017. *The NE Atlantic Region: A Reappraisal of Crustal Structure, Tectonostratigraphy and Magmatic Evolution*. Geological Society, London, Special Publications, **447**, 11–68.
First published online August 11, 2016, updated August 12, 2016, https://doi.org/10.1144/SP447.2

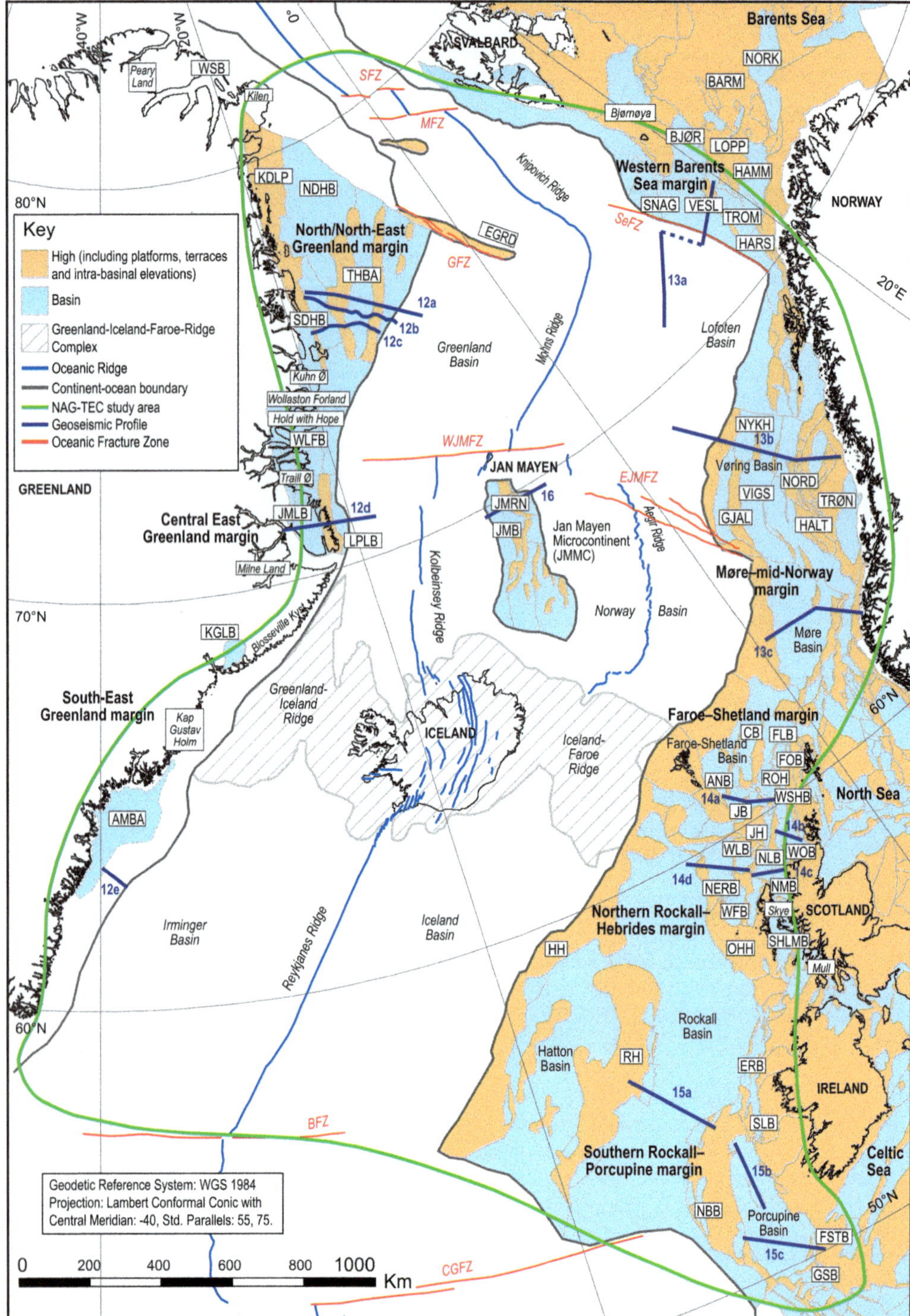

Fig. 1. Map showing location of NAG-TEC study area, with geographical regions, geoseismic profile locations and key structural elements referred to in the text (modified after Hopper *et al.* 2014). See Table 1 for key to structural element abbreviations. Map information: the Geodetic Reference System is WGS 1984; the Projection is Lambert Conformal Conic with Central Meridian of −40, and standard parallels of 55 and 75.

Table 1. *Explanation of structural element abbreviations in Figure 1*

European margin	
ANB	Annika Sub-basin
BARM	Bjarmeland Platform
BJØR	Bjørnøya Basin
CB	Corona Sub-basin
ERB	Erris Basin
FLB	Flett Sub-basin
FOB	Foula Sub-basin
FSTB	Fastnet Basin
GJAL	Gjallar Ridge
GSB	Goban Spur Basin
HALT	Halten Terrace
HAMM	Hammerfest Basin
HARS	Harstad Basin
HH	Hatton High
JB	Judd Sub-basin
JH	Judd High
LOPP	Loppa High
NBB	North Bróna Basin
NERB	NE Rockall Basin
NLB	North Lewis Basin
NMB	North Minch Basin
NORD	Nordland Ridge
NORK	Nordkapp Basin
NYKH	Nyk High
OHH	Outer Hebrides High
RH	Rockall High
ROH	Rona High
SHLMB	Sea of the Hebrides–Little Minch Basin
SLB	Slyne Basin
SNAG	Sørvestsnaget Basin
TROM	Tromsø Basin
TRØN	Trøndelag Platform
VESL	Veslemøy High
VIGS	Vigrid Syncline
WFB	West Flannan Basin
WLB	West Lewis Basin
WOB	West Orkney Basin
WSHB	West Shetland Basin
East Greenland margin and Jan Mayen microcontinent	
AMBA	Ammassalik Basin
EGRD	East Greenland Ridge
JMB	Jan Mayen Basin
JMLB	Jameson Land Basin
JMMC	Jan Mayen Microcontinent
JMRN	Jan Mayen Ridge North
KDLP	Koldewey Platform
KGLB	Kangerlussuaq Basin
LPLB	Liverpool Land Basin
NDHB	North Danmarkshavn Basin
SDHB	South Danmarkshavn Basin
THBA	Thetis Basin
WFLB	Wollaston Forland Basin
WSB	Wandel Sea Basin
Oceanic fracture zones	
BFZ	Bight Fracture Zone
CGFZ	Charlie Gibbs Fracture Zone
EJMFZ	East Jan Mayen Fracture Zone
GFZ	Greenland Fracture Zone
MFZ	Molloy Fracture Zone
SFZ	Spitsbergen Fracture Zone
SeFZ	Senja Fracture Zone
WJMFZ	West Jan Mayen Fracture Zone

Late Silurian Scandic phase of the Caledonian Orogeny: its subsequent Late Palaeozoic incorporation into the Pangaean supercontinent; the prolonged history of extension linked to the Mesozoic break-up of Pangaea; and, ultimately, the early–mid-Cenozoic opening of the NE Atlantic Ocean (Ziegler 1988; Doré *et al.* 1999; Roberts *et al.* 1999; Torsvik *et al.* 2002; Pharaoh *et al.* 2010). The assembly of Pangaea resulted from the convergence and collision of Laurussia with Gondwana (from the south), with final suturing – in the NE Atlantic region – marked by the Late Carboniferous–Early Permian Variscan Orogeny. However, the development of Permian–Triassic rift basins over a large part of the west European and NE Atlantic regions marks the onset of instability shortly after the assemblage of Pangaea. This instability was driven by a number of processes, including the late spreading collapse of the Variscan orogen, the southwards propagation of rifting emanating from the Arctic and the opening of the neo-Tethys (Williams & McKie 2009). This early rift phase instigated a long period of post-Variscan extension that eventually culminated in early–mid-Cenozoic break-up of this wide rifted region.

In general terms, the tectonic evolution of this region has been dominated by several phases of episodic extension, which are increasingly attributed to the interaction of a southwards-propagating 'Arctic' rift and a northwards-propagating 'Atlantic' rift (Doré 1992; Doré *et al.* 1999; Roberts *et al.* 1999; Lundin & Doré 2005*a*; Ellis & Stoker 2014). The break-up phase that led to the formation of the NE Atlantic Ocean was accompanied by magmatic activity, which some authors attribute to an 'Iceland' mantle plume (e.g. White 1988, 1989; White & McKenzie 1989; White & Lovell 1997; Smallwood *et al.* 1999; Skogseid *et al.* 2000; Smallwood & White 2002), whereas others argue that rifting and break-up can be wholly explained by plate tectonic mechanisms, lithospheric thinning and variable decompressive melting along the rifts (e.g. Foulger & Anderson 2005; Foulger *et al.* 2005; Lundin & Doré 2005*a*, *b*; Ellis & Stoker 2014).

Within the NE Atlantic region, Roberts *et al.* (1999) perceived the 'Arctic' rift to extend southward via the Norwegian–Greenland area into the West Shetland area and the North Sea, whereas the 'Atlantic' rift comprises the SE Greenland, Hebrides, Rockall, Porcupine and the western Celtic Sea areas. The overall rift system is generally fairly simple in the Greenland–Norway region, but becomes more complex with different rift arms from the SE Greenland–Faroe–Shetland region southwards. Arguably, this is why Roberts *et al.* (1999) divided the overall system into the Arctic (simple, northern) and the Atlantic (more complex, southern) rifts. Whilst taking into consideration this north–south variation in structural complexity, we prefer to use the term 'NE Atlantic' for the entire rift system – excluding the North Sea – in this paper, which we suggest is a more accurate indicator of the geographical domain that we are focused upon, and does not force or perpetuate this subdivision without a more rigorous analysis of its history of rifting (see below).

According to Roberts *et al.* (1999), these northern and southern sections of the rift system remained largely separate entities until the Late Jurassic or Early Cretaceous; thereafter, the rift system is envisaged to have become a single entity represented by the Rockall, Faroe–Shetland, Møre and Vøring basins. By way of contrast, various palinspastic reconstructions of the NE Atlantic region depict a through-going NE Atlantic rift system along this line of basins since the Permian (e.g. Ziegler 1988; Cope *et al.* 1992; Doré 1992; Knott *et al.* 1993; Torsvik *et al.* 2002; Coward *et al.* 2003; McKie & Williams 2009; Pharaoh *et al.* 2010). Alternatively, Doré *et al.* (1999) have suggested that pre-Cretaceous extensional fault activity within the area between NW Britain/Ireland and SE Greenland was limited and discontinuous, and that the NE Atlantic rift system was not fully linked until the mid-Cretaceous. These authors describe a mosaic-like fragmentation of Pangaea during the Permian–Triassic, replaced by a more systematic, albeit restricted, east–west extension in the Jurassic, with the change to NW–SE extension in the Cretaceous. The latter extension vector was maintained through to plate break-up. The change from east–west Jurassic extension to NW–SE Cretaceous extension is attributed to a switch from a system dominated by north–south Tethyan rift propagation to one dominated by NE–SW rifting driven by northwards-propagating Atlantic spreading (Doré *et al.* 1999).

On the basis of these differing viewpoints, it is clear that the evolution of the NE Atlantic rift systems remains enigmatic. Despite the pioneering work of Ziegler (1988, p. vii), in his own words he described his palinspastic reconstructions as, 'generalised ... they may have serious shortcomings ... and should be regarded as working hypotheses'. Unfortunately, most subsequent authors that have considered the post-Caledonian evolution of the NE Atlantic region have largely continued to rely on the maps of Ziegler (1988), thereby perpetuating all of their inherent limitations. This is particularly evident in that part of the NE Atlantic region between NW Britain/Ireland and SE Greenland – the critical rift tip overlap area in the dual rift model – where the Late Palaeozoic–Mesozoic rock record is very fragmentary and only sparsely known (cf. Ritchie *et al.* 2011*a*; Hitchen *et al.* 2013). Thus, reconstructions that currently show linked Permo-Triassic, Jurassic and Early Cretaceous rift basins across this area are largely without any robust foundation.

In view of the uncertainty relating to the pre-break-up configuration of the NE Atlantic rift system, this paper focuses on the Upper Palaeozoic–Mesozoic (Permian–Cretaceous) rock record preserved on the circum-NE Atlantic continental margins (Fig. 1). The description and distribution of the stratigraphic successions presented in this paper were compiled as part of the NAG-TEC project – a large-scale observational programme designed to better understand the tectonic development of the conjugate NE Atlantic margins and adjacent ocean basin (Hopper *et al.* 2014). The limit of the main NAG-TEC study area is indicated in Figure 1, and extends over 35° of latitude between northern Greenland and the Porcupine Basin. The main objective of the paper is to summarize our current understanding of the Permian–Cretaceous stratigraphy around the NE Atlantic region (within the NAG-TEC study area). We present the first detailed correlation of the various lithostratigraphies developed across this region, and highlight key major unconformities and changes in regional sedimentation patterns. On this basis, we are better able to appraise the history of development of the full NE Atlantic rift system; in particular, to provide a base level of observational data that we hope can be used to test more specific process-orientated studies. The implications of our observations for the putative Late Palaeozoic–Mesozoic palaeogeographical reconstructions will also be briefly discussed.

Data and methods

This stratigraphic summary presents a comprehensive review of the known record of Permian–Cretaceous rocks around the NE Atlantic region, with a focus on distribution, structural setting, stratigraphy, boundary relationships and depositional environments. The stratigraphic architecture is further illustrated in a series of geoseismic profiles from across the study area. The study area has been divided into several sub-areas that include: the North–NE Greenland margin; the western Barents Sea–Svalbard margin; the central East Greenland margin; the Møre–mid-Norway margin; the SE Greenland margin (Cretaceous only); the Faroe–Shetland–northern Rockall–Hebrides margin; and the southern Rockall–Porcupine margin (including the NW Irish Shelf and the western part of the Celtic Sea region) (Fig. 1). The boundary between the NE Greenland and the central East Greenland sub-areas is taken at about 72° N, and between central East Greenland and SE Greenland at about 70° N: however, it should be noted that, for ease of description, the Permian–Triassic rocks between Traill Ø and Hold with Hope (the Wollaston Foreland Basin; 72–74° N) are included within the summary of the central East Greenland sub-area. The Jan Mayen microcontinent (JMMC) is also included as part of this review, and described separately. Although Svalbard lies outside of our main area of study, information from the rocks cropping out on the archipelago has been utilized as a guide to the composition of the adjacent offshore area. Note also that, whilst our stratigraphic distribution maps (Figs 2–4) extend into the northern North Sea, this region is beyond the scope of this study. For details of the North Sea region, the reader is referred to two major atlases that provide a comprehensive review of the stratigraphic framework, and which have been utilized in the production of our maps (Evans *et al.* 2003; Doornenbal & Stevenson 2010).

The decision-making involved in the compilation of the regional stratigraphic distribution maps, the stratigraphic correlation panels and the geoseismic profiles, whilst based in part on published data, was also informed by the intimate expertise and knowledge of the contributors responsible for each sub-area, including the provision and use of unpublished geological survey information – commonly derived from onshore outcrops and offshore boreholes – and released commercial well data. Those boreholes and wells perceived to contain the most significant stratigraphical information have been collated within a stratigraphic database, details of which are presented elsewhere (Hopper *et al.* 2014; Stoker *et al.* 2014). The main processes involved in the creation of the maps and correlation panels are described below.

Stratigraphic distribution maps

Stratigraphic distribution maps have been compiled for the Permian–Triassic, Jurassic and Cretaceous successions (Figs 2–4). These have been compiled largely on the basis of existing published and unpublished geological survey maps, supplemented by correlating chronostratigraphic divisions in key stratigraphic wells and boreholes (cf. Stoker *et al.* 2014), and by cross-checking against published/unpublished seismic data and regional cross-sections. The basic mapping was compiled in ArcGIS, and for each chronostratigraphic division the following polygons have been delivered for each sub-area:

- ‘present’ and ‘not-present’ polygons, indicating the areas where a chronostratigraphic division is present or absent as proven by outcrops, wells and/or seismic interpretation; and
- ‘inferred’ polygons, indicating areas where the presence of a chronostratigraphic division has been assumed based on seismic interpretation or regional maps.

The information from each sub-area was subsequently merged to produce the regional stratigraphic distribution maps illustrated in Figures 2–4.

Stratigraphic correlation panels and geoseismic sections

A series of seven regional stratigraphic correlation panels (Table 2; Figs 5–11) has been constructed, each of which features a number of generalized stratigraphic columns that summarize the Permian–Cretaceous geological history of selected parts of the NE Atlantic region, both onshore and offshore. The details of their geographical extent and the key structural elements/domains that they represent are summarized in Table 2. Each correlation panel is arranged as a general east-to-west conjugate-margin stratigraphic transect.

The correlation panels combine chronostratigraphy with lithostratigraphic (where available) and other information, such as sedimentary environments and magmatic events. Lithostratigraphic information is presented mostly at ‘Group’ level. Each individual offshore stratigraphic column on the correlation panel has been constructed on the basis of information derived from a number of key wells and boreholes located in the area represented by that generalized column (cf. Stoker *et al.* 2014). This type of information is mostly located on the NW European continental margin. On the Greenland margin, the generalized stratigraphic columns (Figs 5–10) are mostly compiled from onshore data (Table 2). All of the generalized stratigraphic columns based on the onshore mapping data and the offshore well and borehole data present a moderate–high reliability in terms of the chronostratigraphy. By way of contrast, the age and composition of the succession in several of the basins offshore north, NE and central East Greenland, as well as the JMMC (Table 2), is inferred on the basis of seismic interpretation and structural position in relation to adjacent established basins. Whereas thick successions are observed on seismic reflection profiles across these basins, the stratigraphic reliability of the infill remains low at the time of this study.

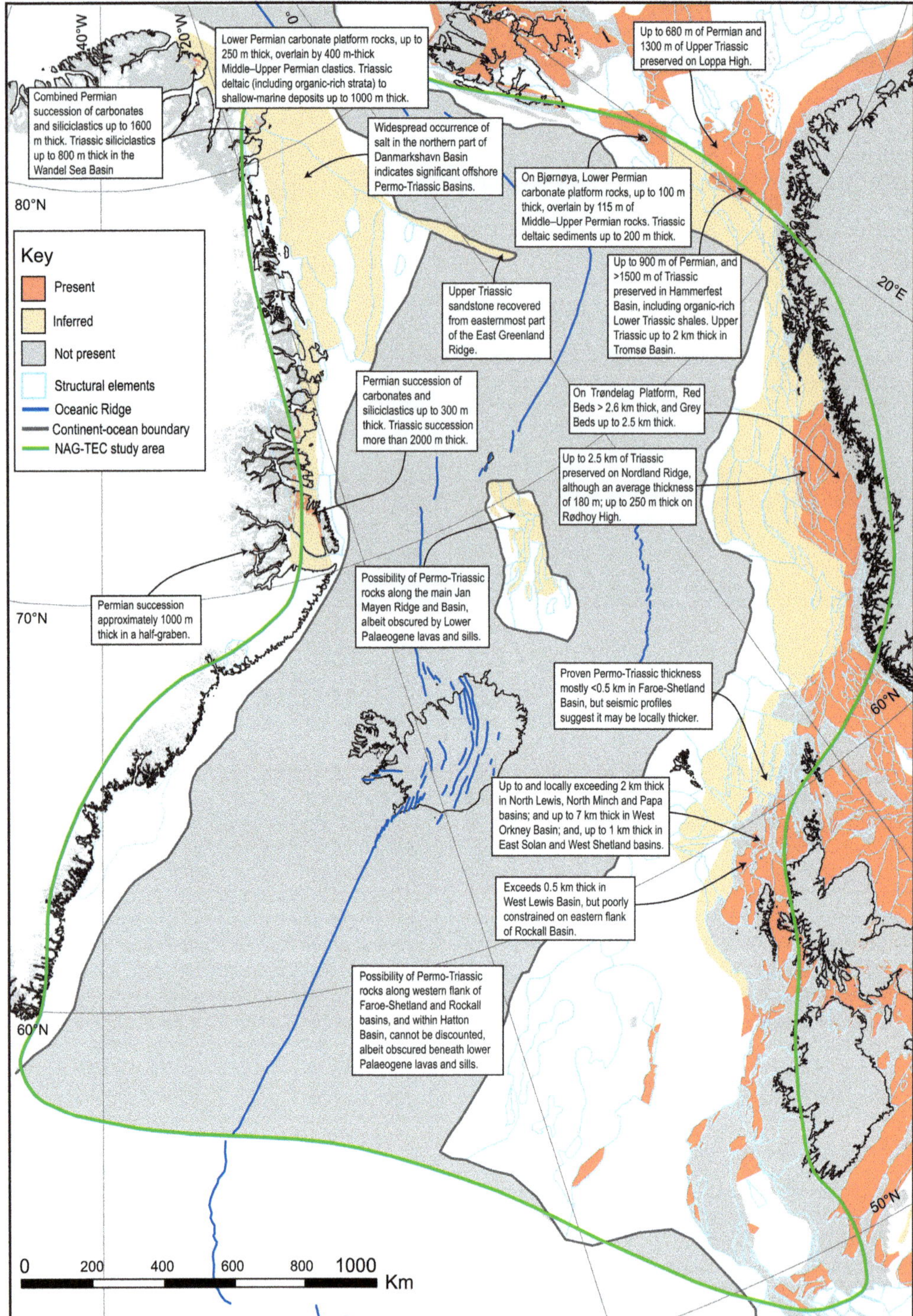

Fig. 2. Stratigraphic distribution of Permian–Triassic rocks (based on the compilation of Stoker *et al.* 2014). Map information: the Geodetic Reference System is WGS 1984; the Projection is Lambert Conformal Conic with a Central Meridian of −40, and Std. Parallels of 55 and 75.

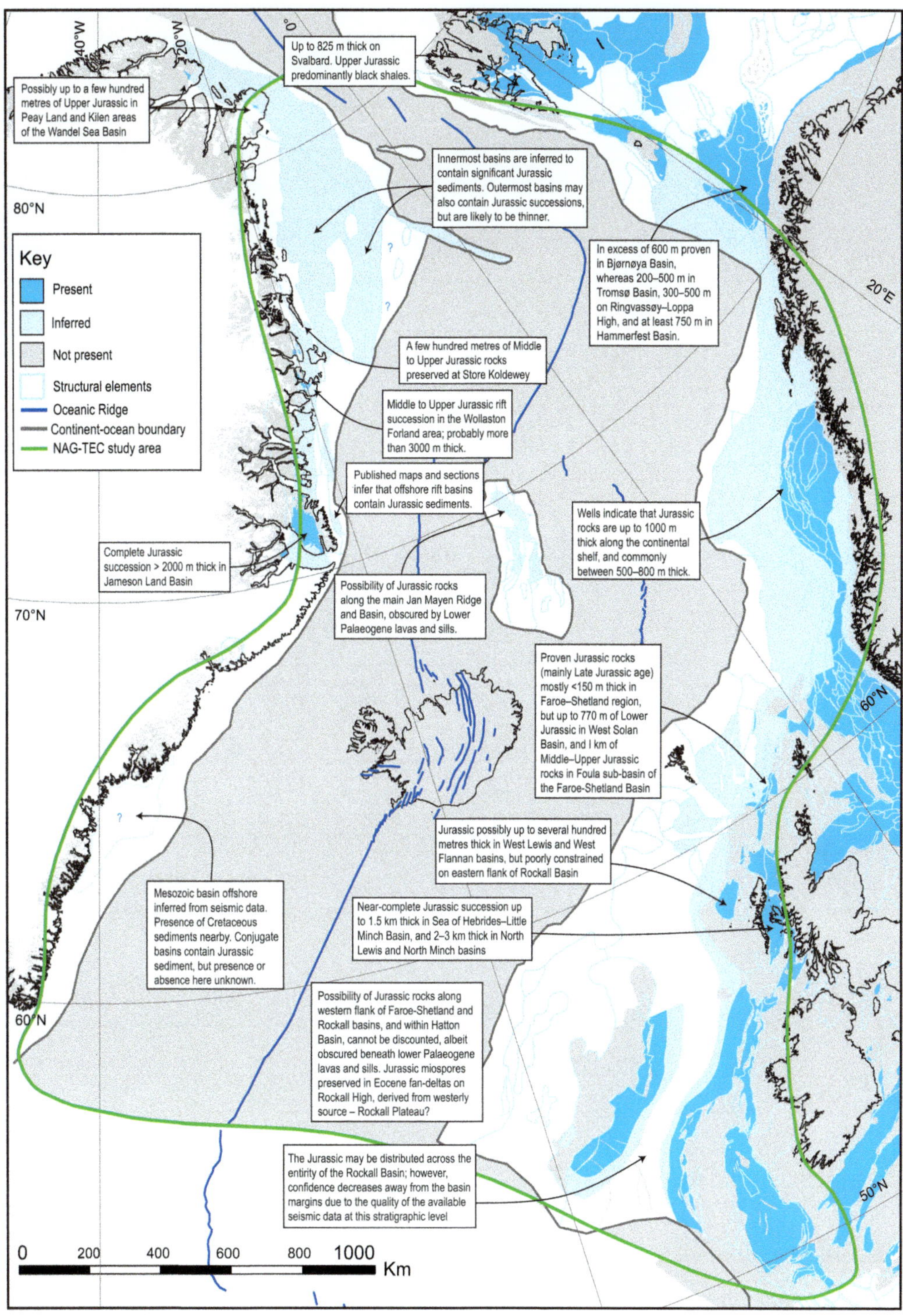

Fig. 3. Stratigraphic distribution of Jurassic rocks (based on the compilation of Stoker *et al.* 2014). Map information: the Geodetic Reference System is WGS 1984; the Projection is Lambert Conformal Conic with Central Meridian of −40, and standard parallels of 55 and 75.

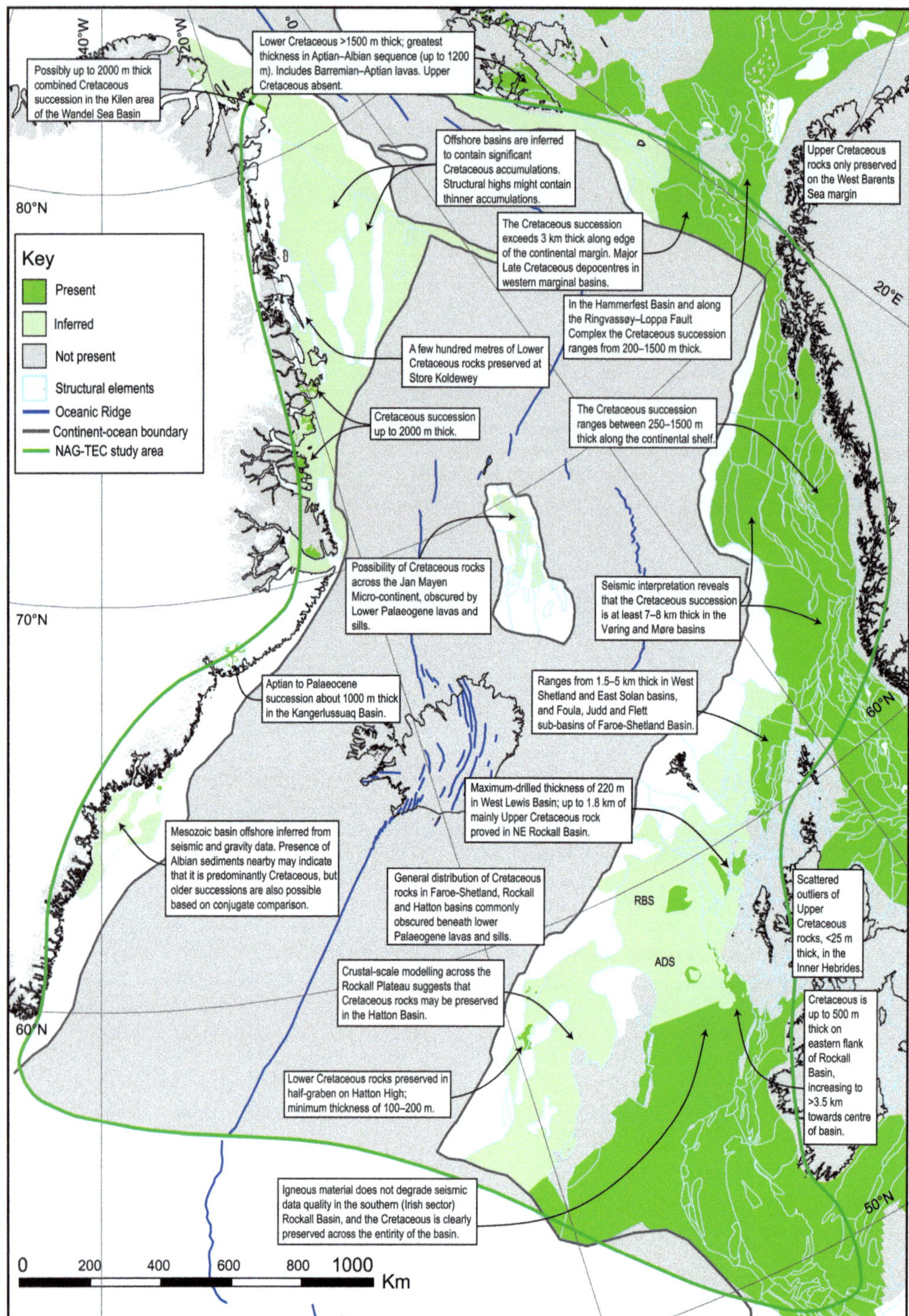

Fig. 4. Stratigraphic distribution of Cretaceous rocks (based on the compilation of Stoker *et al.* 2014). Abbreviations: ADS, Anton Dohrn Seamount; RBS, Rosemary Bank Seamount. Map information: the Geodetic Reference System is WGS 1984; the Projection is Lambert Conformal Conic with Central Meridian of −40, and standard parallels of 55 and 75.

Table 2. *List of regional stratigraphic correlation panels (with figure numbers) arranged as conjugate-margin transects across the NAG-TEC study area*

Fig. No.	Regional correlation panel	Key structural elements/domains
5	North Greenland–Svalbard–western Barents Sea region	Wandel Sea Basin*; northern Danmarkshavn Basin†; Svalbard*; SW Barents Shelf, including the Loppa High and Hammerfest Basin
6	NE Greenland region	Wollaston Foreland–Kuhn Ø*; Koldewey Platform†; southern Danmarkshavn Basin†; Thetis Basin†
7	NE Greenland–Vøring region	Traill Ø–Hold with Hope*; northern and southern Vøring continental shelf
8	Central East Greenland–southern Jan Mayen–Møre region	Jameson Land*; Liverpool Land Basin†; southern Jan Mayen microcontinent†; Møre continental shelf
9	SE Greenland–Faroe–Shetland region	Kangerlussuaq Basin*; Faroe–Shetland Basin; West Shetland marginal basins
10	SE Greenland–northern Rockall Plateau–northern Rockall Basin–Hebrides region	Ammassalik margin; Hatton Basin; northern Rockall Basin; West Lewis Basin; Hebridean basins; Inner Hebrides*
11	Southern Rockall Basin–Porcupine Basin–NW Irish region	Southern Rockall Basin; Porcupine Basin; Erris Basin; Slyne Basin; Goban Spur; Fastnet Basin

*Onshore data.
†Seismic interpretation only.

The stratigraphic architecture across the conjugate continental margins is also displayed in a series of geoseismic profiles, which have been compiled from a variety of sources (see Figs 12–16).

Stratigraphic descriptions

On the basis of this database, the following sections describe the stratigraphy of the Permian–Triassic, Jurassic and Cretaceous successions. These descriptions are presented in order from north to south to illustrate the regional variations, and we have treated each of the sub-areas (outlined above) in a similar manner, addressing key aspects of distribution, structural setting, stratigraphy, boundary relationships and depositional environments.

Permian–Triassic stratigraphy of the NE Atlantic margins

North–NE Greenland margin

Distribution. Permian–Triassic strata are exposed onshore eastern North Greenland in the Wandel Sea Basin (79°–83° N) with an inferred continuation offshore to the north and east (Døssing *et al.* 2010) (Fig. 2). On the NE Greenland margin, the succession is interpreted to be present offshore on the Store Koldewey Platform and in the Danmarkshavn Basin based on seismic data (Hamann *et al.* 2005), and possible Permian–Triassic strata in the Thetis Basin are speculatively suggested by Dinkelman *et al.* (2010) and Jackson *et al.* (2012) (Figs 2 & 12a). Rocks of Late Triassic age have been sampled on the East Greenland Ridge (Nielsen *et al.* 2014) (Fig. 2).

Structural setting. The Late Palaeozoic–Early Mesozoic rift basins trend approximately north–south along the NE Greenland margin. On the North Greenland margin, the Wandel Sea Basin forms a rift basin linking this north–south rift trend with a roughly NW–SE-orientated Palaeozoic rift trend between Greenland and Svalbard (Stemmerik & Håkansson 1991).

Stratigraphy. The Permian in the onshore part of the Wandel Sea Basin is represented by carbonates and mudstones overlain by shale, limestones and sandstones of the Mallemuk Mountain Group, up to about 1600 m thick (Stemmerik *et al.* 1996, 2000) (Figs 2 & 5). This succession is overlain by the Triassic Trolle Land Group of conglomerates, sandstones and mudstones, possibly up to 1 km thick (Håkansson *et al.* 1991; Stemmerik *et al.* 1996) (Fig. 5).

Offshore NE Greenland, a more complete rock record through the Permian–Triassic periods – as suggested by analogy to the conjugate Svalbard and western Barents Sea successions – is inferred for the Danmarkshavn and Thetis basins, and the Koldewey Platform (Figs 5 & 6). An Upper Carboniferous–Triassic succession up to 7 km thick is interpreted for the Danmarkshavn Basin (Hamann *et al.* 2005). The Permian rocks are inferred to comprise platform carbonate, basinal evaporite and marine mudstone deposits overlain by a Triassic

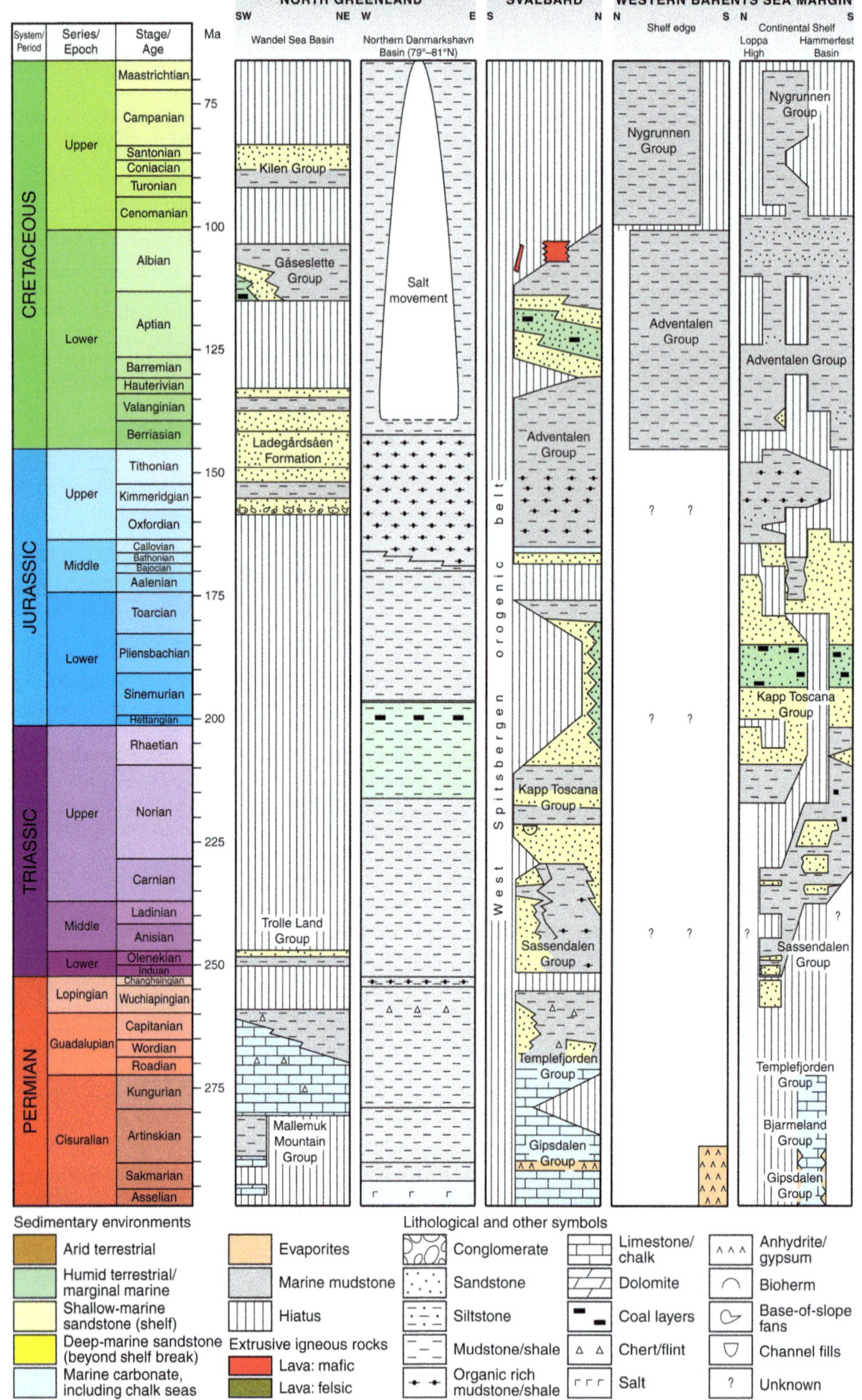
NORTH GREENLAND
SW NE W E
Wandel Sea Basin
Northern Danmarkshavn Basin (79°–81°N)
SVALBARD
S N
WESTERN BARENTS SEA MARGIN
N S N S
Shelf edge
Continental Shelf
Loppa High
Hammerfest Basin
System/ Period
Series/ Epoch
Stage/ Age
Ma
CRETACEOUS
JURASSIC
TRIASSIC
PERMIAN
Upper
Lower
Upper
Middle
Lower
Upper
Middle
Lower
Lopingian
Guadalupian
Cisuralian
Maastrichtian
Campanian
Santonian
Coniacian
Turonian
Cenomanian
Albian
Aptian
Barremian
Hauterivian
Valanginian
Berriasian
Tithonian
Kimmeridgian
Oxfordian
Callovian
Bathonian
Bajocian
Aalenian
Toarcian
Pliensbachian
Sinemurian
Hettangian
Rhaetian
Norian
Carnian
Ladinian
Anisian
Olenekian
Induan
Changhsingian
Wuchiapingian
Capitanian
Wordian
Roadian
Kungurian
Artinskian
Sakmarian
Asselian
75
100
125
150
175
200
225
250
275
Kilen Group
Gåseslette Group
Ladegårdsåen Formation
Trolle Land Group
Mallemuk Mountain Group
Salt movement
West Spitsbergen orogenic belt
Adventalen Group
Kapp Toscana Group
Sassendalen Group
Templefjorden Group
Gipsdalen Group
Nygrunnen Group
Adventalen Group
Nygrunnen Group
Adventalen Group
Kapp Toscana Group
Sassendalen Group
Templefjorden Group
Bjarmeland Group
Gipsdalen Group
?
Sedimentary environments
Arid terrestrial
Humid terrestrial/ marginal marine
Shallow-marine sandstone (shelf)
Deep-marine sandstone (beyond shelf break)
Marine carbonate, including chalk seas
Evaporites
Marine mudstone
Hiatus
Extrusive igneous rocks
Lava: mafic
Lava: felsic
Lithological and other symbols
Conglomerate
Sandstone
Siltstone
Mudstone/shale
Organic rich mudstone/shale
Limestone/ chalk
Dolomite
Coal layers
Chert/flint
Salt
Anhydrite/ gypsum
Bioherm
Base-of-slope fans
Channel fills
Unknown

siliciclastic succession that is dominated by fine-grained marine and marginal-marine sediments (Gautier *et al.* 2011) (Figs 5 & 6). The deposition of halite evaporites is validated by widespread salt diapirism in the northen part of the basin (Fig. 5). A source-rock-prone interval may occur towards the top of the Permian sequence, and the boundary with the Triassic might be, at least locally, an erosional unconformity. A comparable sequence has been inferred for the Koldewey Platform, albeit with coarser clastic facies locally developed along its western margin (Fig. 6). The postulated Permian–Triassic succession in the Thetis Basin remains undivided (Dinkelman *et al.* 2010; Jackson *et al.* 2012). The Upper Triassic rock samples from the East Greenland Ridge consist of sandy mudstones (Nielsen *et al.* 2014).

Boundary relationships with underlying and overlying systems. The onshore part of the Wandel Sea Basin records an unconformity at the Carboniferous–Permian boundary, an intra-Permian unconformity and a major unconformity at the Permian–Triassic transition (Stemmerik *et al.* 1998). A major unconformity separates the Triassic from the Upper Jurassic rocks (Håkansson *et al.* 1991). In the offshore basins, the Carboniferous–Permian transition is inferred to be conformable, whereas an unconformity is inferred at the Permian–Triassic and Triassic–Jurassic boundaries in the southern Danmarkshavn Basin and on the Koldewey Platform.

Depositional environment. The Mallemuk Mountain Group in the Wandel Sea Basin represents the Early Permian shallow-water carbonate shelf. A marked flooding surface forms the base of the overlying succession, with mudstone deposition followed by intercalating carbonates, mudstones and siliciclastics deposited in shelfal environments (Stemmerik & Håkansson 1991). The Triassic Trolle Land Group includes fluvial and shallow-marine sandy to muddy shelf deposition. The basins offshore NE Greenland have an interpreted Upper Carboniferous–Permian marine carbonate platform, with an evaporite succession deposited during a phase of thermal subsidence; this was followed by deposition of marine basinal mudstones during the Triassic, which were succeeded by increasingly marginal-marine and terrestrial deposition during the Late Triassic (Hamann *et al.* 2005).

Western Barents Sea–Svalbard margin

Distribution. Rocks of Permian–Triassic age crop out on Svalbard and Bjørnøya, and have been proved in numerous wells on and adjacent to the southern part of the western Barents Sea margin, and thus are inferred to underlie much of this part of the continental margin (Fig. 2).

Structural setting. A stable carbonate platform persisted from the Late Carboniferous into the Early Permian with only minor uplifts (late-stage Variscan orogenic activity), whereas the Mid–Late Permian was characterized by uplifts and erosional breaks (Steel & Worsley 1984; Brekke *et al.* 2001; Worsley 2008). The latter includes a major Mid–Late Permian unconformity that is widely developed across the Western Barents Sea–Svalbard region (Fig. 5). This is associated with Late Permian–Early Triassic rifting linked to the Uralian Orogeny, which gave rise to normal faulting and rapid subsidence: for example, the western margin of the Loppa High (Brekke *et al.* 2001; Faleide *et al.* 2010). However, for much of the Triassic period, the area was largely one of thermal relaxation (Doré *et al.* 1999; Roberts *et al.* 1999).

Stratigraphy. The Permian Gipsdalen, Bjarmeland and Templefjorden groups comprise a sequence of carbonate, evaporate and shallow-marine clastic rocks across the western Barents Sea–Svalbard margin, including Bjørnøya (Harland 1997; Dallmann 1999; Worsley 2008; Henriksen *et al.* 2011; NORLEX 2014) (Figs 2 & 5). On Svalbard, the Permian succession is up to 650 m thick, albeit punctuated by a late Early Permian unconformity, whereas up to 900 m of Lower Permian strata occur in the Hammerfest Basin, on the SW Barents Sea margin. The shallow-marine clastic succession, which includes mudstone, sandy bioclastic limestone and glauconitic sandstone with minor chert and silicified limestone, becomes increasingly dominant throughout the region in the Mid–Late Permian.

The Triassic succession comprises the Lower–Middle Triassic Sassendalen and the Upper Triassic–Middle Jurassic Kapp Toscana groups (Nøttvedt *et al.* 1993*b*; Harland 1997; Henriksen *et al.* 2011; NORLEX 2014) (Fig. 5). The Triassic rocks are commonly separated from the Permian by an unconformity marking a change from the cherts and siliciclastics of the Templefjorden Group to softer,

Fig. 5. Regional stratigraphic correlation panel showing the generalized Permian–Cretaceous succession preserved in the North Greenland (Wandel Sea Basin)–Svalbard–western Barents Sea region, and inferred (paler colourfill) on the basis of seismic interpretation for the area off NE Greenland (northern Danmarkshavn Basin). Correlation based on the compilation of Stoker *et al.* (2014). Timescale based on Gradstein *et al.* (2012).

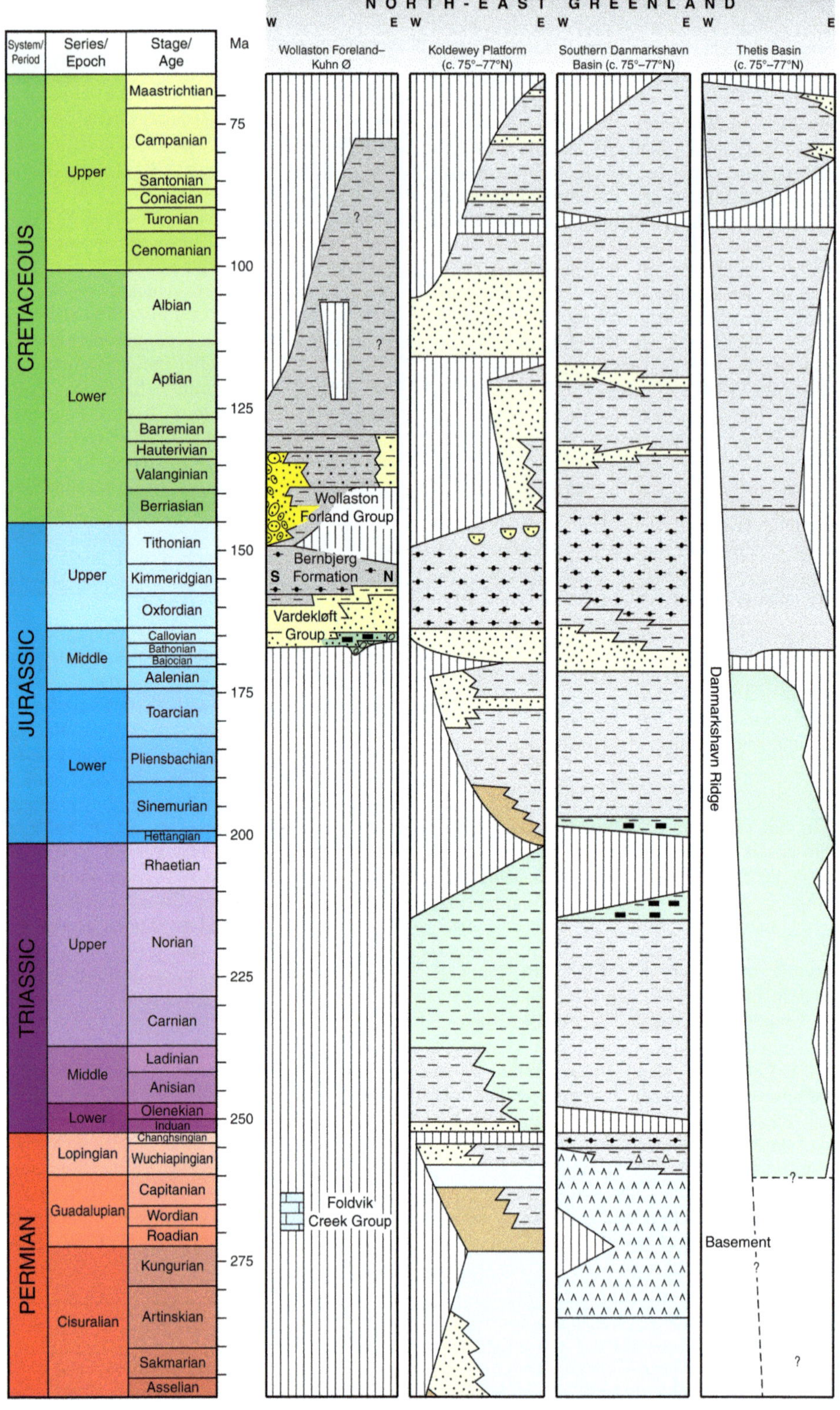

NORTH-EAST GREENLAND
W E W E W E W E
System/Period
Series/Epoch
Stage/Age
Ma
Wollaston Foreland–Kuhn Ø
Koldewey Platform (c. 75°–77°N)
Southern Danmarkshavn Basin (c. 75°–77°N)
Thetis Basin (c. 75°–77°N)
CRETACEOUS
Upper
Lower
Maastrichtian
Campanian
Santonian
Coniacian
Turonian
Cenomanian
Albian
Aptian
Barremian
Hauterivian
Valanginian
Berriasian
JURASSIC
Upper
Middle
Lower
Tithonian
Kimmeridgian
Oxfordian
Callovian
Bathonian
Bajocian
Aalenian
Toarcian
Pliensbachian
Sinemurian
Hettangian
TRIASSIC
Upper
Middle
Lower
Rhaetian
Norian
Carnian
Ladinian
Anisian
Olenekian
Induan
PERMIAN
Lopingian
Guadalupian
Cisuralian
Changhsingian
Wuchiapingian
Capitanian
Wordian
Roadian
Kungurian
Artinskian
Sakmarian
Asselian
75
100
125
150
175
200
225
250
275
Wollaston Forland Group
Bernbjerg Formation
S
N
Vardekløft Group
Foldvik Creek Group
Danmarkshavn Ridge
Basement
?

darker, non-siliceous shales at the base of the Sassendalen Group (Harland 1997; Worsley 2008).

The Sassendalen Group is up to 800 m thick across the western Barents Sea–Svalbard region, and characteristically comprises alternating shale, siltstone and sandstone units. These units are commonly arranged as upwards-coarsening cycles and include a number of organic-rich formations (Dalland *et al.* 1988; Nøttvedt *et al.* 1993*a*, *b*; Harland 1997; Dallmann 1999; Worsley 2008; NORLEX 2014) (Fig. 5).

The Triassic formations of the Kapp Toscana Group are characterized by interbedded mudstone, shale and sandstone with sporadic thin coal beds arranged in small- to large-scale upwards-coarsening units (Dalland *et al.* 1988; Nøttvedt *et al.* 1993*a*, *b*; Dallmann 1999; Worsley 2008; NORLEX 2014). Whereas this succession continues into the Middle Jurassic across much of the region, no rocks younger than Triassic are found on Bjørnøya. The preserved thickness of the Upper Triassic succession is variable: on Svalbard, Bjørnøya and in the Sørvestnaget Basin, the succession is several hundred metres thick; on the southern Loppa High and in the Hammerfest Basin, it exceeds 1000 m in thickness; and in the Tromsø Basin, the Upper Triassic is >2000 m thick (Figs 1 & 2). In the SW Barents Sea region, the Upper Triassic is locally punctuated by unconformities (Fig. 5).

Boundary relationships with underlying and overlying systems. In the western Barents Sea basins and on Svalbard, the Carboniferous–Permian boundary is conformable (Fig. 5). The Triassic–Jurassic boundary is marked by seemingly continuous facies from Rhaetian to Toarcian in Svalbard; elsewhere, a local unconformity may exist (e.g. the Hammerfest Basin).

Depositional environments. The extensive and stable Late Carboniferous–Early Permian carbonate platform was replaced in the Mid–Late Permian by shallow-marine clastic deposits that transgressed the low-lying platform (Harland 1997; Steel & Worsley 1984; Worsley 2008; NORLEX 2014). By the Early Triassic, the deposits of the Sassendalen Group on Svalbard reflect a muddy distal marine shelf with coarser material being sourced from a delta that prograded from the west, possibly sourced by local exposed areas to the west and NW (Dallmann 1999; Riis *et al.* 2008; NORLEX 2014). In the SW Barents Sea, siliciclastic coastal and deltaic sequences fringed a wide shelf. These environments persisted and expanded basinwards through the deposition of the Kapp Toscana Group, in the Mid- and Late Triassic. In the general Barents Sea region, the deltaic system was increasingly sourced from the Uralian Mountains in the east and the Fennoscandian Shield in the south to SE (Riis *et al.* 2008; Worsley 2008; Smelror *et al.* 2009; Henriksen *et al.* 2011). An overall shallowing of the Barents Sea region during the Mid–Late Triassic resulted from the input of large volumes of clastic sediment deposited by the westwards progradation of the deltaic system (Brekke *et al.* 2001; Riis *et al.* 2008; Glørstad-Clark *et al.* 2011; Høy & Lundschien 2011; Anell *et al.* 2014).

Central East Greenland margin

Distribution. Permian rocks are exposed onshore in the Wollaston Forland area and, together with the Triassic, form a combined succession that extends southwards towards, and including, the Jameson Land Basin (Figs 1 & 2). Permian–Triassic strata may also be present in deep fault blocks in the offshore Liverpool Land Basin (Larsen 1990; Hamann *et al.* 2005) (Fig. 12d), but have not been recorded in basins, either onshore or offshore, further south (i.e. in SE Greenland basins, such as the Kangerlussuaq and Ammassalik basins) (Fig. 1).

Structural setting. As with the NE Greenland margin, the Late Palaeozoic–Early Mesozoic rift basins trend approximately north–south along the central East Greenland margin. The NE and central East Greenland margins experienced a Late Palaeozoic rift phase with the formation of half-graben that possibly persisted into the Early Permian (Surlyk *et al.* 1986; Stemmerik *et al.* 1991; Surlyk 1991). Marginal rift blocks in East Greenland were uplifted and peneplained during latest Carboniferous–Early Permian time. The uplift was associated with rotational block-faulting that tilted strata by up to 15° (Surlyk 1990).

The Lower–Upper Permian boundary marks a transition in tectonic style, with the initiation of regional subsidence by thermal contraction in the Late Permian. A Kazanian (Wordian–Capitanian) marine connection was established between the Barents Sea, the Jameson Land Basin and the European Zechstein Basin (Stemmerik 2000) (Figs 6–8).

Fig. 6. Regional stratigraphic correlation panel showing the generalized Permian–Cretaceous succession preserved onshore in NE Greenland (Wollaston Forland–Kuhn Ø), and inferred (paler colourfill) on the basis of seismic interpretation for the adjacent offshore area (Koldewey Platform; southern Danmarkshavn Basin; Thetis Basin). See Figure 5 for the key to colours and symbols. Correlation based on the compilation of Stoker *et al.* (2014). Timescale based on Gradstein *et al.* (2012).

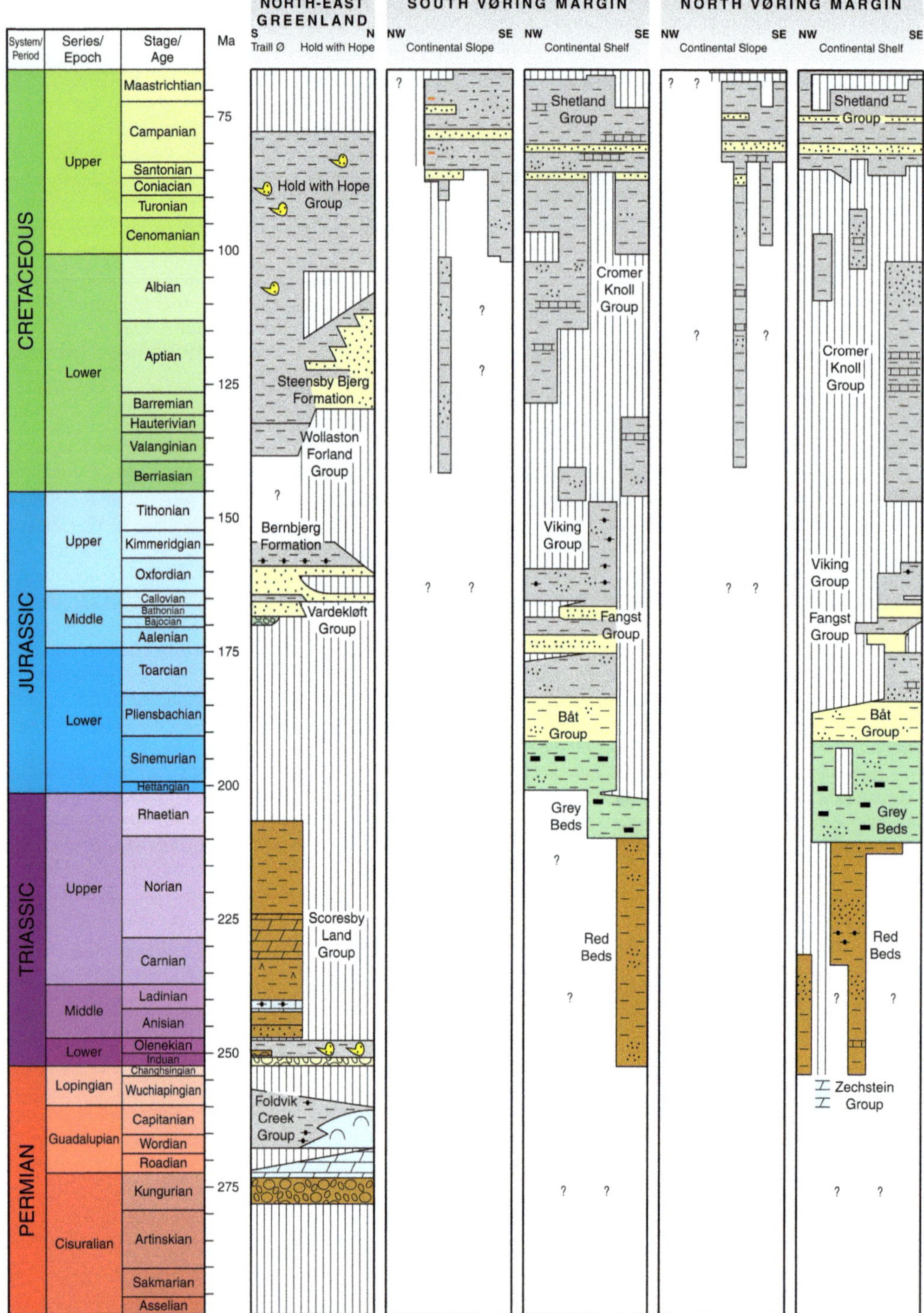

Fig. 7. Regional stratigraphic correlation panel showing the generalized Permian–Cretaceous succession preserved in the Central East Greenland–Vøring region. See Figure 5 for the key to colours and symbols. Correlation based on the compilation of Stoker *et al.* (2014). Timescale based on Gradstein *et al.* (2012).

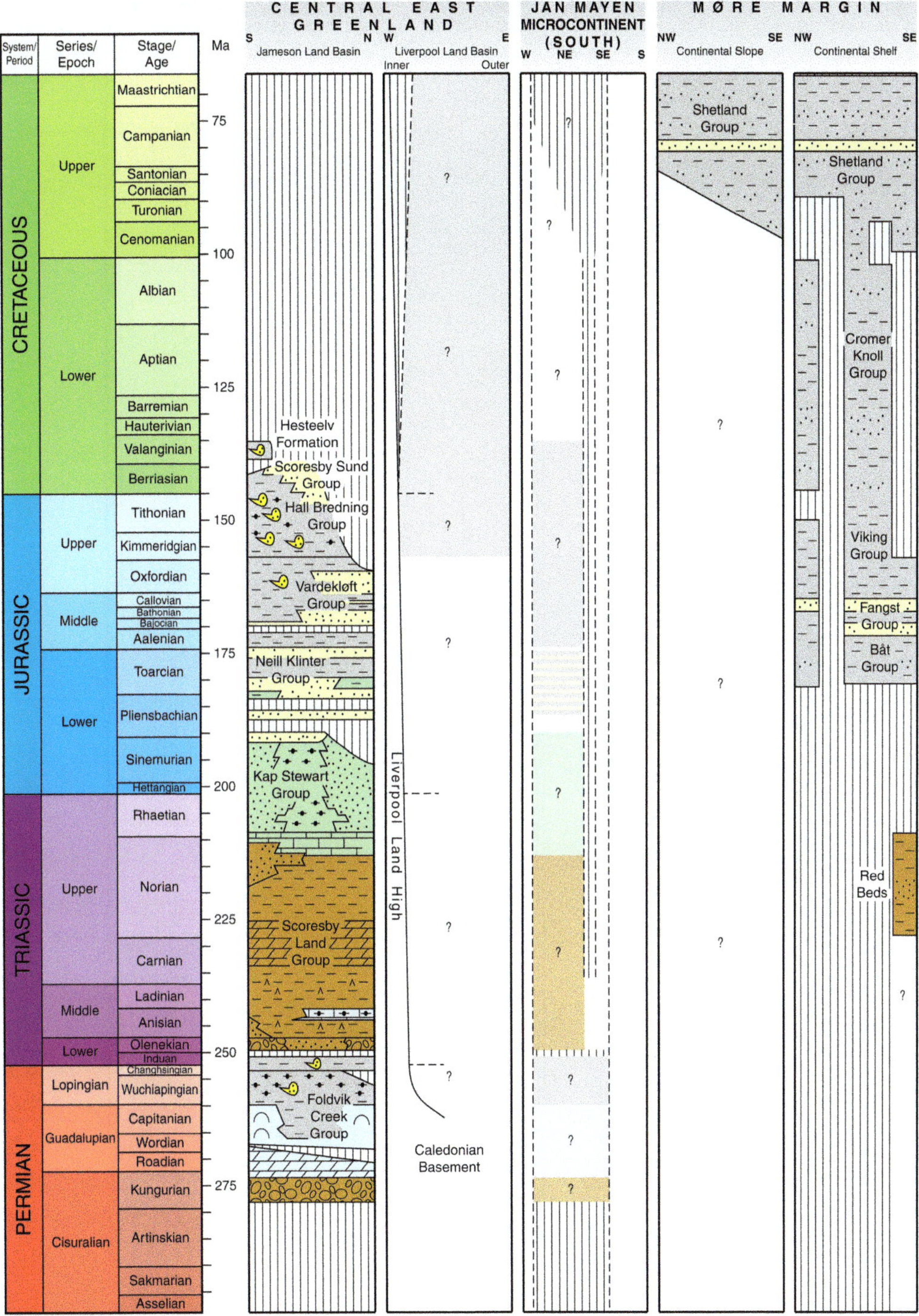

Fig. 8. Regional stratigraphic correlation panel showing the generalized Permian–Cretaceous succession preserved in the Central East Greenland (Jameson Land Basin)–Møre region, and the inferred (paler colourfill) stratigraphic range based on seismic interpretation for the Liverpool Land Basin (offshore Central East Greenland), and the southern Jan Mayen microcontinent. See Figure 5 for the key to colours and symbols. Correlation based on the compilation of Stoker *et al.* (2014). Timescale based on Gradstein *et al.* (2012).

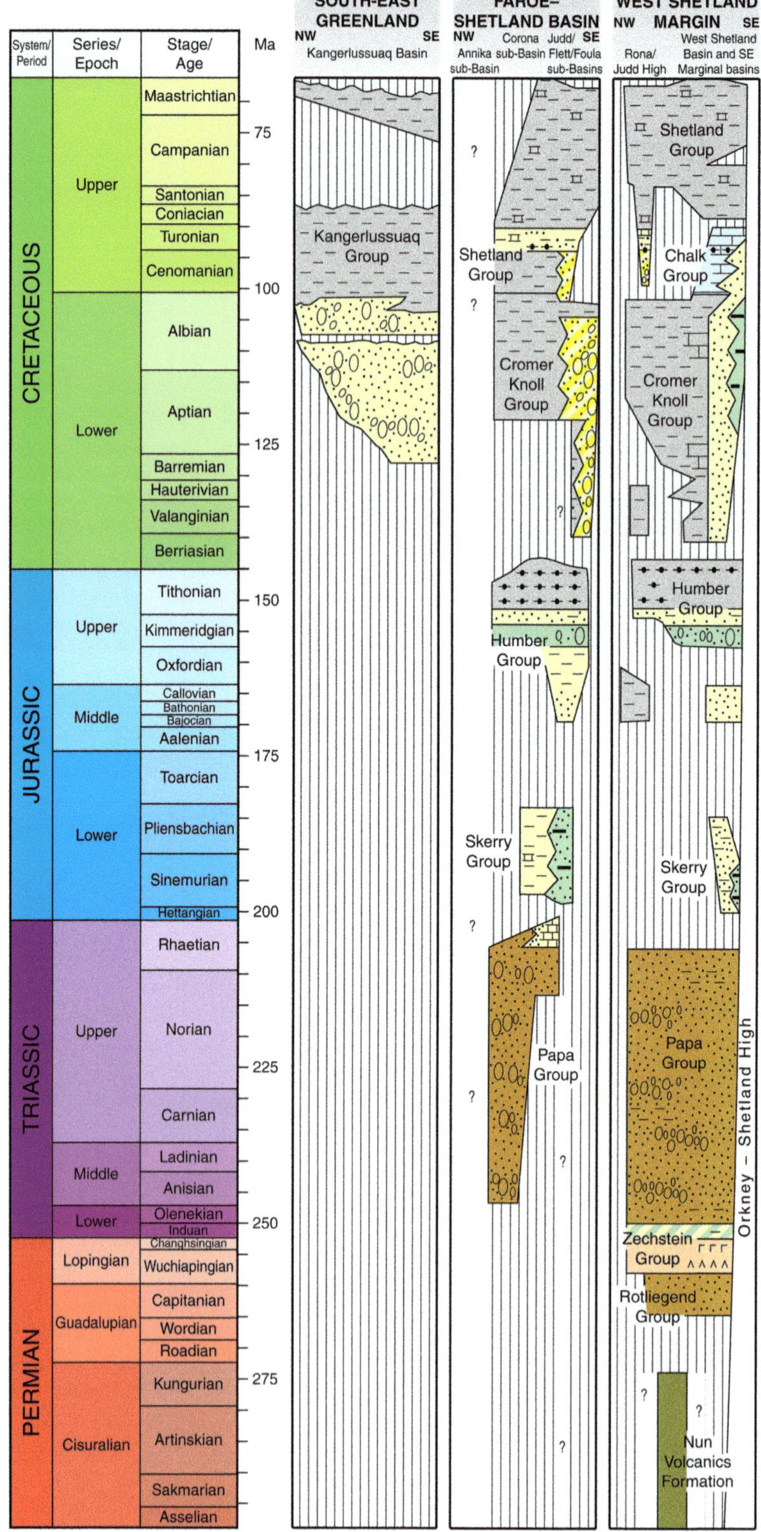

System/ Period
Series/ Epoch
Stage/ Age
Ma
CRETACEOUS
Upper
Lower
Maastrichtian
Campanian
Santonian
Coniacian
Turonian
Cenomanian
Albian
Aptian
Barremian
Hauterivian
Valanginian
Berriasian
JURASSIC
Upper
Middle
Lower
Tithonian
Kimmeridgian
Oxfordian
Callovian
Bathonian
Bajocian
Aalenian
Toarcian
Pliensbachian
Sinemurian
Hettangian
TRIASSIC
Upper
Middle
Lower
Rhaetian
Norian
Carnian
Ladinian
Anisian
Olenekian
Induan
PERMIAN
Lopingian
Guadalupian
Cisuralian
Changhsingian
Wuchiapingian
Capitanian
Wordian
Roadian
Kungurian
Artinskian
Sakmarian
Asselian
75
100
125
150
175
200
225
250
275
SOUTH-EAST GREENLAND
NW SE
Kangerlussuaq Basin
Kangerlussuaq Group
FAROE–SHETLAND BASIN
NW SE
Corona Judd/
Annika sub-Basin Flett/Foula
sub-Basin sub-Basins
Shetland Group
Cromer Knoll Group
Humber Group
Skerry Group
Papa Group
WEST SHETLAND MARGIN
NW SE
Rona/ Judd High
West Shetland Basin and SE Marginal basins
Shetland Group
Chalk Group
Cromer Knoll Group
Humber Group
Skerry Group
Papa Group
Zechstein Group
Rotliegend Group
Nun Volcanics Formation
Orkney – Shetland High

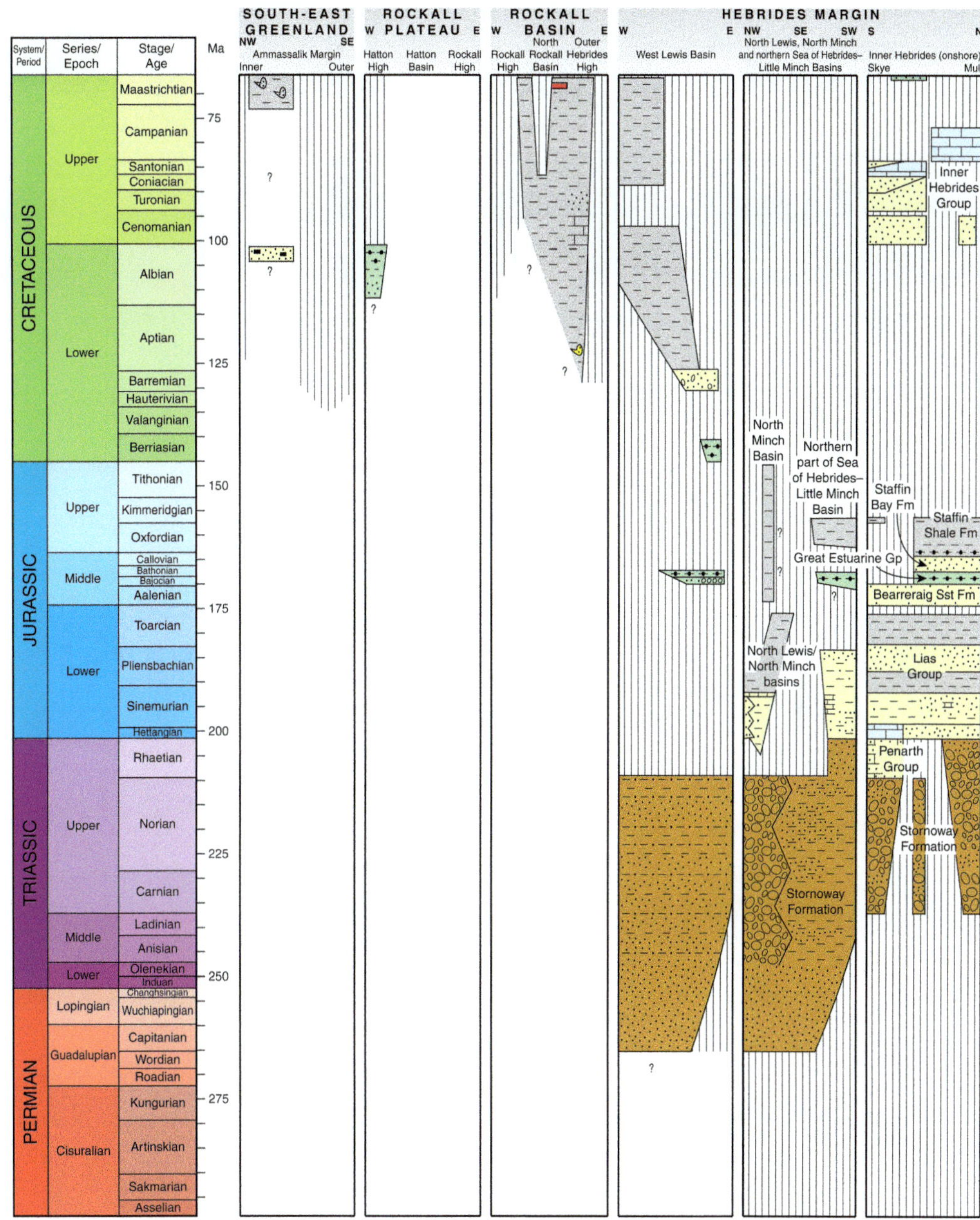

Fig. 10. Regional stratigraphic correlation panel showing the generalized Permian–Cretaceous succession preserved in the SE Greenland–northern Rockall Plateau–northern Rockall Basin–Hebrides region. See Figure 5 for the key to colours and symbols. The sequence in the Ammassalik Basin is based partly on seismic interpretation (Gerlings *et al.*, this volume, in review). Correlation based on the compilation of Stoker *et al.* (2014). Timescale based on Gradstein *et al.* (2012).

Fig. 9. Regional stratigraphic correlation panel showing the generalized Permian–Cretaceous succession preserved in the SE Greenland–Faroe–Shetland region. See Figure 5 for the key to colours and symbols. Correlation based on the compilation of Stoker *et al.* (2014). Timescale based on Gradstein *et al.* (2012).

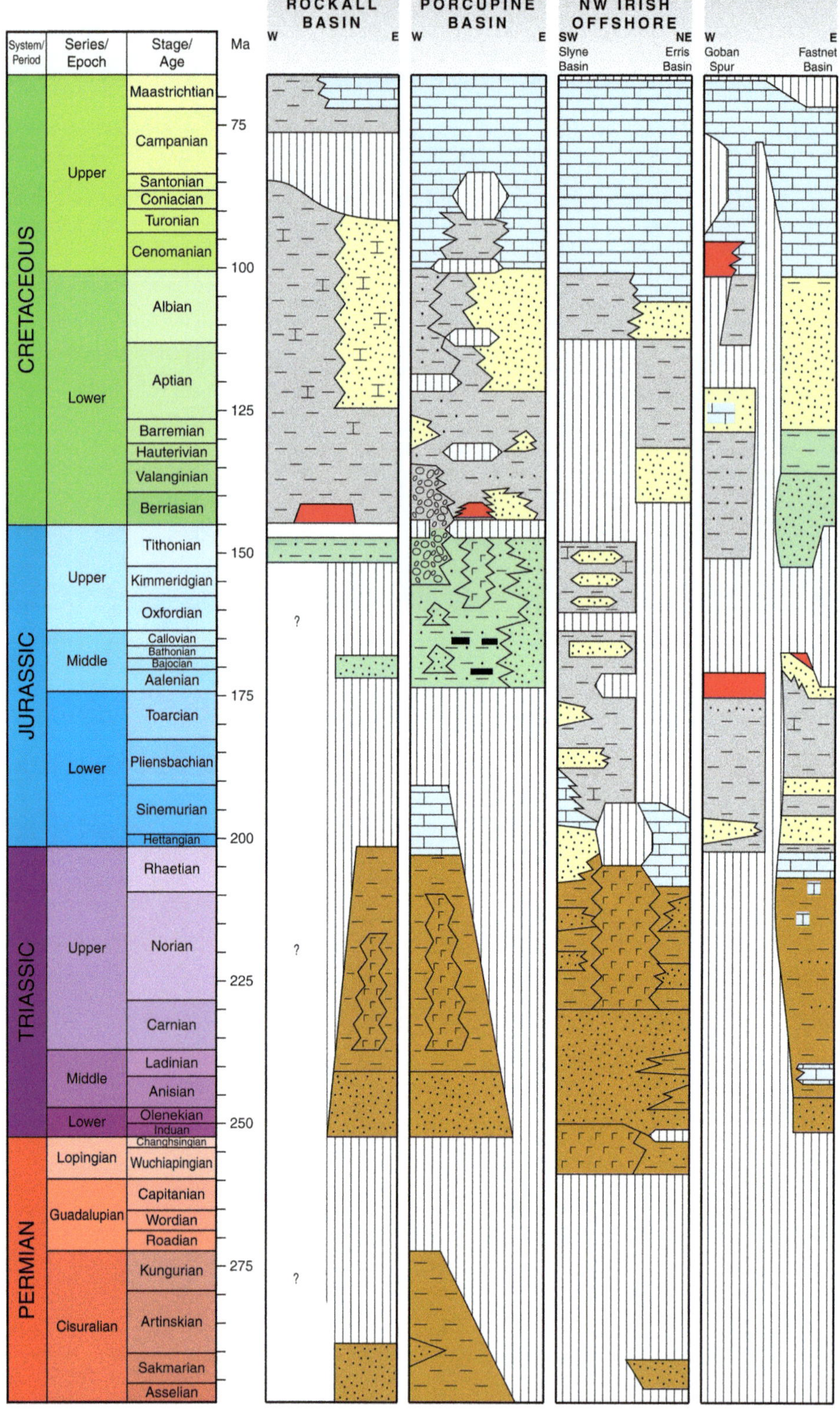
ROCKALL BASIN
W
E
PORCUPINE BASIN
W
E
NW IRISH OFFSHORE
SW
Slyne Basin
NE
Erris Basin
W
Goban Spur
E
Fastnet Basin
System/Period
Series/Epoch
Stage/Age
Ma
CRETACEOUS
Upper
Lower
Maastrichtian
Campanian
Santonian
Coniacian
Turonian
Cenomanian
Albian
Aptian
Barremian
Hauterivian
Valanginian
Berriasian
JURASSIC
Upper
Middle
Lower
Tithonian
Kimmeridgian
Oxfordian
Callovian
Bathonian
Bajocian
Aalenian
Toarcian
Pliensbachian
Sinemurian
Hettangian
TRIASSIC
Upper
Middle
Lower
Rhaetian
Norian
Carnian
Ladinian
Anisian
Olenekian
Induan
PERMIAN
Lopingian
Guadalupian
Cisuralian
Changhsingian
Wuchiapingian
Capitanian
Wordian
Roadian
Kungurian
Artinskian
Sakmarian
Asselian
75
100
125
150
175
200
225
250
275
?

Minor latest Permian–earliest Triassic rifting and rotational block faulting was recorded in an overall east–west extensional regime, and several half-graben were formed in the Early Triassic in East Greenland (Seidler *et al.* 2004). A marine seaway between East Greenland and Norway was connected to the seas covering Svalbard, the Sverdrup Basin and NW Siberia in the Early Triassic (Mørk *et al.* 1992; Seidler *et al.* 2004; Müller *et al.* 2005; Bjerager *et al.* 2006).

Early Triassic rotational block faulting resulted in the formation of angular unconformities in East Greenland (Figs 7 & 8). The main rift event occurred in the late Scythian (Induan–Olenikian transition) along north–south-trending basin-margin faults, with minor rifting also in the Norian. NW–SE-trending transform fault zones probably controlled local differential uplift and subsidence patterns in the basins (Clemmensen 1980*a*). The latest Triassic (Rhaetian)–Mid-Jurassic (early Bajocian) was characterized by thermal subsidence in East Greenland.

Stratigraphy. The stratigraphy of the Permian–Triassic succession is herein described largely from the Wollaston Forland to Jameson Land area (Figs 1 & 2). An isolated Middle–Upper Permian Røde Ø Conglomerate succession, approximately 1 km thick, also occurs in the Rødefjord area, west of Milne Land, in central East Greenland (Stemmerik & Piasecki 2004); in contrast, the stratigraphy of the Liverpool Land Basin remains conjectural (Figs 2, 8 & 12d).

In the combined Jameson Land–Wollaston Forland basins, the Permian Foldvik Creek Group is up to about 300 m thick, and comprises a basal conglomerate that is succeeded by carbonates and evaporates, in turn overlain by carbonate build-ups and organic shales that are capped by siliciclastic deposits (Stemmerik 2001) (Figs 2, 6–8). The top of the Permian consists of basinal bioturbated mudstones and sandy turbidites in the Jameson Land Basin and southern Traill Ø (Kreiner-Møller & Stemmerik 2001). The Upper Permian succession includes a source-rock-prone interval (Christiansen *et al.* 1993).

The Triassic rocks are largely assigned to the Scoresby Land Group. The Lower Triassic sequence is up to approximately 1 km thick in the Traill Ø area, in the deepest part of the combined Jameson Land–Wollaston Forland Basin, and consists of shales, sandstones, conglomerates and minor carbonates (Seidler *et al.* 2004; Bjerager *et al.* 2006) (Figs 7 & 8). These rocks are unconformably overlain by a series of Lower–Middle Triassic basal conglomerates and sandstones, up to 600 m thick, which grade upwards into fine-grained sandstones, mudstones, evaporites and carbonates of Mid–Late Triassic age (Clemmensen 1980*a*). The uppermost Triassic (Rhaetian) rocks in the Traill Ø–Hold with Hope area are missing, whereas in the Jameson Land Basin the Scoresby Land Group is overlain by organic-rich mudstones and sandstone of the Rhaetian–Sinemurian (Lower Jurassic) Kap Stewart Group (see below).

Boundary relationships with underlying and overlying systems. The regional Early–Mid-Permian erosional peneplain of tilted Carboniferous–Devonian strata or Caledonian basement in central East Greenland is overlain by the Foldvik Creek Group (Figs 6–8). A prominent erosional unconformity at the Permian–Triassic boundary is recorded in basin-marginal positions. The associated hiatus expands towards the north with evidence of subaerial exposure, and much of the Upper Permian has been removed by erosion prior to deposition of the Triassic Scoresby Land Group (Surlyk *et al.* 1986) (Fig. 7). In the deeper basinal areas in the south, a complete marine Permian–Triassic boundary interval is recorded in a marine mudstone succession (Stemmerik *et al.* 2001) (Fig. 8). The Triassic succession is conformable with the Jurassic in the Jameson Land Basin, whereas the Upper Triassic is erosionally overlain by Jurassic and/or Cretaceous further north.

Depositional environment. The Foldvik Creek Group shows an overall development from fluvial into shallow-marine and hypersaline carbonate deposition. The successive transgression resulted in deposition of fully marine carbonate build-ups along shelf margins contemporaneously with the deposition of anoxic shales in basinal areas (Stemmerik *et al.* 2001) (Fig. 8). An increased siliciclastic influx is evident in the latest Permian that continued into the Triassic, with the overall upwards-shallowing marine shales and sandy turbidites at the base of the Scoresby Land Group succeeded by shoreface and coastal-plain sandstones and mudstones (Seidler *et al.* 2004; Bjerager *et al.* 2006) (Figs 7 & 8). A fluvial conglomerate was deposited locally in the Traill Ø area as the result of local block tilting.

Marine deposition terminated in the Early Triassic and was followed by the deposition of rift-related

Fig. 11. Regional stratigraphic correlation panel showing the generalized Permian–Cretaceous succession preserved in the southern Rockall Basin–Porcupine Basin–NW Irish region. See Figure 5 for the key to colours and symbols. Correlation based on the compilation of Stoker *et al.* (2014). Timescale based on Gradstein *et al.* (2012).

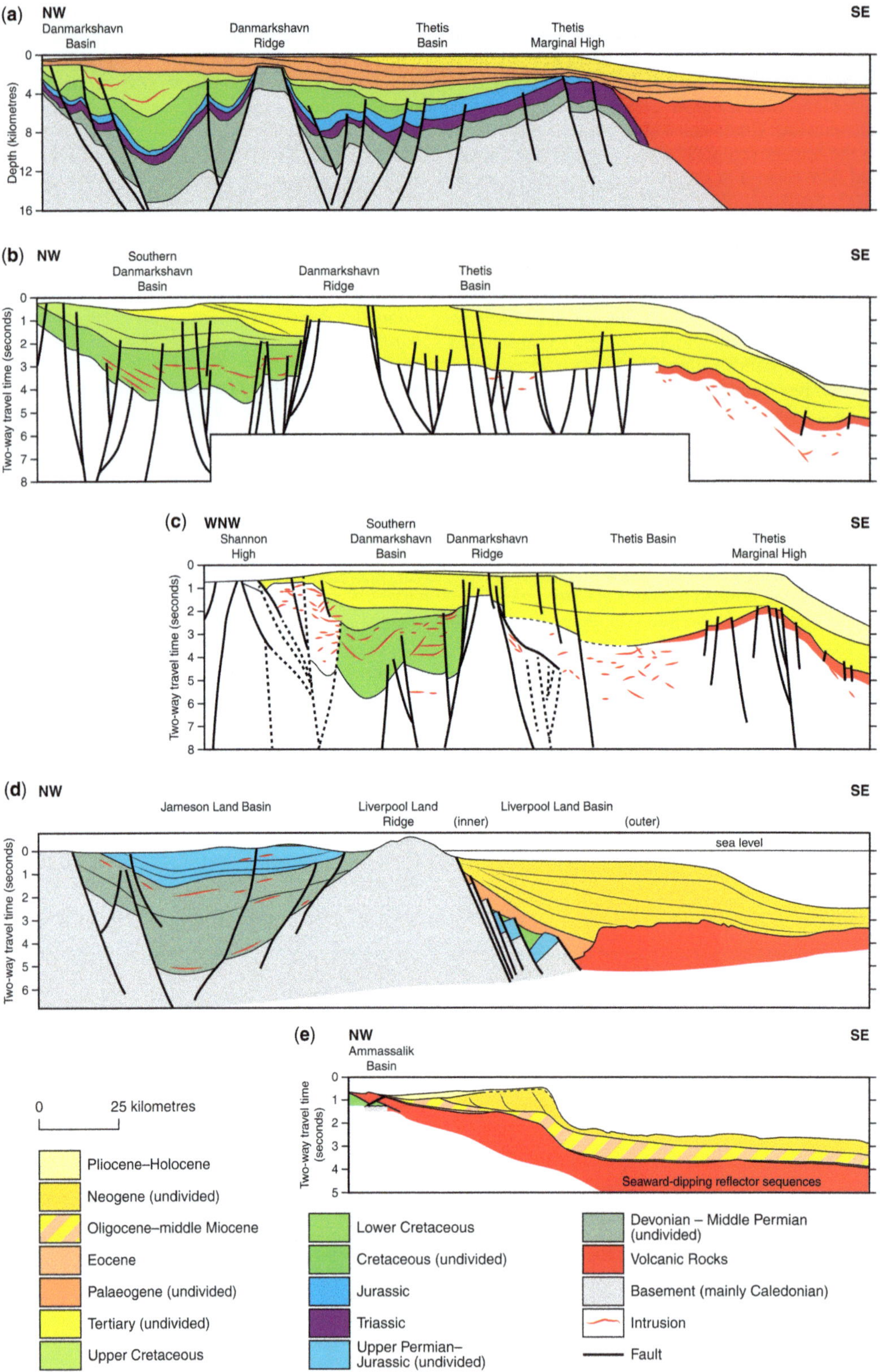
(a) NW
Danmarkshavn Basin
Danmarkshavn Ridge
Thetis Basin
Thetis Marginal High
SE
Depth (kilometres)
(b) NW
Southern Danmarkshavn Basin
Danmarkshavn Ridge
Thetis Basin
SE
Two-way travel time (seconds)
(c) WNW
Shannon High
Southern Danmarkshavn Basin
Danmarkshavn Ridge
Thetis Basin
Thetis Marginal High
SE
Two-way travel time (seconds)
(d) NW
Jameson Land Basin
Liverpool Land Ridge
(inner)
Liverpool Land Basin
(outer)
sea level
SE
Two-way travel time (seconds)
(e) NW
Ammassalik Basin
SE
Two-way travel time (seconds)
Seaward-dipping reflector sequences
0 25 kilometres
Pliocene–Holocene
Neogene (undivided)
Oligocene–middle Miocene
Eocene
Palaeogene (undivided)
Tertiary (undivided)
Upper Cretaceous
Lower Cretaceous
Cretaceous (undivided)
Jurassic
Triassic
Upper Permian–Jurassic (undivided)
Devonian – Middle Permian (undivided)
Volcanic Rocks
Basement (mainly Caledonian)
Intrusion
Fault

alluvial conglomerates and sandstones, which were sourced from the western (Post-Devonian Main Fault area) and the eastern (Liverpool Land) basin margins (Clemmensen 1980*a*). The succession shows a gradual upwards transition into fine-grained floodplain, freshwater, and saline lacustrine and aeolian deposits of Mid–Late Triassic age, although a thin Anisian unit of mudstones and carbonates reflect a shallow-marine interlude (Clemmensen 1980*a*).

Møre–mid-Norway margin

Distribution. Rocks of Permian–Triassic age have been proved on the inner part of the Norwegian margin (Fig. 2), extending from the northern North Sea to the Lofoten archipelago, with the bulk of the rocks sampled in the Møre and Vøring regions being of Triassic age (Figs 7 & 8). However, by analogy with the East Greenland conjugate margin (see below), it has been assumed that the platforms and terraces that underlie the Norwegian margin comprise sedimentary rocks of Permian age atop crystalline basement (Blystad *et al.* 1995; Doré *et al.* 1999; Brekke 2000).

Structural setting. In the Møre and Mid-Norwegian areas, the Late Carboniferous–Early Permian was an active tectonic period, and might have generated a block-faulted terrain beneath the Trøndelag Platform and the Halten Terrace, on the inner Vøring margin (Blystad *et al.* 1995). Further rifting, faulting and fault-block rotation was instigated at the Permian–Triassic boundary and continued during the Early Triassic, whereas a reduction in fault activity and tectonic subsidence characterized the Mid- and Late Triassic (Müller *et al.* 2005).

Stratigraphy. On the Vøring margin, a Permian succession – including carbonates of the Zechstein Group – up to about 250 m thick has been proved in commercial wells (e.g. Nordland Ridge/Rødhoy High) (Figs 1, 2 & 7), whereas no wells on the Møre margin have penetrated Permian strata (Fig. 8). The Triassic has been drilled on both the Møre and Vøring margins, with exploration wells commonly recovering breccia, sandstone and claystone: however, the distribution and thickness of these rocks remains uncertain. The Triassic succession is often referred to as the 'red' and 'grey' beds, which mark an upwards transition from arid to humid conditions in the Rhaetian (Figs 7 & 8). On the Trøndelag Platform, the red beds are >2.6 km thick, whereas the grey beds reach a thickness of up to 2.5 km.

Boundary relationships with underlying and overlying systems. Brekke (2000) reports a regional late Early Permian unconformity in the platform areas – dated by comparison with East Greenland – but the age of the underlying rocks remains ambiguous (NORLEX 2014; Norwegian Petroleum Directorate 2014). On the Vøring margin, the Triassic–Jurassic boundary appears to be largely conformable, whereas a hiatus is noted on the Møre margin. In the deeper parts of the Møre and Vøring basins, the boundary relationship remain unclear (Figs 13b, c).

Depositional environments. By analogy with the East Greenland margin, it has been suggested that alluvial and fluvial facies existed along the Møre–mid-Norway margin in the early Permian (Brekke *et al.* 2001) (Figs 7 & 8). A shallow carbonate platform became established in the East Greenland region in the Late Permian, and equivalent carbonate deposits (dolomite) of the Zechstein Group are recorded in wells in the NW part of the Trøndelag Platform (inner Vøring margin).

The Triassic rocks on the Møre–mid-Norway margin are dominated by continental fluvial and alluvial facies (Brekke *et al.* 2001; Müller *et al.* 2005). The Early Triassic was characterized by marginal-marine and terrestrial environments, with the latter becoming predominant during the Mid-Triassic. In the Late Triassic, evaporites and organic-rich shales formed in arid, isolated, marine sub-basins, which were replaced by fluvio-lacustrine depositional environments and, ultimately, conditions that were more humid in the latest Triassic – the change from red beds to grey beds.

Faroe–Shetland–northern Rockall–Hebrides margin

Distribution. Permian–Triassic rocks are commonly preserved as unconformity bounded deposits within Late Palaeozoic–Mesozoic basins that underlie the Hebrides and West Shetland shelf areas (Fyfe *et al.* 1993; Stoker *et al.* 1993; Quinn &

Fig. 12. Interpreted geoseismic profiles from the East Greenland margin showing the stratigraphic architecture of the Upper Palaeozoic–Mesozoic succession where proven (e.g. the Jameson Land and Ammassalik basins) and where inferred (e.g. the Koldewey Platform and Danmarkshavn, Thetis and Liverpool Land basins). Profile (**a**) based on Dinkelman *et al.* (2010); profiles (**b**) and (**c**) based on Tsikalas *et al.* 2005; profile (**d**) based on Larsen (1990) and Hamann *et al.* (2005); and profile (**e**) based on Larsen *et al.* (1999). The locations of profiles are given in Figure 1.

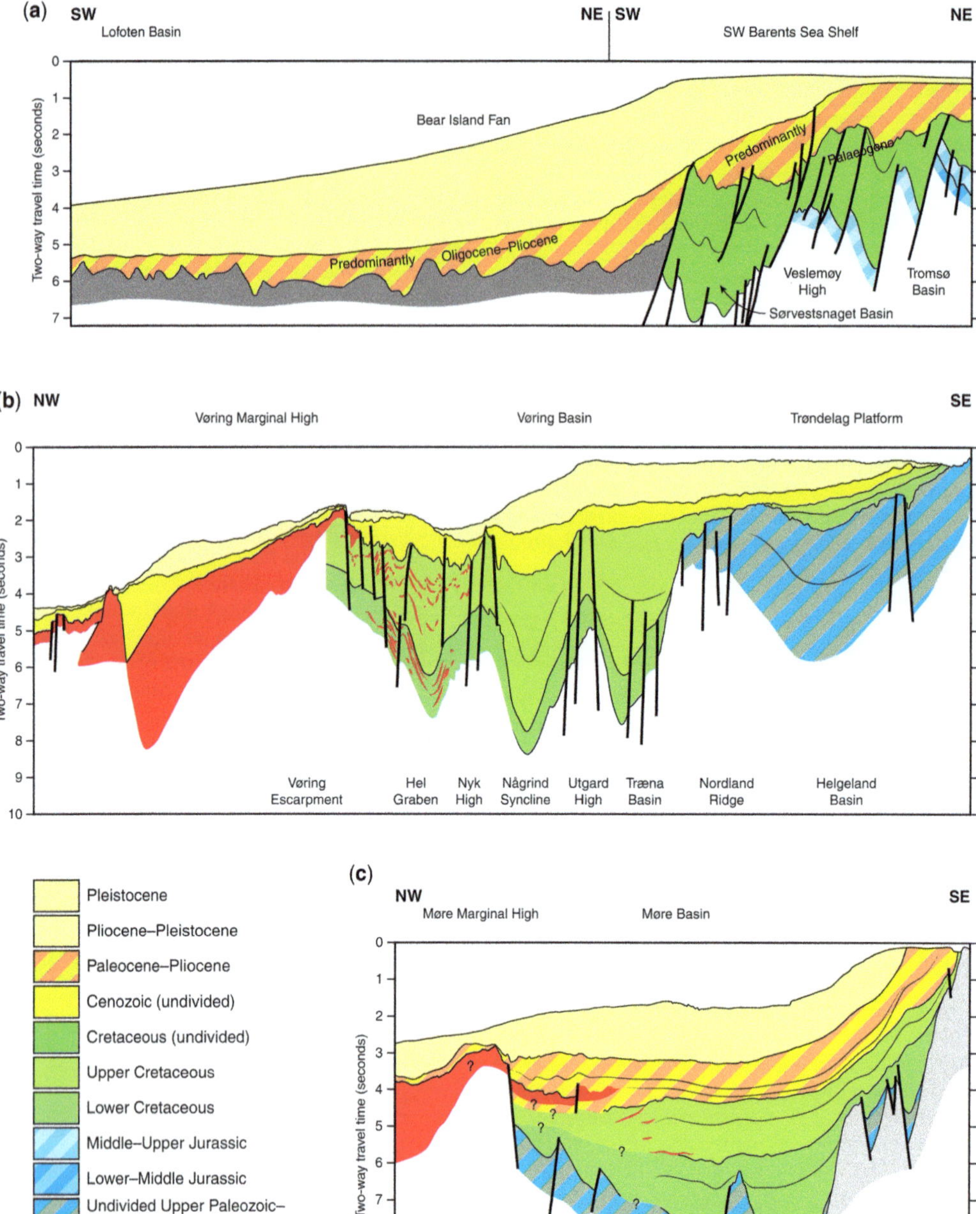

Fig. 13. Interpreted geoseismic profiles from the Mid Norway–SW Barents Sea margin showing the stratigraphic architecture of the Upper Palaeozoic–Mesozoic succession in basins that underlie the SW Barents Sea and mid-Norwegian margin, including the Møre and Vøring basins. Profile (**a**) based on Ryseth *et al.* (2003) and Hjelstuen *et al.* (2007); profiles (**b**) and (**c**) based on Faleide *et al.* (2010) and (**c**) Brekke (2000). The locations of profiles are given in Figure 1.

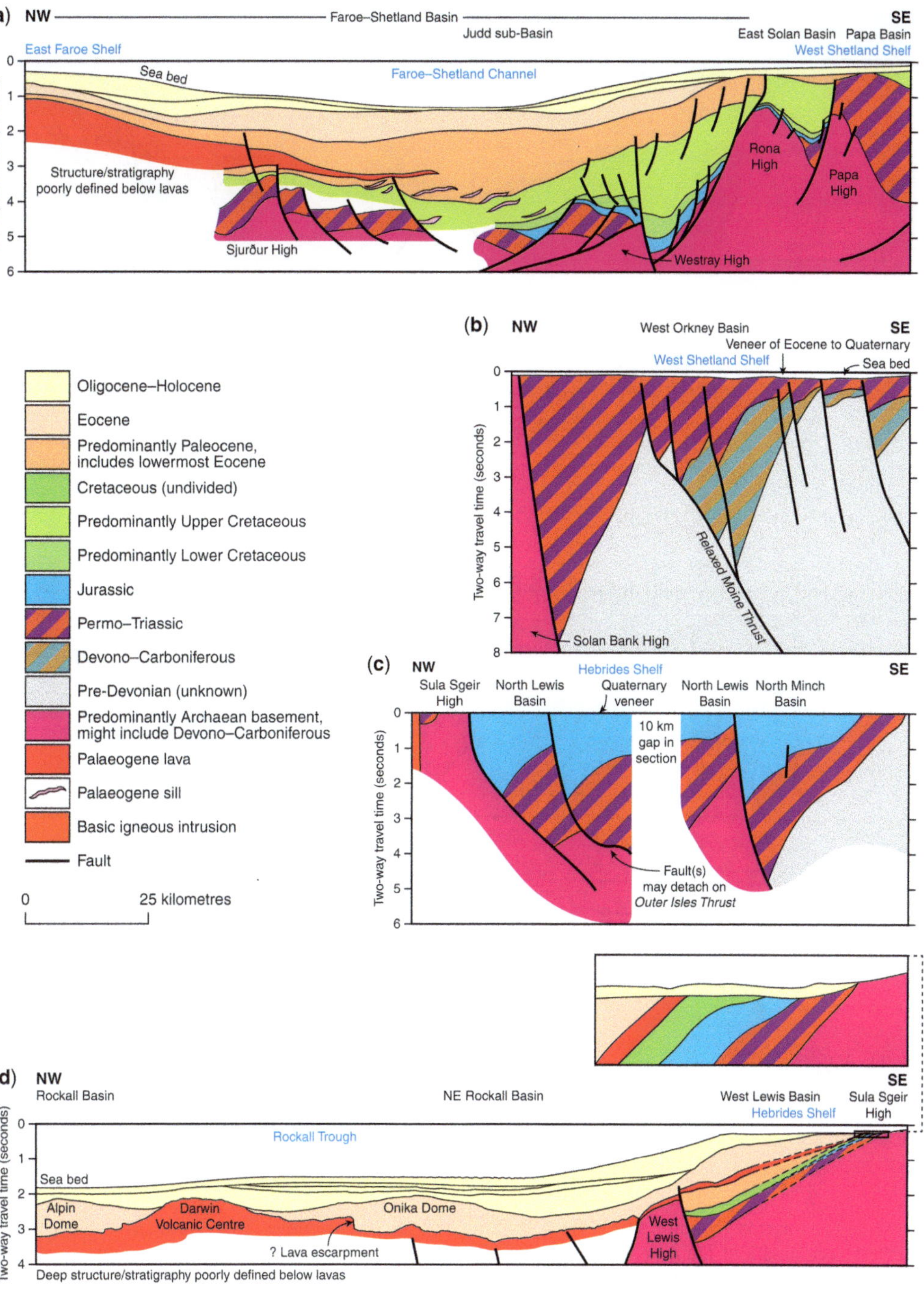

Fig. 14. Interpreted geoseismic profiles from the Faroe–Shetland–northern Rockall–Hebrides margin showing the stratigraphic architecture of the Upper Palaeozoic–Mesozoic succession in basins that underlie the West Shetland and Hebrides shelves, and in the adjacent deep-water Faroe–Shetland and Rockall basins. Profile (**a**) based on Lamers & Carmichael (1999); profile (**b**) based on Earle *et al.* (1989) and Ritchie *et al.* (2011*a*, *b*); profile (**c**) compiled from British Geological Survey (1989, 1990) mapping; and profile (**d**) based on Ritchie *et al.* (2013), with inset showing detail of West Lewis Basin based on Hitchen & Stoker (1993). The locations of profiles are given in Figure 1.

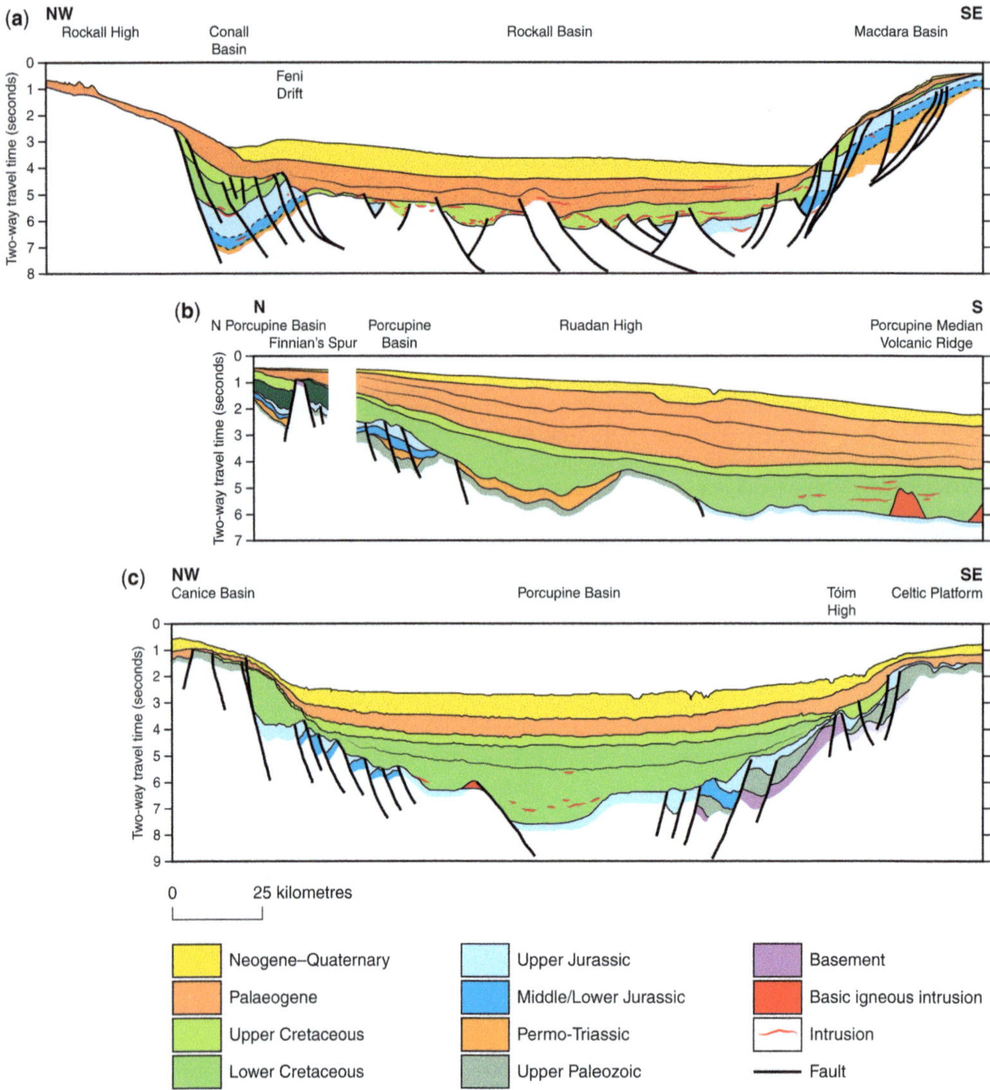

Fig. 15. Interpreted geoseismic profiles from the southern Rockall–Porcupine margin showing the stratigraphic architecture of the Upper Palaeozoic–Mesozoic succession in the Rockall, Porcupine and adjacent marginal basins. Profile (**a**) based on Naylor *et al.* (1999); profiles (**b**) and (**c**) based on Naylor *et al.* (2002). The locations of profiles are given in Figure 1.

Ziska 2011; Johnson & Quinn 2013) (Fig. 2). The thickest accumulations are found in the West Orkney, North Lewis, North Minch, Papa, East Solan and West Shetland basins (Fig. 14). Further west, the distribution is less well constrained. In the Faroe–Shetland Basin, Triassic rocks have been proved in the Judd Sub-basin, and on the Corona and Westray intrabasinal highs. In contrast, no rocks of Permian–Triassic age have been proven to occur in the northern Rockall and Hatton basins.

Structural setting. The Permian–Triassic rocks occupy a series of NE–SW-trending graben and half-graben (Stoker *et al.* 1993; Hitchen *et al.* 1995). The geometry of the preserved successions suggests that the rocks were deposited through a combination of local initial palaeotopographical infill, half-graben synsedimentary fault-controlled geometries and uniformly thick sedimentary fill with no obvious syn-sedimentary control (Kirton & Hitchen 1987; Štolfová & Shannon 2009). In the Hebridean region, the fault-controlled basins (e.g. West Orkney Basin) were largely controlled by listric normal faults downthrowing to the SE: west of Shetland (e.g. West Shetland Basin), however, the faults downthrow to the NW. In some basins (e.g. North Rona Basin), the infill is parallel-bedded with no evidence of thickening towards the present faulted margin. This suggests that this faulting was initiated towards the end of the Permian–Triassic, preserving part of a sequence that originally might have had a greater regional extent, but has been subsequently eroded.

Stratigraphy. Subdivision of the Permian–Triassic succession is poorly constrained over much of the Hebrides–West Shetland shelf areas, where it comprises a largely unfossiliferous clastic assemblage of conglomeratic sandstone, siltstone and mudstone, with only sporadic carbonate, evaporitic and volcanic rocks (Fyfe *et al.* 1993; Stoker *et al.* 1993; Quinn & Ziska 2011; Johnson & Quinn 2013). On the Hebrides Shelf, clastic rocks predominate and are mainly assigned to the Stornoway Formation of Late Permian–Triassic age, although uppermost Triassic (Rhaetian) sandy limestone, preserved locally in the Inner Hebrides, is correlated with the Penarth Group (Storetvedt & Steel 1977; Emeleus & Bell 2005) (Fig. 10). In the North Minch and North Lewis basins, the Stornoway Formation is up to 2 km thick (Fig. 14c). In the West Lewis Basin, an unassigned clastic succession exceeds 0.5 km thick (Fig. 14d).

On the West Shetland Shelf, the Permian–Triassic succession is up to 1 km thick in the East Solan and West Shetland basins, up to 2 km thick in the Papa Basin (Fig. 14a), and up to 7 km thick in the West Orkney Basin (Fig. 14b). In general, a tripartite subdivision has been established for this area: (1) an upper clastic succession assigned to the Triassic Papa Group, which predominates across the region (Booth *et al.* 1993; Ritchie *et al.* 1996; Quinn & Ziska 2011); (2) a middle evaporite succession that has been assigned to the West Orkney Evaporite Formation of the Late Permian Zechstein Group (Ritchie *et al.* 1996); and (3) a lower sequence of sandstone and conglomerate of the Solan Bank Formation, originally assigned by Ritchie *et al.* (1996) to the Zechstein Group, although reassigned by Glennie (2002) to the Upper Rotliegend 2 unit, comparable with the Late Permian sequences in the North Sea Basin (Glennie *et al.* 2003) (Fig. 9).

In the Faroe–Shetland Basin, conglomerate, sandstone and minor mudstone occur in the Judd Sub-basin, and on the Corona and Westray intrabasinal highs, with sporadic limestone beds in the upper part of the succession (Quinn & Ziska 2011) (Fig. 14a). These rocks have been assigned to the Middle–Upper Triassic Foula Formation of the Papa Group (Ritchie *et al.* 1996) (Fig. 9).

By way of contrast, trachyandesitic lavas assigned to the Margarita Volcanics and Nun Volcanics formations occur below the Zechstein Group on the East Shetland High and in the Papa Basin, respectively (Quinn & Ziska 2011). These lavas have been assigned an Early Permian age in common with the age range of most of the Rotliegend volcanics of Scotland (Hitchen *et al.* 1995; Ritchie *et al.* 1996; Glennie 2002) (Fig. 9).

Boundary relationships with underlying and overlying systems. The base of the Permian–Triassic succession is everywhere marked by an unconformity. West of Shetland, the youngest rocks below the basal unconformity are of late Early–early Late Carboniferous in age; in the Hebridean region, the underlying rocks are predominantly pre-Devonian basement. Despite a localized conformity between Triassic and Jurassic strata in the Inner Hebrides, an erosional hiatus separates these systems over the rest of the area.

Depositional environment. The rocks of the Upper Rotliegend 2 unit, Stornoway Group and Papa Group were deposited in a semi-arid, fluvial–lacustrine and/or alluvial braidplain setting (Stoker *et al.* 1993; Quinn & Ziska 2011; Johnson & Quinn 2013). The carbonates and evaporites of the Zechstein Group represent brackish–shallow-marine incursions during the Late Permian, whereas the shallow-marine limestones of the Penarth Group, and in the upper part of the Foula Formation, indicate a Late Triassic marine incursion in both the Inner Hebrides and West Shetland regions.

Southern Rockall–Porcupine margin

Distribution. Permian–Triassic strata are preserved throughout most of the Slyne, Erris and Donegal basins, the eastern part of the southern Rockall Basin, and the northern part of the Porcupine Basin (Naylor & Shannon 2011) (Figs 1, 2 & 15). They are likely to be present in the small half-graben basins lying on the eastern and western shoulders of the southern Rockall Basin, and on the eastern margin of the Hatton Basin (Naylor *et al.* 1999; Štolfová & Shannon 2009). Further south, Triassic and, probably, localized Permian strata occur in the Celtic and Fastnet basins (Robinson *et al.* 1981), and are likely to be present in the Goban Spur (Cook 1987).

Structural setting. The Permian–Triassic rocks throughout most of the region are preserved in NE–SW-orientated half-graben controlled by reactivated Caledonian structures (Shannon 1991) (Fig. 2). The successions in the Slyne and Erris basins occur in narrow, interconnected depocentres that change fault polarity along strike, resulting in the formation of discrete half-graben basins. Croker & Shannon (1987) suggested that Permian–Triassic sediments are preserved in small rift basins along the SW and, possibly, the SE margins of the Porcupine Basin. Naylor *et al.* (1999) interpreted a Permian–Triassic sequence within the group of elongate 'perched' Mesozoic basins along the western margin of the Porcupine High (Fig. 2), and Štolfová & Shannon (2009) interpreted these strata, generally thick and parallel bedded, to have been deposited over a broader region than the current fault-preserved basins. Further south, in the Goban Spur–Fastnet–Celtic Sea region (Figs 1 & 2), Permian–Triassic successions are present in broadly east–west-orientated, fault-controlled basins, with evidence of growth faulting and thickening towards the controlling faults (Naylor *et al.* 2002). Rifting of the Variscan basement in the Goban Spur region probably began in Mid-Triassic time (Cook 1987).

Stratigraphy. Definitive Permian strata have not been encountered by drilling in the Porcupine Basin region. Tentatively dated Autunian mudstones with anhydrite have been drilled on the western margin margin of the Porcupine Basin (Croker & Shannon 1987) (Fig. 11). The Triassic is also little known and poorly dated in the north Porcupine Basin (Naylor & Shannon 2011). Here the Triassic succession, approximately 300 m thick, comprises clastic assemblages of sandstones, siltstones and mudstones overlain by evaporite-bearing shales (Fig. 11).

In the Fastnet–Celtic Sea region, no Permian strata have been proven, although clastic deposits may occur in small, isolated half-graben. Here, the Triassic, typically <600 m thick where drilled, consists of a sandstone-prone Sherwood Sandstone equivalent overlain by evaporitic mudstones of Mercia Mudstone equivalence. A similar Triassic stratigraphic succession is interpreted, on seismic evidence, from the Goban Spur (Fig. 11).

In the Slyne Basin, a massive evaporite sequence, assumed to be Zechstein equivalent, lies beneath the Triassic Sherwood Sandstone that is <400 m thick where drilled (Dancer *et al.* 1999, 2005) (Fig. 11). In the Erris Basin, Zechstein-equivalent marine mudstones, dolomites and anhydrites rest unconformably on the Carboniferous, while similar Upper Permian rocks also occur (Tate & Dobson 1989) resting on red-bed sandstones and shales of possible Early Permian age (Murphy & Croker 1992).

On the eastern margin of the southern Rockall Basin, approximately 180 m of Lower Permian (Asselian) sandstones and mudstones is overlain by almost 50 m of sandy, poorly dated redbeds interpreted as probably Permian–Triassic in age (Tyrrell *et al.* 2010).

Boundary relationships with underlying and overlying systems. The base of the Permian–Triassic succession, which varies in age throughout the region, is everywhere marked by an unconformity. Throughout the Slyne, Erris and Donegal basins, and much of the northern Porcupine Basin, the youngest strata below the basal unconformity are of Upper Carboniferous (Namurian–Westphalian) age, with occasionally Permian–Triassic strata in the north Porcupine Basin resting on pre-Upper Carboniferous strata (Croker & Shannon 1987). Locally in the Celtic Sea region, putative Permian strata are unconformably overlain by the Triassic succession (Smith 1995). The Triassic–Lower Jurassic boundary is typically conformable (Fig. 11). In the eastern part of the southern Rockall Basin, the Permian–Triassic clastic succession is unconformably overlain by probable Middle Jurassic sandstones (Tyrrell *et al.* 2010).

Depositional environment. The Permian, where drilled in the southern Rockall Basin, consists of fluvial channel and aeolian sandstones with lacustrine mudstones, with overlying probably Triassic strata in broadly similar fluvial conglomeratic and sandstone facies. Zechstein-equivalent mudstones, dolomite and evaporitic sequences in the Erris and Donegal basins suggest a narrow marine incursion through these basins (Naylor & Shannon 2011).

Locally, on the western margin of the Porcupine Basin, possible Permian red claystones and siltstones interbedded with sandstones, anhydrites and limestones suggest marginal coastal swamp conditions giving way to brackish and marine conditions, succeeded by lagoonal to playa lake environments.

The Triassic in the Slyne Basin comprises a low-sinuosity braided fluvial sandstone succession with minor sand-flat and playa mudstones. In the Erris Basin, similar facies are recognized with a tuffaceous component seen in some of the sandstones. Within the northern part of the Porcupine Basin, the Triassic was deposited in an alluvial, coastal-plain and shallow-marine environment (Naylor & Shannon 2011).

Jurassic stratigraphy of the NE Atlantic margins

North–NE Greenland margin

Distribution. The Jurassic is exposed onshore in the Wandel Sea Basin and along the NE Greenland margin as far south as the Wollaston Forland (Fig. 3). Offshore, a Jurassic succession has been inferred for the Danmarkshavn and Thetis basins (Hamann *et al.* 2005; Dinkelman *et al.* 2010; Jackson *et al.* 2012) (Figs 3 & 12a).

Structural setting. The most important Mesozoic rifting phase was initiated in the late Bajocian, culminated in the Volgian–early Ryazanian (Tithonian–Berriasian) and waned in the Hauterivian (Early Cretaceous). The Wollaston Forland Basin underwent significant rotational block faulting in the westwards-tilted half-graben with rift-shoulder uplift and deep subsidence along north–south-trending faults (Surlyk 1978, 2003).

Stratigraphy. The Wandel Sea Basin records the middle Oxfordian–Valanginian Ladegårdsåen Formation, about 250 m thick, of sandstone and organic mudstones in the eastern part of Peary Land (Figs 3 & 5), and, further eastwards, a Kimmeridgian–Valanginian (Lower Cretaceous) succession up to 900 m thick is recorded (Håkansson *et al.* 1991; Dypvik *et al.* 2002). The lithostratigraphic subdivision of this region is currently being revised based on new data.

In the Traill Ø–Hold with Hope area, the basal Jurassic rocks of the Wollaston Forland Basin are represented by mainly sandstone, up to 560 m thick, of the Middle–Upper Jurassic Vardekløft Group (Figs 6 & 7). These rocks are overlain by black organic-rich mudstone of the Upper Jurassic Bernbjerg Formation, which shows prominent thickness variations from 10 to 700 m as a consequence of the rift-basin configuration (Surlyk 1977). The Jurassic–Cretaceous boundary in NE Greenland is incorporated within the Wollaston Forland Group, which consists of up to approximately 3 km thickness of conglomerates and sandstones that laterally pass eastwards into mudstones (Surlyk 1978, 2003) (Fig. 6).

In the offshore basins, the Jurassic succession within the Danmarkshavn Basin has an interpreted thickness of up to 3 km, and has been subdivided into three predominantly clastic megasequences bounded by unconformities (Hamann *et al.* 2005). By analogy with the Wollaston Forland and Jameson Land basins (cf. Figs 6–8), Hamann *et al.* (2005) suggested that the lower megasequence (Lower Jurassic) might be comparable to the Kap Stewart and Neill Klinter groups; the middle megasequence – in the southern Danmarkshavn Basin – has a basal unconformity, and is suggested as a sand-dominated equivalent to the Vardekløft Group; whereas the upper megasequence is tentatively correlated with Upper Jurassic source-rock-prone shales of the Bernbjerg Formation. An uppermost Jurassic–Lower Cretaceous succession is suggested to be equivalent to the Wollaston Forland Group in NE Greenland (Hamann *et al.* 2005) (Fig. 6). A comparable Jurassic infill to the Thetis Basin is tentatively inferred (Hamann *et al.* 2005).

Boundary relationships with underlying and overlying systems. In the onshore part of the Wandel Sea Basin, Upper Jurassic strata unconformably overlie Triassic or older rocks, and the upper major unconformity is provisionally placed in the Lower Cretaceous and overlain by upper Lower Cretaceous rocks (Håkansson *et al.* 1991) (Fig. 5).

In the Wollaston Forland Basin, the lower boundary of the Middle Jurassic is represented by an erosional unconformity against the Triassic in the Traill Ø area; against the Permian or Caledonian basement further north in the Wollaston Forland; and against the Caledonian basement at Store Koldewey. The unconformity coincides with the initial transgressional base of the Middle Jurassic Vardekløft Group (Surlyk 2003; Stemmerik & Piasecki 2004). On the Koldewey Platform and in the southern Danmarkshavn Basin, the lower boundary is interpreted to be an erosional unconformity against the Triassic (Fig. 6).

The Jurassic succession is conformably overlain by the Cretaceous in the deep downfaulted parts of the Wollaston Forland Basin, whereas a major unconformity exists against Lower Cretaceous along uplifted footwall crests (Fig. 6). A major unconformity is also interpreted to mark the upper boundary on the Koldewey Platform, whereas the upper boundary in the both the northern and southern parts of the Danmarkshavn Basin is inferred to be conformable (Figs 5 & 6).

Depositional environment. In the Wandel Sea Basin, the basal Jurassic facies reflects an Oxfordian marine transgression in the eastern part of Peary Land (Fig. 2). In the Kilen area, the Kimmeridgian–Valanginian succession comprises restricted marine sediments (Håkansson *et al.* 1991; Dypvik *et al.* 2002).

In the Wollaston Forland Basin, the rift-related Vardekløft Group represents a major westwards and northwards expansion of shallow-marine deposition, with the facies belts moving northwards during the Bajocian–Oxfordian, along en echelon north–south-orientated relay embayments (Surlyk 2003). The Upper Oxfordian–Tithonian interval includes the maximum flooding zone in NE Greenland with deposition of black organic-rich mudstones of the Bernbjerg Formation. The Jurassic–Cretaceous Wollaston Forland Group encompasses the rift-culmination succession of deep-marine proximal gravity-flow conglomerates that pass laterally into more distal mudstones within westwards-tilted fault blocks (Surlyk 1978, 2003).

Comparable depositional environments inferred for the upper part of the Jurassic succession in the

Danmarkshavn Basin were probably preceded by Lower–Middle Jurassic lacustrine, deltaic and marginal-marine facies-equivalents of the Kap Stewart and Neill Klinter groups in the Jameson Land Basin of central East Greenland (see below) (Hamann *et al.* 2005) (Fig. 8).

Western Barents Sea–Svalbard margin

Distribution. Rocks of Jurassic age crop out on Svalbard and have been proven in numerous wells in basins on the southern part of the western Barents Sea margin. It is inferred that Jurassic rocks underlie much of the SW part of the Barents Sea continental margin (Fig. 3).

Structural setting. As for most of the Triassic period, Early–Mid-Jurassic deposition in the western Barents Sea–Svalbard region occurred within a tectonically quiet structural setting that was dominated by passive regional subsidence, and only minor fault activity and uplift (Brekke *et al.* 2001; Henriksen *et al.* 2011). A regional Mid-Jurassic unconformity (Fig. 5) marks the onset of a renewed phase of rifting that caused a marked rejuvenation of the topography into a system of tectonic highs and basins, including the Bjørnøya, Tromsø and Harstad basins (Faleide *et al.* 2010) (Figs 1 & 13a). Intra-Upper Jurassic unconformities are recorded in wells from the Hammerfest Basin and, together with major thickness variations in the Bjørnøya Basin and onshore in Svalbard, indicate further contemporary Late Jurassic tectonism along the western Barents Sea–Svalbard margin (Nøttvedt *et al.* 1993*a*; Doré *et al.* 1999; Worsley 2008; Henriksen *et al.* 2011). This rift phase culminated in the Early Cretaceous (see below).

Stratigraphy. The Upper Triassic–Middle Jurassic Kapp Toscana Group exceeds a thickness of 475 m on Svalbard, and consists of deltaic to shallow-marine shales, siltstones and sandstones (Harland 1997; Dallmann 1999; NORLEX 2014) (Fig. 5). On the western Barents Sea margin, the Kapp Toscana Group is up to 500 m thick (e.g. Hammerfest Basin), and is dominated by a paralic to shallow-marine succession of sandstone with subordinate shale and thin coal beds, which pass upwards into moderately to well sorted and mineralogically mature sandstones (Dalland *et al.* 1988; Dallmann 1999; NORLEX 2014).

The post-Bathonian Middle–Upper Jurassic succession is assigned to the Adventalen Group, which extends into the Early Cretaceous (Nøttvedt *et al.* 1993*a*; Dallmann 1999; NORLEX 2014). On Svalbard, the Jurassic component of this group is up to 350 m thick and is marked at its base by a conglomerate – the Brentskardhaugen Bed – a reworked deposit that is overlain by a poorly sorted upwards-fining basal section of conglomerate, sandstone and siltstone, which pass upwards into dark grey to black organic-rich marine shales with subordinate thin siltstones (Fig. 5). This succession extends into the western Barents Sea where it is up to 470 m thick, and comprises predominantly dark-brown pyritic mudstones, with thin limestones and sporadic thin sandstone beds around basin margins. Thickness variations are observed in some basins (e.g. Hammerfest Basin) as a consequence of contemporary tectonic activity (Dallmann 1999).

Boundary relationships with underlying and overlying systems. In Svalbard, the Triassic–Jurassic boundary is marked, at least locally, by seemingly continuous facies from the Rhaetian to the Toarcian, although elsewhere (e.g. Hammerfest Basin) local unconformities exist (Fig. 5). The Jurassic–Cretaceous boundary is represented by an unconformity in the western Barents Sea, whereas a conformable boundary occurs on Svalbard (Nøttvedt *et al.* 1993*a*; Harland 1997; Dallmann 1999; Henriksen *et al.* 2011).

Depositional environments. The repeated clastic succession of mainly delta-related and coastal and shallow-marine shelf sedimentation that was instigated in the Late Triassic (the Kapp Toscana Group) persisted until the Mid-Jurassic (Bathonian), sourced from multiple provenances. At this time, the palaeogeography of the western Barents Sea–Svalbard region was probably dominated by a low-lying peneplain (Brekke *et al.* 2001). Extensive reworking within this succession has resulted in sandstones that are texturally and mineralogically mature (Worsley 2008; Henriksen *et al.* 2011). Following late Mid–Late Jurassic topographical rejuvenation, a renewed regional transgression led to the submergence of a large part of the western Barents Sea–Svalbard region, and the deposition of deeper-water, muddy marine shelf sediments of the Adventalen Group (Dallmann 1999; Henriksen *et al.* 2011). A fluctuating sea level in combination with the increased tectonic activity resulted in variable oxic and anoxic bottom-water conditions, which gave rise to favourable conditions for the accumulation of black shales (Brekke *et al.* 2001).

Central East Greenland margin

Distribution. Jurassic rocks are exposed in the onshore Jameson Land Basin (Figs 3 & 12d). No Jurassic is exposed onshore south of this basin, although seismic data show that the southwards-dipping Jurassic succession may extend below the volcanic sequence in the Blosseville Coast (Larsen & Marcussen 1992) (Fig. 1). Seismic reflection data suggest that Jurassic rocks might be present in the offshore Liverpool Land Basin (Hamann *et al.* 2005) (Fig. 12d): however, it remains unclear as to whether or not they are present in the Ammassalik Basin, offshore SE Greenland (Gerlings *et al.*, this volume, in review).

Structural setting. The latest Triassic (Rhaetian)–Mid-Jurassic (early Bajocian) was characterized by thermal subsidence in the Jameson Land Basin. The most important Mesozoic rifting phase was initiated in the late Bajocian, culminated in the Volgian–early Ryazanian (Tithonian–early Berriasian (earliest Cretaceous)) and waned in the Hauterivian (Early Cretaceous) (Surlyk 2003). The Jameson Land area only experienced minor tilting during this period, in comparison to NE Greenland (see above) (Surlyk 2003). The Lower Jurassic facies belts follow deep-seated Devonian NW–SE-striking faults in the northern part of Jameson Land (Dam *et al.* 1995).

Stratigraphy. The Jameson Land Basin preserves a near-complete Jurassic succession of up to about 2 km thick; the stratigraphy and basin evolution is comprehensively described by Surlyk (2003). The Rhaetian–Sinemurian Kap Stewart Group is up to 600 m thick, and consists of organic-rich mudstones and sandstones (Dam *et al.* 1995; Surlyk 2003) (Fig. 8). The Kap Stewart Group is overlain by alternating sandstones and mudstones of the Pliensbachian–early Bajocian Neill Klinter Group, which is 300–450 m thick and topped by an erosional unconformity (Dam & Surlyk 1998). This is succeeded by sandstones of the Vardekløft Group (up to about 650 m thick), which are overlain by Oxfordian–Tithonian black organic-rich mudstones and sandstones of the Hall Bredning Group (possibly up to 800 m thick), itself capped by coarse-grained sandstones (a few hundred metres thick) of the Scoresby Sund Group, which extends into the Berriasian (Early Cretaceous) (Surlyk 2003). In the Liverpool Land Basin, the presence of Upper Jurassic marine shales has been speculated upon (Gautier *et al.* 2011) (Fig. 8).

Boundary relationships with underlying and overlying systems. In the Jameson Land Basin, the base of the Kap Stewart Group is associated with a minor unconformity against the Triassic Scoresby Land Group along the SE margin of the Jameson Land Basin. The Jurassic–Cretaceous boundary is a minor erosional unconformity that extends into the Barremian (earliest Cretaceous), and is overlain by Valanginian rocks.

Depositional environment. The Triassic–Jurassic boundary marks a transition towards a temperate and humid climate in the Jameson Land Basin, represented by the deltaic and lacustrine rocks of the Kap Stewart Group (Dam *et al.* 1995; Surlyk 2003) (Fig. 8). The sediments of the Neill Klinter Group represent marine-shoreface, restricted offshore and tidal-embayment environments deposited during an overall Pliesbachian–Early Bajocian transgression (Dam & Surlyk 1998). The Vardekløft Group represents mainly shallow-marine deposition that is succeeded by deposition of the Hall Bredning Group of basinal organic-rich mudstones and mass-flow sandstones during maximum sea level. The southwards progradation of prominent sandy shelf-edge deltas is represented by the sandstones of the Raukelv Formation (Surlyk 2003).

Møre–mid-Norway margin

Distribution. On the Norwegian mainland, Jurassic sedimentary rocks crop out on Andøya, northern Norway, where an approximately 350 m-thick sequence of sandstones, siltstones, shales and thin beds of coal is preserved (Øvrig 1960; Bøe *et al.* 2010). Offshore, Jurassic sediments have been proved in many of the wells drilled on the inner part of the Møre–mid-Norway margin (Figs 3 & 13b, c). On the continental shelf, most wells that prove Jurassic rocks have been drilled on the Trøndelag Platform and Halten Terrace, in the inner Vøring region. Where the wells penetrate the entire preserved Jurassic succession, thicknesses in excess of 1000 m are recorded: more commonly, the drilled succession ranges between 400 and 800 m.

Structural setting. It has been inferred that the mid-Norway margin, as well as the Møre region, was predominantly an area of uplift and erosion throughout the Early and Mid-Jurassic (Brekke *et al.* 1999; Doré *et al.* 1999). During the Late Jurassic–Early Cretaceous, rifting and differential vertical movements created sub-basins and highs in the Møre and Mid-Norwegian region (Faleide *et al.* 2010). Correspondingly, the thickness of the rocks may vary considerably as the sediments were deposited on a series of tilted fault blocks.

Stratigraphy. The Jurassic succession on the continental shelf comprises the Båt (latest Triassic (Rhaetian)–Early Jurassic), Fangst (Mid-Jurassic) and Viking (late Mid-Jurassic–earliest Cretaceous) groups (Dalland *et al.* 1988) (Figs 7 & 8). The Båt Group is best preserved on the inner Vøring margin where it locally exceeds 700 m in thickness, and consists predominantly of alternating sandstone and shale–siltstone units with thin coal beds. The upper part of the Båt Group is progressively truncated towards the crestal region of some of the inner-shelf highs (e.g. the Nordland Ridge), and younger strata within the group onlap directly onto

Precambrian basement. The Fangst Group is up to 250 m thick, and comprises a basal fine- to medium-grained sandstone unit, middle mudstone unit and an upper fine- to coarse-grained sandstone unit (Dalland *et al.* 1988). The Viking Group is best developed in the northern North Sea where it is up to 1 km thick: in contrast, it is more locally developed in the Møre–mid-Norway region, where a succession dominated by shales and mudstones up to 125 m thick has been drilled in commercial wells (Dalland *et al.* 1988). Well data also indicate a variable conformable to unconformable relationship with the underlying Fangst Group, as well as internal unconformities within the Viking Group.

Boundary relationships with underlying and overlying systems. On the inner Møre margin, pre-Toarcian rocks are locally absent, and the Jurassic sequence rests unconformably on Triassic and older strata (Fig. 8). On the inner part of the Vøring margin, a more conformable passage from Triassic to Lower Jurassic deposits is observed (Fig. 9). The boundary with the overlying Lower Cretaceous Cromer Knoll Group is commonly marked by an unconformity throughout the region (Dalland *et al.* 1988; Norwegian Petroleum Directorate 2014).

Depositional environments. The Jurassic rocks on Andøya represent lacustrine–lagoonal to shallow-marine depositional environments (Øvrig 1960; Bøe *et al.* 2010). Deltaic to shallow-marine environments also characterize the latest Triassic to Mid-Jurassic succession on the Møre–mid-Norwegian margin, whereas an open-marine environment existed in the late Mid-Jurassic and Late Jurassic (Dalland *et al.* 1988). The observed variable conformable to unconformable relationship within the Fangst and Viking groups probably reflects the transgressive nature of the upper Middle–Upper Jurassic sequence at a time of transition into an increasingly unstable Cretaceous rifting episode (Blystad *et al.* 1995; Doré *et al.* 1999; Brekke *et al.* 2001).

Faroe–Shetland–northern Rockall–Hebrides margin

Distribution. Most proven occurrences of Jurassic rocks are confined to the basins that underlie the Hebrides and West Shetland shelves (Fyfe *et al.* 1993; Stoker *et al.* 1993; Ritchie & Varming 2011; Evans 2013) (Figs 1 & 3). A near-complete succession is preserved within the Hebridean chain of basins, including the Inner Hebrides, Sea of Hebrides–Little Minch, North Minch and North Lewis basins (Figs 10 & 14c). They have also been encountered in the West Lewis (Fig. 14d) and West Flannan basins. Upper Jurassic rocks are widely present in the North Rona, Papa, West Shetland, West Solan, East Solan (Fig. 14a) and South Solan basins west of Shetland, as well as the Judd, Flett and Foula sub-basins in the Faroe–Shetland Basin (Fig. 9). Their occurrence on the Corona High might indicate a presence in the western half of the Faroe–Shetland Basin, but this remains conjectural. Middle and Lower Jurassic strata have a more restricted distribution across the Faroe–Shetland area. No rocks of Jurassic age have been proven to occur in the northern Rockall and Hatton basins, although the occurrence of Jurassic palynomorphs reworked into Eocene fan-delta deposits on the eastern flank of the Rockall High might reflect a Jurassic source on the Rockall Plateau (Hitchen 2004).

Structural setting. The Jurassic rocks are largely preserved within the same system of NE–SW-trending basins that was active during the Permian–Triassic (Fig. 3). In the Hebrides–West Shetland region, the geometry of the Jurassic basinal successions is asymmetrical and thickest adjacent to the footwall (e.g. North Lewis, North Minch, North Rona and East Solan basins) (Fig. 14) – a pattern of sedimentation that has been linked to pulsed episodes of footwall uplift within this Mesozoic basin system (Harris 1992; Morton 1992; Roberts & Holdsworth 1999). The isolated occurrences of Lower and Middle Jurassic rocks in the Faroe–Shetland Basin make it difficult to determine basin activity at this time. According to both Dean *et al.* (1999) and Doré *et al.* (1999), the general lack of variation in the thickness of the Upper Jurassic strata across the Fasroe–Shetland region (see below) suggests that the effects of any contemporaneous rift activity were negligible.

Stratigraphy. On the Inner Hebridean islands of Skye and Raasay (part of the Sea of Hebrides–Little Minch Basin), the Jurassic succession is up to 1.5 km thick and is divisible into three unconformity bounded sequences: the Lower Jurassic Lias Group, comprising shallow-marine clastic rocks with limestone and sporadic ironstone formations; shallow-water sandstones and partly non-marine organic-rich rocks of the Middle Jurassic Bearreraig Sandstone Formation and Great Estuarine Group, respectively; and shallow-marine Upper Jurassic sandstones and mudstones of the Staffin Bay and Staffin Shale formations (Fyfe *et al.* 1993; Hudson & Trewin 2002) (Fig. 10). A comparable succession of rocks ranging from 2 to 3 km in thickness has been proved in the North Lewis and North Minch basins, including Middle Jurassic organic-rich mudstone (Fyfe *et al.* 1993; Stoker *et al.* 1993; Evans 2013) (Fig. 10). Further west, up to several hundred metres of Middle and Upper Jurassic marginal- to shallow-marine sandstone and mudstone with sporadic thin coal beds might be present in the West Flannan and West Lewis basins (Hitchen & Stoker 1993) (Fig. 14d).

In the Faroe–Shetland region, the Lower Jurassic sequence is separated from the Middle and Upper Jurassic sequences by a major late Early–early Mid-Jurassic (mid-Pliensbachian–Bajocian) hiatus (Ritchie *et al.* 1996; Ritchie & Varming 2011) (Fig. 9). The Lower Jurassic rocks have been grouped collectively within the Skerry Group, which comprises carbonaceous sandstone and mudstone, with maximum-drilled thicknesses of 770 m in the West Solan Basin and 340 m in the Foula Sub-basin (Faroe–Shetland Basin). The upper Middle–Upper Jurassic succession is assigned to the Humber Group, which has a maximum-drilled thickness of 1.05 km in the Foula Sub-basin, although more commonly it is <250 m thick across the Faroe–Shetland region. This group comprises sporadic shallow-marine sandstone and carbonaceous and pyritic marine mudstone of the Bajocian–Oxfordian Heather Formation overlain, locally unconformably, by the Kimmeridge Clay Formation, which is the most extensive Jurassic unit within the Faroe–Shetland region, and extends into the Early Cretaceous (Berriasian). Although the Kimmeridge Clay Formation is typically associated with dark grey to black organic-rich mudstone and minor siltstone, the base of the formation is characterized by sandstone, siltstone and conglomerate.

Boundary relationships with underlying and overlying systems. In the Inner Hebrides there appears to be conformity, at least locally, between the Upper Triassic and Lower Jurassic sequences (Fyfe *et al.* 1993; Hudson & Trewin 2002) (Fig. 10). Over the rest of the area, the base of the Jurassic succession mostly rests unconformably on Triassic and older strata. In the Faroe–Shetland region, the Kimmeridge Clay Formation extends into the early Berriasian (Cretaceous); however, it is almost everywhere separated from the overlying Cromer Knoll Group (Lower Cretaceous) by an unconformity (Ritchie & Varming 2011; Stoker & Ziska 2011). On the Hebrides Shelf, the top of the Jurassic is unconformably overlain by Cretaceous and younger rocks.

Depositional environment. The rocks of the Lias and Skerry groups were deposited in marginal- to shallow-marine environments. In the Hebrides region, interbedded mudstones, siltstones, sandstones and limestones of the Lias Group are arranged in transgressive–regressive facies cycles, whereas thicker, deeper-water mudstone formations are interpreted to reflect contemporaneous subsidence linked to a high clastic input (Fyfe *et al.* 1993; Hesselbo *et al.* 1998; Hudson & Trewin 2002). Following the early Mid-Jurassic hiatus, marginal- to shallow-marine sedimentation was renewed in the Hebrides and was succeeded by fully marine sedimentation in the Late Jurassic, instigated by a major rise in sea level in the Callovian. The Faroe–Shetland region might have remained largely exposed until the widespread deposition of the Kimmeridge Clay Formation. According to Vestralen *et al.* (1995), the Kimmeridge Clay Formation represents the deposition of an overall transgressive succession marked by the transition from subaerial and shallow-marine coarse clastics to organic-rich basinal mudstones. This model contrasts with previous deep-water rift-related models in the Faroe–Shetland area (e.g. Haszeldine *et al.* 1987; Hitchen & Ritchie 1987; Meadows *et al.* 1987).

Southern Rockall–Porcupine margin

Distribution. Jurassic strata have a widespread distribution throughout this region (Figs 3 & 15). Lower Jurassic rocks occur through the Slyne and Erris basins and the Goban Spur, but have a very restricted distribution in the Porcupine region, encountered only in the north Porcupine Basin (Fig. 11). Middle Jurassic strata are widely present in the Slyne, Erris, Porcupine and Goban Spur basins. Upper Jurassic strata are present basinwide through the Porcupine Basin (Croker & Shannon 1987), locally preserved in the Slyne Basin (Corcoran & Mecklenburgh 2005), but generally absent through most of the Erris and Donegal basins. They are very locally present in the Fastnet Basin (Robinson *et al.* 1981) and the Goban Spur due to erosion, although seismic evidence (Cook 1987; Naylor *et al.* 2002) suggests Upper Jurassic strata may be preserved in places. The nature of Jurassic deposition within the southern Rockall and Hatton basins is speculative, although Naylor & Shannon (2005) suggested a widespread presence. Kimmeridgian–Portlandian (late Tithonian–earliest Berriasian (early Cretaceous)) marine sandstones and limestones have been drilled in the north Bróna Basin (Fig. 1) on the western flank of the Porcupine High (Haughton *et al.* 2005). Middle and Upper Jurassic strata are also interpreted from seismic data within the small perched basins along both margins of the southern Rockall Basin (Naylor *et al.* 1999) (Fig. 15a).

Structural setting. Croker & Shannon (1987) suggested that Early Jurassic deposition was limited through the Porcupine and southern Rockall region to isolated fault-controlled basins with a Caledonian trend, mirroring the pattern for the underlying Triassic. These NE–SW-trending basins are larger in the Slyne–Erris region, comprising linked half-graben with a reversal of fault polarity along the basin trend. In the Goban Spur region, Jurassic sedimentation was controlled by extensional reactivation of structures at a high angle to the Caledonian fabric that controlled coeval sedimentation further north along the Porcupine–southern Rockall margin. Middle and Upper Jurassic sedimentation in the Porcupine Basin region was controlled by east–west rifting, resulting in a large fault-bounded basin system which transected the NE–SW caledonoid grain that predominated in the Slyne and Erris region to the north.

Stratigraphy. A fairly complete Lower Jurassic sequence, approximately 600 m thick, is preserved in the Slyne Basin, whereas the upper zones of the Lower Jurassic are missing, through erosion, in wells in the Erris Basin (Fig. 11). Hettangian–Toarcian limestones, siltstones and basal anhydrites are overlain by sandstones, siltstones and capped by up to 100 m of organic-rich shales. The stratigraphy is comparable to the Hebridean region described above (Trueblood & Morton 1991). A broadly similar succession has been encountered in the Goban Spur region.

The Middle Jurassic in the Slyne Basin comprises more than 1300 m of Bajocian–Bathonian claystones, limestones and sandstones with coal interbeds. The Middle and Upper Jurassic succession in the Porcupine Basin comprises coarse clastics in fining-up cycles of Bajocian–early Bathonian age, overlain in the north by sandstones and mudrocks (Croker & Shannon 1987). In the southern Rockall Basin, approximately 120 m of conglomeratic sandstones, overlain by thin marine sandstones and capped by a volcanic unit (Tyrrell *et al.* 2010), are thought to be of Middle Jurassic age. In the Porcupine Basin, the Oxfordian–Tithonian comprises mudstones with thin argillaceous siltstones and sandstones (Croker & Shannon 1987). Thicker sandstones occur towards the base of the succession. There are significant differences in facies and clastic geometries between the northern, central and southern margins of the basin that reflect different depositional settings. Up to 500 m of Upper Jurassic strata are locally preserved in the Slyne Basin, and comprise silty claystones with sandstone, limestone and coal interbeds (Corcoran & Mecklenburgh 2005).

Boundary relationships with underlying and overlying systems. Where the uppermost Triassic succession has been encountered in the region, it is marked by a rapid transition from redbed mudstones to a shallow-marine limestone–shale succession of Rhaetian age. The regionally developed Middle Jurassic sequences in the northern part of the Porcupine Basin (Fig. 11) rest unconformably on older, mostly Upper Carboniferous, strata (Naylor & Shannon 1982; Croker & Shannon 1987). The Upper Jurassic rests conformably on the Middle Jurassic and is regionally unconformably overlain throughout the region by Lower Cretaceous strata.

Depositional environment. The Lower Jurassic succession is largely brackish to marine in nature. In contrast, the Middle Jurassic is predominantly a fluvio-estuarine succession dominated by sandstones and siltstones. In the northern and marginal parts of the Porcupine Basin, the Bajocian–early Bathonian braided river systems prograded southwards, and are succeeded by meandering fluvial and related environments. They pass southwards to shallow-marine sediments (Croker & Shannon 1987). The Late Jurassic was a period of differential subsidence in the basin and the overall facies pattern suggests a northwards encroachment of transgressive facies (Croker & Shannon 1987) along the basin axis, with active movement on marginal bounding faults. Marine influences increased, and nearshore marine conditions became dominant in Kimmeridgian–early Tithonian time. Away from the basin margins, proximal to distal submarine fan sandstones and mudstones are overlain by Tithonian marine shales (Croker & Shannon 1987).

Cretaceous stratigraphy of the NE Atlantic margins

North–NE Greenland margin

Distribution. In north Greenland, the Cretaceous is exposed in Peary Land and at Kilen in the Wandel Sea Basin (Figs 1 & 4). Further south, Cretaceous rocks are exposed along the NE Greenland margin from Store Koldewey to Traill Ø, whereas the offshore Danmarkshavn and Thetis basins have both been interpreted to contain Cretaceous successions, several kilometres thick (Hamann *et al.* 2005).

Structural setting. The Wandel Sea Basin was probably subdivided into several fault-bounded subbasins in the Cretaceous that were increasingly controlled by strike-slip forces (pull-apart basins) (Håkansson *et al.* 1991; Dypvik *et al.* 2002). The Danmarkshavn Basin shows a relatively symmetrical basin configuration in the northern part and an eastwards-tilted half-graben against the Danmarkshavn Ridge in the south (Figs 12a–c). In contrast, the Thetis Basin is interpreted to consist of a westwards-tilted half-graben against the Danmarkshavn Ridge (Fig. 12a). Post-Valanginian basin infills along the NE Greenland margin, including the Wollaston Forland Basin, reflect partly inherited Jurassic–Cretaceous basin configurations. In addition, several episodes of fault reactivations have been suggested to explain the local deposition of coarse-grained units (e.g. Surlyk & Noe-Nygaard 2001).

Stratigraphy. The post-Valanginian Cretaceous siliciclastics in the Wandel Sea Basin comprise the 650 m-thick Gåseslette Group (Aptian–Albian), unconformably overlain by the Kilen Group (Turonian–Santonian), which is probably more than 1900 m thick (Håkansson *et al.* 1991; Dypvik *et al.* 2002).

The post-Valanginian Cretaceous succession exposed from Traill Ø to Store Koldewey is up to a few kilometres thick and is dominated by mudstone, although several coarse-grained units can be identified, including the Barremian–Aptian Steensby Bjerg Formation at Hold with Hope (Whitham *et al.* 1999; Larsen *et al.* 2001), as well as the Albian Rold Bjerge Formation, the Turonian Månedal Formation and the Coniacian Vega Sund Formation at Traill Ø (Surlyk & Noe-Nygaard 2001), all of which are included within the Hold with Hope Group (Fig. 7).

In the offshore basins, the southern part of the Danmarkshavn Basin is interpreted to contain a shale-dominated Cretaceous succession up to about 4 km thick, which is divided by an internal unconformity that roughly separates the Lower and Upper Cretaceous sequences (Tsikalas *et al.* 2005) (Figs 6 & 12a–c). The Cretaceous succession in the northern Danmarkshavn Basin is affected by salt movement (Fig. 5). In the Thetis Basin, a Cretaceous succession several kilometres thick is inferred (Figs 6 & 12a). The succession thickens markedly to the north of the profile (shown in Fig. 12a) and it is possibly analogous to the Vøring Basin offshore mid-Norway (Hamann *et al.* 2005; Dinkelman *et al.* 2010) (Fig. 7).

Boundary relationships with underlying and overlying systems. The onshore part of the Wandel Sea Basin records an unconformity bounded Lower Cretaceous (Aptian–Albian) succession overlying Valanginian or older Mesozoic–Palaeozoic strata (Fig. 5). The Upper Cretaceous comprises unconformity bounded Turonian–Santonian units: Paleocene deposits occur onshore, but the contact relationship with the Cretaceous is not known (Håkansson *et al.* 1991; Dypvik *et al.* 2002).

The post-Valanginian sequences between Traill Ø to Store Koldewey display lower unconformities against Caledonian basement, Carboniferous, Permian, Triassic, Jurassic or lowermost Cretaceous strata. The upper boundaries are controlled by the results of post-Cretaceous uplift events: in some places, the Cretaceous is erosionally overlain by Paleocene sediments and Palaeogene volcanics (e.g. Nøhr-Hansen *et al.* 2011).

The Danmarkshavn and Thetis basins have an interpreted conformable lower boundary against the Jurassic, and an upper boundary that is marked by an angular unconformity: the latter is overlain by Quaternary deposits to the west, and pre-break-up Palaeogene strata to the east along and flanking the Danmarkshavn Ridge (Hamann *et al.* 2005; Tsikalas *et al.* 2005) (Fig. 12a).

Depositional environment. The Ladegårdsåen, Gåseslette and Kilen groups of the Wandel Sea Basin all represent siliciclastic rocks deposited in a restricted marine setting (Håkansson *et al.* 1991; Dypvik *et al.* 2002). Several episodes of rift reactivation have been suggested to explain local deposition of coarse-grained material as preserved by the shallow to deeper-water coarse-grained units, including gravity-flow deposits between Hold with Hope (Whitham *et al.* 1999; Larsen *et al.* 2001) and Traill Ø (Surlyk & Noe-Nygaard 2001) (Figs 6 & 7).

The Danmarkshavn Basin is interpreted to have been a mudstone-dominated marine basin (Figs 5 & 6). Deep-marine gravity-flow sandstones sourced from the west are inferred to have been deposited exclusively in the southern part of the basin during two episodes in the Early and mid-Cretaceous. The Cretaceous succession in the Thetis Basin is similarly interpreted to have been dominated by marine-basinal mudstone deposition (Hamann *et al.* 2005) (Fig. 6).

Western Barents Sea–Svalbard margin

Distribution. Cretaceous rocks crop out in Svalbard and are extensively preserved on the southern part of the western Barents Sea margin: their occurrence west of Svalbard remains uncertain (Figs 4, 5 & 13a). Whereas an accumulation of up to 3 km of Lower–Upper Cretaceous rocks has been identified on the SW Barents Sea margin, most of the greater Barents Sea region, including Svalbard, preserves only Lower Cretaceous rocks (Faleide *et al.* 1993; Breivik *et al.* 1998; Brekke *et al.* 2001).

Structural setting. Increasing tectonic activity throughout the Late Jurassic culminated with rifting in the Early Cretaceous, and the establishment of the present-day structural configuration of basins and highs in the western Barents Sea region (Gabrielsen *et al.* 1990). This included the rapid subsidence and development of deep basins, such as the Harstad, Tromsø, Bjørnøya and Sørvestnaget basins (Figs 1 & 13a), which became decoupled from the rest of the Barents Sea shelf through a series of rift episodes in the Early–mid-Cretaceous (Aptian–Albian, Cenomanian?) (Smelror *et al.* 2009) (Fig. 5). In the Late Cretaceous, much of the NW Barents

Sea region, including Svalbard, was uplifted (Brekke *et al.* 2001), whereas the southern western Barents Sea margin continued to subside (Dallmann 1999; Smelror *et al.* 2009; Faleide *et al.* 2010).

Stratigraphy. The Lower Cretaceous succession on Svalbard and across the western Barents Sea margin is assigned to the upper part of the Adventalen Group (which also includes the Upper Jurassic, see earlier) (Dallmann 1999; NORLEX 2014): Upper Cretaceous rocks on the western Barents Sea margin are assigned to the Nygrunnen Group (Dalland *et al.* 1988) (Fig. 5). An unconformity separates the Adventalen and Nygrunnen groups over much of the western Barents Sea margin (Fig. 5).

On Svalbard, the Cretaceous rocks of the Adventalen Group exceed 1500 m in thickness, and consist predominantly of dark shale with clay-ironstone and siltstone nodules that coarsen upwards into siltstone and sandstone with thin coal beds topped by a wedge-shaped accumulation of interbedded sandstones and shales (Dallmann 1999; Brekke *et al.* 2001; Worsley 2008) (Fig. 5). Between central and southernmost Svalbard, this clastic wedge thickens from 190 m to over 1000 m, increasing further in thickness on the western Barents Sea margin where the rocks mainly include dark marine mudstones with sporadic thin beds of limestone, dolomite, sandstone and siltstone (Dalland *et al.* 1988; Dallmann 1999; NORLEX 2014). In the Tromsø Basin, in excess of 2 km of unconformity bounded Lower Cretaceous rocks have been identified, whereas 700–1000 m of Lower Cretaceous strata have been reported from the Hammerfest Basin. Further west, thicknesses exceeding 2 km can be inferred from seismic data beneath the Veslemøy High and Sørvestsnaget Basin beneath the western Barents Sea shelf (Ryseth *et al.* 2003; Henriksen *et al.* 2011) (Fig. 13a).

Early Cretaceous igneous activity is evidenced by the occurrence of intercalated basalt lava flows of Barremian–Aptian age in Kong Karls Land (eastern Svalbard) (Harland 1997; Dallmann 1999; Henriksen *et al.* 2011; NORLEX 2014) (Fig. 5), with doleritic intrusions within Triassic–Lower Cretaceous shales. Sporadic tuffs are noted in the Hammerfest Basin and on the Senja Ridge.

On the western Barents Sea margin, the Upper Cretaceous Nygrunnen Group exceeds 2 km in thickness in the Tromsø and Sørvestsnaget basins, beneath the outer margin (Henriksen *et al.* 2011) (Fig. 13a). This group consists mostly of greenish-grey to grey shales and mudstones with thin interbeds of limestone, and a tuffaceous component is locally preserved in the Tromsø Basin (Dallmann 1999; NORLEX 2014). These fine-grained clastic deposits pass eastwards into condensed calcareous to sandy units (Dallmann 1999). These strata are commonly truncated to the east (Henriksen *et al.* 2011).

Boundary relationships with underlying and overlying systems. The Jurassic–Cretaceous boundary is represented by a distinct unconformity in the western Barents Sea, whereas continuous sedimentation prevailed across this boundary on Svalbard (Fig. 5). The Cretaceous–Paleocene boundary is characterized by a significant erosional unconformity throughout much of the western Barents Sea–Svalbard region (Henriksen *et al.* 2011).

Depositional environments. On Svalbard, the Lower Cretaceous rocks represent a paralic to shallow-marine shelf succession that records an overall regressive succession deposited under oxic conditions in an open-marine shelf setting. The clastic wedge resulted from a relative sea-level fall and delta progradation from the north in response to a major uplift of the NW Barents Sea area associated with the break-up of the Amerasian Basin (present Arctic Sea) (Nøttvedt *et al.* 1993*a*, *b*; Dallmann 1999; Brekke *et al.* 2001; Worsley 2008). This was accompanied by magmatism in Kong Karls Land and Franz Josef Land (Arctic Ocean, NE of Svalbard) (Steel & Worsley 1984; Harland 1997; Dallmann 1999; Henriksen *et al.* 2011; NORLEX 2014). An Aptian regional sea-level rise cut off most of the coarse clastic supply: however, northern uplift persisted and the southwards-thickening prodelta to distal marine mudstone-dominated wedge was deposited (Dallmann 1999; Brekke *et al.* 2001; Worsley 2008). Large-scale clinoforms prograding from the north are associated with this wedge (Worsley 2008; Henriksen *et al.* 2011).

Thick sequences of Lower Cretaceous rocks in the western Barents Sea were deposited in a partially restricted marine-shelf setting atop rapidly subsiding basins (Dalland *et al.* 1988; Dallmann 1999). In contrast, a sequence of condensed platform carbonate deposits accumulated further east in the Barents Sea (Bjarmeland Platform).

During the Late Cretaceous, much of the Barents Sea shelf was uplifted, and areas to the east were either transgressed only at times of maximum sea level and/or display only condensed calcareous to sandy units (Dallmann 1999). The basins beneath the western Barents Sea margin continued to subside and the Nygrunnen Group was deposited in a well-oxygenated deep-marine setting (Faleide *et al.* 2010).

Central East Greenland margin

Distribution. The lowermost Cretaceous is exposed in the southern part of the Jameson Land Basin and in Milne Land (Surlyk *et al.* 1973; Birkelund *et al.*

1984), and the southwards-dipping strata may continue below the Palaeogene volcanic rocks of the Blosseville Kyst (Larsen & Marcussen 1992) (Figs 1 & 4). Cretaceous rocks are also interpreted to occur within the inner Liverpool Land Basin (Hamann *et al.* 2005) (Fig. 12d).

Structural setting. The structural setting in the Early Cretaceous is considered a continuation from the Jurassic basin configuration in Jameson Land. In the Liverpool Land Basin, the interpreted Cretaceous drapes a westwards-rotated fault-block relief (Fig. 12d).

Stratigraphy. In the Jameson Land Basin, the Tithonian–Berriasian Scorseby Sund Group is unconformably overlain by Valanginian mudstones and sandstones of the 120 m-thick Hesteelv Formation (Surlyk *et al.* 1973) (Fig. 8). To the west, Valanginian and Hauterivian outcrops in Milne Land are up to about 300 m thick and dominated by sandstones of the Hartz Fjeld Formation (Birkelund *et al.* 1984). In the Liverpool Land Basin, Gautier *et al.* (2011) inferred the presence of Cretaceous marine shales.

Boundary relationships with underlying and overlying systems. The base of the Cretaceous succession forms an erosional unconformity against the Upper Jurassic–lowest Cretaceous rocks in the Jameson Land Basin (Surlyk 2003): its upper boundary is a poorly constrained post-Valanginian–Hauterivian erosion surface (Fig. 8). Palaeogene basalts erosionally overlie the Upper Jurassic–Lower Cretaceous rocks in Milne Land (Larsen *et al.* 2003).

Depositional environment. The Jameson Land Basin succession is represented by marginal- to shallow-marine deposits, basinal mudstones and gravity-flow sandstones (Surlyk *et al.* 1973). The Milne Land succession is of shallow-marine origin in its lower part, becoming increasingly marginal marine in the upper part (Birkelund *et al.* 1984).

Møre–mid-Norway margin

Distribution. On the Norwegian mainland, Lower Cretaceous sediments are preserved on Andøya, in northern Norway (e.g. Bøe *et al.* 2010). Offshore, the Cretaceous succession is preserved as thick accumulations within the Møre and Vøring basins (Figs 1, 4 & 13b, c).

Structural setting. In the Early Cretaceous, major rifting occurred along the Møre–mid-Norway continental margin (Doré *et al.* 1999). On seismic profiles, the onset of rifting is marked in both the Møre and Vøring basins by a very pronounced horizon that is observed to onlap the eastern flank of the basins, adjacent platforms and highs, and which is interpreted as the base Cretaceous unconformity (Brekke 2000; Faleide *et al.* 2010) (Fig. 13b, c). The subsequent Cretaceous subsidence was driven by flexuring of the basin flanks rather than faulting, and resulted in exceptionally thick basin fills. Brekke (2000) further describes a 'top Cenomanian' unconformity, which on seismic profiles in the Vøring Basin is identified as a tilted and faulted surface onlapped by younger (post-Cenomanian) strata along both flanks of the basin. This unconformity reflects a phase of tectonism at the end of the Cenomanian, which included compressional deformation and the formation of the Gjallar Ridge (Fig. 1), and eastwards tilting of the westernmost parts of the Vøring Basin (Blystad *et al.* 1995; Lundin & Doré 2011). In the Vøring Basin, the Upper Cretaceous rocks are overprinted by further phases of post-Cenomanian deformation, including the development of anticlines, such as the Nyk and Utgard highs (Figs 1 & 13b), that was instigated in the latest Turonian, and the inversion of the Turonian Vigrid syncline during Maastrichtian–late Paleocene compression (Blystad *et al.* 1995; Brekke 2000; Lundin & Doré 2011). These phases of compression were interrupted by an episode of Campanian extension (Ren *et al.* 2003). By way of contrast, the Møre Basin was tectonically quiet in the Late Cretaceous and underwent passive thermal subsidence (Brekke *et al.* 1999; Brekke 2000).

Stratigraphy. The Cretaceous succession comprises the Cromer Knoll and the Shetland groups, which are separated by the 'top Cenomanian' unconformity. In the Møre–mid-Norway region, the former group is assigned a Berriasian–Turonian age, whereas the latter is designated as Turonian–Maastrichtian (Dalland *et al.* 1988). This contrasts with the North Sea, where the Cromer Knoll and Shetland groups represent the Lower and Upper Cretaceous, respectively.

On Andoya, the Lower Cretaceous sequence is up to 550 m thick. Offshore, in the Møre and Vøring basins, the Cromer Knoll Group is 2–4 km thick, and is composed predominantly of claystones and interbedded marls, carbonates, and sandstones, with the latter becoming more common towards the top of the sequence (Dalland *et al.* 1988; Brekke 2000) (Figs 7 & 8). The overlying Shetland Group is 3–4 km thick, and predominantly comprises claystones with subordinate amounts of carbonate and sandstone. Many wells in the Møre and Vøring basins penetrate into the Upper Cretaceous succession, in which some wells are terminated having penetrating over 3.5 km of Shetland Group claystones. A thinner Cretaceous succession, ranging from 250 to 1500 m, is present outside of the rift basins.

Boundary relationships with underlying and overlying systems. The Cromer Knoll Group rests with a widespread unconformity on Jurassic and older rocks. The boundary between the Upper Cretaceous and the Paleocene successions is marked by a regional erosional unconformity in the Vøring Basin, and along the flanks of the Møre Basin (Brekke 2000) (Figs 7 & 8).

Depositional environments. The Lower Cretaceous Cromer Knoll Group comprises a mixed clastic–carbonate assemblage that was deposited in shallow- to deep-marine environments, and with increased sandstone input towards the top of the sequence, which probably reflects the onset of mid-Cretaceous deformation (Dalland *et al.* 1988; Brekke 2000). The predominance of claystones in the overlying Shetland Group, with only subordinate amounts of carbonate and sandstone, suggests deposition within an open-marine basin, albeit subjected to episodic phases of compression, uplift and extension.

SE Greenland margin

Distribution. Cretaceous rocks are exposed in the Kangerlugssuaq Basin and small outcrops with possible exposed Cretaceous–Palaeogene sediments occur at Kap Gustav Holm further south (Myers *et al.* 1993) (Figs 1, 4 & 9). Offshore, Cretaceous sediments have been sampled in the Ammassalik Basin (Vallier *et al.* 1998; Thy *et al.* 2007) (Figs 4, 10 & 12e).

Structural setting. The Kangerlugssuaq Basin is a fault-bounded Cretaceous–Palaeogene basin juxtaposed with crystalline basement. The structure of the Ammassalik Basin is poorly known, but also appears to be fault-bounded (Gerlings *et al.*, this volume, in review).

Stratigraphy. The Kangerlugssuaq Group represents the 1 km thick Cretaceous–Paleocene succession in the Kangerlugssuaq Basin (Fig. 9). Aptian sandstones and conglomerates are overlain by Albian–Coniacian sandstones and mudstones, which in turn are overlain by Campanian–Maastrichtian mudstones (Larsen *et al.* 2005*b*). The succession in the Ammassalik Basin may be several kilometres thick, although the stratigraphy remains largely unknown. The only samples taken from the basin were recovered in cores from near the top of the succession, and include Lower Cretaceous (Albian?) sandstone with coal flasers and Upper Cretaceous–Lower Paleocene siltstone and sandstone (Vallier *et al.* 1998; Thy *et al.* 2007) (Fig. 10). The contact between these two units, and thus the extent of the stratigraphic gap, has not been established.

Boundary relationships with underlying and overlying systems. The base of the Cretaceous against the crystalline basement is not exposed in detail in the Kangerlugssuaq Basin: the erosional upper boundary is overlain by Danian fluvial conglomerates and Palaeogene volcanics of the Blosseville Group (Larsen *et al.* 1999). Boundary relationships within the Ammassalik Basin remain unknown (Gerlings *et al.*, this volume, in review).

Depositional environment. The Kangerlugssuaq Basin comprises Aptian alluvial units succeeded by Albian–Coniacian shoreface and offshore deposits. The uppermost Cretaceous is represented by marine basinal mudstones and turbidites (Larsen *et al.* 1999). In the Ammassalik Basin, the Albian sandstones are of shallow-marine origin (Thy *et al.* 2007), whereas the Upper Cretaceous–Lower Paleocene rocks represent marine siltstones and turbidites overlain by a thin, sandy fluvial unit (Vallier *et al.* 1998).

Faroe–Shetland–northern Rockall–Hebrides margin

Distribution. Cretaceous strata are probably widespread throughout the Faroe–Shetland–northern Rockall region, although their full extent in the deep-water basins is obscured by Palaeogene volcanic rocks (Figs 4 & 14). The northern Rockall Basin also includes the Late Cretaceous volcanic seamounts of Rosemary Bank and Anton Dohrn (Jones *et al.* 1974; Morton *et al.* 1995) (Fig. 4). The Cretaceous is only sparsely preserved on the Hebrides margin. On the Rockall Plateau, Lower Cretaceous rocks occur in half-graben on the Hatton High (Hitchen 2004) (Fig. 4), but the presence of Cretaceous in the adjacent Hatton Basin remains ambiguous (Shannon *et al.* 1999).

Structural setting. The Cretaceous succession was deposited in a synrift setting (Dean *et al.* 1999). The Lower Cretaceous sequence is largely confined within the various basins and sub-basins, with the thickest accumulations juxtaposed against the footwall of the adjacent basement/intrabasinal high (Dean *et al.* 1999; Grant *et al.* 1999; Goodchild *et al.* 1999; Lamers & Carmichael 1999; Ritchie *et al.* 2011*b*, 2013) (Fig, 14a, d). The Upper Cretaceous sequence is more widely developed and by the end of the Cretaceous period most of the major basement highs, including the Rona High, had been isolated or drowned (Dean *et al.* 1999; Stoker & Ziska 2011). In the Faroe–Shetland region, an angular unconformity separates folded and eroded Turonian and older strata from Coniacian–Maastrichtian rocks in the West Shetland, North Rona, East Solan and West Solan basins, as well as the

Foula Sub-basin (Booth *et al.* 1993; Dean *et al.* 1999; Goodchild *et al.* 1999; Grant *et al.* 1999; Stoker 2016). In other West Shetland Shelf basins, much of the Cenomanian–Turonian section is absent, and the unconformity essentially separates Upper and Lower Cretaceous (Fig. 9). Several other intra-Lower and Upper Cretaceous unconformities occur in the West Shetland Shelf and Hebrides Shelf basins, ranging from the Hauterivian to the Campanian in age (Figs 9 & 10), and indicate the persistence of differential uplift and subsidence throughout the Cretaceous. In the northern Rockall Basin, the intrusion of the Rosemary Bank and Anton Dohrn seamounts is linked to Late Cretaceous extension (Ritchie *et al.* 1999). This pattern of coeval extension and compression is consistent with regional strike-slip associated with transtension and transpression (Roberts *et al.* 1999).

Stratigraphy. A total Cretaceous thickness of 5 and 3.5 km has been estimated for the Faroe–Shetland and northern Rockall basins, respectively (Stoker & Ziska 2011; Smith 2013), the bulk of which is of Late Cretaceous age. In the Faroe–Shetland region, the Lower Cretaceous Cromer Knoll Group is dominated by a punctuated coarse clastic assemblage (Ritchie *et al.* 1996; Stoker & Ziska 2011; Stoker 2016) (Fig. 9). Up to 1 km of sandstones with subordinate conglomerates, mudstones, limestones and argillaceous coal is preserved in the West Shetland Basin, although the sequence is generally much thinner (<0.5 km) in other shelf basins (e.g. the North Rona Basin and the West Solan Basin). In the Faroe–Shetland Basin, the Cromer Knoll Group is predominantly marine mudstone, with sporadic coarse clastic deposits preserved adjacent to the basin margin and intrabasinal highs. A maximum-drilled thickness of 1.5 km occurs in the Foula Sub-basin. The Upper Cretaceous consists of argillaceous limestones and mudstones, with rare organic-rich pyritic mudstone (Chalk Group) of Cenomanian–Turonian age in the North Rona, East Solan and West Shetland basins, which pass northwards and upwards into Cenomanian–Maastrichtian calcareous mudstones (Shetland Group) in the Faroe–Shetland Basin (Ritchie *et al.* 1996; Harker 2002; Stoker & Ziska 2011; Stoker 2016) (Fig. 9). The Chalk Group is generally less than 250 m thick, whereas the Shetland Group possibly exceeds 3.5 km in thickness in the Foula, Judd and Flett sub-basins.

In the Inner Hebrides region, Lower Cretaceous rocks are absent, but a thin (<25 m thick) succession of Upper Cretaceous shallow-marine deposits, including chalk and limestones, comprise the Inner Hebrides Group (Mortimore *et al.* 2001; Hopson 2005; Waters *et al.* 2007) (Fig. 10). The West Flannan Basin contains organic-rich Berriasian mudstones, which are probably equivalent to the Upper Jurassic–lowest Cretaceous Humber Group. In the West Lewis Basin, these rocks are unconformably overlain by thin (<10 m thick) Barremian–Cenomanian shallow-marine sandstones and mudstones, which are, in turn, unconformably overlain by over 200 m of Coniacian–Maastrichtian marine mudstones (Hitchen & Stoker 1993; Smith 2013).

In the NE Rockall Basin, a predominantly Upper Cretaceous mudstone sequence has been proved, with a maximum-drilled thickness of about 1.8 km, although the section is locally intruded by up to 700 m of basic igneous sills of Late Cretaceous or Paleocene age (Archer *et al.* 2005; Smith 2013). On the eastern flank of the central northern Rockall Basin, a 500 m-thick Cretaceous sequence comprises late Barremian–early Aptian to Maastrichtian mudstones (on a possible basal conglomerate) that overlie Archaean basement (Smith 2013). A prominent, largely unfaulted reflector that separates the Lower and Upper Cretaceous sequences is marked by a thick Cenomanian limestone (Musgrove & Mitchener 1996). It is estimated that the Upper Cretaceous is up to 2.5 km thick beneath the central part of the northern Rockall Basin, with the Lower Cretaceous in excess of 1 km (Smith 2013).

The volcanic seamounts of Rosemary Bank and Anton Dohrn have been dated as Late Cretaceous on the basis that upper Maastrichtian limestone with basalt clasts is preserved on the top of Rosemary Bank (Morton *et al.* 1995), whereas unmetamorphosed Maastrichtian chalk is preserved in vesicles in lavas from the Anton Dohrn Seamount (Jones *et al.* 1974).

On the Rockall Plateau, Albian organic-rich mudstones of paralic origin are preserved within half-graben on the Hatton High (Hitchen 2004) (Fig. 10).

Boundary relationships with underlying and overlying systems. The Cretaceous is largely separated from Jurassic and older strata by a widespread unconformity (Figs 9 & 10). On the Hatton High, the base of the sequence has not been sampled. The Cretaceous–Paleocene boundary is sometimes conformable in the deeper parts of basins, but, more commonly, it is marked by a widespread unconformity (Stoker & Ziska 2011; Smith 2013).

Depositional environment. In the basins underlying the West Shetland Shelf, the Cromer Knoll Group was deposited in marginal- to shallow-marine and basinal marine environments that were subject to fluctuating aerobic/anaerobic bottom waters (Ritchie *et al.* 1996; Harker 2002; Stoker & Ziska 2011). Common intraformational unconformities imply contemporary rifting, albeit localized and intermittent (Dean *et al.* 1999; Harker 2002; Larsen *et al.* 2010; Stoker 2016). An expansion of rift activity across the entire Faroe–Shetland–northern Rockall–Hebrides region probably occurred in the late Barremian/Aptian–Albian, with paralic to marine clastic deposition in all basins, including on the Hatton High.

The Chalk and Shetland groups were largely deposited in an aerobic, open-marine, shelf to basinal setting. Sediment thickness and accumulation rate increased during the Late Cretaceous as both the Faroe–Shetland and northern Rockall basins accumulated an expanded Cretaceous sequence (e.g. Fig. 14a). However, the increased subsidence of these basins was periodically interrupted by intermittent compression, uplift and erosion, which persisted throughout the Late Cretaceous (Dean *et al.* 1999; Doré *et al.* 1999; Roberts *et al.* 1999; Larsen *et al.* 2010; Stoker & Ziska 2011; Stoker 2016). In the northern Rockall Basin, the abundant Late Cretaceous–Paleocene sills throughout the Upper Cretaceous sequence, together with the intrusion of the axial volcanic seamounts, are probably associated with the crustal thinning.

Southern Rockall–Porcupine margin

Distribution. Cretaceous strata have a widespread occurrence in the southern Rockall–Porcupine region. Lower Cretaceous clastic-dominant (sandstones, siltstones and mudstones) successions occur in all the basins, with Upper Cretaceous chalk in all the basins and covering the intervening basement highs (Figs 4 & 15).

Structural setting. The Cretaceous succession was deposited in a late synrift to early post-rift setting. The earliest Lower Cretaceous succession in the Porcupine Basin marks the waning phase of major Tithonian rifting (Croker & Shannon 1987; Moore 1992), broadly coeval with North Sea Late Cimmerian events and with a change in regional plate movements. This was succeeded by a tectonically quiescent, post-rift period interrupted locally in the Porcupine, Slyne and Erris basins by an Aptian–Albian rift phase. Early Cretaceous volcanism is suggested in the central part of the Porcupine Basin (Croker & Shannon 1987; Tate & Dobson 1988; Naylor *et al.* 1999) (Fig. 15b, c). The Late Cretaceous saw thermal subsidence and a major sea-level rise resulting in transgression following the commencement of seafloor spreading in the North Atlantic.

Stratigraphy. The total Cretaceous section in the Porcupine Basin is up to 4 km thick. In the northern part of the Porcupine Basin, Valanginian–late Aptian sandstones, siltstones and mudstones occur, with Aptian–Albian sandstones prevalent along parts of the eastern basin margin (Fig. 11). Basinwards, sandy and silty units are encapsulated in shales (Shannon 1993). Lower Cretaceous extrusives and tuffs occur through the Barremian–Albian succession (Tate & Dobson 1988; Naylor *et al.* 2002). The Cenomanian–Danian chalk sequence in the Porcupine Basin ranges from 400 m near the margins of the basin to more than 1000 m in the basin centre (Moore & Shannon 1995).

In the northern Slyne Basin, Albian claystones and glauconitic sandstones pass upwards to sandy claystones with occasional limestones (Dancer *et al.* 2005). In the Erris Basin, late Berriasian–Valanginian claystones and sandstones are overlain by thick (>250 m) late Valanginian–Hauterivian sandstones, followed by claystones and Albian argillaceous limestones (Fig. 11). Upper Cretaceous chalk is present in the northern Slyne Basin (Dancer *et al.* 1999) and throughout the Erris Basin (Chapman *et al.* 1999).

On the eastern margin of the southern Rockall Basin, thin (<40 m) Lower Cretaceous mudstones are overlain by thick Upper Cretaceous mudstones, marls and chalks (Tyrrell *et al.* 2010) (Fig. 11). Further south along the margin, an Early Cretaceous condensed ('brownsand') succession (Haughton *et al.* 2005) is followed by a thin upper Cenomanian–middle Turonian succession, which is overlain unconformably by an upper Cenomanian–upper Maastrichtian mixed clastic–carbonate succession.

In the Fastnet Basin, up to 700 m of Lower Cretaceous strata have been encountered in wells. Berriasian–Hauterivian sandstones and shales are followed by Barremian–Albian sandstones and overlain by latest Albian–Turonian chalk (Robinson *et al.* 1981). To the west, in the Goban Spur, Neocomian sandstones and cherty limestones are overlain by Barremian–lower Aptian limestones and claystones (Cook 1987). Late Aptian–Maastrichtian strata are predominantly chalk (Fig. 11).

An important igneous episode occurred in the Early Cretaceous, represented by the Barra (south Rockall Basin) and Porcupine Volcanic Ridge systems (Naylor & Shannon 2005) (Fig. 11). The Barra system remained upstanding until the Eocene, whereas the Porcupine system was onlapped and buried by the mid-Cretaceous.

Boundary relationships with underlying and overlying systems. Regionally, the Cretaceous rests with angular unconformity on the underlying Jurassic (Figs 11 & 15). Sometimes this is a single unconformity, but locally a set of composite unconformities occurs. Several intra-Lower Cretaceous unconformities occur, with a major Aptian–Albian rift-related unconformity identified in all of the basins (Croker & Shannon 1987; Chapman *et al.* 1999; Dancer *et al.* 2005). The top of the Cretaceous succession is sometimes conformable, but more often is defined by an unconformity or disconformity with variable amounts of erosion (Fig. 11).

Depositional environment. Early Cretaceous deposition in the Porcupine Basin was primarily in a marine shale facies, interrupted locally by deltaic sandstones reflecting minor Aptian–Albian rifting (Fig. 11). Rift-generated deltaic deposits prograded from the northern and SE basin margins, with a series of offshore-bar sandstones rimming the basin. Basinwards, the clastic pulses produced sandy and silty basin-floor fans (Shannon 1993). The Upper Cretaceous (Cenomanian–Danian) chalk sequence in the Porcupine Basin reflects marine deposition with little terrigenous input.

A marine setting dominated the Cretaceous of the Slyne and Erris basins. In the Erris Basin, a thick (>120 m) Valanginian–Hauterivian submarine-fan sandstone complex is preserved (Murphy & Croker 1992; Naylor & Shannon 2011) (Fig. 11). The Upper Cretaceous, like that of the Porcupine Basin, represents a deep-water, non-terrigenous depositional environment.

On the eastern margin of the southern Rockall Basin, the Lower Cretaceous represents a shallow-marine shelf setting (Haughton *et al.* 2005), giving way basinwards to deep-water facies (Tyrrell *et al.* 2010). The Upper Cretaceous on the basin margin is in marine facies, with fluctuations in water depths, with more uniform pelagic deposition in relatively deep water.

The Cretaceous depositional environment in the Goban Spur is predominantly marine with a lower clastic (Lower Cretaceous) and an upper carbonate (Upper Cretaceous) setting reflecting a deepening of the basin. However, in the Fastnet Basin, the Lower Cretaceous comprises a fluvio-deltaic depositional setting, with a shallow-marine transgression in Aptian–Albian times and capped by deep-water Upper Cretaceous chalk facies (Robinson *et al.* 1981) (Fig. 11).

Upper Palaeozoic–Mesozoic stratigraphy of the Jan Mayen microcontinent

Jan Mayen is a volcanic island located at the northern end of the Jan Mayen Ridge (Fig. 1). The latter is a north–south-trending submarine feature that extends about 400 km southwards from Jan Mayen towards Iceland and its insular margin. The Jan Mayen Ridge is a continental remnant derived from the break-up of Norway and Greenland, and was left out in the ocean as a 'microcontinent': the Jan Mayen microcontinent (JMMC) (Blischke *et al.*, this volume, in press).

The JMMC comprises an area larger than the Jan Mayen Ridge (Fig. 1). Its northern limit is ambiguous, and there is uncertainty as to whether it extends beneath Jan Mayen Island, or whether it is located just to the south of the island; its southern limit is also unclear and might extend further towards

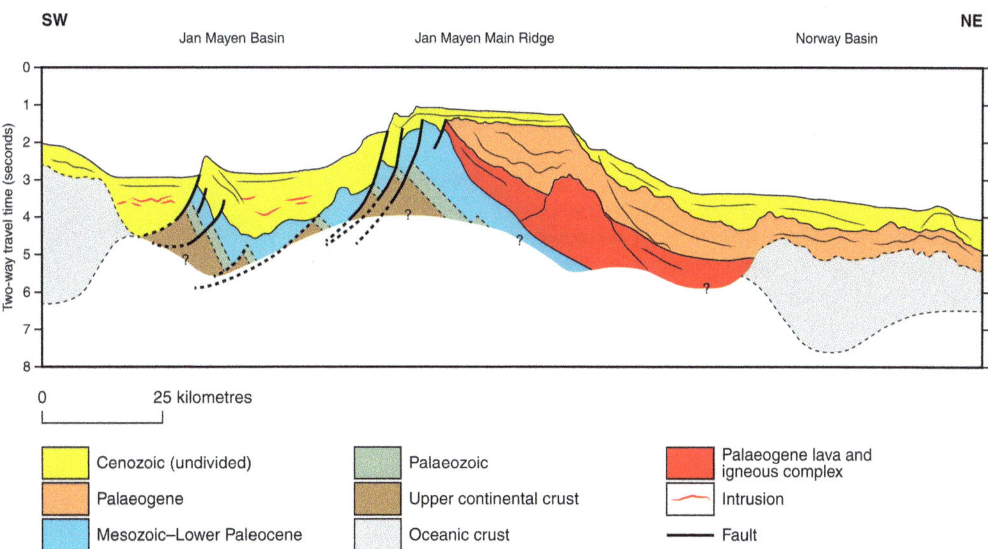

Fig. 16. Interpreted geoseismic profile across the Jan Mayen microcontinent showing the inferred stratigraphic architecture of probable Palaeozoic–Mesozoic rocks beneath the Jan Mayen Ridge and the Jan Mayen Basin. Interpretation based on Peron-Pinvidic *et al.* (2012*a*, *b*). Location of profile given in Figure 1.

Iceland than is shown in Figure 1. Consequently, the stratigraphy and structure of the JMMC is poorly known, particularly at deeper levels: thus, the following description of the Upper Palaeozoic–Mesozoic geology remains preliminary (cf. Blischke *et al.*, this volume, in press).

Distribution. The Upper Palaeozoic–Mesozoic succession is mainly inferred for the JMMC on the basis of its position to its conjugate margins in Norway and Greenland (Figs 2–4). Whereas the possibility of Mesozoic and older Palaeozoic rocks has been interpreted at depth beneath the Jan Mayen Ridge (e.g. Peron-Pinvidic *et al.* 2012*a*, *b*), a major problem in their recognition is caused by lower Palaeogene lavas and sills that commonly obscure the pre-Cenozoic sequences on seismic reflection data (Fig. 16).

The pre-Palaeozoic basement, which is tentatively imaged on seismic reflection data, is herein compared with the granite-intruded Caledonian basement (Gilotti *et al.* 2008; Kalsbeek *et al.* 2008) that is preserved on the Liverpool Land Ridge in East Greenland (Fig. 12d). Seismic refraction and reflection data further suggest that the subbasalt strata might be comparable to the Upper Palaeozoic–Lower Mesozoic successions that are exposed in the Jameson Land Basin and Traill Ø–Hold with Hope area in East Greenland, as well as the Møre and Vøring basins offshore Norway (Figs 7 & 8).

Structural setting. The structural setting of the JMMC is best compared to the structural framework along the central East Greenland coastal area, where the Jameson Land Basin, in particular, provides a potential analogue. This narrow, north–south-trending basin includes a Permian–Lower Cretaceous succession (Fig. 8) that is intruded by Cenozoic igneous rocks (e.g. thick sills) (Larsen & Marcussen 1992; Hald & Tegner 2000). On the JMMC, it is possible that deeper-level structures and rocks, which are visible through windows in the igneous section, might represent comparable Palaeozoic–Mesozoic strata (Blischke *et al.* 2011). As previously noted (see above), the Mesozoic succession in the Jameson Land Basin might continue southwards beneath the Palaeogene volcanic succession in the Blosseville Kyst area of central East Greenland (Larsen *et al.* 2005*a*, 2013), which would have located this structure right across the central part of the JMMC – a scenario previously proposed by Surlyk (1990, 2003). The observation that a thinner and more condensed pre-Cenozoic succession is preserved on the JMMC in comparison to the East Greenland and Møre–mid-Norway areas might imply that the JMMC formed a structurally higher and shallower platform between the adjacent Jameson Land and Møre basins, possibly along strike with the north–south-trending Liverpool Land High (Blischke *et al.*, this volume, in press).

Stratigraphy. The presence and extent of Palaeozoic–Mesozoic rocks along the Jan Mayen Ridge and within the Jan Mayen Basin remains ambiguous on the basis of refraction and reflection seismic interpretations. An inferred stratigraphic range for the southern JMMC is indicated in Figure 8, based on a direct comparison with the conjugate Jameson Land, Møre and Vøring basins: the northern part of the JMMC – formerly juxtaposed with the Vøring margin – has been subjected to intense Palaeogene volcanism that has obscured much of the succession on seismic profile data.

Whereas the recovery of Upper Permian–Lower Triassic and Lower Cretaceous rocks, including limestone, from the Jan Mayen Ridge was initially interpreted by the Norwegian Petroleum Directorate (2012) to indicate the presence of Upper Palaeozoic–Mesozoic rocks on the JMMC, the likelihood is that these samples represent ice-rafted detritus derived and dumped from East Greenland icebergs (Norwegian Petroleum Directorate 2013). Nevertheless, the presence of Cretaceous rocks across the JMMC seems highly probable based on the interpretation of seismic reflection data (Fig. 16), together with evidence from the conjugate Vøring and Møre basins to the east, and the Traill Ø–Hold with Hope area of NE Greenland to the west (Figs 7 & 8).

Boundary relationships with underlying and overlying systems. No boundary relationships can be firmly established: however, potential boundaries indicated in Figure 8 relate to reflecting surfaces observed on seismic data that might correlate with key unconformities preserved in the adjacent conjugate basins.

Depositional environment. In the absence of a proven Late Palaeozoic–Mesozoic stratigraphic record for the JMMC, the conjugate Jameson Land, Møre and Vøring basins are the best providers of information concerning potential depositional environments in this area.

Regional stratigraphic framework

The stratigraphic distribution maps and correlation charts presented in Figures 2–11 provide a strong observational basis for the establishment of a unified Late Palaeozoic–Mesozoic stratigraphic framework for the NE Atlantic region. Any attempt to interpret this framework in terms of the break-up of the Pangaean supercontinent has to take into consideration the general lithostratigraphic pattern and correlation along the entire length of the NE Atlantic rift system. In this section, we address this issue by utilizing a rift-axial correlation chart (Fig. 17) that links basins along the NE Atlantic rift system. In the context of the entire Pangaean supercontinent, this chart represents a transect from the northern (Boreal) marine margin to the centre of the Pangaean plate, as depicted on recent plate-wide reconstructions (e.g. Doré *et al.* 1999; Pharaoh *et al.* 2010; Golonka 2011; Lawver *et al.* 2011). Reference to the southern (Tethyan) marine margin is also included in the text where necessary. The rift system has been arranged into five geographical segments: (1) the North/NE Greenland–Svalbard–western Barents Sea segment; (2) the NE/central East Greenland–mid-Norway–Møre segment; (3) SE Greenland–Faroe–Shetland segment; (4) the southern Greenland–northern Rockall–Hebrides segment; and (5) the Inner Hebrides–NW Irish–Porcupine–southern Rockall segment. This arrangement retains the conjugate placement of basins and margins whilst, at the same time, presenting an along-axis view of the stratigraphic framework. Almost all of the stratigraphic columns shown on this chart represent a proven rock record.

From the stratigraphic information described above, and collated on the rift-axial chart (Fig. 17), we present the regional Late Palaeozoic–Mesozoic stratigraphic framework of the NE Atlantic region as it is currently known from the observed rock record. Even from a cursory inspection of the stratigraphic chart, it is obvious that regional changes in stratigraphic development largely coincide with the Permian–Triassic, Jurassic and Cretaceous successions. In the ensuing discussion, we summarize the large-scale pattern of sedimentation and basin development associated with these successions, as well as the character of the major bounding surfaces. Potential correlation to tectonic events that were affecting the Pangaean plate throughout the Late Palaeozoic–Mesozoic will also be considered, together with established views regarding faunal provinciality (Boreal–Tethyan) across this interval.

Carboniferous–Permian boundary

The northernmost part of the study area is characterized by a largely conformable Carboniferous–Permian boundary. In the North/NE Greenland–Svalbard–western Barents Sea region, a passively subsiding margin formed in the Late Carboniferous and persisted, with only minor interruptions, until the Mid–Late Permian (Stemmerik & Håkansson 1991; Surlyk 1991). A mixed siliciclastic–carbonate shelf sequence of Late Carboniferous–Permian age is preserved in basins throughout this region (see below).

By way of contrast, over most of the study area, the Carboniferous–Permian boundary is marked by a regionally extensive unconformity (Fig. 17). On the western side of the NE/central East Greenland–mid-Norway–Møre region, a regional peneplain formed by Early Permian uplift and erosion is well constrained in outcrops in the Traill Ø–Hold with Hope and Jameson Land areas where rocks of late Early–Mid-Permian age overlie Upper Carboniferous strata (Surlyk *et al.* 1986). The event is also recognized in the North/NE Greenland–Svalbard–western Barents Sea region (‘minor interruptions’), albeit within the overall context of a subsiding margin (Stemmerik 2000; Hamann *et al.* 2005).

Further south, the Permian–Triassic succession is generally more poorly dated, making it difficult to assess the precise nature of the Carboniferous–Permian boundary: however, the duration of the erosional hiatus is generally more expanded. Whereas locally there is evidence (e.g. Porcupine Basin) of a rapid transition from humid to arid conditions and the onset of red-bed deposition, elsewhere (SE Greenland–NW Irish region) there is a widespread unconformity, with rocks ranging in age from Mid-Permian to Early Triassic variably overlying Upper–Lower Carboniferous rocks. In the North Sea Basin, a comparable regional hiatus is known to be an amalgamation of several unconformities rather than a single event (Glennie *et al.* 2003; Gast *et al.* 2010). It has been suggested that the Carboniferous–Permian boundary, particularly in the southern part of the study area, formed through regional uplift and erosion driven by a regional pattern of wrench tectonics and crustal thinning associated with Variscan orogenic collapse (Pharaoh *et al.* 2010).

Permian–Triassic succession

There is a marked north–south contrast in the overall character of the Permian–Triassic succession, with a marine-shelf assemblage of carbonate and siliciclastic rocks preserved in the northernmost (Boreal) part of the study area, whereas a predominantly arid setting prevailed elsewhere (Fig. 17). The most complete Permian–Triassic successions have been proved from Svalbard and the SW Barents Sea. Further south, in the area between NE Greenland/mid-Norway and the Porcupine Basin, Middle/Upper Permian–Triassic sequences predominate, with Triassic strata being the thickest and most widespread unit. In some parts of North and NE Greenland, Permian and/or Triassic rocks are absent, and in other areas, such as SE Greenland, the Rockall Plateau and the north Rockall Basin, no Permian–Triassic rocks have so far been proved.

The bulk of the known Permian rocks are preserved in central East Greenland, North and NE Greenland, Svalbard, and the western Barents Sea. In the Wandel Sea Basin, the Late Carboniferous–Permian marine-shelf succession comprises sandstone, shale and carbonate rocks (Stemmerik & Håkansson 1991; Surlyk 1991). A comparable assemblage is also preserved on Svalbard and in the Hammerfest Basin (SW Barents Sea), and is inferred to be present in the Danmarkshavn Basin (Gautier *et al.* 2011).

In central East Greenland, subsidence was instigated in the Mid-Permian as a consequence of thermal contraction, with minor rifting in mainly westerly tilted half-graben (Surlyk *et al.* 1986). The deposition of an overall transgressive Middle–Upper Permian succession in the Traill Ø–Hold with Hope and Jameson Land areas was initiated by fluvial siliciclastics, followed by marine and hypersaline carbonates and evaporites that are unconformably overlain by fully marine carbonate build-ups along basin margins, together with organic-rich basinal mudstones (Surlyk *et al.* 1986; Surlyk 1990; Stemmerik 2000) (Fig. 17).

Further south, a fragmentary Lower Permian record includes alkaline lavas in the West Shetland region, and ?Lower Permian terrestrial sandstones and mudstones in the Porcupine, southern Rockall and Erris basins, offshore Ireland. The Early Permian volcanism and sedimentation in these areas are linked to tectonism associated with Variscan orogenic collapse.

A significant pulse of rifting affected the NE Atlantic region during the Late Permian–earliest Triassic, which coincided with the final phase of the Uralian Orogeny (Fig. 17) in the eastern Barents Sea region (Smelror *et al.* 2009), to the NE of the study area. Variable large-scale basin architectures are preserved, including half-graben and broad sheet-like geometries. This is interpreted as reflecting a depositional setting controlled by palaeotopographical infill, faulted depocentres (early localized rifts along reactivated Variscan and Caledonian lineaments) and wide-rift extensional processes (e.g. Štolfová & Shannon 2009). Mid–Late Triassic time was generally characterized by post-rift thermal subsidence.

In the northern part of the study area (Greenland–Norway conjugate), the Late Permian–earliest Triassic rifting controlled marine-dominated deposition in mainly westerly tilted half-graben. The Permian–Triassic boundary in this region is a prominent unconformity at basin margins. A coeval rise in sea level resulted in a marine passageway – the Zechstein Sea (cf. Glennie *et al.* 2003) – that transgressed southwards between Greenland and Norway (Seidler *et al.* 2004; Bjerager *et al.* 2006). Upper Permian–Lower Triassic shallow-marine evaporites in the Vøring Basin and in the West Shetland region are most probably linked to this shallow-marine incursion, which extended into the North Sea (Glennie *et al.* 2003), although faunal evidence indicates that there was no connection with the Tethyan realm at this time (Smith 1980; Doré 1992). Equivalent rocks are not present on the Hebrides margin, but are preserved in basins offshore NW Ireland (Fig. 17). However, in the latter area, deposition may have been entirely terrestrial and related to a relative rise of the regional water table rather than marine flooding (Glennie *et al.* 2003).

The Triassic was a time of mainly arid continental deposition with thick coarse clastic successions of red beds preserved in various basins located between NE/central East Greenland–mid-Norway and the Porcupine and southern Rockall basins, offshore Ireland (Fig. 17). Thicknesses in excess of 2 km are preserved in basins offshore NW Scotland, on the inner Vøring margin and onshore central East Greenland (Fig. 2). Recorded Triassic deposits in central East Greenland decrease progressively north of Jameson Land and reappear in the North Greenland Wandel Sea Basin, where Lower Triassic marine siliciclastic rocks are preserved. In contrast, Triassic deposits of presumed marine shale and sandstone are inferred to be present in the Danmarkshavn Basin (Hamann *et al.* 2005), and which may have been sourced from an uplifted western basin margin (Middle and Upper Triassic rocks are absent onshore North and NE Greenland). Triassic deposits are also inferred to be present in the Thetis Basin (Hamann *et al.* 2005; Dinkelman *et al.* 2010). The Triassic siliciclastic succession is essentially restricted to the northern margin of the study area, which was a wide shelf basin at this time. On Svalbard and in the western Barents Sea, a succession of alternating marine, deltaic and coastal facies prograded into this basin (Brekke *et al.* 2001; Riis *et al.* 2008; Worsley 2008; Smelror *et al.* 2009; Henriksen *et al.* 2011). The extensive Mid–Late Triassic westwards progradation of deltaic systems was a consequence of the progressive uplift and erosion in the Uralian highlands, as well as the northern Fennoscandian Shield.

A brief marine incursion from the Boreal Sea in the Mid-Triassic brought shallow-marine mudstones and carbonates to the North/NE and central East Greenland part of the NE Atlantic rift system (Fig. 17). It has been speculated that this ingression might correlate with Tethyan marine incursions into the North Sea (e.g. Muschelkalk Formation) (Jacobsen & van Veen 1984). However, it is stressed that palaeogeographical and faunal evidence do not support the likelihood of a through-going marine connection between the Boreal and Tethyan realms in Mid- and Late Triassic times (Doré 1992).

Triassic–Jurassic boundary

The character of the Triassic–Jurassic boundary is highly variable across the NE Atlantic region. It is marked by a major unconformity in North and NE Greenland, the Møre Basin, and across much of the Faroe–Shetland–Hebrides region. By way of contrast, shallow-marine and paralic sedimentation persisted, at least locally, from the Triassic into the Early Jurassic on Svalbard, in the western Barents Sea, and in the Vøring and Jameson Land basins, whereas, in the Inner Hebrides region, a transition from an arid to a shallow-marine setting, which began in the latest Triassic (Rhaetian) and continued into the Early Jurassic, is locally observed. Further south, the Triassic–Jurassic boundary where observed (e.g. in the northern part of the Porcupine Basin, NW Irish offshore basins and the Fastnet Basin) and interpreted (e.g. Goban Spur, perched basins on the flanks of the southern Rockall Basin) is more widely conformable and marks a rapid transition, through a marine transgression, from continental arid to shallow-marine shelf limestones. The marine transgression commenced in the Rhaetian, and was a regionally extensive event, occurring rapidly and at the same time throughout all of the Irish offshore basins. There is no evidence of any significant tectonism (rifting) at the Triassic–Jurassic boundary in this area.

This variability in the character of the Triassic–Jurassic boundary might be a consequence of events occurring on the flanks of, and within, the Pangaean plate. Along the Boreal margin of the plate, it is suggested that sedimentation in the Svalbard–western Barents Sea region was still controlled by compression and uplift generated by the Uralian Orogeny, as well as a renewal of fault-related tectonic activity at the end of the Triassic (Smelror *et al.* 2009). The latter tectonism is probably linked to an acceleration of rifting and subsidence within the northern part of the NE Atlantic rift system (Greenland–Norway) during the Late Triassic, as it started to propagate southwards towards the Central Atlantic domain and generated contemporaneous uplift and erosion (and, thus, unconformities) on the flanks of the rift system (Pharaoh *et al.* 2010). On the southern edge of the Pangaean plate, incipient ocean-floor spreading was instigated in the Tethys to the SE and in the proto-central Atlantic to the SW (Doré *et al.* 1999) (Fig. 17). According to Doré (1992), marine flooding (?Rhaetian transgression) of the Permian–Triassic rift basins occurred as rifting breached Pangaea.

Jurassic succession

The distribution and stratigraphic range of the Jurassic succession is variable across the NE Atlantic region (Fig. 17). The most complete successions are preserved in Svalbard and the western Barents Sea, the Jameson Land Basin, the inner Vøring margin, the inner Hebrides margin, and offshore NW Ireland (e.g. Slyne Basin), although unconformities are present in these areas, particularly a widespread Mid-Jurassic break. Fragmentary Jurassic sequences are preserved elsewhere (e.g. onshore North and NE Greenland, the Faroe–Shetland Basin, the outer part of the Hebrides margin, the Porcupine Basin, and the southern Rockall Basin), and it was not until the Mid–Late Jurassic that a more regionally extensive succession was deposited. Areas of uncertainty remain, especially the basins offshore NE Greenland, where relatively complete successions are inferred, but remain untested. In some areas, such as SE Greenland, the Rockall Plateau and the northern Rockall Basin, no Jurassic rocks have so far been proved.

In Svalbard, the SW Barents Sea, the Jameson Land Basin and the inner Vøring margin, the repetitive clastic succession of deltaic, paralic and shallow-marine shelf sedimentation initiated in the Triassic, persisted into the Early Jurassic. These areas were characterized by passive thermal subsidence, with only minor fault activity and uplift (Blystad *et al.* 1995; Brekke *et al.* 2001; Surlyk 2003; Henriksen *et al.* 2011). In contrast, the adjacent areas of North and NE Greenland, as well as parts of the Norwegian margin (e.g. Møre region), were areas of more extensive uplift and erosion (Brekke *et al.* 1999; Doré *et al.* 1999). Comparable coastal and marine deposition occurred in the Faroe–Shetland and Inner Hebrides regions, and offshore NW Ireland. Limited extensional fault activity is interpreted to have occurred in the Inner Hebrides region (Morton 1989).

This general pattern of Early Jurassic thermal subsidence and mild extensional tectonism was interrupted in the Mid-Jurassic by widespread uplift and shallowing, which is commonly reflected in the stratigraphic record as an unconformable boundary between Lower and Middle Jurassic rocks. In the North Sea, this boundary is termed the 'Mid-Cimmerian Unconformity', and the uplift has been attributed to a hotspot-related dome (Underhill & Partington 1993; Glennie & Underhill 1998). However, the palaeogeographical extent of the Mid-Jurassic unconformity (Fig. 17) might suggest that this dome was one of a family of uplifts that extended across the European–NE Atlantic region (Doré *et al.* 1999). A major impact of uplift at this time was the lack of marine continuity along what was to become the Atlantic margin. Indeed, the acute faunal provinciality (Callomon 1984) established between the Boreal and Tethyan realms from the late Bajocian onwards is an indication of the existence of a substantial palaeogeographical barrier (Doré 1992).

An intense phase of rifting occurred in some parts of the NE Atlantic region in late Mid–Late Jurassic time, and there is a general trend towards marine conditions. However, continuity of any seaway between NW Europe and the Arctic was probably highly complex, at the very least, as reflected in the variable intensity of the Late Jurassic tectonic activity, as well as the high degree of faunal (ammonite) provinciality even within the Boreal realm (Casey 1971; Callomon & Birkelund 1982; Doré 1992). In central East, NE and North Greenland, major rifting was initiated in the late Bajocian and culminated in the earliest Cretaceous. Middle Jurassic flooding and the northwards propagation of the rifting resulted in the formation of a series of connected, elongate, marine basins in platform and half-graben settings (Surlyk 2003). Whereas the Jameson Land Basin experienced minor tilting during this interval, the Wollaston Forland Basin (Traill Ø–Hold with Hope area) underwent significant rotational block faulting in the westwards-tilted half-graben. In Svalbard and the western Barents Sea, Mid-Jurassic–Early Cretaceous rifting caused a marked rejuvenation of the topography into a system of structural highs and basins (e.g. Bjørnøya Basin, Tromsø Basin), and triggered the development of a marine connection across the Barents Sea shelf (Smelror *et al.* 2009; Faleide *et al.* 2010). In the northern part of this area, some of this tectonic activity may have been related to extension in the Amerasia Basin, as a precursor to the opening of the Arctic Ocean (Smelror *et al.* 2009). Shallow-shelf to deep-marine sedimentation prevailed over a large part of this northern area, including the accumulation of black shales in many of the basins. In the mid-Norway–Møre region, variable thicknesses of marine sediment accumulated on a series of tilted fault blocks, particularly on the Vøring margin (Faleide *et al.* 2010).

Paralic and marginal-marine sedimentation prevailed in the Faroe–Shetland–Hebrides region until latest Jurassic–earliest Cretaceous time when the area was transgressed and marine mudstones, including organic-rich black shales, were deposited. It has been suggested that the general lack of variation in the thickness of the Upper Jurassic rocks across the Faroe–Shetland region is an indication that the effects of any contemporaneous rift activity in this area was negligible (Dean *et al.* 1999; Doré *et al.* 1999).

Offshore Ireland, the Middle Jurassic marks a major change in depositional basin shape and size with the onset of rifting. Some rifting (e.g. NW Irish offshore basins) reactivated existing lineaments (e.g. of Caledonian age), while in other cases (e.g. Porcupine Basin) a new, large north–south basin geometry resulted. The switch from a marine-shelf (Lower Jurassic) to a humid fluvial (Middle Jurassic) environment reflects this tectonic change (i.e. the onset of rifting). Rifting reached its nadir in Late Jurassic times with the formation of tilted fault blocks and major facies variations (Shannon 1991; Sinclair *et al.* 1994). In the central and southern Porcupine Basin, and in the southern Rockall Basin, crustal stretching was extreme (hyperextension) in Late Jurassic time, resulting in significant differential subsidence towards the basin centres. The Upper Jurassic succession is characterized by facies variations, ranging from fluvial to basin-floor sandstones, siltstones and mudstones to shallow-water limestones. This marks a general upwards trend from non-marine to marine conditions.

In the North Sea Basin, a major phase of extensional activity, accelerated subsidence and basin deepening in the Late Jurassic is well established (Glennie & Underhill 1998). Pharaoh *et al.* (2010) have attributed the intensification of the North Sea rift system to a wrench-dominated tectonic regime. Doré *et al.* (1999) suggested that Late Jurassic rifting, in general, throughout NW Europe was driven by an east–west extensional stress regime influenced predominantly by seafloor spreading in the Tethys, and cite close to northerly-trending structural elements, such as the Porcupine Basin, Halten Terrace, and the basins of central East and NE Greenland, as unequivocal Jurassic rift basins.

Jurassic–Cretaceous boundary

In North and NE Greenland, the Jurassic–Cretaceous boundary is transitional and marks the culmination of rifting in this region, with associated basin segmentation, block tilting and footwall uplift. Prominent deep-marine coarse clastic sediments were deposited along rift margins, which pass distally into mudstones (Surlyk 1978). This tectonic activity triggered major salt movements in the northern part of the Danmarkshavn Basin (Hamann *et al.* 2005). A transitional boundary is also observed on Svalbard, whereas an unconformity is developed in the SW Barents Sea.

Over most of the rest of the NE Atlantic region, extending southwards from the central East Greenland–mid-Norway–Møre region, the Jurassic–Cretaceous boundary is predominantly an unconformity, albeit commonly an intra-Berriasian (earliest Cretaceous) event. Significantly, perhaps, this coincides with the most pronounced phase of Ammonite provincialism, with Boreal and Tethyan communities completely cut off from each other during this interval (Rawson & Riley 1982). The unconformity is especially well developed in the

Faroe–Shetland–Hebrides region where it marks the termination of the Upper Jurassic–lowest Cretaceous black shales. Offshore Ireland, the Jurassic–Cretaceous boundary is always marked by one or more unconformities. Unconformities coalesce and individual unconformities do not extend regionally through a basin: however, the overall boundary is characterized by an unconformable nature. Some (diminishing) rifting prevailed across the boundary. In the Faroe–Shetland region, the Upper Jurassic–lowest Cretaceous black shales are locally overlain by paralic deposits, which suggest that there was subaerial exposure, at least locally. In contrast, the Upper Jurassic and lowest Cretaceous strata offshore Ireland is marine in nature, suggesting that there was no regional subaerial exposure in this region.

In the North Sea, the intra-Berriasian unconformity is generally termed the Late Cimmerian Unconformity (Base Cretaceous Unconformity) (Oakman & Partington 1998). This is described as a response to an eustatic sea-level lowstand combined with stress-induced deflection of the lithosphere, which led to earliest Cretaceous emergence and erosion of large parts of western and central Europe (Pharaoh *et al.* 2010). It also marks the approximate end of rifting in the North Sea as the influence of Tethyan spreading waned and Atlantic spreading intensified (Fig. 17) in response to the change in regional (plate-scale) stress directions from east–west in the Jurassic to NW–SE in the Early Cretaceous (Doré *et al.* 1999).

Cretaceous succession

Cretaceous rocks are widespread across the region and, whilst the stratigraphic range of the preserved succession is variable, its thickness is commonly of kilometre-scale (Fig. 4). In several areas, such as the Jameson Land Basin and the inner Hebrides margin, the Cretaceous is largely absent: whereas, in SE Greenland, the Rockall Plateau and the north Rockall Basin, Cretaceous rocks represent the first record of any Upper Palaeozoic–Mesozoic deposition in these areas. The succession mainly comprises marine facies dominated by mudstones, although carbonate rocks become increasingly common in Upper Cretaceous sequences south of the Faroe–Shetland region (Fig. 17), which coincides with the northern limit of the chalk sea in the North Sea region (Oakman & Partington 1998). According to Casey & Rawson (1973), an open-marine seaway had been established by Albian time connecting the Boreal and Tethyan realms via the North Sea. The timing of a fully marine connection with the developing Central Atlantic Ocean via the NW British–SE Greenland region remains equivocal (see below).

As Atlantic spreading propagated northwards, a broad zone of extension and subsidence developed stretching from the southern Rockall Basin to the western Barents Sea (Doré *et al.* 1999; Roberts *et al.* 1999). Along the northernmost edge of the study area, much of the rifting between North Greenland and the Svalbard–western Barents Sea region was progressively taken up by strike-slip movement within the De Geer Zone – a mega-shear system that linked the North Atlantic and Arctic regions prior to break-up (Eldholm *et al.* 2002) – leading to the formation of pull-apart basins in North Greenland and the SW Barents Sea (Håkansson *et al.* 1991; Smelror *et al.* 2009). In North Greenland, the Wandel Sea Basin developed as a pull-apart basin, which rapidly accumulated thick sequences of paralic and marine deposits during discrete, relatively short-lived pulses of rifting in the Early and Late Cretaceous. Activity in this strike-slip zone was terminated close to the Cretaceous–Paleocene boundary by compression associated with the Eurekan Orogeny (Håkansson *et al.* 1991). In the SW Barents Sea, episodic rifting phases led to rapid subsidence and the development of a series of deep basins, such as the Harstad, Tromsø, Bjørnøya and Sørvestsnaget basins, which became decoupled from the rest of the Barents Sea shelf during the rifting. These basins accumulated thick sequences of predominantly marine mudstone.

At the same time, the northern part of the Barents Sea area was uplifted in the Early Cretaceous, which resulted in the deposition of a regressive clastic wedge on Svalbard. This Early Cretaceous uplift was associated with a major volcanic event in the Barents Sea region, including Svalbard, which coincided with the onset of break-up of the Amerasian Basin (Smelror *et al.* 2009). Uplift prevailed in the north, and by the Late Cretaceous large parts of the Barents Sea shelf – including Svalbard – had been uplifted.

The basins of NE and central East Greenland were mainly characterized by thermal subsidence and/or infilling of inherited rift relief, throughout the Cretaceous, accompanied by the deposition of predominantly marine basinal mudstones. This style of sedimentation and basin development was locally and sporadically interrupted by minor fault episodes that generated coarse-grained siliciclastic deposits (Whitham *et al.* 1999; Larsen *et al.* 2001; Surlyk & Noe-Nygaard 2001). Towards the end of the Cretaceous many of the onshore basins were uplifted and eroded, whereas the offshore Danmarkshavn and Thetis basins continued to subside and accumulate sediments (Hamann *et al.* 2005). In the Møre–mid-Norway region, the Vøring and Møre basins also underwent significant subsidence throughout the Cretaceous, and accumulated thick sequences of predominantly marine mudstone. However, whereas the Møre Basin experienced mainly continuous subsidence, the Cretaceous development of the Vøring Basin comprised an early thermal subsidence phase and a post-Cenomanian phase of tectonically driven subsidence involving intermittent phases of normal faulting and compression and folding (Brekke 2000). This deformation has been linked to tectonic activity on structures, such as the Jan Mayen Lineament – the continental extension of the East Jan Mayen Fracture Zone (Blystad *et al.* 1995; Brekke 2000) (Fig. 1). Both the Møre and Vøring basins were uplifted at the end of the Cretaceous.

The Cretaceous development of the basins of the Faroe–Shetland region involved an Early Cretaceous rifting phase that became regionally expansive in the Aptian–Albian, and, as with the Vøring Basin, a post-Cenomanian–Turonian tectonically driven subsidence involving intermittent phases of normal faulting, compression and folding. The accumulation of upper Berriasian–Barremian paralic and shallow-marine deposits in basins on the West Shetland Shelf marks the instigation of rifting: however, it was not until the Aptian–Albian that marine clastic rocks became more widespread across the region as connectivity increased between basins. The development of the Faroe–Shetland Basin as an integrated, wide marine basin occurred in the Turonian–Maastrichtian interval, although the region as a whole was subjected to differential uplift and subsidence, which has been interpreted to be indicative of wrench tectonics (Dean *et al.* 1999; Roberts *et al.* 1999; Stoker 2016). A large part of the region was uplifted and eroded at the end of the Cretaceous, although depositional continuity may have prevailed in the deeper parts of the Faroe–Shetland Basin.

On the Hebrides Shelf, the Cretaceous succession is commonly absent due to uplift and erosion, although a fragmentary Upper Cretaceous shallow-marine clastic and carbonate sequence is preserved in the Inner Hebrides. On the Outer Hebrides Shelf, Barremian–Aptian and younger Cretaceous marine clastic deposits are preserved in the West Lewis Basin, and have also been proved in the adjacent northern Rockall Basin. Comparable sequences of paralic to marine clastic rocks are present in half-graben on the north Hatton High (outer Rockall Plateau), and in the Kangerlussuaq and Ammassalik basins of SE Greenland. In the northern Rockall Basin–Rockall Plateau–SE Greenland region, the Cretaceous rocks represent the first observable indications of rift activity in this area. The folded, faulted and eroded character of the Cretaceous rocks on the north Hatton High suggests that they may have been more extensive prior to the break-up between SE Greenland and the Rockall Plateau.

The Cretaceous succession is extensively preserved offshore Ireland, and is predominantly marine in nature. The Porcupine Basin developed mainly by thermal subsidence, initially following Late Jurassic rifting and, subsequently, in response to Late Cretaceous seafloor spreading in the southernmost North Atlantic. This general pattern of subsidence was interrupted in the Aptian–Albian by a phase of rifting, which also affected the NW Irish basins. The southern Rockall Basin may have undergone a similar development. The Lower Cretaceous succession is predominantly clastic in nature (sandstones and mudstones), and was deposited typically in an outer-shelf to deep-marine setting, with ponded turbidites in residual rift topographical lows and in basin-floor settings. Localized Aptian–Albian deltaic sandstones occur on the flanks of the Porcupine Basin and represent the deposits of minor rifting. The Upper Cretaceous is chalk-dominant, and is present in all basins covering Jurassic footwalls and extending onto basement highs and beyond the edges of the rift basins. Whereas an unconformity separates Cretaceous and Paleocene rocks on the flanks of the Porcupine and southern Rockall basins, the transition in deeper water was more continuous.

It may be no coincidence that the pulse of Aptian–Albian rifting that is widely recorded across the southern part of the study area correlates with the onset of break-up between Newfoundland and Iberia (Doré *et al.* 1999) (Fig. 17). The increasing focus of crustal extension within the Rockall–Faroe–Shetland and Hatton–SE Greenland regions reflects this change in regional stress directions (as described above). This was accompanied by Early Cretaceous volcanism in the central part of the Porcupine Basin and in the southern Rockall Basin (areas of crustal hyperextension according to Lundin & Doré 2011), and extrusives and tuffs occur through the Barremian–Aptian in the Porcupine Basin. The tectonic pulses in Late Cretaceous (post-Cenomanian) times, observed across a large part of the study area, were probably coupled to the opening of the Labrador Sea and the associated anticlockwise rotation of Greenland (Brekke 2000). However, the different structural expressions (faulting, folding and inversion) indicate that the intraplate stress regime changed through time, and the build-up of compressional stresses as part of the evolving Alpine orogen might also have played an important role in modulating stresses throughout the plate (Pharaoh *et al.* 2010). Volcanism continued to prevail in the Rockall Basin with the intrusion of the Late Cretaceous Anton Dohrn and Rosemary Bank igneous centres in the northern part of the basin.

Implications for palaeogeographical reconstruction of the NE Atlantic region

We acknowledge all existing palaeogeographical syntheses that have been established for the NE Atlantic region in terms of their important contribution to the on-going interpretation and discussion of the evolution of this area. Ideas are, however, being constantly revised. As such, the distribution and stratigraphic range of the Permian–Cretaceous rock record that are presented in this paper enable us to test the viability of putative palaeogeographical reconstructions in terms of the accuracy of the reconstructed plate-tectonic base maps, and the history of basin development. Notwithstanding the limitations commonly inherent in plate-tectonic reconstructions (e.g. continental shortening, problems of de-stretching the crust), our basic stratigraphic distribution maps help to illustrate two key problems that are common to many of the existing reconstructions:

- Early Mesozoic plate reconstructions based on palaeomagnetic interpretations (e.g. Ziegler 1988; Coward *et al.* 2003; Pharaoh *et al.* 2010; Torsvik *et al.* 2012) result in a re-fit of the conjugate margins of NE Greenland and Norway that is too tight (i.e. overlapping), leaving little or no space available to accommodate the offshore sedimentary basins that are depicted in Figures 1–4. Whereas the Mid-Jurassic plate reconstruction of Nøttvedt *et al.* (2008) acknowledges the distribution of the Mesozoic basins along the conjugated margins of East/NE Greenland and Norway, the new NAG-TEC data are better constrained and show more detailed basin distributions (Figs 2–4).
- The reconstruction of the area between SE Greenland and NW Britain–Ireland is a long-standing problem. Not only is the amount and timing of extension in this area controversial (e.g. Roberts *et al.* 1988, 1999; Doré 1992; Shannon *et al.* 1995, 1999; Doré *et al.* 1999), the stratigraphic record (as shown in the NAG-TEC data) and, thus, the basin history remain largely unknown (see below) (Figs 2–4 & 17). Despite this, an arrangement of structural highs and basins, as well as inferred rifts, such as the 'Hatton–Greenland rift', comparable to the modern-day structural and bathymetric framework is commonly utilized in all reconstructions back to, and including, the Permian (e.g. Ziegler 1988; Cope *et al.* 1992; Doré 1992; Knott *et al.* 1993; Torsvik *et al.* 2002; Coward *et al.* 2003; McKie & Williams 2009; Pharaoh *et al.* 2010).

Classic plate reconstructions do not address this complexity: thus, existing palaeogeographical maps of this region should be regarded as highly speculative.

In terms of basin history, existing palaeogeographical syntheses have been, and continue to be, updated and modified as more knowledge becomes available. This is especially the case in the North/NE Greenland–western Barents Sea–Svalbard region, where a series of studies (e.g. Doré 1992; Roberts *et al.* 1999; Torsvik *et al.* 2002; Nøttvedt *et al.* 2008; Smelror *et al.* 2009) show a progressive refinement of the palaeogeography. The stratigraphic data presented in this paper should add to this growing database.

By way of contrast, the geological history of the southern part of the study area, especially between SE Greenland and NW Britain, remains poorly understood. For reasons noted above, the structural development of this region remains equivocal and, coupled with the fragmentary nature of the stratigraphic record, challenges the basis of those maps (e.g. Ziegler 1988; Doré 1992; Knott *et al.* 1993; Roberts *et al.* 1999; Torsvik *et al.* 2002; Coward *et al.* 2003; Pharaoh *et al.* 2010) that detail one or several Late Palaeozoic–Early Mesozoic through-going rift systems: that is, along the axis of the Faroe–Shetland and Rockall basins, as well as the inferred 'SE Greenland–Hatton rift'. Inspection of Figures 2–4 indicates that, on the basis of the available evidence (or lack of), the long-term existence of the latter rift is totally without foundation. Currently, the only proven Mesozoic rocks on the Hatton and SE Greenland margins are of Cretaceous age. Moreover, the distribution of Cretaceous strata across a large part of this region remains inferred. Whilst the Cretaceous rocks on the Hatton High have been subjected to later folding, uplift and erosion, which implies a greater former extent, the relatively isolated occurrence of the basins necessitates caution in applying interbasinal connectivity across this region.

The issue of interbasinal connectivity is also pertinent with regard to the validity of a long-lived Faroe–Shetland–Rockall rift. On most reconstructions, the depiction of this proposed rift zone largely mimics the current bathymetric expression of the continental margin off NW Britain and Ireland: however, this morphological expression is largely a late-stage response to Cenozoic subsidence and post-break-up 'passive margin' tectonics (Naylor & Shannon 2005; Praeg *et al.* 2005; Ritchie *et al.* 2011*b*, 2013). It is also increasingly apparent that the Faroe–Shetland and Rockall basins are both segmented by NW-trending lineaments, at least some of which may have been inherited from earlier structures (Kimbell *et al.* 2005). For example, the major offset (and change in trend) between the northern and southern parts of the Rockall Basin (Fig. 1) is related to the Anton Dohrn Lineament

Complex (cf. Kimbell *et al.* 2005), which represents a transfer zone that may have originated as a Precambrian terrane boundary. Doré *et al.* (1999) have suggested that the current offset relates to an Early Cretaceous event; in which case, the Late Palaeozoic–Early Mesozoic development of the southern Rockall Basin might be linked more to basins in the Hebridean region. In this scenario, the status of the northern Rockall Basin, and even the Hatton Basin, remains unclear. Thus, any southern connection with the Faroe–Shetland region might have been via the Hebridean basins, although the latest Triassic–Jurassic stratigraphic record (Fig. 17) indicates sporadic and intermittent deposition of mainly paralic and shallow-marine deposits in both these areas. More specific stratigraphic information includes the near-absence of deposits associated with either the Zechstein or Rhaetian transgressions offshore NW Britain (Fig. 17), which further questions the validity of widespread marine connections in this region, at this time (e.g. Ziegler 1988; Knott *et al.* 1993; Roberts *et al.* 1999; Coward *et al.* 2003; Pharaoh *et al.* 2010).

Of the published plate reconstruction models, the scenario that most accurately complements the proven Late Mesozoic rock record, as presented here, is that of Doré *et al.* (1999), who linked changes in basin development across the NW European and NE Atlantic regions to the change in regional extensional stress directions: that is, from east–west (Jurassic) to NW–SE (Cretaceous) in response to plate break-up. Whilst the possibility of a more extensive Jurassic basin development in the area between NW Britain and SE Greenland remains to be fully tested, the currently available data suggest that it was not until the Late Cretaceous that a substantive rift system linking the Arctic and NE Atlantic regions across the SE Greenland–NW British region was developed. This is consistent with the long-standing view, based on studies of faunal provinciality (as described above), that there was a general lack of continuity between the Boreal and Tethyan realms during much of the Mesozoic.

Conclusions

An overview of the Upper Palaeozoic–Mesozoic rock record around the NE Atlantic region has produced a better understanding of the regional stratigraphic response to the break-up of the Pangaean supercontinent. By combining the distribution and stratigraphic range of the preserved Permian–Triassic, Jurassic and Cretaceous successions we have been able to identify both spatial and temporal changes in the large-scale pattern of sedimentation and basin development throughout the region, and link them to established tectonic events within and on the margins of the Pangaean plate. In particular:

- The Carboniferous–Permian boundary is largely conformable on the northern margin of the study area, where a passively subsiding margin formed in the late Carboniferous and persisted until the Mid–Late Permian. Over most of the rest of the NE Atlantic region, the boundary is a regionally extensive unconformity.
- The Permian–Triassic succession records a northwards transition from an arid, terrestrial facies to a shallow-marine shelf facies that dominates along the northern margin of the study area. Whereas the arid, terrestrial facies is mainly Late Permian–Triassic in age, the shelf facies comprises Permian carbonate rocks that pass upwards into Upper Permian–Triassic siliciclastic deposits. A pulse of rifting affected the entire NE Atlantic region during the Late Permian–earliest Triassic interval that generated variable large-scale basin architectures both in the interior and on the northern margin of the plate.
- The Triassic–Jurassic boundary is variable in character: in the north, paralic to marine clastic deposition persisted locally, although elsewhere the boundary is an unconformity; in the central part of the study area, the boundary is mainly an unconformity; and, in the south, a transitional passage from the arid Triassic into marine Jurassic represents the Rhaetian transgression. This variation is inferred to be a response to tectonic events occurring on the flanks of the Pangaean plate, with a southwards propagation of the rift system between Greenland and Norway, and marine flooding of the southern margin in response to incipient seafloor spreading in the Tethys and the proto-central Atlantic.
- The Jurassic succession preserves a record of Early Jurassic thermal subsidence and mild extensional tectonism, with a fragmentary succession of coastal and shallow-marine deposits predominating across the entire region. This general pattern of development was interrupted in the Mid-Jurassic by widespread uplift and shallowing, and is commonly reflected in the stratigraphic record as an unconformity between Lower and Middle Jurassic rocks. An intense phase of rifting occurred in some (but not all) parts of the NE Atlantic region in late Mid–Late Jurassic time, and there is a general trend towards marine conditions. This phase of rifting may have been driven by seafloor spreading in the Tethys; on the northern margin some of the tectonic activity might also have been related to extension in the Amerasia Basin.
- The Jurassic–Cretaceous boundary is transitional in North and NE Greenland, and marks

the culmination of rifting in this region. Elsewhere, the boundary is predominantly an unconformity, albeit commonly an intra-Berriasian (earliest Cretaceous) event. Its origin has been linked to a eustatic sea-level lowstand combined with stress-induced deflection and uplift of the lithosphere and/or a change in regional stress directions as Tethyan spreading waned and Atlantic spreading intensified.

- The Cretaceous succession is the most widely developed system of rocks proven to occur in the NE Atlantic region. Marine rocks dominate the observed record, with Upper Cretaceous chalks restricted to the southern part of the study area. As Atlantic spreading propagated northwards, a broad zone of extension and subsidence developed stretching from the southern Rockall Basin to the western Barents Sea. Basin development within this zone is variable: strike-slip movement and pull-apart basins characterize the northern margin of the area, whereas thermal subsidence prevailed in a number of basins in the central and southern part of the region. Regional Aptian–Albian rifting included the instigation of Mesozoic basin development in the area between SE Greenland and NW Britain. A post-Cenomanian phase of tectonically driven subsidence involving intermittent phases of normal faulting, compression and folding is recognized, especially in the successions of the Faroe–Shetland region and the Vøring Basin, and has been linked to wrench tectonics within the Pangaean plate. It has been inferred that the intra-plate stress regime at this time may have been modulated by a combination of Atlantic spreading (extension) and the evolving Alpine orogen (compression).

It is clear from the stratigraphic distribution maps that substantially more information remains to be gathered from around the NE Atlantic region: indeed, many areas remain that need to be fully explored. Nevertheless, we feel that there is scope for the development of a new and more comprehensive model of NE Atlantic rifting that takes into account the stratigraphic information presented in this paper. A key question remains: what was the degree of continuity along the proto-NE Atlantic during the Mesozoic? Our stratigraphic data suggest either a lack of continuity or, at least, a degree of complexity to any continuity that is currently not part of any reconstruction.

This work formed part of the NAG-TEC project that developed as part of the Northeast Atlantic Geoscience (NAG) cooperation framework, which comprises the BGR (German Federal Institute for Geosciences and Natural resources), the BGS (British Geological Survey), the GEUS (Geological Survey of Denmark and Greenland), the GSI (Geological Survey of Ireland), the GSNI (Geological Survey of Northern Ireland), ÍSOR (Iceland Geosurvey), JF (Jarðfeingi, Faroe Islands), the NGU (Geological Survey of Norway) and the TNO (Geological Survey of the Netherlands). We also acknowledge the support of the industry sponsors (in alphabetical order): Bayerngas Norge AS; BP Exploration Operating Company Ltd; Bundesanstalt für Geowissenschaften und Rohstoffe (BGR); Chevron East Greenland Exploration A/S; ConocoPhillips Skandinavia AS; DEA Norge AS; Det norske oljeselskap ASA; DONG E&P A/S; E.ON Norge AS; ExxonMobil Exploration and Production Norway AS; Japan Oil, Gas and Metals National Corporation (JOGMEC); Maersk Oil; Nalcor Energy – Oil and Gas Inc.; Nexen Energy ULC; Norwegian Energy Company ASA (Noreco); Repsol Exploration Norge AS; Statoil (UK) Ltd; and Wintershall Holding GmBH. The authors are grateful to Craig Woodward (BGS) for assistance with the figures, and the reviewers – A.G. Doré and S.P. Holford – for their careful and considered reviews of this paper. The contribution of M.S. Stoker and M.A. Stewart is made with the permission of the Executive Director of the British Geological Survey (Natural Environment Research Council).

References

Anell, I., Braathen, A. & Olaussen, S. 2014. The Triassic–Early Jurassic of the northern Barents Shelf: a regional understanding of the Longyearbyen CO_2 reservoir. *Norwegian Journal of Geology*, **94**, 83–98.

Archer, S.G., Bergman, S.C., Iliffe, J., Murphy, C.M. & Thornton, M. 2005. Paleocene igneous rocks reveal new insights into the geodynamic evolution and petroleum potential of the Rockall Trough, NE Atlantic. *Basin Research*, **17**, 171–201.

Birkelund, T., Callomon, J.H. & Fürsich, F.T. 1984. The stratigraphy of the Upper Jurassic and Lower Cretaceous sediments of Milne Land, central East Greenland. *Bulletin Grønlands Geologiske Undersøgelse*, **147**, 56.

Bjerager, M., Seidler, L., Stemmerik, L. & Surlyk, F. 2006. Ammonoid stratigraphy and sedimentary evolution across the Permian–Triassic boundary in East Greenland. *Geological Magazine*, **143**, 635–656.

Blischke, A., Arnarson, T.S. & Gunnarsson, K. 2011. The structural history of the Jan Mayen Micro-Continent (JMMC) and its role during the rift jump between the Aegir to the Kolbeinsey Ridge. *Abstract presented at the Polar Petroleum Potential Conference and Exhibition, American Association of Petroleum Geologists, Halifax*, 30 August – 2 September 2011, Nova Scotia, Canada.

Blischke, A., Peron-Pinvidic, G., Gaina, C., Brandsdóttir, B., Guarnieri, P., Hopper, J.R., Erlendsson, Ö. & Gunnarsson, K. In press. The Jan Mayen Micro-Continent (JMMC): an update of its architecture, structural development, and role during the rift transition from the Aegir Ridge to the Kolbeinsey Ridge. *In*: Péron-Pinvidic, G., Hopper, J., Stoker, M.S., Gaina, C., Doornenbal, H., Funck, T. & Árting, U. (eds) *The NE Atlantic Region: A Reappraisal of Crustal Structure, Tectonostratigraphy and Magmatic*

Evolution. Geological Society, London, Special Publications, **447**, https://doi.org/10.1144/SP447.5

Blystad, P., Brekke, H., Færseth, R.B., Larsen, B.T., Skogseid, J. & Tørudbakken, B. 1995. Structural elements of the Norwegian continental shelf, Part II. The Norwegian Sea Region. *Norwegian Petroleum Directorate Bulletin*, **8**, 45.

Bøe, R., Fossen, H. & Smelror, M. 2010. Mesozoic sediments and structures onshore Norway and in the coastal zone. *Geological Survey of Norway Bulletin*, **450**, 15–32.

Booth, J., Swiecicki, T. & Wilcockson, P. 1993. The tectono-stratigraphy of the Solan Basin, west of Shetland. *In*: Parker, J.R. (ed.) *Petroleum Geology of Northwest Europe: Proceedings of the 4th Conference*. Geological Society, London, 987–998, https://doi.org/10.1144/0040987

Breivik, A.J., Faleide, J.I. & Gudlaugsson, S.T. 1998. Southwestern Barents Sea margin: late Mesozoic sedimentary basins and crustal extension. *Tectonophysics*, **293**, 21–44.

Brekke, H. 2000. The tectonic evolution of the Norwegian Sea continental margin with emphasis on the Vøring and Møre basins. *In*: Nøttvedt, A. (ed.) *Dynamics of the Norwegian Margin*. Geological Society, London, Special Publications, **167**, 327–378, https://doi.org/10.1144/GSL.SP.2000.167.01.13

Brekke, H., Dahlgren, S., Nyland, B. & Magnus, C. 1999. The prospectivity of the Vøring and Møre basins on the Norwegian Sea continental margins. *In*: Fleet, A.J. & Boldy, S.A.R. (eds) *Petroleum Geology of Northwest Europe: Proceedings of the 5th Conference*. Geological Society, London, 261–274, https://doi.org/10.1144/0050261

Brekke, H., Sjulstad, H.I., Magnus, C. & Williams, R.W. 2001. Sedimentary environments offshore Norway – an overview. *In*: Martinsen, O.J. & Dreyer, T. (eds) *Sedimentary Environments Offshore Norway – Palaeozoic to Recent*. Norwegian Petroleum Society, Special Publications, **10**, 7–37.

British Geological Survey 1989. *Sutherland, Sheet 58°N–06°W, Solid Geology, 1:250,000 Series*. British Geological Survey, Nottingham.

British Geological Survey 1990. *Lewis, Sheet 58°N–08°W, Solid Geology, 1:250,000 Series*. British Geological Survey, Nottingham.

Callomon, J.H. 1984. A review of the biostratigraphy of the post-Lower Bajocian Jurassic ammonites of western and northern North America. *In*: Westerman, G.E.G. (ed.) *Jurassic–Cretaceous Biochronology and Palaeogeography of North America*. Geological Association of Canada, Special Papers, **27**, 143–174.

Callomon, J.H. & Birkelund, T. 1982. The ammonite zones of the Boreal Volgian (Upper Jurassic) in East Greenland. *In*: Embry, A.F. & Balkwill, H.R. (eds) *Arctic Geology and Geophysics*. Canadian Society of Petroleum Geologists, Memoirs, **8**, 349–369.

Casey, R. 1971. Facies, Fauna and Tectonics in the Late Jurassic–Early Cretaceous Britain. *In*: Middlemiss, F.A., Rawson, P.F. & Newell, G. (eds) *Faunal Provinces in Space and Time*. Seal House Press, Liverpool, 153–168.

Casey, R. & Rawson, P.F. 1973. A review of the boreal lower cretaceous. *In*: Casey, R. & Rawson, P.F. (eds) *The Boreal Lower Cretaceous. Geological Journal*, Special Issue, **5**, 415–430.

Chapman, T.J., Broks, T.M., Corcoran, D.V., Duncan, L.A. & Dancer, P.N. 1999. The structural evolution of the Erris Trough, offshore northwest Ireland, and implications for hydrocarbon generation. *In*: Fleet, A.J. & Boldy, S.A.R. (eds) *Petroleum Geology of Northwest Europe: Proceedings of the 5th Conference*. Geological Society, London, 455–469, https://doi.org/10.1144/0050455

Christiansen, F.G., Piasecki, S., Stemmerik, L. & Telnaes, N. 1993. Depositional environment and organic geochemistry of the Upper Permian Ravnefjord Formation source rock. *AAPG Bulletin*, **77**, 1519–1537.

Clemmensen, L.B. 1980*a*. Triassic rift sedimentation and palaeogeography of Central East Greenland. *Bulletin Grønlands Geologiske Undersøgelse*, **136**, 72.

Cook, D.R. 1987. The Goban Spur: exploration in a deep-water frontier basin. *In*: Brooks, J. & Glennie, K.W. (eds) *Petroleum Geology of North-West Europe*. Graham & Trotman, London, 623–632.

Cope, J.C.W., Ingham, J.K. & Rawson, P.F. 1992. *Atlas of Palaeogeography and Lithofacies*. Geological Society, London, Memoirs, **13**.

Corcoran, D.V. & Mecklenburgh, R. 2005. Exhumation of the Corrib Gas Field, Slyne Basin, offshore Ireland. *Petroleum Geoscience*, **11**, 239–256, https://doi.org/10.1144/1354-079304-637

Coward, M.P., Dewey, J.F., Hempton, M. & Holroyd, J. 2003. Tectonic evolution. *In*: Evans, D., Graham, C., Armour, A. & Bathurst, P. (eds) *The Millennium Atlas: Petroleum Geology of the Central and Northern North Sea*. Geological Society, London, 17–33.

Croker, P.F. & Shannon, P.M. 1987. The evolution and hydrocarbon prospectivity of the Porcupine Basin, offshore Ireland. *In*: Brooks, J. & Glennie, K.W. (eds) *Petroleum Geology of North West Europe*. Graham & Trotman, London, 633–642.

Dalland, A., Worsley, D. & Ofstad, K. 1988. A lithostratigraphic scheme for the Mesozoic and Cenozoic succession offshore mid- and northern Norway. *Norwegian Petroleum Directorate Bulletin*, **4**, 87.

Dallmann, W.K. 1999. *Lithostratigraphic Lexicon of Svalbard: Review and Recommendations for Nomenclature Use: Upper Palaeozoic to Quaternary Bedrock*. Norwegian Polar Institute Report, **318**.

Dam, G. & Surlyk, F. 1998. Stratigraphy of the Neill Klinter Group; a lower – lower Middle Jurassic tidal embayment succession, Jameson Land, East Greenland. *Geology of Greenland Survey Bulletin*, **175**, 80.

Dam, G., Surlyk, F., Mathiesen, A. & Christiansen, F.G. 1995. Exploration significance of lacustrine forced regressions of the Rhaetian–Sinemurian Kap Stewart Formation, Jameson Land, East Greenland. *In*: Steel, R.J., Felt, V.L., Johannessen, E.P. & Mathieu, C. (eds) *Sequence Stratigraphy: Advances and Application for Exploration and Producing in North West Europe*. Norwegian Petroleum Society, Special Publications, **5**, 509–525.

Dancer, P.N., Algar, S.T. & Wilson, I.R. 1999. Structural evolution of the Slyne Trough. *In*: Fleet, A.J. & Boldy, S.A.R. (eds) *Petroleum Geology of Northwest Europe: Proceedings of the 5th Conference*.

Geological Society, London, 445–453, https://doi.org/10.1144/0050445

Dancer, P.N., Kenyon-Roberts, S.M., Downey, J.W., Baillie, J.M., Meadows, N.S. & Maguire, K. 2005. The Corrib gas field, offshore west of Ireland. *In*: Doré, A.G. & Vining, B.A. (eds) *Petroleum Geology: North-West Europe and Global Perspectives – Proceedings of the 6th Petroleum Geology Conference*. Geological Society, London, 1035–1046, https://doi.org/10.1144/0061035

Dean, K., McLachlan, K. & Chambers, A. 1999. Rifting and the development of the Faeroe–Shetland Basin. *In*: Fleet, A.J. & Boldy, S.A.R. (eds) *Petroleum Geology of Northwest Europe: Proceedings of the 5th Conference*. Geological Society, London, 533–544, https://doi.org/10.1144/0050533

Dinkelman, M.G., Granath, J.W. & Whittaker, R. 2010. The NE Greenland Continental Margin 2010. *GeoExpro*, **7**, 36–40.

Doornenbal, J.C. & Stevenson, A.G. (eds). 2010. *Petroleum Geological Atlas of the Southern Permian Basin Area*. EAGE Publications, Houten.

Doré, A.G. 1992. Synoptic palaeogeography of the Northeast Atlantic Seaway: late Permian to Cretaceous. *In*: Parnell, J. (ed.) *Basins on the Atlantic Seaboard: Petroleum Geology, Sedimentology and Basin Evolution*. Geological Society, London, Special Publications, **62**, 421–446, https://doi.org/10.1144/GSL.SP.1992.062.01.3

Doré, A.G., Lundin, E.R., Jensen, L.N., Birkeland, Ø., Eliassen, P.E. & Fichler, C. 1999. Principal tectonic events in the evolution of the northwest European Atlantic margin. *In*: Fleet, A.J., & Boldy, S.A.R. (eds) *Petroleum Geology of Northwest Europe: Proceedings of the 5th Conference*. Geological Society, London, 41–61, https://doi.org/10.1144/0050041

Døssing, A., Stemmerik, L., Dahl-Jensen, T. & Schlindwein, V. 2010. Segmentation of the eastern North Greenland oblique-shear margin – Regional plate tectonic implications. *Earth and Planetary Science Letters*, **292**, 239–253.

Dypvik, H., Håkansson, E. & Heinberg, C. 2002. Jurassic and Cretaceous palaeogeography and stratigraphic comparisons in the North Greenland–Svalbard region. *Polar Research*, **21**, 91–108.

Earle, M.M., Jankowski, E.J. & Vann, I.R. 1989. Structural and stratigraphic evolution of the Faeroe-Shetland Channel and Northern Rockall Trough. *In*: Tankard, A.J. & Balkwill, H.R. (eds) *Extensional Tectonics and Stratigraphy of the North Atlantic Margins*. American Association of Petroleum Geologists, Memoirs, **46**, 461–469.

Eldholm, O., Tsikalas, F. & Faleide, J.I. 2002. Continental margin off Norway 62–75°N: Palaeogene tectono-magmatic segmentation and sedimentation. *In*: Jolley, D.W. & Bell, B.R. (eds) *The North Atlantic Igneous Province: Stratigraphy, Tectonic, Volcanic and Magmatic Processes*. Geological Society, London, Special Publications, **197**, 39–68, https://doi.org/10.1144/GSL.SP.2002.197.01.03

Ellis, D. & Stoker, M.S. 2014. The Faroe–Shetland Basin: a regional perspective from the Paleocene to the present day and its relationship to the opening of the North Atlantic Ocean. *In*: Cannon, S.J.C. & Ellis, D. (eds) *Hydrocarbon Exploration to Exploitation West of Shetlands*. Geological Society, London, Special Publications, **397**, 11–31, https://doi.org/10.1144/SP397.1

Emeleus, C.H. & Bell, B.R. 2005. *British Regional Geology: The Palaeogene Volcanic Districts of Scotland*. 4th edn. British Geological Survey, Nottingham.

Evans, D. 2013. Jurassic. *In*: Hitchen, K., Johnson, H. & Gatliff, R.W. (eds) *Geology of the Rockall Basin and Adjacent Areas. British Geological Survey Research Report, RR/12/03*. British Geological Survey, Nottingham, 67–70.

Evans, D., Graham, C., Armour, A. & Bathurst, P. (eds). 2003. *The Millennium Atlas: Petroleum Geology of the Central and Northern North Sea*. Geological Society, London.

Faleide, J.I., Vågnes, E. & Gudlaugsson, S.T. 1993. Late Mesozoic-Cenozoic evolution of the southwestern Barents Sea in a regional rift-shear tectonic setting. *Marine and Petroleum Geology*, **10**, 186–214.

Faleide, J.I., Bjørlykke, K. & Gabrielsen, R.H. 2010. Geology of the Norwegian continental shelf. *In*: Bjørlykke, K. (ed.) *Petroleum Geoscience: From Sedimentary Environments to Rock Physics*. Springer, New York, 467–499.

Foulger, G.R. & Anderson, D.L. 2005. A cool model for the Iceland hotspot. *Journal of Volcanology and Geothermal Research*, **141**, 1–22.

Foulger, G.R., Natland, J.H. & Anderson, D.L. 2005. A source for Icelandic magmas in re-melted Iapetus crust. *Journal of Volcanology and Geothermal Research*, **141**, 23–44.

Fyfe, J.A., Long, D. & Evans, D. 1993. *United Kingdom Offshore Regional Report: The Geology of the Malin–Hebrides Sea Area*. HMSO for the British Geological Survey, London.

Gabrielsen, R.H., Færseth, R.B., Jensen, L.N., Kalheim, J.E. & Riis, F. 1990. Structural elements of the Norwegian continental shelf. Part I: the Barents Sea Region. *Norwegian Petroleum Directorate Bulletin*, **6**, 33.

Gast, R., Dusar, M. *et al.* 2010. Rotliegend. *In*: Doornenbal, H. & Stevenson, A.G. (eds) *Petroleum Geological Atlas of the Southern Permian Basin Area*. EAGE Publications, Houten., 101–121.

Gautier, D.L., Stemmerik, L. *et al.* 2011. Assessment of NE Greenland: prototype for development of Circum-Arctic Resource Appraisal methodology. *In*: Spencer, A.M., Embry, A.F., Gautier, D.L., Stoupakova, A.V. & Sørensen, K. (eds) *Arctic Petroleum Geology*. Geological Society, London, Memoirs, **35**, 663–672, https://doi.org/10.1144/M35.43

Gerlings, J., Hopper, J.R., Fyhn, M.B.W. & Frandsen, N. In review. Mesozoic and older rift basins on the SE Greenland shelf near Ammassalik. *In*: Péron-Pinvidic, G., Hopper, J., Stoker, M.S., Gaina, C., Doornenbal, H., Funck, T. & Árting, U. (eds) *The NE Atlantic Region: A Reappraisal of Crustal Structure, Tectonostratigraphy and Magmatic Evolution*. Geological Society, London, Special Publications, **447**.

Gilotti, J.A., Jones, K.A. & Elvevold, S. 2008. Caledonian metamorphic patterns in Greenland. *In*: Higgins, A.K., Gilotti, J.A. & Smith, M.P. (eds) *The Greenland Caledonides: Evolution of the Northeast Margin of Laurentia*. Geological Society of America, Memoirs, **202**, 201–225.

Glennie, K.W. 2002. Permian and triassic. *In*: Trewin, N.H. (ed.) *The Geology of Scotland*. 4th edn. Geological Society, London, 301–322.

Glennie, K.W. & Underhill, J.R. 1998. Origin, development and evolution of structural styles. *In*: Glennie, K.W. (ed.) *Petroleum Geology of the North Sea: Basic Concepts and Recent Advances*. Blackwell Science, Oxford, 42–84.

Glennie, K., Higham, J. & Stemmerik, L. 2003. Permian. *In*: Evans, D., Graham, C.G., Armour, A. & Bathhurst, P. (eds) *Millennium Atlas: Petroleum Geology of the Central and Northern North Sea*. Geological Society, London, 91–103.

Glørstad-Clark, E., Birkelund, E.P., Nystuen, J.P., Faleide, J.I. & Midtkandal, I. 2011. Triassic platform-margin deltas in the western Barents Sea. *Marine and Petroleum Geology*, **28**, 1294–1314.

Golonka, J. 2011. Phanerozoic palaeoenvironment and palaeolithofacies maps of the Arctic region. *In*: Spencer, A.M., Embry, A.F., Gautier, D.L., Stoupakova, A.V. & Sørensen, K. (eds) *Arctic Petroleum Geology*. Geological Society, London, Memoirs, **35**, 79–129, https://doi.org/10.1144/M35.6

Goodchild, M.W., Henry, K.L., Hinkley, R.J. & Imbus, S.W. 1999. The victory gas field, West of Shetland. *In*: Fleet, A.J. & Boldy, S.A.R. (eds) *Petroleum Geology of Northwest Europe: Proceedings of the 5th Conference*. Geological Society, London, 713–724, https://doi.org/10.1144/0050713

Gradstein, F.M., Ogg, J.G., Schmitz, M.D. & Ogg, G.M. 2012. *The Geologic Time Scale 2012*. Elsevier, Amsterdam.

Grant, N., Bouma, A. & McIntyre, A. 1999. The Turonian play in the Faeroe-Shetland Basin. *In*: Fleet, A.J. & Boldy, S.A.R. (eds) *Petroleum Geology of Northwest Europe, Proceedings of the 5th Conference*. Geological Society, London, 661–673, https://doi.org/10.1144/0050661

Håkansson, E., Heinberg, C. & Stemmerik, L. 1991. Mesozoic and Cenozoic history of the Wandel Sea Basin area, North Greenland. *Bulletin Grønlands Geologiske Undersøgelser*, **160**, 153–164.

Hald, N. & Tegner, C. 2000. Composition and age of Tertiary sills and dykes, Jameson Land Basin, East Greenland: relation to regional flood volcanism. *Lithos*, **54**, 207–233.

Hamann, N.E., Wittaker, R.C. & Stemmerik, L. 2005. Geological development of the Northeast Greenland shelf. *In*: Doré, A.G. & Vining, B.A. (eds) *Petroleum Geology: North-West Europe and Global Perspectives – Proceedings of the 6th Petroleum Geology Conference*. Geological Society, London, 887–902, https://doi.org/10.1144/0060887

Harker, S.D. 2002. Cretaceous. *In*: Trewin, N. (ed.) *The Geology of Scotland*. 4th edn. Geological Society, London, 351–360.

Harland, W.B. 1997. *The Geology of Svalbard*. Geological Society, London, Memoirs, **17**.

Harris, J.P. 1992. Mid-Jurassic lagoonal systems in the Hebridean basins: thickness and facies distribution patterns of potential reservoir sandbodies. *In*: Parnell, J. (ed.) *Basins on the Atlantic Seaboard: Petroleum Geology, Sedimentology and Basin Evolution*. Geological Society, London, Special Publications, **62**, 111–144, https://doi.org/10.1144/GSL.SP.1992.062.01.11

Haszeldine, R.S., Ritchie, J.D. & Hitchen, K. 1987. Seismic and well evidence for the early development of the Faroe–Shetland Basin. *Scottish Journal of Geology*, **23**, 283–300.

Haughton, P., Praeg, D. *et al.* 2005. First results from shallow stratigraphic boreholes on the eastern flank of the Rockall Basin, offshore western Ireland. *In*: Doré, A.G. & Vining, B.A. (eds) *Petroleum Geology: North-West Europe and Global Perspectives – Proceedings of the 6th Petroleum Geology Conference*. Geological Society, London, 1077–1094, https://doi.org/10.1144/0061077

Henriksen, E., Ryseth, A.E., Larssen, G.B., Heide, T., Rønning, K., Sollid, K. & Stoupakova, A.V. 2011. Tectonostratigraphy of the greater Barents Sea: implications for petroleum systems. *In*: Spencer, A.M., Embry, A.F., Gautier, D.L., Stoupakova, A.V. & Sørensen, K. (eds) *Arctic Petroleum Geology*. Geological Society, London, Memoirs, **35**, 163–195, https://doi.org/10.1144/M35.10

Hesselbo, S.P., Oates, M.J. & Jenkyns, H.C. 1998. The Lower Lias Group of the Hebrides Basin. *Scottish Journal of Geology*, **34**, 23–60, https://doi.org/10.1144/sjg34010023

Hitchen, K. 2004. The geology of the UK Hatton-Rockall margin. *Marine and Petroleum Geology*, **21**, 993–1012.

Hitchen, K. & Ritchie, J.D. 1987. Geological review of the West Shetland area. *In*: Brooks, J. & Glennie, K.W. (eds) *Petroleum Geology of North West Europe*. Graham & Trotman, London 737–749.

Hitchen, K. & Stoker, M.S. 1993. Mesozoic rocks from the Hebrides Shelf and the implications for hydrocarbon prospectivity in the northern Rockall Trough. *Marine and Petroleum Geology*, **10**, 246–254.

Hitchen, K., Stoker, M.S., Evans, D. & Beddoe-Stephens, B. 1995. Permo-Triassic sedimentary and volcanic rocks in basins to the north and west of Scotland. *In*: Boldy, S.A.R. (ed.) *Permian and Triassic Rifting in Northwest Europe*. Geological Society, London, Special Publications, **91**, 87–102, https://doi.org/10.1144/GSL.SP.1995.091.01.05

Hitchen, K., Johnson, H. & Gatliff, R.W. (eds). 2013. *Geology of the Rockall Basin and Adjacent Areas. British Geological Survey Research Report, RR/12/03*. British Geological Survey, Nottingham.

Hjelstuen, B.O., Eldholm, O. & Faleide, J.I. 2007. Recurrent Pleistocene mega-failures on the SW Barents margin. *Earth and Planetary Science Letters*, **258**, 605–618.

Hopper, J.R., Funck, T., Stoker, M.S., Árting, U., Peron-Pinvidic, G., Doornenbal, H. & Gaina, C. (eds) 2014. *Tectonostratigraphic Atlas of the North-East Atlantic Region*. Geological Survey of Denmark and Greenland (GEUS), Copenhagen.

Hopson, P.M. 2005. A Stratigraphical Framework for the Upper Cretaceous Chalk of England and Scotland with Statements on the Chalk of Northern Ireland and the UK Offshore Sector. British Geological Survey

Research Report, RR/05/01. British Geological Survey, Nottingham.

Høy, T. & Lundschien, B.A. 2011. Triassic deltaic sequences in the northern Barents Sea. *In*: Spencer, A.M., Embry, A.F., Gautier, D.L., Stoupakova, A.V. & Sørensen, K. (eds) *Arctic Petroleum Geology*. Geological Society, London, Memoirs, **35**, 249–260, https://doi.org/10.1144/10.1144/M35.15

Hudson, J.D. & Trewin, N.H. 2002. Jurassic. *In*: Trewin, N.H. (ed.) *The Geology of Scotland*. 4th edn. Geological Society, London, 323–350.

Jackson, D., Protacio, A., Silva, M., Helwig, J.A. & Dinkelman, M.G. 2012. The North East Greenland Danmarkshavn Basin – the sky above, the ice floes, and the Earth Below. *GEO ExPro*, **9**, 60–62.

Jacobsen, V.W. & van Veen, P. 1984. The Triassic offshore Norway north of 62°N. *In*: Spencer, A.M. (ed.) *Petroleum Geology of the North European Margin*. Graham & Trotman, London, 317–328.

Johnson, H. & Quinn, M.F. 2013. Permo-Triassic. *In*: Hitchen, K., Johnson, H. & Gatliff, R.W. (eds) *Geology of the Rockall Basin and Adjacent Areas. British Geological Survey Research Report, RR/12/03*. British Geological Survey, Nottingham, 61–66.

Jones, E.J.W., Siddall, R., Thirlwall, M.F., Chroston, P.N. & Lloyd, A.J. 1974. Anton Dohrn Seamount and the evolution of the Rockall Trough. *Oceanologica Acta*, **17**, 237–247.

Kalsbeek, F., Higgins, A.K., Jepsen, H.F., Frei, R. & Nutman, A.P. 2008. Granites and granites in the East Greenland Caledonides. *In*: Higgins, A.K., Gilotti, J.A. & Smith, M.P. (eds) *The Greenland Caledonides: Evolution of the Northeast Margin of Laurentia*. Geological Society of America, Memoirs, **202**, 227–249.

Kimbell, G.S., Ritchie, J.D., Johnson, H. & Gatliff, R.W. 2005. Controls on the structure and evolution of the NE Atlantic margin revealed by regional potential field imaging and 3D modelling. *In*: Doré, A.G. & Vining, B.A. (eds) *Petroleum Geology: North-West Europe and Global Perspectives – Proceedings of the 6th Conference*. Geological Society, London, 933–945, https://doi.org/10.1144/0060993

Kirton, S.R. & Hitchen, K. 1987. Timing and style of crustal extension of the Scottish mainland. *In*: Coward, M.P., Dewey, J.F. & Hancock, P.L. (eds) *Continental Extensional Tectonics*. Geological Society, London, Special Publications, **28**, 501–510, https://doi.org/10.1144/GSL.SP.1987.028.01.32

Knott, S.D., Burchell, M.T., Jolley, E.J. & Fraser, A.J. 1993. Mesozoic to Cenozoic plate reconstructions of the North Atlantic and hydrocarbon plays of the Atlantic margins. *In*: Parker, J.R. (ed.) *Petroleum Geology of Northwest Europe: Proceedings of the 4th Conference*. Geological Society, London, 953–974, https://doi.org/10.1144/0040953

Kreiner-Møller, M. & Stemmerik, L. 2001. Upper Permian lowstand fans of the Bredehorn Member, Schuchert Dal formation, East Greenland. *In*: Martinsen, O.J. & Dreyer, T. (eds) *Sedimentary Environments Offshore Norway – Paleozoic to Recent*. Norwegian Petroleum Society, Special Publications, **10**, 51–65.

Lamers, E. & Carmichael, S.M.M. 1999. The Paleocene deepwater sandstone play West of Shetland. *In*: Fleet, A.J. & Boldy, S.A.R. (eds) *Petroleum Geology of Northwest Europe, Proceedings of the 5th Conference*. Geological Society, London, 645–659, https://doi.org/10.1144/0050645

Larsen, H.C. 1990. The East Greenland shelf. *In*: Grant, A., Johnson, L. & Sweeney, J.F. (eds) *The Arctic Ocean Region*. The Geology of North America, **L**. Geological Society of America, Boulder, CO, 185–210.

Larsen, H.C. & Marcussen, C. 1992. Sill Intrusion, Flood basalt emplacement and deep crustal structure of the Scoresby Sund region, East Greenland. *In*: Storey, B.C., Alabaster, T. & Pankhurst, R.J. (eds) *Magmatism and Causes of Continental Break-Up*. Geological Society, London, Special Publications, **68**, 365–386, https://doi.org/10.1144/GSL.SP.1992.068.01.23

Larsen, L., Pedersen, A.K., Sørensen, E.V., Watt, W.S. & Duncan, R.A. 2013. Stratigraphy and age of the Eocene Igtertivâ Formation basalts, alkaline pebbles and sediments of the Kap Dalton Group in the graben at Kap Dalton, East Greenland. *Bulletin of the Geological Society of Denmark*, **61**, 1–18.

Larsen, M., Hamberg, L., Olaussen, S., Preuss, T. & Stemmerik, L. 1999. Sandstone wedges of the Cretaceous–Lower Tertiary Kangerlussuaq Basin, East Greenland – outcrop analogues to the offshore North Atlantic. *In*: Fleet, A.J. & Boldy, S.A.R. (eds) *Petroleum Geology of Northwest Europe: Proceedings of the 5th Conference*. Geological Society, London, 337–348, https://doi.org/10.1144/0050337

Larsen, M., Nedkvitne, T. & Olaussen, S. 2001. Lower Cretaceous (Barremian–Albian) deltaic and shallow marine sandstones in North-East Greenland – sedimentology, sequence stratigraphy and regional implications. *In*: Martinsen, O.J. & Dreyer, T. (eds) *Sedimentary Environments Offshore Norway – Paleozoic to Recent*. Norwegian Petroleum Society, Special Publications, **10**, 259–278.

Larsen, M., Piasecki, S. & Surlyk, F. 2003. Stratigraphy and sedimentology of a basement-onlapping shallow marine sandstone succession, the Charcot Bugt Formation, Middle–Upper Jurassic, East Greenland. *In*: Ineson, J.R. & Surlyk, F. (eds) *The Jurassic of Denmark and Greenland*. Geological Survey of Denmark and Greenland, Bulletin, **1**, 893–930.

Larsen, M., Heilmann-Clausen, C., Piasecki, S. & Stemmerik, L. 2005*a*. At the edge of a new ocean: post-volcanic evolution of the Palaeogene Kap Dalton Group, East Greenland. *In*: Doré, A.G. & Vining, B.A. (eds) *Petroleum Geology: North-West Europe and Global Perspectives – Proceedings of the 6th Petroleum Geology Conference*. Geological Society, London, 923–932, https://doi.org/10.1144/0060923

Larsen, M., Nøhr-Hansen, H., Whitham, A.G. & Kelly, S.R.A. 2005*b*. *Stratigraphy of the Pre-Basaltic Sedimentary Succession of the Kangerlussuaq Basin, Volcanic Basin of the North Atlantic*. Final Report for the Sindri Group, September 2005. Danmarks og Grønlands Geologiske Undersøgelse Rapport, **2005/62**.

Larsen, M., Rasmussen, T. & Hjelm, L. 2010. Cretaceous revisited: exploring the syn-rift play of the Faroe–Shetland Basin. *In*: Vining, B.A. & Pickering, S.C. (eds) *Petroleum Geology: From Mature Basins to*

New Frontiers – Proceedings of the 7th Petroleum Geology Conference. Geological Society, London, 953–962, https://doi.org/10.1144/0070953

Lawver, L.A., Gahagan, L.M. & Norton, I. 2011. Palaeogeographic and tectonic evolution of the Arctic region during the Palaeozoic. *In*: Spencer, A.M., Embry, A.F., Gautier, D.L., Stoupakova, A.V. & Sørensen, K. (eds) *Arctic Petroleum Geology*. Geological Society, London, Memoirs, **35**, 61–77, https://doi.org/10.1144/M35.5

Lundin, E.R. & Doré, A.G.D. 2005*a*. NE Atlantic break-up: a re-examination of the Iceland mantle plume model and the Atlantic–Arctic linkage. *In*: Doré, A.G. & Vining, B. (eds) *Petroleum Geology: North West Europe and Global Perspectives: Proceedings of the 6th Conference*. Geological Society, London, 739–754, https://doi.org/10.1144/0060739

Lundin, E.R. & Doré, A.G. 2005*b*. Fixity of the Iceland 'hotspot' on the Mid-Atlantic Ridge: observational evidence, mechanisms and implications for Atlantic volcanic margins. *In*: Foulger, G.R., Anderson, D.L., Natland, J.H. & Presnall, D.C. (eds) *Plates, Plumes and Paradigms*. Geological Society, America, Special Papers, **388**, 627–651.

Lundin, E.R. & Doré, A.G. 2011. Hyperextension, serpentinization, and weakening: a new paradigm for rifted margin compressional deformation. *Geology*, **39**, 347–350.

McKie, T. & Williams, B. 2009. Triassic palaeogeography and fluvial dispersal systems across the northwest European Basins. *Geological Journal*, **44**, 711–741.

Meadows, N.S., Macchi, L., Cubitt, J.M., Bailey, N.J.L. & Johnson, B. 1987. Sedimentology and reservoir potential in the west of Shetland, UK exploration area. *In*: Brooks, J. & Glennie, K.W. (eds) *Petroleum Geology of North West Europe*. Graham & Trotman, London, 723–736.

Moore, J.G. 1992. A syn-rift to post-rift transition sequence in the Main Porcupine Basin, offshore western Ireland. *In*: Parnell, J. (ed.) *Basins on the Atlantic Seaboard: Petroleum Geology, Sedimentology and Basin Evolution*. Geological Society, London, Special Publications, **62**, 333–349, https://doi.org/10.1144/GSL.SP.1992.062.01.26

Moore, J.G. & Shannon, P.M. 1995. The Cretaceous succession in the Porcupine Basin, Offshore Ireland: facies distribution and hydrocarbon potential. *In*: Croker, P.F. & Shannon, P.M. (eds) *The Petroleum Geology of Ireland's Offshore Basins*. Geological Society, London, Special Publications, **93**, 345–370, https://doi.org/10.1144/GSL.SP.1995.093.01.28

Mørk, A., Vigran, J.O., Korchinskaya, M.V., Pchelina, T.M., Fefilova, L.A., Vavilov, M.N. & Weitschat, W. 1992. Triassic rocks in Svalbard, the Artic Soviet Islands and the Barents Shelf: bearing on their correlations. *In*: Vorren, T., Bergsager, E., Dahl-Stamnes, Ø.A., Holter, E., Johansen, B., Lie, E. & Lund, T.B. (eds) *Artic Geology and Petroleum Potential*. Norwegian Petroleum Society, Special Publications, **2**, 457–479.

Mortimore, R., Wood, C. & Gallois, R. 2001. *British Upper Cretaceous Stratigraphy*. Joint Nature Conservation Committee, Peterborough, Geological Conservation Review Series, **23**.

Morton, A.C., Hitchen, K., Ritchie, J.D., Hine, N.M., Whitehouse, M., & Carter, S.G. 1995. Late Cretaceous basalts from Rosemary Bank, Northern Rockall Trough. *Journal of the Geological Society, London*, **152**, 947–952, https://doi.org/10.1144/GSL.JGS.1995.152.01.11

Morton, N. 1989. Jurassic sequence stratigraphy in the Hebrides Basin, NW Scotland. *Marine and Petroleum Geology*, **6**, 243–260.

Morton, N. 1992. Dynamic stratigraphy of the Triassic and Jurassic of the Hebrides Basin, NW Scotland. *In*: Parnell, J. (ed.) *Basins on the Atlantic Seaboard: Petroleum Geology, Sedimentology and Basin Evolution*. Geological Society, London, Special Publications, **62**, 97–110, https://doi.org/10.1144/GSL.SP.1992.062.01.10

Müller, R., Nystuen, J.P., Eide, F. & Lie, H. 2005. Late Permian to Triassic basin infill history and palaeogeography of the Mid-Norwegian shelf–East Greenland Region. *In*: Wandas, B.T.G., Nystuen, J.P., Eide, E. & Gradstein, F. (eds) *Onshore–Offshore Relationships on the North Atlantic Margin*. Norwegian Petroleum Society, Special Publications, **12**, 165–189.

Murphy, N.J. & Croker, P.F. 1992. Many play concepts seen over wide area in Erris, Slyne troughs off Ireland. *Oil and Gas Journal*, **90**, 92–97.

Musgrove, F.W. & Mitchener, B. 1996. Analysis of the pre-Tertiary rifting history of the Rockall Trough. *Petroleum Geoscience*, **2**, 353–360, https://doi.org/10.1144/petgeo.2.4.353

Myers, J.S., Gill, R.C.O., Rex, D.C. & Charnley, N.R. 1993. The Kap Gustav Holm Tertiary plutonic centre, East Greenland. *Journal of the Geological Society, London*, **150**, 259–276, https://doi.org/10.1144/gsjgs.150.2.0259

Naylor, D. & Shannon, P.M. 1982. *The Geology of Offshore Ireland and West Britain*. Graham & Trotman, London.

Naylor, D. & Shannon, P.M. 2005. The structural framework of the Irish Atlantic Margin. *In*: Doré, A.G. & Vining, B.A. (eds) *Petroleum Geology: North-West Europe and Global Perspectives – Proceedings of the 6th Petroleum Geology Conference*. Geological Society, London, 1009–1021, https://doi.org/10.1144/0061009

Naylor, D. & Shannon, P.M. 2011. *Petroleum Geology of Ireland*. Dunedin Academic Press, Edinburgh.

Naylor, D., Shannon, P.M. & Murphy, N. 1999. *Irish Rockall Basin Region – A Standard Structural Nomenclature System*. Petroleum Affairs Division, Dublin, Special Publications, **1/99**.

Naylor, D., Shannon, P.M. & Murphy, N. 2002. *Porcupine–Goban Region – A Standard Structural Nomenclature System*. Petroleum Affairs Division, Dublin, Special Publications, **1/02**.

Nielsen, T., Bjerager, M., Lindström, S., Nøhr-Hansen, H. & Rasmussen, T.L. 2014. Pre-breakup age of East Greenland Ridge strata. *Geophysical Research Abstracts*, **16**, EGU2014–EGU6127.

Nøhr-Hansen, H., Nielsen, L.H., Sheldon, E., Hovokoski, J. & Alsen, P. 2011. Palaeogene deposits in North-East Greenland. *Geological Survey of Denmark and Greenland Bulletin*, **23**, 61–64.

NORLEX 2014. *Norwegian Interactive Offshore Stratigraphic Lexicon*, http://www.nhm2.uio.no/norlex

Norwegian Petroleum Directorate 2012. *Submarine Fieldwork on the Jan Mayen Ridge: Integrated Seismic and ROV-Sampling*. Norwegian Petroleum Directorate, Stavanger, Norway, http://www.npd.no/en/Publications/Presentations/Submarine-fieldwork-on-the-Jan-Mayen-Ridge/

Norwegian Petroleum Directorate 2013. *The Petroleum Resources on the Norwegian Continental Shelf.* Norwegian Petroleum Directorate, Stavanger, Norway, http://www.npd.no/en/Publications/Resource-Reports/2013/

Norwegian Petroleum Directorate 2014. *Lithostratigraphic Chart: Norwegian Sea*. Norwegian Petroleum Directorate, Stavanger, Norway, http://www.npd.no/en/Topics/Geology/

Nøttvedt, A., Cecchi, M. et al. 1993*a*. Svalbard – Barents Sea correlation: a short review. *In*: Vorren, T., Bergsager, E., Dahl-Stamnes, Ø.A., Holter, E., Johansen, B., Lie, E. & Lund, T.B. (eds) *Arctic Geology and Petroleum Potential*. Norwegian Petroleum Society, Special Publications, **2**, 363–375.

Nøttvedt, A., Livbjerg, F., Midbøe, P.S. & Rasmussen, E. 1993*b*. Hydrocarbon potential of the Central Spitsbergen Basin. *In*: Vorren, T., Bergsager, E., Dahl-Stamnes, Ø.A., Holter, E., Johansen, B., Lie, E. & Lund, T.B. (eds) *Arctic Geology and Petroleum Potential*. Norwegian Petroleum Society, Special Publications, **2**, 333–361.

Nøttvedt, A., Johannessen, E.P. & Surlyk, F. 2008. The Mesozoic of Western Scandinavia and East Greenland. *Episodes*, **32**, 59–65.

Oakman, C.D. & Partington, M.A. 1998. Cretaceous. *In*: Glennie, K.W. (ed.) *Petroleum Geology of the North Sea: Basic Concepts and Recent Advances*. Blackwell Science, Oxford, 295–349.

Øvrig, T. 1960. The Jurassic and Cretaceous on Andøya in northern Norway. *In*: Holtedal, O. (ed.) *Geology of Norway*. Geological Survey of Norway, Trondheim, 344–350.

Peron-Pinvidic, G., Gernigon, L., Gaina, C. & Ball, P. 2012*a*. Insights from the Jan Mayen system in the Norwegian-Greenland Sea – I: mapping of a microcontinent. *Geophysical Journal International*, **191**, 385–412.

Peron-Pinvidic, G., Gernigon, L., Gaina, C. & Ball, P. 2012*b*. Insights from the Jan Mayen system in the Norwegian-Greenland Sea – II: architecture of a microcontinent. *Geophysical Journal International*, **191**, 413–435.

Pharaoh, T.C., Dusar, M. et al. 2010. Tectonic evolution. *In*: Doornenbal, J.C. & Stevenson, A.G. (eds) *Petroleum Geological Atlas of the Southern Permian Basin Area*. EAGE Publications, Houten, 25–57.

Praeg, D., Stoker, M.S., Shannon, P.M., Ceramicola, S., Hjelstuen, B., Laberg, J.S. & Mathiesen, A. 2005. Episodic Cenozoic tectonism and the development of the NW European 'passive' continental margin. *Marine and Petroleum Geology*, **22**, 1007–1030.

Quinn, M.F. & Ziska, H. 2011. Permian and Triassic. *In*: Ritchie, J.D., Ziska, H., Johnson, H. & Evans, D. (eds) *Geology of the Faroe–Shetland Basin and Adjacent Areas. BGS Research Report, RR/11/01*. British Geological Survey, Keyworth, Nottingham, 92–102.

Rawson, P.F. & Riley, L.A. 1982. Latest Jurassic-Early Cretaceous Events and the 'Late Cimmerian Unconformity' in the North Sea Area. *American Association of Petroleum Geologists Bulletin*, **66**, 2628–2648.

Ren, S., Faleide, J.I., Eldholm, O., Skogseid, J. & Gradstein, F. 2003. Late Cretaceous-Paleocene tectonic development of the NW Vøring Basin. *Marine and Petroleum Geology*, **20**, 177–206, https://doi.org/10.1016/S0264-8172(03)00005-9

Riis, F., Lundschien, B.A., Høy, T., Mørk, A. & Mørk, M.B.E. 2008. Evolution of the Triassic shelf in the northern Barents Sea region. *Polar Research*, **27**, 318–338.

Ritchie, J.D. & Varming, T. 2011. Jurassic. *In*: Ritchie, J.D., Ziska, H., Johnson, H. & Evans, D. (eds) *Geology of the Faroe–Shetland Basin and Adjacent Areas. BGS Research Report, RR/11/01*. British Geological Survey, Nottingham, 103–122.

Ritchie, J.D., Gatliff, R.W. & Riding, J.B. 1996. *Stratigraphic Nomenclature of the UK North West Margin. 1. Pre-Tertiary Lithostratigraphy*. British Geological Survey, Nottingham.

Ritchie, J.D., Gatliff, R.W. & Richards, P.C. 1999. Early Tertiary magmatism in the offshore NW UK margin and surrounds. *In*: Fleet, A.J. & Boldy, S.A.R. (eds) *Petroleum Geology of Northwest Europe: Proceedings of the 5th Conference*. Geological Society, London, 573–584, https://doi.org/10.1144/0050573

Ritchie, J.D., Ziska, H., Johnson, H. & Evans, D. (eds). 2011*a*. *Geology of the Faroe–Shetland Basin and Adjacent Areas. BGS Research Report, RR/11/01*. British Geological Survey, Nottingham.

Ritchie, J.D., Ziska, H., Kimbell, G., Quinn, M.F. & Chadwick, A. 2011*b*. Structure. *In*: Ritchie, J.D., Ziska, H., Johnson, H. & Evans, D. (eds) *Geology of the Faroe–Shetland Basin and Adjacent Areas. BGS Research Report, RR/11/01*. British Geological Survey, Nottingham, 9–70.

Ritchie, J.D., Johnson, H., Kimbell, G.S. & Quinn, M.F. 2013. Structure. *In*: Hitchen, K., Johnson, H. & Gatliff, R.W. (eds) *Geology of the Rockall Basin and Adjacent Areas. British Geological Survey Research Report, RR/12/03*. British Geological Survey, Nottingham, 10–46.

Roberts, A.M. & Holdsworth, R.E. 1999. Linking onshore and offshore structures: mesozoic extension in the Scottish Highlands. *Journal of the Geological Society, London*, **156**, 1061–1064, https://doi.org/10.1144/gsjgs.156.6.1061

Roberts, D.G., Ginsburg, A., Nunn, K. & McQuillin, R. 1988. The structure of the Rockall Trough from seismic refraction and wide-angle measurements. *Nature*, **332**, 632–635.

Roberts, D.G., Thompson, M., Mitchener, B., Hossack, J., Carmichael, S. & Bjørnseth, H.-M. 1999. Palaeozoic to Tertiary rift and basin dynamics: mid-Norway to the Bay of Biscay – a new context for hydrocarbon prospectivity in the deep water frontier. *In*: Fleet, A.J. & Boldy, S.A.R. (eds) *Petroleum Geology of Northwest Europe: Proceedings of the 5th*

Conference. Geological Society, London, 7–40, https://doi.org/10.1144/0050007

Robinson, K.W., Shannon, P.M. & Young, D.G.G. 1981. The Fastnet Basin: An integrated analysis. *In*: Illing, L.V. & Hobson, G.D. (eds) *The Geology of the Continental Shelf of North-West Europe: Proceedings of the 2nd Conference*. Heyden & Son Ltd, London, 444–454.

Ryseth, A., Augustson, J.H. *et al.* 2003. Cenozoic stratigraphy and evolution of the Sørvestnaget Basin, southwestern Barents Sea. *Norwegian Journal of Geology*, **83**, 107–130.

Seidler, L., Steel, R.J., Stemmerik, L. & Surlyk, F. 2004. North Atlantic marine rifting in the Early Triassic: new evidence from East Greenland. *Journal of the Geological Society, London*, **161**, 583–592, https://doi.org/10.1144/0016-764903-063

Shannon, P.M. 1991. The development of Irish offshore sedimentary basins. *Journal of the Geological Society, London*, **148**, 181–189, https://doi.org/10.1144/gsjgs.148.1.0181

Shannon, P.M. 1993. Submarine fan types in the Porcupine Basin, Ireland. *In*: Spencer, A.M. (ed.) *Generation, Accumulation and Production of Europe's Hydrocarbons*. III. European Association of Petroleum Geoscientists, Special Publications, **3**, 111–120.

Shannon, P.M., Jacob, A.W.B., Makris, J., O'Reilly, B., Hauser, F. & Vogt, U. 1995. Basin development and petroleum prospectivity of the Rockall and Hatton region. *In*: Croker, P.F. & Shannon, P.M. (eds) *The Petroleum Geology of Ireland's Offshore Basins*. Geological Society, London, Special Publications, **93**, 435–457, https://doi.org/10.1144/GSL.SP.1995.093.01.35

Shannon, P.M., Jacob, A.W.B., O'Reilly, B.M., Hauser, F., Readman, P.W. & Makris, J. 1999. Structural setting, geological development and basin modelling in the Rockall Trough. *In*: Fleet, A.J. & Boldy, S.A.R. (eds) *Petroleum Geology of Northwest Europe: Proceedings of the 5th Conference*. Geological Society, London, 421–431, https://doi.org/10.1144/0050421

Sinclair, I.K., Shannon, P.M., Williams, B.P.J., Harker, S.D. & Moore, J.G. 1994. Tectonic control on sedimentary evolution of three North Atlantic borderland Mesozoic basins. *Basin Research*, **6**, 193–217.

Skogseid, J., Planke, S., Faleide, J.I., Pedersen, T., Eldholm, O. & Neverdal, F. 2000. NE Atlantic and volcanic margin formation. *In*: Nøttvedt, A. (eds) *Dynamics of the Norwegian Margin*. Geological Society, London, Special Publications, **167**, 295–326, https://doi.org/10.1144/GSL.SP.2000.167.01.12

Smallwood, J.R. & White, R.S. 2002. Ridge–plume interaction in the North Atlantic and its influence on continental breakup and seafloor spreading. *In*: Jolley, D.W. & Bell, B.R. (eds) *The North Atlantic Igneous Province: Stratigraphy, Tectonic, Volcanic and Magmatic Processes*. Geological Society, London, Special Publications, **197**, 15–37, https://doi.org/10.1144/GSL.SP.2002.197.01.02

Smallwood, J.R., Staples, R.K., Richardson, R.K. & White, R.S. 1999. Crust generated above the Iceland mantle plume: from continental rift to ocean spreading centre. *Journal of Geophysical Research*, **104**, 22,885–22,902.

Smelror, M., Petrov, O.V., Larssen, G.B. & Werner, S. (eds). 2009. *Geological History of the Barents Sea*. Geological Survey of Norway, Trondheim.

Smith, D.B. 1980. The United Kingdom: Permian. *In*: *Geology of the European Countries*. Graham & Trotman, London, 399–402.

Smith, C. 1995. Evolution of the Cockburn Basin: implications for the structural development of the Celtic Sea basins. *In*: Croker, P.F. & Shannon, P.M. (eds) *The Petroleum Geology of Ireland's Offshore Basins*. Geological Society, London, Special Publication, **93**, 279–295.

Smith, K. 2013. Cretaceous. *In*: Hitchen, K., Johnson, H. & Gatliff, R.W. (eds) *Geology of the Rockall Basin and Adjacent Areas. British Geological Survey Research Report, RR/12/03*. British Geological Survey, Nottingham, 71–80.

Steel, R. & Worsley, D. 1984. Svalbard's post-Caledonian strata – an atlas of sedimentological patterns and palaeogeographic evolution. *In*: Spencer, A.M. *et al.* (eds) *Petroleum Geology of the North European Margin*. Graham & Trotman, London, 109–135.

Stemmerik, L. 2000. Late Paleozoic evolution of the North Atlantic margin of Pangea. *Palaeogeography, Palaeoclimatology, Palaeoecology*, **161**, 95–126.

Stemmerik, L. 2001. Sequence stratigraphy of a low productivity carbonate platform succession: the Upper Permian Wegener Halvø Formation, Karstryggen Area, East Greenland. *Sedimentology*, **48**, 79–97.

Stemmerik, L. & Håkansson, E. 1991. Carboniferous and Permian history of the Wandel Sea Basin, North Greenland. *Bulletin Grønlands Geologiske Undersøgelse*, **160**, 141–151.

Stemmerik, L. & Piasecki, S. 2004. Isotope evidence for the age of the Røde Ø Conglomerate, inner Scoresby Sund, East Greenland. *Bulletin of the Geological Society of Denmark*, **51**, 137–140.

Stemmerik, L., Vigran, J.O. & Piasecki, S. 1991. Dating of Late Paleozoic rifting events in the North Alantic: new biostratigraphic data from the uppermost Devonian and carboniferous of East Greenland. *Geology*, **19**, 218–221.

Stemmerik, L., Håkannson, E., Madsen, L., Nilsson, I., Piasecki, S., Pinard, S. & Rasmussen, J.A. 1996. Stratigraphy and depositional evolution of the Upper Palaeozoic sedimentary succession in eastern Peary Land, North Greenland. *Bulletin Grønlands Geologiske Undersøgelse*, **171**, 45–71.

Stemmerik, L., Dalhoff, F., Larsen, B.D., Lyck, J., Mathiesen, A. & Nilsson, I. 1998. Wandel Sea Basin, eastern North Greenland. *Bulletin Grønlands Geologiske Undersøgelse*, **180**, 55–62.

Stemmerik, L., Larsen, B.D. & Dalhoff, F. 2000. Tectono-stratigraphic history of northern Amdrup Land, eastern North Greenland: implications for the northernmost East Greenland shelf. *Geology of Greenland Survey Bulletin*, **187**, 7–19.

Stemmerik, L., Bendix-Almgreen, S.E. & Piasecki, S. 2001. The Permian–Triassic boundary in East Greenland: past and present views. *Bulletin of the Geological Society of Denmark*, **48**, 159–167.

STOKER, M.S. 2016. Cretaceous tectonostratigraphy of the Faroe–Shetland region. *Scottish Journal of Geology*, https://doi.org/10.1144/sjg2016-004

STOKER, M.S. & ZISKA, H. 2011. Cretaceous. *In*: RITCHIE, J.D., ZISKA, H., JOHNSON, H. & EVANS, D. (eds) *Geology of the Faroe–Shetland Basin and Adjacent Areas. BGS Research Report, RR/11/01*. British Geological Survey, Nottingham, 123–150.

STOKER, M.S., HITCHEN, K. & GRAHAM, C.C. 1993. *UK Offshore Regional Report: The Geology of the Hebrides and West Shetland Shelves and Adjacent Deep-Water Areas*. HMSO for the British Geological Survey, London.

STOKER, M.S., DOORNENBAL, H., HOPPER, J.R. & GAINA, C. 2014. Tectonostratigraphy. *In*: HOPPER, J.R., FUNCK, T., STOKER, M.S., ÁRTING, U., PERON-PINVIDIC, G., DOORNENBAL, H. & GAINA, C. (eds) *Tectonostratigraphic Atlas of the North-East Atlantic Region*. Geological Survey of Denmark and Greenland, GEUS, Copenhagen, 129–212.

ŠTOLFOVÁ, K. & SHANNON, P.M. 2009. Permo-Triassic development from Ireland to Norway: basin architecture and regional controls. *Geological Journal*, **44**, 652–767.

STORETVEDT, K.M. & STEEL, R.J. 1977. Palaeomagnetic evidence for the age of the Stornoway Formation. *Scottish Journal of Geology*, **13**, 263–269, https://doi.org/10.1144/sjg13030265

SURLYK, F. 1977. Stratigraphy, tectonics and palaeogeography of the Jurassic sediments of the areas north of Kong Oscars Fjord, East Greenland. *Bulletin Grønlands Geologiske Undersøgelse*, **123**, 56.

SURLYK, F. 1978. Submarine fan sedimentation along fault scarps on tilted fault blocks (Jurassic-Cretaceous boundary, East Greenland). *Bulletin Grønlands Geologiske Undersøgelse*, **128**, 108.

SURLYK, F. 1990. Timing, style and sedimentary evolution of Late Paleozoic-Mesozoic extensional basins of East Greenland. *In*: HARDMAN, R.F.P. & BROOKS, J. (eds) *Tectonic Events Responsible for Britai's Oil and Gas Reserves*. Geological Society, London, Special Publications, **55**, 107–125, https://doi.org/10.1144/GSL.SP.1990.055.01.05

SURLYK, F. 1991. Tectonostratigraphy of North Greenland. *Bulletin Grønlands Geologiske Undersøgelse*, **160**, 25–47.

SURLYK, F. 2003. The Jurassic of East Greenland: a sedimentary record of thermal subsidence, onset and culmination of rifting. *In*: INESON, J.R. & SURLYK, F. (eds) *The Jurassic of Denmark and Greenland*. Geological Survey of Denmark and Greenland Bulletin, **1**, 659–722.

SURLYK, F. & NOE-NYGAARD, N. 2001. Cretaceous faulting and asociated coarse-grained marine gravity flow sedimentation, Traill Ø, East Greenland. *In*: MARTINSEN, O.J. & DREYER, T. (eds) *Sedimentary Environments Offshore Norway – Paleozoic to Recent*. Norwegian Petroleum Society, Special Publications, **10**, 293–319.

SURLYK, F., CALLOMON, J.H., BROMLEY, R.G. & BIRKELUND, T. 1973. Stratigraphy of the Lower Jurassic – Lower Cretaceous sediments of Jameson Land and Scoresby Land, East Greenland. *Bulletin Grønlands Geologiske Undersøgelse*, **105**, 76.

SURLYK, F., HURST, J.M., PIASECKI, S., ROLLE, F., SCHOLLE, P.A., STEMMERIK, L. & THOMSEN, E. 1986. The Permian of the western margin of the Greenland Sea – a future exploration target. *In*: HALBOUTY, M.T. (ed.) *Future Petroleum Provinces of the World*. American Association of Petroleum Geologists, Memoirs, **40**, 629–659.

TATE, M.P. & DOBSON, M.R. 1988. Syn- and post-rift igneous activity in the Porcupine Seabight basin and adjacent continental margin west of Ireland. *In*: MORTON, A.C. & PARSON, L.M. (eds) *Early Tertiary Volcanism and the Opening of the NE Atlantic*. Geological Society, London, Special Publications, **39**, 309–334, https://doi.org/10.1144/GSL.SP.1988.039.01.28

TATE, M.P. & DOBSON, M.R. 1989. Late Permian to early Mesozoic rifting and sedimentation offshore NW Ireland. *Marine and Petroleum Geology*, **6**, 49–59.

THY, P., LESHER, C.E. & LARSEN, H.C. (eds) . 2007. *Proceedings of the Ocean Drilling Program Initial Reports, Volume 163X*. Ocean Drilling Program, College Station, TX, https://doi.org/10.2973/odp.proc.ir.163x.2007

TORSVIK, T.H., CARLOS, D., MOSAR, M., COCKS, L.R.M. & MALME, T. 2002. Global reconstructions and North Atlantic paleogeography 440 Ma to Recent. *In*: EIDE, E.A. (coord.) *BATLAS – Mid Norway Plate Reconstruction Atlas with Global and Atlantic Perspectives*. Geological Survey of Norway, Trondheim, 18–39.

TORSVIK, T.H., VAN DER VOO, R. *ET AL*. 2012. Phanerozoic polar wander, palaeogeography and dynamics. *Earth-Science Reviews*, **114**, 325–368.

TRUEBLOOD, S. & MORTON, N. 1991. Comparative sequence stratigraphy and structural styles of the Slyne Trough and Hebrides Basin. *Journal of the Geological Society, London*, **148**, 197–201, https://doi.org/10.1144/gsjgs.148.1.0197

TSIKALAS, F., FALEIDE, J.I., ELDHOLM, O. & WILSON, J. 2005. Late Mesozoic–Cenozoic structural and stratigraphic correlations between the conjugate mid-Norway and NE Greenland continental margins. *In*: DORÉ, A.G. & VINING, B.A. (eds) *Petroleum Geology: North-West Europe and Global Perspectives – Proceedings of the 6th Petroleum Geology Conference*. Geological Society, London, 785–802, https://doi.org/10.1144/0060785

TYRRELL, S., SOUDERS, A.K., HAUGHTON, P.D.W., DALY, J.S. & SHANNON, P.M. 2010. Sandstone provenance and palaeodrainage on the eastern Rockall Basin margin: evidence from the Pb isotopic composition of detrital K-feldspar. *In*: VINING, B. & PICKERING, S.C. (eds) *From Mature Basins to New Frontiers: Proceedings of the 7th Petroleum Geology Conference*. Geological Society, London, 937–952, https://doi.org/10.1144/0070937

UNDERHILL, J.R. & PARTINGTON, M.A. 1993. Jurassic thermal doming and deflation in the North Sea: implications of the sequence stratigraphic evidence. *In*: PARKER, J.R. (ed.) *Petroleum Geology of Northwest Europe: Proceedings of the 4th Conference*. Geological Society, London, 337–346, https://doi.org/10.1144/0040337

VALLIER, T., CALK, L., STAX, R. & DEMANT, A. 1998. Metamorphosed sedimentary (volcaniclastic?) rocks beneath Paleocene basalt in hole 917a, East Greenland margin. *In*: Saunders, A.D., Larsen, H.C. & Wise, S.W., Jr. (eds) *Proceedings of the Ocean Drilling Program, Scientific Results*. Ocean Drilling Program, College Station, TX, **152**, 129–144.

Vestralen, I., Hartley, A.J. & Hurst, A. 1995. The sedimentology of the Rona Sandstone (Upper Jurassic), West of Shetlands, UK. *In*: Hartley, A.J. & Prosser, D.J. (eds) *Characterisation of Deep-Marine Clastic Systems*. Geological Society, London, Special Publications, **94**, 155–176, https://doi.org/10.1144/GSL.SP.1995.094.01.12

Waters, C.N., Gillespie, M.R. *et al.* 2007. *Stratigraphical Chart of the United Kingdom: Northern Britain*. British Geological Survey, Nottingham.

White, N.J. & Lovell, B. 1997. Measuring the pulse of a plume with the sedimentary record. *Nature*, **387**, 888–891.

White, R. & McKenzie, D. 1989. Magmatism at rift zones: the generation of volcanic continental margins and flood basalts. *Journal of Geophysical Research*, **94**, 7685–7729.

White, R.S. 1988. A hot-spot model for early Tertiary volcanism in the N Atlantic. *In*: Morton, A.C. & Parson, L.M. (eds) *Early Tertiary Volcanism and the Opening of the NE Atlantic*. Geological Society, London, Special Publications, **39**, 3–13, https://doi.org/10.1144/GSL.SP.1988.039.01.02

White, R.S. 1989. Initiation of the Iceland plume and opening of the North Atlantic. *In*: Tankard, A.J. & Balkwill, H.R. (eds) *Extensional Tectonics and Stratigraphy of the North Atlantic Margins*. American Association of Petroleum Geologists, Memoirs, **46**, 149–154.

Whitham, A.G., Price, S.P., Koraini, A.M. & Kelly, S.R.A. 1999. Cretaceous (post-Valanginian) sedimentation and rift events in NE Greenland (71–77°N). *In*: Fleet, A.J. & Boldy, S.A.R. (eds) *Petroleum Geology of Northwest Europe: Proceedings of the 5th Conference*. Geological Society, London, 325–336, https://doi.org/10.1144/0050325

Williams, B.P.J. & McKie, T. 2009. Preface: triassic basins of the Central and North Atlantic Borderlands: models for exploration. *Geological Journal*, **44**, 627–630.

Worsley, D. 2008. The post-Caledonian development of Svalbard and the western Barents Sea. *Polar Research*, **27**, 298–317.

Ziegler, P.A. 1988. *Evolution of the Arctic–North Atlantic and the Western Tethys*. American Association of Petroleum Geologists, Memoir, **43**.

Compilation and appraisal of geochronological data from the North Atlantic Igneous Province (NAIP)

CAMILLA M. WILKINSON[1,2]*, MORGAN GANERØD[1], BART W. H. HENDRIKS[1,3] & ELIZABETH A. EIDE[4]

[1]*Geological Survey of Norway, Post Box 6315 Sluppen, 7491 Trondheim, Norway*

[2]*Present address: Hatfield Marine Science Centre, Newport, Oregon 97365, USA*

[3]*Present address: Statoil Research Centre, Arkitekt Ebbells veg 10, 7053 Ranheim, Norway*

[4]*National Academy of Sciences, 500 Fifth Street NW, Washington, DC 20001, USA*

**Correspondence: camilla.wilkinson@noaa.gov*

Abstract: The North Atlantic Igneous Province (NAIP), composed of volcanic sequences and intrusive rocks, occurs onshore in Greenland, the Faeroe Islands, the UK and Ireland, and offshore surrounding these areas as well as the west coast of Norway. Geochronological data have been published for Cenozoic igneous and volcanic rocks for much of the province, and provide valuable information to analyse the evolution of the province and magmatic processes more broadly. As part of the NE Atlantic Geosciences (NAG) cooperation, we examined approximately 700 dates from over 70 published studies and created a comprehensive database to facilitate ready access to this important information. This includes U–Pb, Rb–Sr, Re–Os, $^{40}Ar/^{39}Ar$ and K–Ar ages presented relative to the Geological Time Scale 2012. $^{40}Ar/^{39}Ar$ and K–Ar ages have been recalculated to a common reference. The complete database includes data that range from approximately 177 to 0.19 Ma. Our evaluation shows that variable sample quality, ambiguous data-handling methods, inadequate data reporting and data interpretation should preclude the use of data for purposes of rigorous geochronological analysis. Through a series of filtering techniques described here, we suggest excluding >500 dates as being of too poor a quality to use in age determinations. Our analysis highlights the need for published geochronological studies to include sufficient information to allow critical assessment of ages and interpretations. We present an 'optimized' dataset containing 130 ages that range from approximately 64 to 13 Ma. The filtered dataset emphasizes the need for firm chronological benchmarks and suggests that some sub-provinces in the NAIP would greatly benefit from renewed research attention.

Supplementary material: The full NAG-TEC Geochronological Database 001 and Data Evaluation 002 are available at https://doi.org/10.6084/m9.figshare.c.3554472

The North Atlantic Igneous Province (NAIP), with its combined onshore and offshore volcanic sequences, is one of the most extended Large Igneous Provinces (LIPs) in the world in terms of areal extent and inferred volume (Saunders *et al.* 1997; Saunders 2016). It is estimated that the entire NAIP covers a total area of 1.3×10^6 km^2. Although the related volume may be difficult to envisage, its original volume has been estimated to reach $5 \times 10^6 – 10 \times 10^6$ km^3 (e.g. Storey *et al.* 2007). This is comparable to some of the best recognized examples of LIPs, such as the Siberian Traps (>4×10^6 km^3: e.g. Ivanov 2007) and the Karoo–Ferrar Province (>2.5×10^6 km^3: cf. Courtillot & Renne 2003), and is larger than others, such as the Deccan Traps (*c.* 1.5×10^6 km^3: e.g. Beane *et al.* 1986) and the Columbia River flood basalt province (e.g. *c.* 2×105 km^3: cf. Reidel *et al.* 2013). The classification of the NAIP as a LIP agrees with the definition of Bryan & Ernst (2008), who stated that LIPs are magmatic provinces with areal extents >1×10^6 km^2, igneous volumes >1×10^6 km^3, maximum lifespans of approximately 50 myr and igneous pulse(s) of short duration (*c.* 1–5 myr), during which a large proportion (>75%) of the total igneous volume is emplaced. The eruption and emplacement of LIPs have been associated with processes such as crustal uplift (Saunders *et al.* 2007) and erosion (Campbell & Griffiths 1990; Nadin *et al.* 1997); continental rifting (White & Mackenzie 1989; Courtillot *et al.* 1999); global environmental change (Svensen *et al.* 2004); and mass extinction (Wignall 2001; Courtillot & Renne 2003).

From: Péron-Pinvidic, G., Hopper, J. R., Stoker, M. S., Gaina, C., Doornenbal, J. C., Funck, T. & Árting, U. E. (eds) 2017. *The NE Atlantic Region: A Reappraisal of Crustal Structure, Tectonostratigraphy and Magmatic Evolution*. Geological Society, London, Special Publications, **447**, 69–103.
First published online November 8, 2016, https://doi.org/10.1144/SP447.10

One of the efforts of the NE Atlantic Geosciences (NAG) cooperation has been the assembly of the *Tectonostratigraphic Atlas* (TEC). This work has involved compiling published geochronological data for Cenozoic igneous rocks from the NAIP in the NAG-TEC Geochronological Database (see Fig. 1b–e). The primary aim of this paper is to examine and assess the quality and geological significance of data contained in that database. As described in the section 'The NAG-TEC Geochronological Database', this database brings all isotope geochronological data together relative to a common age reference system (Geological Time Scale 2012 of Gradstein *et al.* (2012)). This has never before been carried out on this scale for the NAIP. In some cases, insufficient reporting of the analytical protocol has prevented the recalculation of ages (Fitch *et al.* 1969, 1988; Mørk & Duncan 1993; Archer *et al.* 2005). However, full details are provided in the NAG-TEC Geochronological Database. To avoid confusion, it should be noted that other contributions in this Special Publication may refer to data taken directly from the original source without recalculation.

To demonstrate the potential applications of the database, we offer an evaluation of previously reported age data. We concentrate on K–Ar and ^{40}Ar/^{39}Ar data (hereafter referred to as Ar–Ar), with the inclusion of reliable U–Pb data where available. The database offers the possibility of examining the detail of data coverage and provides a reference that can be used for testing models of LIP formation (King & Anderson 1998; Torsvik *et al.* 2001; Foulger 2002; Meyer *et al.* 2007; Ganerød *et al.* 2010). Finally, the database allows identification of the remaining data gaps, which will assist in focusing future research efforts in the NAIP.

The NAIP was divided in two by the opening of the North Atlantic Ocean, and its principal components are now widely distributed from Canada to the British Isles (see Fig. 1a). Although dominantly mafic in composition, the NAIP, like most other LIPs, contains a significant number of ultramafic and silicic components including: the extruded volcanic sequences of Baffin Island and West Greenland (basaltic and picritic in composition); the East Greenland continental flood basalts (basaltic in composition) and intrusions (e.g. central intrusions: sills, and dyke swarms, both mafic and felsic in composition); the seawards-dipping reflector sequences (SDRs) of the East Greenland and NW European rifted margins; the Faeroe Islands onshore and offshore volcanic sequences; and the British–Irish Palaeogene Igneous Province (BPIP), which crops out in Northern Ireland (Antrim), the Inner Hebrides of Scotland (e.g. Isles of Mull and Skye)

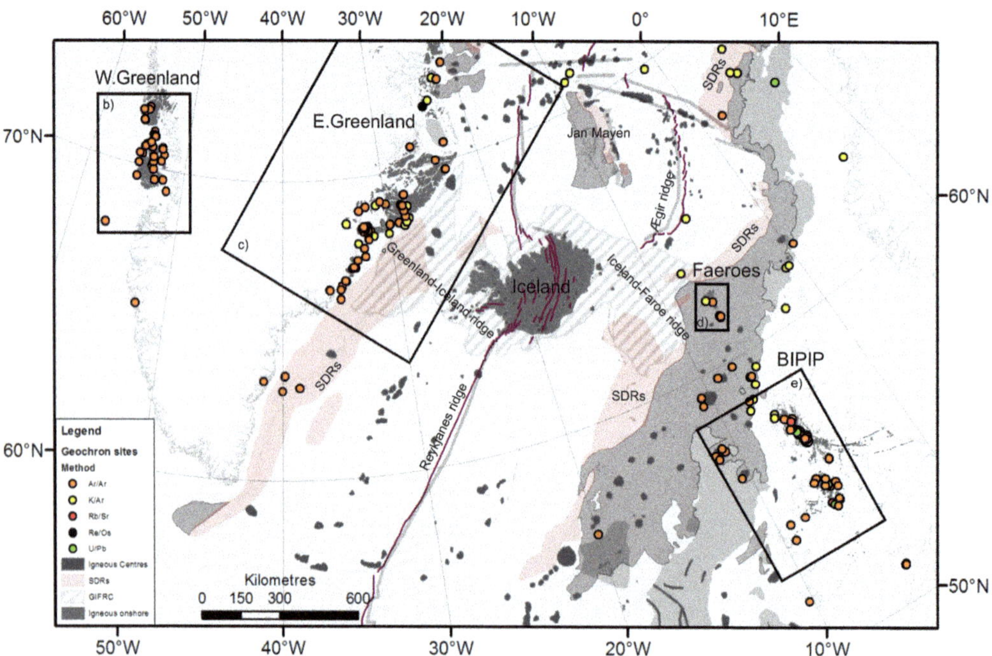

Fig. 1. (**a**) Map of the North Atlantic showing the geographical extent of the North Atlantic Igneous Province (NAIP). Filled circles indicate sampling localities related to geochronological data presented in the unfiltered NAG-TEC Database. Rectangular boxes outline areas shown in Fig. 1(b)–(e).

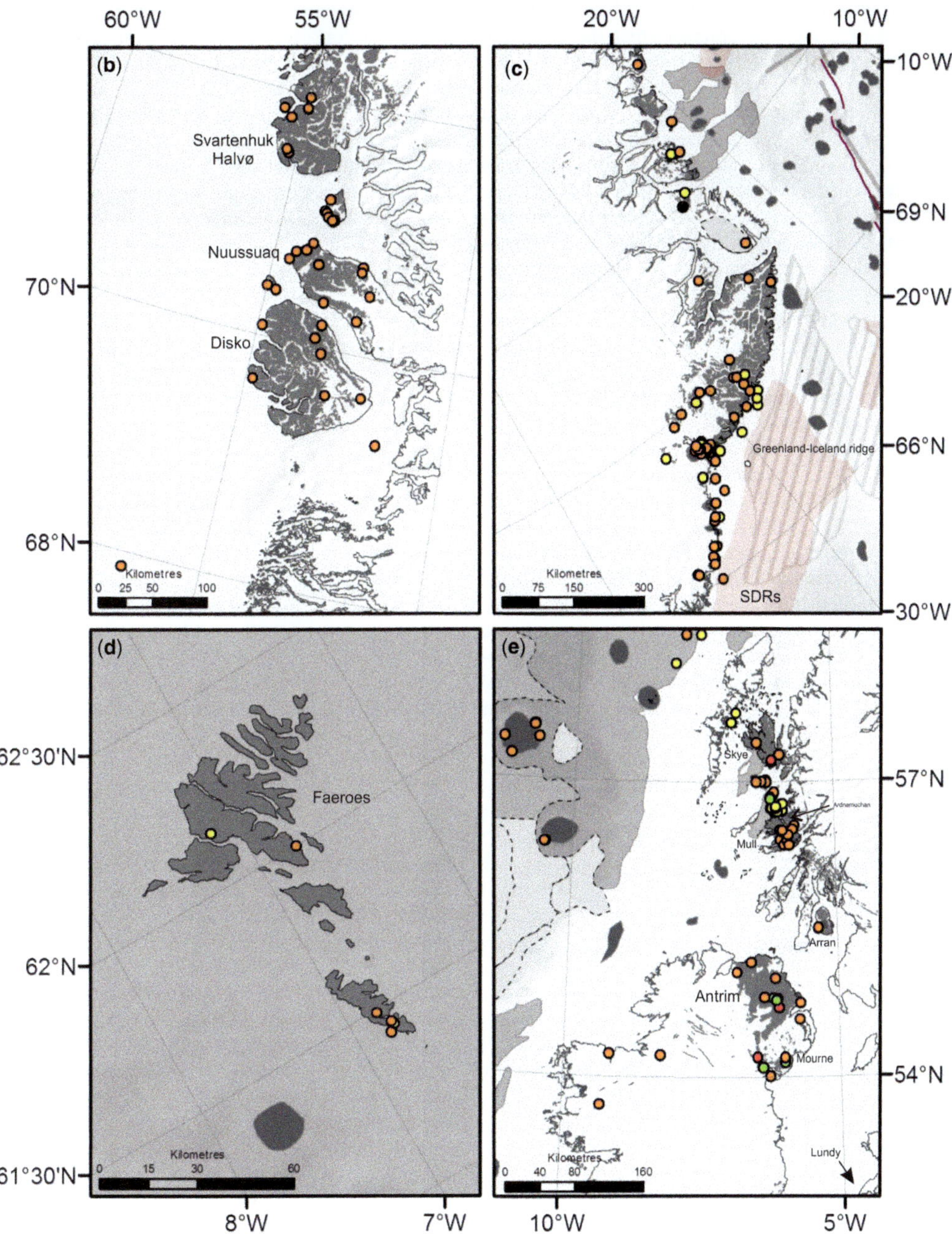

Fig. 1. (**b**) West Greenland; (**c**) East Greenland; (**d**) the Faeroe Islands; and (**e**) the British–Irish Palaeogene Igneous Province (BIPIP). SDRs denote seawards-dipping reflector sequences.

and the Greenland–Iceland–Faeroes Ridge (Saunders *et al.* 1997; Storey *et al.* 1998, 2007 and references therein).

Obtaining precise age determinations for LIPs using various isotopic geochronological techniques is challenging for a number of reasons. In general, magmatic rocks from LIPs erupt at or near the Earth's surface and occasionally in aqueous environments, making them susceptible to chemical alteration and weathering. The application of specific isotopic geochronological methods to these rocks also presents specific challenges. Use of the

U–Pb method, for example, may not be possible due to a lack of suitable minerals on which the technique can be performed (e.g. zircon), coupled with the difficulty in separating such minerals when they are present (e.g. Corfu *et al.* 2013). The K–Ar and Ar–Ar methods require samples that contain sufficient K, which may be problematic for the predominantly basaltic products of LIP magmatism. The Rb–Sr method uses minerals such as mica and feldspar (as does the Ar–Ar method), which are highly susceptible to alteration, resulting in ages that are relatively less precise, difficult to interpret or yield ages that may represent partial or complete overprints of processes other than crystallization (e.g. Dickin 1981).

Despite these difficulties, improvements in analytical capabilities of various isotopic systems and a better understanding of the sources of age variation have, with time, enabled more accurate data to be obtained from individual minerals in LIP rocks and have demonstrated the interpretive value of having accurate ages at the level of individual flows. For example, older geochronological datasets suggested long-lived magmatic systems, with available ages sometimes spreading over tens of millions of years, whereas more recent geochronological data demonstrate eruptions over only a few hundreds of thousands of years, or no more than 1–5 myr (e.g. Deccan LIP: Hofmann *et al.* 2000). For other datasets, the opposite has proven to be the case, with new palaeomagnetic data indicating emplacement of magma over longer durations than previously thought (e.g. the Paraná–Etendeka LIP: Dodd *et al.* 2015).

The North Atlantic Igneous Province (NAIP)

Comprehensive descriptions of the geological setting and geochemistry of the NAIP are provided in several key publications (Upton 1988; Saunders *et al.* 1997; Storey *et al.* 1998, 2004, 2007; Meyer *et al.* 2007; Tegner *et al.* 2008; Larsen *et al.* 2015; Saunders 2016). The following is a summary of the main observations. The Stoker *et al.* (2016) contribution in this volume provides additional, detailed information. Although we exclusively focus on the Cenozoic rift-related volcanism of the NAIP, it should be appreciated that the NAIP is just a recent chapter in the story of the North Atlantic rift system, which experienced *c.* 400 Ma of collisional and extensional tectonics. Older structures are associated with the Caledonian and Appalachian orogenic events, which brought together the early continents of Laurentia and Avalonia as the Iapetus Ocean closed. Contractional events may have started as early as the Ordovician but culminated in Silurian to Devonian times (e.g. Cocks & Torsvik 2011). Devonian rift basins and sedimentary sequences (observed in East Greenland, Norway, Svalbard, and northernmost parts of the UK) are thought to mark the collapse of the Caledonian belts (e.g. Andersen & Jamtveit 1990; Fossen 2010). Rift basins continued to develop through non-continuous stretching of the crust into the Carboniferous, and throughout the Permian and Triassic (e.g. Surlyk 1990; Stemmerik 2000; Müller *et al.* 2005). During the Jurassic and throughout the Cretaceous, major extension and basin formation was occurring in all parts of the proto-North Atlantic (e.g. Norwegian–Greenland Sea and North Sea; Nøttvedt *et al.* 2008; Osmundsen & Ebbing 2008). The separation of Greenland from the Eurasian plate took place at the Paleocene–Eocene boundary and was preceded and accompanied by one of the largest recorded volcanic events on Earth.

It is generally accepted that the formation of the NAIP was initiated during the Palaeogene, in two distinct phases that correspond to pre- and post-rift magmatic events separated by the opening (i.e. syn-rift) of the North Atlantic. Prior to *c.* 63 Ma, there is little evidence to suggest excessive volcanism, although ages of *c.* 80–64 Ma have been reported for parts of the province (e.g. East Greenland; Noble *et al.* 1988); offshore Shetland Isles (K–Ar study of Fitch *et al.* 1988); and the Rockall trough (K–Ar study of Roddick *et al.* 1989). However, since K–Ar studies are unable to identify excess Ar it would be necessary to independently verify an older phase by other means (i.e. Ar–Ar or U–Pb data). The products of the first phase include the emplacement of continental flood basalt sequences (e.g. West Greenland). The ages pertaining to pre-rift volcanism converge on the range 62–58 Ma (Sinton & Duncan 1998; Sinton *et al.* 1998; Storey *et al.* 1998; Hansen *et al.* 2002; Larsen *et al.* 2015), although considered by some to tend towards and beyond the upper limit in this range (e.g. >62 Ma; Ganerød *et al.* 2010). Following a proposed 'gap' or 'hiatus' in widespread volcanism at *c.* 57 Ma (e.g. Saunders *et al.* 1997), a second and more voluminous phase of magmatism, dated at *c.* 56–53 Ma coincides broadly in time with the opening of the North Atlantic (Saunders *et al.* 1997). Products of this second phase are exposed in SDRs in Greenland (Larsen & Saunders 1998; Tegner *et al.* 1998), the Faeroe Islands (e.g. Larsen *et al.* 1999*b*), and Northern Ireland (Meighan *et al.* 1988; Gamble *et al.* 1999; Ganerød *et al.* 2010).

Geochronological methods

The NAG-TEC Geochronological Database contains data that have been determined using a range of isotopic techniques applied to a range of materials, such as basalt and granite (e.g. plagioclase, amphibole, biotite, whole rock and zircon), and rhyolite (e.g. sanidine, zircon, and glass). Samples identified as particularly challenging for dating studies include: dredged seafloor samples (O'Connor *et al.* 2000); rocks later affected by hydrothermal alteration and/or weathering (Mussett 1986; Noble *et al.* 1988; Lenoir *et al.* 2003); rocks containing no zircon (or containing zircon that is difficult to separate; e.g. Corfu *et al.* 2013); rocks with low K content, or very few K-bearing mineral phases (e.g. Tegner & Duncan 1999; Archer *et al.* 2005; Larsen *et al.* 2014); aphyric rocks (Sinton *et al.* 1998), and volcanic glass (e.g. Dickin & Jones 1983).

The majority of ages in the database have been determined by the K–Ar or Ar–Ar methods. These methods are based on the radioactive decay of naturally occurring ^{40}K to stable ^{40}Ar, and so in principle can be applied to any K-bearing rock or mineral. Details of the K–Ar and Ar–Ar dating methods can be found in Dalrymple & Lanphere (1969) and McDougall & Harrison (1999). Ar–Ar ages are calculated relative to a standard mineral (e.g. the Fish Canyon Tuff sanidine: Renne *et al.* 1998, 2010, 2011; Kuiper *et al.* 2008) to allow indirect determination of ^{39}K. Ar–Ar ages may be calculated relative to a number of different standards, as well as to one of three ^{40}K decay constants currently in use (Steiger & Jäger 1977; Min *et al.* 2000; Renne *et al.* 2010, 2011). As described below, we have recalculated the ages relative to a common reference to make age comparisons and interpretations of different datasets possible.

Inaccurate K–Ar and Ar–Ar apparent age determinations will result if data show that the rock or mineral has experienced post-crystallization disturbance, such as weathering, alteration and/or a reheating event. This kind of disturbance may result in episodic diffusive loss of radiogenic ^{40}Ar and partial resetting of the argon clock, redistribution or loss of K, formation of altered minerals that do not retain 100% of Ar over geological time, and the incorporation of an inherited (related to a previous geological event) or 'excess argon' (hereafter referred to as $^{40}Ar_E$) component that contributes additional ^{40}Ar, thus artificially increasing the apparent age.

A small percentage of the data featured in the NAIP database has been determined by other geochronological techniques (i.e. U–Pb and Rb–Sr). U–Pb ages (uranium series decay from ^{238}U to ^{206}Pb) have been considered: however, in some cases, a concordia age has been accepted. It should be noted that a concordia age is often associated with a lower uncertainty compared to the $^{206}Pb/^{238}U$ age and, therefore, it may underestimate the true uncertainty. The majority of Rb–Sr data included in the NAIP database are lower in quality compared to either U–Pb or Ar–Ar data. Rb–Sr ages in the database are often associated with high mean square weighted deviation (MSWD) values (Gibson *et al.* 1987; Meighan *et al.* 1988) or lack comprehensive data reporting (e.g. uncertainties have been changed (increased) to give a lower MSWD value: Dickin 1981).

The NAG-TEC Geochronological Database

The NAG-TEC Geochronological Database incorporates only published data related to the NAIP. It includes age data that range from 0.19 Ma (K–Ar: Jan Mayen) to 177 Ma (K–Ar: offshore UK). Every effort has been taken to make this database as complete as possible, although omissions are unavoidable. We acknowledge that a considerable amount of age data are unpublished (e.g. contained in thesis manuscripts) and these could not be included in the database. In addition, over 550 individual ages related to volcanic activity on Iceland and the present-day position of the Icelandic hotspot are in peer-reviewed literature, but not included in our database. These will form a separate contribution currently in progress.

The compilation of the NAIP database and subsequent data evaluation has relied on the information provided in the publication that originally reported the data. Inadequate data reporting is a problem and hinders thorough evaluation of the data. In some cases, studies may have failed to fully report their analytical protocol or failed to provide raw data in supplementary data tables. We note that the level of detail for analytical reporting of geochronological data 20 years ago or more was different compared to recent times for which a minimum data-reporting standard is now established (Renne *et al.* 2009). In other cases, studies make reference to geochronological data, but not to the original publications in which these data were first presented (e.g. Bell & Williamson 2002). The result is that much of the information on how data were acquired, and how ages were calculated, is ignored or simply lost. Therefore, it has often been impossible, or at best difficult, to verify data quality. In the current compilation only the first published occurrence of the geochronological data is used, and the information that was included there is used to rate the data quality.

Ar–Ar data are calculated relative to an assigned standard (fluence monitor) age and a ^{40}K decay constant, which continue to be the subject of ongoing debate (Renne *et al.* 1998, 2010, 2011; Min *et al.* 2000; Kuiper *et al.* 2008). There are several fluence monitors currently in use by the Ar–Ar community, including sandine (e.g. Fish Canyon, Taylor Creek and Alder Creek), biotite (e.g. GA1550, Tinto and B4) and amphibole (Hb3gr and MMHb-1), and ^{40}K decay constants (e.g. Steiger & Jäger 1977; Min *et al.* 2000; Renne *et al.* 2010, 2011). The ages assigned to fluence monitors and the decay constant for K have changed over time. For example, ages assigned to the Taylor Creek sanidine range from 28.619 to 27.92 Ma. In addition, several ^{40}K decay constants are currently in use, with recent redeterminations designed to reconcile U–Pb and Ar–Ar ages, and to improve precision and accuracy on existing measurements (Renne *et al.* 2010, 2011). This can cause confusion for non-chronologists and the misuse of Ar–Ar age data, which can lead to incorrect geological conclusions. Therefore, in keeping with the Geological Time Scale 2012 (Gradstein *et al.* 2012) chronostratigraphic framework, and based on the values agreed during the EARTHTIME IV Workshop (held in 2009), the entire dataset in the database has been recalculated to the reference system recommended in Kuiper *et al.* (2008).

A summary of the geochronological techniques applied to both onshore and offshore areas of the main geographical regions (divided into sub-provinces) of the NAIP is presented in Table 1. In total, approximately 699 ages exist for the entire NAIP (excluding Iceland). K–Ar and Ar–Ar (290 and 366 ages, respectively) are the most used analytical techniques to date products of extrusive and intrusive activity. Onshore localities, as one would expect, are associated with the greatest density of data, with a focus on Greenland, the UK and Ireland.

Data evaluation

Taking the entire province as a whole and considering all age data ($\pm 2\sigma$ uncertainty), the picture is not a simple one. The cumulative frequency diagram (Fig. 2) shows continuous NAIP volcanism spanning approximately 55 myr. There are no distinguishable pulses or across-province correlated 'breaks or gaps' in volcanism. It is, however, highly unlikely that all 699 dates represent emplacement ages. Therefore, it is necessary to evaluate the data in the database and produce an optimized dataset in order to better understand the time correlation across the province.

Approach to data evaluation

The filtering process comprised selection of those data that are based on tightly constrained isochrons or weighted means (e.g. $^{238}U/^{206}Pb$ ages: Ganerød *et al.* 2011), and ages that conform to criteria defining a statistically robust age (e.g. definition of a valid plateau/isochron for Ar–Ar step heating data: McDougall & Harrison 1999; Baksi 2005*a*), and have well documented analytical procedures. For data that demonstrate an elevated trapped atmospheric ratio ($^{40}Ar/^{36}Ar >$ atmospheric value), and where the use of the inverse isochron age is shown to be justified, the inverse isochron age has been taken as the preferred age. The reproducibility of ages (between separate samples or splits of the same sample) has also been a consideration for accepting data, and in some cases it has been possible to calculate a weighted mean age to further refine the dataset.

U–Pb and Ar–Ar data that are demonstrably variable in quality and K–Ar data have been excluded from the accepted dataset but remain part of the complete NAIP Geochronological Database. Critical examination of Ar–Ar data is commonplace and has been carried out previously by Baksi (2005*a*, *b*, 2007*a*, *b*), with specific reference to Ar–Ar ages related to hotspot activity. Other examples include Jourdan *et al.* (2007, 2009*b*), who reviewed geochronological data related to terrestrial impact structures and the Karoo LIP. For the purpose of this analysis, we conducted a two-stage evaluation and filtering process. The first stage involved critical examination of published data (including a statistical assessment). A second, more rigorous filtering removed Ar–Ar whole-rock and groundmass data, because several studies have now shown that whole-rock/groundmass Ar–Ar

Table 1. *Summary table of published regional geochronological data*

Geographical region	Radiometric technique					Total
	Ar–Ar	K–Ar	U–Pb	Rb–Sr	Re–Os	
West Greenland	49		3	6		58
East Greenland	110	118			2	230
Offshore Greenland	26	7				33
UK + Ireland + offshore UK	164	145	9	21		339
Faeroe Islands + Offshore Faeroes	13	8				21
Offshore Norway + Svalbard	4	12	2			18
Total	366	290	14	27	2	699

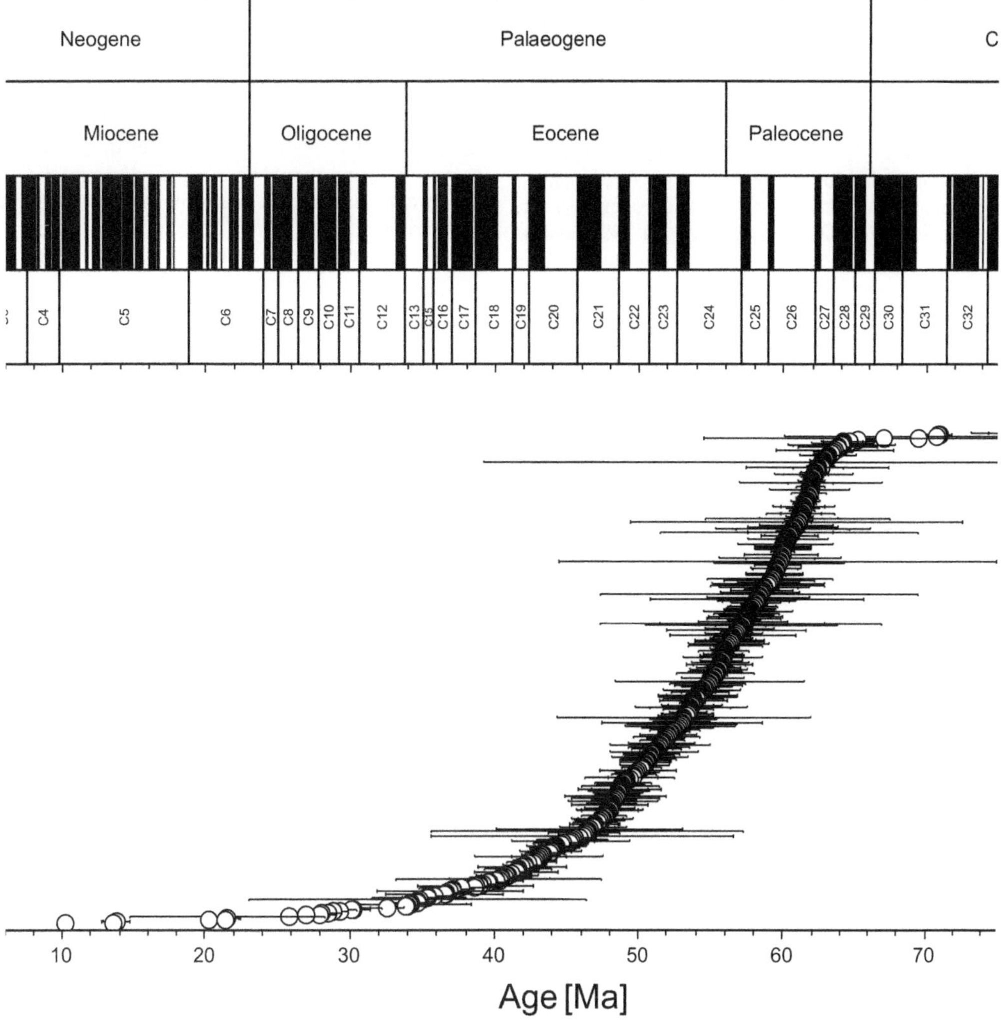

Fig. 2. All geochronological data for the entire North Atlantic Igneous Province (NAIP) shown in terms of their cumulative distribution. Data are presented relative to the geomagnetic polarity timescale of Gradstein *et al.* (2012). Uncertainties are given at the 2σ level. Note: ages older than 90 Ma are not shown.

ages older than 20–30 myr are unreliable (e.g. Hofmann *et al.* 2000). A comparison between the two filtering stages is discussed later.

To allow evaluation of the data, we determined that the following information had to be accessible in the original publication:

- Sample characterization – detailed sample descriptions, including whether the samples are dredged seafloor rocks; characterization of minerals separated for dating (i.e. thin section and/or electron microprobe analysis), or X-ray fluorescence (XRF) data for whole-rock samples (i.e. K_2O content; description of any alteration (secondary minerals, zeolites) and/or inclusions that may be a source of $^{40}Ar_E$, and information regarding additional quantitative methods carried out to assess the alteration state, such as Alteration Index: Baksi 2007*a*, *b*). A description of alteration is critical as altered material may result in partial loss of radiogenic argon ($^{40}Ar^*$) and lead to inaccurate estimates of K–Ar or Ar–Ar crystallization ages, and/or age estimates that are systematically too young.
- Cleaning procedures for whole rock and mineral separates – details of cleaning procedures conducted, or necessary to remove alteration. Examples may include acid leaching (i.e. nitric acid

(HNO_3) for material containing ferromagnesium phases, or hydrofluoric acid (HF) for feldspars: Baksi 2007*a*, *b*).

- Criteria used to define an Ar–Ar 'plateau age' – for the Ar–Ar incremental heating method, it is important to know which criteria were followed to define a 'plateau age'. For example, >50% of the cumulative ^{39}Ar released over three or more consecutive heating steps, which overlap at the 2σ level (McDougall & Harrison 1999). However, 50–70% is considered by some to represent only a 'mini plateau', and a 'plateau age' is one defined by >70% of the cumulative ^{39}Ar (e.g. Hofmann *et al.* 2000; Jourdan *et al.* 2007). The mean square weighted deviation (MSWD) of the plateau and/or inverse isochron age along with the probability value (*P*) should also be given (statistically valid if $P > 0.05$: e.g. Baksi 2003). Jourdan *et al.* (2009*b*) strongly advised that the value of *P* is reported along with the MSWD for any given age, so that readers can estimate for themselves the statistical significance of the data.
- Presentation of full degassing information – for the Ar–Ar incremental heating method, it is important that degassing data (isotopes ^{40}Ar, ^{39}Ar, ^{38}Ar, ^{37}Ar and ^{36}Ar; or ratios ^{40}Ar*/^{39}Ar; ^{40}Ar/^{36}Ar; ^{37}Ar/^{39}Ar and ^{38}Ar/^{39}Ar) for the entire experiment, not only the steps that define the plateau or weighted mean, are included in the paper or provided in a supplementary table. This enables the atmospheric component (through the use of the inverse isochron method), Ca/K and Cl/K ratio of each degassing step to be assessed. These kinds of data can help to identify the presence of a $^{40}Ar_E$ component and the mineral phase(s) being dated. For example, a lower Ca/K than expected for clean and unaltered plagioclase could imply dating of sericite (Verati & Jourdan 2013), and ages should be taken as a minimum estimate of crystallization unless episodic alteration or weathering can be shown to have occurred very soon after crystallization (e.g. Jourdan *et al.* 2009*a*; Polteau *et al.* 2016). The display of degassing spectrums, isochron diagrams (normal and/or inverse) and degassing data (i.e. $^{37}Ar/^{39}Ar = Ca/K$ ratio often presented as K/Ca in publications) must be of a suitable scale so that information can be adequately checked and evaluated (Renne *et al.* 2009; Jourdan *et al.* 2009*b*).
- Sample irradiation information – this information should include correction factors for interfering reactions, the average *J* value; the monitor and monitor age; and the decay constant used for data reduction. Such information allows data from different studies to be recalculated to a common base and compared.

Data presentation by sub-province

As a full review of the accuracy and precision of all 699 ages available for the NAIP is outside the scope of this contribution, our evaluation has focused on onshore data. However, a summary of offshore data is provided. The evaluation of key data and subsequent filtering should be considered as a guideline concerning the current state of geochronological knowledge for the NAIP. For clarity, we present below a general geological and geochronological summary of all data included in the full NAG-TEC database (prior to filtering) with reference to key studies for each sub-province of the NAIP.

The selected dataset (following filtering) for each sub-province is presented in Tables 2–5. Data uncertainties are quoted at the 2σ confidence level. The final optimized dataset is displayed on compilation event charts (radiometric age plotted v. geographical latitude) and summary stratigraphic columns (Fig. 3).

West Greenland

General geology. The Palaeogene volcanic rocks in West Greenland (Fig. 1a) form the second largest exposed basalt succession found in the region, extending from Disko and Nuussuaq in the south, over Ubekendt Ejland, and up to Svartenhuk Halvø in the north (Saunders *et al.* 1997; Storey *et al.* 1998; Larsen & Pedersen 2009; Larsen *et al.* 2015). The flood basalts, which cover an area of approximately 45×10^3 km^2 onshore and continue further offshore (10×10^4 km^2), exhibit varying stratigraphic thickness from approximately 3 km on Disko to approximately 5 km on Ubekendt Ejland and Svartenhuk Halvø. The lava succession in the Disko–Nuussuaq area has been divided into three lithostratigraphical units (Hald & Pedersen 1975), which have then been further subdivided into six formations. From old to young: (1) the Vaigat Formation (present from Disko and northwards to Svartenhuk Halvø), subdivided into three members corresponding to distinct volcanic cycles of lavas and hyaloclastites, mostly of picritic composition; (2) the Maligât Formation (present on Disko and Nuussuaq), subdivided into three major and one minor member, is dominated by feldspar-phyric tholeiitic flood basalts; (3) the Svartenhuk Formation (tholeiitic basalts, corresponding to various members including the Skalø Member); (4) the Naqerloq Formation (tholeiitic basalts corresponding to various members including the Kanísut Member); (5) the Erqua Formation (alkaline basaltic in composition); and (6) the Hareøen Formation (olivine-phyric basaltic in composition). Lavas erupted predominantly in a subaerial environment, but the presence of hyaloclastites, breccias and marine mudstones in the Vaigat Formation and the lower part of the Maligât Formation implies deposition in a subaqueous environment during the earliest stages of activity (Larsen & Pedersen 2009). Eocene volcanic rocks have been identified in the westernmost parts of the onshore areas, the Naqerloq Formation, the Erqua Formation and the Talerua Member of the Hareøen Formation (e.g. Larsen *et al.* 2015). Dykes and sills occur throughout the province, and a central intrusion is situated in Ubekendt Ejland. A swarm of lamprophyre dykes cut the youngest volcanic rocks on Ubekendt Ejland (e.g. Storey *et al.* 1998).

Geochronology. All radiometric age data (Fig. 1b) that exist for West Greenland have been determined using the Ar–Ar technique (whole rock, glass, plagioclase, amphibole and K-feldspar). Ar–Ar ages for this sub-province range from 62 to 27 Ma (Storey *et al.* 1998; Larsen *et al.* 1999*a*, 2009, 2015), and include the Vaigat and Maligât formations (*c.* 62, and *c.* 62–61 Ma, respectively: Storey *et al.* 1998), the Kanisut volcanics (*c.* 52 Ma: Storey *et al.* 1998; and *c.* 56–54 Ma: Larsen *et al.* 2015), and various dyke systems (e.g. Disko, *c.* 58–57 Ma (basaltic): Larsen *et al.* 2015; SW Greenland, *c.* 55–54 Ma (basaltic): Larsen *et al.* 1999*a*, *b*; and Ubekendt Ejland, *c.* 34 Ma (lamprophyre): Storey *et al.* 1998). Until a recent study by Larsen *et al.* (2015), radiometric dating concentrated almost exclusively on the southern parts of the known volcanic successions. The Ar–Ar data of Larsen *et al.* (2015) provide a more complete overview of the age distribution and stratigraphy, and include volcanic successions in the Ubekendt Ejland area (e.g. the Qeqertalik Member, *c.* 59 Ma; the Nûk takisôq Member, *c.* 55 Ma; and the Erqua Formation, *c.* 53 Ma), Hareøen, West Nuussuaq (e.g. the Ifsorisok Member, *c.* 58 Ma; and the Talerua Member of the Hareøen Formation, *c.* 38 Ma) and Svartenhuk Halvø (e.g. the Tunuarsuk Member, *c.* 59 Ma; the Nuuit Member, *c.* 58 Ma; the Skalø Member, *c.* 58 Ma; the Arfertuarsuk trachyte, *c.* 57 Ma; the Naqerloq Formation, *c.* 55–54 Ma; and the Sarqâta qáqâ central intrusion, *c.* 55–57 Ma).

East Greenland

General geology. Voluminous basaltic lavas dominate onshore East Greenland (Fig. 1a), extending from the Gap Gustav Holm region in the south to Shannon Island in the north (Saunders *et al.* 1997). Here we use the Greenland–Iceland Ridge to make the distinction between NE Greenland and SE Greenland. North of the Greenland–Iceland Ridge, thick successions of basalt (exceeding 160×10^3 km^3) are preserved between Kangerlussuaq and Scoresby Sund (divided into the Lower Series Basalts and the Main (or Plateau) Series Basalts). Coastal formations of the Lower Series Basalts (between Kangerlussuaq and Nansen Fjord) include: Vandfalsdalen, Mikis, Jacobsen, Hængefjeldet and the locally identified Nansen Fjord Formation (e.g. Larsen *et al.* 1999*b*; Storey *et al.* 2007). The Main Series Basalts are associated with two major cycles, the first represented by the Milne Land (MLF) and Geikie Plateau formations (GPF), and the second by the Rømer Fjord (RFF) and Skrænterne formations (SF). Sections of the NE Greenland sequence have been temporally correlated with the Middle and Upper Basalt Series of the Faeroe Islands, and the Lower Basalt Series of West Greenland (e.g. Storey *et al.* 2007).

The lavas of the Prinsen af Wales Formation, which overlie the Main Series Basalts (e.g. Nielsen *et al.* 2001), are alkaline in composition, and often strongly olivine and pyroxene-phyric. In contrast, the lavas in the northern coastal areas of Kap Dalton on Blosseville Kyst, south of Scoresby Sund (i.e. the Igtertivâ Formation), are preserved in a small area, and as a result are less well studied (Larsen *et al.* 2013). Inland, in the Prinsen af Wales Bjerge region, the Urbjerget Formation is recognized. The formations notably contain pyroclastic and epiclastic material (Hansen *et al.* 2002; Peate *et al.* 2003), as well as thin picritic lava flows and hyloclastites (e.g. the Hængefjeldet Formation). The overlying Main Series Basalts (between Kangerlussuaq and Scoresby Sund) comprise approximately 300 individual flow units with a combined thickness of approximately 6 km. The eruption of lavas, predominantly tholeiitic in composition, was initially rapid (i.e. <1 Ma: Larsen & Saunders 1998; Larsen & Tegner 2006; Storey *et al.* 2007), then slowed and became punctuated towards the top of the sequence, as implied by intra-basaltic sedimentary rocks and preserved palaeo-soil horizons. Miocene volcanic rocks (e.g. the Vindtop Formation: Storey *et al.* 2004) are also found to conformably overlie the plateau basalts.

Formations in NE Greenland are cut by multiple generations of intrusions (i.e. dykes and sills) and a large number of plutonic centres of diverse composition (e.g. Skaergaard, Kangerlussuaq, Kap Deichman, Kap Boswell, Kræmer Ø syenite intrusion, and Kærven gabbro and granite), which have been linked to continental rifting and early seafloor spreading (e.g. Tegner *et al.* 1998, 2008; Lenoir *et al.* 2003; Larsen *et al.* 2014). Local-scale intrusive networks, associated with igneous centres (e.g. Kangerlussuaq Alkaline Complex), are also common. South of the Greenland–Iceland–Faeroes Ridge, offshore basalts (e.g. Roddick *et al.* 1989), intrusive complexes (e.g. Nualik and Kailineq regions), plutonic centres of diverse composition (e.g. Kruuse Fjord and Imilik layered gabbro intrusions: Tegner *et al.* 1998) and regional-scale dyke swarms, abundant along the coastal margin (Kap Gustav Holm region: Lenoir *et al.* 2003), have been identified.

Table 2. *Selected data for the West Greenland sub-province*

Locality	Unit	Lithology	Method	Mineral	Type	Age (Ma)	±2σ	MSWD (*P*)	References
Disko	Vaigat Formation (Manitdlat Member)	Alkali picrite	Ar–Ar	Glass (264091)	Plateau	61.86	1	–	Storey *et al.* (1998)
Nuussuaq	Vaigat Formation (Naujanguit Member)	Tholeiitic basalt	Ar–Ar	Whole rock (402525)	Plateau	61.55	1	–	Storey *et al.* (1998)
Disko	Maligât Formation (Middle Rinks Dal Member)	Tholeiitic basalt	Ar–Ar	Plagioclase (328406)	Plateau	61.45	0.8	–	Storey *et al.* (1998)
Nuussuaq	Maligât Formation (Lower Rinks Dal Member)	Tholeiitic basalt	Ar–Ar	Plagioclase (400323)	Plateau	61.65	0.8	–	Storey *et al.* (1998)
Hareøen	Hareøen Formation (Talerua Member)	Transitionally alkaline basalt	Ar–Ar	Groundmass (113482)	Plateau	38.9	0.23	2.23 (0.05)	Larsen *et al.* (2015)
Hareøen	Hareøen Formation (Talerua Member)	Transitionally alkaline basalt	Ar–Ar	Groundmass (113482)	Plateau	38.57	0.24	2.27 (0.06)	Larsen *et al.* (2015)
	Hareøen Formation			***Groundmass (113482)***	***Weighted mean (MSWD)***	***38.7***	***2.10 (3.9)***		
Svatenhuk Halvø	Naqerloq Formation	Enriched tholeiitic basalt	Ar–Ar	Plagioclase (251372)	Plateau	54.86	0.44	0.19 (0.99)	Larsen *et al.* (2015)
Svatenhuk Halvø	Naqerloq Formation	Enriched tholeiitic basalt	Ar–Ar	Plagioclase (278596)	Plateau	55.91	0.6	0.06 (1.00)	Larsen *et al.* (2015)
Svatenhuk Halvø	Arfertuarsuk trachyte flow	Trachyte	Ar–Ar	Anorthoclase (1931.1)	Plateau	57.51	0.24	4.15 (1.91)	Larsen *et al.* (2015)
Svatenhuk Halvø	Skalø Member	Tholeiitic basalt	Ar–Ar	Plagioclase (262838)	Plateau	57.98	0.59	0.06 (1.00)	Larsen *et al.* (2015)
Svatenhuk Halvø	Nuuit Member	Tholeiitic basalt	Ar–Ar	Plagioclase (262773)	Plateau	58.05	0.59	0.15 (0.99)	Larsen *et al.* (2015)
Svatenhuk Halvø	Tunuarsuk Member	Tholeiitic basalt	Ar–Ar	Whole rock (278566)	Plateau	59.41	0.61	0.33 (0.90)	Larsen *et al.* (2015)
Ubekendt Eijland	Erqua Formation	Alkali basalt	Ar–Ar	Whole rock (438728)	Plateau	53.47	0.52	0.65 (0.71)	Larsen *et al.* (2015)
Ubekendt Eijland	Nûk takisôq Member	Pitchstone	Ar–Ar	Feldspar (438740)	Plateau	55.94	0.2	1.56 (0.18)	Larsen *et al.* (2015)

Ubekendt Eijland	Qeqertalik Member	Tholeiitic basalt	Ar–Ar	Plagioclase (455809)	Plateau	59.97	0.89	1.26 (0.26)	Larsen *et al.* (2015)
Western Nuussuaq and Hareøen	Upper Kanisut Member	Acid tuff	Ar–Ar	Alkali Feldspar (135152)	Plateau	54.03	0.33	0.48 (0.89)	Larsen *et al.* (2015)
Western Nuussuaq and Hareøen	Middle Kanisut Member	Comendite tuff	Ar–Ar	Sanidine (410140)	Plateau	54.86	0.32	0.91 (0.52)	Larsen *et al.* (2015)
Western Nuussuaq and Hareøen	Lower Kanisut Member	Tuff	Ar–Ar	Plagioclase (456257)	Plateau	56.15	0.41	0.06 (1.00)	Larsen *et al.* (2015)
Western Nuussuaq and Hareøen	Upper Ifsorisok Member	Tuff	Ar–Ar	Alkali Feldspar (489172)	Plateau	58.31	0.3	0.83 (0.58)	Larsen *et al.* (2015)
Western Nuussuaq and Hareøen	Middle Ifsorisok Member	Tuff	Ar–Ar	Alkali Feldspar (489165)	Plateau	58.66	0.34	0.25 (0.99)	Larsen *et al.* (2015)
Hellefisk-1 well	2889.5 m below rotary table	Tholeiitic basalt	Ar–Ar	Whole rock (02A-00-0431)	Plateau	61.34	1.16	1.74 (0.09)	Larsen *et al.* (2015)
Hellefisk-1 well	2938.3 m below rotary table	Tholeiitic basalt	Ar–Ar	Whole rock (02A-00-0447)	Plateau	59.69	1.33	0.16 (0.99)	Larsen *et al.* (2015)
Disko	Dyke	Basalt	Ar–Ar	Plagiocalse (176601)	Plateau	54.62	0.6	–	Storey *et al.* (1998)
Disko (northern)	Dyke	Tholeiitic basalt	Ar–Ar	Whole rock (332904)	Inverse isochron	57.43	2.59	0.30 (0.88)	Larsen *et al.* (2015)
Disko (eastern)	Dyke	Tholeiitic basalt	Ar–Ar	Plagioclase (318800)	Plateau	58.34	0.4	0.12 (1.00)	Larsen *et al.* (2015)
Disko Bugt (SE of Disko)	Gabbro	Tholeiitic gabbro	Ar–Ar	Plagioclase (148D-01)	Plateau	58.96	0.51	1.27 (0.26)	Larsen *et al.* (2015)
Nuussuaq	Dyke	Basalt	Ar–Ar	Plagioclase (340797)	Plateau	55.84	0.8	–	Storey *et al.* (1998)
Nuussuaq (west)	Dyke	Tholeiitic basalt	Ar–Ar	Plagioclase (135129)	Plateau	48.02	2.76	1.1 (0.36)	Larsen *et al.* (2015)
Nuussuaq (north)	Sill	Alkaline	Ar–Ar	Plagioclase (489154)	Plateau	54.52	0.67	0.61 (0.75)	Larsen *et al.* (2015)
Nuussuaq (SE)		Tholeiitic basalt	Ar–Ar	Whole rock (318802)	Plateau	56.42	1.6	0.66 (0.65)	Larsen *et al.* (2009)
Ubekendt Eijland	Dyke	Lamprophyre	Ar–Ar	K-Feldspar (UE500)	Plateau	34.75	0.4	–	Storey *et al.* (1998)

(Continued)

Table 2. *Selected data for the West Greenland sub-province* (*Continued*)

Locality	Unit	Lithology	Method	Mineral	Type	Age (Ma)	$\pm 2\sigma$	MSWD (*P*)	References
Ubekendt Eijland	Sarqâta qáqâ	Granophyre	Ar–Ar	Alkali Feldspar (417747)	Plateau	55.21	0.3	2.27 (0.03)	Larsen *et al.* (2015)
Ubekendt Eijland	Sarqâta qáqâ	Gabbro	Ar–Ar	Plagioclase (455754)	Plateau	56.99	0.49	1.84 (0.09)	Larsen *et al.* (2015)
Godthåbsfjord		Camptonite	Ar–Ar	Plagioclase (201449)	Inverse Isochron	51.46	1.8	1.38 (0.24)	Larsen *et al.* (2009)
Søndre Isortoq		Alkali basalt	Ar–Ar	Plagioclase (KØ17090)	Plateau	58.11	1.2	1.06 (0.38)	Larsen *et al.* (2009)
Søndre Isortoq		Camptonite	Ar–Ar	Whole rock (KØ17154)	Plateau	55.43	1.2	1.94 (0.16)	Larsen *et al.* (2009)
Itilleq	Dyke	Tholeiitic basalt	Ar–Ar	Plagioclase (464627)	Plateau	63.58	2.6	0.28 (0.95)	Larsen *et al.* (2009)
Aasiaat Region	Grønne Ejland Sill	Tholeiitic basalt	Ar–Ar	Plagioclase (455793)	Plateau	60.4	1.6	0.38 (0.82)	Larsen *et al.* (2009)
Aasiaat Region	Globular dyke	Tholeiitic basalt	Ar–Ar	Plagioclase (464536)	Plateau	55.73	0.4	1.49 (0.20)	Larsen *et al.* (2009)

Weighted mean age (recalculated after Larsen *et al.* (2015): $\pm 2\sigma$) and MSWD values (in parentheses) are shown in ***italicized bold*** type. Numbers in parentheses refer to the sample numbers taken directly from the original source. Ages are corrected to Kuiper *et al.* (2008), and the ^{40}K decay constant of Min *et al.* (2000).

Table 3. *Selected data for the East Greenland sub-province*

Locality	Unit	Lithology	Method	Mineral	Type	Age (Ma)	±2σ	MSWD (*P*)	References
Blosseville Kyst	Rømer Fjord Formation	Basalt	Ar–Ar	Plagioclase (436056)	Plateau	55.77	0.5	–	Storey *et al.* (2007)
Blosseville Kyst	Rømer Fjord Formation	Basalt	Ar–Ar	Plagioclase (421522)	Plateau	56.08	0.9	–	Storey *et al.* (2007)
Blosseville Kyst	***Rømer Fjord Formation***	***Basalt***		***Plagioclase***	***Weighted mean (MSWD)***	***55.84***	***0.43 (0.35)***		
Blosseville Kyst	Milne Land Formation	Basalt	Ar–Ar	Plagioclase (404107)	Plateau	56.79	0.5	–	Storey *et al.* (2007)
Blosseville Kyst	Skraenterne Formation (Top)	Basalt	Ar–Ar	Whole rock (412251)	Plateau	55.77	0.5	–	Storey *et al.* (2007)
Blosseville Kyst	Skraenterne Formation	Basalt	Ar–Ar	Plagioclase (436194)	Plateau	55.57	0.9	–	Storey *et al.* (2007)
Blosseville Kyst	Skraenterne Formation	Tephra	Ar–Ar	Sanidine (421564)	Plateau	55.67	0.4	–	Storey *et al.* (2007)
Blosseville Kyst	***Skraenterne Formation***			***Plagioclase + whole rock + Sanidine***	***Weighted mean (MSWD)***	***55.69***	***0.30 (0.1)***		
Blosseville Kyst	Nansen Fjord Formation	Basalt	Ar–Ar	Plagioclase (426296)	Plateau	58.4	0.5	–	Storey *et al.* (2007)
Blosseville Kyst	Nansen Fjord Formation	Basalt	Ar–Ar	Plagioclase (194021)	Inverse isochron	57.49	2.3	–	Storey *et al.* (2007)
Blosseville Kyst	***Nansen Fjord Formation***			***Plagioclase***	***Weighted mean (MSWD)***	***58.36***	***0.48 (0.6)***		
	Vindtop Formation	Basalt	Ar–Ar	Plagioclase (436144)	Plateau	13.56	0.4	–	Storey *et al.* (2004)
Kangerdlugssuaq	Skærgaard	Granophyre	Ar–Ar	Biotite (SG-61 Biotite)	Inverse isochron	55.74	1.44	–	Hirschmann *et al.* (1997)
Kangerdlugssuaq	Skærgaard	Granophyre	Ar–Ar	Hornblende (SG-61 Hornblende)	Inverse isochron	55.82	1.52	–	Hirschmann *et al.* (1997)
Kangerdlugssuaq	***Skærgaard***		***Ar–Ar***		***Weighted mean (MSWD)***	***55.8***	***1.00 (0.01)***		

(*Continued*)

Table 3. *Selected data for the East Greenland sub-province* (*Continued*)

Locality	Unit	Lithology	Method	Mineral	Type	Age (Ma)	$\pm 2\sigma$	MSWD (*P*)	References
Igtutarajik		Gabbroic pegmatite	Ar–Ar	Biotite (I-7)	Plateau	47.9	0.4	–	Tegner *et al.* (1998)
Sorgenfri Gletscher		Diabase sill	Ar–Ar	Plagioclase (413907)	Plateau	57.07	0.8	–	Tegner *et al.* (1998)
Skjoldungen		Basaltic dyke	Ar–Ar	Plagioclase (940211)	Plateau	62.15	1	–	Storey *et al.* (2007)
Tugtilik		Dyke (silico-carbonatite)	Ar–Ar	Biotite (417150)	Plateau	59.01	0.9	–	Storey *et al.* (2007)
Tugtilik		Dyke (silico-carbonatite)	Ar–Ar	Biotite (417151)	Inverse isochron	58.2	1.2	–	Storey *et al.* (2007)
Tugtilik		***Dyke (silico-carbonatite)***			***Weighted mean (MSWD)***	***58.7***	***4.90 (1.17)***		
Kangerlussuaq		Biotite granite	Ar–Ar	Biotite (333132)	Inverse isochron	47.19	0.9	2.40 (0.05)	Tegner *et al.* (2008)
Kangerlussuaq	Kangerlussuaq	Syenite intrusion	Ar–Ar	Biotite (EG4583)	Inverse isochron	51.44	1.1	0.50 (0.81)	Tegner *et al.* (2008)
Kærven		Gabbro	Ar–Ar	Biotite (85659)	Inverse isochron	55.79	1.4	0.80 (0.52)	Tegner *et al.* (2008)
Kærven		Alkali granite	Ar–Ar	Amphibole (40160)	Inverse isochron	53.46	1.3	2.30 (0.08)	Tegner *et al.* (2008)
Kap Dalton	Igtertivâ Formation	Basalt	Ar–Ar	Groundmass (116344)	Plateau	49.09	0.48	0.95 (0.45)	Larsen *et al.* (2013)
Kap Dalton	Igtertivâ Formation	Basalt	Ar–Ar	Plagioclase (475269)	Plateau	43.77	1.08	0.12 (0.99)	Larsen *et al.* (2013)
Lille Pendulum	Lower Plateau Lava Series	Basalt	Ar–Ar	Plagioclase (194233)	Plateau	55.42	0.92	0.07 (1.00)	Larsen *et al.* (2014)
Blasedal (WF)	Lower Plateau Lava Series	Basalt	Ar–Ar	Plagioclase (194187)	Plateau	53.8	0.76	0.43 (0.86)	Larsen *et al.* (2014)
Blæsedal (WF)	Lower Plateau Lava Series	Basalt	Ar–Ar	Groundmass (194194)	Plateau	55.02	0.49	0.17 (0.98)	Larsen *et al.* (2014)
Kap Stosch (HWH)	Lower Plateau Lava Series	Basalt	Ar–Ar	Plagioclase (95346)	Plateau	54.16	0.72	0.53 (0.83)	Larsen *et al.* (2014)
Tværelv (HWH)	Lower Plateau Lava Series	Basalt	Ar–Ar	Groundmass (517303)	Plateau	55.52	0.68	0.12 (1.00)	Larsen *et al.* (2014)
Kap Mackenzie (GSØ)	Lower Plateau Lava Series	Basalt	Ar–Ar	Plagioclase (239531)	Plateau	53.54	0.67	0.04 (1.00)	Larsen *et al.* (2014)
Tobias Dal (HWH)	Upper Plateau Lava Series	Basalt	Ar–Ar	Plagioclase (194150)	Plateau	56.51	0.49	0.39 (0.94)	Larsen *et al.* (2014)

Bontekoe Ø	Upper Plateau Lava Series	Hawaiite	Ar–Ar	Groundmass (1980.274)	Plateau	55.17	0.35	0.18 (0.99)	Larsen *et al.* (2014)
Louise Boyd Land	Inland nunatak zone	Alkaline lava	Ar–Ar	Whole rock (421302)	Inverse isochron	49.8	1.38	1.83 (0.08)	Larsen *et al.* (2014)
Hobbs Land	Inland nunatak zone	Alkaline lava	Ar–Ar	Glass (452434)	Inverse isochron	53.4	1.83	2.20 (0.05)	Larsen *et al.* (2014)
Shannon		Basalt sill	Ar–Ar	Plagioclase (194196)	Plateau	51.85	0.55	0.15 (1.00)	Larsen *et al.* (2014)
Bass Rock (LP)		Basalt sill	Ar–Ar	Plagioclase (194207)	Plateau	53.71	0.62	0.19 (1.00)	Larsen *et al.* (2014)
Kefersteinberg (SØ)		sill	Ar–Ar	Plagioclase (517355)	Plateau	53.98	0.39	0.10 (1.00)	Larsen *et al.* (2014)
Hvalrosø		Pegmatite	Ar–Ar	K-Feldspar (475286)	Plateau	20.31	0.12	0.27 (0.99)	Larsen *et al.* (2014)
Freycinet Bjerg (GSØ)		Basalt sill	Ar–Ar	Plagioclase (239539)	Plateau	52.58	0.66	0.95 (0.48)	Larsen *et al.* (2014)
Traill Ø south coast		Basalt sill	Ar–Ar	Plagioclase (239578)	Plateau	55.14	0.37	0.74 (0.67)	Larsen *et al.* (2014)
Dronning Augusta Dal (WF)		Basalt dyke	Ar–Ar	Plagioclase (475289)	Plateau	53.68	1.59	0.19 (0.94)	Larsen *et al.* (2014)
Bontekoe Ø		Basalt dyke	Ar–Ar	Plagioclase (517310)	Plateau	51.27	1.21	0.05 (1.00)	Larsen *et al.* (2014)
Kap Broer Ruys felsite		Felsite	Ar–Ar	K-Feldspar (228078)	Plateau	48.71	0.51	0.68 (0.82)	Larsen *et al.* (2014)

Weighted mean ages (recalculated after Hirschmann *et al.* (1997) and Storey *et al.* (2007): $\pm 2\sigma$) and MSWD values (in parentheses) are shown in ***italicized bold*** type. Numbers/text in parentheses refers to sample numbers/text taken directly from the original source. Ages are corrected to Kuiper *et al.* (2008), and the ^{40}K decay constant of Min *et al.* (2000).

Table 4. *Selected data for the Faeroe Islands sub-province*

Locality	Unit	Lithology	Method	Mineral	Type	Age (Ma)	$\pm 2\sigma$	MSWD (*P*)	References
Faeroe Islands	Upper Series	Basalt	Ar–Ar	Plagioclase (32/2)	Plateau	55.87	0.70	–	Storey *et al.* (2007)
Faeroe Islands	Middle Series	Basalt	Ar–Ar	Plagioclase (X/2)	Plateau	55.57	0.70	–	Storey *et al.* (2007)
Faeroe Islands	Lower Series	Basalt	Ar–Ar	Plagioclase (89068)	Plateau	57.49	0.60	–	Storey *et al.* (2007)
Faeroe Islands	Lower Series, Lopra Formation	Basalt	Ar–Ar	Plagioclase (L1-0337.5)	Plateau	60.63	0.70	–	Storey *et al.* (2007)
Faeroe Islands	Lower Series, Lopra Formation	Basalt	Ar–Ar	Plagioclase (L1-1923.1)	Plateau	60.83	0.60	–	Storey *et al.* (2007)
Faeroe Islands	***Lower Series, Lopra Formation***	***Basalt***			***Weighted mean (MSWD)***	***60.75***	***0.45 (0.19)***		

Weighted mean age (recalculated after Storey *et al.* (2007): $\pm 2\sigma$) and MSWD value (in parentheses) are shown in ***italicized bold*** type. Numbers/text in parentheses refers to sample numbers/text taken directly from the original source. Ages are corrected to Kuiper *et al.* (2008), and the ^{40}K decay constant of Min *et al.* (2000).

Table 5. *Selected data for the British–Irish Palaeogene Igneous sub-province*

Locality	Unit	Lithology	Method	Mineral	Type	Age (Ma)	$\pm 2\sigma$	MSWD (P)	References*
Rum	Central Complex	Alkaline pegmatite	U–Pb	Zircon (SR314)	Concordia age	60.53	0.08	–	Hamilton *et al.* (1998)
Rum	Central Complex	Alkaline pegmatite	Ar–Ar	Phlogopite (GP387)	Inverse isochron	61.05	1	–	Hamilton *et al.* (1998)
Rum	Early Felsic Rhyodacite	Rhyodacite	Ar–Ar	Plagioclase (UDPNTLH and RDP)	SGF Ideogram	61.08	0.42	–	Troll *et al.* (2008)
Skye	Cuillin	Gabbro pegmatite	U–Pb	Zircon (SC1/2)	Concordia age	58.91	0.06	–	Hamilton *et al.* (1998)
Skye		Trachyte dyke	Ar–Ar	Biotite (GP382)	Inverse isochron	58.51	1.2	–	Hamilton *et al.* (1998)
Mull		Basaltic lava	Ar–Ar	Whole rock (S010)	Plateau	60.71	3.2	–	Mussett (1986)
Mull		Basaltic lava	Ar–Ar	Whole rock (78/5)	Plateau	61.83	2.8	–	Mussett (1986)
Mull		Basaltic lava	Ar–Ar	Whole rock (S023)	Plateau	62.13	2.8	–	Mussett (1986)
Mull		Basaltic lava	Ar–Ar	Whole rock (S007a)	Plateau	60.41	1.8	–	Mussett (1986)
Mull		***Basaltic lava***		***Whole rock***	***Weighted mean (MSWD)***	***61.1***	***1.20 (0.49)***		
Mull	Loch Uisg	Granophyre	Ar–Ar	Whole rock (S043)	Plateau	59.08	3.2	–	Mussett (1986)
Mull	Beinn a' Ghraig	Granophyre	Ar–Ar	Whole rock (S019)	Plateau	57.21	1.6	–	Mussett (1986)
Mull	Loch Bà	Felsite ring dyke	Ar–Ar	Whole rock (S020)	Plateau	56.7	2	–	Mussett (1986)
Mull		Basalt dyke	Ar–Ar	Whole rock (S100Ea)	Plateau	57.71	2.2	–	Mussett (1986)
Mull		***Various intrusives***		***Whole rock***	***Weighted mean (MSWD)***	***57.4***	***1.0 (0.58)***		
Mull		Basalt lava	Ar–Ar	Whole rock (SO17)	Inverse isochron	62.87	0.48	2.02 (0.04)	Chambers & Pringle (2001)
Mull		Dyke	Ar–Ar	Felsic groundmass (MD3)	Plateau	59.29	0.4	1.68 (0.09)	Chambers & Pringle (2001)
Isle of Eigg	Eigg Lava Formation	Tephra	Ar–Ar	Sanidine (Eigg96-1)	Plateau	62.55	1	–	Storey *et al.* (2007)
Isle of Muck	Muck Tuff	Tuff	Ar–Ar	Sanidine (MT2)	Weighted mean	61.95	0.16	–	Chambers *et al.* (2005)

(*Continued*)

Table 5. *Selected data for the British–Irish Palaeogene Igneous sub-province (Continued)*

Locality	Unit	Lithology	Method	Mineral	Type	Age (Ma)	$\pm 2\sigma$	MSWD (*P*)	References*
Isle of Muck	Muck Tuff	Tuff	Ar–Ar	Sanidine (MT3)	Weighted mean	61.74	0.16	–	Chambers *et al.* (2005)
Isle of Muck	Muck Tuff	Tuff	U–Pb	Zircon (Muck Tuff)	Concordia age	61.15	0.5	–	Chambers *et al.* (2005)
Antrim	Upper Basalt Formation	Basalt lava	Ar–Ar	Whole rock (Slimag)	Plateau	59.99	0.64	1.31 (0.25)	Ganerød *et al.* (2010)
Antrim	Lower Basalt Formation	Basalt lava	Ar–Ar	Whole rock (BallyMcillroy 1200)	Plateau	62.61	0.7	0.14 (0.94)	Ganerød *et al.* (2010)
Antrim	Lower Basalt Formation	Basalt lava	Ar–Ar	Whole rock (BallyMcillroy 2320)	Plateau	63.71	0.74	0.16 (0.96)	Ganerød *et al.* (2010)
Antrim	Lower Basalt Formation	Basalt lava	Ar–Ar	Whole rock (BallyMcillroy 2510)	Plateau	64.12	0.68	0.08 (1.00)	Ganerød *et al.* (2010)
Antrim	Lower Basalt Formation	Basalt lava	Ar–Ar	Whole rock (Whitehead 2)	Plateau	63.71	0.68	0.09 (1.00)	Ganerød *et al.* (2010)
Antrim	Lower Basalt Formation	Basalt lava	Ar–Ar	Whole rock (Whitehead 3)	Plateau	63.33	0.66	0.03 (1.00)	Ganerød *et al.* (2010)
Antrim	Lower Basalt Formation	Basalt lava	Ar–Ar	Whole rock (Binvore)	Plateau	63.01	0.68	0.04 (1.00)	Ganerød *et al.* (2010)
Antrim	Lower Basalt Formation	Basalt lava	Ar–Ar	Whole rock (Keady Mt)	Plateau	62.3	0.66	0.03 (1.00)	Ganerød *et al.* (2010)
Antrim	***Lower Basalt Formation***			***Whole rock***	***Weighted mean (MSWD)***	***63.24***	***0.61 (3.7)***		
Antrim	Scrabo Hill	Sill	Ar–Ar	Plagioclase (Scrabo)	Plateau	50.22	0.86	0.36 (0.78)	Ganerød *et al.* (2010)
Antrim	Tardree rhyolite Complex	Rhyolite	Ar–Ar	Sanidine (Sandy Braes)	Plateau	60.9	0.26	–	Ganerød *et al.* (2011)
Antrim	Tardree rhyolite Complex	Rhyolite	Ar–Ar	Sanidine (Tardree Forest)	Plateau	61.02	0.37	0.71 (0.74)	Ganerød *et al.* (2011)
Antrim	***Tardree rhyolite Complex***			***Sanidine***	***Weighted mean (MSWD)***	***60.94***	***0.21 (0.28)***		
Antrim	Tardree rhyolite Complex	Rhyolite	U–Pb	Zircon (Tardree Forest)	206Pb/238U age	61.32	0.05	–	Ganerød *et al.* (2011)
Hebrides shelf	Hebrides terrace	Basalt	Ar–Ar	Whole rock (POS241372DS2)	Plateau	62	0.8	1.80 (0.05)	O'Connor *et al.* (2000)

Hebrides shelf	Hebrides terrace	Basalt	Ar–Ar	Whole rock (POS241374DS4)	Plateau	52.24	0.8	1.60 (0.12)	O'Connor *et al.* (2000)
Rockall Trough seamounts	Anton Dohrn	Basalt	Ar–Ar	Whole rock (POS241377DS8)	Plateau	62.71	0.6	2.30 (0.01)	O'Connor *et al.* (2000)
Rockall Trough seamounts	Anton Dohrn	Basalt	Ar–Ar	Whole rock (POS241366DS3)	Plateau	48.81	1.7	0.20 (0.99)	O'Connor *et al.* (2000)
Rockall Trough seamounts	Anton Dohrn	Basalt	Ar–Ar	Whole rock (POS241364DS3)	Plateau	41.57	0.6	1.20 (0.31)	O'Connor *et al.* (2000)
Rockall Trough seamounts	Rosemary bank	Basalt	Ar–Ar	Whole rock (POS241360DS3)	Plateau	52.64	0.07	0.10 (0.99)	O'Connor *et al.* (2000)
Rockall Trough seamounts	Rosemary bank	Basalt	Ar–Ar	Whole rock (POS241362DS1D)	Plateau	53.75	0.9	1.90 (0.06)	O'Connor *et al.* (2000)
Rockall Trough seamounts	Rosemary bank	Basalt	Ar–Ar	Whole rock (POS241362DS10)	Plateau	53.34	1.5	2.20 (0.02)	O'Connor *et al.* (2000)
Rockall Trough seamounts	Rosemary bank	Basalt	Ar–Ar	Whole rock (VM28-5P452)	Plateau	42.78	0.4	1.00 (0.44)	O'Connor *et al.* (2000)
Lewis	Lewis	Dyke	Ar–Ar	Phlogopite (LR-299)	Plateau	45.91	0.22	–	Faithfull *et al.* (2012)
Isle of Arran	Drumadoon	Dyke (quartz porphyry)	Ar–Ar	Whole rock (Drumadoon Dyke)	Isochron	59.49	1.6	–	Mussett *et al.* (1987)
Isle of Arran	Drumadoon	Porphyritic rhyolite sill	Ar–Ar	K-Feldspar (DD-T)	Plateau	59.42	0.18	–	Meade *et al.* (2009)

Weighted mean ages (recalculated after Mussett (1986), Ganerød *et al.* (2010) and Ganerød *et al.* (2011): $\pm 2\sigma$) and MSWD values (in parentheses) are shown in ***italicized bold*** type. Numbers/text in parentheses refers to sample numbers/text taken directly from the original source. Ages are corrected to Kuiper *et al.* (2008), and the ^{40}K decay constant of Min *et al.* (2000).
SGF, single grain fusion.

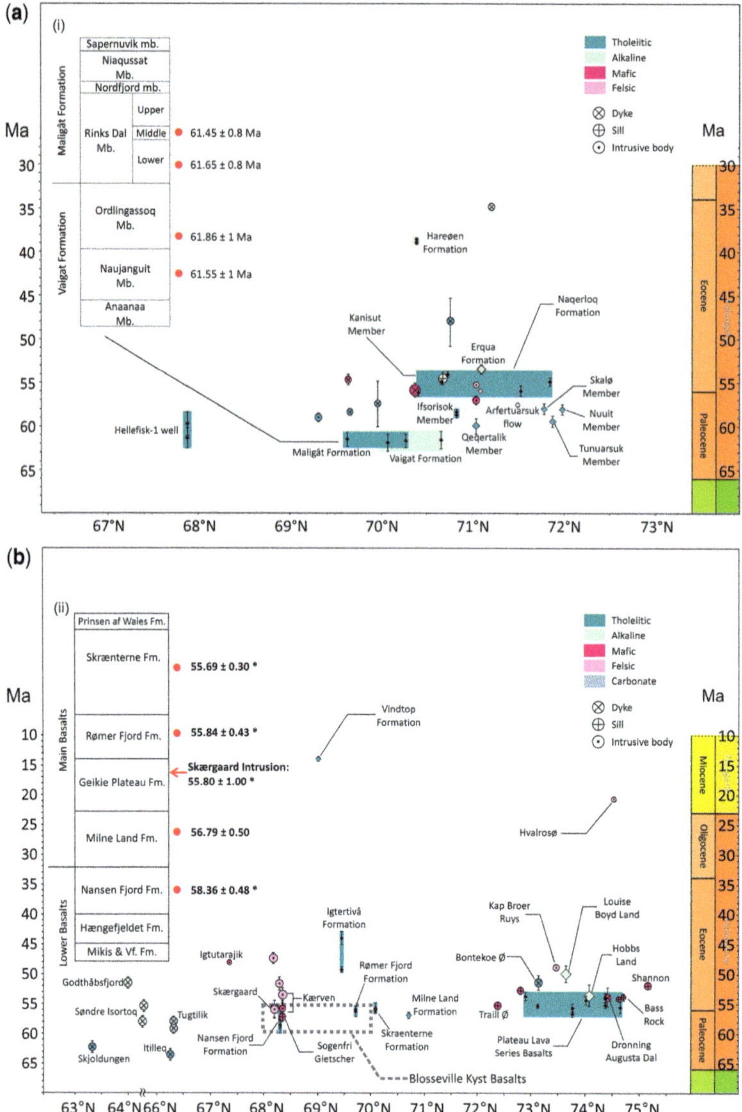

Fig. 3. Timescale (Ma) v. geographical latitude (°N) event charts showing selected age data for the North Atlantic Igneous sub-provinces (*denotes weighted mean age). Data are displayed with the geological timescale of Gradstein *et al.* (2012) and uncertainties are 2σ. Please note the use of a different x-axis and y-axis scale for each diagram: (**a**) West Greenland and (i) summary stratigraphic column displaying selected Ar–Ar data of Storey *et al.* (2007), for various members (Mb.) of the Vaigat and Maligât formations; (**b**) East Greenland (after Larsen *et al.* 2014) and (ii) summary stratigraphic column with Ar–Ar data of Storey *et al.* (2007) for various basalt formations (Fm.), and Hirschmann *et al.* (1997) for the Skærgaard Intrusion (Vf., Vandfaldsdalen Formation).

Geochronology. The chronology of East Greenland (Fig. 1c) has been established using the K–Ar, Ar–Ar and Re–Os techniques (e.g. Noble *et al.* 1988; Roddick *et al.* 1989; Nevle *et al.* 1994; Upton *et al.* 1995; Price *et al.* 1997; Sinton & Duncan 1998; Tegner *et al.* 1998, 2008; Werner *et al.* 1998; Tegner & Duncan 1999; Hald & Tegner 2000; Heister *et al.* 2001; Hansen *et al.* 2002; Lenoir *et al.* 2003; Brooks *et al.* 2004; Storey *et al.* 2007; Larsen *et al.* 2013, 2014). Published age data show the East Greenland volcanic rifted margin is characterized by voluminous flood basalts, which coincide with the two main phases of volcanism. However, ages in excess of 100 Ma have also been

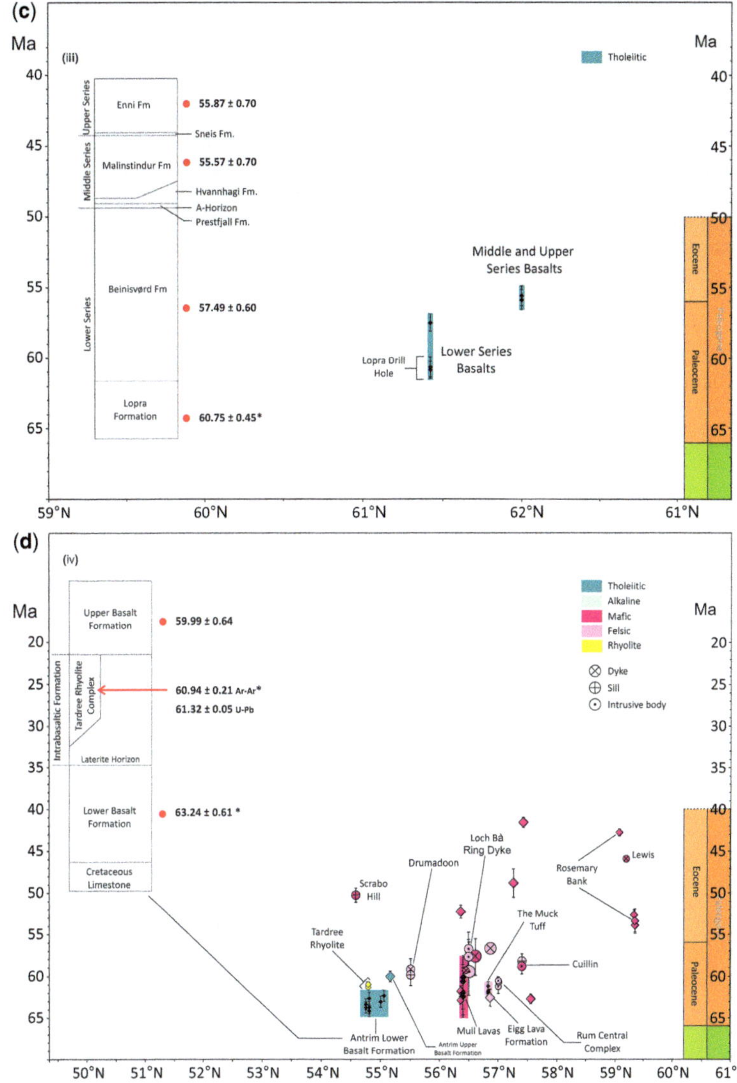

Fig. 3. (**c**) the Faeroe Islands and (iii) summary stratigraphic column displaying the Ar–Ar data of Storey *et al.* (2007) for the Upper, Middle and Lower (Enni; Malinstrindur; Beinisvød and Lopra) formations (Fm.); and (**d**) the British–Irish Palaeogene Igneous Province (BIPIP) with (iv) summary stratigraphic column with Ar–Ar data of Ganerød *et al.* (2010) for the Antrim Upper and Lower Basalt formations (Northern Island). The age of the Intrabasaltic Formation is inferred from the age of the Tardree rhyolite (Ar–Ar and U–Pb: Ganerød *et al.* 2011).

reported (e.g. K–Ar study of Noble *et al.* 1988). The majority of published data presented in the literature are concerned with intrusive rocks (e.g. granitic bodies, mafic sills and dykes). Their emplacement coincides in time with both major phases of volcanism, and continues into the latter part of the Cenozoic (*c.* 53–21 Ma).

Ar–Ar ages date the Nansen Fjord Formation at approximately 59–57 Ma (Storey *et al.* 2007). Older formations of the Lower Basalts have undergone considerable hydrothermal alteration (greenschist grade), resulting in a variable preservation state and attempts to date them have proved unsuccessful. Lavas in the Hold with Hope region are dated at approximately 58 Ma (Upton *et al.* 1995), while the lowermost lavas in the Prinsen af Wales Bjerge inland region are dated at approximately 61–60 Ma (Hansen *et al.* 2002). The Lower Basalts preserved along the coastal margin are thought to be contemporaneous with the lowermost lavas drilled offshore SE Greenland (e.g. Tegner & Duncan 1999), the onshore dykes to the far south (e.g. Storey *et al.* 2007), the Lower Basalt Series of the Faeroe Islands (e.g. Larsen *et al.* 1999*a*) and the lowermost lavas in the inland Urbjerg area (East Greenland). The eruption of the Lower Basalts in East Greenland was contemporaneous with the Paleocene lavas (i.e. the Maligât, Vaigat and Svartenhuk formations) of West Greenland (e.g. Storey *et al.* 1998) and lavas of the small isles of the Inner Hebrides (e.g. Pearson *et al.* 1996). A small number of formations of the Main Basalts have been dated previously. Noble *et al.* (1988) suggested eruption between approximately 53 and 57 Ma, based on the K–Ar dating of basalts along the Blosseville Kyst. The basal flows of the Main Basalts (Milne Land Formation) were later dated at approximately 57 Ma (Ar–Ar; Storey *et al.* 2007), with a possible minimum age of approximately 54–55 Ma, based on an alkaline tephra intercalated at the top of the series (Ar–Ar: Heister *et al.* 2001; Storey *et al.* 2007). The eruption of alkaline lavas of the Prinsen af Wales Bjerge Formation, which may be intercalated with the top of the Main Basalts, are dated at approximately 56–53 Ma (Ar–Ar: Peate *et al.* 2003). In addition, transitional lavas of Miocene age (*c.* 14–13 Ma) overlying the tholeiitic lavas have also been documented (Storey *et al.* 2004). Intrusion emplacement has been recorded between approximately 60 and 20 Ma, with multiple units coinciding with the eruption of the Main Basalts (e.g. Sorgenfrei Gletscher Sill Complex: Tegner *et al.* 1998). Mafic and felsic intrusions emplaced at approximately 55–40 Ma appear to be contemporaneous with continental rifting (e.g. Noble *et al.* 1988; Lenoir *et al.* 2003; Storey *et al.* 2007; Tegner *et al.* 2008), including the well-studied Skaergaard intrusion (*c.* 55 Ma: Hirschmann *et al.* 1997) and the Kangerlussuaq intrusion (*c.* 47 Ma: Tegner *et al.* 2008).

Intrusions that were emplaced following continental rifting at approximately 50–47 Ma (e.g. tholeiitic gabbro complexes at Kap Edvard Holm, Kruuse Fjord and Imilik south of Kangerlussuaq Fjord: Nevle *et al.* 1994; Tegner *et al.* 1998) have been identified; notably alkaline plutons were also emplaced prior to (e.g. the Sulugsut Complex at *c.* 58–59 Ma: Storey *et al.* 2007) and continued after the main phase of continental rifting (e.g. *c.* 33 Ma: Upton *et al.* 1995). Age data related to the East Greenland margin and the outer East Greenland shelf have also been included in the unfiltered NAG-TEC Database. Roddick *et al.* (1989) published whole-rock basalt K–Ar ages that ranged approximately from 71 to 34 Ma. Later studies by Tegner & Duncan (1999), Sinton & Duncan (1998) and Werner *et al.* (1998) applied the Ar–Ar method (whole-rock and plagioclase separates) to correlate sampled offshore units with the upper Lower Series Basalts and Main Series Basalts (e.g. *c.* 58–56, and *c.* 52–48 Ma: Tegner & Duncan 1999). The intercalated basalt and rhyolitic ash layers in the offshore units have been correlated with the base of the Lower Series Basalts (e.g. Sinton & Duncan 1998; Werner *et al.* 1998).

The Faeroe Islands

General geology. The Faeroe Islands are remnants of the widespread subaerial volcanic sequence that completely covered the Faeroe–Rockall Plateau to the north and west of the British Isles in the Palaeogene (Fig. 1a). The plateau was extensively covered with basalts that erupted in connection with the opening of the NE Atlantic. The majority of the basalts were erupted on dry land, but now are almost completely submerged (Waagstein 1988; Boldreel *et al.* 1994), except for land exposures that now exist only in the Faeroe Islands. The Faeroe Island Basalt Group (FIBG) includes: the basaltic lava sequence onshore and its offshore continuation onto the Faeroe Platform, the Faeroe–Shetland Channel, and the banks south of the Faeroe Platform. Covering an estimated area of 120×10^3 km^2, the FIBG has previously been correlated with basalt lava flows in East Greenland (e.g. Storey *et al.* 2007), and represents one of the thickest and stratigraphically most complex volcanic sections within the NAIP (Saunders *et al.* 1997; Waagstein *et al.* 2002; Storey *et al.* 2007; Passey & Jolley 2009).

The formations of the FIBG have an approximate stratigraphic thickness of 6.5 km (Passey & Jolley 2009), and include the Lopra Formation, the Beinisvørð Formation (also referred to as the Lower Basalt Series), the Malinstindur Formation (also referred to as the Middle Basalt Series) and the Enni Formation (also referred to as the Upper Basalt Series). The Lopra Formation, encountered only in the Lopra Drill Hole (Lopra-1A with a depth of 3565 m), consists of volcaniclastic sequences and hyaloclastite later intruded by sills (e.g. Ritchie *et al.* 2011). A shift from the evolved hyaloclastites in the Lopra Formation to a geochemically more primitive sequence marks the horizon between the Lopra and the base of the Beinisvørð Formation, of which approximately 900 m is exposed on the islands of Suðuroy and Mykines (e.g. Ellis *et al.* 2002; Passey 2004). The Prestfjall Formation, a 10 m-thick coal- and clay-bearing sequence is identified between the Beinisvørð and Malinstindur formations, and marks a hiatus between the pre-rift phase of volcanism and the main rifting phase of volcanism in the NAIP. The Prestfjall Formation is followed by the Hvannhagi

Formation (Tuff–agglomerate zone) cut by various and irregular intrusions, which has been interpreted as a second start-up phase of volcanic activity and, more precisely, as the onshore onset of the synrift volcanic phase. The Malinstindur Formation consists of compound flow units, with an upwards evolution from olivine-phyric and aphyric basalts to plagioclase-phyric basalts (e.g. Waagstein 1988). Passey & Jolley (2009) introduced the Sneis Formation, defined by basal reddened volcaniclastic sandstone containing woody material, overlain by beds of volcaniclastic conglomerate. The Enni Formation marks the top of the FIBG and is approximately 900 m thick, although it has been suggested that at least 1 km of the sequence has been removed by erosion (e.g. Waagstein *et al.* 2002). Intrusions (i.e. sills and dykes) are numerous, and include two major sill complexes (Streymoy Sill and Eysturoy Sill), as well as several other minor sill complexes (e.g. described by Hald & Waagstein 1991; Hansen *et al.* 2011).

Geochronology. The Faeroe Island Basalt Group (FIBG) has been the focus of several dating efforts and relative dating studies including palynological and biostratigraphical studies (Lund 1983, 1988; Ellis *et al.* 2002; Jolley & Bell 2002; Passey & Jolley 2009), palaeomagnetic studies (Tarling & Gale 1968; Riisager *et al.* 2002), and K–Ar and Ar–Ar studies (Fig. 1d), which have been applied to basalts of the three main formations, including those sourced from the Lopra Drill Hole (e.g. Noe-Nygaard 1966; Tarling & Gale 1968; Waagstein *et al.* 2002; Storey *et al.* 2007). K–Ar studies, in particular, have highlighted the difficulty in dating the basalts owing to pervasive low-temperature alteration (which often includes low-grade metamorphism, as indicated by the presence of a number of zeolite zones: Waagstein *et al.* 2002; Jørgensen 2006), and resulting in significant ^{40}Ar loss (e.g. Noe-Nygaard 1966; Tarling & Gale 1968; Fitch *et al.* 1978, which re-examined the work of Tarling & Gale 1968). Waagstein *et al.* (2002) published whole-rock mean K–Ar ages (based on multiple samples), yielding an age range of approximately 60–56 Ma. This study also included several whole-rock Ar–Ar plateau ages, which produced a larger age range of approximately 64–55 Ma, with individual ages slightly older, but significantly more precise, than their K–Ar equivalents. A later study conducted by Storey *et al.* (2007) produced Ar–Ar plateau ages for basalt samples of the Lopra Formation, and Lower, Middle and Upper Basalts. To avoid the problems experienced by others, pure mineral separates (plagioclase) were targeted, yielding an age range of approximately 60–55 Ma. Storey *et al.* (2007) reported simple age spectra (clear middle- to high-temperature plateaus), commonly exhibiting Ar-loss from low-temperature heating steps.

The British–Irish Palaeogene Igneous Province (BIPIP)

General geology. The British–Irish sub-province of the NAIP is located on the SE margin of the assumed proto-Icelandic plume head (Fig. 1a, e), which is proposed as being responsible for creating the NAIP (Campbell & Griffiths 1990; Lawver & Müller 1994; Torsvik *et al.* 2001; Ganerød *et al.* 2010). This sub-province has been referred to by several different names in the literature (e.g. the British Tertiary Igneous Province (BTIP) and The Hebridean Province). For this review we have adopted the term the 'British–Irish Palaeogene Igneous Province' (BIPIP: e.g. Meade *et al.* 2009). Main extrusive and intrusive activity is confined to a time window from the Danian to the Ypresian in the Palaeogene period leading up to continental rifting at approximately 56 Ma (Chambers *et al.* 2005; Storey *et al.* 2007). The BIPIP was a site of intense trap-forming volcanic activity of predominantly basaltic composition that covered an estimated area of *ca.* 10 788 km^2: this estimate includes landwards flows documented from present-day remnants.

The main units of the BIPIP now comprise the major lava fields of Skye and Mull in Scotland, and the Antrim plateau in Northern Ireland (Cooper 2004; Emeleus & Bell 2005). Later, significant volumes of silica-rich magmas formed isolated central complexes and granite intrusions at Skye, Rum, Mull, Ardnamurchan, Arran, Mourne, Slieve Gullion, Carlingford and Lundy (Emeleus & Bell 2005). NW–SE- to N–S-trending dyke swarms and arrays of volcanic plugs are characteristic of the BIPIP (Speight *et al.* 1982; Cooper *et al.* 2012), and have been interpreted as feeders for the lava fields (Kerr 1997). The majority of the dykes post-date the lavas, and are in greatest abundance close to the central complexes. The dykes are thought to be contemporaneous with the emplacement of the complexes (Jolly & Sanderson 1995): however, pre-lava dykes have been documented in Northern Ireland (Cooper *et al.* 2012).

Two major lava sequences are preserved and located in the Inner Hebrides: the Skye Lava Group and the Mull Lava Group. The lavas on Canna and NW Rum are considered to belong to the Skye Lava Group (Emeleus 1997), whereas the Eigg Lava Formation (which outcrops at Eigg, Muck and SE Rum) and the lavas of Morvern and Ardnamurchan are interpreted to be part of the Mull Lava Group (Emeleus 1997; Emeleus & Bell 2005). The Skye Lava Group (SLG) as a whole is the most complex sequence in the BIPIP, and includes a variety of formations and members based on distinct lithological lava associations intercalated with sedimentary sequences and red beds. However, the units are not easily correlated across the island as the lava flows are not laterally persistent due to several sets of dissecting faults. Based on the complex nature of the SLG, we direct the reader to the overview presented in Emeleus & Bell (2005). The SLG is intruded by several igneous complexes, collectively termed The Skye Central Complex, including (from oldest to youngest) Cuilli, Srath na Creitheach, and the Western and Eastern Red Hills (Bell & Harris 1986).

The Mull Lava Group (MLG) overlies the Gribun Mudstone Member (e.g. Emeleus & Bell 2005), and includes the tholeiitic basalts of the Staffa Lava Formation, and the predominately olivine-basalts of the Mull Plateau Lava Formation (MPLF: Emeleus & Bell 2005). The entire MLG is cut by three major intrusive centres (which offset each other) comprising gabbros, granophyres and granites, and later ring dykes that cut through centres (Emeleus & Bell 2005). This was followed by extrusion of the olivine-poor pillowed tholeiitic lavas, known as the Mull Central Lavas (MCL) into water-filled calderas (Bailey *et al.* 1924).

Northern Ireland hosts the massive Antrim Lava Group. Volcanism initiated with sporadic explosive eruptions in the north, followed by covering of the karstified early Upper Mastrichtian limestones (Simms 2000; Mitchell 2004) by the Lower Basalt Formation (LBF: Old 1975; Cooper 2004). A time of volcanic dormancy in a wet and humid climate caused a thick weathering profile (>30 m) of laterite to form on the top of the LBF (Hill *et al.* 2000). This period is referred to as the Interbasaltic Formation (IBF) and is laterally extensive across the plateau. The volcanic dormancy was interrupted by the Tardree rhyolite (central Antrim) and the impressive columnar-jointed quartz-tholeiitic Causeway Member in the north. Both the Tardree rhyolite and the Causeway Member are enclosed within the IBF. The subsequent second cycle of voluminous basaltic volcanism is represented by the olivine tholeiite lavas of the Upper Basalt Formation (UBF: Lyle 1979). Boreholes from the Langford Lodge and BallyMcIllroy No. 1 projects have documented thicknesses of 531 and 346 m for the LBF and UBF, respectively (Manning *et al.* 1970; Thompson 1979). Dolerite plugs and several dyke sets cut the ALG and are thought to have functioned as feeders for higher level flows, now removed by erosion (Walker 1959; Cooper *et al.* 2012). The central complexes of the Irish and Northern Ireland section of the BPIP represent the eroded roots of volcanic systems similar to those seen in Scotland. At each centre, erosion and faulting have exposed different levels of the sub-volcanic system from upper-crustal magma chambers (e.g. Mourne), to high-level dyke and sheet intrusions (e.g. Carlingford), and shallow ring-fault systems (e.g. Slieve Gullion) (Cooper & Johnston 2004; Preston 2009).

Geochronology. To date, geochronological studies concerned with the Skye Lava Group (SLG) have been scarce. Hamilton *et al.* (1998) used the Ar–Ar and U–Pb methods to constrain the timing of eruption for the Skye Main Lava succession (SMLS: NW Rum and Canna) at approximately 61 Ma. The study suggests the presence of eroded clasts for all members of the Rum Central Complex within intra-lava sediments of the SMLS; the age of the Rum Central Complex could thus provide a maximum age for the SMLS. Chambers *et al.* (2005) dated the Canna Lava Field, which yielded an Ar–Ar weighted mean age of 60.89 ± 0.23 Ma. A number of dyke swarms and intrusive centres that dissect the lavas have also been dated. Dickin (1981) dated units from the Cuillin (inferred to mark the end of main basaltic activity in the area) and the Western and Eastern Red Hills centres with the Rb–Sr method and obtained ages of 59.3 ± 0.7, 58.7 ± 0.9 and 53.5 ± 0.4 Ma, respectively. Hamilton *et al.* (1998) dated pegmatitic intrusions on Cuillin and Rum using U–Pb zircon geochronology and reported ages of 58.9 ± 0.1 and 60.5 ± 0.1 Ma, respectively. The same study also reported an Ar–Ar age of 60.7 ± 0.5 Ma for a second Rum pegmatite (phlogopite: inverse isochron age). This Ar–Ar age is identical to the U–Pb age for related pegmatites. Hamilton *et al.* (1998) also obtained an age of 58.1 ± 0.6 Ma for a trachyte dyke intruding the upper part of the Skye Lavas (biotite: inverse isochron age).

Basalt lavas of the MLG have previously been dated by the Ar–Ar step heating method. Mussett (1986) presented ages that ranged from approximately 58 to 61 Ma, suggesting that basaltic magmatism lasted for approximately 3 Ma. Gabbros, granophyres and granites that intrude the MLG were dated to approximately 57–56 Ma (Ar–Ar plateau ages: whole rock). A later study carried out by Chambers & Pringle (2001) also dated flows of the MLG using the Ar–Ar method (including the lower lavas), and reported a similar age range (*c.* 62–58 Ma: whole rock) to that of the study by Mussett (1986). The felsic intrusions are themselves cut by various dyke structures (e.g. the Loch Ba ring dyke on Mull), which are themselves cut by the NW–SE-trending regional dyke swarm. This demonstrates the complex nature of volcanism in the BIPIP. The Small Isles of the Inner Hebrides preserve basalt lava flows and intercalated tuffs (e.g. Eigg Lava Formation on the Isle of Muck), and are dated to approximately 63 Ma (e.g. Dagley & Mussett 1986; Pearson *et al.* 1996). A relatively recent study by Chambers *et al.* (2005) dated two exposures of the Muck Tuff, yielding slightly younger Ar–Ar sanidine and U–Pb zircon ages of approximately 61 Ma.

The Lower and Upper Basalt formations (LBF and UBF) of the Antrim Lava Group (ALG) yield an age range of approximately 62–63 and 59 Ma (whole rock: Ar–Ar plateau age), respectively (Ganerød *et al.* 2010). Numerous ages have been reported for the Tardree rhyolite (e.g. Meighan *et al.* 1988; Gamble *et al.* 1999; Ganerød *et al.* 2011), the Mourne granite intrusions at approximately 51–56 Ma (e.g. Gibson *et al.* 1987; Thompson *et al.* 1987; Gamble *et al.* 1999) and Slieve Gullion at approximately 56–57 Ma (e.g. Meighan *et al.* 1988; Gamble *et al.* 1999). Lundy Island is recognized as the most southerly known remnant of the NAIP. It is predominantly granitic in composition (Fitch *et al.* 1969; Thorpe *et al.* 1990) and forms part of a small plutonic complex. Mussett *et al.* (1976, 1988) reported K–Ar ages for dolerite dykes ranging from approximately 54 to 44 Ma.

Offshore studies: a summary. The complete NAG-TEC database contains 116 age dates taken from a range of offshore studies. The majority of published geochronology data relate to basalt, 40 of which have been determined using the K–Ar technique. For example, Roddick *et al.* (1989) presented basalt K–Ar ages from ODP samples (offshore East Greenland); however, the data are described by the author as too variable and imprecise to provide useful age information. K–Ar basalt data also exists for a number of other offshore localities including: the Iceland–Faeroe ridge and outer Vøring plateau (Talwani & Eldholm 1977); mid-Norwegian margin (Bugge *et al.* 1980), and offshore Shetlands and the Hebrides shelf (Hitchen & Ritchie 1993). In addition Fitch *et al.* (1988) presented K–Ar dates based on drill-cored tuff samples. The ages range from approximately 177 Ma to approximately 14 Ma and encompass the full range of ages in the complete database, with the exception of offshore Svalbard and the western Barents Sea margin for which previous studies have offered Pliocene ages (e.g. Talwani & Eldholm 1977; Mørk & Duncan 1993).

The remaining data, obtained using the Ar–Ar technique, cover localities including offshore East Greenland (plagioclase and whole-rock basalt: Sinton & Duncan 1998; Tegner & Duncan 1999), the Vøring margin (glass and plagioclase: Sinton *et al.* 1998), offshore west Greenland (whole-rock basalt: Nelson *et al.* 2015), the Hebrides Shelf (plagioclase and whole-rock basalt: Sinton *et al.* 1998; O'Connor *et al.* 2000) and the Rockall Trough (whole-rock basalt, biotite and plagioclase: O'Connor *et al.* 2000; Archer *et al.* 2005). Svensen *et al.* (2010) provided statistically valid U–Pb zircon concordia ages of 55.6 $\pm$ 0.15 and 56.3 $\pm$ 0.2 Ma for the Utgaard Upper and Lower Sill (Mid-Norwegian margin).

Discussion

An optimized age model for the NAIP

Filtering of geochronological data can be conducted at different levels. For example, filtering criteria could be based on the ability to carry out additional quality control checks (e.g. Baksi 2007*a*, *b*; Verati & Jourdan 2013) or by only accepting data determined on pure mineral separates or determined by the Ar–Ar and U–Pb methods. We have established filtering criteria that attempt to take into account the fact that magmatic rocks of the NAIP are notoriously difficult to date due to very few suitable phases for the U–Pb technique, low potassium contents of the whole rocks (which impacts the usefulness of the Ar–Ar and K–Ar methods), whole rock or groundmass as the only viable materials for dating, and partial to complete alteration of samples during or after crystallization.

In order to systematically check all 699 dates, access to complete datasets, including details of the experimental protocol, is required. For many of the published results, this level of detail is not available. Thus, to avoid automatic exclusion of a large number of data from the database, we have adopted a set of filtering criteria for quality control, which has led to the identification of a set of statistically significant age data. We apply different levels of rigour and share the results with these different filtering approaches.

The final optimized age model for the NAIP (following the first-level filtering) contains 130 ages related to extrusive and intrusive activity covering West Greenland, East Greenland, the Faeroe Islands and the BIPIP. The optimized dataset contains 124 ages determined by the Ar–Ar method and six by the U–Pb (isotope dilution thermal ionization mass spectrometry (ID-TIMS)) method. Figure 3 displays selected radiometric data for each sub-province, and shows a general younging of activity with increasing geographical latitude (in particular for East Greenland, West Greenland and the BIPIP). The precision and accuracy of some selected ages could be challenged further, and would be improved with more thorough statistical analysis. Following a second, more aggressive level of filtering, whereby all whole-rock and groundmass data were also removed because of well-documented problems concerning partial or complete alteration, the age model for the NAIP contains 86 ages, 80 of which are determined by the Ar–Ar method and six by U–Pb.

A review of the two final selected datasets suggests that 86 ages can be considered accurate, 38 are uncertain owing to a lower precision compared to other data (e.g. Meighan *et al.* 1988; Waagstein *et al.* 2002) and six are not well characterized (i.e. poor data reporting: Hirschmann *et al.* 1997; Chambers *et al.* 2005). Where insufficient data reporting has not permitted evaluation of the data at a comparable level to other published results, the data have been acknowledged but are not part of the selected datasets (e.g. Noble *et al.* 1988; Tegner & Duncan 1999).

The current state of geochronological knowledge

The smaller age dataset for the NAIP produced through our filtering process yields a more tightly constrained picture of the current state of geochronological data for the NAIP. The dataset obtained for each sub-province through the data analysis and filtering process described above provides experimentally consistent age estimates for the start and end of magmatic activity. In West Greenland, the Vaigat Formation represents the earliest magmatic event with an age of 61.9 $\pm$ 1.0 Ma (Storey *et al.* 1998). Activity continued until 34.8 $\pm$ 0.4 Ma (dyke, Ubekendt Ejland: Storey *et al.* 1998). For East Greenland, age information relating to the Mikis and Vanfaldsdalen formations is lacking, but the identification of various intrusive events such as the Tugtilik and Skjoldungen dykes, which yield ages of 62.2 $\pm$ 1.0 and 59.0 $\pm$ 0.9 Ma, respectively (Storey *et al.* 2007), allows an inferred age of >61 Ma for the start of the magmatic activity in East Greenland. The Vindtop Formation is dated at 13.6 $\pm$ 0.4 Ma (Storey *et al.* 2004) and marks the youngest basalt formation in East Greenland. For the Faeroe Islands, the Lower Series Basalts (Lopra Formation) is dated at 60.8 $\pm$ 0.5 Ma (weighted mean, recalculated after Storey *et al.* 2007) and represents the start of activity for this sub-province, which continued until 55.9 $\pm$ 0.7 Ma (Upper Series Basalts: Storey *et al.* 2007). The Antrim Lower Basalt Formation yields the oldest age determined for volcanism of the BIPIP, suggesting activity started at approximately 63.2 $\pm$ 0.6 Ma (weighted mean, recalculated after Ganerød *et al.* 2011). Onshore dyke systems (Outer Hebrides, Scotland) extend activity to 45.9 $\pm$ 0.2 Ma (Faithfull *et al.* 2012).

The cumulative frequency of selected data for the NAIP and per sub-province (Fig. 4) is shown to provide a general overview. East Greenland and the BIPIP appear to show activity occurring over a similar duration: however, it could be inferred that magmatic activity in the BIPIP commenced 5 Ma prior to East Greenland (Fig. 4a). Both West Greenland and the BIPIP show high levels of magmatism prior to the break-up of the North Atlantic, whereas the opposite can be seen for East Greenland. There appears to be a shift onshore from predominantly extrusive to intrusive behaviour at approximately 60 Ma for the BIPIP and minor offshore extrusive volcanism recorded after approximately 55 Ma. This type of shift is less obvious for the other sub-provinces: however, it is possible to identify punctuated, younger (<50 Ma) activity across the sub-provinces. Although the Faeroe Islands appear to be the exception, this could simply be a consequence of a small sample dataset bias. The removal of Ar–Ar ages based on whole rock and groundmass has had a significant impact on the distribution of geochronological data for the BIBIP (Fig. 4b). Following the second level of filtering, magmatic activity is primarily focused prior to break-up of the North Atlantic.

Geochronological data have contributed significantly to the discussion about the duration of flood basalt volcanism (Baksi 2005*a*, *b*, 2012; Storey *et al.* 2007; Barry *et al.* 2010, 2012). LIP events with lifespans of more than 15 myr are generally considered to have been multiple-pulsed events (Bryan & Ernst 2008). In the NAIP, a 'gap' or 'hiatus' in widespread volcanism has previously been suggested (e.g. Saunders *et al.* 1997); however, despite carrying out the evaluation and filtering process, we have not been able to recognize a gap or identify multiple 'pulses' in activity. Storey *et al.* (2007) previously suggested the possibility of a short regional hiatus (*c.* 6 Ma) in volcanic activity in East Greenland between a 'first phase' of activity identified at approximately 61–59 Ma, and correlated with palaeomagnetic studies (e.g. Riisager *et al.* 2002), and a 'second phase' of early–middle Eocene age (*c.* 54–34 Ma), correlated with apparent gaps ages of magmatic activity in East Greenland and the Faeroe Islands (Storey *et al.* 1998, 2007). With the recent data presented by Larsen *et al.* (2014, 2015) and the analysis of the suite of published geochronological data from this study, significant pauses in regional volcanism do not seem to have taken place, although breaks in volcanism may have occurred on a local scale. Despite the combination of increased data coverage in some areas and the information revealed by this analysis, gaps in data in other areas, erosion (complete or partial removal of volcanic units) and challenges in correlating stratigraphy across large geographical regions all affect the ability to recognize a cross-province correlated gap.

Recommendations for future work

Earlier studies that contributed new age data and analysed existing age data from the NAIP made significant contributions to understanding the evolution of the province. The current study couples newer,

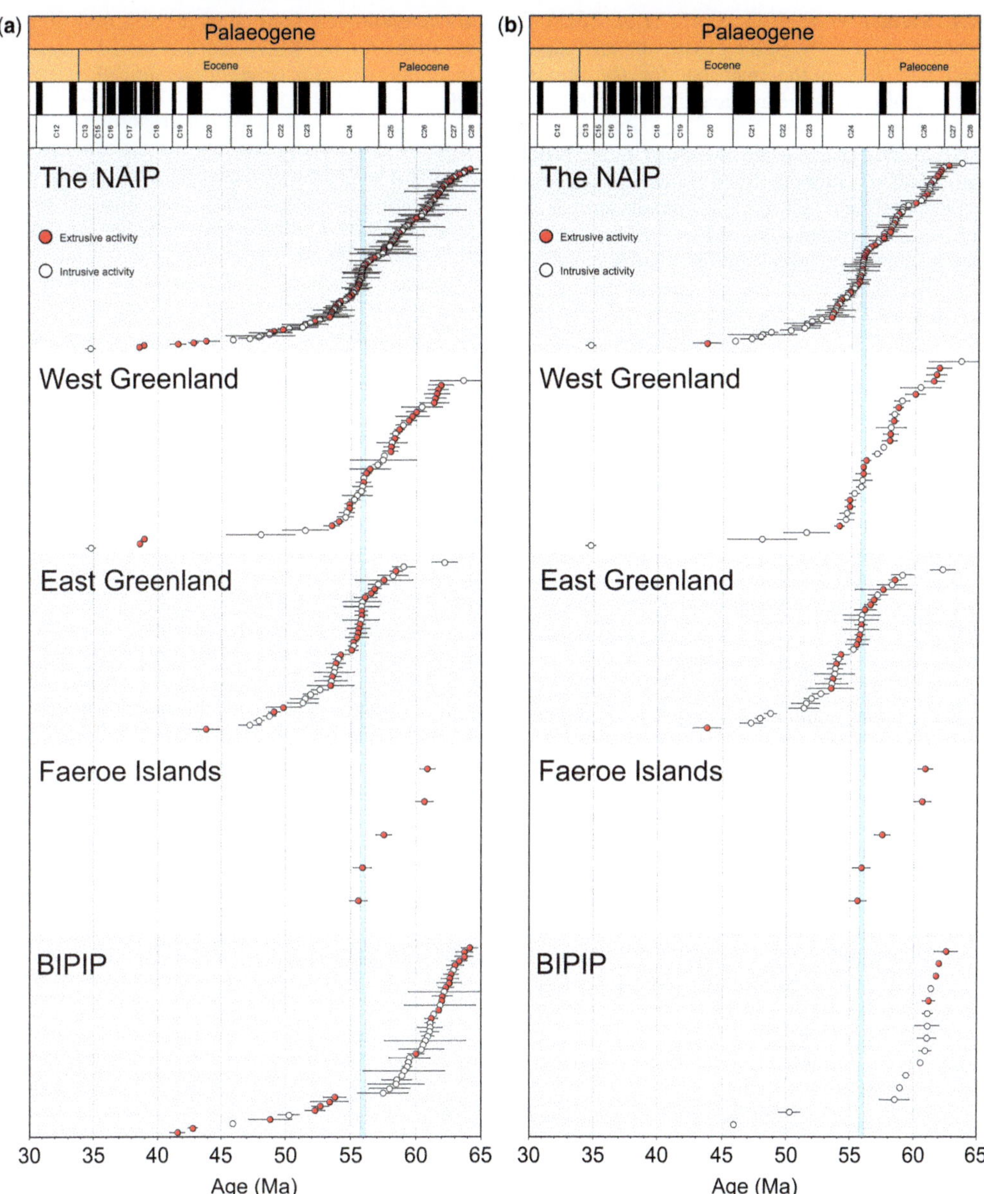

Fig. 4. Selected geochronological data for the NAIP (top), and the sub-provinces of West Greenland, East Greenland, the Faeroe Islands and the British–Irish Palaeogene Igneous Province (BIPIP) shown in terms of their cumulative distribution: (**a**) after the first-level filtering process; and (**b**) after the second-level filtering process. Red and white coloured symbols denote extrusive and intrusive activity, respectively. The vertical light blue line indicates the anomaly C24r at 55.96 Ma (Gradstein *et al.* 2012), which represents the opening of the North Atlantic. Data are presented relative to the geomagnetic polarity timescale after Gradstein *et al.* (2012). Uncertainties are given at the 2σ level.

published results and a more refined, filtered dataset, and emphasizes the need for firm chronological benchmarks and consistent data reporting in order to continue to increase our understanding of the magmatic evolution of the region. For example, it may be worthwhile to consider geochronometers such as the U–Pb method, which has been used with success on mafic rocks from other LIPs (e.g. Svensen *et al.* 2012; Corfu *et al.* 2013), for broader applications in the NAIP, although Ar–Ar data will remain the mainstay for the province because of their broad applicability to the full suite of rocks in the region. The exclusion of over 500 dates in this analysis indicates that many historical NAIP geochronological data do not have high enough precision (due to sample quality and/or analytical and instrumental factors) or lack enough documentation to allow rigorous assessment. The dataset also highlights that the current state of geochronological knowledge for some sub-provinces would greatly benefit from renewed interest and re-sampling. Nonetheless, data coverage will always be an issue in this enormous region. The following are recommendation for future work.

Faeroe Islands. New data could target the Lower, Middle and Upper Series Basalts of the Faeroe Islands, as the selected dataset shows insufficient geochronological data for this sub-province. The lack of suitable or accessible samples may make this effort challenging. However, sill intrusions (cf. Hansen *et al.* 2011) could be targeted as their ages are currently unconstrained.

British–Irish Palaeogene Igneous Province (BIPIP). A systematic sampling strategy of the BIPIP dyke swarms is highly recommended. The curvilinear dyke swarms of the BIPIP (Speight *et al.* 1982; Cooper *et al.* 2012) have received little geochronological attention. Dating the different dyke sets could provide temporal constraints regarding pulses of extension and rifting, magma initiation, and the end of magmatism for higher-level lava flows now removed by erosion. Sampling has been undertaken as part of the Tellus project (lead by the Geological Survey of Northern Ireland) for the Irish–Northern Irish dyke swarms and, if dated, will provide crucial data for geodynamic models of the NAIP.

A recent re-evaluation of the emplacement dynamics of the Ardnamurchan cone sheets indicates that all centres are closely linked in time (Burchardt *et al.* 2013). It may be useful to re-analyse and provide new age determinations for the Ardnamurchan Complex, which would update the low precision K–Ar ages reported by Mitchell & Reen (1973) and further test the model of Burchardt *et al.* (2013). Areas that would benefit from geochronological revision include the lava sequences in Antrim. The whole-rock dates derived by Ganerød *et al.* (2010) should be followed up by mineral separate analysis. The Mull Lava Group (MLG) is not well represented (we excluded most of the data in the study of Chambers & Pringle 2001), and no direct age determinations exist for the Skye Lava Group (SLG) in our selected dataset. New geochronological constraints derived from mineral separates for the MLG and a renewed interest in the units of SLG is strongly encouraged.

Data reporting. Adequate data reporting in future work would greatly ease any further evaluation of data that may be conducted. Critical examination of a number of studies has been hindered by the lack of access to detailed datasets. To that end, it is essential that the publication of data follows the minimum suggested by Renne *et al.* (2009) and Jourdan *et al.* (2009*b*), and the full analytical protocol is described including pre-treatment methods. It is difficult to objectively evaluate age data based only on the information in summary data tables or age plots (i.e. $^{40}Ar/^{39}Ar$ age spectrum and inverse isochron). In addition, all results should be published and not only selected ages (i.e. as supplementary information), including data that are more complex and difficult to interpret. This would benefit geochronologists, as well as non-geochronologists, who are interested in the relevance of a given age and associated geological interpretation.

We recommended that interpretations based on the comparison of Ar–Ar data are carried out once a common reference point has been established and all data have been recalculated. For example, Meade *et al.* (2009) dated K-feldspar from the Drumadoon Sill on the Isle of Arran (BIPIP) and obtained an age of 59.04 ± 0.13 Ma, calculated relative to the Fish Canyon sanidine (FCs) age of 28.02 ± 0.16 Ma. They directly compared their age to Chambers (2000), who obtained a biotite age of 57.85 ± 0.15 Ma for the northern granite body using the Taylor Creek sanidine age of 27.92 Ma (Renne *et al.* 1998). Meade *et al.* (2009) interpreted protracted felsic magmatism based on these two ages being distinguishable at the 95% confidence level. However, if the age of Chambers (2000) is recalculated relative to the FCs age used by Meade *et al.* (2009), the ages are indistinguishable at the 95% confidence level (new biotite age of 58.71 ± 0.15 Ma). It is possible to conduct a quick recalculation of an Ar–Ar age using a tool such as ArArCalibration (i.e. equation (7) in http://www.earth-time.org/ArArReCalc_Explanation.pdf). For a more rigorous recalculation, it is necessary to correctly propagate all uncertainties (see http://earthref.org/ERDA/139/ or Cameron & Hodges 2016).

Summary and conclusions

The North Atlantic Igneous Province (NAIP) is one of the most extensive Large Igneous Provinces (LIPs) in the world, with an original volume estimated to be in the range $5 \times 10^6 - 10 \times 10^6$ km^3. The age and duration of magmatism in the NAIP has been the focus of research for over 50 years.

We have compiled published geochronological data into a single comprehensive database. The unfiltered NAG-TEC Geochronological Database contains 699 dates ranging from 0.19 to 177 Ma from over 60 publications. Over half the ages in the database (366) have been obtained using the Ar–Ar method. We have therefore recalculated all Ar–Ar (and K–Ar) data to a common age reference system following the calibrations in Kuiper *et al.* (2008), in line with the Geological Time Scale 2012 of Gradstein *et al.* (2012).

We have used the NAG-TEC Database to assess the quality and geological significance of published age data. Based on this assessment and following a first round of filtering, a selected dataset is provided containing 130 ages. Removal of Ar–Ar whole-rock and groundmass dates during a second round of filtering has produced a selected dataset containing 86 ages.

The datasets suggest that identified and recorded magmatism associated with the NAIP pre-rift phase started in the early Paleocene (*c.* 64–63 Ma) and continued into the Eocene (*c.* 56 Ma: synrift phase). Episodes of post-rift volcanism took place between approximately 56 and 50 Ma and numerous intrusions extended the activity from around 50 to 35 Ma, with the youngest recorded at approximately 20 Ma. We can use the selected datasets to show the approximate start, periodicity and end of volcanism for the sub-provinces of the NAIP, but the data do not confirm the existence of a province-wide 'gap' in or multiple pulses of volcanic activity.

The construction of the NAG-TEC Geochronological Database and subsequent re-examination of published data suggest targets for future work that might include areas that currently lack reliable, well-constrained ages (e.g. Faeroe Islands). Similarly, targeting some areas for combined U–Pb and Ar–Ar analysis may be warranted both to calibrate the ages across different areas and to help reconcile gaps in data due to lack of exposure, erosion or alteration. Particular attention should be paid to providing geochronological information for complex dyke systems (e.g. BIPIP), and data relating to the Antrim, Mull and Skye Lava groups must be updated.

We advise caution in accepting reported ages without a standard set of accompanying documentation. Studies using the Ar–Ar method should recalculate ages to a common reference to avoid confusion and over-interpretation of data. Although the expectations for data reporting have improved over the past number of years, our examination found that the level of data documentation has not been consistent, making critical evaluation of the results challenging. Publication of future data should be accompanied by full experimental details and should follow recommended data-reporting protocols (e.g. Renne *et al.* 2009).

We would like to thank the following industry partners: Bayerngas Norge AS; BP Exploration Operating Company Ltd; Bundesanstalt für Geowissenschaften und Rohstoffe (BGR); Chevron East Greenland Exploration A/S; ConocoPhillips Skandinavia AS; DEA Norge AS; Det norske oljeselskap ASA; DONG E&P A/S; E.ON Norge AS; ExxonMobil Exploration and Production Norway AS; Japan Oil, Gas and Metals National Corporation (JOGMEC); Maersk Oil; Nalcor Energy – Oil and Gas Inc.; Nexen Energy ULC; Norwegian Energy Company ASA (Noreco); Repsol Exploration Norge AS; Statoil (UK) Ltd; and Wintershall Holding GmBH. CMW would like to thank Susanne Buiter and Per-Terje Osmundsen for thoughtful and constructive comments on earlier versions of this manuscript, and Fred Jourdan and Lotte M. Larsen for thorough, detailed and thought-provoking reviews that have greatly improved this contribution.

References

Andersen, T.B. & Jamtveit, B. 1990. Uplift of deep crust during orogenic extensional collapse – a model based on field studies in the Sogn–Sunnfjord region of western Norway. *Tectonics*, **9**, 1097–1111, https://doi.org/10.1029/Tc009i005p01097

Archer, S.G., Bergman, S.C., Iliffe, J., Murphy, C.M. & Thornton, M. 2005. Palaeogene igneous rocks reveal new insights into the geodynamic evolution and petroleum potential of the Rockall Trough, NE Atlantic Margin. *Basin Research*, **17**, 171–201, https://doi.org/10.1111/j.1365-2117.2005.00260.x

Bailey, E.B., Clough, C.T., Wright, W.B., Richey, J.E. & Wilson, G.V. 1924. *Tertiary and Post-Tertiary Geology of Mull, Loch Aline, and Oban.* Memoirs of the Geological Survey, Scotland.

Baksi, A.K. 2003. Critical evaluation of $^{40}Ar/^{39}Ar$ age for the Central Atlantic Magmatic Province; timing, duration and possible migration of magmatic centers. *In*: Hames, W.E., McHone, J.G., Renne, P.R. & Ruppel, C. (eds) *The Central Atlantic Magmatic Province: Insights from Fragments of Pangea.* American Geophysical Union, Geophysical Monographs, **136**, 77–90.

Baksi, A.K. 2005*a*. Evaluation of radimetric ages. *In*: Foulger, G.R., Natland, J.H., Presnall, D.C. & Anderson, D.L. (eds) *Plates, Plumes and Paradigsm.* Geological Society of America, Special Papers, **388**, 55–70.

Baksi, A.K. 2005*b*. Evaluation of radiometric ages pertaining to rocks hypothesized to have been derived by hotspot activity, in and around the Atlantic, Indian, and Pacific Oceans. *In*: Foulger, G.R., Natland, J.H., Presnall, D.C. & Anderson, D.L. (eds) *Plates, Plumes and Paradigms.* Geological Society of America, Special Papers, **388**, 55–70.

Baksi, A.K. 2007*a*. A quantitative tool for detecting alteration in undisturbed rocks and minerals – I: water, chemical weathering, and atmospheric argon. *In*: Foulger, G.R. & Jurdy, D.M. (eds) *Plates, Plumes and Planetary Processes.* Geological Society of America, Special Papers, **388**, 285–303.

Baksi, A.K. 2007*b*. A quantitative tool for detecting alteration in undisturbed rocks and minerals – II: application to argon ages related to hotspots.

In: Foulger, G.R. & Jurdy, D.M. (eds) *Plates, Plumes and Planetary Processes*. Geological Society of America, Special Papers, **388**, 305–333.

Baksi, A.K. 2012. 'Data reporting norms for $^{40}Ar/^{39}Ar$ geochronology' – comment. *Quaternary Geochronology*, **12**, 50–52, https://doi.org/10.1016/j.quageo.2012.07.004

Barry, T.L., Self, S., Kelley, S.P., Reidel, S., Hooper, P. & Widdowson, M. 2010. New $^{40}Ar/^{39}Ar$ dating of the Grande Ronde lavas, Columbia River Basalts, USA: implications for duration of flood basalt eruption episodes. *Lithos*, **118**, 213–222, https://doi.org/10.1016/j.lithos.2010.03.014

Barry, T.L., Self, S., Kelley, S.P., Reidel, S., Hooper, P. & Widdowson, M. 2012. Response to Baksi, A., 2012, 'New $^{40}Ar/^{39}Ar$ dating of the Grande Ronde lavas, Columbia River Basalts, USA: implications for duration of flood basalt eruption episodes' by Barry *et al.* 2010 – Discussion'. *Lithos*, **146**, 300–303, https://doi.org/10.1016/j.lithos.2012.04.011

Beane, J.E., Turner, C.A., Hooper, P.R., Subbarao, K.V. & Walsh, J.N. 1986. Stratigraphy, composition and form of the Deccan Basalts, Western Ghats, India. *Bulletin of Volcanology*, **48**, 61–83, https://doi.org/10.1007/BF01073513

Bell, B.R. & Harris, J.W. 1986. *An Excursion Guide to the Geology of the Isle of Skye*. Geological Society of Glasgow, Glasgow.

Bell, B.R. & Williamson, I.T. 2002. Tertiary igneous activity. *In*: Trewin, N.H. (ed.) *The Geology of Scotland*. Geological Society, London, 371–407.

Boldreel, L.O., Graversen, O. & Andersen, M.S. 1994. Tertiary development of the Faeroe–Rockall Plateau based on reflection seismic data. *Bulletin of the Geological Society of Denmark*, **41**, 162–180.

Brooks, C.K., Tegner, C., Stein, H. & Thomassen, B. 2004. Re–Os and $^{40}Ar/^{39}Ar$ ages of porphyry molybdenum deposits in the East Greenland volcanic-rifted margin. *Economic Geology*, **99**, 1215–1222.

Bryan, S.E. & Ernst, R.E. 2008. Revised definition of large igneous provinces (LIPs). *Earth-Science Reviews*, **86**, 175–202, https://doi.org/10.1016/j.earscirev.2007.08.008

Bugge, T., Prestvik, T. & Rokoengen, K. 1980. Lower tertiary volcanic rocks off Kristiansund – mid Norway. *Marine Geology*, **35**, 277–286, https://doi.org/10.1016/0025-3227(80)90121-8

Burchardt, S., Troll, V.R., Mathieu, L., Emeleus, H.C. & Donaldson, C.H. 2013. Ardnamurchan 3D cone-sheet architecture explained by a single elongate magma chamber. *Scientific Reports*, **3**, 2891, https://doi.org/10.1038/Srep02891

Campbell, I.H. & Griffiths, R.W. 1990. Implications of mantle plume structure for the evolution of flood basalts. *Earth and Planetary Scientific Letters*, **99**, 79–93.

Cameron, M.M. & Hodges, K.V. 2016. ArAR – a software tool to promote the robust comparison of K–Ar and $^{40}Ar/^{39}Ar$ dates published using different decay, isotopic, and monitor-age parameters. *Chemical Geology*, **440**, 148–163.

Chambers, L.M. 2000. *Age and duration of the British Tertiary Igneous Province: implications for the development of the ancestral Iceland plume*. PhD thesis, University of Edinburgh.

Chambers, L.M. & Pringle, M.S. 2001. Age and duration of activity at the Isle of Mull Tertiary igneous centre, Scotland, and confirmation of the existence of subchrons during Anomaly 26r. *Earth and Planetary Science Letters*, **193**, 333–345.

Chambers, L.M., Pringle, M.S. & Parrish, R.R. 2005. Rapid formation of the Small Isles Teritary centre constrained by precise $^{40}Ar/^{39}Ar$ and U–Pb ages. *Lithos*, **79**, 367–384.

Cocks, L.R.M. & Torsvik, T.H. 2011. The Palaeozoic geography of Laurentia and western Laurussia: a stable craton with mobile margins. *Earth-Science Reviews*, **106**, 1–51, https://doi.org/10.1016/j.earscirev.2011.01.007

Cooper, M.R. 2004. Palaeogene Extrusive Igneous Rocks. *In*: Mitchell, W.I. (ed.) *The Geology of Northern Ireland, Our Natural Foundation*. Geological Survey of Northern Ireland, Belfast, 167–178.

Cooper, M.R. & Johnston, T.P. 2004. Palaeogene intrusive igneous rocks. *In*: Mitchell, W.I. (ed.) *The Geology of Northern Ireland – Our Natural Foundation*. Geological Survey of Northern Ireland, Belfast, 179–198.

Cooper, M.R., Anderson, H., Walsh, J.J., Van Dam, C.L., Young, M.E., Earls, G. & Walker, A. 2012. Palaeogene Alpine tectonics and Icelandic plume-related magmatism and deformation in Northern Ireland. *Journal of the Geological Society, London*, **169**, 29–36, https://doi.org/10.1144/0016-76492010-182

Corfu, F., Polteau, S., Planke, S., Faleide, J.I., Svensen, H., Zayoncheck, A. & Stolbov, N. 2013. U–Pb geochronology of Cretaceous magmatism on Svalbard and Franz Josef Land, Barents Sea Large Igneous Province. *Geological Magazine*, **150**, 1127–1135, https://doi.org/10.1017/S0016756813000162

Courtillot, V., Jaupart, C., Manighetti, I., Tapponnier, P. & Besse, J. 1999. On causal links between flood basalts and continental breakup. *Earth and Planetary Science Letters*, **166**, 177–195.

Courtillot, V.E. & Renne, P.R. 2003. On the ages of flood basalt events. *Comptes Rendus Geoscience*, **335**, 113–140.

Dagley, P. & Mussett, A.E. 1986. Palaeomagnetism and radiometric dating of the British Tertiary igneous province: Muck and Eigg. *Geophysical Journal of the Royal Astronomical Society*, **85**, 221–242.

Dalrymple, G.B. & Lanphere, M.A. 1969. *Potassium–Argon Dating; Principles, Techniques and Applications to Geochronology*. Freeman, San Francisco, CA, USA.

Dickin, A.P. 1981. Isotope geochemistry of Tertiary igneous rocks from the Isle of Skye, N.W. Scotland. *Journal of Petrology*, **22**, 155–189.

Dickin, A.P. & Jones, N.W. 1983. Isotopic evidence for the age and origin of pitchstones and felsites, Isle of Eigg, NW Scotland. *Journal of the Geological Society, London*, **140**, 691–700, https://doi.org/10.1144/gsjgs.140.4.0691

Dodd, S.C., Mac Niocaill, C. & Muxworthy, A.R. 2015. Long duration (>4 Ma) and steady-state volcanic activity in the early Cretaceous Parana–Etendeka large igneous province; new palaeomagnetic data from Namibia. *Earth and Planetary Science Letters*, **414**, 16–29, https://doi.org/10.1016/j.epsl.2015.01.009

Ellis, D., Jolley, D.W., Bell, B.R. & O'Callaghan, M. 2002. The stratigraphy, environment of eruption and age of the Faroes Lava Group, NE Atlantic Ocean. *In*: Jolley, D.W. & Bell, B.R. (eds) *The North Igneous Province: Stratigraphy, Tectonic, Volcanic and Magmatic Processes*. Geological Society, London, Special Publications, **197**, 253–269, https://doi.org/10.1144/GSL.SP.2002.197.01.10

Emeleus, C.H. 1997. *Geology of Rum and the Adjacent Islands: Memoir for the 1:50 000 Geological Sheet 60 (Scotland)*. British Geological Survey, Keyworth, Nottingham.

Emeleus, C.H. & Bell, B.R. 2005. *British Regional Geology: The Palaeogene Volcanic Districts of Scotland*. British Geological Survey, Keyworth, Nottingham.

Faithfull, J.W., Timmerman, M.J., Upton, B.G.J. & Rumsey, M.S. 2012. Mid-Eocene renewal of magmatism in NW Scotland: the Loch Roag Dyke, Outer Hebrides (vol 169, pg 115, 2012). *Journal of the Geological Society, London*, **169**, 363–364, https://doi.org/10.1144/0016-76492011-117Err

Fitch, F.J., Miller, J.A. & Mitchell, J.G. 1969. A new approach to radio-isotopic dating in orogenic belts. *In*: Kent, P.E., Satterthwaite, E. & Spencer, A.M. (eds) *Time and Place in Orogeny*. Geological Society, London, Special Publications, **3**, 157–195, https://doi.org/10.1144/gsl.sp.1969.003.01.09

Fitch, F.J., Hooker, P.J., Miller, J.A. & Brereton, N.R. 1978. Glauconite dating of Palaeocene–Eocene rocks from East Kent and the time-scale of Palaeogene volcanism in the North Atlantic region. *Journal of the Geological Society, London*, **135**, 499–512, https://doi.org/10.1144/gsjgs.135.5.0499

Fitch, F.J., Heard, G.L. & Miller, J.A. 1988. Basaltic magmatism of Late Cretaceous and Palaeogene age recorded in wells NNE of the Shetlands. *In*: Morton, A.C. & Parson, L.M. (eds) *Early Tertiary Volcanism and the Opening of the NE Atlantic*. Geological Society, London, Special Publications, **39**, 253–262, https://doi.org/10.1144/GSL.SP.1988.039.01.23

Fossen, H. 2010. Extensional tectonics in the North Atlantic Caledonides; a regional view. *In*: Law, R.D., Butler, R.W.H., Holdsworth, R.E., Krabbendam, M. & Strachan, R.A. (eds) *Continental Tectonics and Mountain Building: The Legacy of Peach and Horne*. Geological Society, London, Special Publications, **335**, 767–793, https://doi.org/10.1144/SP335.31

Foulger, G.R. 2002. Plumes, or plate tectonics processes? *Astronomy & Geophysics*, **43**, 6.19–16.23.

Gamble, J.A., Wysoczanski, R.J. & Meighan, I.G. 1999. Constraints on the age of the British Tertiary Volcanic Province from ion microprobe U–Pb (SHRIMP) ages for acid igneous rocks from NE Ireland. *Journal of the Geological Society, London*, **156**, 291–299, https://doi.org/10.1144/gsjgs.156.2.0291

Ganerød, M., Smethurst, M.A. *et al.* 2010. The North Atlantic Igneous Province reconstructed and its relation to the Plume Generation Zone: the Antrim Lava Group revisited. *Geophysical Journal International*, **182**, 183–202, https://doi.org/10.1111/j.1365-246X.2010.04620.x

Ganerød, M., Chew, D.M., Smethurst, M.A., Troll, V.R., Corfu, F., Meade, F. & Prestvik, T. 2011. Geochronology of the Tardree rhyolite complex, Northern Ireland; implications for zircon fission-track studies, the North Atlantic igneous province and the age of the Fish Canyon sanidine standard. *Chemical Geology*, **286**, 222–228, https://doi.org/10.1016/j.chemgeo.2011.05.007

Gibson, D., McCormick, A.G., Meighan, I.G. & Halliday, A.N. 1987. The British Tertiary igneous province – young Rb–Sr ages for the Mourne Mountains Granites. *Scottish Journal of Geology*, **23**, 221–225, https://doi.org/10.1144/sjg23020221

Gradstein, F.M., Ogg, J.G., Schmitz, M.D. & Ogg, G.M. 2012. *The Geologic Time Scale 2012*. Elsevier, Amsterdam.

Hald, N. & Pedersen, A.K. 1975. Lithostratigraphy of the early Tertiary volcanic rocks of central West Greenland. *Rapport – Gronlands Geologiske Undersogelse [1964]*, **69**, 17–24.

Hald, N. & Tegner, C. 2000. Composition and age of tertiary sills and dykes, Jameson Land Basin, East Greenland; relation to regional flood volcanism. *Lithos*, **54**, 207–233.

Hald, N. & Waagstein, R. 1991. The dykes and sills of the Early Tertiary Faeroe Island basalt plateau. *Transactions of the Royal Society of Edinburgh: Earth Sciences*, **82**, 373–388.

Hamilton, M.A., Pearson, D.G., Thompson, R.N., Kelley, S.P. & Emeleus, C.H. 1998. Rapid eruption of Skye lavas inferred from precise U–Pb and Ar–Ar dating of the Rum and Cuillin plutonic complexes. *Nature*, **394**, 260–263.

Hansen, J., Jerram, D.A., McCaffrey, K. & Passey, S.R. 2011. Early Cenozoic saucer-shaped sills of the Faroe Islands: an example of intrusive styles in basaltic lava piles. *Journal of the Geological Society, London*, **168**, 159–178, https://doi.org/10.1144/0016-76492010-012

Hansen, H., Pedersen, A.K. *et al.* 2002. Volcanic stratigraphy of the southern Prinsen af Wales Bjerge region, East Greenland. *In*: Jolley, D.W. & Bell, B.R. (eds) *The North Atlantic Igneous Province: Stratigraphy, Tectonic, Volcanic and Magmatic Processes*. Geological Society, London, Special Publications, **197**, 183–218, https://doi.org/10.1144/GSL.SP.2002.197.01.08

Heister, L.E., O'Day, P.A., Brooks, C.K., Neuhoff, P.S. & Bird, D.K. 2001. Pyroclastic deposits within the East Greenland Tertiary flood basalts. *Journal of the Geological Society, London*, **158**, 269–284, https://doi.org/10.1144/jgs.158.2.269

Hill, I.G., Worden, R.H. & Meighan, I.G. 2000. Geochemical evolution of a palaeolaterite: the Interbasaltic Formation, Northern Ireland. *Chemical Geology*, **166**, 65–84.

Hirschmann, M.M., Renne, P.R. & McBirney, A.R. 1997. $^{40}Ar/^{39}Ar$ dating of the Skaergaard intrusion. *Earth and Planetary Science Letters*, **146**, 645–658, https://doi.org/10.1016/S0012-821x(96)00250-6

Hitchen, K. & Ritchie, J.D. 1993. New K–Ar ages, and a provisional chronology, for the offshore part of the British Tertiary Igneous Province. *Scottish Journal of Geology*, **29**, 73–85, https://doi.org/10.1144/sjg29010073

HOFMANN, C., FÉRAUD, G. & COURTILLOT, V. 2000. $^{40}Ar/^{39}Ar$ dating of mineral separates and whole rocks from the Western Ghats lava pile: further constraints on duration and age of the Deccan traps. *Earth and Planetary Science Letters*, **180**, 13–27, https://doi.org/10.1016/S0012-821X(00)00159-X

IVANOV, A.V. 2007. Evaluation of different models for the origin of the Siberian traps. *In*: FOULGER, G.R. & JURDY, D.M. (eds) *Plates, Plumes, and Planetary Processes*. Geological Society of America, Special Paper, **388**, 669–691.

JOLLEY, D.W. & BELL, B.R. 2002. The evolution of the North Atlantic Igneous Province and the opening of the NE Atlantic. *In*: JOLLEY, D.W. & BELL, B.R. (eds) *The North Atlantic Igneous Province: Stratigraphy, Tectonic, Volcanic and Magmatic Processes*. Geological Society, London, Special Publications, **197**, 1–13, https://doi.org/10.1144/GSL.SP.2002.197.01.01

JOLLY, R.J.H. & SANDERSON, D.J. 1995. Variation in the form and distribution of dykes in the Mull swarm, Scotland. *Journal of Structural Geology*, **17**, 1543–1557.

JOURDAN, F., FERAUD, G., BERTRAND, H. & WATKEYS, M.K. 2007. From flood basalts to the inception of oceanization: example from the $^{40}Ar/^{39}Ar$ high-resolution picture of the Karoo large igneous province. *Geochemistry, Geophysics, Geosystems*, **8**, Q02002, https://doi.org/10.1029/2006gc001392

JOURDAN, F., MARZOLI, A. ET AL. 2009*a*. $^{40}Ar/^{39}Ar$ ages of CAMP in North America: implications for the Triassic–Jurassic boundary and the ^{40}K decay constant bias. *Lithos*, **110**, 167–180, https://doi.org/10.1016/j.lithos.2008.12.011

JOURDAN, F., RENNE, P.R. & REIMOLD, W.U. 2009*b*. An appraisal of the ages of terrestrial impact structures. *Earth and Planetary Science Letters*, **286**, 1–13, https://doi.org/10.1016/j.epsl.2009.07.009

JØRGENSEN, O. 2006. The regional distribution of zeolites in the basalts of the Faroe Islands and the significance of zeolites as palaeo-temperature indicators. *Geological Survey of Denmark and Greenland Bulletin*, **9**, 123–156, http://www.geus.dk/publications/bull

KERR, A.C. 1997. The geochemistry and significance of plugs intruding the Tertiary Mull–Movern lava succession, western Scotland. *Scottish Journal of Geology*, **33**, 157–167, https://doi.org/10.1144/sjg33020157

KING, S.D. & ANDERSON, D.L. 1998. Edge-driven convection. *Earth and Planetary Science Letters*, **160**, 289–296.

KUIPER, K.F., DEINO, A., HILGEN, F.J., KRIJGSMAN, W., RENNE, P.R. & WIJBRANS, J.R. 2008. Synchronizing rock clocks of Earth history. *Science*, **320**, 500–504, https://doi.org/10.1126/science.1154339

LARSEN, H.C. & SAUNDERS, A.D. 1998. Tectonism and volcanism at the southeast Greenland rifted margin: a record of plume impact and later continental rupture. *In*: WISE, S.W., JR (ed.) *Proceedings of the Ocean Drilling Program, Scientific Results, Volume 152*. Ocean Drilling Program, College Station, TX, 503–533.

LARSEN, L.M. & PEDERSEN, A.K. 2009. Petrology of the Paleocene Picrites and Flood Basalts on Disko and Nuussuaq, West Greenland. *Journal of Petrology*, **50**, 1667–1711, https://doi.org/10.1093/petrology/egp048

LARSEN, L.M., REX, D.C., WATT, W.S. & GUISE, P.G. 1999*a*. $^{40}Ar/^{39}Ar$ dating of alkali basaltic dykes along the southwest coast of Greenland; Cretaceous and Tertiary igneous activity along the eastern margin of the Labrador Sea. *Geology of Greenland Survey Bulletin*, **184**, 19–29.

LARSEN, L.M., WAAGSTEIN, R., PEDERSEN, A.K. & STOREY, M. 1999*b*. Trans-Atlantic correlation of the Palaeogene volcanic successions in the Faeroe Islands and East Greenland. *Journal of the Geological Society, London*, **156**, 1081–1095, https://doi.org/10.1144/gsjgs.156.6.1081

LARSEN, L.M., HEAMAN, L.M., CREASER, R.A., DUNCAN, R.A., FREI, R. & HUTCHINSON, M. 2009. Tectonomagnetic events during stretching and basin formation in the Labrador Sea and the Davis Strait: evidence from age and composition of Mesozoic to Palaeogene dyke swarms in West Greenland. *Journal of the Geological Society, London*, **166**, 999–1012, https://doi.org/10.1144/0016-76492009-038

LARSEN, L.M., PEDERSEN, A.K., SORENSEN, E.V., WATT, W.S. & DUNCAN, R.A. 2013. Stratigraphy and age of the Eocene Igtertiva Formation basalts, alkaline pebbles and sediments of the Kap Dalton Group in the graben at Kap Dalton, East Greenland. *Bulletin of the Geological Society of Denmark*, **61**, 1–18.

LARSEN, L.M., PEDERSEN, A.K., TEGNER, C. & DUNCAN, R.A. 2014. Eocene to Miocene igneous activity in NE Greenland: northward younging of magmatism along the East Greenland margin. *Journal of the Geological Society, London*, **171**, 539–553, https://doi.org/10.1144/jgs2013-118

LARSEN, L.M., PEDERSEN, A.K., TEGNER, C., DUNCAN, A.R., HALD, N. & LARSEN, J.G. 2015. Age of Tertiary volcanic rocks on the West Greenland continental margin: volcanic evolution and event correlation to other parts of the North Atlantic Igneous Province. *Geological Magazine*, **153**, 487–511, https://doi.org/10.1017/S0016756815000515

LARSEN, R.B. & TEGNER, C. 2006. Pressure conditions for the solidification of the Skaergaard intrusion: eruption of East Greenland flood basalts in less than 300 000 years. *Lithos*, **92**, 181–197, https://doi.org/10.1016/j.lithos.2006.03.032

LAWVER, L.A. & MÜLLER, R.D. 1994. Iceland hotspot track. *Geology*, **22**, 311–314.

LENOIR, X., FÉRAUDC, G. & GEOFFROY, L. 2003. High-rate flexure of the East Greenland volcanic margin: constraints from $^{40}Ar/^{39}Ar$ dating of basaltic dykes. *Earth and Planetary Science Letters*, **214**, 515–528.

LUND, J. 1983. Biostratigraphy of interbasaltic coals from the Faroe Islands. *In*: BOTT, M.H.P., SAXOV, S., TALWANI, M. & THIEDE, J. (eds) *NATO Conference Series, Series IV*. Marine Sciences, **8**. Plenum Press, New York, 417–423.

LUND, J. 1988. A late Paleocene non-marine microflora from the interbasaltic coals of the Faeroe Islands, North Atlantic. *Bulletin of the Geological Society of Denmark*, **37**, 181–203.

LYLE, P. 1979. A petrological and geochemical study of the Tertiary basaltic rocks of northeast Ireland. *Journal of Earth Sciences, Royal Irish Academy*, **2**, 137–152.

MANNING, P.I., ROBBIE, J.A. & WILSON, H.E. 1970. *Geology of Belfast and the Lagan Valley: (One-Inch*

Geological Sheet 36). 2nd edn. Memoirs of the Geological Survey, Northern Ireland.

McDougall, I. & Harrison, T.M. 1999. *Geochronology and Thermochronology by the $^{40}Ar/^{39}Ar$ Method.* 2nd edn. Oxford University Press, New York.

Meade, F.C., Chew, D.M., Troll, V.R., Ellam, R.M. & Page, L.M. 2009. Magma Ascent along a Major Terrane Boundary: crustal contamination and Magma Mixing at the Drumadoon Intrusive Complex, Isle of Arran, Scotland. *Journal of Petrology*, **50**, 2345–2374.

Meighan, I.G., McCormick, A.G., Gibson, D., Gamble, J.A. & Graham, I.J. 1988. Rb–Sr isotopic determinations and the timing of Tertiary central complex magmatism in NE Ireland. *In*: Morton, A.C. & Parson, L.M. (eds) *Early Tertiary Volcanism and the Opening of the North-East Atlantic*. Geological Society, London, Special Publications, **39**, 349–360, https://doi.org/10.1144/GSL.SP.1988.039.01.30

Meyer, R., van Wijk, J. & Gernigon, L. 2007. The North Atlantic Igneous Province: a review of models for its formation. *In*: Foulger, G.R. & Jurdy, D.M. (eds) *Plates, Plumes, and Planetary Processes*. Geological Society of America, Special Papers, **430**, 525–552.

Min, K.W., Mundil, R., Renne, P.R. & Ludwig, K.R. 2000. A test for systematic errors in $^{40}Ar/^{39}Ar$ geochronology through comparison with U/Pb analysis of a 1.1-Ga rhyolite. *Geochimica et Cosmochimica Acta*, **64**, 73–98, https://doi.org/10.1016/S0016-7037(99)00204-5

Mitchell, J.G. & Reen, K.P. 1973. Potassium–argon ages from the Tertiary Ring Complexes of the Ardnamurchan Peninsula, Western Scotland. *Geological Magazine*, **110**, 331–340.

Mitchell, W.I. 2004. Cretaceous. *In*: Mitchell, W.I. (ed.) *The Geology of Northern Ireland, Our Natural Foundation*. Geological Survey of Northern Ireland, Belfast, 149–160.

Mussett, A.E. 1986. $^{40}Ar–^{39}Ar$ step heating ages of the Tertiary igneous rocks of Mull, Scotland. *Journal of the Geological Society, London*, **143**, 887–896, https://doi.org/10.1144/gsjgs.143.6.0887

Mussett, A.E., Dagley, P. & Eckford, M. 1976. The British Tertiary igneous province; palaeomagnetism and ages of dykes, Lundy Island, Bristol Channel. *Geophysical Journal of the Royal Astronomical Society*, **46**, 595–603.

Mussett, A.E., Dagley, P., Hodgson, B. & Skelhorn, R.R. 1987. Palaeomagnetism and age of the quartz-porphyry intrusions, Isle of Arran. *Scottish Journal of Geology*, **23**, 9–22, https://doi.org/10.1144/sjg230 10009

Mussett, A.E., Dagley, P. & Skelhorn, R.R. 1988. Time and duration of the British Tertiary Igneous Province. *In*: Morton, A.C. & Parson, L.M. (eds) *Early Tertiary Volcanism and the Opening of the North-East Atlantic*. Geological Society, London, Special Publications, **39**, 337–348, https://doi.org/10.1144/GSL.SP.1988.039.01.29

Müller, R., Nystuen, J.P., Eide, F. & Lie, H. 2005. Late Permian to Triassic basin infill history and palaeogeography of the Mid-Norwegian shelf–East Greenland region. *In*: Wandås, B.T.G., Nystuen, J.P., Eide, E. & Gradstein, F.M. (eds) *Onshore–Offshore Relationships on the North Atlantic Margin.* Norwegian Petroleum Society, Special Publications, **12**, 155–189.

Mørk, M.B.E. & Duncan, R.A. 1993. Late Pliocene basaltic volcanism on the western Barents Shelf margin; implications from petrology and $^{40}Ar/^{39}Ar$ dating of volcaniclastic debris from a shallow drill core. *Norsk Geologisk Tidsskrift*, **73**, 209–225.

Nadin, P.A., Kusznir, N.J. & Cheadle, M.J. 1997. Early Tertiary plume uplift of the North Sea and Faeroe–Shetland basins. *Earth and Planetary Science Letters*, **148**, 109–127.

Nevle, R.J., Brandriss, M.E., Bird, D.K., McWilliams, M.O. & Oneil, J.R. 1994. Tertiary Plutons Monitor Climate-Change in East Greenland. *Geology*, **22**, 775–778, https://doi.org/10.1130/0091-7613(1994)022<0775:Tpmcci>2.3.Co;2

Nelson, C.E., Jerram, D.A., Clayburn, J.A.P., Halton, A.M. & Roberge, J. 2015. Eocene volcanism in offshore southern Baffin Bay. *Marine and Petroleum Geology*, **67**, 678–691, https://doi.org/10.1016/j.marpetgeo.2015.06.002

Nielsen, T.F.D., Hansen, H., Brooks, C.K., Lesher, C.E. & Parties, A.F. 2001. The East Greenland continental margin, the Prinsen af Wales Bjerge and new Skaergaard intrusion initiatives. *Geology of Greenland Survey Bulletin*, **189**, 83–98.

Noble, R.H., Macintyre, R.M. & Brown, P.E. 1988. Age constraints on Atlantic evolution: timing of magmatic activity along the E Greenland continental margin. *In*: Morton, A.C. & Parson, L.M. (eds) *Early Tertiary Volcanism and the Opening of the NE Atlantic*. Geological Society, London, Special Publications, **39**, 201–214, https://doi.org/10.1144/GSL.SP.1988.039.01.19

Noe-Nygaard, A. 1966. The invisible part of the Faroes. *Meddelelser fra Dansk Geologisk Forening*, **16**, 191–195.

Nøttvedt, A., Gee, D.G., Johannessen, E.P., Surlyk, F. & Ladenberger, A. 2008. The Mesozoic of western Scandinavia and East Greenland. *Episodes*, **31**, 59–65.

O'Connor, J.M., Stoffers, P., Wijbrans, J.R., Shannon, P.M. & Morrissey, T. 2000. Evidence from episodic seamount volcanism for pulsing of the Iceland plume in the past 70 Myr. *Nature*, **408**, 954–958, https://doi.org/10.1038/35050066

Old, R.A. 1975. The age and field relationships of the Tardree Tertiary rhyolite complex, County Antrim, N. Ireland. *Bulletin of the Geological Survey of Great Britain*, **51**, 21–40.

Osmundsen, P.T. & Ebbing, J. 2008. Styles of extension offshore mid-Norway and implications for mechanisms of crustal thinning at passive margins. *Tectonics*, **27**, https://doi.org/10.1029/2007tc002242

Passey, S.R. 2004. *The Volcanic and Sedimentary Evolution of the Faeroe Plateau Lava Group, Faeroe Islands and the Faeroe–Shetland Basin, NE Atlantic*. PhD, University of Glasgow.

Passey, S.R. & Jolley, D.W. 2009. A revised lithostratigraphic nomenclature for the Palaeogene Faroe Island Basalt Group, NE Atlantic Ocean. *Earth and Environmental Science Transactions of the Royal Society of Edinburgh*, **99**, 127–158.

Pearson, D.G., Emeleus, C.H. & Kelley, S.P. 1996. Precise $^{40}Ar/^{39}Ar$ age for the initiation of Palaeogene volcanism in the Inner Hebrides and its regional significance. *Journal of the Geological Society, London*, **153**, 815–818, https://doi.org/10.1144/gsjgs.153.6.0815

Peate, D.W., Baker, J.A. *et al.* 2003. The Prinsen af Wales Bjerge formation lavas, East Greenland: the transition from tholeiitic to alkalic magmatism during Palaeogene continental break-up. *Journal of Petrology*, **44**, 279–304, https://doi.org/10.1093/petrology/44.2.279

Polteau, S., Hendriks, B.W.H. *et al.* 2016. The Early Cretaceous Barents Sea Sill Complex: distribution, $^{40}Ar/^{39}Ar$ geochronology, and implications for carbon gas formation. *Palaeogeography, Palaeoclimatology, Palaeoecology*, **441**, 83–95, https://doi.org/10.1016/j.palaeo.2015.07.007

Preston, J. 2009. Tertiary igneous activity. *In*: Holland, C.H. & Saunders, I.S. (eds) *The Geology of Ireland*. Dunedin Academic Press, Edinburgh, 353–373.

Price, S., Brodie, J., Whitham, A. & Kent, R. 1997. Mid-Tertiary rifting and magamatism in the Traill Ø region, East Greenland. *Journal of the Geological Society, London*, **154**, 419–434, https://doi.org/10.1144/gsjgs.154.3.0419

Reidel, S.P., Camp, V.E., Tolan, T.L., Martin, B.S., Ross, M.E., Wolff, J.A. & Wells, R.E. 2013. The Columbia River flood basalt province; stratigraphy, areal extent, volume, and physical volcanology. *In*: Reidel, S.P., Camp, V.E., Ross, M.E., Wolff, J.A., Martin, B.S., Tolan, T.L. & Wells, R.E. (eds) *The Columbia River Flood Basalt Province*. Geological Society of America, Special Papers, **497**, 1–43, https://doi.org/10.1130/2013.2497(01)

Renne, P.R., Swisher, C.C., Deino, A.L., Karner, D.B., Owens, T.L. & DePaolo, D.J. 1998. Intercalibration of standards, absolute ages and uncertainties in $^{40}Ar/^{39}Ar$ dating. *Chemical Geology*, **145**, 117–152, https://doi.org/10.1016/S0009-2541(97)00159-9

Renne, P.R., Deino, A.L. *et al.* 2009. Data reporting norms for $^{40}Ar/^{39}Ar$ geochronology. *Quaternary Geochronology*, **4**, 346–352, https://doi.org/10.1016/j.quageo.2009.06.005

Renne, P.R., Mundil, R., Balco, G., Min, K.W. & Ludwig, K.R. 2010. Joint determination of ^{40}K decay constants and $^{40}Ar^*/^{40}K$ for the Fish Canyon sanidine standard, and improved accuracy for $^{40}Ar/^{39}Ar$ geochronology. *Geochimica et Cosmochimica Acta*, **74**, 5349–5367, https://doi.org/10.1016/j.gca.2010.06.017

Renne, P.R., Balco, G., Ludwig, K.R., Mundil, R. & Min, K. 2011. Response to the comment by W.H. Schwarz *et al.* on 'Joint determination of ^{40}K decay constants and $^{40}Ar^*/^{40}K$ for the Fish Canyon sanidine standard, and improved accuracy for $^{40}Ar/^{39}Ar$ geochronology' by PR Renne *et al.* (2010). *Geochimica et Cosmochimica Acta*, **75**, 5097–5100, https://doi.org/10.1016/j.gca.2011.06.021

Riisager, P., Riisager, J., Abrahamsen, N. & Wagstein, R. 2002. New paleomagnetic pole and magnetostratigraphy of Faroe Islands flood volcanics, North Atlantic igneous province. *Earth and Planetary Science Letters*, **201**, 261–276.

Ritchie, D.K., Ziska, H., Johnson, H. & Evans, D. (eds) 2011. *Geology of the Faroe–Shetland Basin and Adjacent Areas*. BGS Research Report RR/11/01. British Geological Survey, Keyworth, Nottingham.

Roddick, J.C., Srivastava, S.P. *et al.* 1989. K–Ar dating of basalts from Site 647, ODP Leg 105. *In*: *Proceedings of the Ocean Drilling Program, Scientific Results, Volume 105*. Ocean Drilling Program, College Station, TX, 885–887.

Saunders, A.D. 2016. Two LIPs and two Earth-system crises: the impact of the North Atlantic Igneous Province and the Siberian Traps on the Earth-surface carbon cycle. *Geological Magazine*, **153**, 201–222, https://doi.org/10.1017/S0016756815000175

Saunders, A.D., Fitton, J.G., Kerr, A.C., Norry, M.J. & Kent, R.W. 1997. The North Atlantic Igneous Province. *In*: Mahoney, J.J. & Coffin, M.F. (eds) *Large Igneous Provinces*. American Geophysical Union, Geophysical Monograph, **100**, 45–93.

Saunders, A.D., Jones, S.M., Morgan, L.A., Pierce, K.L., Widdowson, M. & Xu, Y.G. 2007. Regional uplift associated with continental large igneous provinces: the roles of mantle plumes and the lithosphere. *Chemical Geology*, **241**, 282–318, https://doi.org/10.1016/j.chemgeo.2007.01.017

Simms, M.J. 2000. The sub-basaltic surface in northeast Ireland and its significance for interpreting the Tertiary history of the region. *Proceedings of the Geologists' Association*, **111**, 321–336.

Sinton, C.W. & Duncan, R.A. 1998. $^{40}Ar/^{39}Ar$ ages of lavas from the Southeast Greenland margin, ODP Leg 152 and the Rockall Plateau, DSDP Leg 81. *In*: Saunders, A.D., Larsen, H.C. & Wise, W. (eds) *Proceedings of the Ocean Drilling Program, Scientific Results, Volume 152*. Ocean Drilling Program, College Station, TX, 387–402.

Sinton, C.W., Kitchen, K. & Duncan, R.A. 1998. $^{40}Ar/^{39}Ar$ geochronology of silicic and basic volcanic rocks on the margins of the North Atlantic. *Geological Magazine*, **135**, 161–170.

Speight, J.M., Skelhorn, R.R., Sloan, T. & Knapp, R.J. 1982. The dyke swarms of Scotland. *In*: Sutherland, D.S. (ed.) *Igneous Rocks of the British Isles*. John Wiley & Son, Chichester, 449–459.

Steiger, R.H. & Jäger, E. 1977. Subcommission on geochronology: convention on the use of decay constants in geo- and cosmochronology. *Earth and Planetary Science Letters*, **36**, 359–362.

Stemmerik, L. 2000. Late Palaeozoic evolution of the North Atlantic margin of Pangea. *Palaeogeography, Palaeoclimatology, Palaeoecology*, **161**, 95–126.

Stoker, M.S., Stewart, M.A. *et al.* 2016. An overview of the Upper Palaeozoic–Mesozoic stratigraphy of the NE Atlantic region. *In*: Peron-Pinvidic, G., Hopper, J.R., Stoker, M.S., Gaina, C., Doornenbal, J.C., Funck, T. & Arting, U.E. (eds) *The NE Atlantic Region: A Reappraisal of Crustal Structure, Tectonostratigraphy and Magmatic Evolution*. Geological Society, London, Special Publications, **447**. First published online August 11, 2016, https://doi.org/10.1144/SP447.2

Storey, M., Duncan, R.A., Pedersen, A.K., Larsen, L.M. & Larsen, H.C. 1998. $^{40}Ar/^{39}Ar$ geochronology of the West Greenland Tertiary volcanic province. *Earth and Planetary Science Letters*, **160**, 569–586.

STOREY, M., PEDERSEN, A.K. *ET AL.* 2004. Long-lived postbreakup magmatism along the East Greenland margin: evidence for shallow-mantle metasomatism by the Iceland plume. *Geology*, **32**, 173–176, https://doi.org/10.1130/G19889.1

STOREY, M., DUNCAN, R.A. & TEGNER, C. 2007. Timing and duration of volcanism in the North Atlantic Igneous Province: implications for geodynamics and links to the Iceland Hotspot. *Chemical Geology*, **241**, 264–281.

SURLYK, F. 1990. Timing, style and sedimentary evolution of late Palaeozoic–Mesozoic extensional basins of East Greenland. *In*: HARDMAN, R.F.P. & BROOKS, J. (eds) *Tectonic Events Responsible for Britain's Oil and Gas Reserves*. Geological Society, London, Special Publications, **55**, 107–125, https://doi.org/10.1144/GSL.SP.1990.055.01.05

SVENSEN, H., PLANKE, S., MALTHE-SØRENSSEN, A., JAMTVEIT, B., MYKLEBUST, R. & RASMUSSEN EIDEM, T. 2004. Release of methane from a volcanic basin as a mechanism for initial Eocene global warming. *Nature*, **429**, 542–545.

SVENSEN, H., PLANKE, S. & CORFU, F. 2010. Zircon dating ties NE Atlantic sill emplacement to initial Eocene global warming. *Journal of the Geological Society, London*, **167**, 433–436, https://doi.org/10.1144/0016-76492009-125

SVENSEN, H., CORFU, F., POLTEAU, S., HAMMER, O. & PLANKE, S. 2012. Rapid magma emplacement in the Karoo Large Igneous Province. *Earth and Planetary Science Letters*, **325**, 1–9, https://doi.org/10.1016/j.epsl.2012.01.015

TALWANI, M. & ELDHOLM, O. 1977. Evolution of the Norwegian-Greenland sea. *Geological Society of America Bulletin*, **88**, 969–999, https://doi.org/10.1130/0016-7606(1977)88<969:EOTNS>2.0.CO;2

TARLING, D.H. & GALE, N.H. 1968. Isotopic dating and palaeomagnetic polarity in the Faeroe islands. *Nature [London]*, **218**, 1043–1044.

TEGNER, C. & DUNCAN, R.A. 1999. A $^{40}Ar/^{39}Ar$ chronology for the Volcanic history of the southeast Greenland rifted margin, Leg 163. *In*: LARSEN, H.C., DUNCAN, R.A., ALLAN, J.F. & BROOKS, K. (eds) *Proceedings of the Ocean Drilling Program, Scientific Results, Volume 163*. Ocean Drilling Program, College Station, TX, 53–62.

TEGNER, C., DUNCAN, R.A., BERNSTEIN, S., BROOKS, C.K., BIRD, D.K. & STOREY, M. 1998. $^{40}Ar/^{39}Ar$ geochronology of Tertiary mafic intrusions along the East Greenland rifted margin: relation to flood basalts and the Iceland hotspot track. *Earth and Planetary Science Letters*, **156**, 75–88.

TEGNER, C., BROOKS, C.K., DUNCAN, R.A., HEISTER, L.E. & BERNSTEIN, S. 2008. Ar-40–Ar-39 ages of intrusions in East Greenland: rift-to-drift transition over the Iceland hotspot. *Lithos*, **101**, 480–500, https://doi.org/10.1016/j.lithos.2007.09.001

THOMPSON, P., MUSSETT, A.E. & DAGLEY, P. 1987. Revised $^{40}Ar/^{39}Ar$ age for Granites of the Mourne Mountains, Ireland. *Scottish Journal of Geology*, **23**, 215–220, https://doi.org/10.1144/sjg23020215

THOMPSON, S.J. 1979. *Preliminary Report on the Ballymacilroy No. 1 Borehole, Co. Antrim, Ahoghill.* Geological Survey of Northern Ireland, Open-file Report, **63**.

THORPE, R.S., TINDLE, A.G. & GLEDHILL, A. 1990. The petrology and origin of the Tertiary Lundy Granite (Bristol Channel, UK). *Journal of Petrology*, **31**, 1379–1406, https://doi.org/10.1093/petrology/31.6.1379

TORSVIK, T.H., MOSAR, J. & EIDE, E.A. 2001. Cretaceous–Tertiary geodynamics: a North Atlantic exercise. *Geophysical Journal International*, **146**, 850–866.

TROLL, V.R., NICOLL, G.R., DONALDSON, C.H. & EMELEUS, H.C. 2008. Dating the onset of volcanism at the Rum Igneous Centre, NW Scotland. *Journal of the Geological Society, London*, **165**, 651–659, https://doi.org/10.1144/0016-76492006-190

UPTON, B.G.J. 1988. History of Tertiary igneous activity in the N Atlantic borderlands. *In*: MORTON, A.C. & PARSON, L.M. (eds) *Early Tertiary Volcanism and the Opening of the NE Atlantic*. Geological Society, London, Special Publications, **39**, 429–453, https://doi.org/10.1144/GSL.SP.1988.039.01.38

UPTON, B.G.J., EMELEUS, C.H., REX, D.C. & THIRLWALL, M.F. 1995. Early Tertiary magamatism in NE Greenland. *Journal of the Geological Society, London*, **152**, 959–964, https://doi.org/10.1144/GSL.JGS.1995.152.01.13

VERATI, C. & JOURDAN, F. 2013. Modelling effect of sericitization of plagioclase on the $^{40}K/^{40}Ar$ and $^{40}Ar/^{39}Ar$ chronometers; implication for dating basaltic rocks and mineral deposits. *Geological Society, London, Special Publications*, **378**, https://doi.org/10.1144/sp378.14

WAAGSTEIN, R. 1988. Structure, composition and age of the Faeroe basalt plateau. *In*: MORTON, A.C. & PARSON, L.M. (eds) *Early Tertiary Volcanism and the Opening of the North-East Atlantic*. Geological Society, London, Special Publications, **39**, 225–238, https://doi.org/10.1144/GSL.SP.1988.039.01.21

WAAGSTEIN, R., GUISE, P. & REX, D. 2002. K/Ar and $^{40}Ar/^{39}Ar$ whole-rock dating of zeolite facies metamorphosed flood basalts: the Upper Paleocene basalts of the Faeroe Islands, NE Atlantic. *In*: JOLLEY, D.W. & BELL, B.R. (eds) *The North Atlantic Igneous Province: Stratigraphy, Tectonic, Volcanic and Magmatic Processes*. Geological Society, London, Special Publications, **197**, 219–252, https://doi.org/10.1144/GSL.SP.2002.197.01.09

WALKER, G.P.L. 1959. Some observationa on the Antrim basalts and associated dolerite intrusions. *Proceedings of the Geologists' Association*, **70**, 179–205.

WERNER, R., BOGAARD, P.V.d., LACASSE, C. & SCHMINCKE, H.U. 1998. Chemical composition, age, and sources of volcaniclastic sediments from Sites 917 and 918. *In*: SAUNDERS, A.D., LARSEN, H.C. & WISE, S.W., JR (eds) *Proceedings of the Ocean Drilling Program, Scientific Results, Volume 152*. Ocean Drilling Program, College Station, TX.

WHITE, R. & MACKENZIE, D. 1989. Magmatism at rift zones: the generation of volcanic continental margins and flood basalts. *Journal of Geophysical Research*, **94**, 7685–7729.

WIGNALL, P.B. 2001. Large igneous provinces and mass extinctions. *Earth-Science Reviews*, **53**, 1–33.

Regional distribution of volcanism within the North Atlantic Igneous Province

JIM Á HORNI[1*], JOHN R. HOPPER[2], ANETT BLISCHKE[3], WOLFRAM H. GEISLER[4], MARGARET STEWART[5], KENNETH MCDERMOTT[6,7], MARIA JUDGE[8], ÖGMUNDUR ERLENDSSON[3] & UNI ÁRTING[1]

[1]*Jarðfeingi – Faroese Geological Survey, Jóannesar Paturssonargøta 32-34, Postbox 3059, FO-110 Tórshavn, Faroe Islands*

[2]*Geological Survey of Denmark and Greenland, Øster Voldgade 10, DK1350 Copenhagen K, Denmark*

[3]*Iceland GeoSurvey, Branch at Akureyri, Rangárvöllum, 602 Akureyri, Iceland*

[4]*Alfred Wegener Institute, Helmholtz Centre for Polar and Marine Research, Am Alten Hafen 26, 27568 Bremerhaven, Germany*

[5]*British Geological Survey, Lyell Centre, Research Avenue South, Edinburgh EH14 4AP, UK*

[6]*University College Dublin, Stillorgan Rd, Belfield, Dublin 4, Ireland*

[7]*Present address: ION, 31 Windsor Street, Chertsey, Surrey KT16 8AT, UK*

[8]*Geological Survey of Ireland, Beggars Bush, Haddington Road, Dublin D04 K7X4, Ireland*

**Correspondence: jim.a.horni@jardfeingi.fo*

Abstract: An overview of the distribution of volcanic facies units was compiled over the North Atlantic region. The new maps establish the pattern of volcanism associated with breakup and the initiation of seafloor spreading over the main part of the North Atlantic Igneous Province (NAIP). The maps include new analysis of the Faroe–Shetlands region that allows for a consistent volcanic facies map to be constructed over the entire eastern margin of the North Atlantic for the first time. A key result is that the various conjugate margin segments show a number of asymmetric patterns that are interpreted to result in part from pre-existing crustal and lithospheric structures. The compilation further shows that while the lateral extent of volcanism extends equally far to the south of the Iceland hot spot as it does to the north, the volume of material emplaced to the south is nearly double of that to the north. This suggests that a possible southward deflection of the Iceland mantle plume is a long-lived phenomenon originating during or shortly after impact of the plume.

Pre-breakup volcanism from the North Atlantic Igneous Province (NAIP) extends over a pre-drift east–west distance of approximately 2000 km from Baffin Island, Canada to Lundy Island, Bristol Channel, UK (Saunders *et al.* 1997) while breakup volcanism extends over a NE–SW distance of nearly 3000 km from Lofoten/East Greenland Ridge in the north to Cape Farewell/Edoras Bank in the south (Fig. 1). Although detailed mapping of the NAIP has been published for smaller local areas (Planke & Alvestad 1999; Planke *et al.* 2000; Berndt *et al.* 2001) and simple regional maps have been published (Eldholm & Grue 1994; Saunders *et al.* 1997; Brooks 2011), the vast area covered by volcanism is a challenge for developing consistent mapping of various volcanic features in a systematic way (Fig. 2).

Nevertheless, mapping of the volcanic facies and understanding their emplacement environment and magmatic processes play an important role in the understanding of rifting and tectonic processes associated with the generation of large igneous provinces. Indications for subaerial v. subaqueous extrusion and indicators for water depth provide information regarding palaeogeography during breakup and initial rifted margin evolution (e.g. Wright *et al.* 2011). In addition, detailed mapping of key units is the first step towards establishing the volume of material extruded and intruded, which when combined with accurate geochronology

From: Péron-Pinvidic, G., Hopper, J. R., Stoker, M. S., Gaina, C., Doornenbal, J. C., Funck, T. & Árting, U. E. (eds) 2017. *The NE Atlantic Region: A Reappraisal of Crustal Structure, Tectonostratigraphy and Magmatic Evolution.* Geological Society, London, Special Publications, **447**, 105–125.
First published online July 11, 2017, https://doi.org/10.1144/SP447.18

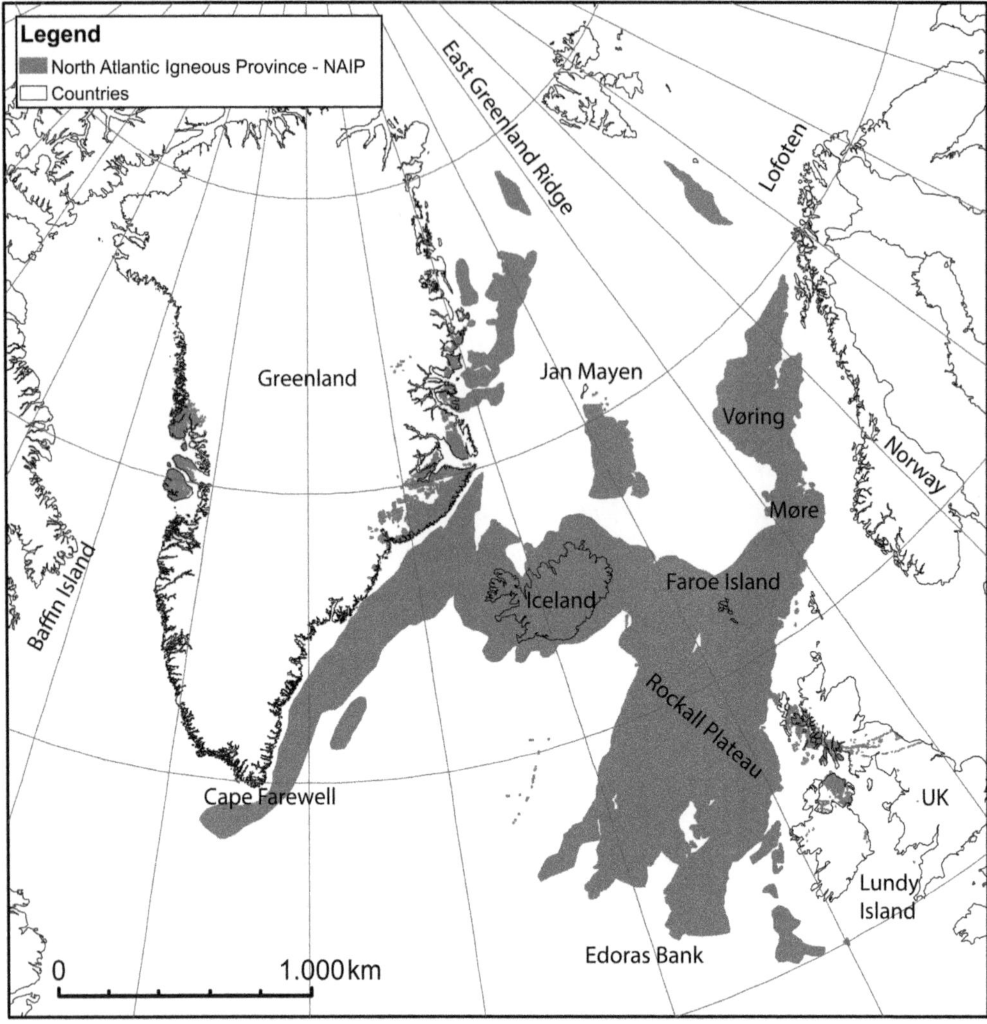

Fig. 1. Overview map of the North Atlantic showing the outline of the NAIP based on this work.

can constrain productivity rates. Together with geochemistry, the combined information places constraints on the mantle thermal and chemical structure associated with volcanic margin formation.

It is increasingly clear that NAIP volcanism is associated with a mantle thermal anomaly (e.g. Nielsen & Hopper 2002, 2004; Brown & Lesher 2014). Thus volcanism is typically interpreted in terms of mantle plume dynamics. The role of the overlying crust and lithosphere is generally disregarded. Several papers, however, have suggested that the lithosphere may play a key role in determining when and where volcanism occurs when a mantle plume impacts the base of the lithosphere (e.g. Ebinger & Sleep 1998; Nielsen *et al.* 2002; Sleep 2002). Regional mapping of volcanism and volcanic facies provides information on the patterns and distribution of volcanism that may reflect pre-existing lithospheric and crustal structure.

This contribution presents a regionally consistent interpretation of the main volcanic facies that can be mapped along the North Atlantic margins from approximately the Bight Fracture zone in the south up to the Fram Strait and SW Barents Sea margin to the north. Several areas that have not been considered in previous work are included: NE Greenland margin, the Jan Mayen microcontinent, the Greenland–Iceland–Faroe Ridge Complex (GIFRC), and Faroe–Shetlands platform. The former three regions are discussed in Blischke *et al.* (2016), Geissler *et al.* (2016) and Hjartason

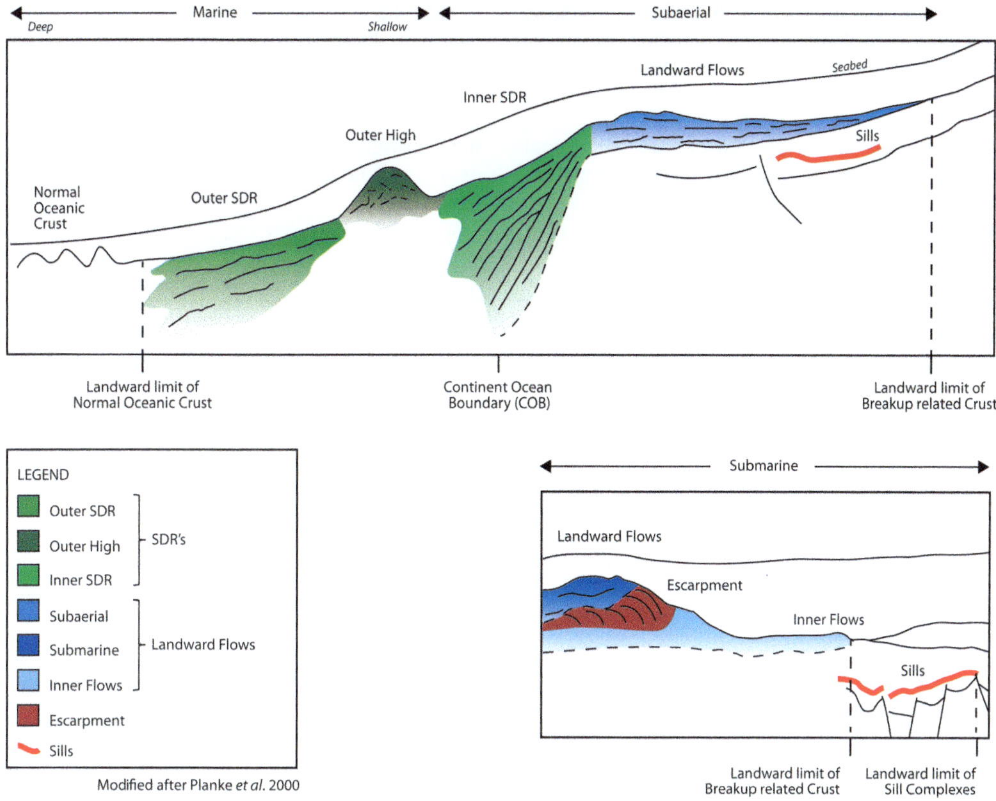

Fig. 2. Schematic model showing the different facies. The typical transition from subaerial to marine environment is illustrated in the upper panel. The lower panel shows the typical transition from landward flow to inner flow divided by an escarpment.

et al. (2017) and therefore are not discussed here. The new mapping around the Faroes is discussed in more detail and provides a regionally consistent view of the entire eastern margin from the Hatton margin up to the Lofoten margin.

The new mapping highlights a number of regional asymmetries in volcanism that characterize the NAIP. We suggest that these asymmetries are the result of the pre-existing lithospheric and crustal architecture of the region, which had an important influence on determining the lateral flow of hot material and played a first-order role in determining the pattern of surface volcanism observed in the region.

Seismic facies units

The mapping is based mainly on seismic volcanostratigraphy; however, gravity and magnetic anomalies provide supplementary information about possible distribution of volcanic rocks in areas with sparse seismic data.

Seismic volcanostratigraphy is the study of the nature and geological history of volcanic rocks and their emplacement environment from seismic data. Volcanic seismic facies units are based on their shape, reflections patterns and boundary reflections (Planke *et al.* 2000; á Horni *et al.* 2014). In areas well covered by seismic data and wells, it is possible to construct detailed maps with many facies. Here, however, we focus on a simplified scheme to provide a basic regional overview of the distribution of volcanism. The scheme is based on the work of Planke *et al.* (2000) (Fig. 2). Five primary units are considered and described briefly below: landward flows, escarpments, seaward dipping reflector (SDR) sequences, sills, and intrusions and igneous centres. Along the eastern margins, a subdivision of some units is included as outlined below.

Landward flows

Landward flows are composed of both subaerial and submarine lavas. Subaerial lavas are erupted onshore and are largely the products of fissure or vent eruptions. They are often seen to flow downslope for several hundreds of kilometres. Submarine lavas are lavas of the same source but emplaced on, or close to, the coast in a submarine environment (Self *et al.* 1997).

Subaerial landward flows are commonly identified on seismic sections as a strong, fairly smooth, top reflector (Fig. 3a). The external shape is sheet-like, whereas internal reflections are disrupted or hummocky and sub-parallel. The submarine landward flows typically have a rougher top reflector than the subaerial landward flows. The landward-flow unit frequently wedges out in the landward direction and terminates at a regional escarpment or merges with prograding reflections (Fig. 3b).

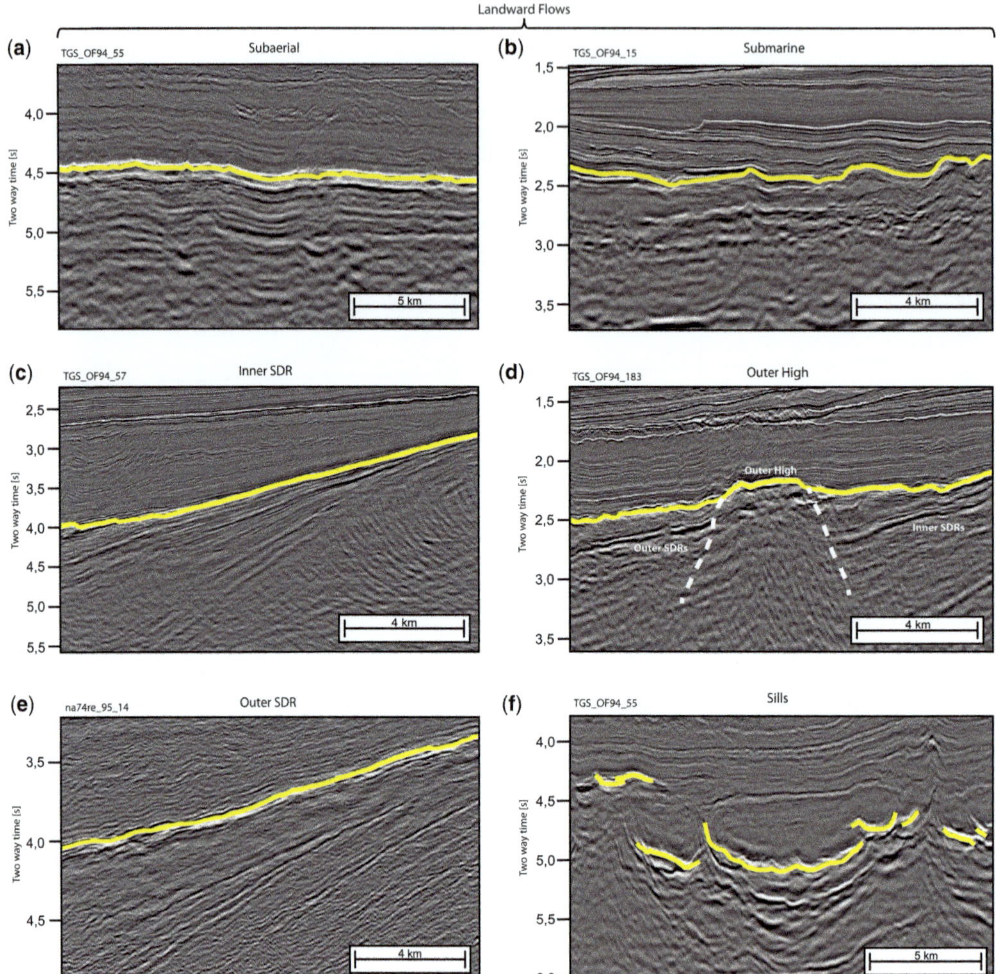

Fig. 3. Examples of seismic sections showing the different volcanic facies. The location of the profiles is not displayed for confidentiality reasons. (**a**) Subaerial landward flows typically are indicated by a strong, smooth and laterally continuous top reflection. (**b**) Submarine landward flows are also indicated by a strong top reflection, but it is rougher and laterally discontinuous. (**c**) Inner SDRs are observed as a wedge-shaped unit with strong internal, dipping reflections capped by a continuous, smooth top reflection. (**d**) An outer high is a mounded feature characterized by a fairly strong top reflector, chaotic internal reflection and is located near the seaward termination of the inner SDRs. (**e**) Outer SDRs are similar to inner SDRs, but are less well developed and are located seaward of an outer high (when present). The top reflection of outer SDRs trends is smooth, but is somewhat less continuous. (**f**) Sills emplaced in sedimentary sections are very distinct on seismic sections because of the significant acoustic impedance contrast between the intrusion and the sedimentary host rock. They frequently have a typical saucer shape.

Escarpments

An escarpment is a feature with steep relief. In volcanic systems, escarpments form typically from eruptive and emplacement processes. Changes in the deposition rate, the type of material erupted, or the available accumulation space are key factors that can control the development of escarpments (Smythe *et al.* 1983; Kiørboe 1999). In a recent review of the Vøring escarpment, Abdelmalak *et al.* (2015) show that major volcanic escarpments form in areas where lava flows encroach into basins. They are thus often seen in association with lava delta fronts and provide a key marker for the palaeoshoreline (Moore *et al.* 1989).

Seaward dipping reflectors (SDRs)

One of the characteristic features of a volcanic rifted margin is presence of well-defined seaward dipping reflections below the top basalt reflection (Fig. 3c–e). SDRs have smooth to hummocky geometries and are interpreted to be subaerial lava flows erupted during an early stage of seafloor spreading. Laterally extensive continuous internal reflections are interpreted as large sheet flows, which are common in subaerial extrusions having a large magma supply that can continuously feed individual lava flows. In areas where they are particularly well developed, they have a concave-down and/or wedge appearance (e.g. Fig. 3e). SDRs form by sequential loading of older flows as rifting and seafloor spreading initiates. They are considered the most important indicator for anomalously large volcanism along rifted margins (e.g. Hinz 1981; Mutter *et al.* 1982).

Along many margins, two distinct sets of SDRs are often observed. The inner, landward set typically terminates at an outer high, followed by a second, outer set. The outer high is a mounded feature characterized by a fairly strong top reflection and chaotic internal reflections (Fig. 3d) (Planke *et al.* 2000). The outer high is interpreted to represent a transition from a subaerial to subaqueous environment. Flows erupted in shallow water can be more explosive and form hyaloclastites, preventing the formation of sheet flows that produce SDRs (e.g. Gregg & Fornari 1998). The outer SDRs are interpreted to indicate a transition from shallow to deep water, where the water pressure prevents degassing and large sheets flows can form, similar to the subaerial eruptions (Gregg & Fornari 1998; Planke *et al.* 2000; Hopper *et al.* 2003). The seismic characteristics of both SDR units are similar, however (Fig. 3c, e).

Sills and intrusions

Sills and intrusions are emplaced in existing host rock of any lithology. In this study, shallow intrusions are typically observed as sills intruded into sedimentary sections (Fig. 3f) (Planke *et al.* 2000, 2005; Smallwood & Maresh 2002). These are very distinct on seismic sections because of the significant acoustic impedance contrast between the intrusion and the sedimentary host rock (Fig. 3f).

Magmatic intrusions emplaced into basaltic sections have not been identified on seismic data, but studies from exposures and wells from the Faroese area shows that they are present within the basaltic province (Varming 2009; Hansen *et al.* 2011). However, they are difficult to image on seismic sections because of the similarities in physical parameters between the host rock and the intrusions. Similarly, sills can be intruded into crystalline continental crust, although this is difficult to observe on seismic sections.

Igneous centres

The igneous centres have been divided into six subdivisions: offshore seamounts, igneous complexes, inactive calderas, active calderas, onshore inactive central volcanoes and active central volcanoes (á Horni *et al.* 2014).

Igneous centres represent a persistent volcanic eruptive or vent area that has built a complex combination of volcanic forms over time. These eruptions may be associated with faults or fissures. Igneous centres are identified on seismic sections as a mound or a bank feature with dipping sides and commonly erosional features on the top. In areas with no seismic data, it is possible to infer additional igneous centres from circular gravity anomalies (Passey & Hitchen 2011; á Horni *et al.* 2014).

Results

Figure 4 shows the regional map of main volcanic features and seismic facies. The map presents a regional overview of NAIP volcanism, including an assessment of several areas that have not been considered in detail before. In particular, new mapping of the Faroese platform is included that provides a consistent interpretation connecting the well-mapped Norwegian margin (Berndt *et al.* 2001; Planke *et al.* 2005) to the Rockall–Hatton margins and the UK and Irish margins (Elliott & Parson 2008). The map also shows several aspects of the regional distribution patterns that are discussed further below – including a comparison of the breakup-related volcanic volumes between different margin segments – and a brief discussion on the northern extent of the NAIP.

Faroe–Rockall–Hatton detailed mapping

Figure 5 illustrates the details of the main volcanic facies in the Faroese–Rockall–Hatton area and is

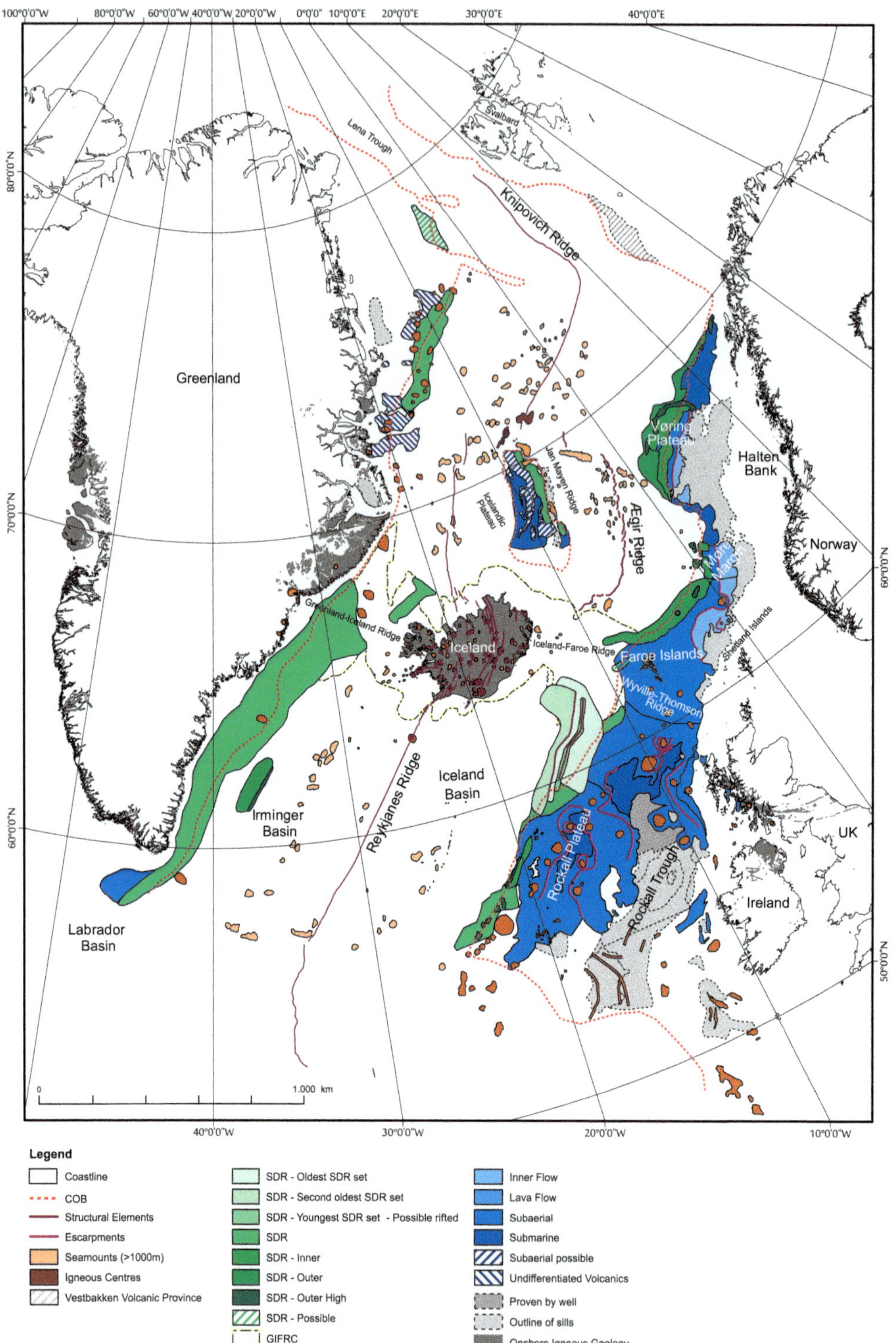

Fig. 4. Volcanic facies map of the NAIP. The COB (Funck *et al.* 2016) is shown as are the main volcanic escarpments.

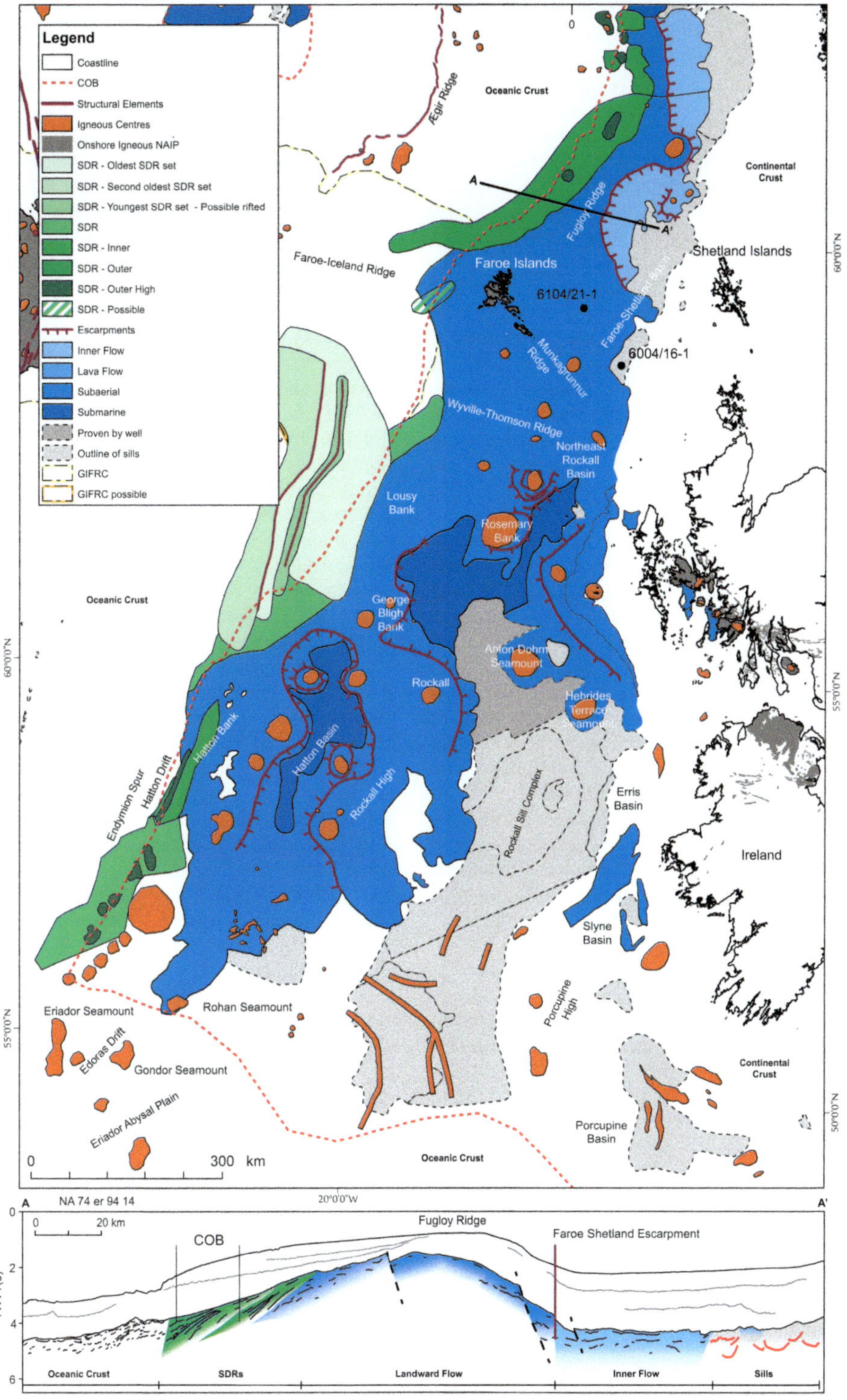

Fig. 5. Map of the volcanic facies in the Faroe–Rockall area and a cross section showing an example of the volcanic facies on a geo-seismic profile.

mainly based on a compilation of previously published maps (á Horni *et al.* 2014) with updated seismic interpretation in several areas.

The area is dominated by basaltic volcanic rocks derived from many different sources, including numerous volcanic centres (Hitchen 2004; Hitchen & Johnson 2013). In terms of seismic data, the top surface of the lavas produces an easily mappable prominent reflection, which shows considerable variation in depth across the margin. The top reflection occurs at or near to the seabed on structural highs and shelves, but is at depths greater than 3 km below the seabed in the northern part of the Rockall Basin. Most of the lavas are interpreted to be subaerial (Hitchen & Johnson 2013). Thus in the Rockall Basin, significant post-early Eocene subsidence has occurred. However, it should be noted that lavas interpreted to be of submarine origin have also been described from parts of the basin (á Horni *et al.* 2014).

The Faroe Island basalt group (FIBG) is mainly composed of landward flows, which are an expression of large-scale volcanic activity which covered an area of approximately 120 000 km^2 (Figs 4 & 5). The FIBG is estimated to have a total stratigraphic thickness of 6.6 km (Passey & Jolley 2009). In offshore commercial wells, wireline logs indicate that the subaerial lavas have a thickness of 795 m and the hyaloclastites a thickness of 778 m (Brugdan well – 6104/21-1) (Fig. 5). Well data further shows that the volcanic material thins towards the SE. In the Marjun well (6004/16-1), there are no subaerial lava flows, only tuff and some intrusions (Fig. 5). It has been suggested that approximately 2000 m of the FIBG have been eroded in the western part of the islands and around 200–300 m in the southern part of the islands (Waagstein 1988; Andersen *et al.* 2002; Jørgensen 2006).

SDR sequences are observed oceanward at the transition into the Norway Basin and Iceland Basin to the NW and to the SE, respectively (Fig. 4). SDRs, however, are not identified on seismic data to the west and NW where the Faroese platform connects to the Faroe–Iceland Ridge. SDRs typically form near the continent–ocean boundary (COB) and are associated with rapid subsidence as spreading progresses. The lack of well-imaged SDRs towards the Faroe–Iceland Ridge may indicate that along the ridge, little subsidence occurred as spreading was established. On the conjugate side, the main SDRs along SE Greenland become poorly defined towards the Greenland–Iceland Ridge. Both of these aseismic ridges are characterized by extremely thick igneous crust that is comparable to Iceland today (Smallwood *et al.* 1999; Holbrook *et al.* 2001). This suggests that well-developed SDRs form primarily along margin segments where the melt supply decreases as spreading becomes established, resulting in a rapid reduction in crustal thickness.

The Faroe–Shetland Basin, which is to the east of the main landward flows, is highly intruded by sills and dykes (Figs 4 & 5). This intrusive complex is referred to as the Faroe–Shetland sill complex and its mapped extent covers an area of approximately 22 500 km^2 (Passey & Hitchen 2011). The sill complex, however, covers a much greater area as has been proven by well data showing that sills are found below the landward flows within the Faroese area (á Horni *et al.* 2015). Sills are also found onshore on the Faroe Islands (Hansen *et al.* 2011).

Close to the eastward edge of the landward-flow lavas, a major continuous escarpment has been mapped crossing the border to both the Norwegian and UK areas and is referred to as the Faroe–Shetland escarpment (Figs 4 & 5) (Smythe 1983). A vertical relief of up to 1000 m is observed across the escarpment. Regional escarpments are only well mapped along the eastern margins of the North Atlantic. Along SE Greenland and around Jan Mayen, no escarpments have been mapped. Along NE Greenland, poorly developed escarpments are indicated, but show distinct differences compared to the conjugate margins – see Geissler *et al.* (2016) for a discussion. The escarpments are found between the inner flows and landward lava flows (Figs 2 & 5) and are interpreted to represent a period of volcanism associated with subsidence below sea level, generating prograding lava deltas into shallow water, thus marking a palaeoshoreline (e.g. Wright *et al.* 2011). The age of the escarpment is not known from direct dating. Stratigraphically, it is within the uppermost Beinisvørð Formation close to the break with the Malinstindur Formation of the FIBG (Smythe *et al.* 1983; Kiørboe 1999; Passey & Hitchen 2011). This break probably correlates to a volcanic hiatus observed throughout the NAIP from 58 to 56 Ma (Graham *et al.* 1998; Storey *et al.* 2007; Larsen *et al.* 2015). Thus, the escarpment may indicate a period of subsidence and submergence prior to the main flood volcanism and final breakup at 56 Ma.

Spatial distribution of volcanism

The regional volcanic facies maps present a new view of the spatial patterns of NAIP extrusive volcanism that provide insight into the development of volcanism prior to and during breakup. In this section, the spatial pattern of pre-breakup volcanism is briefly described; the volumes and production rates of breakup volcanism are then analysed. The results are compared to previous work regarding volume and productivity estimates along with early observations of asymmetric patterns.

The implications of these patterns of volcanism in terms of lithosphere plume interaction are then discussed.

Pre-breakup volcanism. Pre-breakup to breakup volcanism is best shown by the mapped landward lava flows and inner flows, which cover much of the Faroese, UK and Irish margins over a broad area. Landward flows are also observed to a lesser extent along the Greenland, Jan Mayen and Norwegian margins. It is difficult to estimate the volume of pre-breakup volcanism since generally the base of the basalt is not mappable and so thicknesses are poorly constrained.

In addition, it is difficult in these regions to separate pre-breakup volcanism from breakup- volcanism. Where sampled along the Rockall–Hatton margins, dating of volcanic rocks from the landward and inner flows shows that these units include both pre-breakup and breakup volcanic rocks and that the pre-breakup phase is less extensive (Hitchen & Johnson 2013). It should be noted that onshore areas in West Greenland, East Greenland, the Faroe Islands, and parts of Scotland and Northern Ireland also include pre-breakup volcanic rocks. These areas are marked by dark grey in Figure 4. The maps show that pre-breakup volcanism affects only local areas except for the region around the Rockall Basin and the Faroes platform. Pre-breakup to breakup volcanism affected a total area of approximately 6×10^5 km^2, of which more than half (3.2×10^5 km^2) is along the eastern margins south of the Wyville Thompson Ridge.

Breakup volcanism. To estimate the volumes and production rates of breakup-related volcanism, several key observations are required: the area covered by erupted basalt together with their intrusive counterparts; thickness estimates based on seismic reflection and refraction data; and estimates of timing based on both geochronology of basalt samples and interpretation of magnetic spreading anomalies.

Eldholm & Grue (1994) considered the areal coverage of flood basalts together with thickness determinations to estimate the volume of extrusive basalt of the NAIP. By assuming a distribution of high velocity lower crust based on limited refraction profiles, the total volcanic volume was constrained. Combined with available timing constraints, they then estimated the average eruptive rates for the entire province. While the distribution of extrusive rocks is fairly straightforward to constrain, a significant uncertainty arises from assumptions regarding the distribution of high velocity lower crust along-strike, discussed further below. Holbrook *et al.* (2001) and Breivik *et al.* (2009) considered igneous thickness variations interpreted along well-resolved 2D seismic refraction profiles to estimate volumes per km along-strike. The work of Holbrook *et al.* (2001) is based on profiles on the SE Greenland margin and provides constraints on the region south of Iceland, whereas Breivik *et al.* (2009) analysed a profile along the northern Vøring margin and obtained an estimate that applies to the Vøring margin. These latter two approaches provide accurate estimates locally, but interpolating and extrapolating the crustal structure along-strike presents problems in many key areas.

The approach here is to use the mapped SDR sequences to obtain an estimate of the volumes and production rates. In general, the inner SDRs mark the main part of the volcanic transition zone along a volcanic rifted margin. In terms of volumetric significance, they mark the location where the bulk of the breakup-related volcanism occurred. Previous work on volcanic rifted margins shows that the transition from rifting to full seafloor spreading is closely associated with the formation of the main, inner SDRs (e.g. Hinz 1981; Mutter *et al.* 1982). Along the North Atlantic margins, they straddle the COB and the first identifiable seafloor spreading anomalies are typically within this volcanic facies unit. For example, along both the Vøring and SE Greenland margins, magnetic anomaly C24 is well identified and within the mapped inner SDRs (Hopper *et al.* 2003; Breivik *et al.* 2009; respectively). The COB is typically located near to where the thickest new igneous crust is observed, since oceanward the entire crust is volcanic whereas continentward, only a portion of the crust is volcanic.

In addition, the inner SDRs are closely associated with the high velocity lower crust, where P-wave velocities range up to 7.5–7.6 km s^{-1} (e.g. Mjelde *et al.* 2007). High velocity lower crust along volcanic rifted margins is normally interpreted as magmatic underplating, although there remains some discussion on precisely what this means and how an underplate is emplaced. As noted by White *et al.* (2008), the term 'underplate' implies that the entire high velocity body represents an igneous cumulate and thus is 100 per cent volcanic. They show, however, that along a seismic profile near the Faroe Islands, the high velocity region most likely consists of sills intruded into the continental crust. Thus the high velocity region landward of the COB may include less breakup-related igneous material than assumed by the pure underplate model.

High velocities in continental lower crust, it should be noted, are not necessarily diagnostic for magmatic underplating. For example, the lower crust beneath Lake Baikal – an intracontinental rift in central Russia – is characterized by a lower crust with velocities from 7.05 to 7.4 km s^{-1}, which ten Brink & Taylor (2002) attribute to an unthinned part of the Siberian Platform. In the case of volcanic margins, proximity to a major and anomalous volcanic event is a key argument to support the underplating model. Recently, the suggestion that parts of the high velocity lower crust on some margins may be related to hyperextension has raised the possibility that some high velocities may represent serpentinized mantle below very thin continental crust (e.g. Osmundsen & Ebbing 2008). This complicates the interpretation of rifted margins that may have undergone a phase of protracted hyperextension prior to volcanism and final breakup as suggested for the North Atlantic (e.g. White *et al.* 2008; Lundin & Doré 2011). Lundin & Doré (2011) propose that the anomalously wide zone of high velocity lower crust below the Vøring margin can be explained if the innermost part of the margin is underlain by serpentized mantle and only the outer part of the margin includes intruded lower crust.

Given the large uncertainties over how much of the high velocity lower crust along the Atlantic margins is related to volcanism, and, if it is indeed related, how much of the lower crust is composed of new igneous intrusions, the approach here is to not include mapping of the lower crust directly in the volume calculations. Nevertheless, the bulk of the high velocity lower crust is included in an ad hoc way since much of it falls within the assessment polygons used to estimate volcanic volume. This is explained further below in the description of the assessment polygons using an example from the SE Greenland margin.

To assess the volume and productivity of breakup volcanism, the North Atlantic is broken down into distinct conjugate margin segments to better quantify the differences between the various areas. Four distinct segment pairs are recognized: SE Greenland and Hatton–Rockall margins; East Greenland and the Faroe Islands along the GIFRC; the Jan Mayen microcontinent and the Møre margin; and NE Greenland and the Vøring–Lofoten margins. Holbrook *et al.* (2001) note a dramatic difference in igneous crustal thickness that develops over the GIFRC compared to other areas that is attributed to proximity to the Iceland plume track. They divide the North Atlantic into proximal margins and distal margins based on the influence of active upwelling from the Iceland plume. In this analysis, conjugate comparisons are focused on the distal margins for the purposes of examining transient volcanic margin formation away from the plume track.

Except for the GIFRC, which is described separately, assessment polygons are constructed for each margin segment based on the mapped inner SDRs and magnetic Chron C23n.2no (51.826 Ma; (Ogg 2012; Gaina *et al.* 2016)). The polygons constructed are shown in Figure 6. Along each margin segment, the COB based on Funck *et al.* (2014) falls within the SDRs, consistent with the general understanding that they are closely associated with the initiation of seafloor spreading. Seaward of the COB, the entire crust up to Chron C23n.2no is newly accreted intrusive and extrusive volcanic material and the volume can be calculated directly from the area of the oceanward assessment polygon combined with the regional crustal thickness derived from wide-angle seismic data (Funck *et al.* 2016). In these areas, all of the high velocity lower crust is included in the estimate.

Landward of the COB, the total volume of the volcanic rocks is more difficult to estimate and several simplifying assumptions must be made. In general, the amount of volcanism decreases rapidly landward and the extrusive layer of SDRs thins before pinching out entirely. In general, the high velocity lower crust also thins landward. An example from SE Greenland is shown in Figure 7. This example is based on the seismic reflection and refraction profile of Hopper *et al.* (2003). The inner SDRs span a distance of over 100 km and the seawardmost extent are emplaced within unambiquous oceanic crust (Chron C24, see Hopper *et al.* (2003)). At the landward end, the SDRs thin and pinch out. A high velocity lower crust is observed that also thins and pinches out. The location of the COB along this profile is poorly constrained. The COB – based on the seismic velocity structure – is far landward, approximately 50 km further landward than the first clear spreading anomaly (Hopper *et al.* 2003). From this point and landward, the transition zone comprises both continental crust and igneous intrusive and extrusive rocks associated with breakup. The volume of crust from the seismic COB to the pinchout of the extrusive layer is *c.* 930 $km^3\ km^{-1}$, of which *c.* 130 $km^3\ km^{-1}$ is the extrusive basalts. The question remains how much of the high velocity lower crust is intrusive v. continental crust. Assuming a pure underplate from 7.0 $km\ s^{-1}$ to the Moho, the volume of intrusives is *c.* 320 $km^3\ km^{-1}$. Thus, approximately half of the transition zone crust is accreted volcanic rock. The ratio of intrusive to extrusive rock is 2.4:1, very similar to typical oceanic crust (White *et al.* 1992), but much higher than expected if the lower crust is sill intruded continental crust (e.g. White *et al.* 2008).

Based on the SE Greenland example, it is assumed that half of the volcanic transitional crust, defined as the area between the COB and the pinch-out of the inner SDR, is newly accreted material associated with breakup. This approach has slightly different implications for each margin segment. However, given the previously noted uncertainties in constraining volcanic volumes, this assumption

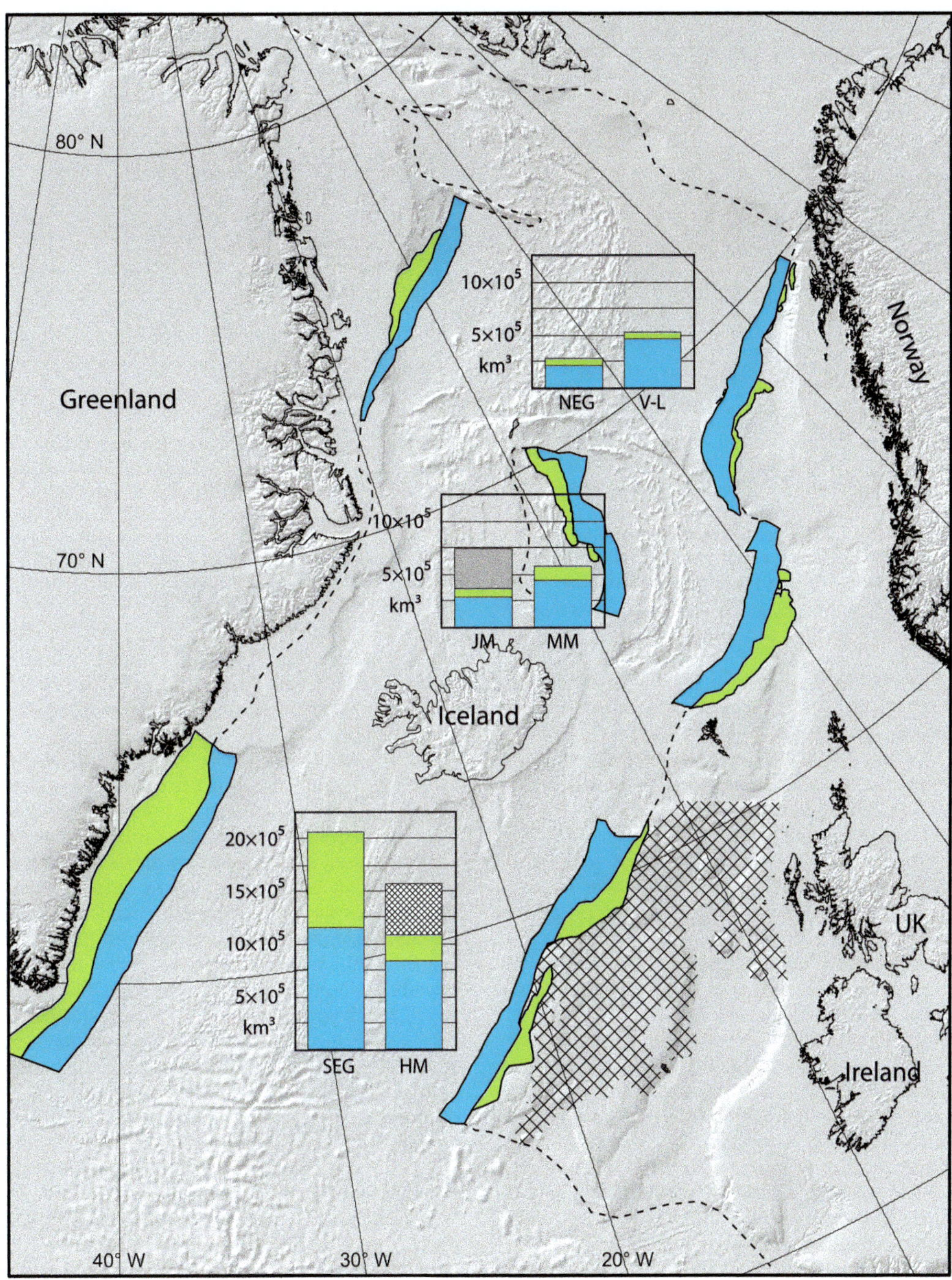

Fig. 6. Breakup-related volcanism around the North Atlantic. Blue polygons are oceanic crust from the COB (*c.* 56 Ma, dashed line) to magnetic Chron 23r (*c.* 52 Ma); green polygons are mapped SDR sequences located over continental crust. Hashed areas are landward and inner flows along the Rockall–Hatton margins south of the Wyville Thompson Ridge. Histograms show the calculated volumes of igneous accretion. The additional grey component for Jan Mayen is the estimated volume of onshore flood basalts from East Greenland. See text for details. NEG, NE Greenland margin; V-L, Vøring-Lofoten margin; JM, Jan Mayen microcontinent; MM, Møre margin; SEG, SE Greenland margin; HM, Hatton margin.

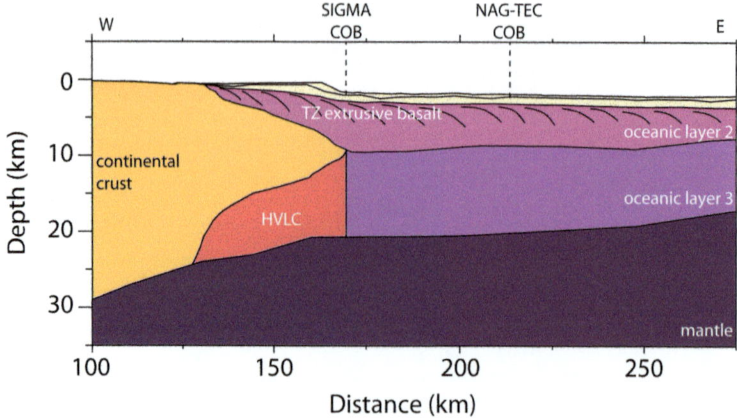

Fig. 7. Schematic cross section along the SIGMA III profile along the SE Greenland margin highlighting the key aspects of a typical volcanic margin. The volcanic transition zone (TZ) is defined as the area where both continental crust and breakup-related volcanic rocks are observed landward of the COB. The SIGMA COB is from Hopper *et al.* (2003) and the NAG-TEC COB is from the NAG-TEC Atlas (Funck *et al.* 2014). Magnetic Chron C24r is located at km 230. Oceanic layer 3 merges into the high velocity lower crust (HVLC) below continental crust. The HVLC thins and pinches out landward approximately where the extrusive basalts also pinch out. From km 125 to the SIGMA COB, about half the crust is volcanic assuming that the HVLC is magmatic underplate. See text for additional discussion.

is broadly consistent with previous estimates and provides a basis for relative comparisons between margin segments. Two consequences of this assumption are: (1) volcanic volumes associated the landward flows and other volcanic units landward of the SDRs are missing; and (2) any intrusive rocks landward of the SDR pinchout are missing. Intrusive rocks associated with the bulk of the high velocity lower crust are included in an ad hoc fashion as described above. In addition, the misplacement of the COB will lead to further errors. If placed too far landward, the volcanic volume may well be overestimated since the entire crust is assumed volcanic. In contrast, if placed too far seaward, the volcanic volume will be underestimated.

As previously discussed, landward flows are locally important, but only the Rockall–Hatton margins show significant areal coverage that potentially has a large impact on the volume estimate. Estimating volumes of these units is complicated by the fact that in most areas, the thickness of the landward flows is unconstrained except where drilled, and there is no straightforward way to estimate sill volumes. In addition, some portion of these units includes pre-breakup volcanism. On all margins except for Rockall– Hatton, these units are therefore not included. Where drilled along Rockall, Hatton and the Faroes, the thickness of landward lava flows is commonly more than 1 km (Varming 2009), and on the Outer Hebrides High, 2–2.5 km of lava flows are estimated (Stoker *et al.* 2012). For this exercise, a 1.5 km-thick layer of basalt is included for the Rockall– Hatton margins where landward flows and sills are mapped (see Figs 4 and 6).

The issues of COB placement and high velocity lower crust potentially affect two areas of investigation. In general, the COB is poorly constrained along much of the East Greenland margin. Discrepancies between the placement of the COB based on magnetic anomaly interpretations v. seismic velocity structure have already been noted along the SE Greenland margin. To avoid overestimating the volcanic volume and productivity here, the COB has been placed 50 km seaward from that interpreted by Hopper *et al.* (2003). This better reconciles misfits to reconstructions based on magnetic spreading anomalies (Gaina *et al.* 2016), and takes into account the possibility that hyperextended continental crust underlies part of the Greenland margin as proposed by White *et al.* (2008). Along NE Greenland, Voss *et al.* (2009) suggest that the NE Greenland margin in the vicinity of the Jan Mayen fracture zone broke up much later than further north. A clear Chron C24 anomaly to the north becomes indistinct to the south, and the first clear anomaly in the southern area is C21. They place the COB further seaward than plate reconstructions suggest (Gaina *et al.* 2009). The entire high velocity crust in their interpretation forms transitional crust. Here, the COB is placed 60 km landward from estimates by Voss *et al.* (2009) and 25 km seaward from those by Gaina *et al.* (2009) as a compromise between the seismic evidence and the magnetic anomaly evidence. Thus, in terms of the Voss *et al.* (2009) interpretation, the volumes here are overestimated, since a significant portion of transition zone crust is assumed 100 per cent volcanic, whereas in the Gaina *et al.* (2009) interpretation, the volume is underestimated.

The East Greenland margin north of Iceland is further complicated by the double breakup associated with the formation of the Jan Mayen microcontinent (see Blischke *et al.* 2016). High velocity lower crust and anomalously thick oceanic crust is observed off East Greenland conjugate to the Jan Mayen microcontinent (e.g. Weigel *et al.* 1995; Voss *et al.* 2009). However, the oceanic crust is associated with the second breakup event (see age grids and magnetic spreading anomalies of Gaina *et al.* (2016)). Onshore, tholeiitic lavas have been dated to 56–53 Ma and are thus clearly related to the first breakup event and separation of the eastern Jan Mayen microcontinent from the Møre margin. Additional volcanism continues until the Miocene (Larsen *et al.* 2014). Given the protracted history of volcanism and possibility that much of the high velocity lower crust is associated with the second breakup event (e.g. Voss *et al.* 2009), this region is considered separately in the volume estimation.

Finally, to allow for comparison to previous work, the volumes and productivity of the proximal region over the plume track are estimated to constrain the breakup-related volcanism of the entire province. The main complication here is that spreading anomalies are not clearly identifiable along the GIFRC. This is in part a result of repeated ridge relocation in response to the relative motion between the mid-ocean ridge spreading system and the Iceland mantle plume. In addition, along the GIFRC the location of the COB is unconstrained on the Greenland side, and poorly constrained on the Faroes side, so constructing a reasonable assessment polygon as above for the distal margin segments is not possible. Instead, a 400 km-wide plume track is assumed (e.g. Holbrook *et al.* 2001), and the initial opening rates based on plate reconstructions are used to estimate the total volume of new crust. At the time of breakup, the spreading rate between Greenland and Europe was relatively fast before decreasing to the slow rates observed today. Half rates as high as 45 mm a^{-1} have been estimated off East Greenland (Larsen & Saunders 1998) and regional reconstructions suggest values around 30 mm a^{-1} (e.g. Gaina 2014). Richardson *et al.* (1998), Smallwood *et al.* (1999) and Holbrook *et al.* (2001) show that the crustal thickness of the GIFRC is on average approximately 30 km thick. From breakup to anomaly C23n.2no, which spans 4.2 million years, *c.* 3.1×10^6 km^3 of volcanic rock was emplaced along the GIFRC. This does not include the landward flood basalts, which Larsen *et al.* (1999) estimate to have a total volume of 0.25×10^6 km^3, including both the East Greenland and Faroese lava flows. Assuming an intrusive to extrusive ratio of 2:1, an additional 0.75×10^6 km^3 is added to the total volume for the GIFRC.

The results are summarized in Figure 6 and Table 1. The average productivity assumes that the duration of volcanism spans from breakup at 56 Ma to anomaly C23n.2no at 51.8 Ma. All three of the margin segments distal to the Iceland plume track show asymmetry, with one margin showing nearly double the volume of volcanic rock as the other. Given the problems with estimating the amount of volcanism over the transition zone, the plots are further broken down into the oceanic and continental areas. Significantly, the earliest oceanic crust produced shows the same volumetric conjugate asymmetry as the full margin estimate along all three conjugate segment pairs.

Conjugate asymmetry. Between SE Greenland and the Hatton Bank, asymmetry in early accretion has been noted previously (Hopper *et al.* 2003). While there remains some uncertainty over the placement of the COB along Greenland, here it is placed significantly further seaward than Hopper *et al.* (2003). This should account for the possibility that portions of the margin are underlain by hyper-extended crust as proposed by White *et al.* (2008). Despite this, there still appears to be more breakup-related volcanism along East Greenland, even when taking into account possible volumes of landward flows and sills covering the Hatton and Rockall basins. Thus, it seems clear from the analysis here that significant volcanic asymmetry between SE Greenland and the Hatton–Rockall margins is a robust observation.

North of Iceland, the sense of asymmetry is the opposite between NE Greenland and the Vøring–Lofoten margins. Here too, there are large uncertainties regarding the COB placement, especially along the Greenland margin where data are sparse. Assuming the Voss *et al.* (2009) placement of the COB, the volume along NE Greenland is potentially even less and would only make the observed asymmetry more dramatic. If instead, the COB is moved landward to where Gaina *et al.* (2009) place it, the volcanic volume would increase to approximately 0.36×10^6 km^3, still significantly less than the Norwegian margins.

For Jan Mayen, it remains somewhat unclear how much volcanism along East Greenland should be included as part of the GIFRC or as part of the Jan Mayen microcontinent. Assuming that half of the estimated volume in Larsen *et al.* (1999) is from the Greenland side, more symmetric volcanism may be indicated (Table 1). In addition, the sense of asymmetry could be the opposite if a

Table 1. *Volcanic volumes and production rates of North Atlantic margins*

Margin segment	Volume (km^3)	Per cent of pair	Average productivity ($km^3\ km^{-1}\ Ma^{-1}$)	Previous productivity estimates[5]
SE Greenland margin	2.06×10^6	65%	490	576[2]
Hatton–Rockall margins	1.09×10^6	35%	258	125[2]
Jan Mayen MC[6]	0.38×10^6 (0.76×10^6)	39% (56%)	199	
Møre margin	0.59×10^6	61% (44%)	312	
NE Greenland margin	0.29×10^6	36%	107	
Vøring/Lofoten margins	0.53×10^6	64%	195	
Vøring only	0.42×10^6		366	428[3]
North of Iceland	1.79×10^6	36%	388	
South of Iceland	3.14×10^6	64%	748	700[2]
Greenland–Iceland–Faroe Ridge Complex[4]	3.85×10^6		2246	1800[2]
Total NAIP Atlantic margins	8.78×10^6		864	830[1], 800–1000[2]

[1]Eldholm & Grue (1994).
[2]Holbrook *et al.* (2001).
[3]Breivik *et al.* (2009): note that in terms of productivity, the estimate for the Vøring is less than in Breivik *et al.* (2009). However, their estimate is based on a single profile along a relatively magma-rich portion of the Vøring margin, whereas here, the productivity has been averaged across the entire margin and includes magmatically less robust areas.
[4]Includes volumes estimated by Larsen *et al.* (1999) for the onshore East Greenland basalts and the Faroe Island basalts.
[5]It should be noted that productivity estimates are less easy to compare than total volumes, since different authors use different timescales. For example, Eldholm & Grue (1994) consider the volume of crust from breakup to anomaly C23, similar to here. However, they assume 3 myr for this interval compared to 4.2 myr here based on the Ogg (2012) timescale and a breakup time of 56 Ma.
[6]The values in parentheses are the volumes if the Blosseville Coast volcanism is included as part of the Jan Mayen volcanism.

substantial volume of early breakup volcanism occurred along East Greenland in the Jameson Land Basin. From Scoresby Sund up to the western Jan Mayen fracture zone, the areal extent of this region is approximately 27 500 km^2. Mathiesen *et al.* (2000) estimate that 2 km of basaltic lava flows covered the Jameson Land Basin and were subsequently eroded away. High velocity lower crust observed by Weigel *et al.* (1995) could be the intrusive counterpart to the eroded basalts. In that case, 5–6 km of intrusive rocks are unaccounted for in the volume estimates here. As noted earlier, volcanism along this part of the margin continued into the Miocene (Storey *et al.* 2004) and much of the volcanism may be younger. Thus, it remains unclear from the available data if this segment is asymmetric, and if so, which side has more volcanism.

For both margin segments north of Iceland, there are significant uncertainties for many aspects of the estimates here. Thus, while the mapping suggests possible asymmetries, it is clear that additional work and more systematic conjugate analysis are required to investigate this further. A useful exercise would be to re-analyse key conjugate seismic refraction profiles to establish a consistent interpretation of the velocity structure and implications for crustal types and COB placement.

Asymmetric spreading is known from many studies of mid-ocean ridges and back-arc spreading systems (Hayes 1976; Stein *et al.* 1977; Barker & Hill 1980). This is thought to be a response to the migration of the lithosphere over the asthenosphere, skewing the thermal field and leading to differences in the shear traction at the base of the plate that causes the ridge to migrate. In these models, faster accretion is predicted over the cooler plate (Hayes 1976; Barker & Hill 1980). Müller *et al.* (1998) noticed that many areas of asymmetric accretion are associated with proximity to mantle plumes. They showed that the ridge axis migrates towards the warmer, plume-affected mantle resulting in a deficit of accretion on the plate over the plume and a surplus of accretion of the ‘cooler’ plate. Hopper *et al.* (2003) suggested that a similar thermal effect could explain asymmetric accretion between SE Greenland and the Hatton Bank since the Greenland margin is bounded by thick, and presumably cooler, Archean lithosphere. Applying this idea to the north, proximity of the Norwegian margin to the thick, cooler lithosphere of the Baltic shield could result in a similar effect, resulting in a crustal deficit along East Greenland. The kinematic modelling, however, indicates an accretional crustal deficit along the magmatically more robust margin, calling into question how important early ridge migration might be during volcanic margin formation. Along the Norwegian margins, it appears that the excess volumes are primarily expressed as anomalously thick crust along the Vøring margins compared to NE Greenland (Funck *et al.* 2016; Haase *et al.* 2016). Nevertheless, the possibility that the magmatically rich half of the conjugate pairs seems to be the one closest to the stable cratons and thick lithosphere is intriguing and suggests pre-existing lithospheric structure may play a role in asymmetric volcanic margin development.

North–south asymmetry. The variation in volcanism away from Iceland along the North Atlantic margins has been considered previously in a number of studies. Barton & White (1995) noted an apparent symmetric distribution of estimated melt thickness away from the Faroe–Iceland Ridge by comparing profiles from Edoras Bank, Hatton Bank, the Vøring Plateau and the Lofoten Basin margin. This was explained as a decrease in asthenospheric potential temperature away from the plume centre. In contrast, Eldholm & Grue (1994) noted that the areal coverage of basaltic lavas over continental crust is greatest south of Iceland. They suggest decreased melt production to the north as a response to a propagating rift away from a plume impact site beneath central East Greenland. This mechanism also reflects decreasing temperature away from the mantle plume as heat is rapidly advected out of the system by melting after plume impact. Why the same effect does not propagate away from the plume to the south is not explained, however. Similar to Eldholm & Grue (1994), Voss *et al.* (2009) noted that the northward decrease in volcanism is more pronounced than the southward decrease in volcanism away from Iceland. They further note that to north, the volumes indicated by high velocity lower crust are very different than to south, but note the complication with the two breakup events and the later separation of Jan Mayen from East Greenland.

A comparison between the volume of volcanism north of Iceland v. south of Iceland is shown in Figure 8. The results here confirm the observations of Eldholm & Grue (1994) and Voss *et al.* (2009) that there is significant north–south asymmetry in early NAIP volcanism. This is the case for both pre-breakup volcanism, indicated by the mapped landward flows, as well as the main breakup volcanism, indicated by the SDR sequences and the development of the main part of the volcanic margins.

It is well established that the present day influence of the Iceland plume is highly asymmetric (Howell *et al.* 2014). A strong influence to the south along the Reykjanes Ridge is well documented whereas the influence to north is significantly reduced (e.g. Vogt 1971; Ito 2001; Delorey *et al.* 2007). This is typically attributed to a deflection of the Iceland mantle plume to the south, causing stronger plume–ridge interaction. A key unanswered question is when this stronger influence to the south began, and why there is a deflection to south. Based on the patterns of both the pre-breakup and breakup volcanism, it is suggested that this southward-biased influence has existed for the entire history of the NAIP and that this pattern probably reflects pre-existing lithosphere structure.

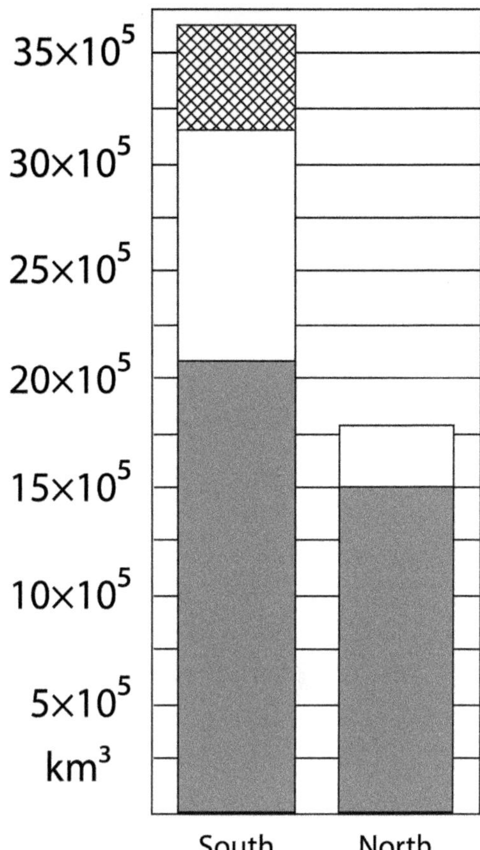

Fig. 8. Histogram summarizing the total volume of material erupted at breakup to the north and to the south of the GIFRC.

Following Nielsen *et al.* (2002), it is assumed that the Iceland plume impinged beneath central Greenland in the Paleocene and is preferentially channelled into thin spots. It is well documented that the proto-North Atlantic experienced a long period of extension from the Late Palaeozoic and throughout the Mesozoic. Figure 9 shows the Cretaceous stratigraphic distribution (Stoker *et al.* 2016) reconstructed to 80 Ma, which highlights the pattern of lithospheric thinning prior to the arrival of the plume beneath Greenland. To the north, rifting follows the Iapetus suture but splits into two distinct branches towards the south, one along the Rockall–Hatton region, and one towards the North Sea. Closer proximity of the Rockall–Hatton region to

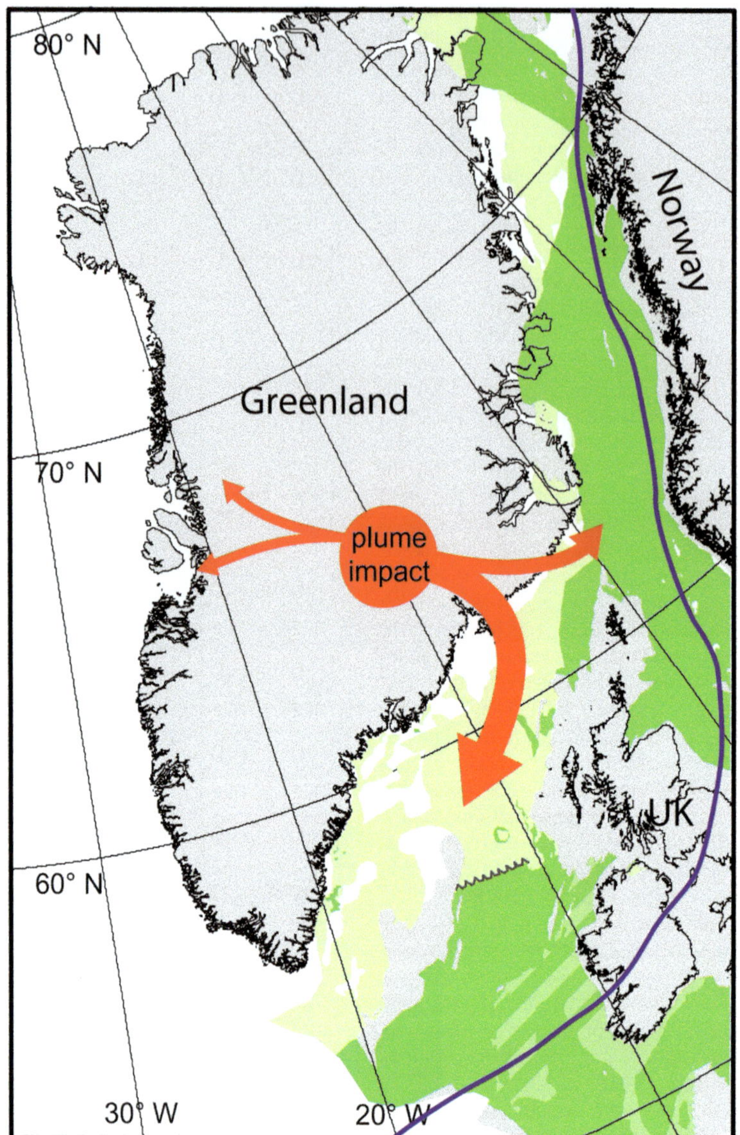

Fig. 9. Distribution of Cretaceous basins reconstructed at 80 Ma, approximately 20 myr prior to plume impact in the Paleocene. Dark green are areas with proven Cretaceous, light green are areas of inferred Cretaceous. Dark blue line is the Iapetus suture. Plume impact location proposed by Nielsen *et al.* (2002) is shown, and arrows indicate lateral flow into lithospheric thin spots. Material is preferentially channelled into the Rockall–Hatton margins as a consequence of the pre-existing lithospheric configuration.

the site of the impacting plume probably led to it being the preferred region for lateral flow of plume material.

The northern limit of NAIP volcanism. In the discussion so far, one feature mapped to the north has been ignored. Along the SW Barents margin, a volcanic unit referred to as the Vestbakken Volcanic Province is shown on most maps of the NAIP (e.g. Abdelmalak *et al.* 2015). Faleide *et al.* (1988) interpret seismic reflection data as showing evidence for a volcanic basement along the shear margin. Volcanism associated with the shear margin is further indicated in exploration well 7316/5-1,where sills intruded into Eocene sediments were encountered (http://factpages.npd.no [Last accessed 26 June

2017]). The sills themselves, however, have not been dated. At another exploration well, 7216/11-1S, tuffs and other volcanic debris are found in Upper Cretaceous through Eocene sediments (Ryseth *et al.* 2003). Additional evidence for volcanism comes from basalts recovered in shallow cores from south of Bjørnøya (Mørk & Duncan 1993). These were dated to Late Pliocene and thus are unrelated to earlier breakup events of early shear margin evolution.

Eldholm *et al.* (2002) note that there is no evidence for SDR units along the shear margin. Berndt *et al.* (2001) show a strong decrease in magmatism to the north along the Norwegian margins. This is also the case along the NE Greenland margin from the Jan Mayen fracture zone towards the East Greenland Ridge (Geissler *et al.* 2016). Further north between the East Greenland Ridge and the Hovgaard Ridge, poorly developed SDRs (or better lava flows) may be identified (Geissler *et al.* in press). These are fairly minor compared to typical development of SDRs in the very magma-rich areas such as the SE Greenland and Vøring margins. In addition, much of the Boreas Basin is interpreted to be underlain by very thin crust with indications for serpentinized mantle. Within the interpreted oceanic areas, the crustal thickness of the region shows normal to thin crust north of Lofoten and the East Greenland Ridge. Overall, the evidence for significant Paleocene to Eocene volcanism is generally lacking in this region. Thus, we question if the Vestbakken Volcanic Province should be considered part of the NAIP. Geochemical data showing an Icelandic signature would be needed to firmly establish a connection between the NAIP and Vestbakken volcanic rocks, but no data exist to determine this.

Conclusion

New mapping of the NAIP establishes the pattern of volcanism associated with breakup and the initiation of seafloor spreading of the region. The new map provides a consistent view of the distribution of volcanism along the eastern margins and presents some of the first maps available for the entire East Greenland margin and Jan Mayen microcontinent. A comparison of conjugate margin pairs shows that the NE Greenland/Vøring–Lofoten margins and the SE Greenland/Rockall–Hatton margins are highly asymmetric, with the bulk of the volume of new igneous crust emplaced on one side. An explanation for this observation remains elusive, but we note that in both cases, the relatively magma-rich margin is in close proximity to thicker lithosphere associated with the stable cratons. Finally, the amount of pre-breakup and breakup volcanism is strongly asymmetric from south to north, with significantly more volcanism south of the GIFRC. Today, the Iceland plume shows a much stronger interaction to the south along the Reykjanes Ridge compared to the Kolbeinsey Ridge to the north. We suggest that this southward bias has persisted since pre-breakup. After impact of the Iceland plume in the Paleocene, lateral flow of ponded plume material was preferentially channelled along the proto-Hatton margin, resulting in the bulk of the magmatism being emplaced to the south. The patterns of volcanism thus show that pre-existing lithospheric structure plays a first-order role in the development of the North Atlantic volcanic margins. Finally, we note that the northern limit of the NAIP appears to be the northernmost Lofoten margin and NE Greenland margin up to the East Greenland Ridge. The Vestbakken Volcanic Province and its NE Greenland counterpart are suggested to be related to local tectonics associated with shear margin development and thus we question if it is part of the NAIP.

This work was part of the NAG-TEC project and was sponsored by: Bayerngas Norge AS; BP Exploration Operating Company Limited, Bundesanstalt für Geowissenschaften und Rohstoffe (BGR); Capricorn Norge A/S; Chevron East Greenland Exploration A/S; ConocoPhillips Skandinavia AS; DEA Norge AS; Det norske oljeselskap ASA; DONG E&P A/S; E.ON Norge AS; ExxonMobil Exploration and Production Norway AS; Japan Oil, Gas and Metals National Corporation (JOGMEC); Maersk Oil; Nalcor Energy – Oil and Gas Inc.; Nexen Energy ULC; Norwegian Energy Company ASA (Noreco); Repsol Exploration Norge AS; Statoil (UK) Limited; and Wintershall Holding GmBH.

References

Abdelmalak, M.M., Andersen, T.B. *et al.* 2015. The ocean-continent transition in the mid-Norwegian margin: insight from seismic data and an onshore Caledonian field analogue. *Geology*, **43**, 1011–1014, https://doi.org/10.1130/G37086.1

Á Horni, J., Geissler, W.H. *et al.* 2014. Offshore volcanic facies. *In*: Hopper, J.R., Funck, T., Stoker, M.S., Árting, U., Peron-Pinvidic, G., Doornenbal, H. & Gaina, C. (eds) *Tectonostratigraphic Atlas of the North-East Atlantic Region.* Geological Survey of Denmark and Greenland (GEUS), Copenhagen, Denmark, 235–253.

Á Horni, J., Boldreel, L.O. & Larsen, M. 2015. 3D mapping of an intrusive complex in the Faroe-Shetland Basin. *In*: Eidesgaard, Ó.R. (ed.) *Faroe Islands Exploration Conference Proceedings of the 4th Conference*. Fróðskapur – Faroe University Press, Tórshavn, 57–73.

Andersen, M.S., Sørensen, A.B., Boldreel, L.O. & Nielsen, T. 2002. Cenozoic evolution of the Faroe Platform: comparing denudation and deposition. *In*: Doré, A.G., Cartwright, J.A., Stoker, M.S., Turner, J.P. & White, N. (eds) *Exhumation of the*

North Atlantic Margin: Timing, Mechanisms and Implications for Petroleum Exploration. Geological Society, London, Special Publications, **196**, 291–311, https://doi.org/10.1144/GSL.SP.2002.196.01.16

Barker, P.F. & Hill, I.A. 1980. Asymmetric spreading in back-arc basins. *Nature*, **285**, 652–654, https://doi.org/10.1038/285652a0

Barton, A.J. & White, R.S. 1995. The Edoras Bank margin: continental break-up in the presence of a mantle plume. *Journal of the Geological Society, London*, **152**, 971–974, https://doi.org/10.1144/GSL.JGS.1995.152.01.15

Berndt, C., Planke, S., Alvestad, E., Tsikalas, F. & Rasmussen, T. 2001. Seismic volcanostratigraphy of the Norwegian Margin: constraints on tectonomagmatic break-up processes. *Journal of the Geological Society, London*, **158**, 413–426, https://doi.org/10.1144/jgs.158.3.413

Blischke, A., Gaina, C. *et al.* 2016. The Jan Mayen microcontinent: an update of its architecture, structural development and role during the transition from the Ægir Ridge to the mid-oceanic Kolbeinsey Ridge. *In*: Péron-Pinvidic, G., Hopper, J.R., Stoker, M.S., Gaina, C., Doornenbal, J.C., Funck, T. & Árting, U.E. (eds) *The NE Atlantic Region: A Reappraisal of Crustal Structure, Tectonostratigraphy and Magmatic Evolution*, Geological Society, London, Special Publications. First published online September 8, 2016, https://doi.org/10.1144/SP447.5

Breivik, A.J., Faleide, J.I., Mjelde, R. & Flueh, E.R. 2009. Magma productivity and early seafloor spreading rate correlation on the northern Vøring Margin, Norway – constraints on mantle melting. *Tectonophysics*, **468**, 206–223, https://doi.org/10.1016/j.tecto.2008.09.020

Brooks, C. 2011. The East Greenland rifted volcanic margin. *Geological Survey of Denmark and Greenland Bulletin*, **24**, 1–96.

Brown, E.L. & Lesher, C.E. 2014. North Atlantic magmatism controlled by temperature, mantle composition and buoyancy. *Nature Geoscience*, **7**, 820–824, https://doi.org/10.1038/NGEO2264

Delorey, A.A., Dunn, R.A. & Gaherty, J.B. 2007. Surface wave tomography of the upper mantle beneath the Reykjanes Ridge with implications for ridge–hot spot interaction. *Journal of Geophysical Research*, **112**, B08313, https://doi.org/10.1029/2006JB004785

Ebinger, C.J. & Sleep, N.H. 1998. Cenozoic magmatism throughout east Africa resulting from impact of a single plume. *Nature*, **395**, 788–791, https://doi.org/10.1038/27417

Eldholm, O. & Grue, K. 1994. North Atlantic volcanic margins: dimensions and production rates. *Journal of Geophysical Research*, **99**, 2955–2968, https://doi.org/10.1029/93jb02879

Eldholm, O., Tsikalas, F. & Faleide, J.I. 2002. Continental margin off Norway 62–75° N: palaeogene tectono-magmatic segmentation and sedimentation. *In*: Jolley, D.W. & Bell, B.R. (eds) *The North Atlantic Igneous Province: Stratigraphy, Tectonic, Volcanic and Magmatic Processes*. Geological Society, London, Special Publications, **197**, 39–68, https://doi.org/10.1144/GSL.SP.2002.197.01.03

Elliott, G.M. & Parson, L.M. 2008. Influence of margin segmentation upon the break-up of the Hatton Bank rifted margin, NE Atlantic. *Tectonophysics*, **457**, 161–176, https://doi.org/10.1016/j.tecto.2008.06.008

Faleide, J.I., Myhre, A.M. & Eldholm, O. 1988. Early Tertiary volcanism at the western Barents Sea margin. *In*: Morten, A.C. & Parrson, L.M. (eds) *Early Tertiary volcanism and the opening of the NE Atlantic*. Geological Society, London, Special Publications, **39**, 135–146, https://doi.org/10.1144/GSL.SP.1988.039.01.13

Funck, T., Hopper, J.R. *et al.* 2014. Crustal structure. *In*: Hopper, J.R., Funck, T., Stoker, M.S., Árting, U., Peron-Pinvidic, G., Doornenbal, H. & Gaina, C. (eds) *Tectonostratigraphic Atlas of the North-East Atlantic Region*. Geological Survey of Denmark and Greenland (GEUS), Copenhagen, Denmark, 69–126.

Funck, T., Geissler, W.H., Kimbell, G.S., Gradmann, S., Erlendsson, Ö., McDermott, K. & Petersen, U.K. 2016. Moho and basement depth in the NE Atlantic Ocean based on seismic refraction data and receiver functions. *In*: Péron-Pinvidic, G., Hopper, J.R., Stoker, M.S., Gaina, C., Doornenbal, J.C., Funck, T. & Árting, U.E. (eds) *The NE Atlantic Region: A Reappraisal of Crustal Structure, Tectonostratigraphy and Magmatic Evolution*, Geological Society, London, Special Publications. First published online July 13, 2016, https://doi.org/10.1144/SP447.1

Gaina, C. 2014. Plate reconstructions and regional kinematics. *In*: Hopper, J.R., Funck, T., Stoker, M.S., Árting, U., Peron-Pinvidic, G., Doornenbal, H. & Gaina, C. (eds) *Tectonostratigraphic Atlas of the North-East Atlantic Region*. Geological Survey of Denmark and Greenland (GEUS), Copenhagen, Denmark, 53–68.

Gaina, C., Gernigon, L. & Ball, P. 2009. Paleocene-Recent Plate Boundaries in the NE Atlantic and the formation of Jan Mayen microcontinent. *Journal of the Geological Society, London*, **166**, 601–616, https://doi.org/10.1144/0016-76492008-112

Gaina, C., Blischke, A., Geissler, W.H., Kimbell, G.S. & Erlendsson, Ö. 2016. Seamounts and oceanic igneous features in the NE Atlantic: a link between plate motions and mantle dynamics. *In*: Péron-Pinvidic, G., Hopper, J.R., Stoker, M.S., Gaina, C., Doornenbal, J.C., Funck, T. & Árting, U.E. (eds) *The NE Atlantic Region: A Reappraisal of Crustal Structure, Tectonostratigraphy and Magmatic Evolution*, Geological Society, London, Special Publications. First published online September 8, 2016, https://doi.org/10.1144/SP447.6

Geissler, W.H., Gaina, C. *et al.* 2016. Seismic volcanostratigraphy of the NE Greenland continental margin. *In*: Péron-Pinvidic, G., Hopper, J.R., Stoker, M.S., Gaina, C., Doornenbal, J.C., Funck, T. & Árting, U.E. (eds) *The NE Atlantic Region: A Reappraisal of Crustal Structure, Tectonostratigraphy and Magmatic Evolution*, Geological Society, London, Special Publications. First published online December 14, 2016, https://doi.org/10.1144/SP447.11

Graham, D.W., Larsen, L.M., Hanan, B.B., Storey, M., Pedersen, A.K. & Lupton, J.E. 1998. Helium isotope composition of the early Iceland mantle plume inferred from the Tertiary picrites of West Greenland. *Earth and Planetary Science Letters*, **160**, 241–255, https://doi.org/10.1016/S0012-821X(98)00083-1

GREGG, T.K.P. & FORNARI, D.J. 1998. Long submarine lava flows: observations and results from numerical modeling. *Journal of Geophysical Research*, **103**, 27517–27531, https://doi.org/10.1029/98jb02465

HAASE, C., EBBING, J. & FUNCK, T. 2016. A 3D regional crustal model of the NE Atlantic based on seismic and gravity data. *In*: PÉRON-PINVIDIC, G., HOPPER, J.R., STOKER, M.S., GAINA, C., DOORNENBAL, J.C., FUNCK, T. & ÁRTING, U.E. (eds) *The NE Atlantic Region: A Reappraisal of Crustal Structure, Tectonostratigraphy and Magmatic Evolution*, Geological Society, London, Special Publications. First published online October 12, 2016, https://doi.org/10.1144/SP447.8

HANSEN, J., JERRAM, D.A., MCCAFFREY, K. & PASSEY, S.R. 2011. Early Cenozoic saucer-shaped sills of the Faroe Islands: an example of intrusive styles in basaltic lava piles. *Journal of the Geological Society, London*, **168**, 159–178, https://doi.org/10.1144/0016-76492010-012

HAYES, D.E. 1976. Nature, implications of asymmetric sea-floor spreading – 'Different rates for different plates'. *Geological Society of America Bulletin*, **87**, 994–1002, https://doi.org/10.1130/0016-7606(1976)87<994:NAIOAS>2.0.CO;2

HINZ, K. 1981. A hypothesis on terrestrial catastrophes: wedges of very thick oceanward dipping layers beneath passive continental margins. *Geologisches Jahrbuch*, **E22**, 3–28.

HITCHEN, K. 2004. The geology of the UK Hatton-Rockall margin. *Marine and Petroleum Geology*, **21**, 993–1012, https://doi.org/10.1016/j.marpetgeo.2004.05.004

HITCHEN, K. & JOHNSON, H. 2013. *Geology of the Rockall Basin and Adjacent Areas*, Report RR/12/03, British Geological Survey, Keyworth, Nottingham, UK.

HJARTASON, A., ERLENDSSON, O. & BLISCHKE, A. 2017. The Greenland-Iceland-Faroe Ridge Complex. *In*: PERON-PINVIDIC, G., HOPPER, J.R., STOKER, M., GAINA, C., FUNCK, T., ARTING, U. & DOORNENBAL, J.C. (eds) *The North-East Atlantic Region: A Reappraisal of Crustal Structure, Tectono-Stratigraphy and Magmatic Evolution*. Geological Society, London, Special Publications. First published online April 19, 2017, https://doi.org/10.1144/SP447.14

HOLBROOK, W.S., LARSEN, H.C. *ET AL*. 2001. Mantle thermal structure and active upwelling during continental breakup in the North Atlantic. *Earth and Planetary Science Letters*, **190**, 251–266, https://doi.org/10.1016/S0012-821X(01)00392-2

HOPPER, J.R., DAHL-JENSEN, T. *ET AL*. 2003. Structure of the SE Greenland margin from seismic reflection and refraction data: Implications for nascent spreading center subsidence and asymmetric crustal accretion during North Atlantic opening. *Journal of Geophysical Research: Solid Earth*, **108**, 2269, https://doi.org/10.1029/2002jb001996

HOWELL, S.M., ITO, G. *ET AL*. 2014. The origin of the asymmetry in the Iceland hotspot along the Mid-Atlantic Ridge from continental breakup to present-day. *Earth and Planetary Science Letters*, **392**, 143–153, https://doi.org/10.1016/j.epsl.2014.02.020

ITO, G. 2001. Reykjanes 'V'-shaped ridges originating from a pulsing and dehydrating mantle plume. *Nature*, **411**, 681–684, https://doi.org/10.1038/35079561

JØRGENSEN, O. 2006. The regional distribution of zeolites in the basalts of the Faroe Islands and the significance of zeolites as palaeotemperature indicators. *Geological Survey of Denmark and Greenland Bulletin*, **9**, 123–156.

KIØRBOE, L. 1999. Stratigraphic relationships of the Lower Tertiary of the Faeroe Basalt Plateau and the Faeroe-Shetland Basin. *In*: FLEET, A.J. & BOLDY, S.A.R. (eds) *Petroleum Geology of Northwest Europe: Proceedings of the 5th Conference*. Geological Society, London, 559–572.

LARSEN, H.C. & SAUNDERS, A.D. 1998. Tectonism and volcanism at the southeast Greenland rifted margin: a record of plume impact and later continental rupture. *Proceedings of the Ocean Drilling Program, Scientific Results*, **152**, 503–534, https://doi.org/10.2973/odp.proc.sr.152.240.1998

LARSEN, L.M., WAAGSTEIN, R., PEDERSEN, A.K. & STOREY, M. 1999. Trans-Atlantic correlation of the Paleogene volcanic successions in the Faroe Islands and East Greenland. *Journal of the Geological Society, London*, **156**, 1081–1095, https://doi.org/10.1144/gsjgs.156.6.1081

LARSEN, L.M., PEDERSEN, A.K., TEGNER, C. & DUNCAN, R.A. 2014. Eocene to Miocene igneous activity in NE Greenland: northward younging of magmatism along the East Greenland margin. *Journal of the Geological Society, London*, **171**, 539–553, https://doi.org/10.1144/jgs2013-118

LARSEN, L.M., PEDERSEN, A.K., TEGNER, C., DUNCAN, R.A., HALD, N. & LARSEN, J.G. 2015. Age of Tertiary volcanic rocks on the West Greenland continental margin: volcanic evolution and event correlation to other parts of the North Atlantic Igneous Province. *Geological Magazine*, 153, 1–25, https://doi.org/10.1017/S0016756815000515

LUNDIN, E.R. & DORÉ, A.G. 2011. Hyperextension, serpentinization, and weakening: a new paradigm for rifted margin compressional deformation. *Geology*, **39**, 347–350, https://doi.org/10.1130/g31499.1

MATHIESEN, A., BIDSTRUP, T. & CHRISTIANSEN, F.G. 2000. Denudation and uplift history of the Jameson Land basin, East Greenland – constrained from maturity and apatite fission track data. *Global and Planetary Change*, **24**, 275–301, http://www.sciencedirect.com/science/article/pii/S0921818100000138

MJELDE, R., RAUM, T., MURAI, Y. & TAKANAMI, T. 2007. Continent–ocean-transitions: Review, and a new tectono-magmatic model of the Vøring Plateau, NE Atlantic. *Journal of Geodynamics*, **43**, 374–392, https://doi.org/10.1016/j.jog.2006.09.013

MOORE, J.G., CLAGUE, D.A., HOLCOMB, R.T., LIPMAN, P.W., NORMARK, W.R. & TORRESAN, M.E. 1989. Prodigious submarine landslides on the Hawaiian Ridge. *Journal of Geophysical Research: Solid Earth*, **94**, 17465–17484, https://doi.org/10.1029/JB094iB12p17465

MØRK, M.B. & DUNCAN, R.A. 1993. Late Pliocene basaltic volcanism on the Western Barents Shelf margin: implications from petrology and ^{40}Ar-^{39}Ar dating of volcaniclastic debris from a shallow drill core. *Norsk Geologisk Tidsskrift*, **73**, 209–225.

MÜLLER, R.D., ROEST, W.R. & ROYER, J.-Y. 1998. Asymmetric sea-floor spreading caused by ridge–plume interactions. *Nature*, **396**, 455–459, https://doi.org/10.1038/24850

Mutter, J.C., Talwani, M. & Stoffa, P.L. 1982. Origin of seaward-dipping reflectors in oceanic crust off the Norwegian margin by subaerial sea-floor spreading. *Geology*, **10**, 353–357, https://doi.org/10.1130/0091-7613(1982)10<353:OOSRIO>2.0.CO;2

Nielsen, T.K. & Hopper, J.R. 2002. Formation of volcanic rifted margins: are temperature anomalies required? *Geophysical Research Letters*, **29**, 18–11, https://doi.org/10.1029/2002GL015681

Nielsen, T.K. & Hopper, J.R. 2004. From rift to drift: mantle melting during continental breakup. *Geochemistry, Geophysics, Geosystems*, **5**, Q07003, https://doi.org/10.1029/2003GC000662

Nielsen, T.K., Larsen, H.C. & Hopper, J.R. 2002. Contrasting rifted margin styles south of Greenland: implications for mantle plume dynamics. *Earth and Planetary Science Letters*, **200**, 271–286, https://doi.org/10.1016/S0012-821X(02)00616-7

Ogg, J.G. 2012. The geomagnetic polarity timescale. *In*: Gradstein, F.M., Ogg, J.G., Schmitz, M.D. & Ogg, G. (eds) *The Geologic Time Scale 2012*. Elsevier, Amsterdam, The Netherlands, **2**, 85–115.

Osmundsen, P.T. & Ebbing, J. 2008. Styles of extension offshore mid-Norway and implications for mechanisms of crustal thinning at passive margins. *Tectonics*, **27**, TC6016, https://doi.org/10.1029/2007tc002242

Passey, S.R. & Hitchen, K. 2011. Cenozoic (igneous). *In*: Ritchie, J.D., Ziska, H., Johnson, H. & Evans, D. (eds) *The Geology of the Faroe-Shetland Basin and Adjacent Areas*. British Geological Survey, Edinburgh and Jarðfeingi, Tórshavn, 209–228.

Passey, S.R. & Jolley, D.W. 2009. A revised lithostratigraphic nomenclature for the Palaeogen Faroe Island Basalt Group, NE Atlantic Ocean. *Earth and Environmental Science Transactions of the Royal Society of Edinburgh*, **99**, 127–158.

Planke, S. & Alvestad, E. 1999. Seismic volcanostratigraphy of the extrusive breakup complexes in the northeast Atlantic: implications from ODP/DSDP drilling. *In*: Larsen, H.C., Duncan, R.A., Allan, J.F. & Brooks, K. (eds) *Proceedings of the Ocean Drilling Program, Scientific Results*. Ocean Drilling Program, College Station, Texas, **163**, 3–16.

Planke, S., Symonds, P.A., Alvestad, E. & Skogseid, J. 2000. Seismic volcanostratigraphy of large-volume basaltic extrusive complexes on rifted margins. *Journal of Geophysical Research*, **105**, 19335–19351, https://doi.org/10.1029/1999jb900005

Planke, S., Rasmussen, T., Rey, S.S. & Myklebust, R. 2005. Seismic characteristics and distribution of volcanic intrusions and hydrothermal vent complexes in the Vøring and Møre basins. *In*: Doré, A.G. & Vining, B.A. (eds) *Petroleum Geology: North-West Europe and Global Perspectives – Proceedings of the 6th Petroleum Geology Conference*. Geological Society, London, 833–844.

Richardson, K.R., Smallwood, J.R., White, R.S., Snyder, D.B. & Maguire, P.K.H. 1998. Crustal structure beneath the Faroe Islands and the Faroe-Iceland Ridge. *Tectonophysics*, **300**, 159–180, https://doi.org/10.1016/s0040-1951(98)00239-x

Ryseth, A., Augustson, J.H. *et al.* 2003. Cenozoic stratigraphy and evolution of the Sørvestsnaget Basin, southwestern Barents Sea. *Norwegian Journal of Geology*, **83**, 107–130.

Saunders, A.D., Fitton, J.G., Kerr, A.C., Norry, M.J. & Kent, R.W. 1997. The North Atlantic Igneous Province. *In*: Mahoney, J.J. & Coffin, M.L. (eds) *Large Igneous Provinces: Continental, Oceanic, and Planetary Flood Volcanism*. American Geophysical Union, Washington, D.C., Geophysical Monographs, **100**, 45–93.

Self, S., Thordarson, T. & Keszthelyi, L. 1997. Emplacement of continental flood basalt lava flows. *In*: Mahoney, J.J. & Coffin, M.L. (eds) *Large Igneous Provinces: Continental, Oceanic, and Planetary Flood Volcanism*. American Geophysical Union, Washington, D.C., Geophysical Monographs, **100**, 381–410.

Sleep, N.H. 2002. Local lithospheric relief associated with fracture zones and ponded plume material. *Geochemistry, Geophysics, Geosystems*, **3**, 8506, https://doi.org/10.1029/2002GC000376

Smallwood, J.R. & Maresh, J. 2002. The properties, morphology and distribution of igneous sills: modelling, borehole data and 3D seismic from the Faroe-Shetland area. *In*: Jolley, D.W. & Bell, B.R. (eds) *The North Atlantic Igneous Province. Stratigraphy, Tectonic, Volcanic and Magmatic Processes*. Geological Society, London, Special Publications, **197**, 271–206, https://doi.org/10.1144/GSL.SP.2002.197.01.11

Smallwood, J.R., Staples, R.K., Richardson, K.R. & White, R.S. 1999. Crust generated above the Iceland mantle plume: from continental rift to oceanic spreading center. *Journal of Geophysical Research: Solid Earth*, **104**, 22885–22902, https://doi.org/10.1029/1999JB900176

Smythe, D.K. 1983. Faeroe-Shetland Escarpment and continental margin north of the Faeroes. *In*: Bott, M.H.P., Saxov, S., Talwani, M. & Thiede, J. (eds) *Structure and Development of the Greenland-Scotland Ridge. New Methods and Concepts*. Plenum Press, New York, 109–119.

Smythe, D.K., Chalmers, J.A., Skuce, A.G., Dobinson, A. & Mould, A.S. 1983. Early opening history of the North Atlantic – I. Structure and origin of the Faeroe-Shetland Escarpment. *Geophysical Journal of the Royal Astronomical Society*, **72**, 373–398.

Stein, S., Melosh, H.J. & Minster, J.B. 1977. Ridge migration and asymmetric sea-floor spreading. *Earth and Planetary Science Letters*, **36**, 51–62, https://doi.org/10.1016/0012-821X(77)90187-X

Stoker, M.S., Kimbell, G.S., McInroy, D.B. & Morton, A.C. 2012. Eocene post-rift tectonostratigraphy of the Rockall Plateau, Atlantic margin of NW Britain: linking early spreading tectonics and passive margin response. *Marine and Petroleum Geology*, **30**, 98–125, https://doi.org/10.1016/j.marpetgeo.2011.09.007

Stoker, M.S., Stewart, M.A. *et al.* 2016. An overview of the Upper Palaeozoic–Mesozoic stratigraphy of the NE Atlantic region. *In*: Péron-Pinvidic, G., Hopper, J.R., Stoker, M.S., Gaina, C., Doornenbal, J.C., Funck, T. & Árting, U.E. (eds) *The NE Atlantic*

Region: A Reappraisal of Crustal Structure, Tectonostratigraphy and Magmatic Evolution, Geological Society, London, Special Publications. First published online August 11, 2016, https://doi.org/10.1144/SP447.2

STOREY, M., PEDERSEN, A.K. ET AL. 2004. Long-lived postbreakup magmatism along the East Greenland margin; evidence for shallow-mantle metasomatism by the Iceland Plume. *Geology*, **32**, 173–176, https://doi.org/10.1130/g19889.1

STOREY, M., DUNCAN, R.A. & TEGNER, C. 2007. Timing and duration of volcanism in the North Atlantic Igneous Province: implications for geodynamics and links to the Iceland hotspot. *Chemical Geology*, **241**, 264–281, https://doi.org/10.1016/j.chemgeo.2007.01.016

TEN BRINK, U.S. & TAYLOR, M.H. 2002. Crustal structure of central Lake Baikal: insights into intracontinental rifting. *Journal of Geophysical Research*, **107**, 2132, https://doi.org/10.1029/2001JB000300

VARMING, T. 2009. Results from the drilling of the 1st license round wells in the Faroese part of the Judd Basin. *In*: VARMING, T. & ZISKA, H. (eds) *Faroe Islands Exploration Conference Proceedings of the 2nd Conference*. Føroya Fróðskaparsetur, Tórshavn, 346–363.

VOGT, P.R. 1971. Asthenosphere motion recorded by the ocean floor south of Iceland. *Earth and Planetary Science Letters*, **13**, 153–160, https://doi.org/10.1016/0012-821X(71)90118-X

VOSS, M., SCHMIDT-AURSCH, M.C. & JOKAT, W. 2009. Variations in magmatic processes along the East Greenland volcanic margin. *Geophysical Journal International*, **177**, 755–782, https://doi.org/10.1111/j.1365-246X.2009.04077.x

WAAGSTEIN, R. 1988. Structure, composition and age of the Faeroe basalt plateau. *In*: MORTON, A.C. & PARSON, L.M. (eds) *Early Tertiary Volcanism and the Opening of the NE Atlantic*. Geological Society, London, Special Publications, **39**, 225–238, https://doi.org/10.1144/GSL.SP.1988.039.01.21

WEIGEL, W., FLÜH, E.R. ET AL. 1995. Investigations of the East Greenland continental margin between 70° and 72° N by deep seismic sounding and gravity studies. *Marine Geophysical Researches*, **17**, 167–199, https://doi.org/10.1007/BF01203425

WHITE, R.S., MCKENZIE, D.P. & O'NIONS, R.K. 1992. Oceanic crustal thickness from seismic measurements and rare-earth element inversions. *Journal of Geophysical Research*, **97**, 19683–19715, https://doi.org/10.1029/92jb01749

WHITE, R.S., SMITH, L.K., ROBERTS, A.W., CHRISTIE, P.A.F. & KUSZNIR, N.J. & TEAM, I. 2008. Lower-crustal intrusion on the North Atlantic continental margin. *Nature*, **452**, 460–464, https://doi.org/10.1038/nature06687

WRIGHT, K.A., DAVIES, R.J., JERRAM, D.A., MORRIS, J. & FLETCHER, R. 2011. Application of seismic and sequence stratigraphic concepts to a lava-fed delta system in the Faroe-Shetland Basin, UK and Faroes. *Basin Research*, **24**, 91–106, https://doi.org/10.1111/j.1365-2117.2011.00513.x

The Greenland–Iceland–Faroe Ridge Complex

ÁRNI HJARTARSON[1]*, ÖGMUNDUR ERLENDSSON[1] & ANETT BLISCHKE[2]

[1]*Iceland GeoSurvey, Grensasvegi 9, 108 Reykjavik, Iceland*

[2]*Iceland GeoSurvey, Branch at Akureyri, Rangarvöllum, 602 Akureyri, Iceland*

**Correspondence: ah@isor.is*

Abstract: The Greenland–Iceland–Faroe Ridge Complex (GIFRC) has been forming since the opening of the NE Atlantic (<55 Ma), standing out as a prominent feature on all geoscientific datasets. Our interpretations have revealed several new potential abandoned rift centres, mapped as syncline and anticline structures. The synclines are suggested to be manifestations of former rift axes that were abandoned by rift jumps. These appear to be more common inside the GIFRC region than in the adjacent ocean basins, and can be confirmed by observations of cumulative crustal accretion data through time. A major post-40 Ma unconformity is proposed across the East Iceland Shelf, forming a distinct 16–20 myr-long hiatus that is covered by a thick, younger sedimentary section.

Several seamounts were identified on multibeam datasets at around 1200 m water depth in the Vesturdjúp Basin, just south of the Greenland–Iceland Ridge. These seamounts appear to be younger in formation time than the surrounding ocean floor, possibly indicating a still active intraplate volcanic zone. Young tectonic features, such as faults, graben and transverse ridges, characterize the area and present a good example of the complexity of the GIFRC in comparison to the adjacent abyssal plain.

The purpose of this study is to review the structural segmentation and links to chronostratigraphic processes affecting the areas of the Greenland–Iceland Ridge (GIR), the Iceland Plateau and the Iceland–Faroe Ridge (IFR) (Figs 1 & 2; Table 1). These ridges and plateau are summarized here as the Greenland–Iceland–Faroe Ridge Complex (GIFRC) and part of the North Atlantic Igneous Province (NAIP) (Fig. 1), one of the largest igneous provinces in the world (Saunders *et al.* 1997). This review addresses the initiation of the GIFRC, its extent, defines rift jump areas within the complex, and addresses the Iceland-type central volcanoes and seamounts in their offshore regions. In addition, crustal thickness variations are compared to structural and geochronological variations within the GIFRC based on potential field, seismic and geological sections, and surface geological field data (Table 2).

Abandoned rift systems have previously been mapped for the onshore region of Iceland for the last 16 myr (Hjartarson 2003; Jóhannesson & Sæmundsson 2009) (Table 3), and in offshore areas around Iceland by Vogt (1971), Talwani & Eldholm (1977), Harðarson *et al.* (1997), Vogt & Jung (2009) and Erlendsson & Blischke (2013). These extinct rift systems show a distinct dip of subaerial lava-flow formation sequences from both sides towards the old rift centre (Böðvarsson & Walker 1964), which can be observed in surface geological maps (Jóhannesson & Sæmundsson 2009) (Table 4), and on seismic reflection data for offshore areas as synclines and anticlines.

Field and refraction data studies have previously defined the term Icelandic-type crust, which has been used to describe the crust beneath the Greenland–Iceland–Faroe Ridge, as it differs fundamentally from both oceanic and continental crust (Foulger *et al.* 2003). Oceanic crust is divided into three distinct layers: Layer 1 is referred to as sediments; Layer 2 is composed of extrusive volcanic eruptive rock, mostly pillow lava; and Layer 3 is equivalent to gabbro and cumulates ultramafic rocks. Icelandic-type crust can be divided in a similar way but pillow lavas in Layer 2 are replaced by subaerial lava flows. The seismic velocities of Icelandic-type crust are generally equivalent to those of normal oceanic crust, but have a more variable Layer 3 of overall much thicker crust (Brandsdóttir & Menke 2008). This atypical build-up of thick oceanic crust is of interest, as multichannel seismic (MCS) reflection data interpretations might shed light on the internal structures up to 15 km in depth that cannot be imaged by seismic refraction or potential field data.

The GIFRC area has long been regarded as a hotspot track, mainly formed by subaerial igneous activity (Bjarnason 2008). Its structure is, however,

From: Péron-Pinvidic, G., Hopper, J. R., Stoker, M. S., Gaina, C., Doornenbal, J. C., Funck, T. & Árting, U. E. (eds) 2017. *The NE Atlantic Region: A Reappraisal of Crustal Structure, Tectonostratigraphy and Magmatic Evolution*. Geological Society, London, Special Publications, **447**, 127–148.
First published online April 19, 2017, https://doi.org/10.1144/SP447.14

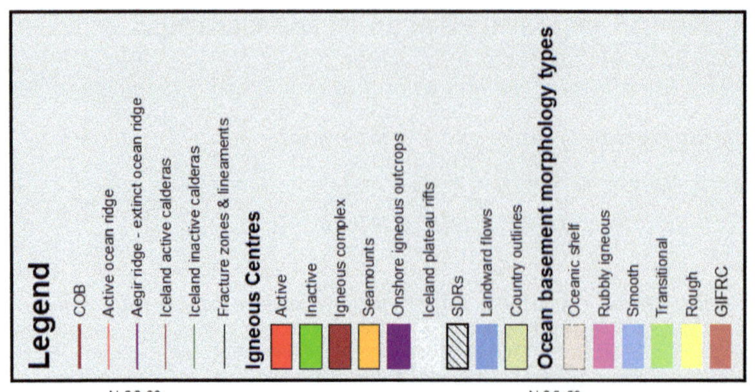

Legend
COB
Active ocean ridge
Aegir ridge - extinct ocean ridge
Iceland active calderas
Iceland inactive calderas
Fracture zones & lineaments
Igneous Centres
Active
Inactive
Igneous complex
Seamounts
Onshore igneous outcrops
Iceland plateau rifts
SDRs
Landward flows
Country outlines
Ocean basement morphology types
Oceanic shelf
Rubbly igneous
Smooth
Transitional
Rough
GIFRC
65°0'0"N
60°0'0"N
55°0'0"N
15°0'0"W
30°0'0"W
VB
MB
UK
NB
FI
ÆR
EJMFZ
IFFZ
IFR
IPR
JMMC
JMI
MR
WJMFZ
KR
NVZ
EVZ
WVZ
TFZ
MIB
SISZ
SVB
RR
GIR
CEG
Iceland basin
Irminger basin
?
0 65 130 260 Kilometers

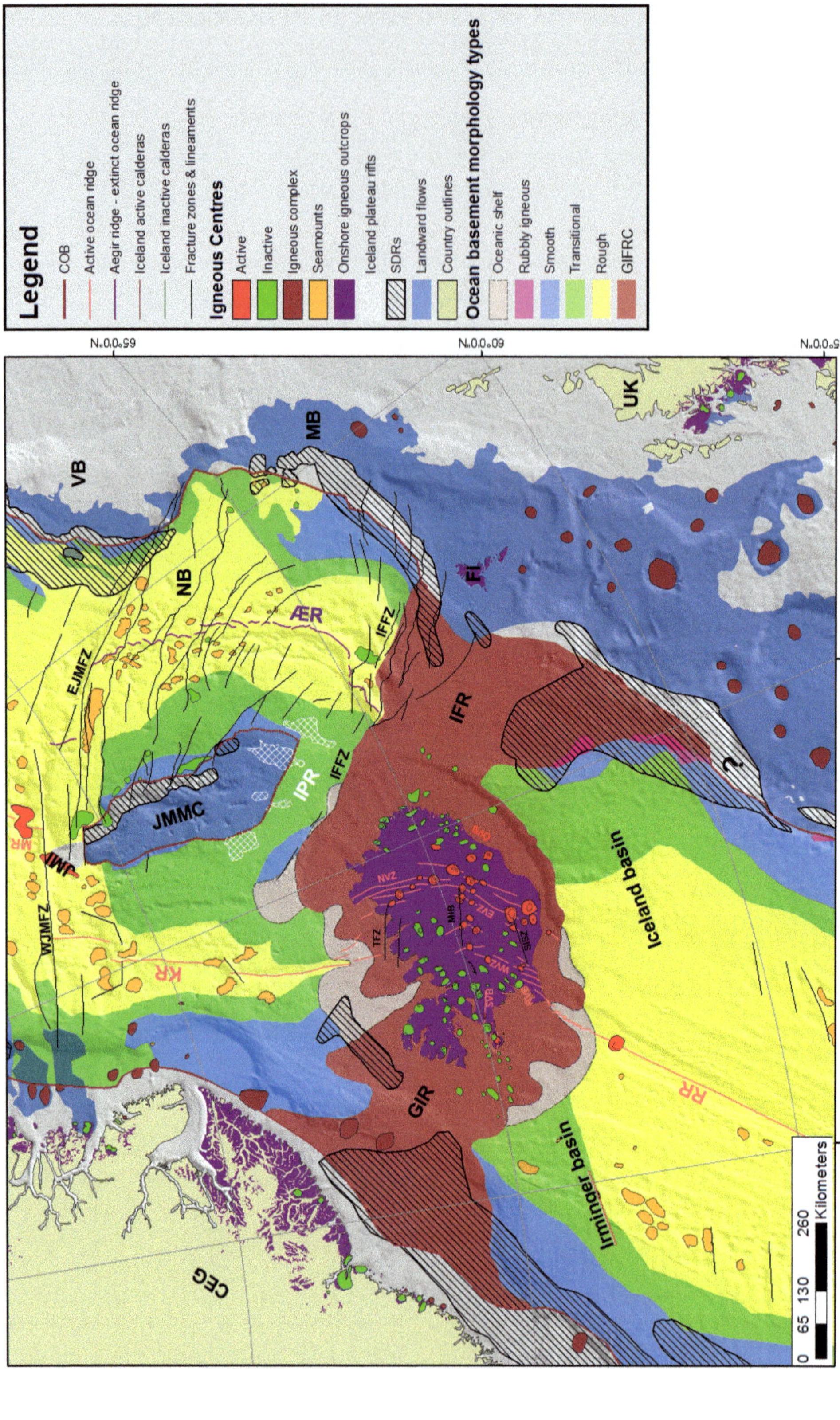

poorly understood due to the lack of geological profiles, dated rock samples from the offshore areas, and difficulties in interpreting consistent magnetic chron data across the region (Gaina *et al.* 2017) (Fig. 3a).

It can be debated whether or not the Greenland–Iceland–Faroe Ridge and Iceland itself belong to the NAIP. The definition of a large igneous province (LIP) defines an area that has undergone extremely large accumulation of igneous rocks in a short time. In that sense, the formation of the NAIP ended by the opening of the ocean, when break-up volcanism turned into drift volcanism and concentrated around the Iceland hotspot and the Ægir, Reykjanes and Kolbeinsey mid-ocean ridges (MORs). Most published work, however, includes the GIFRC and Iceland itself in that province (e.g. Saunders *et al.* 1997). In that sense, the NAIP is still expanding and the duration of its formation exceeds 60 myr.

Tectonic setting

The NAIP started to form 60–63 myr ago during the pre-break-up phase of the NE Atlantic, when a deep-seated mantle plume reached the lower crust below central Greenland (Morgan 1971; Brooks 1973; White & McKenzie 1989). This caused intensive volcanism in East and West Greenland and northern Canada, as well as in the Faroe and British islands (Ganerød *et al.* 2010). The NE Atlantic break-up phase took place during the Late Paleocene (57–55 Ma), forming magma-rich margins across the area before the final opening of the North Atlantic rift system along structurally weak areas within the crust (Larsen & Saunders 1998; Gaina *et al.* 2009; Blischke *et al.* 2016). The central NE Atlantic was covered with extensive lava flows on adjacent elevated margins, also referred to as plateau basalts (Larsen & Watt 1985; Larsen *et al.* 1989; Søager & Holm 2009). In the Early Eocene (55–54 Ma), the lithosphere finally ruptured, initiating the post-break-up phase and marking the onset of seafloor spreading in the NE Atlantic. Initially, three MOR segments formed: the Ægir, Mohn's, and Reykjanes ridges. The initiation sequences are present in seismic reflection data as distinct seawards-dipping reflectors formations (SDR) along the break-up margins (e.g. Talwani & Eldholm 1977; Hinz 1981; Mutter *et al.* 1982; Larsen & Jakobsdóttir 1988; Larsen & Saunders 1998; Elliott & Parson 2008; Blischke *et al.* 2016; Geissler *et al.* 2016). The Late Paleocene and Early Eocene plateau basalts and SDR sequences that form the bulk of the NAIP were split up by the opening of the North Atlantic, and are now widely distributed and exposed along both margins of the NE Atlantic Ocean.

The GIFRC covers 480 000 km^2 of a thick volcanic crust that stretches 1150 km across the central NE Atlantic Ocean between the central East Greenland and the NW European margins (Fig. 1). It incorporates the Iceland Plateau, the aseismic GIR and the IFR. The GIFRC appears as a prominent phenomenon with respect to bathymetry, ocean basement morphology, gravity, palaeomagnetism, crustal thickness, geochemistry and petrology characteristics, with direct influence from the mantle plume (Figs 1–3) (e.g. Jakobsson 1972; Fitton *et al.* 1997; Thirlwall *et al.* 2004; Kokfelt *et al.* 2006; Thordarson & Larsen 2007; Parnell-Turner *et al.* 2014). Seismic refraction studies have confirmed a crustal thickness variation from 20 to 40 km within the area, accounting for at least a 3–4 times thicker crust than observed for the average oceanic crust (Funck *et al.* 2014) (Fig. 3c).

The western border of the GIFRC corresponds to the continent–ocean boundary (COB) of central East Greenland, and its eastern border corresponds to the continent–ocean boundary west of the Faroe Islands (Hopper *et al.* 2014). The GIFRC reaches up to 2100 m above sea level in SE Iceland, whereas the bathymetrically deepest points of the

Fig. 1. Regional settings map showing the main volcanic facies elements (á Horni *et al.* 2014), major structural lineaments (Funck *et al.* 2014), the Iceland Plateau Rift system (Blischke *et al.* 2016), ocean basin morphology type map (Funck *et al.* 2014; Gaina *et al.* 2016; Geissler *et al.* 2016) and bathymetry data (Hopper *et al.* 2014) of the GIFRC, surrounding oceanic basins and continental margins. This compilation illustrates the complex segmentation of the central NE Atlantic region with its active and extinct volcanic systems (e.g. SDRs, igneous complexes and rift systems, and major boundary structural elements, such as fracture zones). *Abbreviations*: CEG, central East Greenland; GIR, Greenland–Iceland Ridge; IFR, Iceland–Faroe Ridge; GIFRC, Greenland–Iceland–Faroe Ridge Complex; FI, Faroe Islands; VB, Vøring Basin; MB, Møre Basin; KR, Kolbeinsey mid-oceanic ridge (MOR); ÆR, Ægir MOR; RR, Reykjanes MOR; MR, Mohn's MOR; JMMC, Jan Mayen microcontinent; JMI, Jan Mayen Island Volcanic Complex; EJMFZ, East Jan Mayen Fracture Zone; WJMFZ, West Jan Mayen Fracture Zone; IFFZ, Iceland–Faroe Fracture Zone; TFZ, Tjörnes Fracture Zone; MIB, Mid-Iceland Volcanic Belt; SISZ, South Iceland Seismic Zone; NVZ, North Iceland Volcanic Zone; EVZ, East Iceland Volcanic Zone; WVZ, West Iceland Volcanic Zone; RVB, Reykjanes Volcanic Belt; SVB, Snæfellsnes Volcanic Belt; and ÖVB, Öræfajökull Volcanic Belt (Einarsson 2008; Gaina *et al.* 2009; Gernigon *et al.* 2015; Blischke *et al.* 2016).

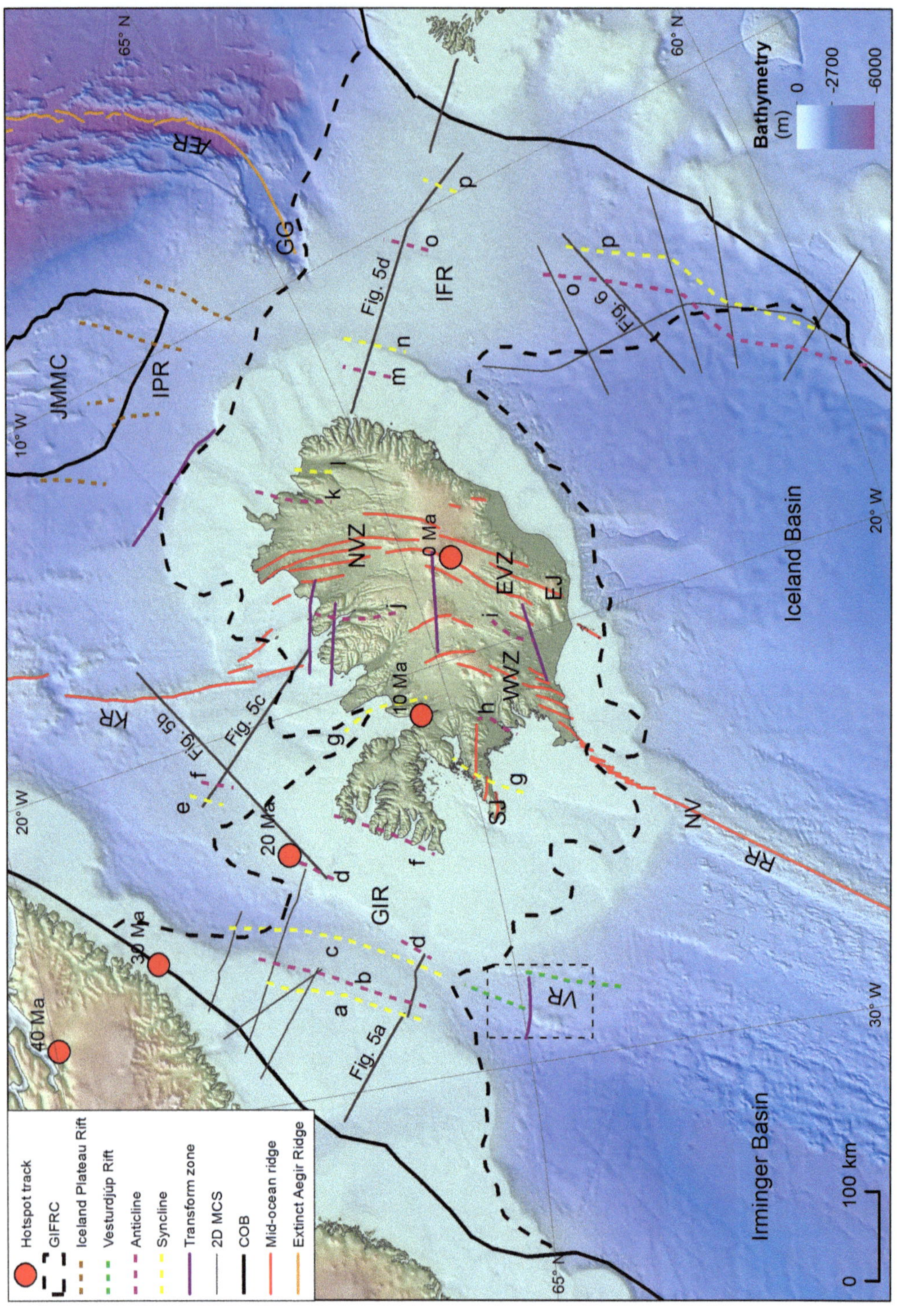
Hotspot track
GIFRC
Iceland Plateau Rift
Vesturdjúp Rift
Anticline
Syncline
Transform zone
2D MCS
COB
Mid-ocean ridge
Extinct Aegir Ridge
Bathymetry
(m)
0
-2700
-6000
0
100 km
Irminger Basin
Iceland Basin
40 Ma
30 Ma
20 Ma
10 Ma
0 Ma
GIR
IFR
IPR
JMMC
AER
GG
KR
RR
NVZ
EVZ
WVZ
EJ
SJ
VR
Fig. 5a
Fig. 5b
Fig. 5c
Fig. 5d
Fig. 6
65° N
60° N
10° W
20° W
30° W

Table 1. *Milestones in the history of the North Atlantic Igneous Province*

Ma	Event	NB
62–63	First volcanism of the NAIP	Saunders *et al.* (1997)
62.6	The Antrim Lava Group, Ireland (Chron 27r)	Ganerød *et al.* (2010)
62	West Greenland	Saunders *et al.* (1997)
62	Baffin Island	Saunders *et al.* (1997)
60–62	East Greenland (Mikis and Vandfaldsdalen formations)	Brooks (2011)
60.94	Faroe Islands (Lopra Formation)	Storey *et al.* (2007)
54–56	Commencement of the break-up and seafloor spreading	Gaina *et al.* (2009)
54–56	Formation of GIFRC starts	This study
52–56	SDRs	á Horni *et al.* (2014)
51–52	End of regional volcanism (ODP dating)	á Horni *et al.* (2014)
26–30	Ægir Ridge became extinct	Gaina *et al.* (2009)
24–33	Ægir–Kolbeinsey rift jump	Gaina *et al.* (2009)
25	Proto-Iceland	Harðarson *et al.* (2008)
20	Complete separation of the JMMC	Gaina *et al.*(2009)

Table 2. *GIFRC databases that have been reviewed and used in this study*

Acquisition date	Survey ID	Streamer length	No. of channels/ grid spacing	Data repository	Survey type
1976	BGR-76	2400 m	48	BGR	2D MCS
1978	WGC-78	2400 m	48	Western Geco	2D MCS
1980	GGU1980	3000 m	60	GEUS	2D MCS
1981	GGU1981	2400 m	48	GEUS	2D MCS
1994	FIRE	6000 m	240	2D MCS	
2000	KRISE 2000	20 m mini-hydrophone streamer	8	UiB	2D MCS
2003	IFR-2003	6000 m	480	NEA/ISOR	2D MCS
2009/2012	Vesturdjúp		100 m	HAFRO	Multibeam data

complex lie at 600 m below sea level (bsl) between Iceland and Greenland, within the Denmark Strait, and at approximately 500 m bsl between Iceland and the Faroe Islands. The ocean basins north and south of the ridge are over 2000 m deep. The submarine areas of the GIFRC are initially believed to have formed subaerially, but have been subsiding below sea level due to erosion and cooling of the crust (Lundin & Doré 2004; Denk *et al.* 2011). The subsidence process has resulted in an approximately 1500 m difference in elevation along the GIFRC crest (Fig. 2).

Data

Primary data control is affected by potential field and section data, such as bathymetry, magnetic and gravity anomaly compilations, MCS reflection, seismic refraction, and multibeam data from the Vesturdjúp area (Funck *et al.* 2014; Gaina 2014; Haase & Ebbing 2014; Hopper *et al.* 2014; Nasuti & Olesen 2014) (Figs 3a–c & 4; Tables 1 & 2). Analogue studies and surface geology map data also proved vital for a comprehensive understanding of the area (Sæmundsson 1974, 1979; Talwani &

Fig. 2. Map showing the distribution of synclines and anticlines observed, based on interpretation from seismic reflection data and onshore geological data of Iceland. Names of synclines and anticlines are given in Table 3. Tracing of plume positions back to 40 Ma relative to Greenland are based on Gaina *et al.* (2009). Suggested synclines buried below the volcanic zones of Iceland are not shown. See Figure 5 for a cross-section of the seismic reflection profile of a complex rift area, and Figures 7 and 8 for the Vesturdjúp multibeam data and interpretation maps (VR). Bathymetry and elevation map from Hopper *et al.* (2014). *Abbreviations*: GIR, Greenland–Iceland Ridge; SJ, Snæfellsnes Jökull; EJ, Eyjafjallajökull; WVZ, Western Volcanic Zone; EVZ, Eastern Volcanic Zone; NVZ, North Volcanic Zone; KR, Kolbeinsey Ridge; JMMC, Jan Mayen microcontinent; ÆR, Ægir Ridge; GG, Gridar Gorge; IFR, Iceland–Faroe Ridge; NV, Njörður volcano; RR, Reykjanes Ridge.

Table 3. *Rift zones and rift relocations*

Rift zone	Duration (Ma)	References
XR52Ma (Rockall and Vøring margins)	54–52	Torsvik *et al.* (2015)
XR33Ma (Greenland–Iceland Ridge)	54–33	Torsvik *et al.* (2015)
Ægir Ridge	52–26	Gernigon *et al.* (2015)
XR40Ma (East Iceland insular margin)	52–40	Torsvik *et al.* (2015)
Iceland Plateau Rift	49–252	Brandsdóttir *et al.* (2015), Blischke *et al.* (2016)
Kolbeinsey Ridge	25–0	Talwani & Eldholm (1977)
Vesturdjúp Rift	<20	This study
NW Iceland Rift Zone	25–15	Harðarson *et al.* (1997, 2008)
Snæfellsnes–Húnaflói Rift Zone	15–7	Harðarson *et al.* (1997, 2008)
Reykjanes–Langjökull–North Iceland Rift Zone	7–0	Sæmundsson (1979)
Eastern Rift Zone	3–0	Sæmundsson (1979)
Skagafjörður Rift Zone	1.7–0.5	Hjartarson (2003)

Eldholm 1977; Harðarson *et al.* 1997, 2008; Hjartarson 2003; Gaina *et al.* 2009; Jóhannesson & Sæmundsson 2009; Erlendsson & Blischke 2013; Hjartarson & Sæmundsson 2014; Brandsdóttir *et al.* 2015; Gernigon *et al.* 2015; Torsvik *et al.* 2015; Blischke *et al.* 2016) (Tables 3 & 4). These datasets combined allowed a detailed analysis of the GIFRC structure, its development and its geological history.

Comparisons of the potential field data maps were used to define the outlines and extent of the GIFRC (Figs 2 & 3a–c). The resulting GIFRC outline is a compromise between bathymetry, gravity anomaly, magnetic anomaly and crustal thickness data, along with the geochron model, ocean-floor type map, and seismic reflection and refraction profile data across the region (Figs 1–3a–c).

Table 4. *Synclines and anticlines*

Label*	Syncline/anticline	Comment	References
a	Strede Bank Syncline	Same as XR33Ma? (Torsvik *et al.* 2015)	This study
b	Strede Bank Anticline		This study
c	Vesturdjúp Syncline	Connected with the VR Rift	This study
d	Djúpáll Anticline		This study
e	NW Syncline	NW Iceland Rift Zone	This study
f	NW Iceland Anticline	Unconformity and hiatus at the shore	Harðarson *et al.* (1997), Jóhannesson & Sæmundsson (2009)
g	Snæfellsnes–Húnaflói Syncline	Inland Iceland and off Húnaflói	Jóhannesson & Sæmundsson (2009)
h	Borgarnes Anticline	Onland Iceland	Jóhannesson & Sæmundsson (2009)
i	Hreppar Anticline	Onland Iceland	Jóhannesson & Sæmundsson 2009
j	Eyjafjörður Anticline	Onland Iceland	Based on strike and dip: Jóhannesson & Sæmundsson (2009)
k	Vopnafjörður Anticline	Onland Iceland	Based on strike and dip: Jóhannesson & Sæmundsson (2009)
l	Héraðsflói Syncline	Onland Iceland	Jóhannesson & Sæmundsson (2009)
m	Austurbrún Anticline		This study
n	Austurbrún Syncline	Same as XR40 Ma? (Torsvik *et al.* 2015)	This study
o	Iceland Basin Anticline		Erlendsson & Blischke (2013)
p	Iceland Basin Syncline	Same as XR52 Ma? (Torsvik *et al.* 2015)	Erlendsson & Blischke (2013)

*For the locations, see Figure 2.

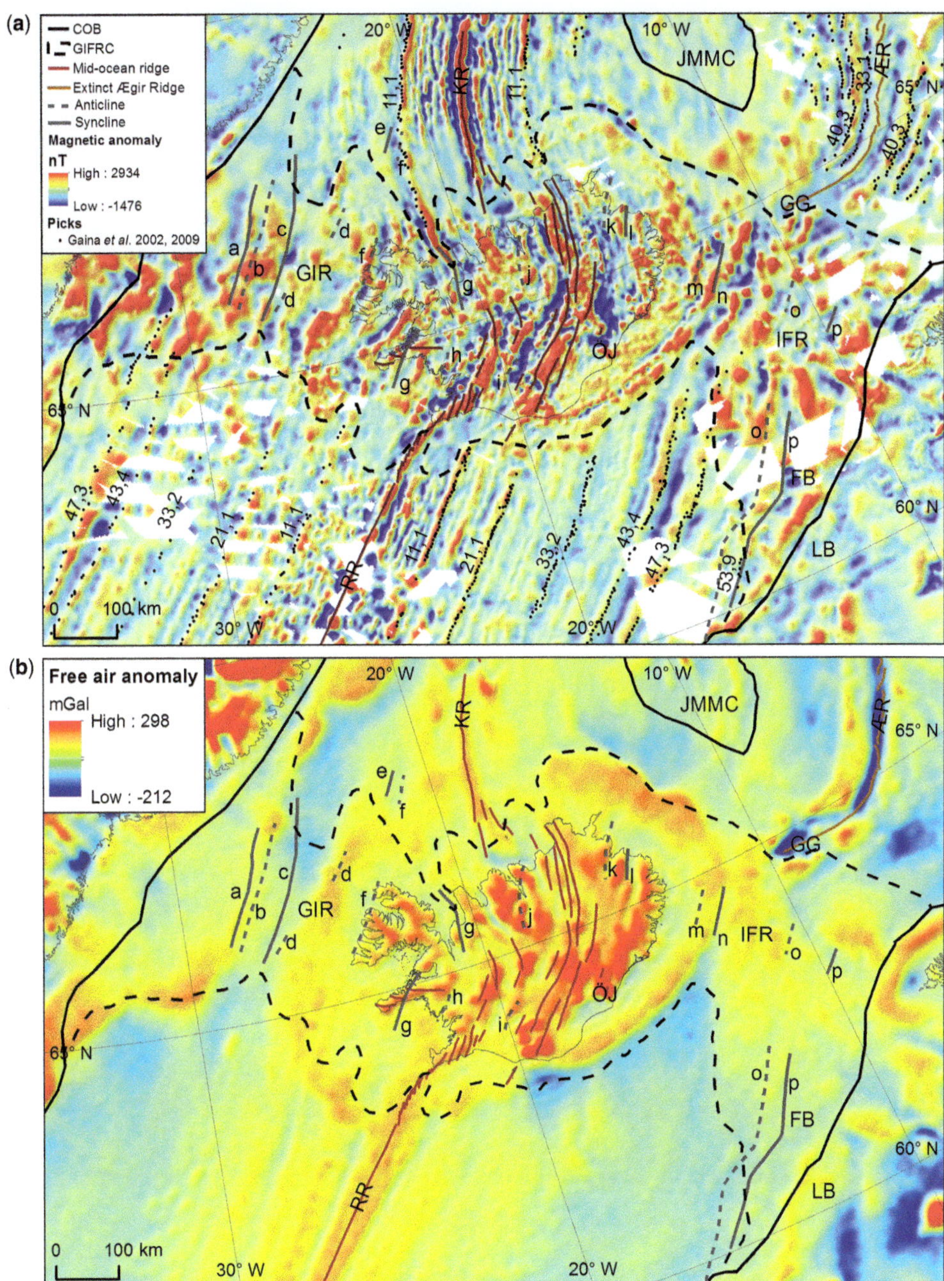

Fig. 3. (**a**) Magnetic intensity anomaly map (Nasuti & Olesen 2014). A dotted line indicates the outline of the GIFRC and the solid thick black line indicates the continent–ocean boundary. Magnetic anomaly data appear as regular time-parallel strip patterns within the oceanic crust domain, but within the GIFRC are irregular and chaotic. This is thought to be the result of subaerial volcanism and repeatedly active areas through time in the same areas, overprinting the first-order magnetic response (Nunns *et al.* 1983). Synclines and anticlines are shown on the map with a name key given in Table 3. (**b**) Free-air gravity anomaly map (Haase & Ebbing 2014) providing an insight into the subsurface features and the main outline of the GIFRC. Gravity data are also the basis for the Moho interface and crustal thickness distribution inversions for crustal thickness estimates. *Abbreviations*: GIR, Greenland–Iceland Ridge; KR, Kolbeinsey MOR; JMMC, Jan Mayen microcontinent; ÆR, Ægir MOR; GG, Gridar Gorge; IFR, Iceland–Faroe Ridge; FB, Faroe Basin; LB, Lousy Bank; RR, Reykjanes MOR.

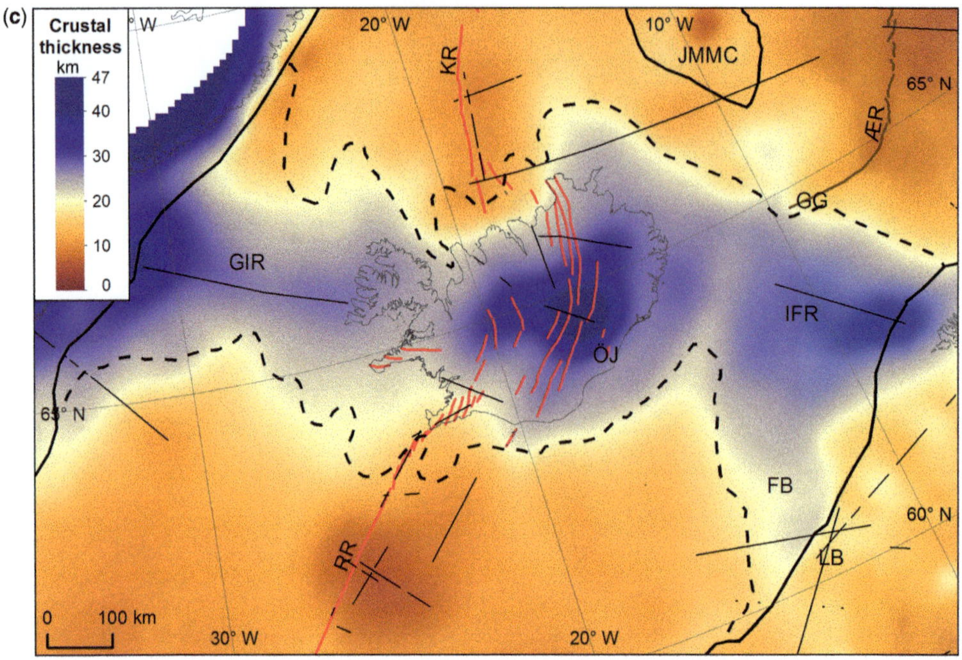

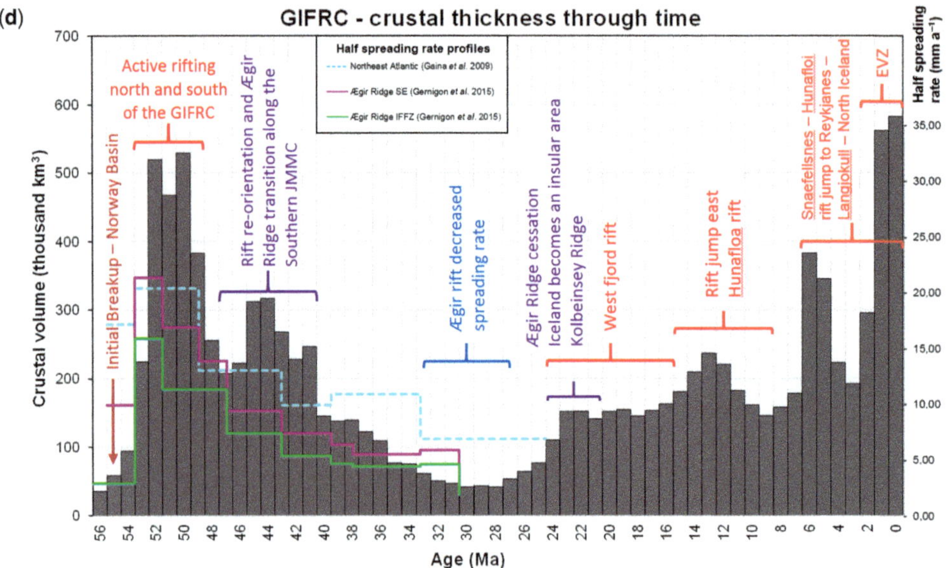

Fig. 3. (*Continued*) (**c**) Crustal thickness map based on gravity inversion of the Moho interface (Funck *et al.* 2014). The thin black lines show the locations of the seismically constrained Moho (SCM) points from seismic refraction data (Funck *et al.* 2014). The GIFRC area shows clear crustal thickness variations, and consists of a much thicker crust than the oceanic crust south and north of the ridge complex. (**d**) Crustal volume–GIFRC crustal thickness estimates through time, by comparing age grid data (Gaina 2014) to crustal thickness data, summarized as cumulative datasets for 1 Ma increments since break-up. This crustal accretion v. time plot is then compared with half spreading rate profiles for the NE Atlantic by Gaina *et al.* (2009), and for the southern part of the Norway Basin by Gernigon *et al.* (2015). It has been shown that the GIFRC follows the NE Atlantic opening process until the rift reorganization and transfer from the Ægir MOR across to the Iceland Plateau Rift to the Kolbeinsey MOR. After the cessation of the Ægir Ridge system, the GIFRC shows a good correlation to the idea that the mantle plume linked to the Icelandic MOR system caused the rift jumps from the Westfjord to the Snæfellsnes–Húnaflói Rift and from there to the Western Volcanic Zone. Finally, can we observe the still active rift transfer to the East Iceland Volcanic Zone (EVZ).

Submarine volcanic complexes were mapped offshore to localize rift centres mainly using high-resolution bathymetry and multibeam data, gravity-magnetic anomalies data, and seismic reflection data, where applicable.

Mapping data compilations and methods

The combination of various data observations was vital to validate areas of rift jumps and the formation of complex igneous structures across the GIFRC. These were obtained by accessing the palaeomagnetic arrangement, and comparing the area to the surrounding oceanic basement morphology and crustal thickness variations through time.

Rift jumps, synclines and anticlines

Relocation of the magmatic focus through rift-jumping is a prominent process in the evolution of Iceland, and appears to apply to the GIFRC in general. Rift jumps generate unconformities that have been observed across Iceland, marked as time hiatuses, often accompanied by distinct sedimentary horizons containing plant remains between lava formations (Denk *et al.* 2011). The exposed Iceland crust contains evidence of several rift jumps during the last 16 myr (Sæmundsson 1974, 1979; Jóhannesson 1980; Pálmason 1981; Harðarson *et al.* 1997, 2008; Hjartarson 2003). Here, the mid-Atlantic ridge axis approached the mantle plume from the SE, fully connected to it at approximately 40–35 Ma (Gaina *et al.* 2017) and proceeded with a WNW drift with respect to the plume at present. At the same time, the active rift axis attempted to maintain its position near the centre of the plume by repeatedly relocating its rift system, leaving behind extinct axial rift zones. Such extinct rift zones have been identified as synclinal structures within the volcanic formations across Iceland (Sæmundsson 1979; Jóhannesson 1980; Harðarson *et al.* 2008) ('g'–'l' in Fig. 2 & Table 4). Our investigations and interpretation of seismic reflection data profiles across the GIFRC, specifically east and west of Iceland, have revealed several potential synclines and anticlines ('a'–'f' and 'm'–'q' in Figs 2, 4–6).

A seismic reflection study along the southern flank of the IFR has revealed a complex system of rifts and volcanic facies of different, but unknown, ages (Figs 5 & 6) (Erlendsson & Blischke 2013). The seismic section in Figure 6 extends along the SW slope of the IFR, with a SW–NE orientation (location given in Fig. 2). This section shows clearly the internal complex structures of two rift relocations, where syncline, anticline and SDR structures are interlayered along the IFR and Lousy Bank margins. As no core or seafloor samples are available, it is only possible to link to magnetic data interpretations to obtain an approximate time frame for these ridge jumps. The interpreted syncline may be defined as an extinct volcanic zone or rift system, possibly a remainder from first break-up events in 55–53 Ma between Europe and Greenland (Fig. 6) (Gaina *et al.* 2009). The syncline has clear inwards-sloping reflectors and a clearly visible anticline is observed along the western slope of that syncline. This anticline may have developed due to a westwards ridge jump, when the old volcanic zone or ridge system within the interpreted syncline became extinct. Further evolution of a new rift system can be seen as sets of SDR sequences, dipping westwards and forming the Proto-Reykjanes MOR located west of the described anticline. An extinct stratovolcano near the eastern end of the section, covering partially the previous dipping reflector, indicates a reactivation of the old rift located within the above described syncline. Remains of the youngest volcanic activity are observed within the sediments as sill intrusions, volcanic ridges and cones at the seafloor.

Igneous complexes

Igneous centres are subdivided into four groups in the onshore and offshore areas, such as igneous complexes, inactive and active central volcanoes, and seamounts (Fig. 1), which are described in detail by Hopper *et al.* (2014) and Gaina *et al.* (2016). These submarine complexes are compared to active igneous centres along the rift zones across Iceland, as potential present-day geological analogues (Fig. 4) that are formed due to persistent volcanic eruptive or vent systems along fault/fissure swarms (Sæmundsson 1979), building complex volcanic structures over time. Igneous centres can be seen on seismic reflection data as mounds with steeply dipping flanks or as flat-topped structures due to erosion.

Palaeomagnetism arrangement, unconformities and age distribution

Magnetic anomaly data and geochron model data (Gaina 2014; Nasuti & Olesen 2014) are key to defining the outlines of the GIFRC (Fig. 3a). The regular pattern of the magnetic anomalies on the ocean floor south and north of Iceland, that are used for age determinations, is mixed up and shows a quasi-chaotic pattern across the GIR and the IFR, making it very hard or impossible to cross-interpolate the magnetically derived age interpretations. The patchy magnetic pattern of the Greenland–Iceland–Faroe Ridge is thought to be a result of subaerial lava extrusions (Nunns *et al.* 1983), interacting with the topography of pre-existing flows, and

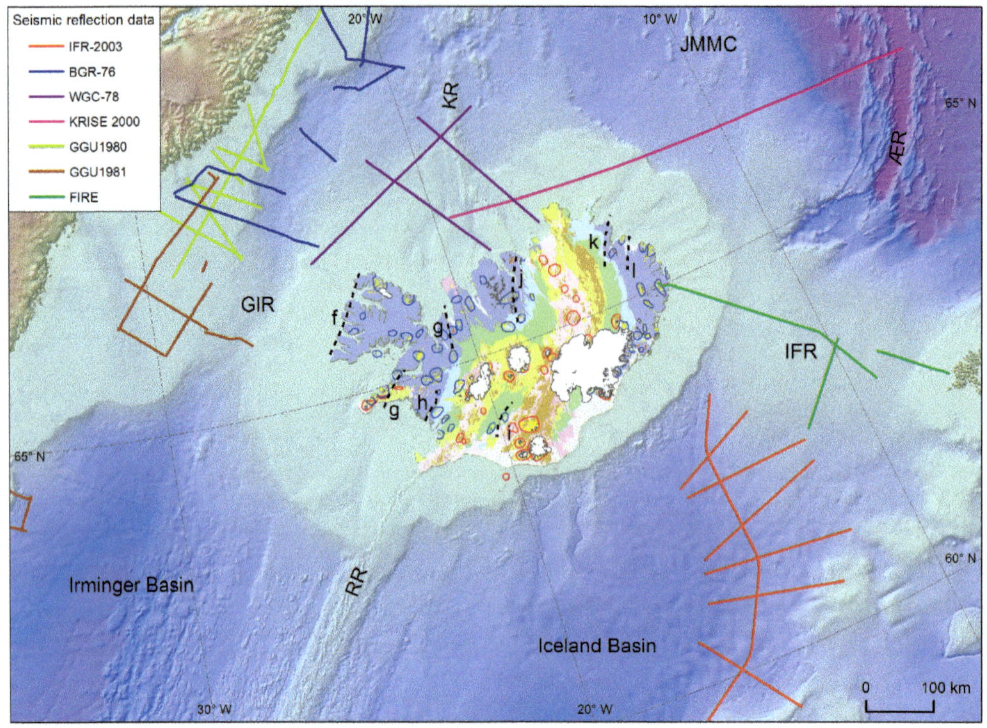

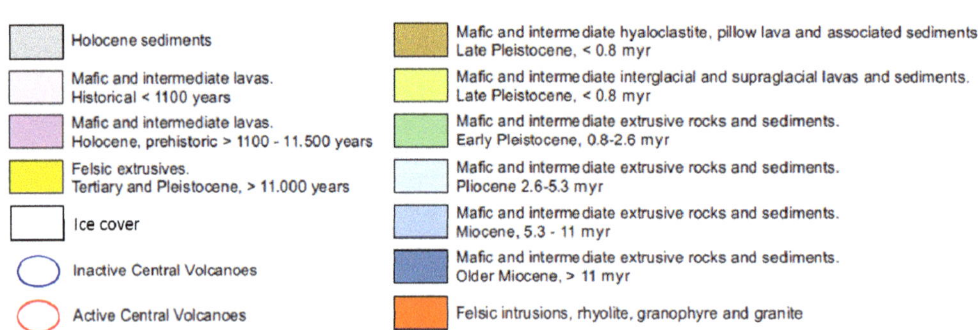

Fig. 4. Regional map showing the seismic reflection lines (see Table 2) and the geological map of Iceland (Hjartarson & Sæmundsson 2014). Black dotted lines indicate the onshore syncline and anticline from Jóhannesson & Sæmundsson 2009 (see Table 4).

being further complicated by frequent rift jumps and erosion. Very few direct age measurements have been carried out on the offshore part of the GIFRC, and all ODP and DSDP site boreholes are outside the area. Therefore, the magnetic anomaly interpretation has been extrapolated across the ridge complex (Gaina *et al.* 2017).

Magnetic anomalies were compared to unconformity observations, geochron age grid interpretation (Gaina *et al.* 2017), localized syncline and anticline observations, and crustal thickness and ocean-floor morphology.

Unconformities are known from onshore records since 16 Ma (Denk *et al.* 2011; Stoker *et al.* 2014), but older unconformities offshore are visible on seismic reflection data as horizons, where the younger strata lie discordantly on an older erosion surface (Figs 5 & 6).

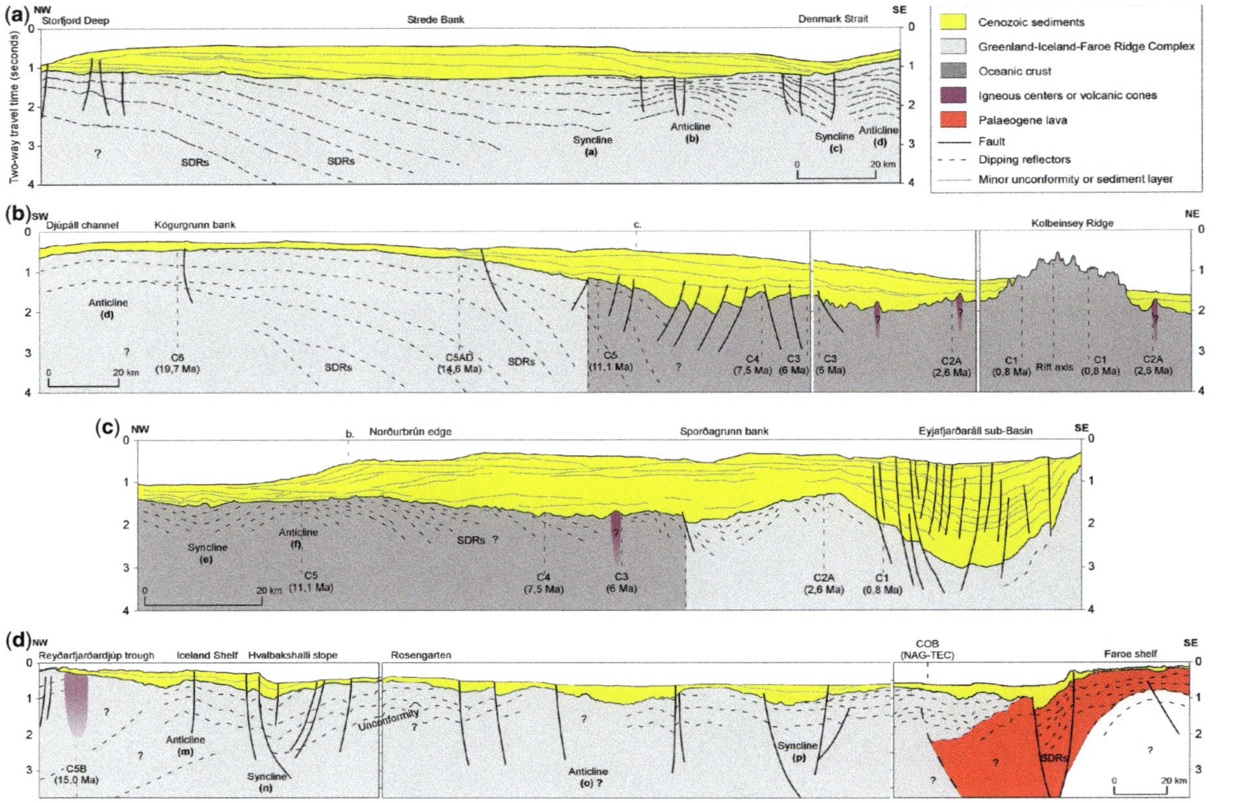

Fig. 5. Geoseismic profiles from the Greenland–Iceland Ridge, North Iceland Shelf and Iceland–Faroe Ridge. Key structural features to note include: faulted and folded layering (anticlines and synclines) within the basement is observed in all sections. Profile locations are shown in Figure 2. (**a**) shows SDRs along the eastern Greenland margin, and synclines and anticlines in the SE part of the section (**b**) is the western part of the section, which is smooth and shows only indistinct sub-parallel internal layering of the crust. But further east eastwards-dipping reflectors appear; they become steeper and are clearly analogous to the SDRs found on the volcanic continental margins. This part of the profile is interpreted to be a part of thicker oceanic crust defined as the Greenland–Iceland–Faroe Ridge Complex. The structure of the basement east of the SDRs becomes rougher and more faulted, where the crust slightly changes to a typical oceanic crust. (**c**) shows the correlation between the infill of the Eyjafjörður Sub-basin and the outer shelf succession. A rather clear syncline and an anticline are observed in the SW part of the section. (**d**) is a composite section showing well-defined structures of the basaltic basement and lava sequences. Both margins appear to be affected by large normal faults, and SDR structures are observed along the Faroe Shelf, covering a nicely shaped anticline that might be interpreted as a possible crystalline basement. (Ages are based on Gaina 2014.)

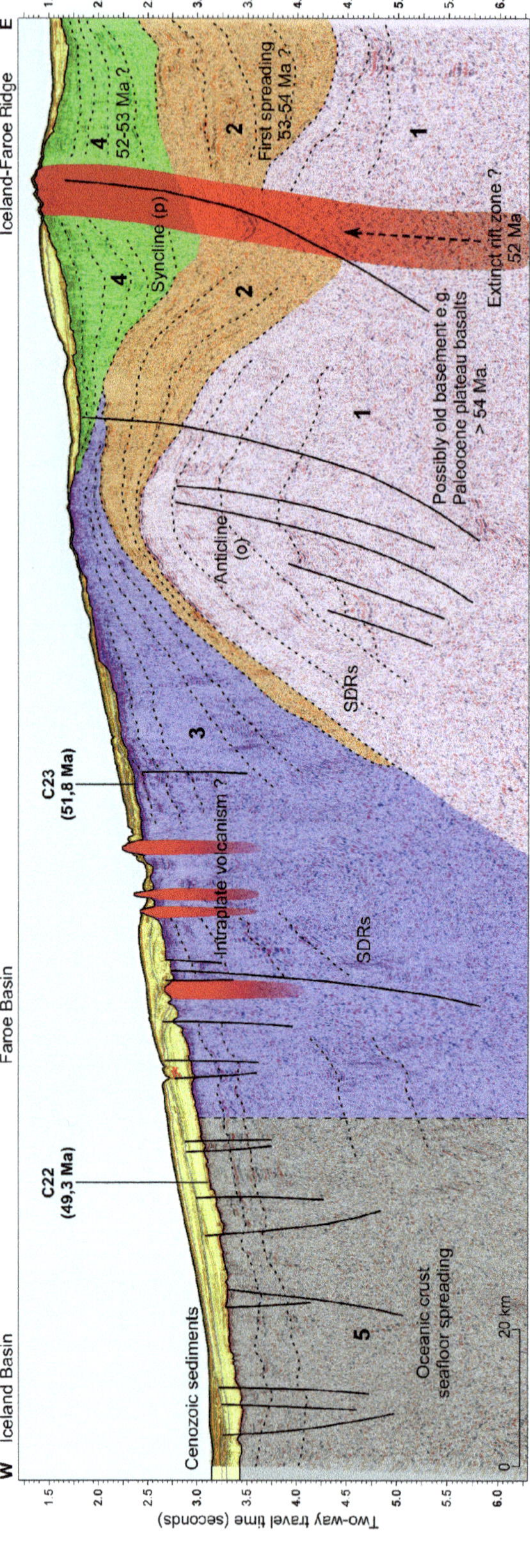

Fig. 6. Geoseismic profile based on seismic reflection data, extending from the Iceland Basin along the SW slope of the Iceland–Faroe Ridge (see Fig. 2 for location and Table 4). The section shows the internal structures of the basaltic basement, where syncline, anticline and SDRs are visible, forming a complex rift propagation structure that is specific for the GIFRC. The numbers and different colours display different chronologically volcanic formations. (Ages are based on Gaina 2014.)

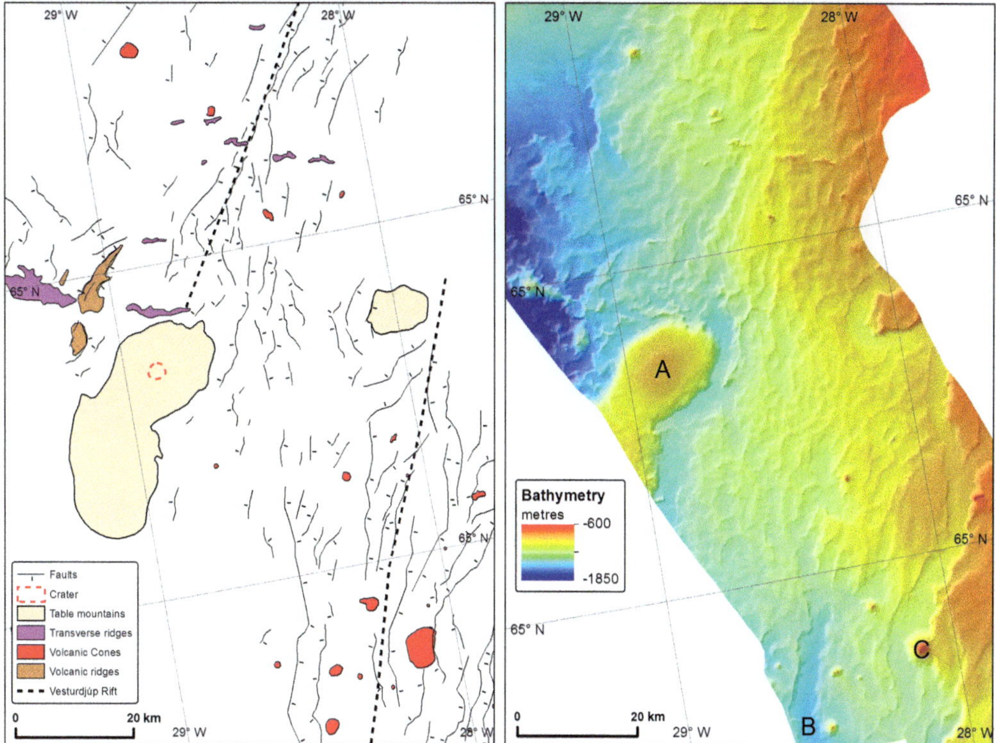

Fig. 7. Vesturdjúp multibeam data from the Icelandic Marine Research Institute (multibeam data available from the Icelandic Marine Research Institute website: http://www.hafro.is). Caris and Cfloor software was used to clean, processes and grid the data (100 × 100 m). Features such as volcanic cones, crater, table mountains, traverse and volcanic ridges, faults, and rift valleys were outlined and interpreted on the multibeam data and differentiated as shown in the legend (see Fig. 2 for the location).

Oceanic basement morphology

The oceanic basement morphology is primarily controlled by an interplay between tectonic and magmatic processes (e.g. Carbotte & Macdonald 1994; Ito & Behn 2008), and whether an eruption takes place subaerially or subaqueously (Fig. 1). Thus, water depth has a significant influence on the morphological characteristics of lava flows. This determines the ocean-floor surface characteristics, as they are different for sheet flows, pillow basalts and hyaloclastite formations, directly linked to volcanic seismic facies occurring along rifted margins (e.g. Planke *et al.* 2000; Hopper *et al.* 2003; Elliott & Parson 2008; Gaina *et al.* 2016; Geissler *et al.* 2016). The different morphology types (Fig. 1) are subdivided into (according to Funck *et al.* 2014): (1) smooth basement of strong subplanar continuous, high-amplitude, seismic reflections that are very common in the areas of SDRs; (2) transitional basement, where seismic reflections are discontinuous or absent, and the ocean floor is structurally affected and broken up by smaller-scale faults, creating a moderately irregular surface; (3) rough basement, indicating areas with significant basement relief and frequent occurrences of gravity anomalies, indicating volcanic and structural processes affecting the area, most commonly closer to the present-day MORs; (4) rubbly basement that is characterized by a seismically chaotic reflector package with limited energy penetration beneath, mostly close to the break-up margins (Elliott & Parson 2008); (5) volcanic cones and seamounts that form isolated, high-relief features of typically small peaks above the surrounding seafloor, often clustered in areas of increased magmatic activity, such as the GIFRC (Figs 6–8); and (6) the Greenland–Iceland–Faroe Ridge Complex that is a build-up of complex, overlapping rift, igneous and structural systems, thus forming an anomalously thick crust across the complex.

Crustal thickness of GIFRC

Crustal thickness derived from gravity and seismic refraction data (Haase *et al.* 2016) is one of the

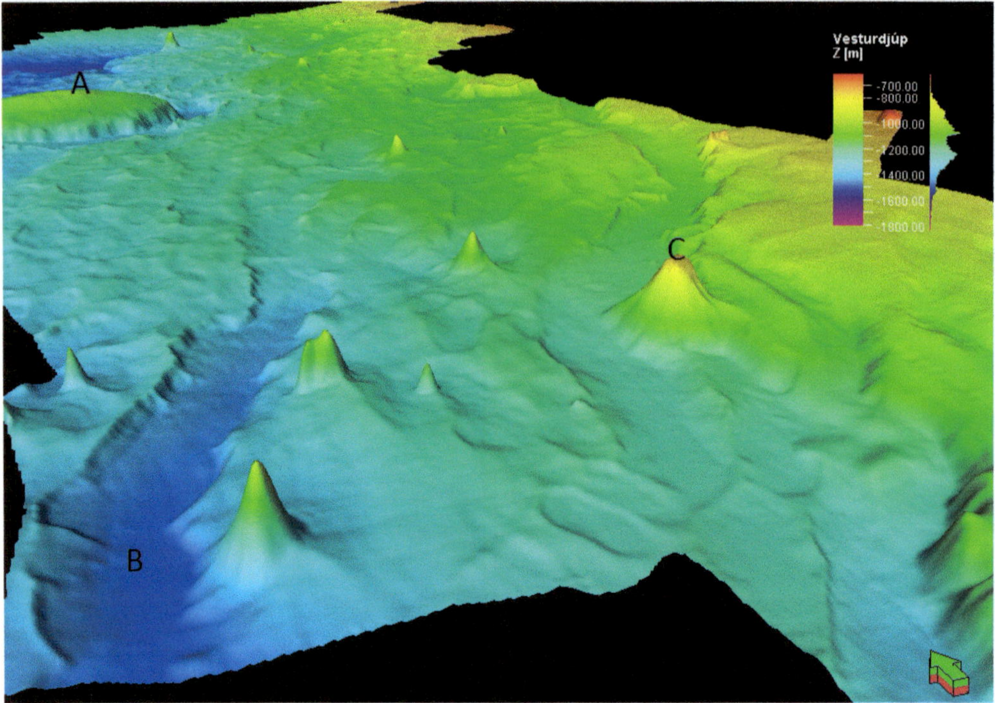

Fig. 8. 3D view of the Vesturdjúp multibeam data. Cone-shaped seamounts, faults and graben structures can clearly be seen in the foreground. Part of two table mountains may be seen in the distance. The geographical location is shown in Figure 2.

main factors that was used to define the outline of the GIFRC. The anomalously thick crust of the GIFRC is a result of the production of the interaction between the MOR and the Iceland plume (Fig. 3c) (White & Lovell 1997; Funck *et al.* 2014). The thickest section of up to 40 km-thick crust of the GIFRC is located in SE central Iceland, right next to the postulated centre of the Iceland plume area, and next to the COB boundaries at its eastern and western extensions (Darbyshire *et al.* 2000; Allen *et al.* 2002; Funck *et al.* 2014). The average crustal thickness along the FIR and GIR is estimated to be about 30 km (Fig. 3c) (Richardson *et al.* 1998; Holbrook *et al.* 2001). As the GIFRC area is clearly a region with a thick oceanic crust (Fig. 3c), the crustal thickness grid data was compared to the age grid data presented by Gaina (2014). All crustal thickness grid points within a grid cell of 4000×4000 m were assigned a value of the overlapping age grid within the GIFRC area. All thickness values were summarized in 1 myr steps since break-up time. A cumulative graph was generated showing the cumulative amount of crustal thickness for each 1 myr step (Fig. 3d).

Crustal thickness variations between 15 and 40 km are observed beneath Iceland (Fig. 3c). This could be related to changes in magmatic activity of the MOR–Iceland plume system through time (Darbyshire *et al.* 2000) (Fig. 3d). Alternatively, these could be related to pre-existing complex structural settings within the subsurface (Fig. 6). Foulger & Anderson (2005) proposed an explanation for the increased crustal thickness in SE central Iceland, suggesting that microplates of older oceanic crust are submerged beneath that region. Recently, Torsvik *et al.* (2015) also proposed that geochemical data indicate that fragments of continental crust are present beneath the SE coast of Iceland, suggesting the presence of a SW extension of the Jan Mayen microcontinent, with deeply buried fragments under volcanic rocks. However, there is no evidence of a distinct lateral velocity anomaly seen in refraction data for east Iceland (Funck *et al.* 2014).

The review of seismic reflection profiles in Figures 5 and 6 show that the IFR is structurally complex, with multiple examples of old crustal blocks buried beneath younger subaerial volcanic sections. This vertical crustal build-up is not just due to increased magmatism from the ridge–plume system, but also due to complex structural events in that area (Blischke *et al.* 2016). However, changes in magmatic production and crustal accretion can be seen through time by estimating the cumulative crustal thickness for each 1 myr increment across the GIFRC (Fig. 3d). Here, it can be shown that regional events from break-up at around 55 Ma through to the present are reflected in the crustal volume inferred. These include initial break-up, regional and localized rift jumps, decreased spreading activity,g and ridge cessation.

GIFRC rift centres and rift relocations

The complexity of the GIFRC appears to be closely connected to frequent rift jumps. Several rift jumps are known and documented (Table 3). One specific example is the Ægir Ridge system that formed during the initial opening of the NE Atlantic, with an initial break-up phase between 55 and 53 Ma, and was fully established around 50 Ma (Gaina *et al.* 2009; Gernigon *et al.* 2015). It propagated from north to south, spanning the distance between the IFR and the Jan Mayen Transform Zone (Blischke *et al.* 2016). Spreading on the Iceland Plateau rift took place simultaneously with that on the Ægir Ridge from 49 to 25 Ma along the Iceland–Faroe Fracture Zone, before the complete rift transfer to the Kolbeinsey Ridge system, connecting the Reykjanes Ridge directly and separating the Jan Mayen microcontinent from the central East Greenland coast (Brandsdóttir *et al.* 2015; Blischke *et al.* 2016).

Cessation of seafloor spreading and extinction of the Ægir MOR system occurred in the Early Oligocene, around 24–21 Ma, coinciding with the activation of the Kolbeinsey Ridge around anomaly C6b, or 22–21 Ma (Gernigon *et al.* 2015), and Iceland becoming an insular shelf probably due to the plume–ridge activity (Fig. 3d). Intensive volcanism and a high lava production rate accompanying the initiation of the Kolbeinsey MOR led to the formation of the Iceland Shelf as a volcanic region within the GIFRC.

The NW Rift Zone is thought to have formed some 24 myr ago west of the NW peninsula of Iceland (Harðarson *et al.* 1997) (Table 4). It was most likely to have been a direct continuation of the Kolbeinsey Ridge, forming its oldest and southernmost part. It was active for 8–10 myr until approximately 15 Ma. Its exact location has always been speculative, as its manifestations have never been clear in geophysical potential field data. However, in seismic reflection data profiles, syncline structures can be seen (syncline ‘e’ in Figs 2 & 5c). The site is just north of the GIR and is approximately parallel to the 15 Ma time line according to the geochron model of Gaina (2014). We suggest that this hypothetical NW Rift Zone, thought to be found somewhere on the insular shelf off the NW peninsula of Iceland (Harðarson *et al.* 1997), correlates with syncline ‘e’, thus confirming this ancient spreading axis by geophysical data. The seawards-dipping formations of the NW Rift Zone are nearly totally submerged below the seafloor, except for their easternmost extensions, exposed along the outermost coast of the Icelandic Westfjords, where they form the anticline structure ‘f’ shown in Figures 2, 5c and 9b. A lignite horizon overlies these formations, representing a 1–1.5 myr hiatus before the next rift jump and before the onset of the Snæfellsnes–Húnaflói Zone took place (Riishuus *et al.* 2013) (Figs 2 & 3d).

The Snæfellsnes–Húnaflói Rift Zone formed approximately 14–15 myr ago by an eastwards spreading centre relocation from the NW Rift Zone, and was active for about 8–10 myr (Harðarson *et al.* 2008). This rift zone is present onshore west Iceland as regional dipping formations that form a distinct syncline centre line of that rift zone (Fig. 2). There are two segments of the rift zone located in Snæfellsnes and Húnaflói, respectively, which may have been connected by a transform fault system. The majority of the Icelandic subaerial Miocene volcanic strata was formed in this rift zone.

The present-day rift zones formed approximately 6 myr ago by relocating active spreading from the Snæfellsnes–Húnaflói Rift Zone to its present location, forming the Western Volcanic Zone (WVZ) and the Northern Volcanic Zone (NVZ) (Sæmundsson 1974, 1979; Jóhannesson 1980) (Figs 2 & 4). However, the rift zone remained active until about 5 myr ago (Pringle *et al.* 1997). Separately, the most recent rift relocation to the Skagafjörður Rift Zone took place in north Iceland, becoming activated at 1.7 Ma and forming a temporary rift axis for about 1 myr (Hjartarson 2003). The East Iceland Volcanic Zone (EVZ) in south Iceland appears to be an evolving spreading system (Sæmundsson 1979) (Figs 1 & 3d). This zone was initiated 2–3 myr ago, and is slowly propagating to the SE from the WVZ towards the EVZ, forming a dual-zone rift system (Einarsson 2008).

Central volcanoes and seamounts on the GIFRC

Central volcanoes play an important role in the build-up and structure of the Icelandic volcanic strata and have been studied intensively (e.g. Sæmundsson 1979; Harðarson *et al.* 2008). Very little is known about their existence and role within the submarine areas of the GIFRC. The central volcanoes of Iceland can be divided into rift zone and off-rift central volcanoes. The off-rift central volcanoes often form high and prominent stratovolcanoes (e.g. Snæfellsjökull and Eyjafjallajökull) (Fig. 2),

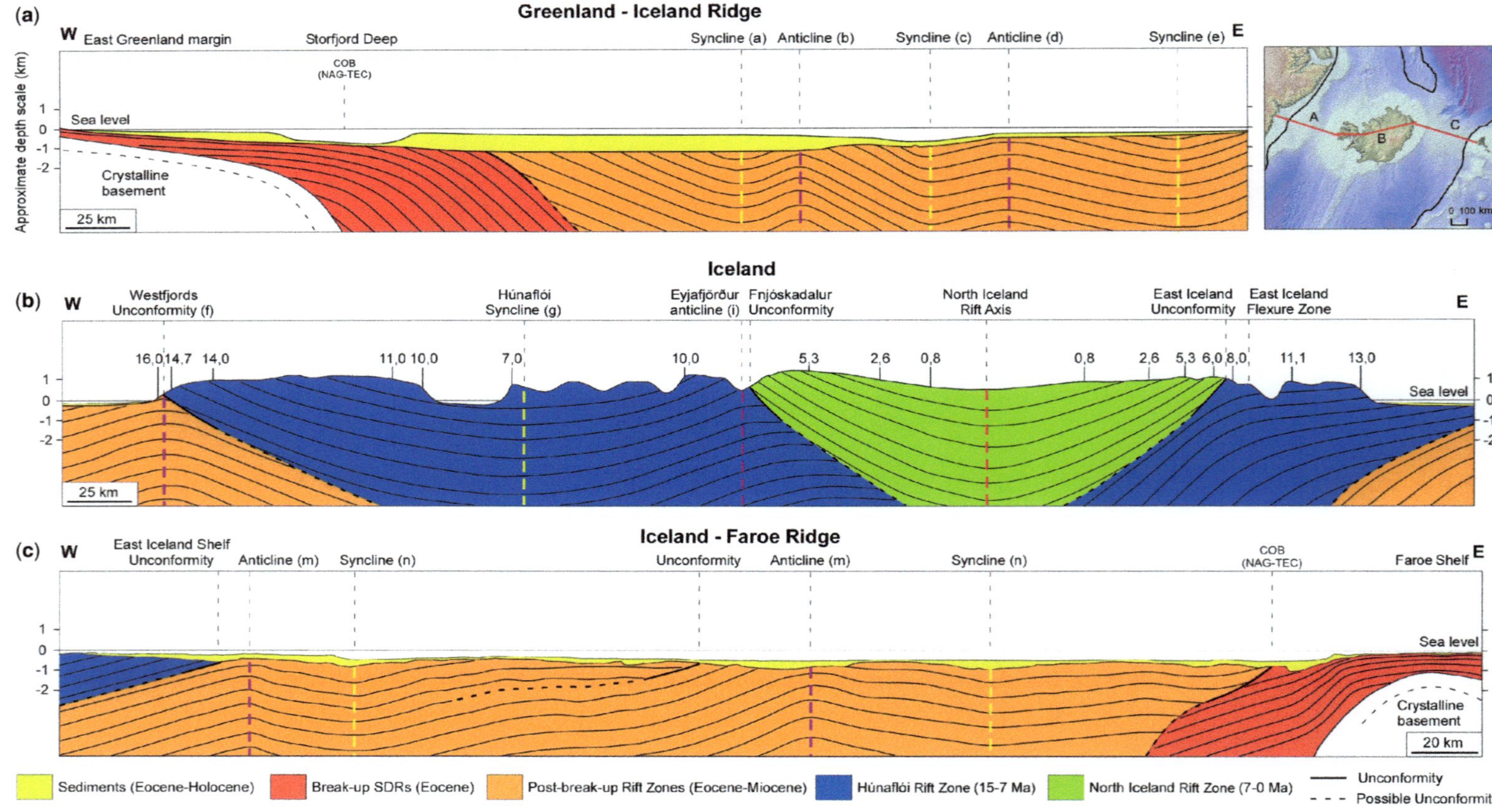

Fig. 9. Schematic cross-section based on an interpreted seismic section and the geological map of Iceland. This section extends from the East Greenland margin, across the Greenland–Iceland Ridge, Iceland and Iceland–Faroe Ridge and towards the Faroe Shelf. The main purpose of this section is to illustrate in a schematic way the development, basement dip, and synclines and anticlines across GIFRC based on seismic reflection data (for Section (a) and Section (b) see Fig. 5), surface geology mapping and data observations (for Section (b) see Fig. 4 and Hjartarson & Sæmundsson 2014), and data modelling (Böðvarsson & Walker 1964).

in contrast to rift zone central volcanoes that are lower and more irregular in shape, and many of them have formed calderas. The lifespan of an individual central volcano, until it cools down, has been found to vary from 300 kyr to over 1 myr (Sæmundsson 1979; Harðarson *et al.* 2008). The only known example of an active submarine central volcano today is the Njörður volcano situated on the Reykjanes Ridge close to the Icelandic shelf (Höskuldsson *et al.* 2013).

More than 40 inactive former rift zone central volcanoes are known in the Neogene formations of Iceland. They are often deeply eroded, and represented by acid and intermediate rocks, local cone sheets swarms, and swarms of regional dykes and faults (Sæmundsson 1979).

Central volcanoes also exist on the shelf area all around Iceland (Figs 1, 7 & 8). Several central volcanoes have been inferred from potential field data east and west of Iceland, and in some cases confirmed by dredging (Kristjánsson 1976; Jónsson & Kristjánsson 1997). They are thought to have formed subaerially, but have been submerged as their emplacement area cooled down, while drifting away from the spreading axis.

Seamounts are defined as isolated topographical features of volcanic origin, rising from the ocean floor that did not rise high enough to break through the sea level and turn into islands. Their height differs from some hundred metres up to 4000 m. They follow a distinctive pattern of growth, activity and cessation, and are generally formed near mid-ocean spreading ridges, over upwelling mantle plumes (hotspots) and in island-arc convergent settings (Staudigel & Clague 2010). Thus, they can give important clues as to where old rift systems might have been located in magmatically inactive areas.

Seamount features have been mapped across the GIFRC area, and north and south of Iceland (Funck *et al.* 2014; Gaina *et al.* 2016). They are mostly situated on the deep ocean floor on both sides of the Reykjanes and Kolbeinsey ridges (Fig. 1). Some are near to the GIFRC but very few on the ridge complex itself. Possibly the igneous complexes on the GIFRC were formed subaerially, partially eroded after cessation and submerged due to thermal cooling. However, shapely seamounts have been found close to the ridge complex, specifically on the flanks of east and west Iceland's offshore areas (Figs 1, 7 & 8). In the Vesturdjúp Basin, west of Iceland and just south of the GIR at around 1200 m depth, a group of small seamounts has been identified (Figs 7 & 8). Helgadóttir (2012) has described and discussed these as volcanoes or mud volcanoes. Most of them are cone-shaped ridges, but eroded table-like mountains are also found. Because of these various types of igneous-complex-like structures, conventional volcanism appears to be the most likely cause for these features. The largest cone rises to around 500 m above the surroundings, with a diameter of 5000 m. These seamounts appear to be much less eroded and younger than the neighbouring ocean floor, and possibly indicate a flank igneous system or intraplate volcanism, accompanied by young tectonism with faults, graben and transverse ridges characterizing this area (Fig. 8).

Discussion

In order to assess the development of the GIFRC as an igneous complex within the NAIP, this section addresses individual key stages that affected the GIFRC since the break-up of the North Atlantic (Fig. 3d).

Initial break-up

The GIFRC started to form along with the continental break-up between Greenland and Eurasia, and the initiation of the seafloor spreading, 55–53 and 36 Ma, north and south of the GIFRC (Gaina *et al.* 2009; Gernigon *et al.* 2015), but not affecting the GIFRC to a great extent (Fig. 3d). During the Eocene, the Eurasian and North American continental margins were located very close to each other and the plume situated below Greenland, sustaining a subaerial connection between the two continents, and forming a land bridge between Greenland and the Faroes and onwards to the European continent. This is supported by geoseismic investigations (Parnell-Turner *et al.* 2014), as well as palaeobotanical evidence (Denk *et al.* 2011).

Active rifting north and south of the GIFRC

The first rifting phase (53.36–49 Ma) of the NE Atlantic after break-up affected the GIFRC area heavily, with the emplacement of large volumes of extrusive and intrusive magmatic material building up the oldest part of the complex (Fig. 3d). During this phase, overlapping rift systems were active (Figs 5, 6 & 9) that overlaid older crustal segments and led to very thick crustal formation from east Iceland to the IFR region (Fig. 3c).

Rift orientation and Ægir Ridge transition

Continuous spreading was active in the Reykjanes MOR system to the south, but rift transfer started to form from the Ægir Ridge system along the Iceland Plateau Rift (IPR) corridor south of the JMMC between approximately 49 and 40 Ma. This is reflected in crustal accretion of the GIFRC, with increased magmatic activity between the IPR system and the Iceland–Faroe Fracture Zone (IFFZ) (Blischke *et al.* 2016) (Figs 1 & 3d). Recent reconstruction work of the region indicates that the East Iceland Shelf edge is parallel to the proto-Reykjanes Ridge location at anomaly C19n (40.32 Ma) (Gaina 2014; Blischke *et al.* 2016), cutting into the older crust of the IFR area (Fig. 3c, d).

Decrease in spreading rate along the Ægir MOR system

The Greenland–Eurasian plate system moved NW relative to the mantle plume, which was situated at 35–30 Ma below the Greenland margin (Fig. 1) (Torsvik *et al.* 2015). Oceanic spreading was active along the Reykjanes and the Ægir MORs, but was gradually slowing down to ultra-slow-spreading past 30 Ma within the Ægir MOR system (Gernigon *et al.* 2015). This also directly affects the GIFRC, where much lower crustal accretion volume through time can be observed (Fig. 3d) in connection with the slowing down of the rift systems. Rift jumps possibly took place in a westwards direction that can be seen in the seismic reflection record in the form of synclines and anticlines along the Greenland–Iceland and Iceland–Faroe ridges (Figs 2, 6 & 9).

Ægir Ridge cessation: Kolbeinsey Ridge insular shelf

Spreading activity along the Ægir Ridge ceased around 22 Ma (Gernigon *et al.* 2015) (Fig. 3d) and seafloor spreading concentrated only along the Kolbeinsey Ridge from 24 Ma onwards. The process of establishing a new plate boundary and ultimately the Kolbeinsey Ridge resulted in the extension of the Greenland margin situated immediately north of the Reykjanes Ridge, which led to the final detachment of the Jan Mayen microcontinent (JMMC) from the central East Greenland margin (Blischke *et al.* 2016). The initiation of the Kolbeinsey Ridge, and the interaction between the Iceland plume and the newly formed MOR system, led to an increase in magmatic and volcanic activity along the GIFRC area, as well as further north along the western to SW margin of the JMMC.

Recent age models based on palaeomagnetic chron interpretations of the ocean floor around Iceland indicate a major hiatus crossing the insular shelf near the eastern to SE coast (Gaina *et al.* 2017) that may possibly be identified on seismic reflection data interpretations (Fig. 5). We suggest that this hiatus is related to increased magmatic accretion (Fig. 3d) of the central Iceland region, with extrusive rock discordantly overlying older igneous formation and crust, forming a much thicker crust in that area. The subcrop of this unconformity boundary is buried beneath thick layers of sediments, 8–10 km inside the bathymetric shelf break, according to Jónsson & Kristjánsson (1997), but can be seen near the anticline 'm' in Figures 2 and 5d. The age of the volcanic rocks at the edge of the East Iceland Shelf might be 20–24 Ma, which would correlate to the original opening time of the Kolbeinsey Ridge system during its initiation. The age of the underlying volcanic basement for that section might be around 40 Ma according to the age model by Gaina *et al.* (2017) (Fig. 3a), corresponding to a hiatus time span of approximately 16–20 myr.

This increase in magmatic activity, and most probably thermal uplift of the GIFRC area, created the insular shelf of proto-Iceland; thus creating a major hiatus and related unconformity between the young formations of the Kolbeinsey Ridge and the older volcanic basement of the IFR and GIR (Figs 5d & 9c). This early stage of the subaerial insular shelf region is today mostly submerged, but forms large areas of the insular shelf in the east, west and north of Iceland.

Miocene: Pliocene rift jumps on Iceland

In the Early Miocene, the mid-Atlantic ridge axis approached and crossed the location of the Iceland plume (Harðarson *et al.* 2008). Since then, the spreading ridge systems in Iceland have remained linked to the plume. As the spreading axis moves away from the central plume location, the rift centres are periodically recaptured by the plume through rift-jumping. It has been proposed that a complete rift cycle for Iceland lasts for at least 12 myr, from initial propagation to extinction (Harðarson *et al.* 1997, 2008). The control on rift-jumping is clearly related to the interaction of the static mantle plume with the overlying NW-migrating plate of Eurasia (Gaina *et al.* 2017). Relocation of active magmatism towards the plume may simply be a response to this migration.

As the North Atlantic Ocean widened, the marginal eastern and western parts of the land bridge cooled, partially eroded and gradually submerged (Denk *et al.* 2011), as can be seen as a base Cenozoic sediment horizon in Figures 5d and 9. This first took place along the eastern area of the IFR, followed by the western area due to its proximity to the mantle plume. Palaeobotanical observations indicate that the latest evidence for plant migration on land between Europe and Iceland is dated at around 9 myr ago, and between Greenland and Iceland around 6 myr ago (Denk *et al.* 2011). The age of the GIR and the IFR as submarine areas is therefore less than 10 myr, and the age of Iceland as an isolated island is approximately 6 myr.

Conclusions

The Greenland–Iceland–Faroe Ridge Complex (GIFRC) has been in development since the opening of the NE Atlantic at around 55 Ma. It appears as a prominent feature in all geological and geophysical datasets. Its shape and size can be drawn in slightly different ways according to the various data sources available. Despite a small areal outline compromise between the different datasets, comprising bathymetry, gravity, magnetic and crustal thickness maps along with seismic profiles over the region, all show the area as an anomalous feature within the oceanic crustal fabric of the NE Atlantic.

Published synclines and anticlines have been summarized (Table 3), and several new synclines and anticlines that were revealed in seismic reflection data across the GIFRC east, west and north of Iceland (Figs 2, 5, 6 & 9). Specifically, the offshore anticlines and synclines may be related to old rift systems prior the forming of Iceland as an insular shelf region (>24 Ma). Synclines are suggested to be manifestations of former rift axes that have been abandoned by rift jumps. These rift jumps appear to be more common inside the GIFRC region than in the ocean basins south and north of the area, and can also be confirmed by the observation of cumulative crustal accretion through time (Fig. 3d).

Thus, the GIFRC represents a complex region of crustal accretion in three dimensions due to overlapping rift systems, complex interlinked rift and transform zones, and several unconformities that suggest a variable uplift and subsidence history for the ridge complex. An excellent example to visualize such processes of vertical crustal accretion and rift jumps is seen in seismic reflection data that extends along the SW slope of the Iceland–Faroe Ridge (IFR). They clearly display the internal structures of basement blocks, separated by a syncline and younger rift system, and the formation of an anticline across the deeply buried basement blocks that are overlain by SDR structures (Fig. 6).

We suggest a major hiatus (from 40 to 24–20 Ma) and a related unconformity at the boundary of the volcanic insular shelf edge of east Iceland and the Faroe Ridge, buried beneath thick layers of sediments, 8–10 km inside the bathymetric shelf break (Fig. 5d).

We suggest that the hypothetical NW Rift Zone, thought to be somewhere on the insular shelf off the NW peninsula of Iceland (Harðarson *et al.* 1997), correlates with syncline 'e' just north of the Greenland–Iceland Ridge (GIR) (Figs 2 & 5), parallel to age line 15 Ma of the geochron model by Gaina *et al.* (2017), and thus confirms this ancient spreading axis by geophysical data.

Several seamounts were observed on multibeam datasets from the Vesturdjúp Basin west of Iceland, just south of the GIR at around 1200 m depth (Figs 2, 7 & 8). Most of them are cone shaped, but ridges and table mountains are also found. These seamounts appear to be much less eroded and younger than the neighbouring ocean floor, and might indicate a still active flank or intraplate volcanic zone. Young tectonism with faults, graben and transverse ridges also characterize the area, and most of the volcanic cones are located along fault plans or/and within the graben of the Vesturdjúp (Fig. 8), giving a good example of the complexity of the GIFRC in comparison to simple ocean-floor areas.

This work is part of the Northeast Atlantic Tectonostratigraphic Atlas (NAG-TEC) consortium, with data permission provided by the National Energy Authority of Iceland (Orkustofnun), the Iceland GeoSurvey and GEUS (Geological Survey of Denmark and Greenland). Support from the following industry sponsors of NAG-TEC is gratefully acknowledged (in alphabetical order): Bayerngas Norge AS; BP Exploration Operating Company Ltd; Bundesanstalt für Geowissenschaften und Rohstoffe (BGR); Chevron East Greenland Exploration A/S; ConocoPhillips Skandinavia AS; DEA Norge AS; Det norske oljeselskap ASA; DONG E&P A/S; E.ON Norge AS; ExxonMobil Exploration and Production Norway AS; Japan Oil, Gas and Metals National Corporation (JOG-MEC); Maersk Oil; Nalcor Energy – Oil and Gas Inc.; Nexen Energy ULC, Norwegian Energy Company ASA (Noreco); Repsol Exploration Norge AS; Statoil (UK) Ltd; and Wintershall Holding GmBH. In addition, we would like to thank Bryndís Brandsdóttír (HÍ), Dr Bjarni Gautason (ÍSOR) and Dr Halldór Ármannsson (ÍSOR) for their help and advice in this project.

References

ALLEN, R.M., NOLET, G. *ET AL.* 2002. Plume-driven plumbing and crustal formation in Iceland. *Journal of Geophysical Research*, **107**, ESE 4-1–ESE 4-19, https://doi.org/10.1029/2001JB000584

Á HORNI, J., GEISSLER, W.H. *ET AL.* 2014. Offshore volcanic facies. *In*: HOPPER, J.R., FUNCK, T., STOKER, M., ÁRTING, U., PERON-PINVIDIC, G., DOORNENBAL, H. & GAINA, C. (eds) *Tectonostratigraphic Atlas of the North-East Atlantic Region*. GEUS, Copenhagen, Denmark, 235–253.

BJARNASON, I.Þ. 2008. An Iceland hotspot saga. *Jökull*, **58**, 3–16.

BLISCHKE, A., GAINA, C. *ET AL.* 2016. The Jan Mayen microcontinent: an update of its architecture, structural development, and role during the transition from the Ægir Ridge to the mid-oceanic Kolbeinsey Ridge. *In*: PÉRON-PINVIDIC, G., HOPPER, J.R., STOKER, T., GAINA, C., DOORNEBAL, H., FUNCK, T. & ÁRTING, U. (eds) *The NE Atlantic Region: A Reappraisal of Crustal Structure, Tectonostratigraphy and Magmatic Evolution*. Geological Society, London, Special Publications, **447**. First published online September 8, 2016, https://doi.org/10.1144/SP447.5

Brandsdóttir, B. & Menke, W.H. 2008. The seismic structure of Iceland. *Jökull*, **58**, 17–35.

Brandsdóttir, B., Hooft, E.E. & Murai, Y. 2015. Origin and evolution of the Kolbeinsey Ridge and Iceland Plateau, N-Atlantic. *Geochemistry, Geophysics, Geosystems*, **16**, 612–634, https://doi.org/10.1002/2014GC005540

Brooks, C.K. 1973. Rifting and Doming in Southern East Greenland. *Nature Physical Science*, **244**, 3.

Brooks, C.K. 2011. The East Greenland rifted volcanic margin. *Geological Survey of Denmark and Greenland Bulletin*, **24**, 96.

Böðvarsson, G. & Walker, G.P.L. 1964. Crustal drift in Iceland. *Geophysical Journal of the Royal Astronomical Society*, **8**, 285–300, https://doi.org/10.1111/j.1365-246X.1964.tb06295.x

Carbotte, S.M. & Macdonald, K.C. 1994. Comparison of seafloor tectonic fabric at intermediate, fast, and super-fast spreading ridges: influence of spreading rate, plate motions, and ridge segmentation on fault patterns. *Journal of Geophysical Research*, **99**, 13609–13631.

Darbyshire, F.A., White, R.S. & Priestley, K.F. 2000. Structure of the crust and uppermost mantle of Iceland from a combined seismic and gravity study. *Earth and Planetary Science Letters*, **181**, 409–428.

Denk, T., Grimsson, F., Zetter, R. & Simonarson, L. 2011. *Late Cainozoic Floras of Iceland. 15 Million Years of Vegetation and Climate History in the Northern North Atlantic*. Topics in Geology, **35**. Springer, Dordrecht.

Einarsson, P. 2008. Plate boundaries, rifts and transforms in Iceland. *Jökull*, **58**, 35–58.

Elliott, G.M. & Parson, L.M. 2008. Influence of margin segmentation upon the break-up of the Hatton Bank rifted margin, NE Atlantic. *Tectonophysics*, **457**, 161–176.

Erlendsson, Ö. & Blischke, A. 2013. *Haffréttarmál: hljóðendurvarps- og bylgjubrots-mælingar: skýrsla um stöðu mála á úrvinnslu og túlkun gagna [Law of the Sea. Seismic Reflection and Refraction Datasets. Status Report on the Processing and Interpretation work]*. ÍSOR-2013/067. Iceland GeoSurvey (ÍSOR), Reykjavík, Iceland.

Fitton, J.G., Saunders, A.D., Norry, M.J., Hardarson, B.S. & Taylor, R.S. 1997. Thermal and chemical structure of the Iceland plume. *Earth and Planetary Science Letters*, **153**, 197–208.

Foulger, G.R. & Anderson, D.L. 2005. A cool model for the Iceland hotspot. *Journal of Volcanology and Geothermal Research*, **141**, 1–22.

Foulger, G.R., Du, Z. & Julian, R.R. 2003. Icelandic-type crust. *Geophysical Journal International*, **155**, 567–590.

Funck,T., Hopper, J.R. *et al.* 2014. Crustal structure. *In*: Hopper, J.R., Funck, T., Stoker, M., Árting, U., Peron-Pinvidic, G., Doornenbal, H. & Gaina, C. (eds) *Tectonostratigraphic Atlas of the North-East Atlantic Region*. Geological Survey of Denmark and Greenland (GEUS), Copenhagen, Denmark, 69–128.

Gaina, C. 2014. Plate reconstructions and regional kinematics. *In*: Hopper, J.R., Funck, T., Stoker, M., Árting, U., Peron-Pinvidic, G., Doornenbal, H. & Gaina, C. (eds) *Tectonostratigraphic Atlas of the North-East Atlantic Region*. Geological Survey of Denmark and Greenland (GEUS), Copenhagen, Denmark, 53–68.

Gaina, C., Gernigon, L. & Ball, P. 2009. Palaeocene–Recent plate boundaries in the NE Atlantic and the formation of the Jan Mayen microcontinent. *Journal of the Geological Society, London*, **166**, 601–616, https://doi.org/10.1144/0016-76492008-112

Gaina, C., Blischke, A., Geissler, W., Kimbell, G.S. & Erlendsson, Ö. 2016. Seamounts and oceanic igneous features in NE Atlantic: a link between plate motions and mantle dynamics. *In*: Péron-Pinvidic, G., Hopper, J.R., Stoker, T., Gaina, C., Doornebal, H., Funck, T. & Árting, U. (eds) *The NE Atlantic Region: A Reappraisal of Crustal Structure, Tectonostratigraphy and Magmatic Evolution*. Geological Society, London, Special Publications, **447**. First published online September 8, 2016, https://doi.org/10.1144/SP447.6

Gaina, C., Nasuti, A., Kimbell, G.S. & Blischke, A. 2017. Break-up and seafloor spreading domains in the NE Atlantic. *In*: Péron-Pinvidic, G., Hopper, J.R., Stoker, T., Gaina, C., Doornebal, H., Funck, T. & Árting, U. (eds) *The NE Atlantic Region: A Reappraisal of Crustal Structure, Tectonostratigraphy and Magmatic Evolution*. Geological Society, London, Special Publications, **447**. First published online February 3, 2017, https://doi.org/10.1144/SP447.12

Ganerød, M., Smethurst, M.A. *et al.* 2010. The North Atlantic Igneous Province reconstructed and its relation to the Plume Generation Zone: the Antrim Lava Group revisited. *Geophysical Journal International*, **182**, 183–202.

Geissler, H.W., Gaina, C. *et al.* 2016. Seismic volcanostratigraphy of the NE Greenland continental margin. *In*: Péron-Pinvidic, G., Hopper, J.R., Stoker, T., Gaina, C., Doornebal, H., Funck, T. & Árting, U. (eds) *The NE Atlantic Region: A Reappraisal of Crustal Structure, Tectonostratigraphy and Magmatic Evolution*. Geological Society, London, Special Publications, **447**. First published online December 14, 2016, https://doi.org/10.1144/SP447.11

Gernigon, L., Blischke, A., Nasuti, A. & Sand, M. 2015. Conjugate volcanic rifted margins, seafloor spreading, and microcontinent: Insights from new high-resolution aeromagnetic surveys in the Norway Basin. *Tectonics*, **34**, 907–933.

Haase, C. & Ebbing, J. 2014. Gravity data. *In*: Hopper, J.R., Funck, T., Stoker, M., Árting, U., Peron-Pinvidic, G., Doornenbal, H. & Gaina, C. (eds) *Tectonostratigraphic Atlas of the North-East Atlantic Region*. Geological Survey of Denmark and Greenland (GEUS), Copenhagen, Denmark, 29–40.

Haase, C., Ebbing, J. & Funck, T. 2016. A 3D crustal model of the NE Atlantic based on seismic and gravity data. *In*: Péron-Pinvidic, G., Hopper, J.R., Stoker, T., Gaina, C., Doornebal, H., Funck, T. & Árting, U. (eds) *The NE Atlantic Region: A Reappraisal of Crustal Structure, Tectonostratigraphy and Magmatic Evolution*. Geological Society, London, Special Publications, **447**. First published online October 12, 2016, https://doi.org/10.1144/SP447.8

Harðarson, B.S., Fitton, J.G., Ellam, R.M. & Pringle, M.S. 1997. Rift relocation – a geochemical and geochronological investigation of a palaeo-rift in northwest Iceland. *Earth and Planetary Science Letters*, **153**, 181–196.

Harðarson, B.S., Fitton, J.G. & Hjartarson, Á. 2008. Tertiary volcanism in Iceland. *Jökull*, **58**, 161–178.

Helgadóttir, G. 2012. Mud volcanoes west of Iceland. *Hafrannsóknir*, **162**, 36–39.

Hinz, K. 1981. A hypothesis on terrestrial catastrophes: wedges of very thick oceanward dipping layers beneath passive continental margins. *Geologisches Jahrbuch*, **E22**, 3–28.

Hjartarson, Á. 2003. *The Skagafjördur Unconformity North Iceland and its geological history*. PhD thesis, University of Copenhagen, Copehagen, Denmark.

Hjartarson, Á. & Sæmundsson, K. 2014. *Geological Map of Iceland. Bedrock. 1:600 000*. Iceland GeoSurvey (ÍSOR), Reykjavik, Iceland.

Holbrook, W.S., Larsen, H.C. *et al.* 2001. Mantel thermal structure and active upwelling during continental breakup in the North Atlantic. *Earth and Planetary Science Letters*, **190**, 250–261.

Hopper, J.R., Dahl-Jensen, T. *et al.* 2003. Structure of the SE Greenland margin from seismic reflection and refraction data: Implications for nascent spreading center subsidence and asymmetric crustal accretion during North Atlantic opening. *Journal of Geophysical Research*, **108**, 2269.

Hopper, J.R., Funck, T., Stoker, M., Árting, U., Peron-Pinvidic, G., Doornenbal, H. & Gaina, C. (eds) . 2014. *Tectonostratigraphic Atlas of the North-East Atlantic Region*. Geological Survey of Denmark and Greenland (GEUS), Copenhagen, Denmark.

Höskuldsson, Á., Vésteinsson, Á., Steinþórsson, S. & Sigurðsson, O. 2013. Eldstöðvar í sjó. *In*: Sólnes, J., Sigmundsson, F. & Bessason, B. (eds) *Náttúruvá á Íslandi* [Natural Hazards in Iceland – Eruptions and Earthquakes]. Viðlagatrygging Íslands and Háskólaútgáfan, Reykjavík, 403–425.

Ito, G. & Behn, M.D. 2008. Magmatic and tectonic extension at mid-ocean ridges: 2. Origin of axial morphology. *Geochemistry, Geophysics, Geosystems*, **9**, 20.

Jakobsson, S.P. 1972. *The Geology and Petrography of Vestmanna Islands. A Preliminary Report*. University of Copenhagen, Copenhagen, Denmark.

Jóhannesson, H. 1980. Jarðlagaskipan og þróun rekbelta á Vesturlandi [Evolution of rift zones in western Iceland]. *Náttúrufræðingurinn*, **50**, 13–31.

Jóhannesson, H. & Sæmundsson, K. 2009. *Geological Map of Iceland. Tectonics. 1:600 000*. Iceland GeoSurvey (ÍSOR), Reykjavik, Iceland.

Jónsson, G. & Kristjánsson, L. 1997. Magnetic surveys south and southeast of Iceland. *Journal of Geodynamics*, **26**, 45–56.

Kristjánsson, L. 1976. Central volcanoes on the Western Icelandic shelf. *Marine Geophysical Researches*, **2**, 285–289.

Kokfelt, T.F., Hoernle, K., Hauff, F., Fiebig, J., Werner, R. & Garbe-Schõnberg, D. 2006. Combined trace element and Pb–Nd–Sr–O isotope evidence for recycled oceanic crust (upper and lower) in the Iceland mantle plume. *Journal of Petrology*, **47**, 1705–1749.

Larsen, H.C. & Jakobsdóttir, S. 1988. Distribution, crustal properties and significance of seawards-dipping sub-basement reflectors off East Greenland. *In*: Morton, A.C. & Parson, L.M. (eds) *Early Tertiary Volcanism and the Opening of the Northeast Atlantic*. Geological Society of London, Special Publications, **39**, 95–114, https://doi.org/10.1144/GSL.SP.1988.039.01.10

Larsen, H.C. & Saunders, A.D. 1998. Tectonism and volcanism at the southeast Greenland rifted margin: a record of plume impact and later continental rupture. *In*: *Proceedings of the Ocean Drilling Program, Scientific Results, Volume 152*. Ocean Drilling Program, College Station, TX, USA, 503–534, https://doi.org/10.2973/odp.proc.sr.152.240.1998

Larsen, L.M. & Watt, W.S. 1985. Episodic volcanism during break-up of the North Atlantic: evidence from the East Greenland plateau basalts. *Earth and Planetary Science Letters*, **73**, 105–116, https://doi.org/10.1016/0012-821X(85)90038-X

Larsen, L.M., Watt, W.S. & Watt, M. 1989. Geology and petrology of the Lower Tertiary plateau basalts of the Scoresby Sund region, East Greenland. *Bulletin Grønlands Geologiske Undersøgelse*, **157**, 164.

Lundin, E. & Doré, T. 2004. The Iceland 'anomaly' – an outcome of plate tectonics. www.mantleplumes.org, http://www.mantleplumes.org/Iceland2.html

Morgan, W.J. 1971. Convection plumes in the lower mantle. *Nature*, **230**, 42–43.

Mutter, J.C., Talwani, M. & Stoffa, P.L. 1982. Origin of seaward-dipping reflectors in oceanic crust off the Norwegian margin by subaerial sea-floor spreading. *Geology*, **10**, 353–357, https://doi.org/10.1130/0091-7613(1982)102.0.CO;2

Nasuti, A. & Olesen, O. 2014. Magnetic data. *In*: Hopper, J.R., Funck, T., Stoker, M.S., Árting, U., Peron-Pinvidic, G., Doornenbal, H. & Gaina, C. (eds) *Tectonostratigraphic Atlas of the North-East Atlantic Region*. GEUS, Copenhagen, Denmark, 41–52.

Nunns, A.G., Talwani, M. *et al.* 1983. Magnetic anomalies over Iceland and surrounding seas *In*: Bott, M.H.P., Saxov, S., Talwani, M. & Thiede, J. (eds) *Structure and Development of the Greenland–Scotland Ridge*. Plenum, New York, 661–678.

Pálmason, G. 1981. Crustal rifting and related thermomechanical processes in the lithosphere beneath Iceland. *Geologische Rundschau*, **70**, 244–260.

Parnell-Turner, R., White, N., Henstock, T., Murton, B., Maclennan, J. & Jones, S.M. 2014. A continuous 55-million-year record of transient mantle plume activity beneath Iceland. *Nature Geosience*, **7**, 914–919.

Planke, S., Symonds, P.A., Alvestad, E. & Skogseid, J. 2000. Seismic volcanostratigraphy of large-volume basaltic extrusive complexes on rifted margins. *Journal of Geophysical Research*, **105**, 19 335–19 351.

Pringle, M.S., Hardarson, B.S., Kristjansson, L. & Johannesson, H. 1997. Decrease in crustal production rate from established and dying rift: examples from western Iceland. *Eos, Transactions of the American Geophysical Union*, **78**, 656.

Richardson, K.R., Smallwood, J.R., White, R.S., Snyder, B.D. & Maguire, P.K.H. 1998. Crustal structure beneath Faroe Islands and the Faroe–Iceland Ridge. *Tectonophysics*, **300**, 159–180.

Riishuus, M.S., Kristjansson, L. & Duncan, R. 2013. Revised $^{40}Ar/^{39}Ar$ geochronology and magnetostratigraphy of Vestfirðir. The Autumn Conference of the Geological Society of Iceland. Abstract volume, 21.

Saunders, A.D., Fitton, J.G., Kerr, A.C., Norry, M.J. & Kent, R.W. 1997. The North Atlantic igneous province. *In*: Mahoney, J.J. & Coffin, M.F. (eds) *Large Igneous Provinces*. American Geophysical Union, Geophysical Monograph, **100**, 45–93.

Sæmundsson, K. 1974. Evolution of the Axial Rifting Zone in Northern Iceland and the Tjörnes Fracture Zone. *Geological Society of America Bulletin*, **85**, 495–504.

Sæmundsson, K. 1979. Outline of the geology of Iceland. *Jökull*, **29**, 7–28.

Søager, N. & Holm, P.M. 2009. Extended correlation of the Paleogene Faroe Islands and East Greenland plateau basalts. *Lithos*, **107**, 205–215, https://doi.org/10.1016/j.lithos.2008.10.002

Staudigel, H. & Clague, D.A. 2010. The geological history of deep sea volcanoes. *Oceanography*, **23**, 59–71.

Stoker, M., Doornenbal, H., Hopper, J.R. & Gaina, C. 2014. Tectonostratigraphy. *In*: Hopper, J.R., Funck, T., Stoker, M., Árting, U., Peron-Pinvidic, G., Doornenbal, H. & Gaina, C. (eds) *Tectonostratigraphic Atlas of the North-East Atlantic Region*. Geological Survey of Denmark and Greenland (GEUS), Copenhagen, Denmark, 129–212.

Storey, M., Duncan, R.A. & Tegner, C. 2007. Timing and duration of volcanism in the North Atlantic Igneous Province: Implications for geodynamics and links to the Iceland hotspot. *Chemical Geology*, **241**, 264–281, https://doi.org/10.1016/j.chemgeo.2007.01.016

Talwani, M. & Eldholm, O. 1977. Evolution of the Norwegian–Greenland Sea. *Geological Society of America Bulletin*, **88**, 969–999.

Thirlwall, M.F., Gee, M.A.M., Taylor, R.N. & Murton, B.J. 2004. Mantle components in Iceland and adjacent ridges investigated using double-spike Pb isotope ratios. *Geochimica et Cosmochimica Acta*, **68**, 361–386.

Thordarson, T. & Larsen, G. 2007. Volcanism in Iceland in historical time: Volcano types, eruption styles and eruptive history. *Journal of Geodynamics*, **43**, 118–152.

Torsvik, T.H., Amundsen, H.E.F. *et al.* 2015. Continental crust beneath southeast Iceland. *Proceedings of the National Academy of Sciences of the United States of America*, **112**, 1818–1827.

Vogt, P.R. 1971. Asthenosphere motion recorded by the ocean floor south of Iceland. *Earth and Planetary Science Letters*, **13**, 153–160.

Vogt, P.R. & Jung, W.Y. 2009. Treitel Ridge: a unique inside corner hogback on the west flank of extinct Aegir spreading ridge, Norway basin. *Marine Geology*, **267**, 86–100.

White, N. & Lovell, B. 1997. Measuring the pulse of a plume with the sedimentary record. *Nature*, **387**, 888–891, https://doi.org/10.1038/43151

White, R. & McKenzie, D. 1989. Magmatism at rift zones: the generation of volcanic continental margins and flood basalts. *Journal of Geophysical Research*, **94**, 7685–7729.

Seismic volcanostratigraphy of the NE Greenland continental margin

WOLFRAM H. GEISSLER[1]*, CARMEN GAINA[2], JOHN R. HOPPER[3], THOMAS FUNCK[3], ANETT BLISCHKE[4], UNI ARTING[5], JIM Á HORNI[5], GWENN PÉRON-PINVIDIC[6] & MANSOUR M. ABDELMALAK[2]

[1]*Alfred Wegener Institute Helmholtz Centre for Polar and Marine Research, Am Alten Hafen 26, 27568 Bremerhaven, Germany*

[2]*Centre for Earth Evolution and Dynamics (CEED), University of Oslo, Sem Sælands vei 24, PO Box 1048, Blindern, NO-0316 Oslo, Norway*

[3]*Geological Survey of Denmark and Greenland, Øster Voldgade 10, 1350 Copenhagen K, Denmark*

[4]*Iceland GeoSurvey, Branch at Akureyri, Rangárvöllum, 602 Akureyri, Iceland*

[5]*Jarðfeingi, Faroese Earth and Energy Directorate, Brekkutún 1, 110 Tórshavn, Faroe Islands*

[6]*Geological Survey of Norway, Leiv Eirikssons vei 39, 7040 Trondheim, Norway*

**Correspondence: wolfram.geissler@awi.de*

Abstract: The Early Eocene continental break-up between the NE Greenland and the mid-Norwegian–SW Barents Sea margins was associated with voluminous magmatism and led to the emplacement of massive volcanic complexes including wedges of seawards-dipping reflections (SDR). We study the distribution of these break-up-related volcanic rocks along the NE Greenland margin by revisiting existing seismic reflection data and comparing our observations to better-studied segments of the conjugate margin. Seismic facies types match between the conjugate margins and show strong lateral variations. Seaward-dipping wedges are mapped offshore East Greenland, the conjugate to the Vøring continental margin. The geophysical signature of the SDRs becomes less visible towards the north, as it does along the conjugate Lofoten–Vesterålen margin. We suggest that the Traill Ø volcanic ridge is a result of plume–ridge interactions formed between approximately 54 and 47 Ma. North of the East Greenland Ridge, strong basement reflections conjugate to the Vestbakken Volcanic Province are interpreted as lava flows or 'spurious' SDRs. We discuss our findings in conjunction with results from seismic wide-angle experiments, gravity and magnetic data. We focus on the spatial and temporal relationships of the break-up volcanic rocks, and their structural setting in a late rift and initial oceanic drift stage.

Supplementary material: The figures show the original seismic data used as the base for the interpretations shown in this paper. The seismic profiles are marked on Figure 1 (in the paper) as numbers 1 to 10 and are available at https://doi.org/10.6084/m9.figshare.c.3593780

The onset of continental break-up in the NE Atlantic region marked a culmination of an approximately 350 myr period of predominantly extensional deformation and intermediate cooling events subsequent to the Caledonian Orogeny (Ziegler 1988; Doré *et al.* 1999; Skogseid *et al.* 2000; Tsikalas *et al.* 2008; Faleide *et al.* 2010). Through the late Palaeozoic and Mesozoic, lithospheric thinning resulted in large sedimentary sag-basins controlled by regional detachment faults. Some of the early rift events were associated with magmatism, others were not. The final continental break-up occurred in the Early Eocene (at *c.* 54 Ma – magnetic Chron C24: e.g. Ziegler 1988; Doré *et al.* 1997) and was accompanied by massive magmatism that caused the formation of the North Atlantic Igneous Province (NAIP: Saunders *et al.* 1997), which extends into the NW Atlantic west of Greenland (Baffin Bay/Davis Strait) and extensively into the NE Atlantic from the Charlie Gibbs Fracture Zone south of Greenland to the Senja Fracture Zone in the north. Igneous rocks emplaced around the time of

From: Péron-Pinvidic, G., Hopper, J. R., Stoker, M. S., Gaina, C., Doornenbal, J. C., Funck, T. & Árting, U. E. (eds) 2017. *The NE Atlantic Region: A Reappraisal of Crustal Structure, Tectonostratigraphy and Magmatic Evolution.* Geological Society, London, Special Publications, **447**, 149–170.
First published online December 14, 2016, https://doi.org/10.1144/SP447.11

break-up are mainly associated with voluminous extrusive complexes, thick initial oceanic crust or transitional crust with magmatically underplated material (high-velocity lower-crustal bodies) and intrusives. According to Eldholm & Grue (1994), a volume of 1.8×10^6 km^3 of flood basalts was emplaced onshore and offshore over an area of approximately 1.3×10^6 km^2. á Horni *et al.* (In review, this volume) estimated that the total volcanic volume including deeper intrusives, is approximately 9×10^6 km^3. Some authors favour a mantle plume contribution for these large volumes (e.g., Storey *et al.* 2007).

Hinz *et al.* (1987) provided the first detailed seismic study of the East Greenland continental margin, and summarized important findings published earlier by Hinz (1981), Mutter *et al.* (1982) and Talwani *et al.* (1983). These studies reported the presence of wedge-shaped bodies bounded at the top by a prominent seismic unconformity and generally exhibiting an internal pattern of divergent reflectors with ubiquitous seawards dips (the so-called 'SDRs' – seawards-dipping reflectors). Drilling during DSDP Leg 38 (Eldholm *et al.* 1986) proved that these wedges are composed of extrusive basaltic rocks, distinctly different from that of oceanic layer 2, of much greater thickness and erupted in a subaerial to shallow-marine environment (Hinz *et al.* 1987). However, a strong scientific debate about the structural and temporal setting of emplacements has remained. One model preferred a late rift stage with dyke intrusions, implying pre-break-up emplacement (Hinz 1981); the other preferred an origin by incipient seafloor spreading, implying post-break-up emplacement (Mutter *et al.* 1982). In consequence, the crust beneath the SDRs differs in both models: modified (transitional) continental crust is assumed in the late rift-stage model, and oceanic crust produced by seafloor spreading in the assumed post-break-up model.

Hinz *et al.* (1987) found a high degree of symmetry on the conjugate East Greenland and mid-Norwegian margins. Both along-strike and across-dip variations in the structure of the wedges are comparable, as well as the relationship to magnetic lineation patterns. Since the northernmost SDRs are landwards of lineation C24 at both margins, a late rift phase might be plausible (Hinz *et al.* 1987). Further south, SDRs are imaged at lineation C23 (disturbed), which would support a true seafloor-spreading origin of the SDRs. From these early observations, Hinz *et al.* (1987) concluded that initial subaerial seafloor spreading took place partly on highly extended continental crust (as evidenced by the geochemistry of basalts drilled in ODP 642: see Viereck *et al.* 1988). Recent geochemical analysis combined with C, Pb, Sr and Nd isotope compositions of drill-core ODP 642 samples indicate the interaction of mid-ocean ridge basalt (MORB)-type melts with continental material, partly including highly radiogenic pelagic sediments rich in organic carbon deposits (Meyer *et al.* 2009; Abdelmalak *et al.* 2016*a*).

Later, Tsikalas *et al.* (2005, 2012) and Berger & Jokat (2008) provided new seismic data and details on the internal structure of the margin and its evolution, but did not focus on the magmatic activity and its variation along the margin. Voss & Jokat (2007) and Voss *et al.* (2009) studied the deep structure of the margin and the extent and volume of the magmatic underplate by means of wide-angle seismic data, and demonstrated the impact of break-up-related magmatism on the margin structure.

Results of comprehensive geophysical and geological studies along the SE Greenland volcanic margin are summarized in Larsen & Saunders (1998) and Hopper *et al.* (2003). Massive well-developed packages of SDRs could be mapped across the shelf and far offshore (e.g. Holbrook *et al.* 2001; Hopper *et al.* 2003).

While Hinz *et al.* (1987) argued that there is a well-developed symmetry between the NE Greenland and mid-Norwegian margins to the north of the Jan Mayen Fracture Zone (JMFZ), Elliott & Parson (2008) showed clear asymmetries between the Hatton Bank and the SE Greenland rifted margins. We re-evaluate the distribution and characteristics of break-up-related volcanic rocks along the NE Greenland continental margin (Fig. 1), applying the recent concept of seismic volcanostratigraphy proposed by Elliott & Parson (2008). This will allow for a detailed characterization of the temporal and lateral variations in the SDRs along the margin to better understand the volcanic activity before, during and after break-up.

Seismic volcanostratigraphy

Based on seismic observations along the North Atlantic and Western Australian rifted margins, Planke *et al.* (2000) developed a concept of seismic volcanostratigraphy to characterize volcanic deposits along passive continental margins based on seismic facies. Six seismic facies units had originally been defined: landward flows, lava delta, inner flows, inner SDR (seaward-dipping reflector), outer high and, finally, outer SDR. These units were interpreted in terms of a five-stage tectonomagmatic margin evolution model. The widely used concept was applied to various volcanic margins and underwent further modifications (e.g. Elliott & Parson 2008; Calvès *et al.* 2011).

Elliott & Parson (2008) adapted the Planke *et al.* (2000) model to characterize the Hatton Bank rifted

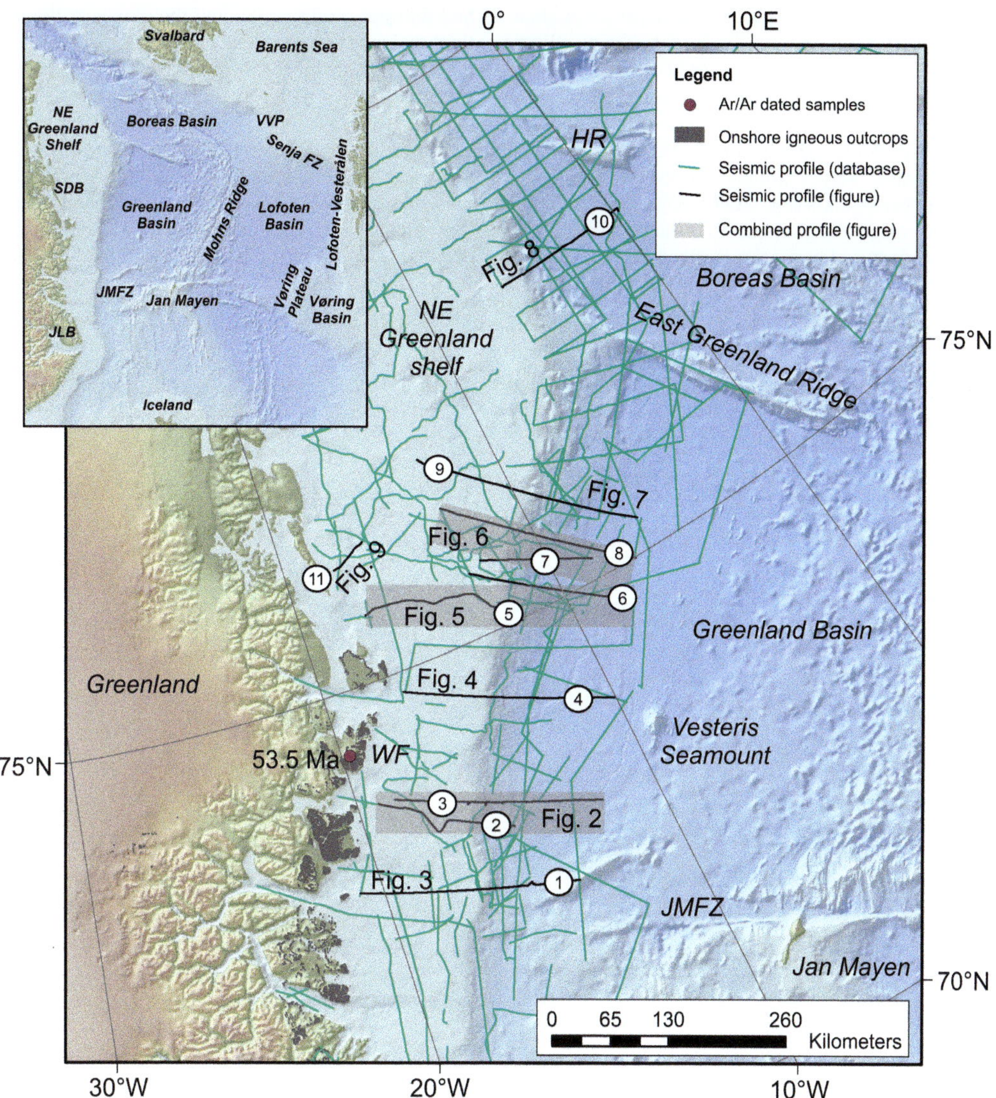

Fig. 1. Seismic reflection data (turquoise lines) along the NE Greenland continental margin used in this study. Profiles marked in black are shown in the figures. Profiles are numbered as follows: 1, AWI-20030550; 2, KAN91-05A; 3, AWI-20030560; 4, AWI-20030350; 5, KAN91-21; 6, AWI-20030370; 7, KAN95-11; 8, AWI-20030380; 9, AWI-20030390; 10, AWI-20020655; 11, KAN94-07. Dated magmatic outcrops (see Ganerød *et al.* 2014). HR: Hovgard Ridge; JLB: Jameson Land Basin; JMFZ: Jan Mayen Fracture Zone; SDB: southern Danmarkshavn Basin; VVP: Vestbakken Volcanic Province; WF: Wollaston Foreland.

margin, conjugate to the SE Greenland margin. Besides the main seismic facies units of Planke *et al.* (2000), Elliott & Parson (2008) slightly modified and added more characteristic features along volcanic rifted margins, such as rubbly basement and volcanic cones, but also included the smoothness or roughness of the nearby true oceanic basement (smooth, transitional, rough: see Funck *et al.* 2014; Gaina *et al.* 2016). The original outer SDRs were classified as oceanic SDRs by Elliott & Parson (2008).

Data

Our re-evaluation of the NE Greenland margin is based on a combined set of existing and publically available seismic reflection data (Fig. 1). We reviewed data previously used in separate publications by Hinz *et al.* (1987), Tsikalas *et al.* (2005) and Berger & Jokat (2008, 2009). Since the data were acquired over a period of more than 40 years using different seismic acquisition systems and processing protocols, the quality of the seismic reflection images varies greatly. However, the major seismic facies units (e.g. SDRs, outer high/volcanic mound) originally proposed by Planke *et al.* (2000) can be identified and characterized. In our description and interpretation, we follow the work of Elliott & Parson (2008), which additionally includes the concept of oceanic basement morphology.

The different seismic facies types were interpreted as shown on selected sections (Figs 2–8) and mapped along strike; and the extent of the distinct volcanic types were interpolated in-between the sparse seismic profiles based on gravity and magnetic data (Haase & Ebbing 2014; Nasuti & Olesen 2014). In addition, we incorporated interpretations from regional wide-angle seismic profiles (Funck *et al.* 2014; Funck *et al.* 2016 and references therein) to account for magmatic additions to the base of the crust (as high-velocity lower crust and oceanic layer 3B).

Observations

In the following, we describe observations along selected seismic reflection profiles that illustrate

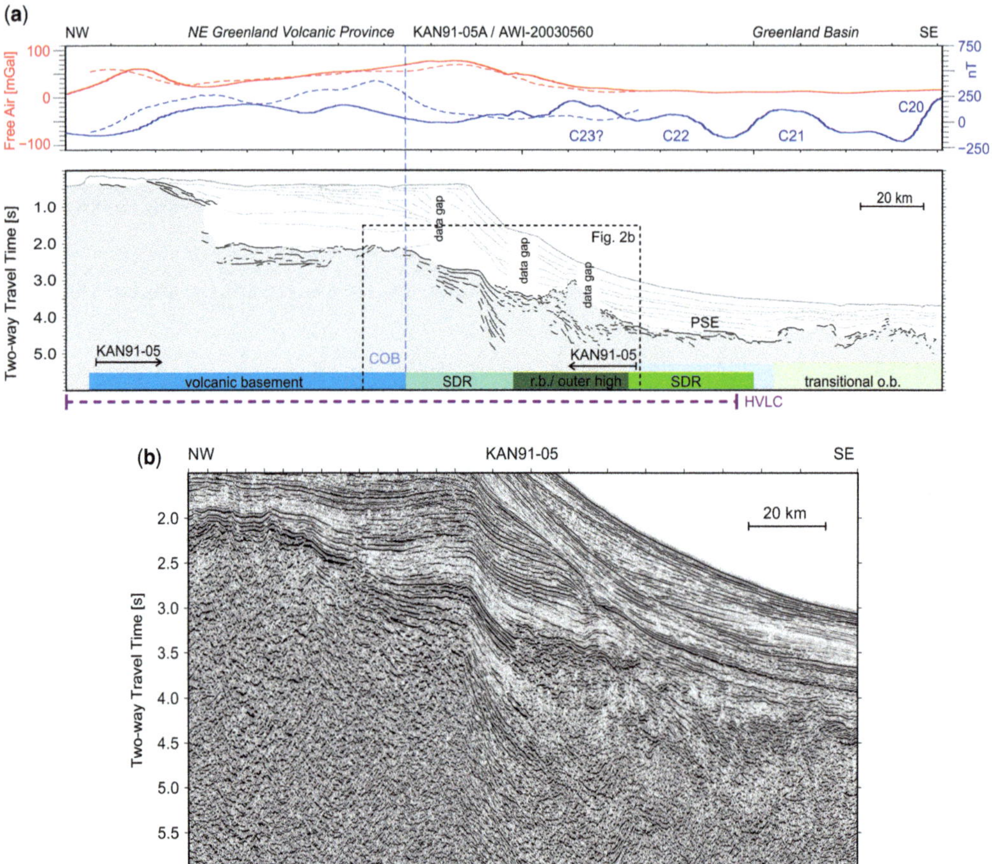

Fig. 2. (**a**) Interpretation of combined profiles KAN91-05A and AWI-20030560, illustrating the various volcanic seismic facies types. Helicopter starts and landings caused data gaps along line AWI-20030560 during the acquisition. The grey-shaded area marks the acoustic basement below the Cenozoic sedimentary cover. The continent–ocean boundary (COB) is as defined in Funck *et al.* (2014). The extent of the high-velocity lower crust (HVLC) (layer 3B in the oceanic part) as modelled by Voss & Jokat (2007) is indicated at the bottom. Gravity (Haase & Ebbing 2014) and magnetic (Nasuti & Olesen 2014) data are plotted on top. (**b**) Reflection profile KAN91-05A showing the interpreted seawards-dipping reflection (SDR) sequences in detail. o.b., oceanic basement; PSE, pseudo-escarpments; r.b., rubbly basement.

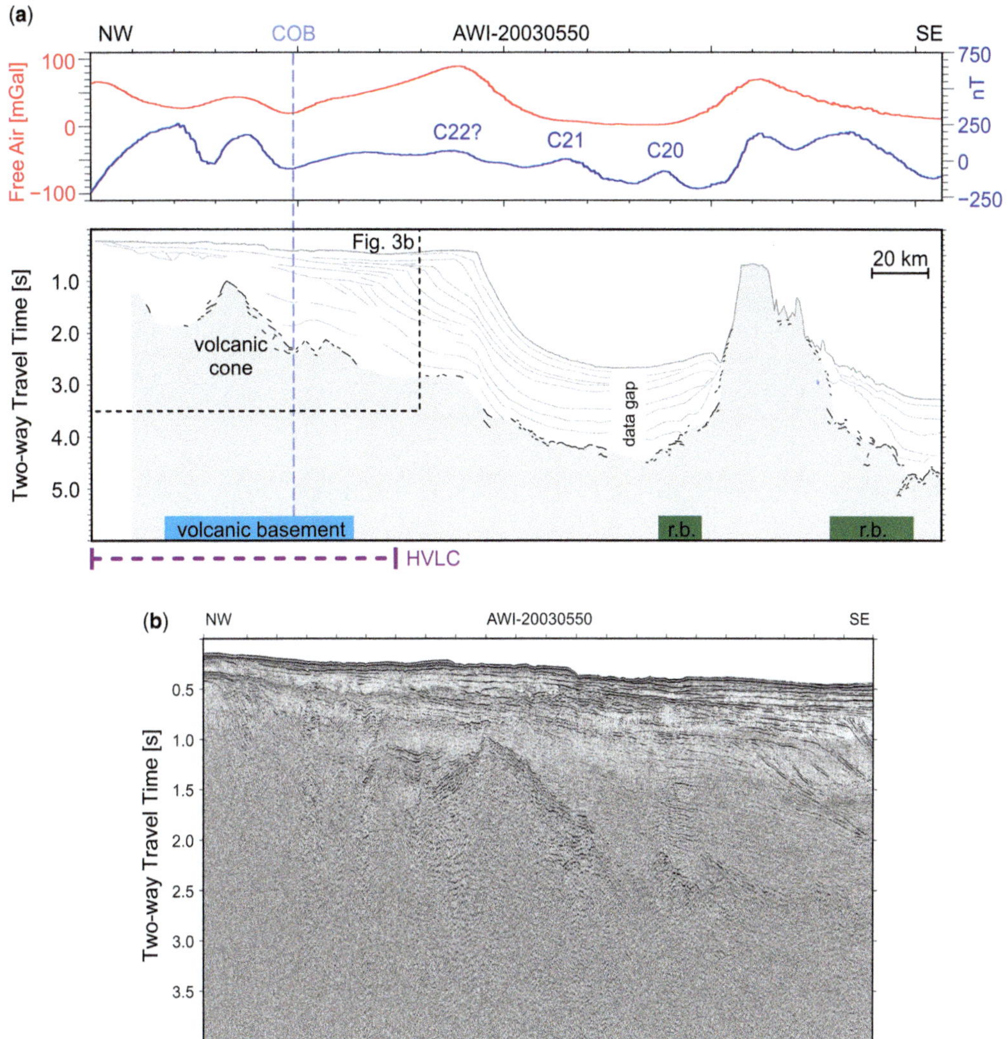

Fig. 3. (**a**) Interpretation of profile AWI-20030550 across the shelf just to the north of the JMFZ. The grey-shaded area marks the acoustic basement below the Cenozoic sedimentary cover. The continent–ocean boundary (COB) is as defined in Funck *et al.* (2014). The extent of the high-velocity lower crust (HVLC) (layer 3B in the oceanic part) as modelled by Voss & Jokat (2007) is indicated at the bottom. Gravity (Haase & Ebbing 2014) and magnetic (Nasuti & Olesen 2014) data are plotted on top. Line intervals of rubbly basement (r.b.) are marked in dark green. (**b**) Detail of profile AWI-20030550 showing the volcanic cone or ridge buried beneath the Cenozoic sediments.

the general quality of the data and the distribution of the magmatic features along the NE Greenland margin, from the JMFZ in the south to the NE Greenland sheared margin in the north.

The combined profiles KAN91-05A and AWI-20030650 image the shelf and the adjacent ocean basin at about 73.5° N (Fig. 2). This transect is the best example to illustrate the general volcano-stratigraphic model as originally proposed by Planke *et al.* (2000) for the conjugate Vøring plateau–margin. High-amplitude continuous reflections, close to the seafloor at the western end of the transect, indicate the presence of lava flows beneath a thin sedimentary veneer that might be tied to nearby basalt outcrops along the East Greenland coast at Wollaston Foreland (age *c.* 53.5 Ma: Price *et al.* 1997) (see Fig. 1). A major scarp offsets these lava flows (volcanic basement) beneath the shelf. From the data, it is not fully understood whether this scarp represents a major fault. At the

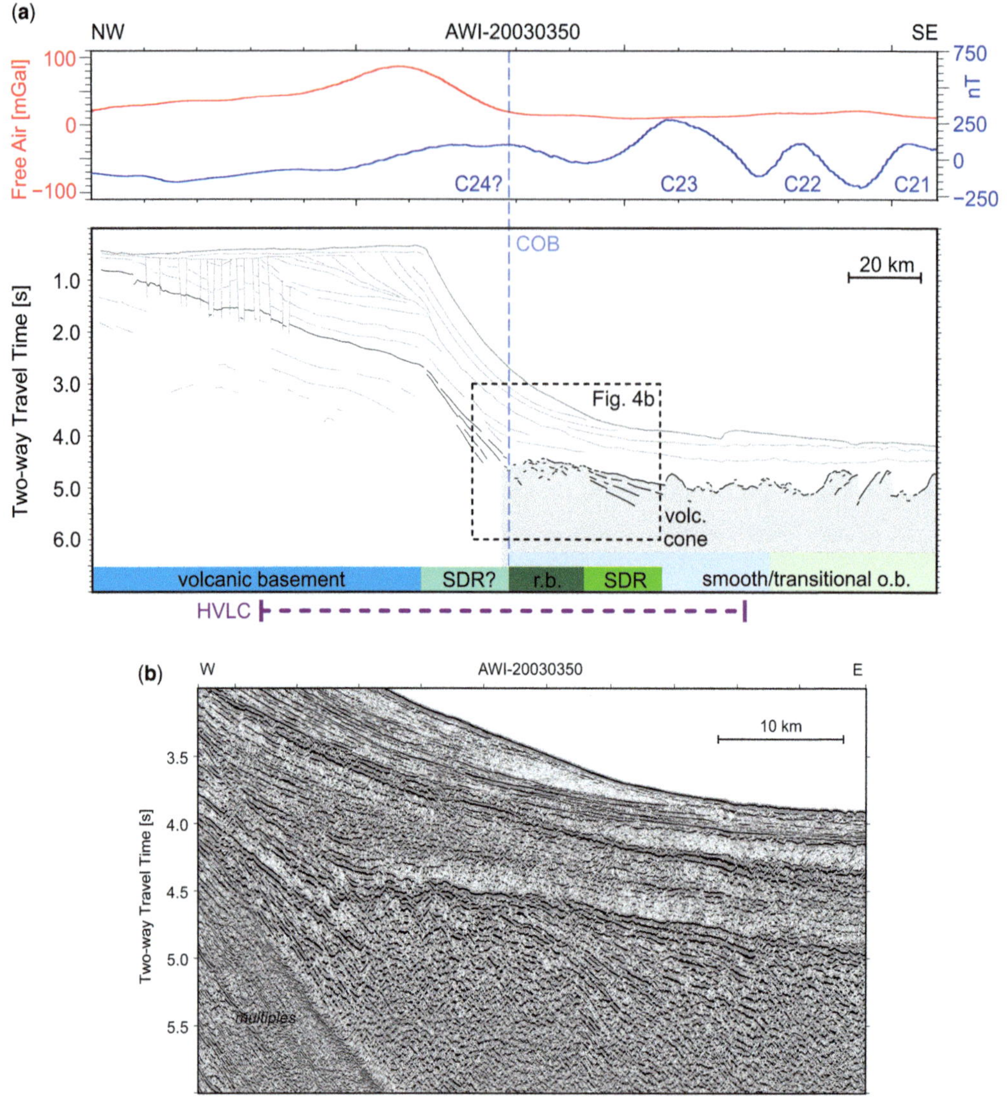

Fig. 4. (**a**) Interpretation of profile AWI-20030350. The grey-shaded area marks the acoustic basement below the Cenozoic sedimentary cover. The continent–ocean boundary (COB) is as defined in Funck *et al.* (2014). The extent of the high-velocity lower crust (HVLC) (layer 3B in the oceanic part) as modelled by Voss *et al.* (2009) is indicated at the bottom. Gravity (Haase & Ebbing 2014) and magnetic (Nasuti & Olesen 2014) data are plotted on top. (**b**) Detail of profile AWI-20030350 showing the rubbly basement. o.b., oceanic basement; r.b., rubbly basement; SDR, seawards-dipping reflection; volc., volcanic.

position of the shelf break, the reflection pattern of the acoustic basement (a wedge of high-amplitude reflections) resembles the inner SDRs of Planke *et al.* (2000). SDRs also continue beneath the shelf slope along profile KAN91-05A (see Fig. 2b), while the flanks of an outer high (magmatic centre) are imaged on profile AWI-20030560. Beneath the shelf-rise sediments, lava flows (high-amplitude continuous reflections) with pseudo-escarpments *sensu* Larsen (Larsen 1990) are imaged, which represent the outer SDRs of Planke *et al.* (2000) and oceanic SDRs of Elliott & Parson (2008). Further eastwards in the Greenland Basin, transitional (regarding roughness) oceanic basement is imaged. Voss & Jokat (2007) map magmatic underplating (high-velocity lower crust–oceanic layer 3B) up to

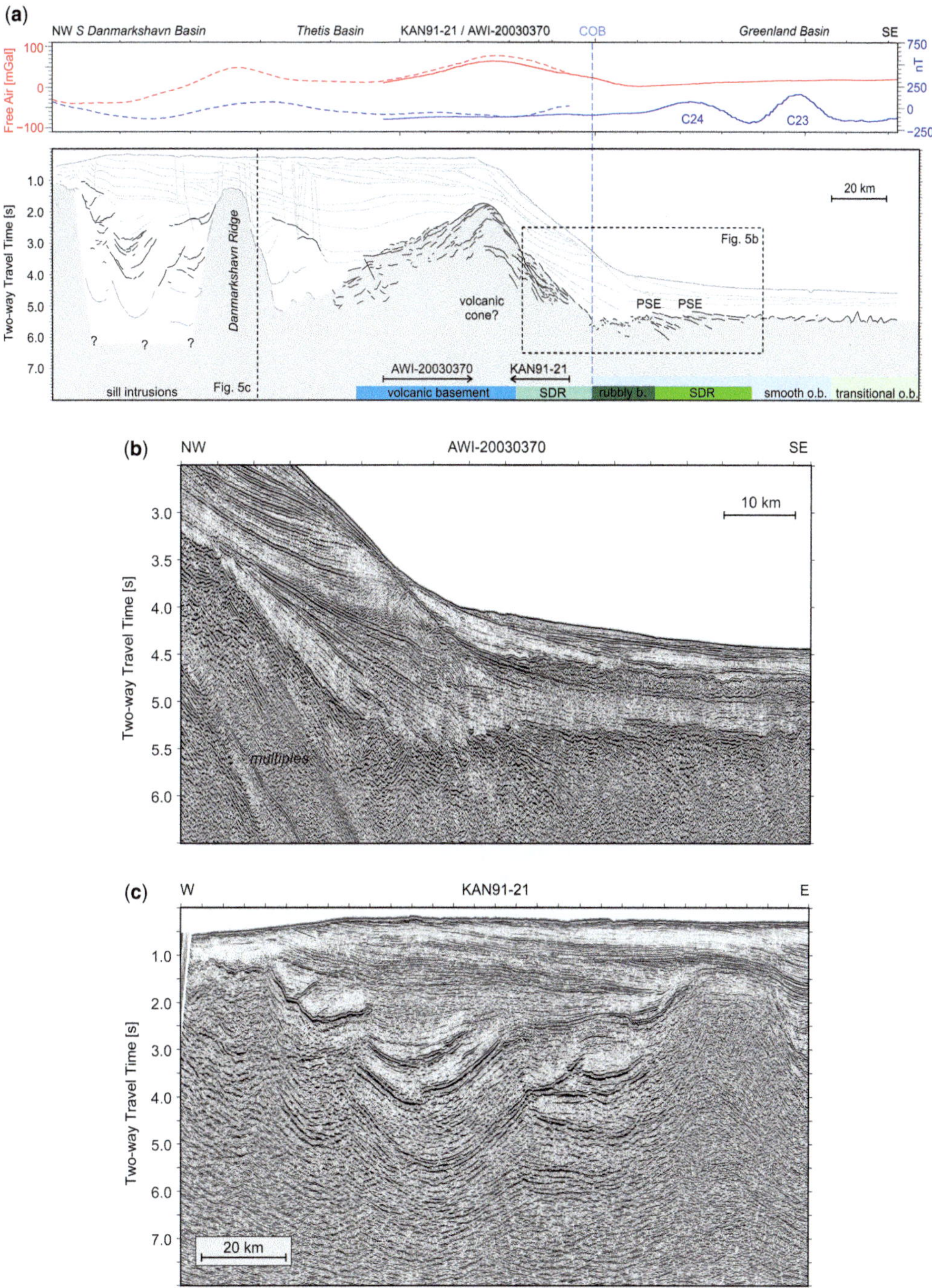

Fig. 5. (**a**) Interpretation of combined profile KAN91-21 and AWI-20030370 spanning from the East Greenland coast towards the Greenland Basin. Continent–ocean boundary (COB) is as in Funck *et al.* (2014). Gravity (Haase & Ebbing 2014) and magnetic (Nasuti & Olesen 2014) data are plotted on top. Sill complexes are imaged in the southern Danmarkshavn Basin. (**b**) Seismic section of profile AWI-20030370 illustrating the rubbly basement and the outer (oceanic) SDRs with its pseudo-escarpments. (**c**) Seismic section of profile KAN91-21 illustrating the sill complexes in the southern Danmarkshavn Basin. o.b., oceanic basement; PSE, pseudo-escarpments; rubbly b., rubbly basement; SDR, seawards-dipping reflection.

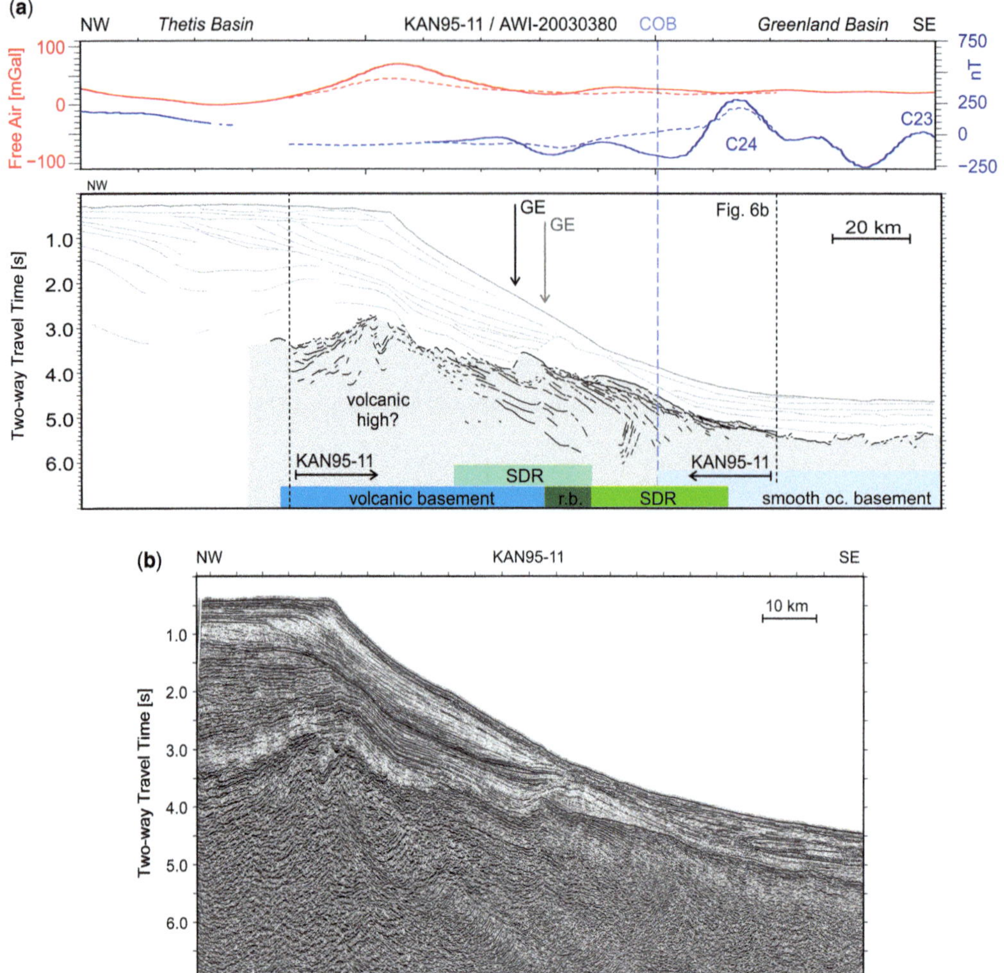

Fig. 6. (**a**) Interpretation of combined profiles KAN95-11 and AWI-20030380 imaging the magmatic complexes at the continent–ocean boundary (COB). The COB is as in Funck *et al.* (2014). Gravity (Haase & Ebbing 2014) and magnetic (Nasuti & Olesen 2014) data are plotted on top. (**b**) Data of profile KAN95-11 clearly showing the seawards-dipping reflections (SDRs) beneath the continental slope and rise. GE, Greenland Escarpment; oc., oceanic; r.b., rubbly basement.

Chron C22 and slightly eastwards within an area where SDRs and a smooth oceanic basement are mapped.

Seismic profile AWI-20030550 images the shelf at about 73° N and extends into the SW corner of the oceanic Greenland Basin just to the north of the JMFZ (Fig. 3). Owing to the thick sedimentary cover, the seismic resolution of the acoustic/basaltic basement is limited. The profile crosses the area, where the nature of the underlying crust and the position of the continent–ocean boundary (COB) is highly debated (e.g. Voss & Jokat 2007). The profile shows a mountain-like feature buried beneath the inner shelf sediments. The gravity data and, in particular, the tilt derivative (Haase & Ebbing 2014) indicate that this structure may represent a ridge. The magnetic data (Nasuti & Olesen 2014) show strong anomalies in that area. So an interpretation suggesting a volcanic cone or ridge is plausible. Close to the shelf break, there are sparse reflections on top of the acoustic basement, which could be interpreted as lava flows or the top of the SDRs. Transitional oceanic basement is mapped between the shelf and the mountain-like feature to the east in an area where Voss & Jokat (2007) did not map magmatic underplating or oceanic layer 3B. The

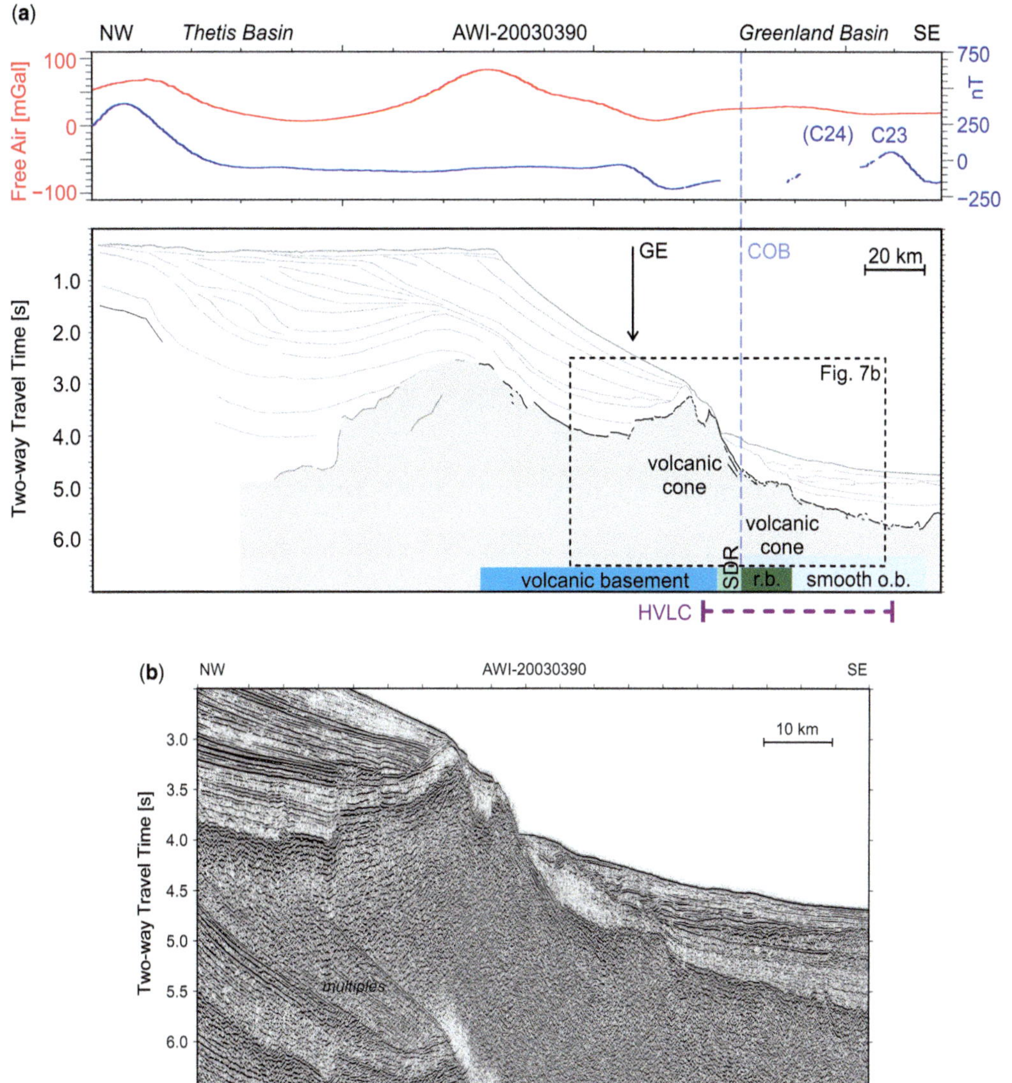

Fig. 7. (**a**) Interpretation of profile AWI-20030390. The continent–ocean boundary (COB) is as in Funck *et al.* (2014). The extent of the high-velocity lower crust (HVLC) (layer 3B in the oceanic part) as modelled by Voss *et al.* (2009) is indicated at the bottom. Gravity (Haase & Ebbing 2014) and magnetic (Nasuti & Olesen 2014) data are plotted on top. (**b**) Seismic section of profile AWI-20030390 showing details of the COB. GE, Greenland Escarpment; o.b., oceanic basement; r.b., rubbly basement; SDR, seawards-dipping reflection.

mountain-like feature imaged to the east (Fig. 3a) might be a volcano or volcanic ridge emplaced after the break-up of the Greenland Basin, and therefore is not further discussed below.

The area between 73.7 and 74.4° N is not well covered by our data (Fig. 1). However, thick sediments can be imaged along sparse lines on top of most probably basaltic flows/crust, as indicated by aeromagnetic data (Voss & Jokat 2007; Nasuti & Olesen 2014). Line AWI-20030350 crosses the shelf at 74.5° N (Fig. 4). Here the thick sediments beneath the shelf did not allow imaging of the acoustic basement. However, strong continuous reflections indicate the presence of lava flows beneath the Cenozoic sediments. Beneath the shelf slope, seawards-dipping basement reflections are interpreted as magmatic SDRs. Further basinwards, rubbly basement (chaotic reflections: see Fig. 4b) is

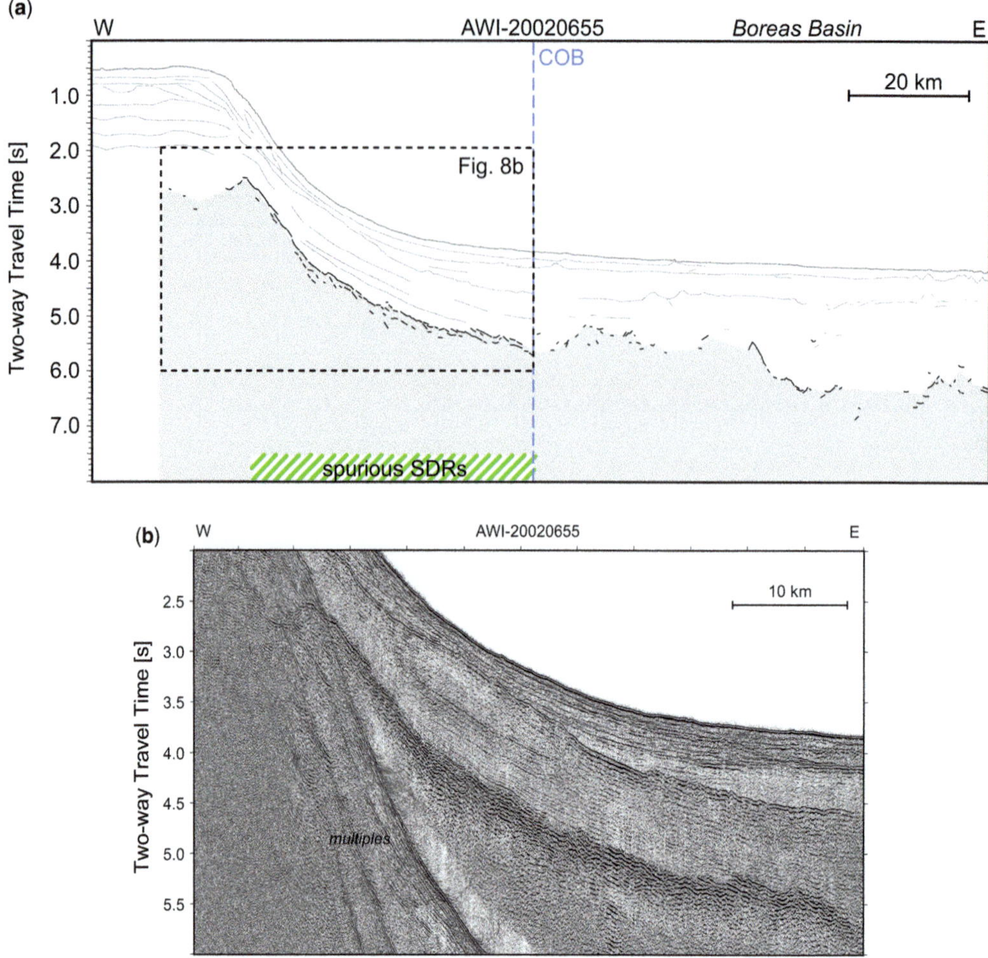

Fig. 8. (**a**) Interpretation of profile AWI-20020655 at the western flank of the Boreas Basin. The continent–ocean boundary (COB) is as in Funck *et al.* (2014). (**b**) Detail showing the spurious seawards-dipping reflections (SDRs), which might actually be simple lava or hyaloclastite flows and not SDR wedges. Note that there are multiples at the western part of the profile that mask the primary reflections.

imaged at a position where the outer high is expected, following the concept of Planke *et al.* (2000). Eastwards, a small wedge of outer SDRs could be imaged, adjacent to a small cone, which we interpret as an extinct volcano. This cone could be the potential vent for lavas forming the outer SDR wedge. Transitional oceanic basement is imaged towards the centre of the basin. Along this transect, the various published COBs are in good agreement with only minor differences (see Funck *et al.* 2014). There is evidence from seismic wide-angle data (line AWI-20030300, Voss *et al.* 2009) of small-scale magmatic underplating close to the COB, and of an oceanic layer 3B beneath the mapped outer SDRs and smooth oceanic basement.

Profile AWI-20030370 images the volcanic facies of the acoustic basement around 75.5° N (Fig. 5). The profile's reflectivity pattern of the acoustic basement implies the existence of basalt flows beneath the outer shelf. However, beneath the shelf slope, no clear SDRs can be imaged and rubbly basement can be observed beneath the shelf rise, instead of the outer high of the Planke *et al.* (2000) conceptual model. Further seawards, outer SDRs with pseudo-escarpments are identified (Fig. 5b), followed by transitional oceanic crust. This transect is extended towards the East Greenland coast by profile KAN91-21 (Fig. 5), which crosses the Danmarkshavn Basin and the Danmarkshavn Ridge. Strong, upwards-concave reflections

(Fig. 5c) can be interpreted as magmatic sill intrusions, well known from the Jameson Land Basin in East Greenland (e.g. Larsen & Marcussen 1992) and the Vøring Basin (e.g. Berndt *et al.* 2000). The structural high beneath the shelf break imaged on profile KAN91-21 (Fig. 5) lies over a cone-like structure that appears to be covered by lava flows. The existing data do not make it possible to clearly differentiate whether this cone-like structure is of igneous origin or marks the outer marginal high of the Thetis Basin (see Bjerager *et al.* 2014), as postulated by Haase & Ebbing (2014) based on gravity data.

The combined section KAN95-11 and AWI-20030380 images the margin at about 75.7° N (Fig. 6). The seismic energy on line AWI-20030380 was not sufficient to clearly characterize the acoustic basement beneath the shelf. Profiles KAN95-11 and AWI-20030380 are very close to each other. Both profiles clearly image magmatic features close to the COB. The SDRs are more pronounced along profile KAN95-11 (see Fig. 6b) than along other lines in that area. We therefore suggest that the profile images a regional magmatic centre, with rubbly basement extending landwards, which is marked as 'volcanic basement' in Figure 6. This indicates that the magma might have been emplaced in a shallow-marine environment, forming hyaloclastites instead of smooth far-extending lava flows. Lava flows and a major pseudo-escarpment (the formerly called Greenland Escarpment) exist beneath the slope. Recently, this structure has been recognized as a lava delta front similar to the Vøring and Faroe–Shetland escarpments, and was named as Thetis Escarpment (see Abdelmalak *et al.* 2016*b*). Seawards of the SDRs, a smooth oceanic basement suggests the existence of laterally extending lava flows on top of the oceanic crust. At the seawards end of the profile, the roughness of the oceanic basement becomes more transitional, indicating less magmatic activity along the palaeo-mid-ocean ridge.

Profile AWI-20030390 images the volcanic basement at about 76.0° N (Fig. 7). Similar to the previously described profile, the signal penetration does not allow clear characterization of the acoustic basement beneath the shelf. Strong basement reflections indicate the presence of lava flows beneath the slope. However, these lava flows are probably not as thick as observed further south. At the continental rise, a basement high could represent a volcanic edifice with an escarpment at its western flank, previously called the Greenland Escarpment. Seawards of this basement high, there are spatially very limited SDRs bounded by rubbly basement and smooth oceanic basement further seawards, suggesting minor magmatic activity during break-up. Voss *et al.* (2009) reported minor underplating westwards of the COB and a thin oceanic layer 3B along nearby seismic line AWI-20030200, where we map weak SDRs and smooth oceanic basement.

Profile AWI-20020655 images the margin at about 77.9° N (Fig. 8), which is located to the north of the East Greenland Ridge. The extent of the profile does not allow the basement beneath the shelf to be characterized, but strong and laterally coherent reflection at the base of sediments beneath the slope indicates the presence of lava flows on most probably transitional or highly extended continental crust. However, the geometry of the reflectors does not allow these reflections to be classified as true SDRs. Therefore we call them 'spurious'

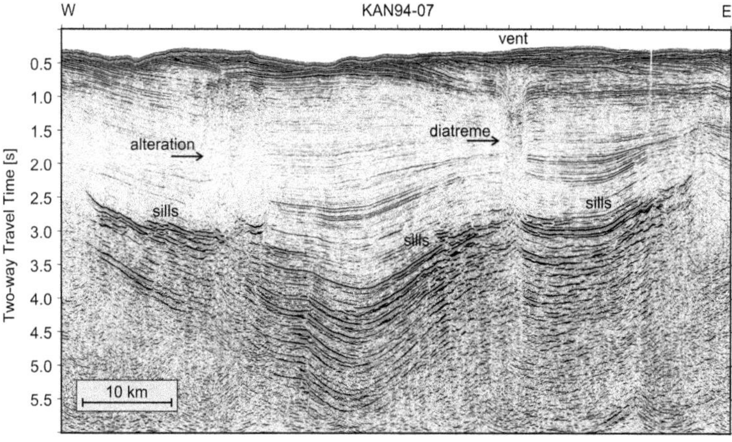

Fig. 9. Profile KAN94-07 showing sill intrusions and a volcanic–hydrothermal diatreme–vent system in the southern Danmarkshavn Basin.

SDRs. Further seawards, the data show a very rough oceanic basement, which indicates a more tectonic than magmatic imprint during the formation of the oceanic crust. This part of the continental margin is conjugate to the Vestbakken Volcanic Province on the SW Barents Sea continental margin, where Faleide *et al.* (1988, 1991) found indications of Early Eocene lava flows.

Figure 9 shows a section along the southern Danmarkshavn Basin. This part of the early rift basin is highly intruded by magmatic sills, and can be regarded as the northwards extension of the sill provinces in the Vøring and Jameson Land basins, as stated earlier. Some of the sills are connected to a palaeo-surface (most probably late Paleocene or Early Eocene in age) by diatreme-vent structures resembling hydrothermal vent complexes well known from the Vøring Basin (Planke *et al.* 2005).

Discussion

Figure 10 shows the compilation of volcanic facies along the East Greenland margin based on the seismic interpretations along all seismic lines and illustrated above for selected lines. Additional constraints for the spatial distribution of the different seismic volcanic facies units are obtained from aeromagnetic (Nasuti & Olesen 2014) (Fig. 11) and gravity data (Haase & Ebbing 2014), which are used to interpolate between the seismic profiles. Figure 12 shows plate kinematic reconstructions for the study area for time slices at 52 and 49 Ma. In Figure 12, we concentrate only on the area between the JMFZ in the south and the Greenland and Senja fracture zones in the north.

There are four major topics that will be discussed in detail: (1) the distribution of break-up volcanics along the NE Greenland margin (including the margin bordering the Boreas Basin); (2) the comparison with the conjugate mid-Norwegian margin; (3) the sill complexes and hydrothermal vents in the southern Danmarkshavn Basin; and (4) the East Greenland Shelf Magmatic Province including the Traill Ø–Vøring magmatic complex (or Traill Ø–Vøring Igneous Province – TVIP).

Break-up volcanics along the NE Greenland continental margin

Outer SDRs are commonly assumed to be emplaced in a late rift stage (e.g. Voss & Jokat 2007) or during an early oceanic stage, when the continental margin was already subsided below the sea surface (Elliott & Parson 2008). As has already been stated by Hinz *et al.* (1987) and Voss & Jokat (2007), and confirmed by the new interpretation (Fig. 10), the age of the outer SDRs becomes younger to the south, towards the JMFZ. Between latitudes 76° and 74.8° N, the outer SDRs are located close to the interpreted magnetic Chron C24 (*c.* 54 Ma); around latitude 73.2° N, however, they are placed close to Chron C22 (*c.* 49 Ma). Bastow & Keir (2011) proposed that SDRs could be the product of rapid stretching and associated fast decompression melting of an already partially molten mantle during the final rupture. The production of SDRs during a late rift stage or initial seafloor spreading stage is proposed by various authors who studied the western Afar Depression region (the Erta Ale and Tat'Ali ridges: e.g. Bastow & Keir 2011; Bosworth *et al.* 2012; Pagli *et al.* 2012).

In our case, most of the outer SDRs (with the exception of the northernmost SDRs) are located seawards of the COB as defined by Funck *et al.* (2014). That means that, along the NE Greenland continental margin, the outer SDRs define the landwards boundary of 'normal' oceanic crust accretion. The juvenescence of the outer SDRs to the south is also in accordance with the distribution of the rough, intermediate and smooth oceanic crust. The transition to rough morphology is close to magnetic Chron C20o (43.4 Ma: Gaina *et al.* 2002) in the south, but close to magnetic Chron C22 (49.3 Ma: Gaina *et al.* 2002) in the vicinity of the East Greenland Ridge. According to Voss *et al.* (2009), there is evidence of a magma-starved break-up and a rapidly thinning oceanic crust until Chron C21 (*c.* 47 Ma) in the northernmost part of the NE Atlantic oceanic system. The transition from smooth to transitional oceanic crust morphology is also wider to the south, indicating a continuing magma supply in the evolving southern Greenland–Lofoten Basin. But the smooth oceanic basement might be also a result of a faster spreading rate.

The lateral extent of the outer SDRs and the smooth oceanic basement correlate partially with the imaged oceanic layer 3B extent along the NE Greenland margin (see Voss & Jokat 2007; Voss *et al.* 2009) (see Figs 2, 4 & 7). Therefore, the smooth oceanic basement could generally be used to map the lateral distribution of the oceanic layer 3B away from the seismic wide-angle profiles. All the above-mentioned observations point to a high magma supply during the early drift phase. This is in accordance with the fact that the Early Eocene spreading was the fastest in the history of the Norwegian–Greenland Sea (Tsikalas *et al.* 2002; Gaina *et al.* In press, this volume).

Close to the COB, we mapped a volcanic facies that we named 'rubbly basement' and which forms a structure more similar to an 'outer low' (Fig. 5), and its location within the volcanic sequence fits the 'outer high' definition in the Planke *et al.* (2000) model. According to that model, the outer highs are emplaced in a shallow-marine environment as

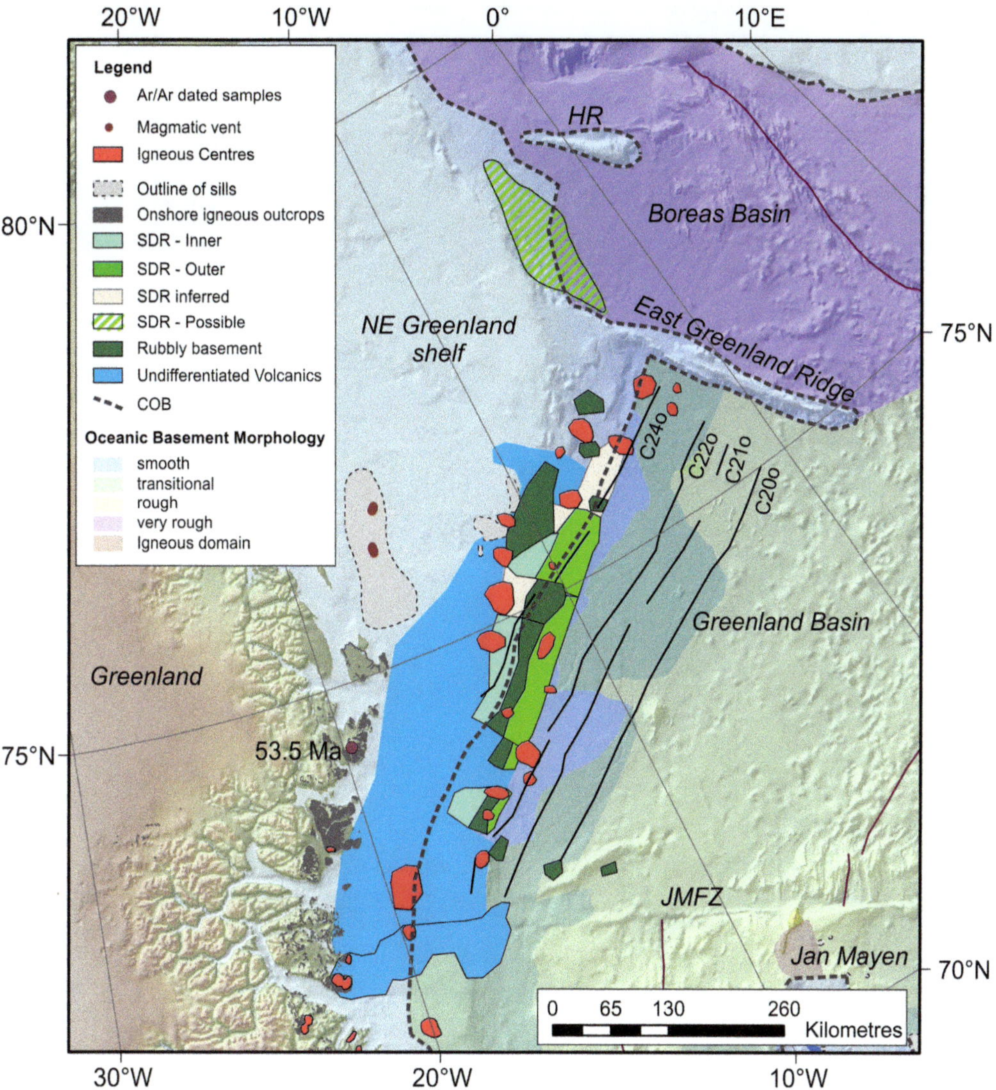

Fig. 10. Volcanic facies map of the NE Greenland margin shown together with interpreted magnetic lineations from Gaina *et al.* (2009) and dated magmatic outcrops (Ganerød *et al.* 2014). The volcanic facies for Jan Mayen is taken from á Horni *et al.* (2014). The facies 'SDR inferred' marks areas where we do not have clear seismic evidence for the existence of SDRs. The dashed grey line marks the position of the continent–ocean boundary (COB) (Funck *et al.* 2014). HR, Hovgard Ridge; JMFZ, Jan Mayen Fracture Zone; SDR, seawards-dipping reflection.

hyaloclastite ridges and represent the deepening stage of the margin. The area with the rubbly basement along the NE Greenland continental margin is about 10–30 km wide (Fig. 10), quite similar to the width of the outer highs along the Hatton Bank continental margin (7–14 km: Elliott & Parson 2008). Roberts *et al.* (1984) proved at DSDP Site 554 that the outer highs are actually composed of hyaloclastites and fractured pillow basalts. The rubbly basement is located more seawards in the southern part of the study area, in accordance with the outer SDRs. Interpreting the rubbly basement as being related to volcanism in a shallow-marine environment, we conclude that the area close to the JMFZ was above sea level for a few millions years more than the northern part of the Greenland Basin, and that the (proto-) spreading axis deepened with time from north to south. In that sense, the outer SDRs do not mark the general onset of oceanic crustal accretion as interpreted by Voss & Jokat

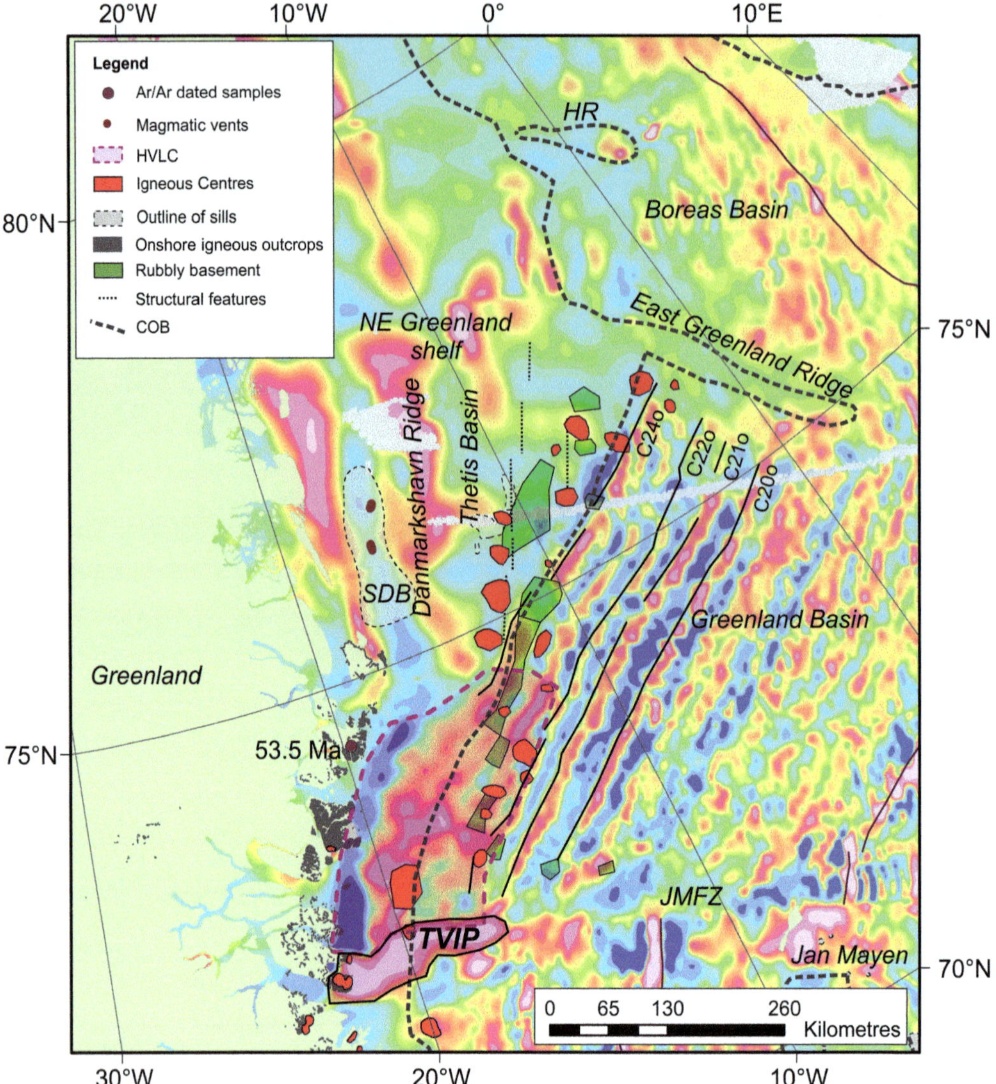

Fig. 11. Magnetic map of the NE Greenland margin (Nasuti & Olesen 2014) shown together with the magnetic lineations of Gaina *et al.* (2009) and dated magmatic outcrops (Ganerød *et al.* 2014). Selected volcanic facies types (rubbly basement, igneous centres, magmatic sills and vents) are overlain. Black dashed lines indicate the trend of Thetis Basin marginal highs. The continent–ocean boundary (COB) is as in Funck *et al.* (2014). The extent of the high-velocity lower crust (HVLC) (layer 3B in the oceanic part) as modelled by Voss & Jokat (2007) and the western part of the Traill Ø–Vøring Igneous Complex are outlined. HR, Hovgard Ridge; JMFZ, Jan Mayen Fracture Zone; SDB, southern Danmarkshavn Basin; SDR, seawards-dipping reflections; TVIP, Traill Ø–Vøring Igneous Complex.

(2007) but, rather, the oceanic crustal accretion at a submerged mid-ocean ridge. Therefore, the outer SDRs and the rubbly basement alone cannot be taken as a stand-alone justification to map the COB.

The inner SDRs along the NE Greenland margin cannot be mapped with the same confidence as the outer ones. This could be due to data coverage and quality at the often sea-ice-covered shelf. However, it could also be the case that there were no large continuous fissures producing thick and coherent magmatic wedges. Most probably, the lava flows originate from local magmatic (igneous) centres and spread radially. Possibly, the magma supply was not yet as high and laterally focused as it was

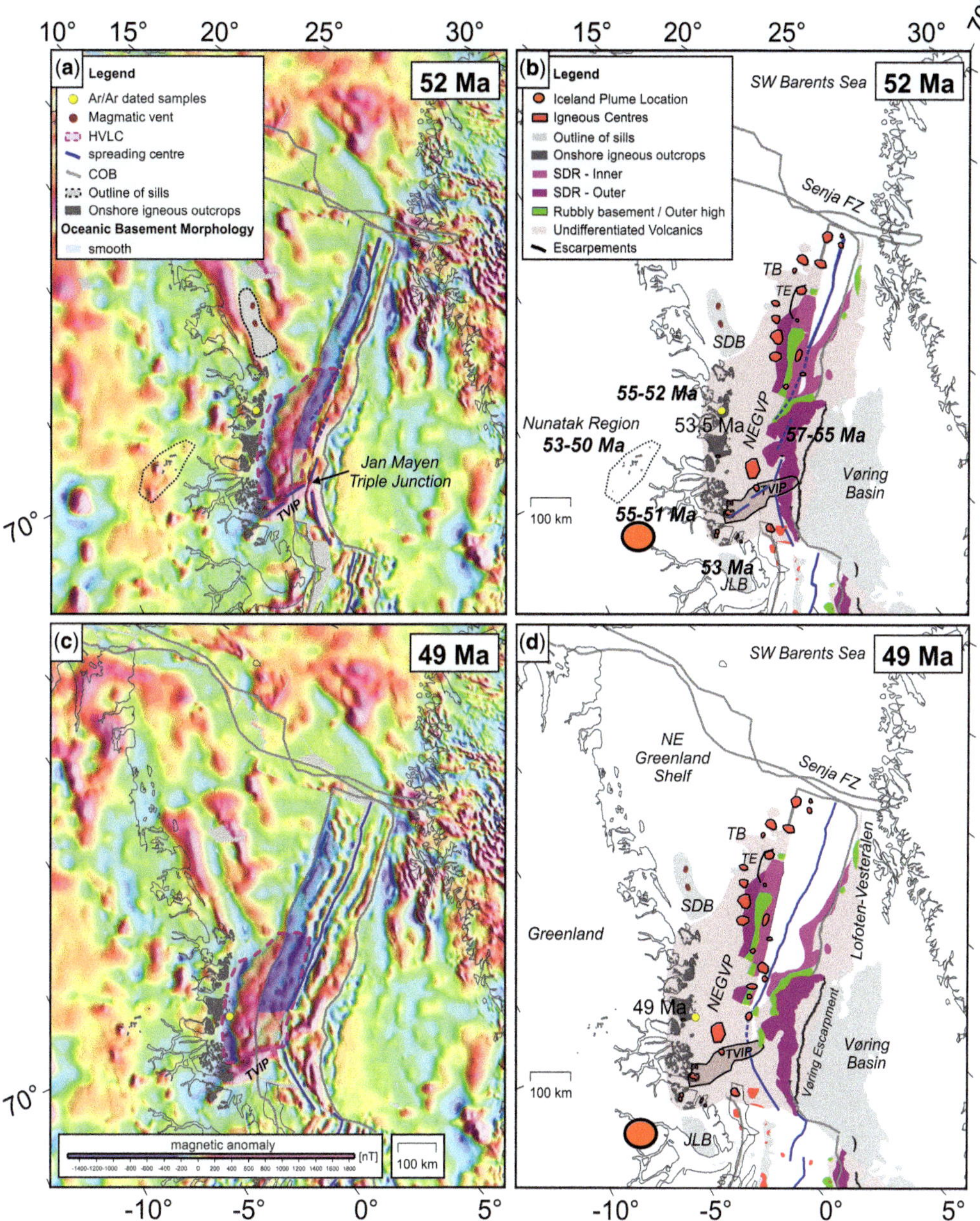

Fig. 12. Plate kinematic reconstructions for 52 and 49 Ma (in a fixed European Plate frame). For reconstruction parameters see Gaina *et al.* In press, this volume). (**a**) & (**c**) show a reconstructed magnetic anomaly grid (Gaina *et al.* In press, this volume); and (**b**) & (**d**) show interpreted volcanic facies types from this study and from Abdelmalak *et al.* (2016*b*) for the conjugate mid-Norwegian margin. Ar/Ar ages of magmatic rocks are indicated after Larsen *et al.* (2014) (in bold type) and Ganerød *et al.* (2014). The reconstructed location of the Iceland plume is based on the Doubrovine *et al.* (2012) model. COB, continent–ocean boundary; FZ, Fracture Zone; HVLC, high-velocity lower crust; JLB, Jameson Land Basin; NEGVP, NE Greenland Volcanic Province; SDB, southern Danmarkshavn Basin; SDR, seawards-dipping reflections; TB, Thetis Basin; TE, Thetis Escarpment; TVIP, Traill Ø–Vøring Igneous Complex.

the case during the emplacement of the outer SDRs along the proto mid-ocean ridge. This also concurs with the uneven distribution of landwards flows (volcanic basement). No voluminous volcanism appears to have been active along the evolving continental margin north of 75° N. This is also evident in seismic wide-angle data, showing only a thin and narrow high-velocity lower-crustal body (magmatic underplating/oceanic layer 3B) close to the line of final break-up (Voss *et al.* 2009; Funck *et al.* 2016).

Most of the mapped igneous/magmatic centres are placed in close spatial relationship to the SDRs, but do not form long continuous ridges. Based on the magnetic field and gravity data, we assume that they are more like elongated central volcanoes or rift flank volcanoes similar to those found today onshore Iceland (e.g. Sæmundsson *et al.* 1980; Harðarson *et al.* 2008; Óskarsson & Riishuus 2013), as well as in the western Afar depression and the North Ethiopian Rift (e.g. Ebinger & Casey 2001; Keir *et al.* 2013). The seismic data also show that there is no evidence of a single far-reaching continuous Greenland Escarpment. Igneous centres mapped along the Thetis Basin marginal high might not all be true magmatic centres, as stated above. Some might represent existing rift-related topography covered by lava flows during the break-up magmatic pulse. However, the positions of our volcanic highs or magmatic centres beneath the shelf are controlled by the early rift architecture rather than the line of final break-up.

The margin north of 77° N bounding the Boreas Basin shows only weak indications for break-up-related magmatism. The basement reflections landwards of the COB indicate the presence of basalt flows or, perhaps, hyaloclastites. However, there is no wedge-like structure resembling real SDRs. Instead, the spurious SDRs indicate that there was minor volcanism during the Early Eocene rifting similar to that observed in the Vestbakken Volcanic Province at the conjugate SW Barents Sea continental margin (e.g. Faleide *et al.* 2008).

Comparison with the conjugate Norwegian continental margin

The conjugate Norwegian continental margin was extensively studied, especially the segment comprising the Vøring marginal plateau and the Vøring Basin. Besides seismic investigations, this area was also the target of several scientific DSDP and ODP legs, as well as commercial drilling campaigns. Eldholm *et al.* (2002) and Tsikalas *et al.* (2002) discussed the segmentation of the margin, which is interpreted to have influenced the abundance of magmatism during break-up. This along-margin segmentation stems most probably from ancient pre-rift-inherited lithospheric structures and later Mesozoic rift transfer or accommodation zones (Eldholm *et al.* 2002; Tsikalas *et al.* 2002).

Prominent SDR wedges could be mapped on the Vøring marginal plateau (most recent works by Abdelmalak *et al.* 2015, 2016*a*, *b*), conjugate to the East Greenland Magmatic Province *sensu* Hamann *et al.* (2005) (see Fig. 12). These SDRs originally provided the basis for the concept of seismic volcanostratigraphy as published by Planke *et al.* (2000). In addition to the primary facies types, seismic data reveal cross-cutting dyke swarms and sill intrusions beneath the SDRs, as well as lower series lava flows along the Vøring marginal plateau (Abdelmalak *et al.* 2015, 2016*a*). A model discussing the deeper structure of the volcanic margins based on a comparison between along the Norwegian margin and outcrops from the Swedish Caledonides suggests that the conjugate East Greenland Magmatic–Vøring Plateau province was a site of intensive magmatic activity shortly before, during and also after the final break-up.

Along the Lofoten–Vesterålen margin, break-up lavas cover almost the entire continental slope and terminate in the vicinity of the shelf edge against older fault scarps along the Lofoten–Vesterålen margin (Tsikalas *et al.* 2001, 2002). However, only sparse original seismic data are published for the northernmost Lofoten–Vesterålen margin, showing no or only weakly developed SDR wedges west of the COB (Tsikalas *et al.* 2002). Eldholm *et al.* (1984) interpreted smooth basement dipping reflectors at the Lofoten margin, but did not identify SDR-like structures further north at the Vesterålen margin. Generally, only a single outer (continuous) SDR wedge is mapped seawards of a basaltic flow unit (e.g. Abdelmalak *et al.* 2015, 2016*b*) (see Fig. 12). The data from the NE Greenland margin do not support a continuous magmatically active zone along this segment of the break-up line. However, there are magmatic centres with characteristics similar to the central volcanoes in the recent Iceland rift zone (Óskarsson & Riishuus 2013), where clear evidence of two SDR wedges (inner and outer SDRs) can be observed in the seismic data. As the data coverage in that part of the NE Greenland margin is rather sparse, this interpretation needs to be substantiated by further studies.

The break-up volcanic sequences in the mid-Norwegian margin include two prominent escarpments: the Vøring Escarpment on the Vøring margin; and the Faroe–Shetland Escarpment on the Møre and UK margins further south. The Vøring Escarpment consists of an approximately 350 km prominent feature along the Vøring margin with a height ranging between 200 and 1600 m (Abdelmalak *et al.* 2016*b*). It separates the Vøring Marginal High to the west from the Vøring Basin to the east. Further north, the Vøring Escarpment does not continue into the Lofoten–Vesterålen margin. There, the escarpment seems to represent a volcanic build-up or simply a flow front. On the NE Greenland conjugate margin, a similar escarpment (Greenland or Thetis Escarpment) facing to the west is identified. The Greenland Escarpment presents a limited extent compared to the Vøring and Faroe–Shetland examples (Abdelmalak *et al.* 2016*b*). This indicates that most of the NE Greenland margin was subaerially exposed, and the lava front reached the water-filled rift basin to the north (Thetis Rift) and SE (Vøring Rift).

Seismic data along the sheared and rifted SW Barents Sea margin are sparser. No SDRs and seafloor-spreading anomalies have been identified at the marginal high in the Vestbakken Volcanic Province (see Faleide *et al.* 1991). Smooth opaque basement reflections are interpreted as lava extruded during the break-up (Faleide *et al.* 1988, 1991). Tsikalas *et al.* (2002) assumed that the complete Vestbakken Volcanic Province was captured on the Eurasian side of the rift after Chron C21n (47 Ma). Our data show comparable reflections along the sheared NE Greenland margin. Hence, we think that the rifting and later drifting were symmetrical without any early rift jump. Tsikalas *et al.* (2002) and references therein also report renewed magmatism near the Eocene–Oligocene transition and during the late Pliocene.

Sill complexes and hydrothermal vents

Sill complexes are imaged by numerous high-amplitude reflections in the southern Danmarkshavn Basin, mainly within the Cretaceous successions (Figs 5 & 9) (see Hamann *et al.* 2005). These are similar in style and position to the sills observed in the Cretaceous–Paleocene sediments of the Vøring and Jameson Land rift basins to the south (e.g. Skogseid & Eldholm 1989; Skogseid *et al.* 1992, Larsen & Marcussen 1992). Given the age of the Utgaard Upper Sill in the Vøring Basin (55.6 Ma: Svensen *et al.* 2010), the sills most probably represent a late rift phase (pre-break-up) event. Most of the magma seems to be intrusive; possibly, the buoyancy of the magma was not large enough to reach the palaeo-surface (Neumann *et al.* 2013). Only a few explosive vents could be detected in the southern Danmarkshavn Basin (Figs 9 & 10) but, perhaps, this observation is biased by the scarce seismic coverage of the area. These structures are similar to well-described and widespread vents in the Vøring Basin (e.g. Berndt *et al.* 2000; Planke *et al.* 2005).

NE Greenland Volcanic Province (NEGVP)

Hamann *et al.* (2005) defined the NE Greenland magmatic province dominated by Palaeogene plateau basalts beneath the shelf between about 72°30′ and 75° N. Unfortunately, our seismic reflection data base covers this part of the margin only sparsely. Along most of the existing profiles, the acoustic basement is not well imaged. However, two of the seismic sections prove the existence of a volcanic basement composed of possibly landwards lava flows and volcanic ridges and mountains (Figs 2 & 3). This interpretation is supported by magnetic and gravity data (Haase & Ebbing 2014; Nasuti & Olesen 2014). Furthermore, wide-angle seismic data image an up to 15 km-thick high-velocity lower crust beneath the wide shelf, which was interpreted as Palaeogene voluminous magmatic underplating by Voss & Jokat (2007) and Voss *et al.* (2009). Figures 10 and 11 show the outline of this magmatic underplating, which has a width of 120–130 km.

The NEGVP is a key region in pre-break-up reconstructions of the NE Atlantic, but there is some disagreement over the location of the COB (e.g. Voss & Jokat 2007; Funck *et al.* 2014). Clear magnetic anomalies indicate that, at least in the northern part, there was some kind of initial seafloor spreading starting close to anomaly C24 (*c.* 54 Ma). However, in the southern part close to the JMFZ, the magmatic activity seems to have been less focused at that time. This could mean that this part was an area of long-lasting Iceland- or Afar-type rifting and initial subaerial oceanic spreading close to the Jan Mayen triple junction (Gaina *et al.* 2009) (see Fig. 12) as a result of increased magma supply and multiple rift zones. The position of the rubbly basement/outer high indicates that the volcanism was subaerial until about 49 Ma (Chron C22) in that area (Fig. 12).

A compelling feature of the NE Greenland margin is the Traill Ø–Vøring Igneous Complex (Olesen *et al.* 2007; Gernigon *et al.* 2009). It clearly stands out as a continuous magmatic feature in the magnetic data (Fig. 11). Unfortunately, it is not well covered by seismic data. Olesen *et al.* (2007) associated this igneous complex with magmatism at the time of Chron C22 (49–50 Ma) and related it to contemporaneous magmatic intrusions in East Greenland. There seems to be a link to the initial magmatic lineament proposed by Nielsen (1987) and Larsen (1988) between the Kangerlussuaq Intrusion (50 Ma, Noble *et al.* 1988) and Traill Ø. Gernigon *et al.* (2009) proposed anomalous melt production and igneous activity during the early spreading history of the Norwegian–Greenland Sea in the Early Eocene (Ypresian) by linking the TVIP to mapped magnetic anomalies. In their model, the western JMFZ represents a leaky fracture zone similar to the Azores triple junction today. Døssing & Funck (2012) suggested that the change in plate motion around C22 (as postulated by Gaina *et al.* 2009) could have had an influence on the activation or deactivation of this leaky fracture zone.

However, the magnetic anomaly pattern (e.g. Gernigon *et al.* 2009) suggests that the Traill Ø–Vøring continuous magnetic anomaly may have already partly existed before Chron C21. Talwani & Eldholm (1972) had speculated that they might represent original geological structures of Palaeozoic or even Precambrian age. The basalts drilled in ODP Site 642E1 on the Vøring Plateau are dated to 54.3 Ma (Sinton *et al.* 1998). This age implies a magmatic phase close to the initial seafloor spreading (chrons C24, 53.5 or C24o; 54 Ma; according to Gaina *et al.* 2002) in the Greenland–Lofoten Basin.

According to Eldholm *et al.* (2002) and references therein, the upper series drilled in the Vøring Plateau represent transitional, mid-oceanic tholeiitic basalts and altered, interbedded, basaltic vitric tuffs emplaced at about 55 Ma. There is contemporaneous magmatic activity on the conjugate East Greenland Shelf, as indicated by a dated basalt sample in the Wollaston Foreland with an age of 53.5 Ma (Price *et al.* 1997) and recent age data from a wider area along the coast (55–51 Ma: see Fig. 12) (Larsen *et al.* 2014). At the same time or shortly after, outer SDRs were emplaced in the northern Greenland–Lofoten Basin (see above).

A comparison of the Early Cenozoic evolution of the NEGVP with recent activity in the Afar depression and the north Ethiopian rift shows a number of similarities. The late rifting and early oceanic spreading in the NE Atlantic Ocean might have taken place in an en echelon pattern and with partial overlap of two or multiple rift/spreading zones (cf. Ebinger & Casey 2001; Keir *et al.* 2013). In this case, the Traill Ø–Vøring Igneous Complex could be an abandoned magmatically active detachment (cf. Ebinger & Casey 2001). As discussed by Wolfenden *et al.* (2005), there is little evidence of final break-up along such crustal- or lithospheric-scale detachments. The location of 'magma injection zones' seems to define finally the line of break-up and seems to establish a regime similar to spreading along normal mid-ocean ridges (Wolfenden *et al.* 2005). In our case, the line of final break-up seems to be associated with the position of the early (inner) SDRs along the northern part of the rifted NE Greenland margin south of the East Greenland Ridge. Rubbly basement, outer highs and outer oceanic SDRs then indicate the submergence of the evolving magma injection zone (spreading centre) below sea level, as proposed by Planke *et al.* (2000).

Alternatively, Traill Ø volcanic ridge is the result of a plume–ridge interaction, as the postulated (pre) Iceland plume position was very close to the East Greenland margin at early seafloor spreading times (see Gaina *et al.* In press, this volume). A similar volcanic, elongated feature has been described in the Indian Ocean (the Rodriguez Ridge) and was formed over 8 myr (from *c.* 11 to 3 Ma) due to the interaction between the Central Indian Ridge and the Reunion Plume (e.g. Morgan 1978) or elsewhere in the oceanic realm (for a review, see Dyment *et al.* 2007).

Conclusions

Revisiting the seismic volcanostratigraphy along the NE Greenland continental margin, we confirm the previously reported strong variations in seismically imaged magmatic and volcanic structures. There are no clear indications for break-up-related magmatic structures in the region where the JMFZ meets the East Greenland Shelf. This could be partly caused by the masking effect of the thick sedimentary cover and partly due to the poor quality of data, but more probably it could represent the structural setting involving the pre-break-up Traill Ø–Vøring Igneous Complex. Along most sections of the margin between the JMFZ and the East Greenland Ridge, we image SDRs and related seismic facies types similar to the conjugate Vøring continental margin. The visibility of the SDR sequences on seismic data decreases towards the north, as it does along the conjugate Lofoten–Vesterålen margin and also towards the TVIP in the south. North of the East Greenland Ridge, strong basement reflections conjugated to the Vestbakken Volcanic Province are interpreted as lava flows or 'spurious' SDRs. At the northernmost sheared margin, there are no indications for the emplacement of break-up-related volcanic rocks. Sill and vent complexes in the southern Danmarkshavn Basin show spatial and temporal relationships to the counterparts in the Jameson Land and Vøring basins. As proposed earlier, the positions of the outer highs or rubbly basement are evidence of margin subsidence, but are not diagnostic for the location of the COB or onset of oceanic crustal accretion. The seawards extent of the outer (oceanic) SDRs and the smooth oceanic basement correlate partially with the extent of initially formed oceanic layer 3B along the NE Greenland margin.

Our seismic volcanostratigraphic interpretations along the rifted NE Greenland margin and the conjugate Norwegian margin, together with potential field and seismic wide-angle data, confirm that the initial oceanic rifting occurred subaerially in the area immediately north of the JMFZ until about 49 Ma, when the spreading axis submerged below the sea level and normal oceanic crust was accreted. We also postulate that the Traill Ø volcanic ridge could have been formed as a result of plume–ridge interactions in the early seafloor-spreading time period.

This work was part of the NAG-TEC project and was sponsored by Bayerngas Norge AS; BP Exploration Operating Company Ltd; Bundesanstalt für Geowissenschaften und Rohstoffe (BGR); Chevron East Greenland Exploration A/S; ConocoPhillips Skandinavia AS; DEA Norge AS; Det norske oljeselskap ASA; DONG E&P A/S; E.ON Norge AS; ExxonMobil Exploration and Production Norway AS; Japan Oil, Gas and Metals National Corporation (JOGMEC); Maersk Oil; Nalcor Energy – Oil and Gas Inc.; Nexen Energy ULC, Norwegian Energy Company ASA (Noreco); Repsol Exploration Norge AS; Statoil (UK) Ltd; and Wintershall Holding GmBH. CG acknowledges support from the Research Council of Norway's Centres of Excellence project No. 223272.

We thank two anonymous reviewers for their helpful comments.

References

Abdelmalak, M.M., Andersen, T.B. *et al.* 2015. The ocean–continent transition in the mid-Norwegian margin: Insight from seismic data and an onshore Caledonian field analogue. *Geology*, **43**, 1011–1014, https://doi.org/10.1130/g37086.1

Abdelmalak, M.M., Meyer, R. *et al.* 2016*a*. Pre-breakup magmatism on the Vøring Margin: Insight from new sub-basalt imaging and results from Ocean Drilling Program Hole 642E. *Tectonophysics*, **675**, 258–274.

Abdelmalak, M.M., Planke, S., Faleide, J.I., Jerram, D.A., Zastrozhnov, D., Eide, S. & Myklebust, R. 2016*b*. The development of volcanic sequences at rifted margins: New insights from the structure and morphology of the Vøring Escarpment, mid-Norwegian Margin. *Journal of Geophysical Research: Solid Earth*, **121**, 5212–5236, https://doi.org/10.1002/2015JB012788

Á Horni, J., Geissler, W.H. *et al.* 2014. Offshore volcanic facies. *In*: Hopper, J.R., Funck, T., Stoker, M., Artling, U., Peron-Pindivic, G., Gaina, C. & Doornenbal, H. (eds) *NAG-TEC Atlas: Tectonostratigraphic Atlas of the North-East Atlantic Region*. Geological Survey of Denmark and Greenland (GEUS), Copenhagen, 235–253.

Á Horni, J., Blischke, A. *et al.* In review. Seismic volcanostratigraphy of the North Atlantic Igneous Province: the volcanic facies units. *In*: Péron-Pinvidic, G., Hopper, J.R., Stoker, M.S., Gaina, C., Doornenbal, J.C., Funck, T. & Árting, U.E. (eds) *The NE Atlantic Region: A Reappraisal of Crustal Structure, Tectonostratigraphy and Magmatic Evolution*. Geological Society, London, Special Publications, **447**.

Bastow, I.D. & Keir, D. 2011. The protracted development of the continent–ocean transition in Afar. *Nature Geoscience*, **4**, 248–250, https://doi.org/10.1038/ngeo1095

Berger, D. & Jokat, W. 2008. A seismic study along the East Greenland margin from 72°N to 77°N. *Geophysical Journal International*, **174**, 733–748.

Berger, D. & Jokat, W. 2009. Sediment deposition in the northern basins of the North Atlantic and characteristic variations in shelf sedimentation along the East Greenland margin. *Marine and Petroleum Geology*, **26**, 1321–1337, https://doi.org/10.1016/j.marpetgeo.2009.04.005

Berndt, C., Skogly, O.P., Planke, S. & Eldholm, O. 2000. High-velocity breakup-related sills in the Voring Basin, off Norway. *Journal of Geophysical Research*, **105**, 28,443–28,454.

Bjerager, M., Nielsen, T. & Guarnieri, P. 2014. 7.5 The East Greenland margin. *In*: Hopper, J.R., Funck, T., Stoker, M., Artling, U., Peron-Pindivic, G., Gaina, C. & Doornenbal, H. (eds) *NAG-TEC Atlas: Tectonostratigraphic Atlas of the North-East Atlantic Region*. Geological Survey of Denmark and Greenland (GEUS), Copenhagen, 178–184.

Bosworth, W., Huchon, P. & McClay, K. 2012. The Red Sea and Gulf of Aden basins. *In*: Roberts, D.G. & Bally, A.W. (eds) *Regional Geology and Tectonics: Phanerozoic Passive Margins, Cratonic Basins and Global Tectonic Maps, Volume 1A*. Elsevier, Amsterdam, 62–139, https://doi.org/10.1016/b978-0-444-56357-6.00003-2

Calvès, G., Schwab, A.M. *et al.* 2011. Seismic volcanostratigraphy of the western Indian rifted margin: the pre-Deccan igneous province. *Journal of Geophysical Research*, **116**, B01101, https://doi.org/10.1029/2010jb000862

Doré, A.G., Lundin, E.R., Birkeland, O., Eliassen, P.E. & Jensen, L.N. 1997. The NE Atlantic Margin: implications of late Mesozoic and Cenozoic events for hydrocarbon prospectivity. *Petroleum Geoscience*, **3**, 117–131, https://doi.org/10.1144/petgeo.3.2.117

Doré, A.G., Lundin, E.R., Jensen, L.N., Birklamd, Ø., Eliassen, P.E. & Fichler, C. 1999. Principal tectonic events in the evolution of the northwest European Atlantic margin. *In*: Fleet, A.J. & Boldy, S.A.R. (eds) *Petroleum Geology of Northwest Europe: Proceedings of the 5th Conference*. Geological Society, London, Petroleum Geology Conference series, **5**, 41–61, https://doi.org/10.1144/0050041

Døssing, A. & Funck, T. 2012. Greenland Fracture Zone–East Greenland Ridge(s) revisited: indications of a C22-change in plate motion? *Journal of Geophysical Research: Solid Earth*, **117**, B01103, https://doi.org/10.1029/2011JB008393

Doubrovine, P.V., Steinberger, B. & Torsvik, T.H. 2012. Absolute plate motions in a reference frame defined by moving hot spots in the Pacific, Atlantic, and Indian oceans. *Journal of Geophysical Research: Solid Earth*, **117**, B09101, https://doi.org/10.1029/2011JB009072

Dyment, J., Lin, J. & Baker, E.T. 2007. Ridge-hotspot interactions – what mid-Ocean ridges tell us about deep earth processes. *Oceanography*, **20**, 102–115, https://doi.org/10.5670/oceanog.2007.84

Ebinger, C.J. & Casey, M. 2001. Continental breakup in magmatic provinces: an Ethiopian example. *Geology*, **29**, 527–530, https://doi.org/10.1130/0091-7613(2001)029<0527:CBIMPA>2.0.CO;2

Eldholm, O. & Grue, K. 1994. North Atlantic volcanic margins: dimensions and production rates. *Journal of Geophysical Research*, **99**, 2955–2968, https://doi.org/10.1029/93JB02879

Eldholm, O., Sundvor, E., Myhre, A.M. & Faleide, J.I. 1984. Cenozoic evolution of the continental margin off Norway and western Svalbard. *In*: Spencer, A.M. (ed.)

Petroleum Geology of the North European Margin. Graham and Trotman, London, 3–28.

Eldholm, O. & Thiede, J. & Leg 104 Shipboard Scientific Party. 1986. Ocean Drilling at the Vøring Plateau in the Norwegian Sea. *Nature*, **319**, 360–361.

Eldholm, O., Tsikalas, F. & Faleide, J.I. 2002. The continental margin off Norway 62–75°N: Palaeogene tectono-magmatic segmentation and sedimentation. *In*: Jolley, D.W. & Bell, B.R. (eds) *The North Atlantic Igneous Province: Stratigraphy, Tectonics, Volcanic and Magmatic Processes*. Geological Society, London, Special Publications, **197**, 39–68, https://doi.org/10.1144/GSL.SP.2002.197.01.03

Elliott, G. & Parson, L. 2008. Influence of margin segmentation upon the break-up of the Hatton Bank rifted margin, NE Atlantic. *Tectonophysics*, **457**, 161–176, https://doi.org/10.1016/j.tecto.2008.06.008

Faleide, J.I., Myhre, A.M. & Eldholm, O. 1988. Early Tertiary volcanism at the western Barents Sea margin. *In*: Morton, A. & Parson, L.M. (eds) *Early Tertiary Volcanism and the Opening of the NE Atlantic*. Geological Society, London, Special Publications, **39**, 135–146, https://doi.org/10.1144/GSL.SP.1988.039.01.13

Faleide, J.I., Gudlaugsson, S.T., Eldholm, O., Myhre, A.M. & Jackson, H.R. 1991. Deep seismic transects across the sheared western Barents Sea–Svalbard continental margin. *Tectonophysics*, **189**, 73–89.

Faleide, J.I., Tsikalas, F. *et al.* 2008. Structure and evolution of the continental margin off Norway and the Barents Sea. *Episodes*, **31**, 82–91.

Faleide, J., Bjørlykke, K. & Gabrielsen, R. 2010. *Geology of the Norwegian Continental Shelf. Petroleum Geoscience*. Springer, Berlin.

Funck, T., Hopper, J.R. *et al.* 2014. Crustal structure. *In*: Hopper, J.R., Funck, T., Stoker, M., Artling, U., Peron-Pindivic, G., Gaina, C. & Doornenbal, H. (eds) *NAG-TEC Atlas: Tectonostratigraphic Atlas of the North-East Atlantic Region*. Geological Survey of Denmark and Greenland (GEUS), Copenhagen, 69–126.

Funck, T., Erlendsson, Ö., Geissler, W.H., Gradmann, S., Kimbell, G.S., McDermott, K. & Petersen, U.K. 2016. A review of the NE Atlantic conjugate margins based on seismic refraction data. *In*: Péron-Pinvidic, G., Hopper, J.R., Stoker, M.S., Gaina, C., Doornenbal, J.C., Funck, T. & Árting, U.E. (eds) *The NE Atlantic Region: A Reappraisal of Crustal Structure, Tectonostratigraphy and Magmatic Evolution*. Geological Society, London, Special Publications, **447**. First published on October 12, 2016, https://doi.org/10.1144/SP447.9

Gaina, C., Roest, W.R. & Müller, R.D. 2002. Late Cretaceous–Cenozoic deformation of northeast Asia. *Earth and Planetary Science Letters*, **197**, 273–286, https://doi.org/10.1016/S0012-821X(02)00499-5

Gaina, C., Gernigon, L. & Ball, P. 2009. Palaeocene–Recent plate boundaries in the NE Atlantic and the formation of the Jan Mayen microcontinent. *Journal of the Geological Society, London*, **166**, 601–616, https://doi.org/10.1144/0016-76492008-112

Gaina, C., Nasuti, A., Kimbell, G.S. & Blischke, A. In press. Break-up and seafloor spreading domains in the NE Atlantic. *In*: Péron-Pinvidic, G., Hopper, J.R., Stoker, M.S., Gaina, C., Doornenbal, J.C., Funck, T. & Árting, U.E. (eds) *The NE Atlantic Region: A Reappraisal of Crustal Structure, Tectonostratigraphy and Magmatic Evolution*. Geological Society, London, Special Publications, **447**, https://doi.org/10.1144/SP447.12.

Gaina, C., Blischke, A., Geissler, W.H., Kimbell, G.S. & Erlendsson, Ö. 2016. Seamounts and volcanic edifices in the NE Atlantic oceanic basins: a link between plate motions and mantle dynamics. *In*: Péron-Pinvidic, G., Hopper, J.R., Stoker, M.S., Gaina, C., Doornenbal, J.C., Funck, T. & Árting, U.E. (eds) *The NE Atlantic Region: A Reappraisal of Crustal Structure, Tectonostratigraphy and Magmatic Evolution*. Geological Society, London, Special Publications, **447**. First published on September 8, 2016, https://doi.org/10.1144/SP447.6

Ganerød, M., Wilkinson, C.M. & Hendriks, B. 2014. Geochronology. *In*: Hopper, J.R., Funck, T., Stoker, M., Artling, U., Peron-Pindivic, G., Gaina, C. & Doornenbal, H. (eds) *NAG-TEC Atlas: Tectonostratigraphic Atlas of the North-East Atlantic Region*. Geological Survey of Denmark and Greenland (GEUS), Copenhagen, 253–261.

Gernigon, L., Olesen, O. *et al.* 2009. Geophysical insights and early spreading history in the vicinity of the Jan Mayen Fracture Zone, Norwegian–Greenland Sea. *Tectonophysics*, **468**, 185–205, https://doi.org/10.1016/j.tecto.2008.04.025

Haase, C. & Ebbing, J. 2014. Gravity data. *In*: Hopper, J.R., Funck, T., Stoker, M., Artling, U., Peron-Pindivic, G., Gaina, C. & Doornenbal, H. (eds) *NAG-TEC Atlas: Tectonostratigraphic Atlas of the North-East Atlantic Region*. Geological Survey of Denmark and Greenland (GEUS), Copenhagen, 29–39.

Hamann, N.E., Whittaker, R.C. & Stemmerink, L. 2005. Geological development of the Northeast Greenland Shelf. *In*: Doré, A.G. & Vinning, B.A. (eds) *Petroleum Geology: North-West Europe and Global Perspectives – Proceedings of the 6th Petroleum Geology Conference*. Geological Society, London, 887–902, https://doi.org/10.1144/0060887

Harðarson, B.S., Fitton, J.G. & Hjartarson, Á. 2008. Tertiary volcanism in Iceland. *Jökull*, **58**, 161–178.

Hinz, K. 1981. A hypothesis on terrestrial catastrophes: Wedges of very thick oceanward dipping layers beneath passive margins – Their origin and paleoenvironment significance. *Geologische Jahrbuch*, **E22**, 345–363.

Hinz, K., Mutter, J.C. & Zehnder, C.M. 1987. Symmetric conjugation of continent-ocean boundary structures along the Norwegian and East Greenland Margins. *Marine and Petroleum Geology*, **4**, 166–187, https://doi.org/10.1016/0264-8172(87)90043-2

Holbrook, W.S., Larsen, H.C. *et al.* 2001. Mantle thermal structure and active upwelling during continental breakup in the North Atlantic. *Earth and Planetary Science Letters*, 190, 251–266.

Hopper, J.R., Dahl-Jensen, T. *et al.* 2003. Structure of the SE Greenland margin from seismic reflection and refraction data: Implications for nascent spreading center subsidence and asymmetric crustal accretion during North Atlantic opening. *Journal of Geophysical Research: Solid Earth*, **108**, 2269, https://doi.org/10.1029/2002jb001996

Keir, D., Bastow, I.D., Pagli, C. & Chambers, E.L. 2013. The development of extension and magmatism in the Red Sea rift of Afar. *Tectonophysics*, **607**, 98–114, https://doi.org/10.1016/j.tecto.2012.10.015

Larsen, H.C. 1988. A multiple and propagating rift model for the NE Atlantic. *In*: Morton, A.C. & Parson, L.M. (eds) *Early Tertiary Volcanism and the Opening of the NE Atlantic*. Geological Society, London, Special Publications, **39**, 157–158, https://doi.org/10.1144/GSL.SP.1988.039.01.15

Larsen, H.C. 1990. The East Greenland shelf. *In*: Grantz, A., Johnson, L. & Sweeney, J.F. (eds) *The Arctic Ocean Region*. The Geology of North America, **L**. Geological Society of America, Boulder, CO, 185–210.

Larsen, H.C. & Marcussen, C. 1992. Sill-intrusion, flood basalt emplacement and deep crustal structure of the Scoresby Sund region, east Greenland. *In*: Storey, B.C., Alabaster, T. & Pankhurst, R.J. (eds) *Magmatism and the Causes of Continental Break-up*. Geological Society, London, Special Publications, **68**, 365- 386, https://doi.org/10.1144/GSL.SP.1992.068.01.23

Larsen, H.C. & Saunders, A.D. 1998. 41. Tectonism and volcanism at the Southeast Greenland rifted margin: a record of plume impact and later continental rupture. *In*: Saunders, A.D., Larsen, H.C. & Wise, S.W., Jr. (eds) *Proceedings of the Ocean Drilling Program, Scientific Results, Volume 152*. Ocean Drilling Program, College Station, TX, 503–533, https://doi.org/10.2973/odp.proc.sr.152.240.1998

Larsen, L.M., Pedersen, A.K., Tegner, C. & Duncan, R.A. 2014. Eocene to Miocene igneous activity in NE Greenland: northward younging of magmatism along the East Greenland margin. *Journal of the Geological Society, London*, **171**, 539–553, https://doi.org/10.1144/jgs2013-118

Meyer, R., Hertogen, J., Pedersen, R.B., Viereck-Götte, L. & Abratis, M. 2009. Interaction of mantle derived melts with crust during the emplacement of the Vøring Plateau, N.E. Atlantic. *Marine Geology*, **261**, 3–16.

Morgan, W.J. 1978. Rodriguez, Darwin, Amsterdam, . . ., A second type of Hotspot Island. *Journal of Geophysical Research*, **83**, 5355–5360, https://doi.org/10.1029/JB083iB11p05355

Mutter, J.C., Talwani, M. & Stoffa, P.L. 1982. Origin of tseaward-dipping reflectors in oceanic crust off the Norwegian margin by 'sub-aerial seafloor spreading'. *Geology*, **10**, 353–357.

Nasuti, A. & Olesen, O. 2014. Magnetic data. *In*: Hopper, J.R., Funck, T., Stoker, M., Artling, U., Peron-Pindivic, G., Gaina, C. & Doornenbal, H. (eds) *NAG-TEC Atlas: Tectonostratigraphic Atlas of the North-East Atlantic Region*. Geological Survey of Denmark and Greenland (GEUS), Copenhagen, 41–51.

Neumann, E.-R., Svensen, H., Tegner, C., Planke, S., Thirlwall, M. & Jarvis, K.E. 2013. Sill and lava geochemistry of the mid-Norway and NE Greenland conjugate margins. *Geochemistry, Geophysics, Geosystems*, **14**, 3666–3690, https://doi.org/10.1002/ggge.20224

Nielsen, T.F.D. 1987. Tertiary alkaline magmatism in East Greenland: a review. *In*: Fitton, J.G. & Upton, B.G.J. (eds) *Alkaline Igneous Rocks*. Geological Society, London, Special Publications, **30**, 489–515, https://doi.org/10.1144/GSL.SP.1987.030.01.24

Noble, R.H., Macintyre, R.M. & Brown, P.E. 1988. Age constraints on Atlantic evolution: timing of magmatic activity along the E Greenland continental margin. *In*: Morton, A.C. & Parson, L.M. (eds) *Early Tertiary Volcanism and the Opening of the NE Atlantic*. Geological Society, London, Special Publications, **39**, 201–214, https://doi.org/10.1144/GSL.SP.1988.039.01.19

Olesen, O., Ebbing, J. et al. 2007. An improved tectonic model for the Eocene opening of the Norwegian–Greenland Sea: Use of modern magnetic data. *Marine and Petroleum Geology*, **24**, 53–66, https://doi.org/10.1016/j.marpetgeo.2006.10.008

Óskarsson, B.V. & Riishuus, M.S. 2013. The mode of emplacement of Neogene flood basalts in Eastern Iceland: Facies architecture and structure of the Hólmar and Grjótá olivine basalt groups. *Journal of Volcanology and Geothermal Research*, **267**, 92–118, https://doi.org/10.1016/j.jvolgeores.2013.09.010

Pagli, C., Wright, T.J., Ebinger, C.J., Yun, S.-H., Cann, J.R., Barnie, T. & Ayele, A. 2012. Shallow axial magma chamber at the slow-spreading Erta Ale Ridge. *Nature Geoscience*, **5**, 284–288, https://doi.org/10.1038/ngeo1414

Planke, S., Symonds, P.A., Alvestad, E. & Skogseid, J. 2000. Seismic volcanostratigraphy of large-volume basaltic extrusive complexes on rifted margins. *Journal of Geophysical Research*, **105**, 19,335–19,351.

Planke, S., Rasmussen, T., Rey, S.S. & Myklebust, R. 2005. Seismic characteristics and distribution of volcanic intrusions and hydrothermal complexes in the Vøring and Møre basins. *In*: Doré, A.G. & Vining, B.A. (eds) *Petroleum Geology: North-West Europe and Global Perspectives – Proceedings of the 6th Petroleum Geology Conference*. Geological Society, London, 833–844, https://doi.org/10.1144/0060833

Price, S., Brodie, J., Whitham, A. & Kent, R. 1997. Mid-Tertiary rifting and magmatism in the Traill Ø region, east Greenland. *Journal of the Geological Society, London*, **154**, 419–434, https://doi.org/10.1144/gsjgs.154.3.0419

Roberts, D.G., Backman, J., Morton, A.C., Murray, J.W. & Keane, J.B. 1984. Evolution of volcanic rifted margins: synthesis of Leg 81 results on the west margin of Rockall Plateau. *In*: Roberts, D.G., Schnitker, D. et al. (eds) *Initial Reports of the Deep Sea Drilling Project, Volume 81*. Ocean Drilling Program, College Station, TX, 883–911.

Sæmundsson, K., Kristjánsson, L., McDougall, I. & Watkins, N.D. 1980. K-Ar dating, geological and paleomagnetic study of a 5-km lava succession in Northern Iceland. *Journal of Geophysical Research*, **85**, 3628–3646.

Saunders, A.D., Fitton, J.G., Kerr, A.C., Norry, M.J. & Kent, R.W. 1997. The North Atlantic igneous province. *In*: Coffin, M.F. & Mahoney, J.J. (eds) *Large Igneous Provinces: Continental. Oceanic, and Planetary Flood Volcanism*. American Geophysical Union, Geophysical Monographs, **100**, 45–93.

Sinton, C.W., Hitchen, K. & Duncan, R.A. 1998. $^{40}Ar-^{39}Ar$ geochronology of silic and basic volcanic rocks on the margins of the North Atlantic. *Geological Magazine*, **135**, 161–170.

Skogseid, J. & Eldholm, O. 1989. 50. Voring plateau continental margin: seismic interpretation, stratigraphy, and vertical movements. *In*: Eldholm, O., Thiede, J., Taylor, E. et al. (eds) *Proceedings of the Ocean Drilling Program, Scientific Results*. Ocean Drilling Program, College Station, TX, 993–1029.

Skogseid, J., Pedersen, T. & Larsen, V.B. 1992. Vøring Basin: subsidence and tectonic evolution. *In*: Larsen, R.M., Brekke, H., Larsen, B.T. & Talleras, E. (eds) *Structural and Tectonic Modelling and its Application to Petroleum Geology*. Norwegian Petroleum Society, Special Publications, **1**, 55–82.

Skogseid, J., Planke, S., Faleide, J.I., Pedersen, T., Eldholm, O. & Neverdal, F. 2000. NE Atlantic continental rifting and volcanic margin formation. *In*: Nottvedt, A. (eds) *Dynamics of the Norwegian Margin*. Geological Society, London, Special Publications, **167**, 295–326, https://doi.org/10.1144/GSL.SP.2000.167.01.12

Storey, M., Duncan, R.A. & Tegner, C. 2007. Timing and duration of volcanism in the North Atlantic Igneous Province: Implications for geodynamics and links to the Iceland hotspot. *Chemical Geology*, **241**, 264–281, https://doi.org/10.1016/j.chemgeo.2007.01.016.

Svensen, H., Planke, S. & Corfu, F. 2010. Zircon dating ties NE Atlantic sill emplacement to initial Eocene global warming. *Journal of the Geological Society, London*, **167**, 433–436, https://doi.org/10.1144/0016-76492009-125

Talwani, M. & Eldholm, O. 1972. The continental margin off Norway: A geophysical study. *Geological Society of America Bulletin*, **83**, 3573–3606.

Talwani, M., Mutter, J.C. & Hinz, K. 1983. Ocean–continent boundary under the Norwegian continental margin. *In*: Bott, M.H.P., Saxov, S., Talwani, M. & Thiede, J. (eds) *Structure and Development of the Greenland–Scotland Ridge*. NATO Conference Series IV: Marine Sciences, **8**. Plenum Press, New York, 121–131.

Tsikalas, F., Faleide, J.I. & Eldholm, O. 2001. Lateral variations in tectono-magmatic style along the Lofoten–Vesteralen volcanic margin off Norway. *Marine and Petroleum Geology*, **18**, 807–832.

Tsikalas, F., Eldholm, O. & Faleide, J.I. 2002. Early Eocene sea floor spreading and continent–ocean boundary between Jan Mayen and Senja fracture zones in the Norwegian–Greenland Sea. *Marine Geophysical Researches*, **23**, 247–270.

Tsikalas, F., Eldholm, O. & Faleide, J.I. 2005. Crustal structure of the Lofoten–Vesterålen continental margin, off Norway. *Tectonophysics*, **404**, 151–174, https://doi.org/10.1016/j.tecto.2005.04.002

Tsikalas, F., Faleide, J.I. & Kuznir, N.J. 2008. Along-strike variations in rifted margin crustal architecture and lithosphere thinning between northern Vøring and Lofoten margin segments off mid-Norway. *Tectonophysics*, **458**, 68–81.

Tsikalas, F., Faleide, J.I., Eldholm, O. & Blaich, O.A. 2012. The NE Atlantic conjugate margins. *In*: Roberts, D.G. & Bally, A.W. (eds) *Regional Geology and Tectonics: Phanerozoic Passive Margins, Cratonic Basins and Global Tectonic Maps, Volume 1A*. Elsevier, Amsterdam, 140–201, https://doi.org/10.1016/b978-0-444-56357-6.00004-4

Viereck, L.G., Taylor, P.N., Parson, L.M., Morton, A.C., Hertogen, J. & Gibson, I.L. the ODP Leg 104 Scientific Party 1988. Origin of the Palaeogene Vøring Plateau volcanic sequence. *In*: Morton, A. & Parson, L.M. (eds) *Early Tertiary Volcanism and the Opening of the NE Atlantic*. Geological Society, London, Special Publications, **39**, 69–83, https://doi.org/10.1144/GSL.SP.1988.039.01.08

Voss, M. & Jokat, W. 2007. Continent–ocean transition and voluminous magmatic underplating derived from P-wave velocity modelling of the East Greenland continental margin. *Geophysical Journal International*, **170**, 580–604, https://doi.org/10.1111/j.1365-246X.2007.03438.x

Voss, M., Schmidt-Aursch, M.C. & Jokat, W. 2009. Variations in magmatic processes along the East Greenland volcanic margin. *Geophysical Journal International*, **177**, 755–782, https://doi.org/10.1111/j.1365-246X.2009.04077.x

Wolfenden, E., Ebinger, C., Yirgu, G., Renne, P.R. & Kelley, S.P. 2005. Evolution of a volcanic rifted margin: Southern Red Sea, Ethiopia. *Geological Society of America Bulletin*, **117**, 846, https://doi.org/10.1130/b25516.1

Ziegler, P.A. 1988. *Evolution of the Arctic–North Atlantic and the Western Tethys*. American Association of Petroleum, Memoirs, **43**.

A review of the NE Atlantic conjugate margins based on seismic refraction data

THOMAS FUNCK[1]*, ÖGMUNDUR ERLENDSSON[2], WOLFRAM H. GEISSLER[3], SOFIE GRADMANN[4], GEOFFREY S. KIMBELL[5], KENNETH MCDERMOTT[6] & UNI K. PETERSEN[7]

[1]*Geological Survey of Denmark and Greenland, Øster Voldgade 10, 1350 Copenhagen K, Denmark*

[2]*Iceland Geosurvey, Grensasvegi 9, 108 Reykjavík, Iceland*

[3]*Alfred Wegener Institute Helmholtz Centre for Polar and Marine Research, Am Alten Hafen 26, 27568 Bremerhaven, Germany*

[4]*Geological Survey of Norway, Leiv Eirikssons vei 39, 7040 Trondheim, Norway*

[5]*British Geological Survey, Keyworth, Nottingham NG12 5GG, UK*

[6]*UCD School of Geological Sciences, University College Dublin, Belfield, Dublin 4, Ireland*

[7]*Faroese Earth and Energy Directorate, Brekkutún 1, 110 Tórshavn, Faroe Islands*

**Correspondence: tf@geus.dk*

Abstract: The NE Atlantic region evolved through several rift episodes, leading to break-up in the Eocene that was associated with voluminous magmatism along the conjugate margins of East Greenland and NW Europe. Existing seismic refraction data provide good constraints on the overall tectonic development of the margins, despite data gaps at the NE Greenland shear margin and the southern Jan Mayen microcontinent. The maximum thickness of the initial oceanic crust is 40 km at the Greenland–Iceland–Faroe Ridge, but decreases with increasing distance to the Iceland plume. High-velocity lower crust interpreted as magmatic underplating or sill intrusions is observed along most margins but disappears north of the East Greenland Ridge and the Lofoten margin, with the exception of the Vestbakken Volcanic Province at the SW Barents Sea margin. South of the narrow Lofoten margin, the European side is characterized by wide margins. The opposite trend is seen in Greenland, with a wide margin in the NE and narrow margins elsewhere. The thin crust beneath the basins is generally underlain by rocks with velocities of >7 km s^{-1} interpreted as serpentinized mantle in the Porcupine and southern Rockall basins; while off Norway, alternative interpretations such as eclogite bodies and underplating are also discussed.

The opening of the NE Atlantic Ocean between East Greenland and NW Europe created mainly magma-rich margins, with exception of the northernmost area at the shear margin between Svalbard/SW Barents Sea and NE Greenland (Fig. 1). The magmatism is associated with the North Atlantic Igneous Province (Coffin & Eldholm 1994; Saunders *et al.* 1997), as evidenced by onshore flood basalts, basalt flows, seaward-dipping reflections (SDRs), volcanic centres and sills. The main phase of pre-break-up volcanism started at 61 Ma (Storey *et al.* 2007 and references therein). An increase in magma production rate at 56 Ma marks the transition to the syn-break-up volcanism (Storey *et al.* 2007).

Our knowledge of magma-poor rifted margins has significantly advanced over the last two decades, in particular as a result of the detailed studies along the Newfoundland–Iberia conjugate margin pair that involved the drilling of a number of scientific wells (see summary in Tucholke *et al.* 2007). In contrast, the understanding of magma-rich margins is still not satisfactory (cf. Quirk *et al.* 2014) as rift structures are overprinted by magmatism and seismic imaging problems are common. Seismic refraction data provide information on the large-scale crustal velocity structure. In particular, continental, transitional and oceanic crust can be distinguished, the amount of crustal thinning can be determined,

From: Péron-Pinvidic, G., Hopper, J. R., Stoker, M. S., Gaina, C., Doornenbal, J. C., Funck, T. & Árting, U. E. (eds) 2017. *The NE Atlantic Region: A Reappraisal of Crustal Structure, Tectonostratigraphy and Magmatic Evolution*. Geological Society, London, Special Publications, **447**, 171–205.
First published online October 12, 2016, updated October 19, 2016, https://doi.org/10.1144/SP447.9

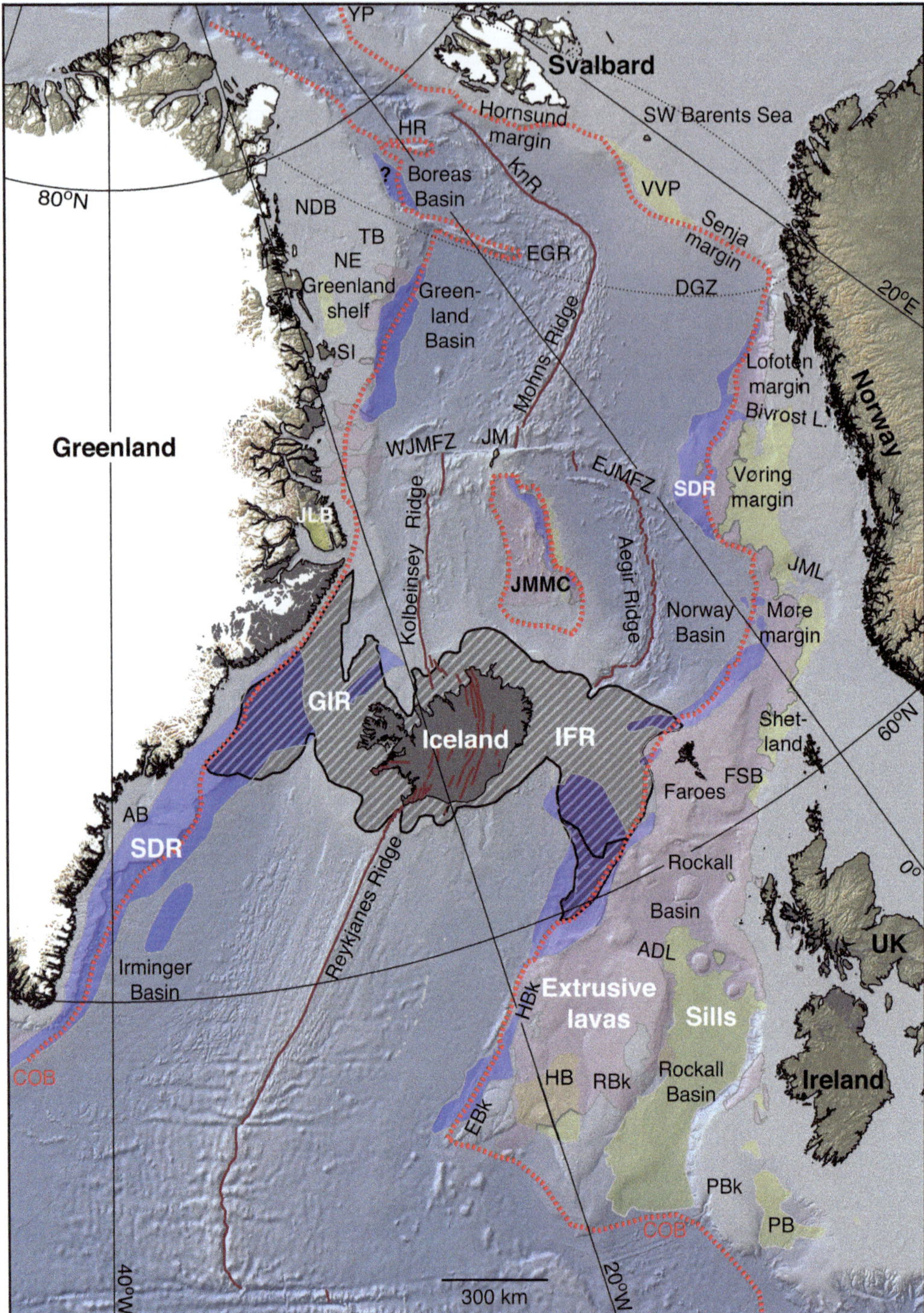

Fig. 1. Physiographical map (after Hopper *et al.*, this volume, in prep) of the NE Atlantic Ocean with the extent of the North Atlantic Igneous Province. The continent–ocean boundary (COB: dashed red line) is taken from Funck *et al.* (2014). Transparent overlays mark SDRs (blue), extrusive lavas (purple/pink), sills (green), the Greenland–Iceland–Faroe Ridge complex (hachured) and the onshore volcanic rocks (grey): all after á Horni *et al.* (2014), see also á Horni *et al.* (this volume, in review). Red lines mark the location of active and extinct spreading centres.

and magmatic additions can be recognized. However, when velocity models from conjugate margins in the NE Atlantic are compared, there are often inconsistencies. For example, the initial seafloor spreading between SE Greenland and the Hatton margin seems to be highly asymmetrical, with a higher spreading rate on the Greenland side (White & Smith 2009). If this asymmetry is real or just perceived depends largely on the correct identification of the continent–ocean boundary (COB). Funck *et al.* (2014) noticed that the various published COBs in the NE Atlantic deviate by as much as 150 km.

Despite a large number of published seismic refraction lines in the NE Atlantic realm, there is a lack of truly conjugate profiles (as derived from plate reconstructions such as the one in Gaina 2014 and Gaina *et al.*, this volume, in review). Often the lines are too short to image the entire margin from the proximal to the distal zone. When velocity models are stitched together, misfits at the intersection are very common and can result from differences in the modelling technique (e.g. forward modelling v. tomography), data density and quality, availability of coincident seismic reflection data, and the general concepts of margin evolution, which changed over time and vary between the working groups performing the modelling and interpretation. In addition, the experience of the modeller can also have some influence on the modelling results, as the proper identification of seismic phases is essential. Misfits at line intersections are easy to spot but become more of an issue when lines at conjugate margins are compared and conclusions on the margin development are drawn.

This paper reviews the seismic refraction data along the continental margins bordering the NE Atlantic Ocean and summarizes the current knowledge on the crustal structure. Conjugate margin transects are compiled to present the full rift system and to check for any inconsistencies in the models or interpretations. A full listing of the available seismic refraction lines in the study region is given in Funck *et al.* (2016, their fig. 2 and table 1). After a brief summary of the opening history of the NE Atlantic, this paper provides a more detailed account on the margin structure in three different subregions: the area south of Iceland; the Greenland–Iceland–Faroe Ridge; and the margins north of Iceland. In each section, the individual margin segments are described based on available seismic refraction lines, followed by a compilation of conjugate margin transects. The overall scope is to summarize the current knowledge of the crustal structure of the margins and to provide some ideas of where additional data would be beneficial to address some of the noticed inconsistencies in the reviewed datasets.

Geological setting

The NE Atlantic Ocean evolved from the Palaeogene break-up of an amalgamated landmass composed of Avalonia, Baltica and Laurentia that had formed the vast Caledonian mountain chain (Ziegler 1988). Prior to break-up, the landmass was subject to post-orogenic collapse in Devonian times, followed by several episodes of extension and rifting starting in the Carboniferous (Ziegler 1988; Doré *et al.* 1999). Magmatism associated with the North Atlantic Igneous Province began at 61 Ma and continued throughout the break-up phase (Storey *et al.* 2007), which resulted in magma-rich (volcanic-style) continental margins extending over a length of 2000 km on either side of the ocean (Fig. 1). Storey *et al.* (2007) estimated that continental break-up in the NE Atlantic occurred in the Eocene at 55.5 $\pm$0.3 Ma.

The margins are divided into a number of segments that correlate to a large degree with the segmentation of the present-day Mid-Atlantic Ridge (Fig. 1). South of Iceland, the SE Greenland continental margin is narrow and characterized by the absence of major rift basins, with exception of the poorly studied Ammassalik Basin (Hopper *et al.* 1998). On the conjugate Faroe–Hatton–Rockall margin, the Hatton and Rockall basins are the most prominent rift basins that together span a width of 600 km. The Hatton Basin terminates at the Anton Dohrn Lineament, which Kimbell *et al.* (2005) interpreted as a transfer zone. Across this zone, the axis of the Rockall Basin is shifted by 200 km.

Iceland is part of the Greenland–Iceland–Faroe Ridge (Fig. 1) where thick igneous crust in excess of 30 km is observed (Darbyshire *et al.* 1998; Richardson *et al.* 1998; Holbrook *et al.* 2001). The crustal thickness is the result of enhanced melting in the Iceland mantle plume (White & McKenzie 1995).

The opening history of the Norwegian–Greenland Sea between Iceland and the Jan Mayen

Fig. 1. (*Continued*) The dotted line indicates the region affected by the De Geer Zone (DGZ) megashear system. Abbreviations: AB, Ammassalik Basin; ADL, Anton Dohrn Lineament; EBk, Edoras Bank; EGR, East Greenland Ridge; EJMFZ, East Jan Mayen Fracture Zone; FSB, Faroe–Shetland Basin; GIR, Greenland–Iceland Ridge; HB, Hatton Basin; HBk, Hatton Bank; HR, Hovgaard Ridge; IFR, Iceland–Faroe Ridge; JLB, Jameson Land Basin; JM, Jan Mayen Island; JML, Jan Mayen Lineament; JMMC, Jan Mayen microcontinent; KnR, Knipovich Ridge; L., Lineament; NDB, North Danmarkshavn Basin; PB, Porcupine Basin; PBk, Porcupine Bank; RBk, Rockall Bank; SDR, seaward-dipping reflection; SI, Shannon Island; TB, Thetis Basin; VVP, Vestbakken Volcanic Province; WJMFZ, West Jan Mayen Fracture Zone; YP, Yermak Plateau.

Fracture Zone is characterized by numerous plate-boundary relocations that formed the highly or super-extended Jan Mayen microcontinent (JMMC) (Gaina *et al.* 2009; Gernigon *et al.* 2015). The initial opening occurred along the Aegir Ridge between the JMMC and the mid-Norwegian Møre Margin, and lasted until at least 30 Ma when the ridge became extinct (Jung & Vogt 1997; Gaina *et al.* 2009; Gernigon *et al.* 2015 and references therein). Gaina *et al.* (2009) related the fragmented character of the JMMC to several failed ridge-propagation attempts of the Kolbeinsey Ridge. By Chrons C6–C7 (25–20 Ma), stable seafloor spreading was established at the Kolbeinsey Ridge between East Greenland and the JMMC (Gaina *et al.* 2009).

Between the Jan Mayen Fracture Zone and the Greenland/Senja fracture zones, the East Greenland shelf widens northwards (Hamann *et al.* 2005), while the shelf off Norway with the Vøring and Lofoten margins narrows (Fig. 1). These two margins are separated by the Bivrost Lineament (Mjelde *et al.* 2003, 2005; Tsikalas *et al.* 2005).

Rifting in the NE Atlantic Ocean was linked to the Eurasia Basin in the Arctic by the De Geer Zone megashear system, encompassing the NE part of Greenland and the westernmost Barents Sea (Harland 1969; Mosar *et al.* 2002*b*). Off Norway, this region includes a transform zone off western Svalbard, and the sheared Hornsund and Senja margins (Fig. 1) (Eldholm *et al.* 1987; Engen *et al.* 2008). The strike-slip shear margin did not develop into a fully passive margin before the Oligocene (Faleide *et al.* 1996). The East Greenland Ridge is a sliver of continental crust that was sheared off the NE Greenland margin (Døssing *et al.* 2008; Døssing & Funck 2012). Another bathymetric high in this shear region is the Hovgaard Ridge, the crustal affinity of which is not yet fully resolved (Engen *et al.* 2008). North of the Greenland and Senja fracture zones, little volcanism is observed and restricted to the southernmost part (á Horni *et al.* 2014, this volume, in review). This includes the Vestbakken Volcanic Province on the Norwegian side of the margin, where sills intruded into Eocene sediments (Faleide *et al.* 1988; Ryseth *et al.* 2003). Conjugate to Vestbakken, some strong reflections observed in seismic data may indicate a volcanic cover (á Horni *et al.* 2014; cf. á Horni *et al.*, this volume, in review).

Continent–ocean boundary

Unequivocal oceanic and continental crust are separated by a continent–ocean transition (COT) zone, the composition and internal structure of which is often complex and ambiguous. At magma-poor margins, the nature of the basement in the transition zone is controversial, with models ranging from thinned and disrupted continental crust to exhumed mantle or ultra-slow spreading oceanic crust (cf. Peron-Pinvidic *et al.* 2013). At magma-rich margins, the interpretation of the COT becomes even more complicated as remnants of thinned continental crust may become indistinguishable from oceanic crust due to the break-up-related magmatic overprint. In the paper by White & Smith (2009), the COT was defined as the zone where a seaward increase in the average lower-crustal velocity is observed, indicating magmatic intrusions into older, seismically slower crust. With a dense spacing of seismic receivers, it is possible to determine the lower-crustal velocity distribution with sufficient resolution. However, many older seismic refraction lines do not have the necessary resolution to map the COT this way. At magma-rich margins, the initial oceanic crust generally has a higher lower-crustal velocity than normal oceanic crust, which White & Smith (2009) attributed to increased mantle temperatures; in the case of the NE Atlantic, the higher temperatures have been linked to the influence of the Iceland mantle plume.

It is often useful to define a distinct continent–ocean boundary (COB), outboard of which pure oceanic crust is present. This boundary is needed, for example, for plate kinematic reconstructions. In the following presentation of crustal velocity models along the NE Atlantic margins, the COB of Funck *et al.* (2014) is used (Fig. 1). As a starting point for the construction of the COB, Funck *et al.* (2014) mapped the landward limit of clear oceanic crust on seismic refraction lines. Initial refinements were then implemented on the basis of potential field data and other mapped structural features. Final adjustments were made after compiling plate reconstructions (Gaina 2014; cf. Gaina *et al.*, this volume, in review) for the time of break-up. In this paper, the COB is regarded as the seaward limit of the COT.

The plate reconstruction used in this paper is based on the identification of magnetic anomalies and fracture zones (Gaina 2014; cf. Gaina *et al.*, this volume, in review). In this reconstruction, the misfit of the COBs of conjugate plates at the time of closure is variable and displays some of the uncertainty in defining the COB. North of the Jan Mayen Fracture Zone up to the Senja margin (Fig. 1), the conjugate COBs of Funck *et al.* (2014) match within 20 km upon closure. The COBs of the northern JMMC and the Møre margin fit within 10 km, while the largest misfit of up to 70 km is observed south of Iceland.

Continental margins south of Iceland

This section starts with a review of the seismic refraction lines along the SE Greenland continental margin, followed by a description of the margins and rift basins in the Faroe–Hatton–Rockall region. Finally, three conjugate transects are presented and discussed with a focus on the observed asymmetries in crustal structure.

SE Greenland continental margin

The SE Greenland continental margin was studied in the SIGMA experiment, which was carried out in 1996 and consisted of four seismic refraction lines (Holbrook *et al.* 2001). They are all dip lines and three of them (SIGMA lines 2–4) are located south of the Greenland–Iceland Ridge (Fig. 2). The occurrence of thick igneous crust, including a high-velocity lower crust (HVLC: velocities well above 7.0 km s^{-1}) and a thick extrusive layer with systematically SDRs (Fig. 1), confirms that the SE Greenland margin is magma-rich along its entire length (Holbrook *et al.* 2001). The lateral changes along the margin are primarily related to the distance to the Iceland plume.

SIGMA line 2 (Korenaga *et al.* 2000) is located just to the south of the Greenland–Iceland Ridge and is characterized by 27 km-thick igneous crust at the location of break-up (Fig. 2a). The line is close to what Holbrook *et al.* (2001) defined as the proximal zone of the Iceland hotspot, which explains the increased crustal thickness. The oceanic crust remains relatively thick for another 50 km seaward before a noticeable thinning of the crust occurs. However, nowhere is the oceanic crust thinner than 8 km, which is more than the 7 km global average thickness of oceanic crust (White *et al.* 1992).

SIGMA line 3 (Fig. 2b) (Hopper *et al.* 2003) is at a more distal position to the Iceland plume and this is reflected by a further decrease in the oceanic crustal thickness at the break-up location (16 km). If the lower-crustal velocities are used as criteria for the definition of the COB (White & Smith 2009; see above), the COB may be located 60 km more landward than indicated by the COB of Funck *et al.* (2014). Irrespective of the location of the COB, the thinning of the oceanic crust is more gradual than along SIGMA line 2. Landward of the COB, the crust thickens and displays a maximum Moho depth of 33 km and a HVLC is observed.

SIGMA line 4 (Holbrook *et al.* 2001), at the southern tip of Greenland, is the most distal profile to the Iceland plume (Fig. 2c). The initial thickness of the oceanic crust at the break-up location (16 km) is not too dissimilar to what is observed on line 3. However, seaward, the crust thins to 8–9 km over only 70 km, which is narrower than what is observed on line 3 and indicates that the period of excess magmatism was shorter. Landward of the COB, the Moho deepens to 28 km and the crust displays a HVLC with a thickness of up to 8 km. Velocities of 6.7–7.3 km s^{-1} in the HVLC are not too well constrained, but the seismic records show prominent reflections from both the top and the base of the HVLC (J.R. Hopper pers. comm.). Extrusive volcanic rocks with velocities of >4.0 km s^{-1} can be correlated from the COB to the NW end of line 4, with the exception of a basement high south of Cape Farvel. Looking at the igneous crustal thickness along the SE Greenland margin, Holbrook *et al.* (2001) concluded that the initially wide melting anomaly (>1200 km) became restricted to the Greenland–Iceland Ridge in only 6 myr following the break-up.

SIGMA lines 2–4 did not detect any appreciable rift basins beneath the extrusive volcanics in the COT zone. The poorly defined Ammassalik Basin (Hopper *et al.* 1998; Gerlings *et al.*, this volume, in review) seems to be restricted to the area between SIGMA lines 2 and 3 (Funck *et al.* 2016). Line 3 is located along legs 152 (Saunders *et al.* 1998) and 163 (Larsen *et al.* 1999) of the Ocean Drilling Program (ODP). Here some meta-sedimentary rocks of unknown age were drilled at ODP site 917 (Larsen *et al.* 1998).

Faroe–Hatton–Rockall region

Similar to the SE Greenland continental margin, the Faroe–Hatton–Rockall region (Fig. 1) is characterized by high levels of pre-, syn- and post-break-up magmatism, as evidenced by SDRs, lava flows and sills (Hopper *et al.* 2014; á Horni *et al.*, this volume, in review). The Hatton, Rockall and Porcupine basins form the main rift basins in this area. Along RAPIDS line 4 in the Porcupine Basin (Fig. 3d), an up to 12 km-thick sedimentary sequence is underlain by continental crust that is extremely thin, in some places less than 2 km (O'Reilly *et al.* 2006). The three sedimentary layers are interpreted as Cretaceous and Cenozoic strata on top of predominantly Jurassic syn-rift deposits (O'Reilly *et al.* 2006). Mantle velocities beneath the basin centre are as low as 7.2 km s^{-1} and indicate a partial serpentinization of the mantle rock (O'Reilly *et al.* 2006). These velocities would be also compatible with magmatic underplating but O'Reilly *et al.* (2006) argued against this alternative for several reasons, such as the lack of a double reflection from the top and base of an underplated body. Beneath the adjacent Porcupine Bank and the Irish Shelf, the Moho deepens to 30 km, which is similar to full-thickness continental crust beneath Ireland and the UK (Funck *et al.* 2016). The crustal thinning to either side of the basin is asymmetrical, with a narrower necking zone in the west. O'Reilly *et al.* (2006) interpreted this to be the result of simple shear along low-angle westwards-dipping detachment surfaces formed during later stages of extension.

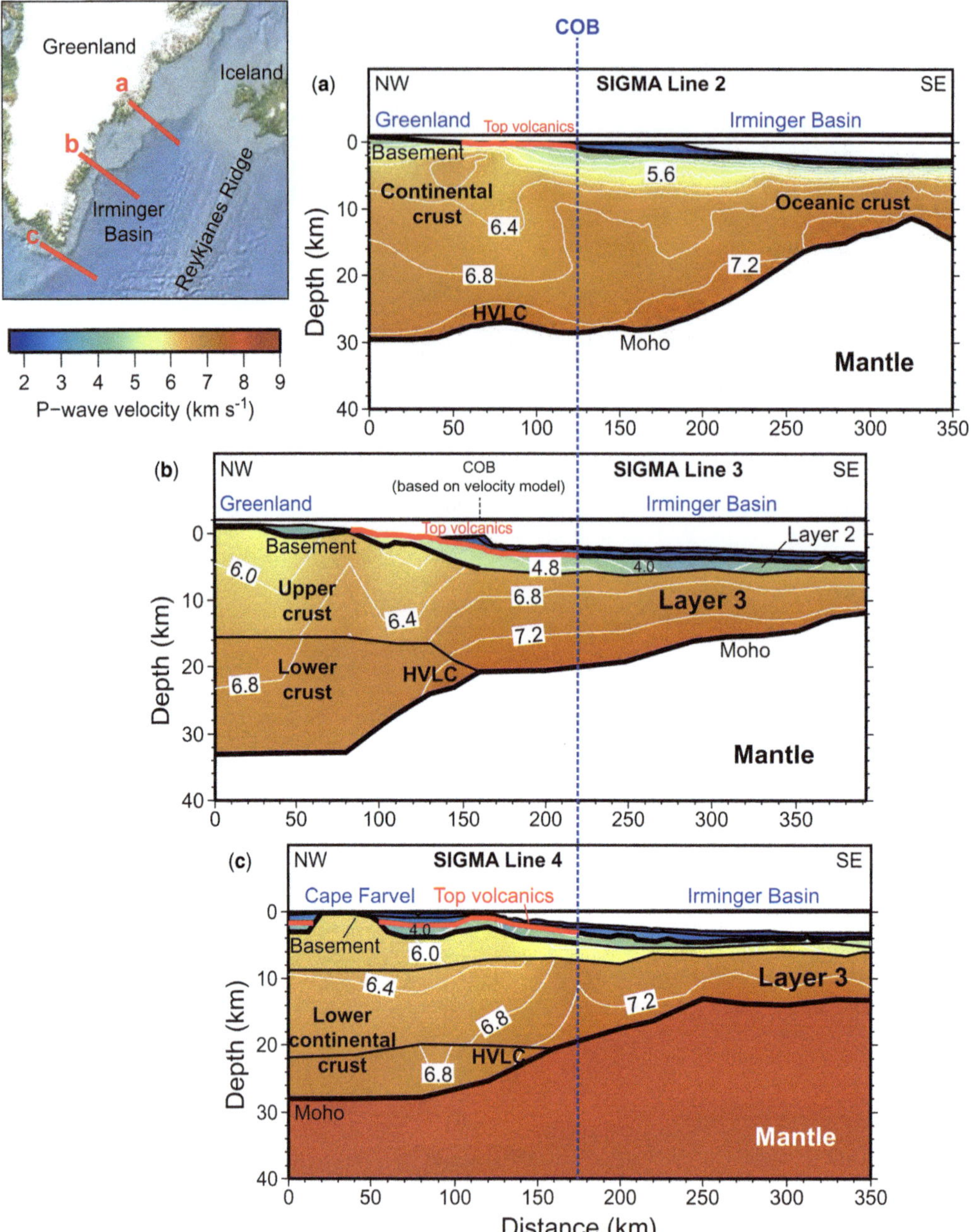

Fig. 2. P-wave velocity models at the SE Greenland margin: (**a**) SIGMA line 2 (after Korenaga *et al.* 2000); (**b**) SIGMA line 3 (after Hopper *et al.* 2003); and (**c**) SIGMA line 4 (after Holbrook *et al.* 2001). Velocities are given in km s^{-1} and the contour interval (white lines) is 0.4 km s^{-1}. Black lines mark layer boundaries in the velocity models. Abbreviations: COB, continent–ocean boundary; HVLC, high-velocity lower crust.

Serpentinized mantle is also interpreted beneath the southern Rockall Basin: for example, along RAPIDS line 1 (Fig. 3c) (Shannon *et al.* 1999). However, mantle velocities do not decrease below 7.5 km s^{-1}, indicating a lower degree of serpentinization (Christensen 2004) when compared to the Porcupine Basin. This is consistent with a thicker continental crust beneath the Rockall Basin. Along

RAPIDS line 1, the continental crust is stretched to a thickness of 5–7 km in a 200 km-wide zone (Shannon *et al.* 1999).

The amount of crustal thinning in the Rockall Basin decreases northwards. On AMP line E (Fig. 3a) (Klingelhöfer *et al.* 2005) in the northern Rockall Basin, the minimum crustal thickness is 12 km, twice as much as in the south. No mantle refractions are observed on AMP line E and there is no indication of mantle serpentinization on the cross-line (AMP line D) (Klingelhöfer *et al.* 2005). The sedimentary cover sequence on AMP line E is about 5 km thick, including up to 1.5 km of Palaeogene volcanic rocks. The sedimentary succession in the southern Rockall Basin (Fig. 3c) is typically 4.5 km thick and reaches a maximum of 7 km (Mackenzie *et al.* 2002).

To the west, the southern Rockall Basin is bounded by the Rockall Bank (Fig. 3c) that is characterized by a three-layered continental crust with a total thickness of 28–30 km (Bunch 1979; Shannon *et al.* 1999). The Hatton and Rockall banks are separated by the Hatton Basin. Available refraction data (Fig. 3c) indicate that the basin is underlain by continental crust with a thickness of about 15 km (Vogt *et al.* 1998; White & Smith 2009). The basin contains 1–3 km of post-break-up sedimentary rocks, but the deeper basin-fill is poorly understood because of the masking effects of the Cenozoic lavas. However, it appears likely that these lavas are underlain by Mesozoic rifts that contain rocks with relatively high velocities, because of either overcompaction or a high concentration of sills. Magnetic anomalies over the Hatton Basin suggest substantial volumes of magmatic rocks (Kimbell *et al.* 2010).

The outer continental margin extends from the Faroe Islands to Edoras Bank, linking a series of bathymetric highs (Fig. 1). The thickness of the crystalline crust beneath these highs is generally between 20 and 25 km (Funck *et al.* 2008; White & Smith 2009), but is slightly less (17 km) beneath Edoras Bank (Barton & White 1997). All these bathymetric highs are capped by several kilometres of volcanic rocks (Funck *et al.* 2008) (see also Fig. 3a). In places, low-velocity zones were detected between the volcanic strata and the underlying basement (Funck *et al.* 2008; Eccles *et al.* 2009), which may indicate the presence of sedimentary rocks predating the volcanism. The volcanic strata and SDRs are the result of voluminous pre-, syn- and post-break-up magmatism. Intrusive rocks and a HVLC are observed in a 40–50 km-wide transition zone (e.g. AMP line E and iSIMM Hatton dip line in Fig. 3a, b). The initial thickness of the oceanic crust is high, but decreases from north to south. At Lousy Bank, AMP line E (Fig. 3a) indicates that initial oceanic crustal thickness may be as great as 28 km (Klingelhöfer *et al.* 2005), further south at Hatton Bank it is 17 km (iSIMM Hatton dip line in Fig. 3b) (White & Smith 2009) and at Edoras Bank it is about 14 km (Fig. 4c) (Barton & White 1997).

North of Hatton Bank, the published seismic refraction lines are not long enough to fully map the decrease in oceanic crustal thickness over time. The best dataset available for the margin is the iSIMM Hatton line (Fig. 3b), where the oceanic crust thins from 17 to 9 km over a distance of 70 km to remain at that thickness to the end of the line (Parkin & White 2008; White & Smith 2009).

Conjugate transects south of Iceland

For all three dip lines (SIGMA lines 2–4) covering the SE Greenland continental margin, seismic refraction profiles could be found at the Faroe–Hatton–Edoras margin that are conjugate within 100 km (Fig. 4). The southernmost transect (Fig. 4c) is composed of SIGMA line 4 (Holbrook *et al.* 2001) and line CAM77 (Barton & White 1997). The two lines are joined at their position at magnetic Chron C22N (49 Ma), where both basement and Moho depth match each other. Up to a distance of 120 km away from Chron C22N, the Moho depth is fairly symmetrical. However, when the COB of Funck *et al.* (2014) is used, the zone of oceanic crust is roughly 20 km wider off Greenland. The initial oceanic crustal thickness on SIGMA line 4 is 16 km, while it is 14 km on line CAM77. While this deviation can be explained within the modelling uncertainties, the symmetry in Moho depth may, instead, argue for an incorrect position of the COB. Gaina *et al.* (2009) used a COB that is 35 km more landward at the Edoras Bank compared to the COB of Funck *et al.* (2014) used here. This indicates that the uncertainty in the position of the COB may well remove the apparent asymmetry in crustal accretion along this transect. However, it should be noted that basalts drilled at DSDP site 552 show evidence of contamination by either ancient U-depleted continental crust or subcontinental lithosphere (Merriman *et al.* 1988). This well is located about 10 km seaward of the COB proposed by Gaina *et al.* (2009).

Line CAM77 is too short to image the full width of the margin including the Hatton and Rockall basins. Therefore, RAPIDS line 1 (Vogt *et al.* 1998; Shannon *et al.* 1999) is shown in Figure 4c as a landward extension of line CAM77. The two lines are separated by 140 km, which may explain the difference in the crustal thickness where the models are joined. However, there is no isostatic balance between the Hatton Bank and the Hatton Basin for the model of RAPIDS line 1, which may

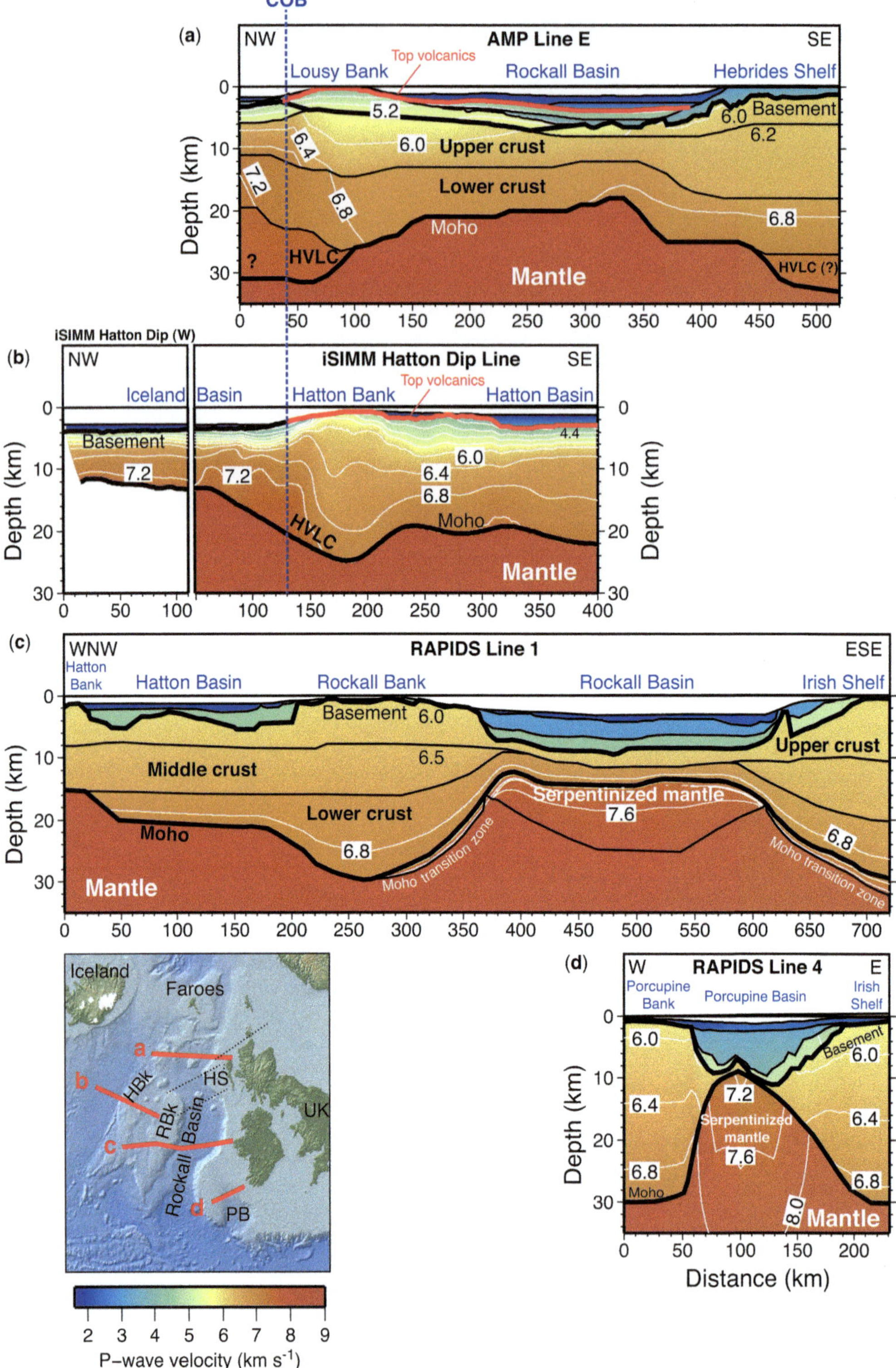
COB
(a)
NW
AMP Line E
SE
Top volcanics
Lousy Bank
Rockall Basin
Hebrides Shelf
Basement
Upper crust
Lower crust
Moho
HVLC
HVLC (?)
Mantle
Depth (km)
iSIMM Hatton Dip (W)
(b)
iSIMM Hatton Dip Line
Iceland Basin
Hatton Bank
Hatton Basin
(c)
WNW
RAPIDS Line 1
ESE
Hatton Bank
Hatton Basin
Rockall Bank
Irish Shelf
Middle crust
Serpentinized mantle
Moho transition zone
Iceland
Faroes
HBk
HS
RBk
Rockall Basin
UK
PB
(d)
W
RAPIDS Line 4
E
Porcupine Bank
Porcupine Basin
Serpentinized mantle
Distance (km)
P–wave velocity (km s^{-1})

indicate potential issues with the velocity model. The full transect shows three rift axes located within the Rockall Basin, the Hatton Basin and at the later break-up position. The Rockall Basin is fairly symmetrical, with similar necking profiles beneath the adjacent Rockall Bank and the Irish Shelf. Crustal stretching factors of 5–6 have thinned the crust sufficiently to allow for a partial serpentinization of the underlying mantle rock (O'Reilly *et al.* 1996). In the narrower Hatton Basin, the stretching factor is around 2. Final break-up occurred further to the west and is associated with a massive volcanic overprint of the crustal structure.

SIGMA line 2 (Korenaga *et al.* 2000) and AMP line E (Klingelhöfer *et al.* 2005) form a conjugate transect just to the south of the Greenland–Iceland–Faroe Ridge (Fig. 4a). The two lines are conjugate within 100 km and are stitched together at magnetic Chron C24N (53 Ma). AMP line E was primarily designed to determine the crustal velocity structure beneath the northern Rockall Basin, which is why the data coverage in the oceanic domain is limited. The width of the zone with inferred oceanic crust is, again, wider on the Greenland side than on the conjugate NW European margin. Given the deviation in the published COBs at either margin (cf. Funck *et al.* 2014), it is difficult to judge whether this asymmetry is real or not. In addition, a potential ridge jump in this region prior to Chron C21 (Kimbell *et al.* 2005) may further complicate this issue.

The initial thickness of the oceanic crust matches well, with 27 and 29 km on SIGMA line 2 and AMP line E, respectively (Fig. 4a). Volcanic rocks and HVLC are interpreted in the COT of either line. The transect covers the entire width of the conjugate margin pair, extending from onshore Greenland to full-thickness continental crust beneath the Hebrides Shelf. The crustal velocity structure between the two lines is difficult to compare owing to the different modelling approaches. The model of SIGMA line 2 is based on tomography, in which both the crust and any overlying volcanic rocks are modelled as one continuous layer (Korenaga *et al.* 2000). In contrast, the model for AMP line E is divided into several layers (Klingelhöfer *et al.* 2005), which distinguishes both a volcanic cover and up to four layers in the crystalline crust. However, both lines show a HVLC in the COT. Beneath the Hebrides Shelf, another 5 km-thick HVLC is indicated in the model based on an interpreted double reflection from the top and base of this layer. However, other lines in this region (their location is indicated by dotted lines on the map in Fig. 3) do not corroborate the presence of a HVLC (Powell & Sinha 1987; Roberts *et al.* 1988; Morgan *et al.* 2000). Volcanic rocks can be correlated from the top of Lousy Bank through to the east side of the Rockall Basin, where they become unrecognizable. However, some isolated lavas are known on the Hebrides Shelf (á Horni *et al.* 2014, this volume, in review).

The last conjugate transect along this margin segment is composed of SIGMA line 3 (Hopper *et al.* 2003) and the iSIMM Hatton dip line (White & Smith 2009) (Fig. 4b). The two lines are conjugate within 60 km and are joined at Chron C22N (49 Ma). Similar to the previous transect, the two velocity models were developed by different techniques. On the Hatton line, the volcanics and the crust were modelled as one continuous layer in the tomographical model, while the crust at the Greenland margin is divided into sublayers constrained by forward and inverse modelling. Despite these differences, the crustal thickness and velocity range match fairly well where the lines join. The crustal thinning on the European side of the transect is distributed over a much wider zone than on the Greenland side, as already discussed for the previous transects. The thickness of the continental crust beneath Hatton Bank (<24 km) is less than the full-thickness crust of Greenland (32 km). While these asymmetries are not surprising given that the margin was subject to several rift episodes preceding the final break-up (Ziegler 1988; Doré *et al.* 1999), the model shows a rather pronounced asymmetry in the accretion of oceanic crust.

The zone between Chron C22 and the COB interpretation of Funck *et al.* (2014) is 65 km wide at the Hatton margin, while it is 140 km wide at the SE Greenland margin. When the original interpretation of the SIGMA line 3 velocity model is used, the COB on the Greenland side may be even 60 km further landward (Hopper *et al.* 2003), which would increase the asymmetry even more. To avoid some of the problems associated with such a gross asymmetry at a spreading centre, White & Smith (2009) looked into possible alternative interpretations of the COB. In particular, they noted that some rather weak magnetic anomalies off Greenland that are

Fig. 3. P-wave velocity models of selected seismic refraction lines in the Faroe–Hatton–Rockall region: (**a**) AMP line E (after Klingelhöfer *et al.* 2005); (**b**) iSIMM Hatton dip line (W) (after Parkin & White 2008) and iSIMM Hatton dip line (after White & Smith 2009); (**c**) RAPIDS line 1 (after Vogt *et al.* 1998; Shannon *et al.* 1999); and (**d**) RAPIDS line 4 (after O'Reilly *et al.* 2006). Dotted lines indicate the location of the seismic refraction profiles on the Hebrides shelf (HS) with no evidence of the presence of a HVLC (Powell & Sinha 1987; Roberts *et al.* 1988; Morgan *et al.* 2000). Abbreviations: COB, continent-ocean boundary; HBk, Hatton Bank; HS, Hebrides Shelf; HVLC, high-velocity lower crust; PB, Porcupine Basin; RBk, Rockall Bank.

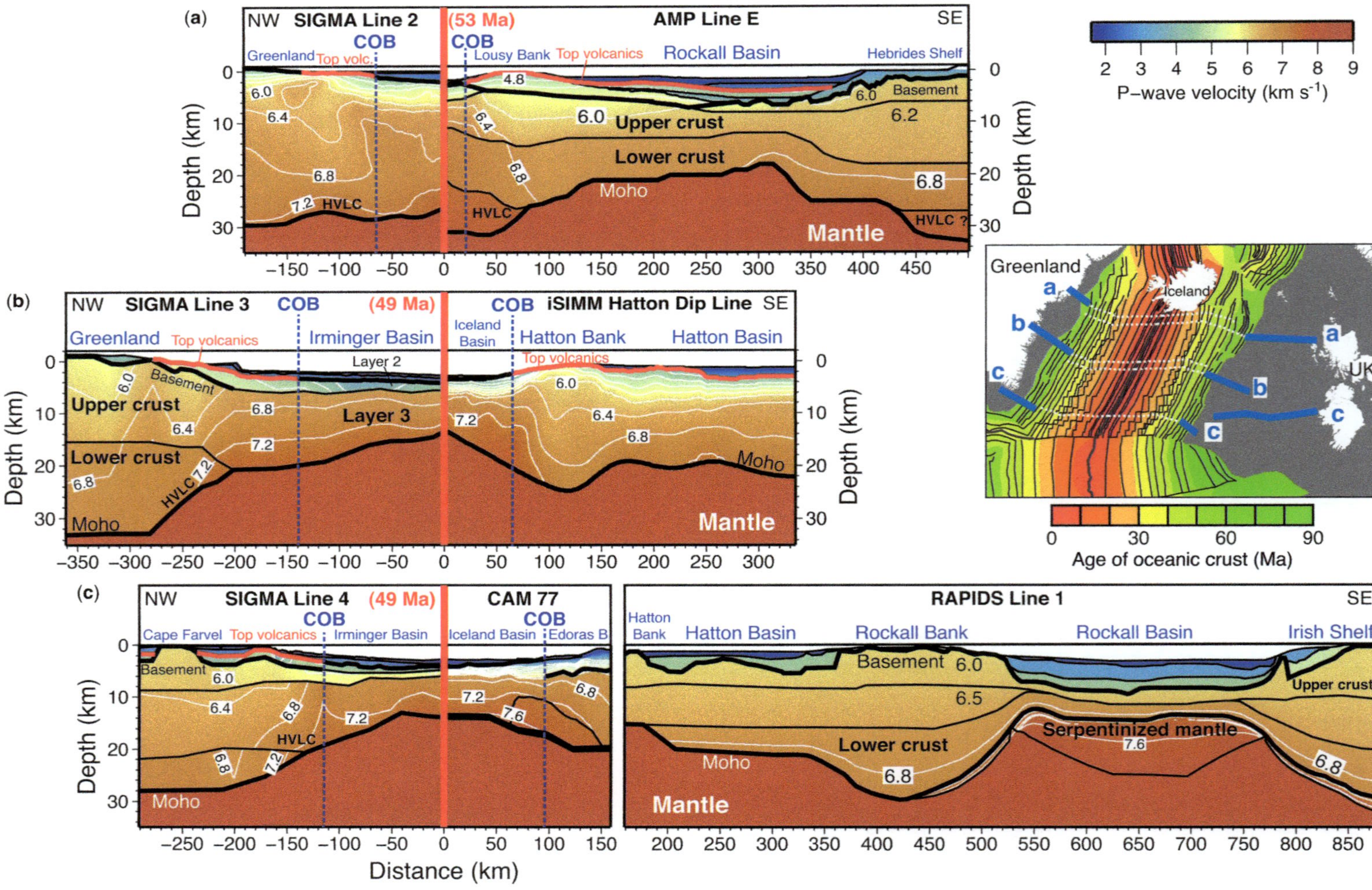
(a)
NW SIGMA Line 2
(53 Ma)
AMP Line E
SE
COB
COB
Greenland
Top volc.
Lousy Bank
Top volcanics
Rockall Basin
Hebrides Shelf
Basement
Upper crust
Lower crust
Moho
Mantle
HVLC
HVLC ?
Depth (km)
P-wave velocity (km s-1)
(b)
NW SIGMA Line 3
COB
(49 Ma)
COB
iSIMM Hatton Dip Line
SE
Greenland
Top volcanics
Irminger Basin
Iceland Basin
Hatton Bank
Hatton Basin
Layer 2
Basement
Upper crust
Layer 3
Lower crust
HVLC
Moho
Mantle
Greenland
Iceland
UK
Age of oceanic crust (Ma)
(c)
NW SIGMA Line 4
(49 Ma)
CAM 77
COB
COB
Cape Farvel
Top volcanics
Irminger Basin
Iceland Basin
Edoras B
Basement
HVLC
Moho
Distance (km)
RAPIDS Line 1
SE
Hatton Bank
Hatton Basin
Rockall Bank
Rockall Basin
Irish Shelf
Basement
Upper crust
Serpentinized mantle
Lower crust
Moho
Mantle

interpreted as magnetic cryptochrons 24.1n–24.11n (Larsen *et al.* 1994; Larsen & Saunders 1998) might not be true seafloor spreading anomalies. Instead, White & Smith (2009) proposed that these anomalies could represent the edges of subhorizontal lava flows. If this model is true, the COB would be located further seaward than indicated by Funck *et al.* (2014). A remodelling of the SIGMA line 3 dataset may provide some more definitive answer as to where the COB is located if a similar modelling approach is used, as for the iSIMM Hatton dip line. The Monte Carlo tomography used on the Hatton line is better suited to map the lateral velocity variation across the COT than the forward modelling applied to the SIGMA line.

Some support for the reinterpretation of the weak magnetic anomalies on SIGMA line 3 by White & Smith (2009) may come from the conceptual break-up model at volcanic margins proposed by Quirk *et al.* (2014). This model is based on deep seismic reflection data from the NE Greenland margin and appears to provide a mechanism for generating asymmetry during the early stages of break-up. In this model, lava-filled half-graben develop on either side of an axial horst and there is no reason why these have to be symmetrical. In addition to the asymmetry, the model provides a means of generating broad zones of subaerial SDRs above crust that looks more oceanic than continental. Eventually, the rising asthenosphere causes the horst to split and spreading is accommodated by a sheeted dyke system. At this stage, an outer high is predicted, and there is a change to submarine conditions and a more symmetrical spreading. Applied to SIGMA line 3 (Fig. 2b), there is no prominent outer high. However, a zone with rough basement separates subaerial SDRs from submarine SDRs according to the interpretation of Hopper *et al.* (2003). The seaward limit of the subaerial SDRs occurs some 50 km seaward of the COB proposed by Funck *et al.* (2014). Moving the COB by this amount would greatly reduce the asymmetry in the initial seafloor spreading (Fig. 4b).

Greenland–Iceland–Faroe Ridge

The Greenland–Iceland–Faroe Ridge (GIFR) is composed of thick igneous crust and extends from the eastern coast of Greenland to the Faroe Islands (Fig. 5). The 25–35 km-thick crust of the GIFR is the product of the interaction between the Mid-Atlantic Ridge and the Iceland mantle plume (White 1997). White (1997) showed that an increase in mantle temperature of as little as 50°C causes an increase of 30% in crustal thickness. The three main tectonic elements of the GIFR are the Greenland–Iceland Ridge, Iceland and the Iceland–Faroe Ridge. The two seismic refraction lines on the crest of the Greenland–Iceland Ridge and the Iceland–Faroe Ridge (Fig. 5a, c) show a fairly constant Moho depth of around 30 km (Richardson *et al.* 1998; Holbrook *et al.* 2001), but there is some indication in the data that the Moho deepens close to the shore, although this is only poorly constrained on the Greenland side. The velocity model of SIGMA line 1 (Fig. 5a) indicates a fairly homogeneous velocity structure along the entire length of the mapped Greenland–Iceland Ridge with no clear hint as to where the COB could be located. The COB of Funck *et al.* (2014) is located at km 82. Other authors place the COB close to the coast of Greenland (e.g. Escher & Pulvertaft 1995; Mosar *et al.* 2002*a*), which would be more compatible with the velocity model of SIGMA line 1.

Similarly, the FIRE offshore line (Fig. 5c) provides only limited resolution of the lower-crustal velocity structure close to the Faroe Islands. Richardson *et al.* (1998) presented two models, one with and one without a high-velocity lower-crustal layer. Both models fit the seismic observations, and the deepening of the Moho towards the Faroes is robust. However, the maximum Moho depth is 4 km shallower in the model without the high-velocity layer. Richardson *et al.* (1998) preferred the model with the high-velocity lower crust (this is the model shown in Fig. 5c) based mainly on gravity modelling and on the amplitude characteristics of the Moho reflection. This model would also be consistent with nearby lines that show a high-velocity lower crust (e.g. the iSIMM Faroes line: Roberts *et al.* 2009). Crustal velocities gradually decrease towards the Faroes, starting some 100 km away from the islands. This argues for the presence of continental crust beneath the Faroes (Richardson *et al.* 1998). However, the resolution of the data is not sufficient to determine the COB based on the lower-crustal velocity distribution. Published

Fig. 4. P-wave velocity models along conjugate transects south of Iceland: (**a**) SIGMA line 2 (after Korenaga *et al.* 2000) and AMP line E (after Klingelhöfer *et al.* 2005); (**b**) SIGMA line 3 (after Hopper *et al.* 2003) and iSIMM Hatton dip line (after White & Smith 2009); and (**c**) SIGMA line 4 (after Holbrook *et al.* 2001), line CAM77 (after Barton & White 1997) and RAPIDS line 1 (after Vogt *et al.* 1998; Shannon *et al.* 1999). The inset map displays the age of the oceanic crust (after Gaina 2014; cf. Gaina *et al.*, this volume, in review), together with the line locations (blue lines) and flow lines (dashed white lines). Thin solid lines mark selected isochrons. Abbreviations: B., Bank; COB, continent–ocean boundary; HVLC, high-velocity lower crust; volc., volcanics.

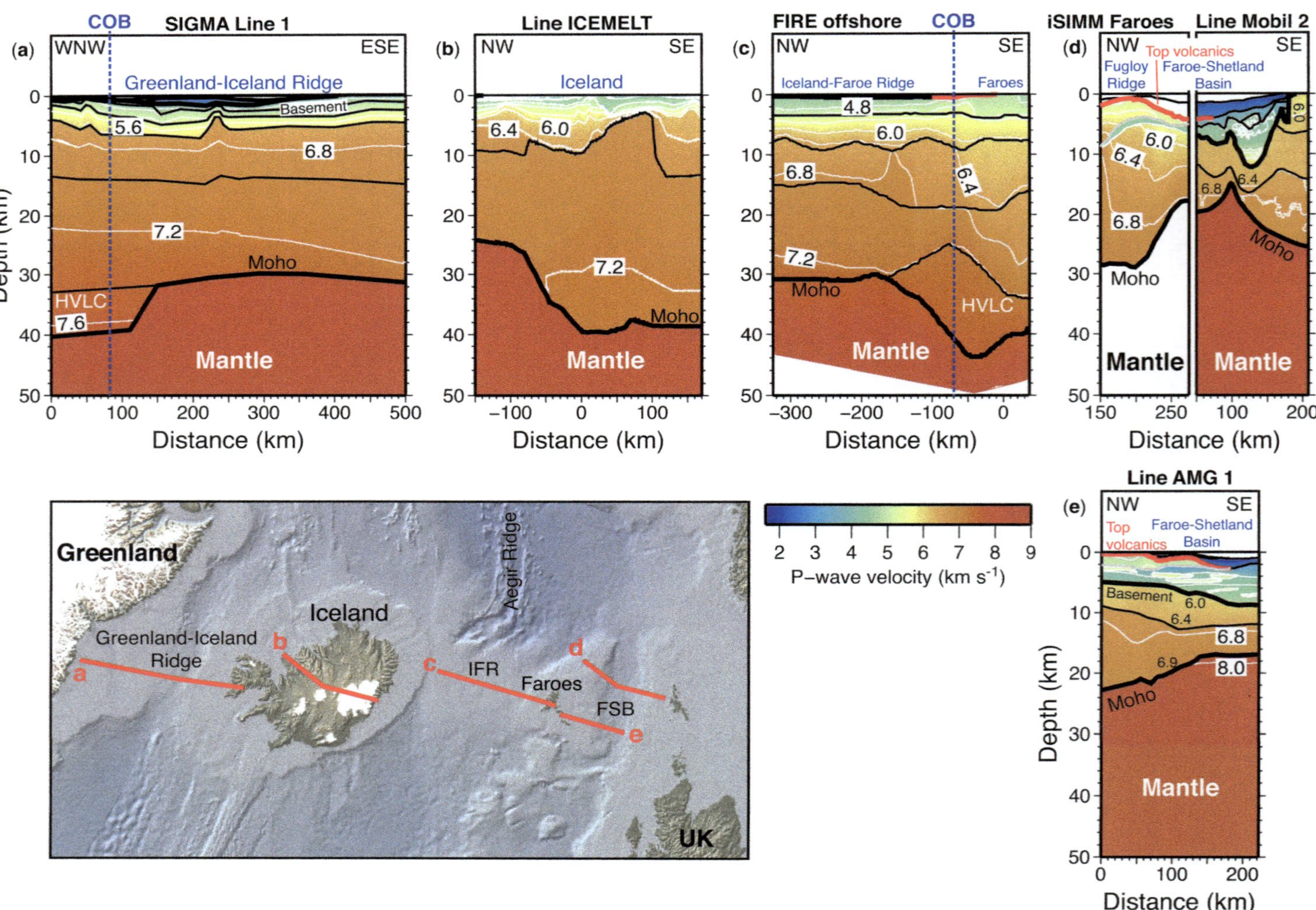
(a)
COB
SIGMA Line 1
WNW
ESE
Greenland-Iceland Ridge
Basement
5.6
6.8
7.2
Moho
HVLC
7.6
Mantle
Depth (km)
Distance (km)
(b)
Line ICEMELT
NW
SE
Iceland
6.4
6.0
7.2
Moho
Mantle
Distance (km)
(c)
FIRE offshore
COB
NW
SE
Iceland-Faroe Ridge
Faroes
4.8
6.0
6.8
6.4
7.2
Moho
HVLC
Mantle
Distance (km)
(d)
iSIMM Faroes
Line Mobil 2
NW
SE
Top volcanics
Fugloy Ridge
Faroe-Shetland Basin
6.0
6.4
6.8
Moho
Mantle
Distance (km)
Greenland
Iceland
Aegir Ridge
Greenland-Iceland Ridge
IFR
Faroes
FSB
UK
a
b
c
d
e
P-wave velocity (km s-1)
(e)
Line AMG 1
NW
SE
Top volcanics
Faroe-Shetland Basin
Basement
6.0
6.4
6.8
6.9
8.0
Moho
Mantle
Depth (km)
Distance (km)

COBs in this area vary by as much as 50 km (cf. Funck *et al.* 2014).

Beneath Iceland, the Moho depth varies between 15 and 40 km (Funck *et al.* 2016). The maximum value is found on the ICEMELT line (Fig. 5b) (Darbyshire *et al.* 1998) in central SE Iceland above the postulated centre of the Iceland plume (Darbyshire *et al.* 2000). However, there is an alternative explanation for the presence of thick crust beneath that region. Foulger & Anderson (2005) proposed the presence of a microplate that may contain oceanic crust submerged beneath younger lavas. Recently, Torsvik *et al.* (2015) suggested the presence of a sliver of continental crust beneath SE Iceland that links with the JMMC. This interpretation is based on geochemistry data. The postulated continental fragment is covered by the eastern end of the ICEMELT line (Fig. 5b) (Darbyshire *et al.* 1998) and the onshore portion of the FIRE profile (Staples *et al.* 1997). Neither of these lines shows a distinct lateral velocity anomaly that could point to the presence of continental crust. Given that the continental crust may have been thinned significantly, followed by substantial magmatic addition, its velocity structure might be indistinguishable from Iceland-type igneous crust. This often-used term comprises the abnormally thick crust formed above the centre of the Iceland mantle plume, the surface of which was originally subaerial (White 1997).

The break-up-related basaltic rocks produced at the rift zone between Greenland and the Faroes reach a thickness of 5 km and display velocities of approximately 4.4–5.2 km s^{-1} (Richardson *et al.* 1998). Where the palaeotopography adjacent to the rift was either flat-lying or formed by gently dipping sediments, the lavas were able to flow long distances uninterrupted by topographical barriers (Fliedner & White 2003). For example, east of the Faroe Islands, lava flows extend 150 km away from the islands and cover older sediments (Fliedner & White 2003). The feather edge of these basalt flows is mapped on a number of seismic refraction lines, such as Mobil line 2 (Fig. 5d) (Makris *et al.* 2009) or AMG line 1 (Fig. 5e) (Raum *et al.* 2005). As discussed in Petersen & Funck (2016), the velocity models of the seismic refraction lines in the Faroe–Shetland Basin match fairly well down to the top of the basalt cover, but can show large differences further below. AMG line 1 (Raum *et al.* 2005) terminates close to the Faroe Islands where it has a Moho depth of 22 km (Fig. 5e), which is significantly less than the depths of >30 km observed on the FIRE offshore line (Fig. 5c) (Richardson *et al.* 1998) in an extension of AMG line 1. Receiver function analysis on the Faroe Islands provides an estimate of 29–32 km for the depth to Moho (Harland *et al.* 2009). These values are compatible with what is observed on the Fugloy Ridge (iSIMM Faroes line in Fig. 5d) (Roberts *et al.* 2009), a little to the north of the main axis of the GIFR.

Continental margins north of Iceland

In this section, individual margin segments are described in a clockwise direction starting with the NE Greenland continental margin to the north of the GIFR and finishing with the margins bordering the JMMC. After the review of the margins, five conjugate transects are discussed, one of which is based on gravity inversion (Haase *et al.*, this volume, in press) and not on seismic refraction data, as for the others.

NE Greenland continental margin

The NE Greenland continental margin (Fig. 1) can be divided into three main segments. The southernmost segment comprises the area between the GIFR and the West Jan Mayen Fracture Zone (WJMFZ). The central segment spans the region between the WJMFZ and the East Greenland Ridge, while the northernmost region developed as a shear margin in the De Geer Zone megashear system linking the Atlantic and Arctic spreading systems (Doré *et al.* 2015).

Between the GIFR and the WJMFZ. Gernigon *et al.* (2015) recognized a Mid-Eocene kinematic event at around magnetic Chron C21r (48 Ma) in the Norway Basin that coincides with the onset of dyking and increasing rifting activity between the proto-JMMC and the East Greenland margin. Separation of the JMMC from Greenland started at approximately 30 Ma (Gaina *et al.* 2009). The southern JMMC was completely detached from Greenland by magnetic Chron C6 (20 Ma) (Gaina *et al.* 2009), leading to the accretion of oceanic crust along the Kolbeinsey Ridge.

Hermann & Jokat (2016) published a composite velocity model across this margin segment using lines AWI 94340 and AWI 20090100 (Fig. 6e). The maximum constrained Moho depth on the line

Fig. 5. P-wave velocity models along the Greenland-Iceland-Faroe Ridge and the Faroe-Shetland Basin (FSB). (**a**) SIGMA line 1 (after Holbrook *et al.* 2001); (**b**) ICEMELT line (after Darbyshire *et al.* 1998); (**c**) FIRE offshore line (after Richardson *et al.* 1998); (**d**) iSIMM Faroe line (after Roberts *et al.* 2009), Mobil line 2 (after Makris *et al.* 2009); and (**e**) AMG line 1 (after Raum *et al.* 2005). Abbreviations: COB, continent-ocean boundary; HVLC, high-velocity lower crust; IFR, Iceland-Faroe Ridge.

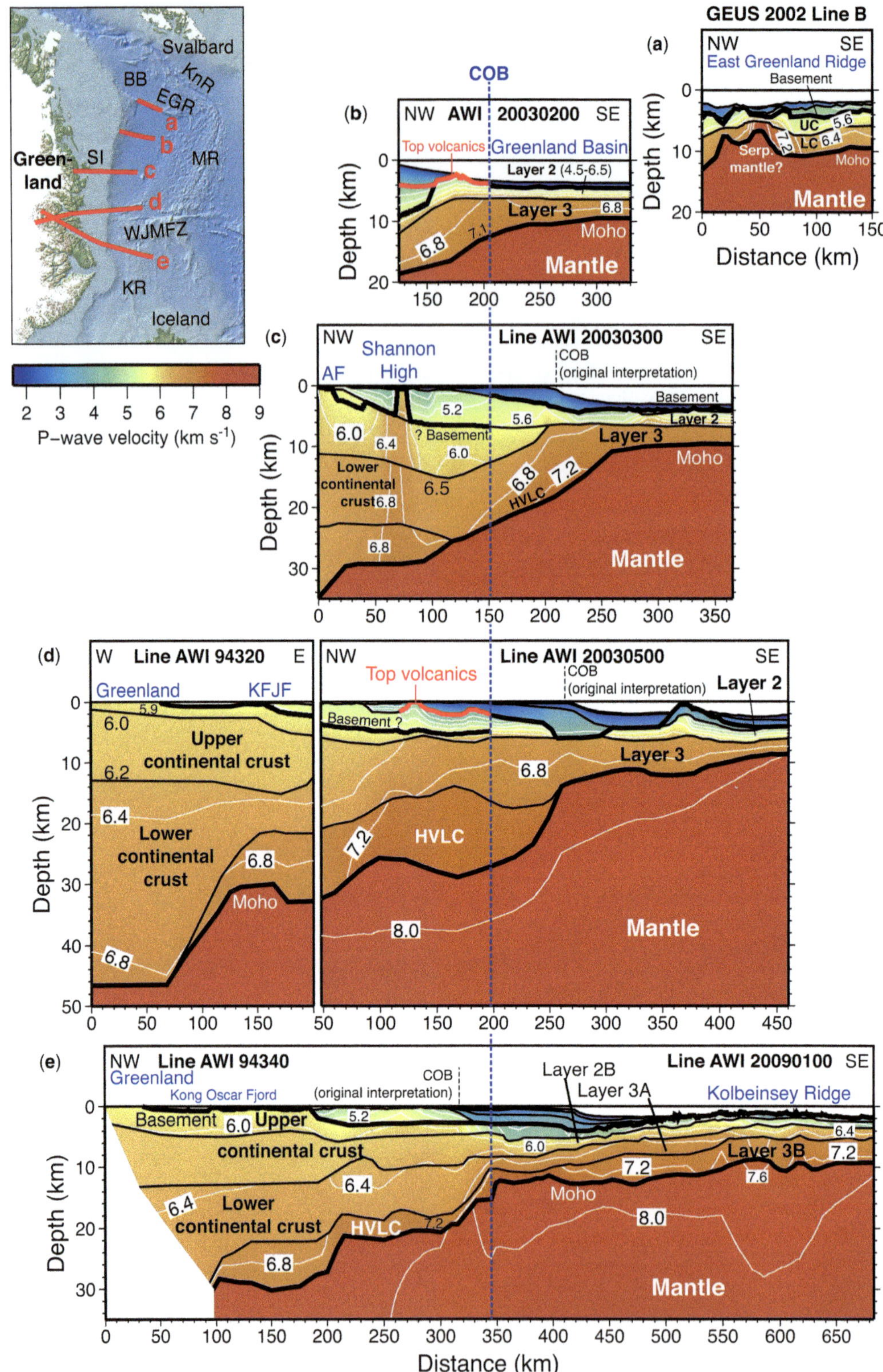

Svalbard
BB
KnR
EGR
a
b
Green-
land
SI
c
MR
d
WJMFZ
e
KR
Iceland
2 3 4 5 6 7 8 9
P-wave velocity (km s-1)
GEUS 2002 Line B
(a)
NW SE
East Greenland Ridge
Basement
UC 5.6
7.2
LC 6.4
Serp. mantle?
Moho
Mantle
Depth (km)
Distance (km)
COB
(b)
NW AWI 20030200 SE
Top volcanics
Greenland Basin
Layer 2 (4.5-6.5)
Layer 3
6.8
7.1
Moho
Mantle
(c)
NW
Shannon High
AF
Line AWI 20030300 SE
COB (original interpretation)
Basement
5.2
5.6
Layer 2
6.0
6.4
? Basement
Layer 3
Lower continental crust
6.5
HVLC
7.2
Mantle
(d)
W Line AWI 94320 E
Greenland
KFJF
5.9
6.0
Upper continental crust
6.2
6.4
Lower continental crust
6.8
Moho
NW Line AWI 20030500 SE
Top volcanics
COB (original interpretation)
Layer 2
Basement ?
Layer 3
HVLC
8.0
Mantle
(e)
NW Line AWI 94340
Greenland
Kong Oscar Fjord
COB (original interpretation)
Layer 2B
Layer 3A
Line AWI 20090100 SE
Kolbeinsey Ridge
Basement
6.0
Upper continental crust
5.2
6.4
Layer 3B
7.2
7.6
Moho
Lower continental crust
HVLC
8.0
6.8
Mantle
Distance (km)

is 30 km, and the thinning of the continental crust occurs in two steps around km 210 and km 320 in the model. A 150 km-wide and up to 3 km-thick HVLC marks the base of the crust in the COT. The HVLC is relatively poorly constrained (Hermann & Jokat 2016), but its thickness is substantially less than is found in the COT north of the WJMFZ where the high-velocity lower-crustal body displays a thickness of 15 km (Fig. 6d). Hermann & Jokat (2016) argued that the high-velocity body on lines AWI 94340 and 20090100 (Fig. 6e) is the product of excess magma production that was focused along the WJMFZ during the break-up of the JMMC from Greenland. It should be noted that no such HVLC is observed on the conjugate western side of the JMMC (Kodaira *et al.* 1998*b*), which might be expected if the magma production were related to break-up. This could either suggest that the rather poorly constrained HVLC on the Greenland side is not real or that such a layer was missed in the model of the western JMMC. On the Greenland side, there is at least one other line in support for a thin HVLC (Weigel *et al.* 1995), but the wide-angle seismic constraints are also very poor. At the western margin of the JMMC, 4 km-thick continental crust lies adjacent to 9 km-thick oceanic crust (Kodaira *et al.* 1998*a*). The thin continental crust is constrained by only two ocean bottom seismometers (OBSs) 45 km apart from each other, which may not be sufficient to map a thin HVLC.

Along the transect AWI 94340 and AWI 20090100 (Fig. 6e), the COB of Funck *et al.* (2014) fits fairly well with the lower-crustal velocity distribution, even though Hermann & Jokat (2016) suggested a location 30 km further landward. The oceanic crust between the COB and the present Kolbeinsey Ridge has a fairly constant thickness of about 9 km.

The sedimentary succession along lines AWI 94340 and 20090100 (Fig. 6e) has a maximum thickness of 3 km (Hermann & Jokat 2016). Thicker sedimentary basins along this margin segment are found onshore Greenland where seismic refraction data indicate a depth of 15 km for the Jameson Land Basin (Weigel *et al.* 1995). Larsen & Marcussen (1992) suggested that the basin infill there might even be up to 18 km thick. Following the post-Caledonian extension with fault-controlled Devonian basins, late Jurassic–early Cretaceous rifting resulted in the deposition of marine sediments over the Devonian sediments in Jameson Land (Surlyk 1990).

Between the WJMFZ and the East Greenland Ridge. The margin segment between the WJMFZ and the East Greenland Ridge is characterized by massive break-up-related Cenozoic volcanism, especially in the southern part. Three seismic refraction transects are shown in Figure 6b–d, illustrating the northwards decrease of magmatic addition to the margin. In the south, on line AWI 20030500 (Fig. 6d), the magmatism is manifested in a 15 km-thick high-velocity lower-crustal body and an up to 5 km-thick series of volcanic rocks at the top of the crust (Voss & Jokat 2007). At the northernmost available line (AWI 20030200: Fig. 6b), hardly any signs of break-up-related volcanism are left. Maximum velocities in the lower crust close to the COB are 7.1 km s^{-1} and the total crustal thickness there is only 7 km (Voss *et al.* 2009). Landward of the COB, seismic reflection data indicate the presence of basalts forming an outer high (Voss *et al.* 2009). Line AWI 20030200 does not image the proximal part of the margin due to sea ice that inhibited the seismic data acquisition there.

Line AWI 20030300 (Fig. 6c) at the centre of the margin segment displays a more pronounced HVLC than the line discussed previously. In the continental domain, Voss *et al.* (2009) noticed a positive velocity anomaly beneath the Shannon High. East of the high, the thickness of the basalts and the depth to basement are poorly resolved due to a lack of velocity contrasts. The problem is that the velocities in the deeper Mesozoic and Palaeozoic sediments are similar to the ones expected in crystalline crust, which makes it difficult to distinguish between the two. Voss *et al.* (2009) suggested that the rift-related basin infill might be up to 15 km thick. In Figure 6c, the basement is indicated by the 5.7 km s^{-1} contour, similar to what is done along the lines in the north and south (Fig. 6b, d). This leaves a substantial uncertainty in the thickness of the rift-related basins along this margin segment. In the northern part of the margin, no seismic refraction data are available. However, seismic reflection data from the KANUMAS group released in 2014 can be used to compile a sediment thickness map for the NE Greenland Shelf (Hopper *et al.*, this

Fig. 6. P-wave velocity models at the NE Greenland margin: (**a**) GEUS2002 line B (after Døssing & Funck 2012); (**b**) line AWI 20030200 (after Voss *et al.* 2009); (**c**) line AWI 20030300 (after Voss *et al.* 2009); (**d**) lines AWI 94320 (after Schlindwein & Jokat 1999) and AWI 20030500 (after Voss & Jokat 2007); and (**e**) lines AWI 94340 and 20090100 (after Hermann & Jokat 2016). Abbreviations: AF, Ardencaple Fjord; BB, Boreas Basin; COB, continent–ocean boundary; EGR, East Greenland Ridge; HVLC, high-velocity lower crust; KFJF, Kejser Franz Joseph Fjord; LC, lower crust; Serp., Serpentinized; SI, Shannon Island; UC, upper crust; WJMFZ, West Jan Mayen Fracture Zone.

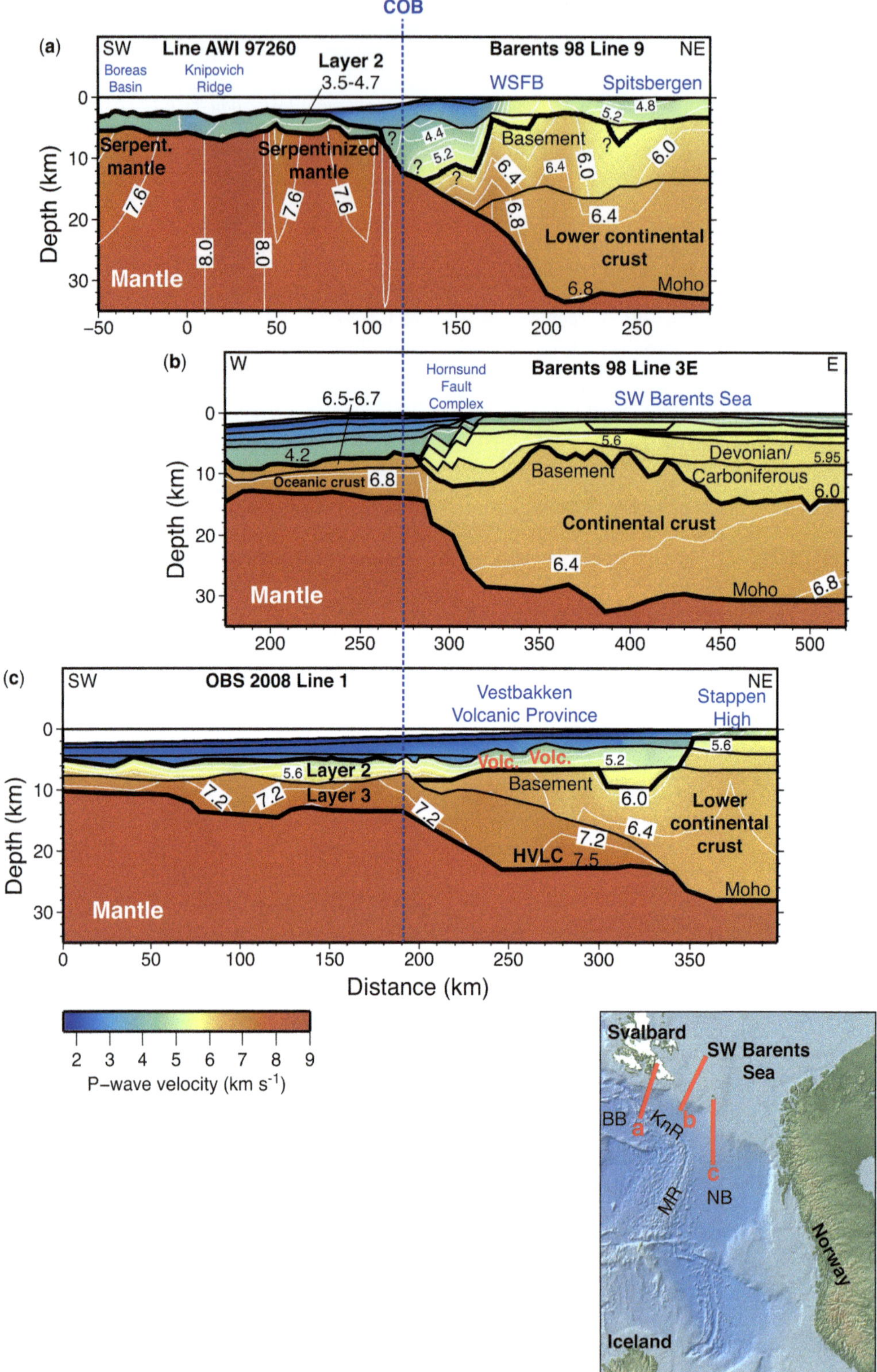
COB
(a)
SW
Line AWI 97260
Barents 98 Line 9
NE
Boreas Basin
Knipovich Ridge
Layer 2
3.5-4.7
WSFB
Spitsbergen
Serpent. mantle
Serpentinized mantle
Basement
Lower continental crust
Mantle
Moho
Depth (km)
(b)
W
Barents 98 Line 3E
E
Hornsund Fault Complex
SW Barents Sea
6.5-6.7
Oceanic crust
Devonian/ Carboniferous
Basement
Continental crust
Mantle
Moho
Depth (km)
(c)
SW
OBS 2008 Line 1
NE
Vestbakken Volcanic Province
Stappen High
Layer 2
Layer 3
Volc.
Volc.
Basement
Lower continental crust
HVLC
Mantle
Moho
Depth (km)
Distance (km)
P-wave velocity (km s^{-1})
Svalbard
SW Barents Sea
BB
KnR
MR
NB
Iceland
Norway

volume, in prep). This map indicates less than 5 km of sediments to the east of Shannon Island. Thicker sedimentary basins are observed in the north, where the sedimentary pile is up to 17 km thick in the North Danmarkshavn Basin.

Another point of interest is the location of the COB that is subject to some controversy. In the original publications, with the velocity models of lines AWI 20030300 and 20030500 (Fig. 6c, d), the COB is located approximately 60 km seaward of the one proposed by Funck *et al.* (2014) that is, to a large degree, determined by plate reconstructions (Gaina 2014; cf. Gaina *et al.*, this volume, in review) using the arguably better-defined location of the COB at the conjugate mid-Norwegian margin. The originally proposed COB along line AWI 20030300 (Fig. 6c) (Voss *et al.* 2009) fits well with the modelled lower-crustal velocity variations, while the velocities on line AWI 20030500 (Fig. 6d) would, instead, support the more landward COB of Funck *et al.* (2014).

North of the East Greenland Ridge. The margin north of the East Greenland Ridge (EGR) developed as a shear margin and is the least-studied margin in the NE Atlantic owing to the year-round cover with sea ice. The continental domain of the margin is not studied with seismic refraction lines, with the exception of the EGR that developed along the Greenland Fracture Zone and protrudes from the NE Greenland shelf into the oceanic basins (Fig. 1). The ridge (Fig. 6a) is a 250 km-long and up to 50 km-wide bathymetric high that is composed of continental crust thinned to 2–6 km and locally underlain by partially serpentinized mantle (Døssing *et al.* 2008; Døssing & Funck 2012; Gerlings *et al.* 2014; Funck *et al.* 2015). The EGR was sheared from the shelf and has a complex internal structure with two overstepping main ridge segments (Døssing & Funck 2012).

Immediately to the north of the EGR in the SW Boreas Basin, seismic refraction and coincident reflection data indicate the presence of an extremely thin and faulted transitional crust (Døssing *et al.* 2008). The exact extent of this crust is unknown, but can be correlated at least 40 km to the north of the central part of the EGR (Døssing *et al.* 2008). Discontinuous and often weak magnetic lineations characterize the Boreas Basin (cf. Engen *et al.* 2008). Hermann & Jokat (2013) showed the presence of thin (generally <3 km) oceanic crust in the basin.

SW Barents Sea margin

The SW Barents Sea margin is a transform margin that extends from northern Norway to Svalbard (Fig. 1). The Barents Sea is a wide rifted continental shelf domain composed of numerous fault-bounded interconnected and segmented basins that are linked both to the Atlantic and Arctic rift systems. The formation of the western Barents Sea began in the Carboniferous, followed by two additional rift phases in Middle Jurassic–Early Cretaceous times and in the early Cenozoic (Faleide *et al.* 1993, 2008). The basins in the Barents Sea locally exceed 18 km in depth (Ritzmann *et al.* 2007).

Late Cretaceous rifting, subsequent break-up and initial seafloor spreading in the Norwegian–Greenland Sea was linked to the Eurasia Basin in the Arctic by the regional De Geer Zone megashear system (Faleide *et al.* 2008). The southern limit of this predominantly sheared margin system in the SW Barents Sea is marked by the Senja Fracture Zone, the conjugate to the Greenland Fracture Zone. The margin is divided into two large shear segments, the Hornsund and Senja margins (Fig. 1), and a central rifted segment with volcanism, the Vestbakken margin. At Svalbard, initial shearing was followed by rifting, while north of Svalbard and up to the Yermak Plateau, the margin developed as a complex sheared and rifted margin (Faleide *et al.* 2008).

Line 3E of the Barents 98 survey (Breivik *et al.* 2003) is a dip line at the Hornsund margin (Fig. 7b) and illustrates the shear-margin setting. The Moho shallows from 29 to 14 km over a distance of only 35 km and the transition from continental to oceanic crust can be resolved to lie within a narrow zone of less than 5 km in width (Breivik *et al.* 2003). The upper part of the continental crust in the vicinity of the Hornsund fault complex is dominated by two large, rotated downfaulted blocks with throws of 2–3 km on each fault, apparently formed during the transform margin development (Breivik *et al.* 2003). Beneath the shelf, the top of the crystalline basement is as deep as 16 km and is primarily constrained by a coincident seismic reflection line. Velocities in the interpreted Devonian–Carboniferous sedimentary section vary between 5.6 and 6.0 km s^{-1}, which would be difficult to distinguish from crystalline basement using seismic refraction data alone. The oceanic crust adjacent to the COB has a thickness of 7 km, but thins to as little as 4 km further to the west. The oceanic crust lacks

Fig. 7. P-wave velocity models at the SW Barents Sea margin: (**a**) lines AWI 97260 and Barents 98 line 9 (after Ritzmann *et al.* 2002); (**b**) Barents 98 line 3E (after Breivik *et al.* 2003); and (**c**) line 1 of the OBS 2008 survey (Libak *et al.* 2012). Abbreviations: BB, Boreas Basin; COB, continent–ocean boundary; HVLC, high-velocity lower crust; KnR, Knipovich Ridge; MR, Mohns Ridge; NB, Norway Basin; Serpent., Serpentinized; Volc., Volcano; WSFB, West Spitsbergen Fold Belt.

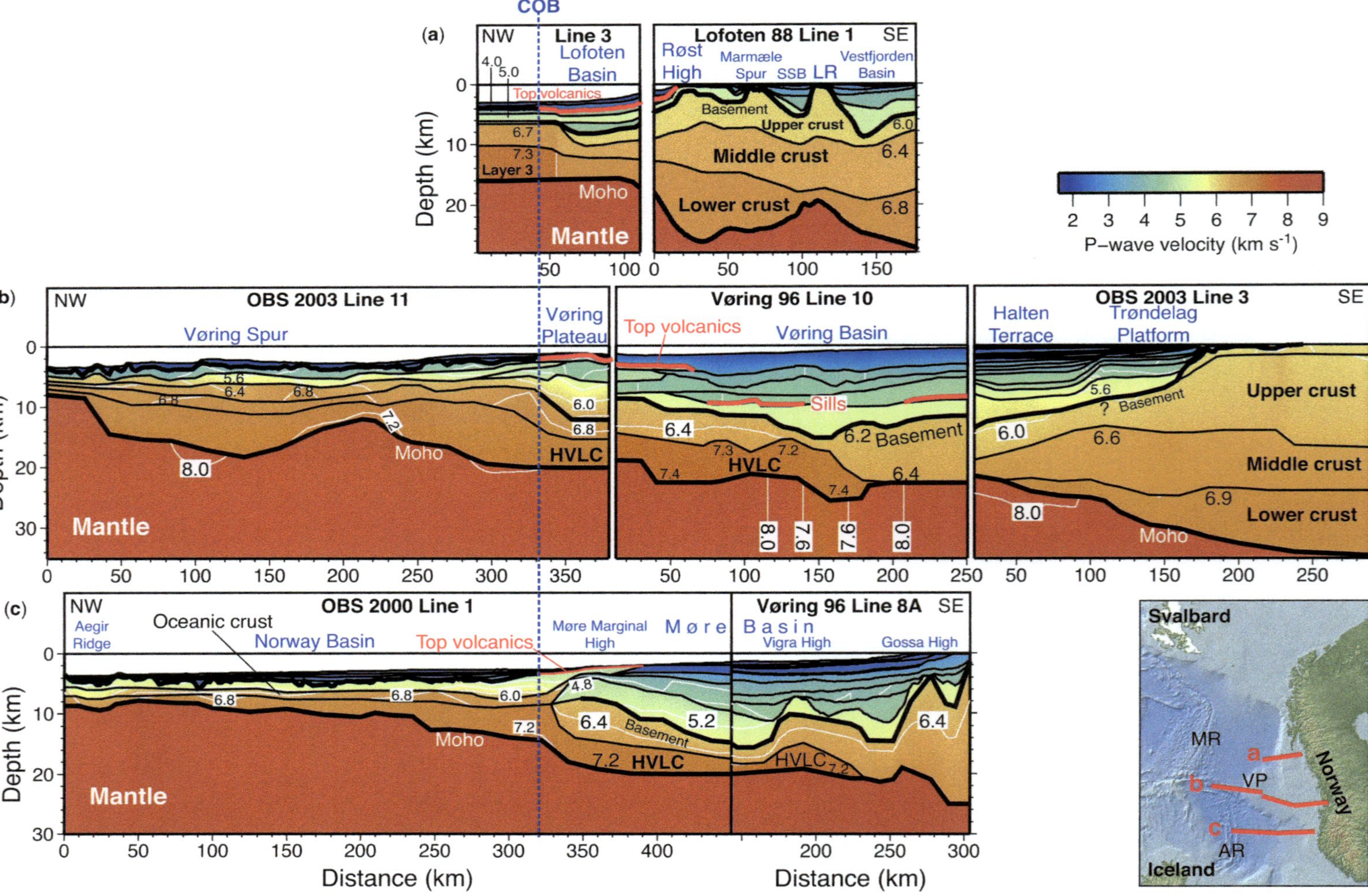

COB
(a)
Depth (km)
NW
Line 3
Lofoten Basin
Top volcanics
Layer 3
Moho
Mantle
Lofoten 88 Line 1
SE
Røst High
Marmæle Spur
SSB
LR
Vestfjorden Basin
Basement
Upper crust
Middle crust
Lower crust
P–wave velocity (km s-1)
(b)
NW
OBS 2003 Line 11
Vøring Spur
Vøring Plateau
Moho
Mantle
HVLC
Vøring 96 Line 10
Top volcanics
Vøring Basin
Sills
Basement
HVLC
OBS 2003 Line 3
SE
Halten Terrace
Trøndelag Platform
Basement
Upper crust
Middle crust
Lower crust
Moho
(c)
NW
Aegir Ridge
Oceanic crust
OBS 2000 Line 1
Norway Basin
Top volcanics
Møre Marginal High
Møre Basin
Basement
HVLC
Moho
Mantle
Vøring 96 Line 8A
SE
Vigra High
Gossa High
HVLC
Distance (km)
Distance (km)
Svalbard
MR
VP
AR
Norway
Iceland
a
b
c

velocities typical for oceanic layer 2, but is characterized by basement velocities of 6.6 km s^{-1}. Breivik *et al.* (2003) saw this as a response to mineral infilling and the closure of cracks, fissures and voids in layer 2 induced by the 7–8 km-thick sedimentary overburden.

The rifted Vestbakken margin segment is illustrated by line 1 of the OBS 2008 experiment (Fig. 7c) (Libak *et al.* 2012). In contrast to the rapid crustal thinning observed at the Hornsund margin, the shallowing of the Moho is more gradual from a depth of 28 to 13 km over a 170 km-wide zone. Landward of the COB, increased velocities of up to 7.5 km s^{-1} are observed in the lower crust, which are interpreted as intrusions (Libak *et al.* 2012). Within the region of the HVLC, volcanoes are observed (their position is indicated in Fig. 7c) and Libak *et al.* (2012) interpreted the upwarping of velocity contours in the lowermost sedimentary layer (velocities of 4.4–5.2 km s^{-1}) as marking feeder dykes. Prominent volcanoes, as well as sill intrusions, are interpreted in seismic reflection data at the outer margin at the Vestbakken Volcanic Province with two phases of volcanism in the early Eocene and early Oligocene (Faleide *et al.* 1988). The oceanic crust along this profile has a general thickness of 7 km, with variations of between 5 and 9 km. Velocities in the interpreted oceanic layer 3 (6.8–7.7 km s^{-1}) are rather high and could also be compatible with partially serpentinized mantle, a possibility that was not ruled out by Libak *et al.* (2012). In this case, the thickness of the oceanic crust would be overestimated.

The Hornsund margin is presented by a seismic refraction transect (Fig. 7a) that uses data from two experiments (line AWI 97260 and Barents 98 line 9). On this profile, the oceanic crust is characterized by a 3.5 km-thick layer 2 with velocities of between 3.5 and 5.7 km s^{-1}, while a layer 3 could not be mapped (Ritzmann *et al.* 2002). Velocities of 7.3–8.0 km s^{-1} beneath the oceanic crust indicate a partial serpentinization of the mantle rocks. The thinning of the continental crust occurs over an 80 km-wide zone extending from the West Spitsbergen Fold Belt seaward. In this zone, the Moho shallows from 33 to 7 km. Between model km 150 and 180, a region with increased lower-crustal velocities (>7.2 km s^{-1}) is observed, although the seismic resolution is low. Ritzmann *et al.* (2002) related these velocities to magmatic intrusions of unknown origin. The nature of the deep sedimentary basin landward of the COB is not clear and the depth of the crystalline basement is not resolved by the data. The crystalline crust beneath Spitsbergen has a thickness of 30 km, with velocities ranging from 5.5 to 6.8 km s^{-1} and is overlain by a 3–7 km-thick sedimentary sequence.

Mid-Norwegian continental margin

The mid-Norwegian continental margin is divided into three main segments, referred to as the Møre, Vøring, and Lofoten margins (Fig. 1). These segments are delimited by regional transfer zones. The Vøring margin is bounded by the Bivrost and Jan Mayen lineaments to the north and south, respectively. At the Jan Mayen Lineament, Olesen *et al.* (2007) notice only a weak expression in basement structure while the related East Jan Mayen Fracture Zone is associated with a major shift in the COB. In contrast, the Bivrost Lineament is well expressed in the basement structure but lacks an outboard fracture zone according to the interpretation of Olesen *et al.* (2007), who relate a previously proposed Bivrost Fracture Zone to data artefacts. Doré *et al.* (1997) suggest that the transfer zones are likely linked to the structural heritage of the region.

A first extensional phase took place during Devonian–Carboniferous times and was possibly related to the gravitational collapse of the Caledonian orogen (Seranne & Seguret 1987; Andersen & Jamtveit 1990; Osmundsen & Andersen 2001 and references therein). Early rift basins formed most likely during Carboniferous, Permian and Middle Triassic times in the proximal setting (e.g. the Trøndelag Platform), while renewed extension in the Middle Jurassic and in the Late Jurassic to Early/middle-Cretaceous shaped the distal basins (e.g. the Lofoten, Vøring, and Møre basins) (Doré *et al.* 1999; Brekke 2000). The complex outer ridges were affected by the last rifting phase of Late Cretaceous–Paleocene age (Gernigon *et al.* 2003; Ren *et al.* 2003).

The mid-Norwegian margin is densely sampled by seismic refraction lines (see figure 2 in Funck *et al.* 2016) but individual profiles are rather short and do not cover the entire margin from the proximal to the distal domain. Hence, composite profiles were constructed to illustrate the crustal structure of the three margin segments (Fig. 8).

Fig. 8. P-wave velocity models at the mid-Norwegian margins: (**a**) the Lofoten margin with lines 1 (after Mjelde *et al.* 1993) and 3 (after Mjelde *et al.* 1992) of the Lofoten 88 survey; (**b**) the Vøring margin with lines 3 (after Breivik *et al.* 2011) and 11 (after Breivik *et al.* 2008) of the OBS 2003 survey and Vøring 96 line 10 (after Raum *et al.* 2002); and (**c**) the Møre margin with line 1 of the OBS 2000 survey (after Breivik *et al.* 2006) and Vøring 96 line 8A (after Raum 2000). Abbreviations: AR, Aegir Ridge; COB, continent–ocean boundary; HVLC, high-velocity lower crust; LR, Lofoten Ridge; MR, Mohns Ridge; SSB, Skomvær Sub-basin; VP, Vøring Plateau.

Lofoten margin. The transect across the Lofoten margin is composed of lines 1 and 3 (Fig. 8a) of the Lofoten 88 experiment (Mjelde *et al.* 1992, 1993). The imaged continental crust has a maximum thickness of 25 km beneath the Røst High. Landward of the high, the transect crosses the Marmæle Spur, the Lofoten Ridge and the Vestfjorden Basin, resulting in a variable basement depth of between 1 and 9 km that is consistent with a highly faulted basement, as indicated in the interpretation of the coincident seismic reflection data (Mjelde *et al.* 1993). Volcanics extend from the western flank of the Røst High onto the oceanic crust. The seismic resolution in the COT is limited as the spacing of OBSs was rather high (*c.* 30 km). The velocity model indicates a sharp transition between the continental and oceanic crust, which is different from modern experiments that observe a smoother velocity transition in the COT of magma-rich margins (e.g. White & Smith 2009). This may explain why the COB as interpreted by Funck *et al.* (2014) is further seaward than the velocity model of Mjelde *et al.* (1993) may suggest. In any case, velocities in the lower oceanic crust are modelled at 7.3 km s^{-1}, which is compatible with high-velocity lower crust commonly observed at magma-rich margins. However, landward of the COB, no distinct HVLC is mapped. This distinguishes the Lofoten margin from the Vøring and Møre margins to the south. The thickness of the initial oceanic crust is 12 km, including the overlying flood basalts.

Vøring margin. The Vøring margin is much wider than the Lofoten margin. The chosen transect (Fig. 8b) is composed of lines 3 and 11 of the OBS 2003 experiment (Breivik *et al.* 2008, 2011) and line 10 of the Vøring 96 survey (Raum *et al.* 2002). There are some slight deviations in the models where the lines join, which is of no concern here as we look at large-scale structures. At the landward end of the transect, the maximum Moho depth constrained by the data is 32 km. The basement deepens seaward to a depth of 13 km beneath the Halten Terrace. However, beneath the terrace and the Trøndelag Platform there is some uncertainty about the exact location of the basement, as velocities of 5.3–5.7 km s^{-1} are observed that are intermediate between typical crystalline crust and Mesozoic sedimentary strata (Breivik *et al.* 2011).

In the Vøring Basin, the sedimentary section is up to 14 km thick (Fig. 8b), including sills and inner flows (Raum *et al.* 2002). The basin is underlain by thinned continental crust with a thickness of 2–11 km, not including the HVLC (7.2–7.4 km s^{-1}) beneath the western part of the basin that has a maximum thickness of 8 km. Raum *et al.* (2002) interpreted the HVLC as a magmatic underplated body. However, this interpretation is controversial. Under the Rån Ridge at the outer Vøring margin, velocities locally exceed 8.5 km s^{-1} and have been mapped in a tectonically complex setting where densities of 3500 kg m^{-3} are used for gravity modelling (Raum *et al.* 2006). This HVLC is accordingly explained as eclogite because the velocities are too high for both crustal crystalline rocks and mantle peridotite. Other studies use much lower densities for gravity modelling (Rouzo *et al.* 2006; Reynisson *et al.* 2010) and the 3D complexity of the area might suggest that 2D seismic refraction profiles did not adequately map the deep crustal structures in sufficient detail. In contrast, velocities between 7.0 and 7.8 km s^{-1} could represent both mafic intrusions (including magmatic underplating) and partially serpentinized mantle peridotite (Miller & Christensen 1997; Christensen 2004). Lundin & Doré (2011) proposed that the thinned crust beneath the Vøring Basin presents hyper-extended crust that is underlain by partially serpentinized upper mantle. Mjelde *et al.* (2009) favoured a model that relates the HVLC to mafic intrusions, as there is a spatial relationship between the HVLC and sill intrusions. Gernigon *et al.* (2003, 2004) showed, based on high-resolution reflection seismic data and gravity modelling, that the HVLCs of the outer Vøring margin correlate with structural highs and that crustal thinning occurred adjacent to them. Thus, the HVLCs were already in place prior to break-up and probably represent high-pressure metamorphic rocks and not serpentinized mantle.

Beneath the Vøring Plateau, crustal velocities increase seaward up to the position of the COB (Fig. 8b). At the onset of seafloor spreading, the oceanic crustal thickness is 18 km, but it thins to 9 km over a distance of 115 km. However, beneath the Vøring Spur, a renewed crustal thickening is observed to a maximum of 15 km. Breivik *et al.* (2008) relate this to magmatic underplating beneath oceanic crust.

Møre margin. Compared to the Vøring margin, the Møre margin is significantly narrower. The transect shown in Figure 8c is composed of line 1 of the OBS 2000 experiment in the NW (Breivik *et al.* 2006), while the SE part is taken from line 8A of the Vøring 96 survey (Raum 2000). The full-thickness continental crust is not really covered by this transect as the shelf is relatively narrow and no land stations were deployed. However, other lines nearby obtain Moho depths of 29 km beneath the shelf and 37 km onshore (Kvarven *et al.* 2014). Faulted continental crust is observed inboard of the Gossa High, and HVLC has been mapped here in a number of seismic reflection and refraction studies (Olafsson *et al.* 1992; Kvarven *et al.* 2014; Nirrengarten

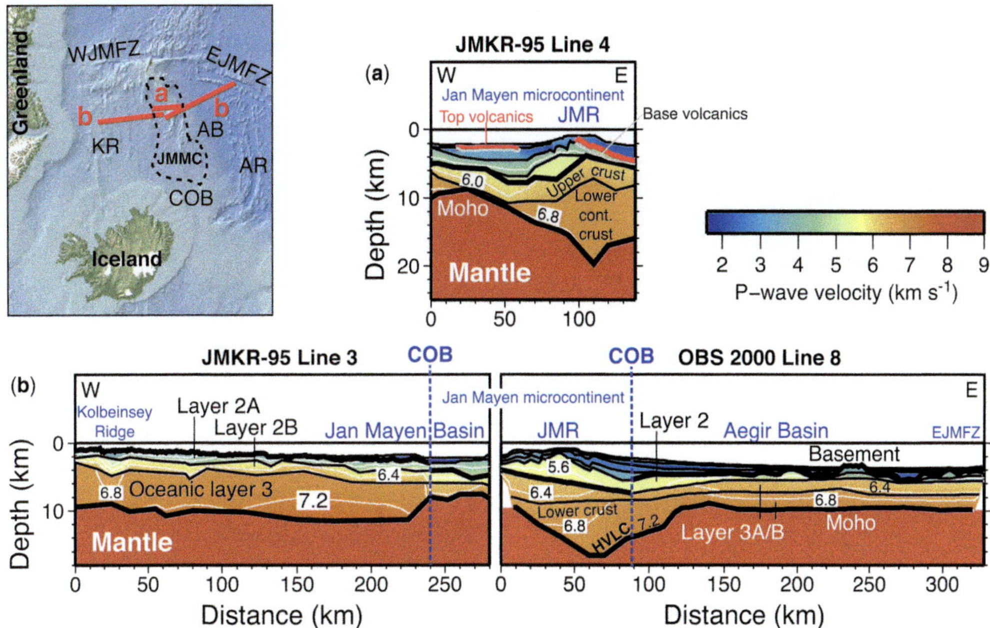

Fig. 9. P-wave velocity models at the Jan Mayen microcontinent: (**a**) JMKR-95 line 4 (after Kodaira *et al.* 1998*b*); and (**b**) JMKR-95 line 3 (after Kodaira *et al.* 1998*a*) and line 8 of the OBS 2000 survey (after Mjelde *et al.* 2008). Abbreviations: AB, Aegir Basin; AR, Aegir Ridge; COB, continent–ocean boundary; cont., continental; EJMFZ, East Jan Mayen Fracture Zone; HVLC, high-velocity lower crust; JMMC, Jan Mayen microcontinent; JMR, Jan Mayen Ridge; KR, Kolbeinsey Ridge; WJMFZ, West Jan Mayen Fracture Zone.

et al. 2014), as well as inferred from isostatic modelling (Gradmann *et al.*, this volume, in review). Seaward, the continental crust is thinned to 3–7 km, overlain by the up to 14 km-thick sedimentary sequence of the Møre Basin and underlain by an up to 4 km-thick HVLC with velocities of around 7.2 km s^{-1}. Similar to the Vøring margin, the nature of the HVLC is also under debate at the Møre margin. Lundin & Doré (2011) again advocated partially serpentinized mantle peridotite. Kvarven *et al.* (2014) noticed velocity variations within the HVLC, with higher velocities in the west (7.6–7.7 km s^{-1}) than in the east (7.2 km s^{-1}), although not very well constrained. They proposed that the higher velocities could indicate the presence of a partially eclogized body, while the lower velocities are interpreted as magmatic underplating. In a recent review of the HVLC beneath the Møre margin, Nirrengarten *et al.* (2014) concluded that the HVLC in the proximal part of the margin is most likely to represent inherited crustal bodies and not rift-related serpentinized mantle. For the distal part of the margin, their preferred interpretation is that the HVLC is made of boudins of hyper-extended, pre-rift lower-continental crustal rocks more or less intruded by Early Tertiary magmatic material.

The Møre Marginal High is covered by an up to 4 km-thick volcanic sequence (Fig. 8c). The initial oceanic crustal thickness is 10 km, which is similar to the Lofoten margin but substantially less than at the Vøring margin (18 km). Further seaward, a relatively homogeneous oceanic crust, with a thickness of 5–7 km, is observed beneath the Norway Basin.

Margins of the Jan Mayen microcontinent

The Jan Mayen microcontinent (JMMC) is a fragment of continental crust extending from north of Iceland up to the East Jan Mayen Fracture Zone (EJMFZ: Fig. 1). Jan Mayen Island itself is not considered part of the JMMC as geochemical data provide no evidence of continental contamination of the rocks on the island (Svellingen & Pedersen 2003). Gaina *et al.* (2009) proposed that the major Oligocene plate-boundary reorganization and microcontinent formation might have been preceded by various ridge propagations and/or short-lived triple junctions NE and, possibly, SW of the JMMC from the initiation of seafloor spreading (54 Ma) to Chron C18 (40 Ma). This resulted in the formation of a highly extended or even fragmented JMMC, and subsequent deformation of its margins and surrounding regions.

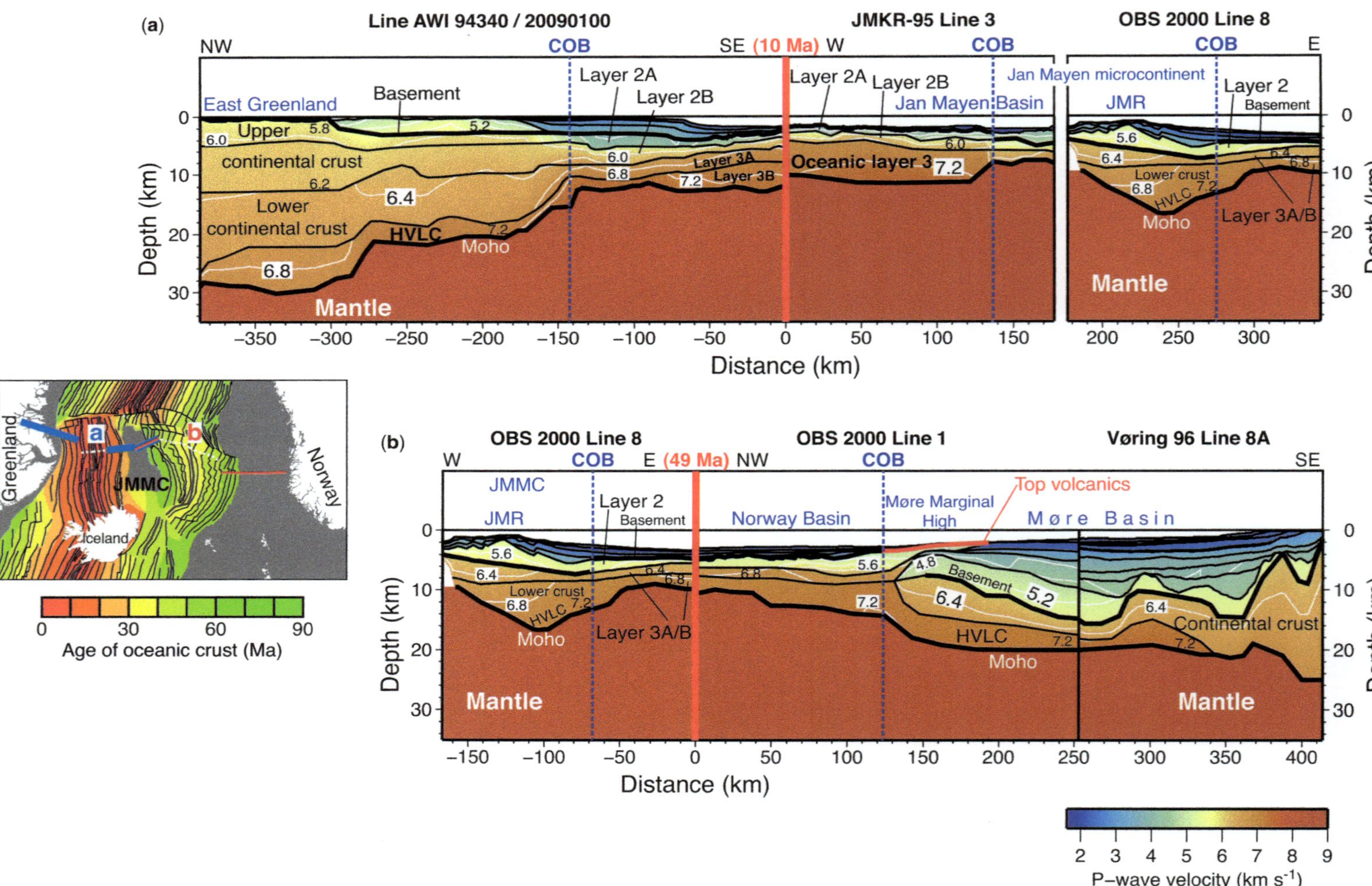
(a)
Line AWI 94340 / 20090100
JMKR-95 Line 3
OBS 2000 Line 8
NW
COB
SE (10 Ma) W
COB
COB
E
East Greenland
Basement
Layer 2A
Layer 2B
Layer 2A
Layer 2B
Jan Mayen Basin
Jan Mayen microcontinent
JMR
Layer 2
Basement
Upper
continental crust
Lower
continental crust
HVLC
Moho
Mantle
Layer 3A
Layer 3B
Oceanic layer 3
Lower crust
HVLC
Moho
Layer 3A/B
Mantle
Depth (km)
Distance (km)
(b)
OBS 2000 Line 8
OBS 2000 Line 1
Vøring 96 Line 8A
W
COB
E (49 Ma) NW
COB
SE
JMMC
JMR
Layer 2
Basement
Norway Basin
Møre Marginal
High
Top volcanics
M ø r e B a s i n
Lower crust
HVLC
Moho
Layer 3A/B
Basement
HVLC
Moho
Continental crust
Mantle
Mantle
Depth (km)
Distance (km)
Greenland
a
b
JMMC
Iceland
Norway
Age of oceanic crust (Ma)
P–wave velocity (km s-1)

During Paleocene rifting, the JMMC was still part of Greenland (Gaina *et al.* 2009). Break-up in the Early Eocene occurred between the JMMC and the mid-Norway and Faroe margins along the Aegir Ridge, and is imaged by regional SDRs along the eastern margin of the JMMC (Peron-Pinvidic *et al.* 2012). At 30 Ma, the Aegir Ridge became extinct and the separation of the JMMC from East Greenland was completed at 20 Ma (Gaina *et al.* 2009) when a seafloor spreading system developed along the Kolbeinsey Ridge. Gaina *et al.* (2009) suggested that the southernmost extended, fragmented character of the southernmost JMMC is a product of several failed ridge-propagation attempts of the Kolbeinsey Ridge. This is also the reason for some uncertainty in defining the COB in the southern JMMC (cf. Funck *et al.* 2014).

The combination of JMKR-95 line 3 (Kodaira *et al.* 1998*a*) and OBS 2000 line 8 (Mjelde *et al.* 2008) presents a seismic refraction transect from the Kolbeinsey Ridge across the JMMC and into the Aegir Basin (Fig. 9b). The western and eastern margins of the JMMC look distinctively different. In the west, a 40 km-wide zone with thin continental crust (thickness of 2.5–4 km) is observed with no indication of a HVLC. In contrast, the eastern margin is characterized by up to 10 km-thick crystalline crust with a high-velocity lower-crustal body (up to $7.2\ \mathrm{km\ s^{-1}}$) at the base. A nearby line (JMKR-95 line 4: Fig. 9a) indicates the presence of basalts extending from the top of the Jan Mayen Ridge eastwards (Kodaira *et al.* 1998*b*). This line also displays a layer with velocities of 4.6–$5.0\ \mathrm{km\ s^{-1}}$ in the Jan Mayen Basin interpreted as basalts that were erupted at 30 Ma (Kodaira *et al.* 1998*b*). In the area of the northern JMMC, the sedimentary column has a maximum thickness of 6 km observed on the western flank of the Jan Mayen Ridge (Fig. 9a).

Moho depth beneath the JMMC is 17–19 km on the lines shown in Figure 9, but close to the northern limit of the JMMC; Kandilarov *et al.* (2012) reported a depth of 27 km. The initial oceanic crust at the eastern margin of the JMMC has a thickness of 9 km, but thins to 5 km over a distance of 30 km (Fig. 9b). Further eastwards, the oceanic crustal thickness varies between 5 and 6 km. The oceanic crust at the western COB of the JMMC has a thickness of 5 km, but increases to 9 km just 15 km to the east. The remainder of JMKR-95 line 3 displays a fairly constant and unusually high oceanic crustal thickness of 9 km.

Conjugate transects north of Iceland

In this section, conjugate transects involving the JMMC are presented first, followed by three transects north of the Jan Mayen Fracture Zone. The western flank of the JMMC (Fig. 10a) is constrained by JMKR-95 line 3 (Kodaira *et al.* 1998*a*) and OBS 2000 line 8 (Mjelde *et al.* 2008). On the Greenland side, line ARK 1988-3 (Weigel *et al.* 1995) is perfectly conjugate to these two lines, but has a rather poor resolution owing to a curved line geometry and a limited number of receivers in the offshore segment of the line. Hence, lines AWI 94340 and 20090100 (Hermann & Jokat 2016) are chosen instead to illustrate this conjugate margin pair (Fig. 10a). There is an offset of about 85 km between the opposite lines. In the oceanic domain, the velocity structure is fairly similar on either side, with 9–10 km of crust. A difference is the division of the lower oceanic crust into a layer 3A and 3B on the Greenland side, while it is modelled as a single layer on the JMMC side.

While the Moho deepens immediately landward of the COB on the Greenland side, the Moho stays almost horizontal for another 40 km at the JMMC before crustal thickening is observed (Fig. 10a). Another difference is the presence of a 2–3 km-thick high-velocity body at the base of the East Greenland crust, while such a lower-crustal body is not indicated at the western margin of the JMMC. As mentioned earlier, the seismic constraints in the HVLC are limited off East Greenland and there is also the possibility that a HVLC at the western margin of the JMMC was missed in the dataset. Hence, the asymmetry with respect to the HVLC may just be perceived. Given the thicker than normal oceanic crust that formed along the Kolbeinsey Ridge, the system seems to be magma-rich rather than magma-poor. If both velocity models are correct, the asymmetry may relate to the focusing of magma production along the WJMFZ during break-up of the JMMC from Greenland, as suggested by Hermann & Jokat (2016). The absence of the HVLC on the JMMC profile can then be attributed to its

Fig. 10. P-wave velocity models of conjugate transects at the JMMC: (**a**) Combined lines AWI 94340 and 20090100 (after Hermann & Jokat 2016), JMKR-95 line 3 (after Kodaira *et al.* 1998*a*), and line 8 of the OBS 2000 survey (after Mjelde *et al.* 2008); and (**b**) lines 8 (after Mjelde *et al.* 2008) and 1 (Breivik *et al.* 2006) of the OBS 2000 survey, and Vøring 96 line 8a (after Raum 2000). The inset map displays the age of the oceanic crust (after Gaina 2014; cf. Gaina *et al.*, this volume, in review), together with the line locations (red and blue lines) and flow lines (dashed white lines). Thin solid lines mark selected isochrons. Abbreviations: COB, continent–ocean boundary; HVLC, high-velocity lower crust; JMMC, Jan Mayen microcontinent; JMR, Jan Mayen Ridge.

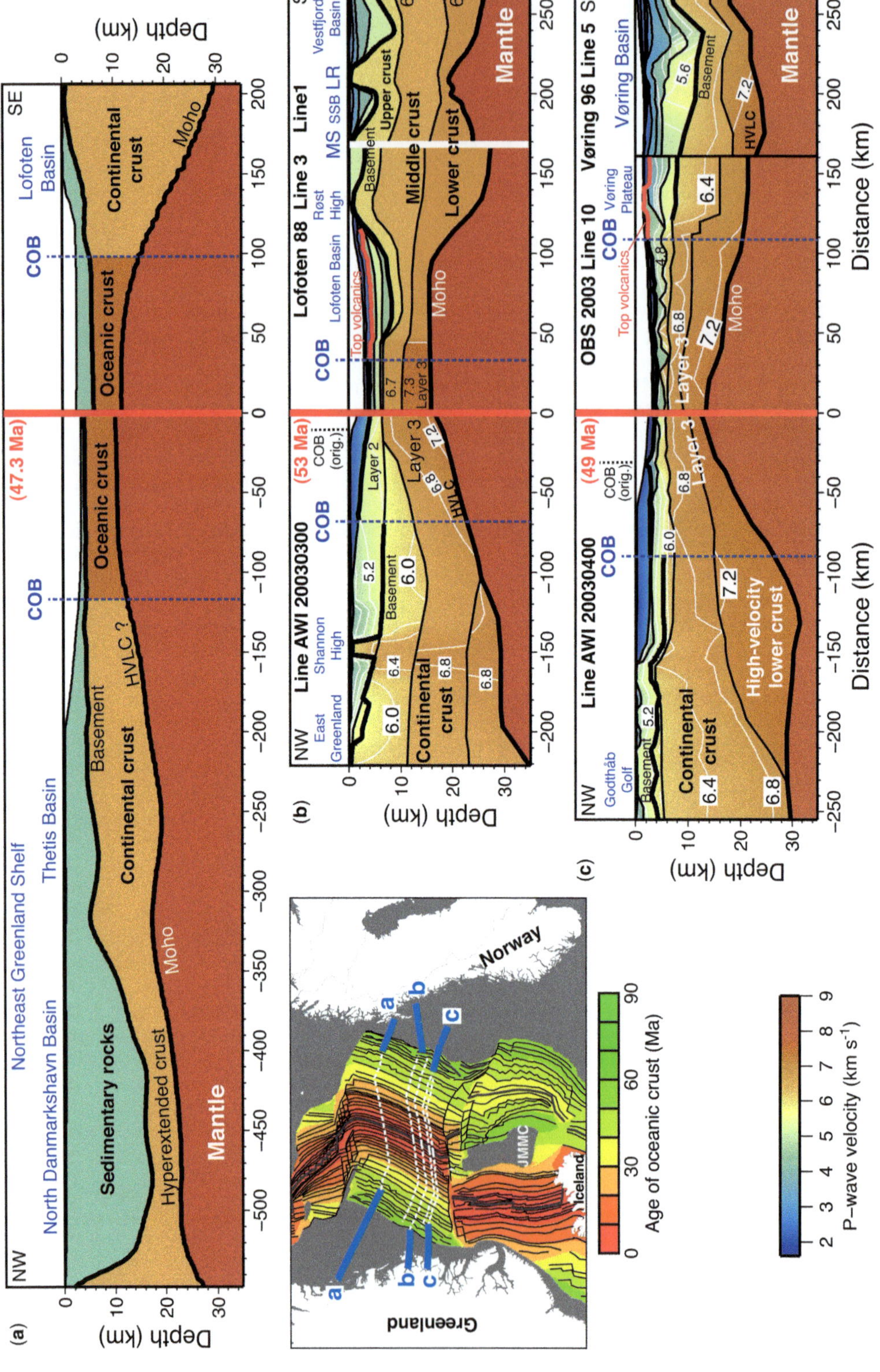
(a)
NW
SE
North Danmarkshavn Basin
Northeast Greenland Shelf
Thetis Basin
COB
(47.3 Ma)
COB
Lofoten Basin
Sedimentary rocks
Hyperextended crust
Basement
Continental crust
HVLC ?
Oceanic crust
Oceanic crust
Continental crust
Moho
Moho
Mantle
Depth (km)
(b)
Line AWI 20030300
(53 Ma)
Lofoten 88 Line 3 Line1
East Greenland
Shannon High
COB
COB (orig.)
Lofoten Basin
Røst High
MS SSB LR
Vestfjorden Basin
Top volcanics
Layer 2
Layer 3
Basement
Upper crust
Middle crust
Lower crust
Continental crust
HVLC
Moho
Mantle
Depth (km)
(c)
Line AWI 20030400
(49 Ma)
OBS 2003 Line 10
Vøring 96 Line 5
Godthåb Golf
COB
COB (orig.)
Top volcanics
Vøring Plateau
Vøring Basin
Basement
Continental crust
High-velocity lower crust
Layer 3
HVLC
Moho
Mantle
Depth (km)
Distance (km)
Greenland
Norway
Iceland
JMMC
Age of oceanic crust (Ma)
P-wave velocity (km s-1)

more distal position to the WJMFZ when compared to the East Greenland line.

At the western margin of the JMMC, the continental crust is only 3–4 km thick in a 40 km-wide zone adjacent to the COB (Fig. 10a). Mantle velocities beneath this hyper-extended crust are 8.0 km s^{-1} (Kodaira *et al.* 1998*a*) and, hence, do not support the presence of partially serpentinized mantle rock. This is, to some degree, surprising as the entire crust becomes brittle at stretching factors of 3–5 (Pérez-Gussinyé & Reston 2001), which would then provide pathways for water to enter the mantle and enable the serpentinization of the peridotite. In analogy to the previous discussion on the absence of a HVLC at that margin, reduced mantle velocities may not have been resolved with the set-up of the experiment.

In the centre of the JMMC, crustal velocities vary between 6.2 and 6.8 km s^{-1} (Mjelde *et al.* 2008). Velocities of 5.8–6.2 km s^{-1} that are observed in the upper crust of Greenland (Hermann & Jokat 2016) are not present. This may indicate that the upper crust may have been completely removed during the rifting. However, some lower velocities of 6.0 km s^{-1} are reported on line 4 of the JMKR-95 survey (Kodaira *et al.* 1998*b*) a little further to the north on the JMMC (Fig. 9a). These velocities would be more compatible with the upper-crustal velocities of central East Greenland.

The transect compiled for the eastern JMMC and Møre margins (Fig. 10b) is conjugate within 75 km, and the reconstruction is for 49 Ma corresponding to Chron C22. The oceanic crust at that age has a similar thickness (5 km) and velocity structure on either side. However, some asymmetry in the spreading rate becomes visible with a wider zone of oceanic crust on the Møre side. In addition, the oceanic crust on the Møre side becomes thicker towards the COB when compared to the JMMC side. The initial oceanic crust has a thickness of 8 and 11 km at the eastern JMMC and Møre margins, respectively. These thickness differences can probably be attributed to deviations from the exact conjugacy. Landward of the COB, a high-velocity lower-crustal body is observed on either side. Beneath the Møre marginal high, the HVLC is 4 km thick: however, the velocity model for the JMMC does not show a sharp boundary between regular lower crust and a HVLC, but, rather, combines them into a single layer with a maximum thickness of 7 km. The extended continental crust beneath the Møre Basin is modelled as a single layer, the thickness of which is less than 7 km in a 240 km-wide zone landward of the COB. Crustal velocities range from 6.1 to 6.7 km s^{-1}, which is compatible with the JMMC, where the maximum crustal thickness is 7 km outside the zone with the HVLC. While the crustal extension at this margin pair is distributed over a wide zone, the break-up occurred close to the western edge of this zone. The HVLC is observed beneath the entire length of the hyper-extended crust, but disappears beneath the western part of the JMMC that was probably much thicker at the time of the Early Eocene break-up between the JMMC and Norway. This thicker crust could then have prevented magma spreading westwards.

The first conjugate transect to the north of the Jan Mayen Fracture Zone is composed of lines AWI 20030400 (Voss & Jokat 2007) off East Greenland, and a combination of line 10 of the OBS 2003 survey (Breivik *et al.* 2009) and line 5 of the Vøring 96 experiment (Mjelde *et al.* 1998) at the Vøring margin (Fig. 11c). The lines are reconstructed for Chron C22 (49 Ma) and they are conjugate within 30 km. Unfortunately, the two Norwegian profiles cover only the outer part of the margin, and do not extend across the Trøndelag Platform and further to the coast. At Chron C22, there is a good match of the velocities and thickness (10 km) of the oceanic crust on either side of the transect.

Problems arise when the COB is looked at in more detail (Fig. 11c). At the Vøring margin, the initial oceanic crustal thickness is 18 km, compared to 24 km on the Greenland side when the COB of Funck *et al.* (2014) is used. Employing the original COB interpretation of Voss & Jokat (2007) on the Greenland side would make for an extreme asymmetry of oceanic crustal accretion, with a 30 km-wide zone of oceanic crust accreted prior to C22 on the Greenland side compared to more than 100 km off Norway. The oceanic crustal thickness at the original COB of Voss & Jokat (2007) is 14 km, which does not fit with the conjugate

Fig. 11. (**a**) Conjugate transect between NE Greenland and the Lofoten margin constructed from results of gravity inversion (Haase *et al.*, this volume, in press) and the sediment thickness compilation of Hopper *et al.* (this volume, in prep). P-wave velocities are not known along the transect. (**b**) & (**c**) P-wave velocity models along conjugate transects between East Greenland and the Vøring margin. Line AWI 20030300 (after Voss *et al.* 2009), lines 3 (after Mjelde *et al.* 1992) and 1 (after Mjelde *et al.* 1993) of the Lofoten 88 survey, line AWI 20030400 (Voss & Jokat 2007), line 10 of the OBS 2003 survey (after Breivik *et al.* 2009), and Vøring 96 line 5 (after Mjelde *et al.* 1998). The inset map displays the age of the oceanic crust (after Gaina 2014; cf. Gaina *et al.*, this volume, in review), together with the line locations (blue lines) and flow lines (dashed white lines). Thin solid lines mark selected isochrons. Abbreviations: COB, continent–ocean boundary; HVLC, high-velocity lower crust; JMMC, Jan Mayen microcontinent; LR, Lofoten Ridge; orig., original interpretation; SSB Skomvær Sub-basin.

margin either unless there was magmatic addition after the accretion of the crust, which could account for the difference. The mismatch between line AWI 20030400 and the conjugate Vøring margin was also discussed by Voss & Jokat (2007). They argued that anomaly C22 is the oldest true seafloor spreading anomaly along their Greenland line. Magnetic anomalies landward of C22 are interpreted to relate to intrusions into stretched continental crust. Furthermore, Voss & Jokat (2007) argued that their magnetic data show that anomalies C24A–C21 terminate against the East Greenland margin. This would be compatible with a north–south propagation of rifting between Shannon Island and the Jan Mayen Fracture Zone. If this model holds, the COB on the Norwegian margin would need to be moved further seaward and the oldest identified magnetic anomalies be questioned. Given the wealth of data on the Vøring margin, it is easier to attribute problems in plate reconstructions to a lack of sufficient data at the conjugate Greenland margin. However, it is worthwhile to mention that magnetic anomalies C24B–C23 become progressively more diffuse to the south at the Vøring margin (Olesen *et al.* 2010), not too dissimilar to what is observed off Greenland. Based on the velocity models alone, it is difficult to pinpoint the exact location of the first true oceanic crust on either of the lines, which is why the real question might be whether or not the diffuse magnetic anomalies could be highly stretched and intruded continental crust or associated extrusive volcanic rocks that may overlie such crust.

Both lines display a HVLC (Fig. 11c), although it is much thicker on the Greenland side (up to 16 km) than beneath the Vøring margin (up to 9 km but mostly *c.* 6 km). Most of the Mesozoic crustal extension is taken up by the Vøring margin. At the eastern end of the transect, some 160 km landward of the COB, the continental crust excluding the HVLC is only 8 km thick, while the Greenland crust has a thickness of 25 km at a similar distance.

Moving northwards, line AWI 20030300 (Voss *et al.* 2009) is conjugate within 60 km of lines 1 (Mjelde *et al.* 1993) and 3 (Mjelde *et al.* 1992) of the Lofoten 88 survey (Fig. 11b). While the two Norwegian lines are located northwards of the Bivrost Lineament and, hence, are part of the Lofoten margin, line AWI 20030300 would reconstruct to a position just to the south of the lineament. Hence, along-margin variations across the lineament may affect the conclusions drawn from this conjugate transect. At 53 Ma, the oceanic crust has a thickness of 12 km on either profile. While the Moho deepens landward on the Greenland side, its depth does not change over a distance of 90 km at the Lofoten margin. The original interpretation of the COB on line AWI 20030300 by Voss *et al.* (2009) seems to be in better agreement with the conjugate Lofoten margin than the revised COB by Funck *et al.* (2014). Since the lower-crustal velocity increases steadily from west to east between km −90 and km −10 (Fig. 11b), this zone fits better the characteristics of a COT than those of oceanic crust. Some HVLC is observed in this transition zone, while it is less clear whether there is such a high-velocity lower-crustal body on the Lofoten lines. Instead of a gradual velocity increase, there is a sharp velocity change some 15 km landward of the COB. The Moho deepens gradually on the Greenland side. In contrast, some local shallowing of the Moho is indicated beneath the Lofoten Ridge. The ridge separates the Skomvær Subbasin from the Vestfjorden Basin with a maximum sediment thickness of 7 km. The depth of the sedimentary basins on the Greenland margin is poorly resolved, as explained earlier. They are possibly deeper than the 5 km indicated in Figure 11b and might be as deep as 14 km, as suggested in the original interpretation of the velocity model (Voss *et al.* 2009).

Given the lack of seismic refraction data covering the deep and wide basins beneath the NE Greenland Shelf, alternative data have to be used to illustrate the conjugate Lofoten–NE Greenland margins in their full width. Figure 11a shows such a transect, which is based on three datasets: the sediment thickness map of Hopper *et al.* (this volume, in prep); the Moho depth obtained from gravity inversion (Haase *et al.*, this volume, in press); and the COB of Funck *et al.* (2014). While this transect does not allow verification of crustal features based on their seismic velocities, some general remarks on the crustal architecture are possible. The conjugates are reconstructed for 47.3 Ma (C21), where the oceanic crust has a thickness of 6 km. At that position, the basement is slightly deeper on the Lofoten profile, which is related to a thicker sediment load. The oceanic crust thickens towards the COB, where it has a thickness of 8 and 10 km at the Greenland and Lofoten margins, respectively. This thickening indicates a larger magma supply following break-up and would be compatible with some HVLC in the adjacent COT. This is also consistent with the HVLC observed on line AWI 20030200 (Fig. 6b) that lies within 5 km of the transect.

The NE Greenland margin is characterized by a 410 km-wide zone with extended continental crust (Fig. 11a). At the Lofoten margin, the continental crust thickens from 10 to 30 km over a distance of just 100 km. In contrast to the Vøring margin further to the south, the polarity of the margin has changed from a wide margin on the Norwegian side (Fig. 8c) and a narrow margin off East Greenland (Fig. 6d) to the opposite configuration (Fig. 11a). Beneath the NE Greenland Shelf, two deep basins are encountered: the 7 km-deep Thetis Basin and the 17 km-deep North Danmarkshavn Basin (Fig. 11a). While the crust beneath the Thetis Basin still has a thickness of 11 km, it thins to 6 km beneath the North Danmarkshavn Basin and has to be considered as hyper-extended. Unfortunately, there is no information on the crustal and mantle velocities beneath the basin that could provide information on the existence of a HVLC or partially serpentinized mantle. This could provide additional constraints on the nature of the HVLC at the Vøring margin, where its composition is controversial (cf. Mjelde *et al.* 2009).

Conclusions

The review of the seismic refraction data has shown that there is a wealth of profiles that provide information on the continental margins bordering the NE Atlantic Ocean. The margins are, in large parts, characterized by excess magmatism associated with the volcanic break-up in the Early Eocene. This magmatism is evidenced by SDRs on seismic reflection sections, while refraction data image mafic intrusions that are accompanied by an increase in crustal velocities. A HVLC is also often observed, which is commonly explained by magmatic underplating or as a result of sill intrusions into the lower crust (White *et al.* 2008) at the distal parts of the margins. However, high velocities are also observed below more proximal parts of the margins, such as beneath the southern Rockall Basin and the Porcupine Basin, where they are interpreted as partially serpentinized mantle (Shannon *et al.* 1999; O'Reilly *et al.* 2006). At the mid-Norwegian margin, interpretations vary from magmatic underplating to eclogite bodies and partially serpentinized mantle (cf. Mjelde *et al.* 2009; Lundin & Doré 2011). Crustal thinning down to a few kilometres (e.g. Fig. 8b) is observed beneath the rift basins that are generally underlain by material with seismic velocities in excess of 7.0 km s^{-1}. As stated above, the lithological interpretation of these velocities is often controversial.

The Iceland plume with its associated excess magmatism has shaped the continental margins bordering the NE Atlantic. Along the Greenland–Iceland–Faroe Ridge (Fig. 5), the crustal thickness is generally greater than 30 km (Richardson *et al.* 1998; Holbrook *et al.* 2001). The thickness of the initial oceanic crust formed at break-up decreases southwards. However, even some 1200 km away from the axis of the Greenland–Iceland Ridge, the igneous crust at the SE Greenland margin is still 16 km thick (Holbrook *et al.* 2001) (Fig. 2c). Magmatism to the north of Iceland was more restrictive when compared to the south. At the central Møre margin, oceanic crust at the time of break-up is only 11 km thick and thins to 5 km towards the extinct Aegir Ridge (Fig. 8c) (Breivik *et al.* 2006). Later, when the spreading moved from the Aegir Ridge to the Kolbeinsey Ridge, the thickness of the oceanic crust increased again to 9 km to remain fairly constant around this value (Figs 6e & 9b) (Kodaira *et al.* 1998*a*; Hermann & Jokat 2016). A significant increase in the amount of magmatic addition is noticed north of the Jan Mayen Fracture Zone with an initial oceanic crustal thickness of 18 km at the Vøring margin (Fig. 8b) (Breivik *et al.* 2008) or even up to 25 km on the conjugate NE Greenland margin (Fig. 6d) (Voss & Jokat 2007). Similar to the area south of Iceland, the amount of magmatism decreases with distance to the plume. On the Greenland side, the northernmost evidence of a HVLC is found just to the south of the East Greenland Ridge (Fig. 6b) (Voss *et al.* 2009). On the conjugate Norwegian side, there is no evidence of a HVLC at the Lofoten margin, even though the region was still affected by extrusive magmatism, including SDRs (Fig. 1).

The overall architecture of the rifted margins in the NE Atlantic Ocean reveals a pronounced asymmetry of the conjugate margin segments. South of the Bivrost Lineament, the wide margins with deep rift basins are located on the European side, while the conjugate East Greenland margin is comparatively narrow. This pattern changes north of the Bivrost Lineament, where the Lofoten margin is narrow, while the conjugate NE Greenland Shelf with the deep Danmarkshavn and Thetis basins widens northwards.

This review has shown that there are still many questions that the presently available data cannot answer. In particular, there is a shortage of truly conjugate transects that can help to fully understand the complex rifting history in this part of the Atlantic. There is a lack of continuous profiles extending over the entire width of, in particular, the wide margins (the Hatton–Rockall area, and the Vøring and NE Greenland margins) from the proximal to the distal zones into truly oceanic crust. The mid-Norwegian margin is studied by a dense net of seismic refraction lines that are generally short. Even if they can be merged into a continuous profile, there are often substantial deviations at the line intersections (e.g. Figs 8b & 11c). In NE Greenland, seismic data acquisition is impeded by the presence of sea ice, and north of the East Greenland Ridge there are no refraction lines to image the margin there. The mid-Norwegian margin is generally considered as well studied, while the coverage on the conjugate East Greenland margin is less dense. However, some modern refraction lines are available for East Greenland and cover that margin in its full width (Voss & Jokat 2007; Voss *et al.* 2009; Hermann & Jokat 2016). In contrast, the most recent lines published on the Lofoten margin were acquired in 1988 (Mjelde *et al.* 1992, 1993) with a poor resolution in the COT due to a wide receiver spacing. Velocity models there indicate a sharp transition from oceanic to continental crust, in contrast to wider COTs observed at magma-rich margins studied with more receivers, such as the Hatton margin (White & Smith 2009) or the NE Greenland margin (Voss & Jokat 2007, 2009; Voss *et al.* 2009). As reflection data and potential field data alone are not necessarily sufficient to define the COB, some modern OBS experiments would be desirable to confirm or refine the COB at the Lofoten margin and also at the Vøring margin, at the transition from the Vøring to the Møre margin, and probably many more places.

Similarly, at the SE Greenland margin, there remains uncertainty about the location of the COB, as White & Smith (2009) questioned the interpretation of some weak magnetic anomalies as true seafloor spreading anomalies (Larsen *et al.* 1994; Larsen & Saunders 1998). Using the conceptual model of Quirk *et al.* (2014) for the break-up of volcanic margins, some of the previously interpreted oceanic crust at the SE Greenland margin may, indeed, be thinned and intruded continental crust that is overlain by a broad zone of subaerial SDRs. This may create a velocity structure that looks more oceanic than continental, and would reduce the asymmetry in the initial accretion of oceanic crust. However, a verification of the model in Quirk *et al.* (2014) would require seismic reflection data imaging down to the Moho, preferentially on both the SE Greenland and the conjugate Hatton margins. Poorly defined COBs may be the reason for the apparent asymmetry in the production of early oceanic crust in many parts of the NE Atlantic.

The margins of the JMMC are only poorly sampled so far. Some refraction profiles are available for the northern JMMC but leave some questions open, in particular about the western margin. A recently published line at the East Greenland margin (Hermann & Jokat 2016) conjugate to the JMMC shows some indication for the presence of a HVLC with no equivalent at the JMMC (Kodaira *et al.* 1998*a*). Since the two lines are not completely conjugate, the differences could relate to along-strike variations at the margin. Alternatively, there could be problems in the velocity models. The southern limit of the JMMC is not defined with any confidence and there is no seismic refraction data available to provide information on this highly segmented part of the microcontinent.

Another major data gap is the sheared NE Greenland margin to the north of the East Greenland Ridge. This area is ice-covered year-round in contrast to the much better studied conjugate SW Barents Sea margin. Of particular interest here would be a conjugate to the Vestbakken Volcanic Province that is characterized by volcanic rocks and a HVLC (Fig. 7c) (Libak *et al.* 2012). Conjugate to this province, seismic reflection data show some strong, but discontinuous, reflectors that might be SDRs (cf. Geissler *et al.*, this volume, in review). Refraction data could provide information on the deeper crustal velocity structure and verify whether or not the margin is similar to the Vestbakken Volcanic Province.

Another issue with many seismic refraction lines crossing the deep sedimentary basins off NE Greenland, mid-Norway and in the SW Barents Sea is the clear identification of the basement, as velocities of Mesozoic and Palaeozoic sedimentary rocks can be similar to crystalline basement when deeply buried. This can leave some uncertainty with respect to crustal thinning factors, which are important for basin modelling and deformable plate reconstructions. Some of the ambiguity can be reduced by having coincident deep seismic reflection data of high quality available for the modelling of the refraction data. This may require some collaboration between industry and academia, as there are excellent industry seismic reflection data acquired along the margins that would benefit from adding refraction data. A better integration of potential field and seismic reflection data could help to resolve issues related to sparse seismic refraction data and difficulties in imaging deep structures.

While magma-poor margins are fairly well understood by now, thanks to extensive studies including drilling at the conjugate Iberia–Newfoundland margin pair (cf. Peron-Pinvidic *et al.* 2013), there has been less focus on the rifting processes at magma-rich margins (Quirk *et al.* 2014). The NE Atlantic with the North Atlantic Igneous Province is a natural laboratory to study magma-rich margins. However, despite decades of seismic and geophysical data acquisition, there is still a lot of uncertainty on first-order structures, such as the location of the COB. This is why some key conjugate refraction transects should be acquired in the NE Atlantic. These should be truly conjugate, cover the entire width of the margins, have good coincident reflection data available, use a similar instrumentation and experimental set-up on either side, and would be analysed in a consistent way with the same techniques, ideally by the same person.

This work was part of the NAG-TEC project that was supported by the following industry sponsors (in alphabetic order): BayernGas Norge AS; BP Exploration Operating Company Ltd; Bundesanstalt für Geowissenschaften und Rohstoffe (BGR); Chevron East Greenland Exploration A/S; ConocoPhillips Skandinavia AS; DEA Norge AS; Det norske oljeselskap ASA; DONG E&P A/S; E.ON Norge AS; ExxonMobil Exploration and Production

Norway AS; Japan Oil, Gas, and Metals National Corporation (JOGMEC); Maersk Oil; Nalcor Energy – Oil and Gas Inc.; Nexen Energy ULC; Norwegian Energy Company ASA (Noreco); Repsol Exploration Norge AS; Statoil (UK) Ltd; and Wintershall Holding GmbH. Many thanks go to Carmen Gaina for assistance with the plate reconstructions. Reviews from two anonymous referees helped to improve the manuscript. The work of Kenneth McDermott was supported by the Geological Survey of Ireland. Geoffrey S. Kimbell publishes with permission of the Executive Director of the British Geological Survey (Natural Environment Research Council).

References

Á Horni, J., Geissler, W.H. *et al.* 2014. Offshore volcanic facies. *In*: Hopper, J.R., Funck, T., Stoker, M., Árting, U., Peron-Pinvidic, G., Doornenbal, H. & Gaina, C. (eds) *NAG-TEC Atlas: Tectonostratigraphic Atlas of the North-East Atlantic Region*. Geological Survey of Denmark and Greenland (GEUS), Copenhagen, Denmark, 235–253.

Á Horni, J., Blischke, A. *et al.* In review. Seismic volcanostratigraphy of the North Atlantic Igneous Province: the volcanic facies units. *In*: Péron-Pinvidic, G., Hopper, J.R., Stoker, M.S., Gaina, C., Doornenbal, J.C. & Funck, T. & Árting, U.E. (eds) *The NE Atlantic Region: A Reappraisal of Crustal Structure, Tectonostratigraphy and Magmatic Evolution*. Geological Society, London, Special Publications, **447**.

Andersen, T.B. & Jamtveit, B. 1990. Uplift of deep crust during orogenic extensional collapse: a model based on field studies in the Sogn–Sunnfjord region of western Norway. *Tectonics*, **9**, 1097–1111, https://doi.org/10.1029/TC009i005p01097

Barton, A.J. & White, R.S. 1997. Crustal structure of Edoras Bank continental margin and mantle thermal anomalies beneath the North Atlantic. *Journal of Geophysical Research*, **102**, 3109–3129, https://doi.org/10.1029/96jb03387

Breivik, A.J., Mjelde, R., Grogan, P., Shimamura, H., Murai, Y. & Nishimura, Y. 2003. Crustal structure and transform margin development south of Svalbard based on ocean bottom seismometer data. *Tectonophysics*, **369**, 37–70, https://doi.org/10.1016/s0040-1951(03)00131-8

Breivik, A.J., Mjelde, R., Faleide, J.I. & Murai, Y. 2006. Rates of continental breakup magmatism and seafloor spreading in the Norway Basin-Iceland plume interaction. *Journal of Geophysical Research*, **111**, B07102, https://doi.org/10.1029/2005jb004004

Breivik, A.J., Faleide, J.I. & Mjelde, R. 2008. Neogene magmatism northeast of the Aegir and Kolbeinsey ridges, NE Atlantic: Spreading ridge–mantle plume interaction? *Geochemistry Geophysics Geosystems*, **9**, Q02004, https://doi.org/10.1029/2007gc001750

Breivik, A.J., Faleide, J.I., Mjelde, R. & Flueh, E.R. 2009. Magma productivity and early seafloor spreading rate correlation on the northern Vøring Margin, Norway – Constraints on mantle melting. *Tectonophysics*, **468**, 206–223, https://doi.org/10.1016/j.tecto.2008.09.020

Breivik, A.J., Mjelde, R., Raum, T., Faleide, J.I., Murai, Y. & Flueh, E.R. 2011. Crustal structure beneath the Trøndelag Platform and adjacent areas of the Mid-Norwegian margin, as derived from wide-angle seismic and potential field data. *Norwegian Journal of Geology*, **90**, 141–161.

Brekke, H. 2000. The tectonic evolution of the Norwegian Sea continental margin with emphasis on the Vøring and Møre basins. *In*: Nøttvedt, A. (ed.) *Dynamics of the Norwegian Margin*. Geological Society, London, Special Publications, **167**, 327–378, https://doi.org/10.1144/GSL.SP.2000.167.01.13

Bunch, A.W.H. 1979. A detailed seismic structure of Rockall Bank (55°N, 15°W) – A synthetic seismogram analysis. *Earth and Planetary Science Letters*, **45**, 453–463, https://doi.org/10.1016/0012-821x(79)90144-4

Christensen, N.I. 2004. Serpentinites, peridotites, and seismology. *International Geology Review*, **46**, 795–816, https://doi.org/10.2747/0020-6814.46.9.795

Coffin, M.F. & Eldholm, O. 1994. Large igneous provinces: crustal structure, dimensions, and external consequences. *Reviews of Geophysics*, **32**, 1–36, https://doi.org/10.1029/93rg02508

Darbyshire, F.A., Bjarnason, I.T., White, R.S. & Flóvenz, Ó.G. 1998. Crustal structure above the Iceland mantle plume imaged by the ICEMELT refraction profile. *Geophysical Journal International*, **135**, 1131–1149, https://doi.org/10.1046/j.1365-246X.1998.00701.x

Darbyshire, F.A., White, R.S. & Priestley, K.F. 2000. Structure of the crust and uppermost mantle of Iceland from a combined seismic and gravity study. *Earth and Planetary Science Letters*, **181**, 409–428, https://doi.org/10.1016/s0012-821x(00)00206-5

Doré, A.G., Lundin, E.R., Birkeland, O., Eliassen, P.E. & Jensen, L.N. 1997. The NE Atlantic margin; implications of late Mesozoic and Cenozoic events for hydrocarbon prospectivity. *Petroleum Geoscience*, **3**, 117–131, https://doi.org/10.1144/petgeo.3.2.117

Doré, A.G., Lundin, E.R., Jensen, L.N., Birkeland, Ø., Eliassen, P.E. & Fichler, C. 1999. Principal tectonic events in the evolution of the northwest European Atlantic margin. *In*: Fleet, A.J. & Boldy, S.A.R. (eds) *Petroleum Geology of Northwest Europe: Proceedings of the 5th Conference*. Geological Society, London, 41–61, https://doi.org/10.1144/0050041

Doré, A.G., Lundin, E.R., Gibbons, A., Sømme, T.O. & Tørudbakken, B.O. 2015. Transform margins of the Arctic: a synthesis and re-evaluation. *In*: Nemčok, M., Rybár, S., Sinha, S.T. & Hermeston, S.A. & Ledvényiová, L. (eds) *Transform Margins: Development, Controls and Petroleum Systems*. Geological Society, London, Special Publications, **431**, https://doi.org/10.1144/SP431.8

Døssing, A. & Funck, T. 2012. Greenland Fracture Zone–East Greenland Ridge(s) revisited: indications of a C22-change in plate motion? *Journal of Geophysical Research*, **117**, B01103, https://doi.org/10.1029/2011jb008393

Døssing, A., Dahl-Jensen, T., Thybo, H., Mjelde, R. & Nishimura, Y. 2008. East Greenland Ridge in the North Atlantic Ocean: an integrated geophysical study of a continental sliver in a boundary transform fault setting. *Journal of Geophysical Research*, **113**, B10107, https://doi.org/10.1029/2007jb005536

Eccles, J.D., White, R.S. & Christie, P.A.F. 2009. Identification and inversion of converted shear waves: case studies from the European North Atlantic continental margins. *Geophysical Journal International*, **179**, 381–400, https://doi.org/10.1111/j.1365-246X.2009.04290.x

Eldholm, O., Faleide, J.I. & Myhre, A.M. 1987. Continent–ocean transition at the western Barents Sea/Svalbard continental margin. *Geology*, **15**, 1118–1122, https://doi.org/10.1130/0091-7613(1987)15<1118:CTATWB>2.0.CO;2

Engen, Ø., Faleide, J.I. & Dyreng, T.K. 2008. Opening of the Fram Strait gateway: a review of plate tectonic constraints. *Tectonophysics*, **450**, 51–69, https://doi.org/10.1016/j.tecto.2008.01.002

Escher, J.C. & Pulvertaft, T.C.R. 1995. *Geological Map of Greenland, Scale 1:2,500,000.* Geological Survey of Greenland, Copenhagen.

Faleide, J.I., Myhre, A.M. & Eldholm, O. 1988. Early Tertiary volcanism at the western Barents Sea margin. *In*: Morton, A.C. & Parson, L.M. (eds) *Early Tertiary Volcanism and the Opening of the NE Atlantic.* Geological Society, London, Special Publications, **39**, 135–146, https://doi.org/10.1144/GSL.SP.1988.039.01.13

Faleide, J.I., Vågnes, E. & Gudlaugsson, S.T. 1993. Late Mesozoic-Cenozoic evolution of the southwestern Barents Sea in a regional rift-shear tectonic setting. *Marine and Petroleum Geology*, **10**, 186–214, https://doi.org/10.1016/0264-8172(93)90104-Z

Faleide, J.I., Solheim, A., Fiedler, A., Hjelstuen, B.O., Andersen, E.S. & Vanneste, K. 1996. Late Cenozoic evolution of the western Barents Sea–Svalbard continental margin. *Global and Planetary Change*, **12**, 53–74, https://doi.org/10.1016/0921-8181(95)00012-7

Faleide, J.I., Tsikalas, F. *et al.* 2008. Structure and evolution of the continental margin off Norway and the Barents Sea. *Episodes*, **31**, 82–91.

Fliedner, M.M. & White, R.S. 2003. Depth imaging of basalt flows in the Faeroe–Shetland Basin. *Geophysical Journal International*, **152**, 353–371, https://doi.org/10.1046/j.1365-246X.2003.01833.x

Foulger, G.R. & Anderson, D.L. 2005. A cool model for the Iceland hotspot. *Journal of Volcanology and Geothermal Research*, **141**, 1–22, https://doi.org/10.1016/j.jvolgeores.2004.10.007

Funck, T., Andersen, M.S., Keser Neish, J. & Dahl-Jensen, T. 2008. A refraction seismic transect from the Faroe Islands to the Hatton–Rockall Basin. *Journal of Geophysical Research*, **113**, B12405, https://doi.org/10.1029/2008jb005675

Funck, T., Hopper, J.R. *et al.* 2014. Crustal structure. *In*: Hopper, J.R., Funck, T., Stoker, M., Árting, U., Peron-Pinvidic, G., Doornenbal, H. & Gaina, C. (eds) *NAG-TEC Atlas: Tectonostratigraphic Atlas of the North-East Atlantic Region.* Geological Survey of Denmark and Greenland (GEUS), Copenhagen, Fenmark, 69–126.

Funck, T., Gerlings, J., Hopper, J.R., Castro, C.F., Marcussen, C., Nielsen, T. & Døssing, A. 2015. The East Greenland Ridge: Geophysical mapping and geological sampling reveal a highly segmented continental sliver. *In*: Smelror, M. (ed.) *7th International Conference on Arctic Margins – ICAM 2015. NGU Report 2015.032.* Geological Survey of Norway, Trondheim, Denmark, 45.

Funck, T., Geissler, W.H., Kimbell, G.S., Gradmann, S., Erlendsson, Ö., McDermott, K. & Petersen, U.K. 2016. Moho and basement depth in the NE Atlantic Ocean based on seismic refraction data and receiver functions. *In*: Péron-Pinvidic, G., Hopper, J.R., Stoker, M.S., Gaina, C., Doornenbal, J.C., Funck, T. & Árting, U.E. (eds) *The NE Atlantic Region: A Reappraisal of Crustal Structure, Tectonostratigraphy and Magmatic Evolution.* Geological Society, London, Special Publications, **447**. First published online July 13, 2016, https://doi.org/10.1144/SP447.1

Gaina, C. 2014. Plate reconstructions and regional kinematics. *In*: Hopper, J.R., Funck, T., Stoker, M., Árting, U., Peron-Pinvidic, G., Doornenbal, H. & Gaina, C. (eds) *NAG-TEC Atlas: Tectonostratigraphic Atlas of the North-East Atlantic Region.* Geological Survey of Denmark and Greenland (GEUS), Copenhagen, 53–66.

Gaina, C., Gernigon, L. & Ball, P. 2009. Palaeocene–Recent plate boundaries in the NE Atlantic and the formation of the Jan Mayen microcontinent. *Journal of the Geological Society*, **166**, 601–616, https://doi.org/10.1144/0016-76492008-112

Gaina, C., Nasuti, A., Kimbell, G.S. & Blischke, A. In review. Break-up and seafloor spreading domains in the NE Atlantic. *In*: Péron-Pinvidic, G., Hopper, J.R., Stoker, M.S., Gaina, C., Doornenbal, J.C., Funck, T. & Árting, U.E. (eds) *The NE Atlantic Region: A Reappraisal of Crustal Structure, Tectonostratigraphy and Magmatic Evolution.* Geological Society, London, Special Publications, **447**.

Geissler,W.H., Hopper, J.R. *et al.* In review. Seismic volcano-stratigraphy of the Northeast Greenland continental margin revisited. *In*: Péron-Pinvidic, G., Hopper, J.R., Stoker, M.S., Gaina, C., Doornenbal, J.C., Funck, T. & Árting, U.E. (eds) *The NE Atlantic Region: A Reappraisal of Crustal Structure, Tectonostratigraphy and Magmatic Evolution.* Geological Society, London, Special Publications, **447**.

Gerlings, J., Funck, T., Castro, C.F. & Hopper, J.R. 2014. The East Greenland Ridge – a continental sliver along the Greenland Fracture Zone. *Geophysical Research Abstracts*, **16**, EGU2014–EGU2879.

Gerlings, J., Hopper, J.R., Fyhn, M.B.W. & Frandsen, N.K. In review. Mesozoic and older rift basins on the SE Greenland shelf near Ammassalik. *In*: Péron-Pinvidic, G., Hopper, J.R., Stoker, M.S., Gaina, C., Doornenbal, J.C., Funck, T. & Árting, U.E. (eds) *The NE Atlantic Region: A Reappraisal of Crustal Structure, Tectonostratigraphy and Magmatic Evolution.* Geological Society, London, Special Publications, **447**.

Gernigon, L., Ringenbach, J.C., Planke, S., Le Gall, B. & Jonquet-Kolstø, H. 2003. Extension, crustal structure and magmatism at the outer Vøring Basin, Norwegian margin. *Journal of the Geological Society, London*, **160**, 197–208, https://doi.org/10.1144/0016-764902-055

GERNIGON, L., RINGENBACH, J.-C., PLANKE, S. & LE GALL, B. 2004. Deep structures and breakup along volcanic rifted margins: insights from integrated studies along the outer Vøring Basin (Norway). *Marine and Petroleum Geology*, **21**, 363–372, https://doi.org/10.1016/j.marpetgeo.2004.01.005

GERNIGON, L., BLISCHKE, A., NASUTI, A. & SAND, M. 2015. Conjugate volcanic rifted margins, seafloor spreading, and microcontinent: insights from new high-resolution aeromagnetic surveys in the Norway Basin. *Tectonics*, **34**, 907–933, https://doi.org/10.1002/2014tc003717

GRADMANN, S., HAASE, C. & EBBING, J. In review. Isostasy as a tool to validate interpretations of regional geophysical data sets – application to the mid-Norwegian continental margin. *In*: PÉRON-PINVIDIC, G., HOPPER, J.R., STOKER, M.S., GAINA, C., DOORNENBAL, J.C., FUNCK, T. & ÁRTING, U.E. (eds) *The NE Atlantic Region: A Reappraisal of Crustal Structure, Tectonostratigraphy and Magmatic Evolution*. Geological Society, London, Special Publications, **447**.

HAASE, C., EBBING, J. & FUNCK, T. In press. A 3D regional crustal model of the NE Atlantic based on seismic and gravity data. *In*: PÉRON-PINVIDIC, G., HOPPER, J.R., STOKER, M.S., GAINA, C., DOORNENBAL, J.C., FUNCK, T. & ÁRTING, U.E. (eds) *The NE Atlantic Region: A Reappraisal of Crustal Structure, Tectonostratigraphy and Magmatic Evolution*. Geological Society, London, Special Publications, **447**, https://doi.org/10.1144/SP447.8

HAMANN, N.E., WHITTAKER, R.C. & STEMMERIK, L. 2005. Geological development of the Northeast Greenland Shelf. *In*: DORÉ, A.G. & VINING, B.A. (eds) *Petroleum Geology: North-West Europe and Global Perspectives – Proceedings of the 6th Petroleum Geology Conference*. Geological Society, London, 887–902, https://doi.org/10.1144/0060887

HARLAND, W.B. 1969. Contribution of Spitsbergen to understanding of tectonic evolution of the North Atlantic region. *In*: KAY, M. (ed.) *North Atlantic – Geology and Continental Drift*. American Association of Petroleum Geologists, Memoirs, **12**, 817–851.

HARLAND, K.E., WHITE, R.S. & SOOSALU, H. 2009. Crustal structure beneath the Faroe Islands from teleseismic receiver functions. *Geophysical Journal International*, **177**, 115–124, https://doi.org/10.1111/j.1365-246X.2008.04018.x

HERMANN, T. & JOKAT, W. 2013. Crustal structures of the Boreas Basin and the Knipovich Ridge, North Atlantic. *Geophysical Journal International*, **193**, 1399–1414, https://doi.org/10.1093/gji/ggt048

HERMANN, T. & JOKAT, W. 2016. Crustal structure off Kong Oscar Fjord, East Greenland: evidence for focused melt supply along the Jan Mayen Fracture Zone. *Tectonophysics*, https://doi.org/10.1016/j.tecto. 2015.12.005

HOLBROOK, W.S., LARSEN, H.C. *ET AL*. 2001. Mantle thermal structure and active upwelling during continental breakup in the North Atlantic. *Earth and Planetary Science Letters*, **190**, 251–266, https://doi.org/10.1016/s0012-821x(01)00392-2

HOPPER, J.R., LIZARRALDE, D. & LARSEN, H.C. 1998. Seismic investigations offshore South-East Greenland. *Geology of Greenland Survey Bulletin*, **180**, 145–151.

HOPPER, J.R., DAHL-JENSEN, T. *ET AL*. 2003. Structure of the SE Greenland margin from seismic reflection and refraction data: implications for nascent spreading center subsidence and asymmetric crustal accretion during North Atlantic opening. *Journal of Geophysical Research*, **108**, 2269, https://doi.org/10.1029/2002jb001996

HOPPER, J.R., FUNCK, T., STOKER, M., ÁRTING, U., PERON-PINVIDIC, G. & DOORNENBAL, H. & GAINA, C. (eds) 2014. *NAG-TEC Atlas: Tectonostratigraphic Atlas of the North-East Atlantic Region*. Geological Survey of Denmark and Greenland (GEUS), Copenhagen, Denmark.

HOPPER, J.R., ABDUL FATTAH, R., GAINA, C., GEISSLER, W., DOORNENBAL, H., FUNCK, T. & KIMBELL, G.S. In prep. Sediment thickness and residual topography of the North Atlantic: estimating dynamic topography around Iceland. *In*: PÉRON-PINVIDIC, G., HOPPER, J.R., STOKER, M.S., GAINA, C., DOORNENBAL, J.C., FUNCK, T. & ÁRTING, U.E. (eds) *The NE Atlantic Region: A Reappraisal of Crustal Structure, Tectonostratigraphy and Magmatic Evolution*. Geological Society, London, Special Publications, **447**.

JUNG, W.-Y. & VOGT, P.R. 1997. A gravity and magnetic anomaly study of the extinct Aegir Ridge, Norwegian Sea. *Journal of Geophysical Research*, **102**, 5065–5089, https://doi.org/10.1029/96jb03915

KANDILAROV, A., MJELDE, R. *ET AL*. 2012. The northern boundary of the Jan Mayen microcontinent, North Atlantic determined from ocean bottom seismic, multichannel seismic, and gravity data. *Marine Geophysical Research*, **33**, 55–76, https://doi.org/10.1007/s11001-012-9146-4

KIMBELL, G.S., RITCHIE, J.D., JOHNSON, H. & GATLIFF, R.W. 2005. Controls on the structure and evolution of the NE Atlantic margin revealed by regional potential field imaging and 3D modelling. *In*: DORÉ, A.G. & VINING, B.A. (eds) *Petroleum Geology: North-West Europe and Global Perspectives – Proceedings of the 6th Petroleum Geology Conference*. Geological Society, London, 933–945, https://doi.org/10.1144/0060933

KIMBELL, G.S., RITCHIE, J.D. & HENDERSON, A.F. 2010. Three-dimensional gravity and magnetic modelling of the Irish sector of the NE Atlantic margin. *Tectonophysics*, **486**, 36–54, https://doi.org/10.1016/j.tecto.2010.02.007

KLINGELHÖFER, F., EDWARDS, R.A., HOBBS, R.W. & ENGLAND, R.W. 2005. Crustal structure of the NE Rockall Trough from wide-angle seismic data modeling. *Journal of Geophysical Research*, **110**, B11105, https://doi.org/10.1029/2005jb003763

KODAIRA, S., MJELDE, R., GUNNARSSON, K., SHIOBARA, H. & SHIMAMURA, H. 1998*a*. Evolution of oceanic crust on the Kolbeinsey Ridge, north of Iceland, over the past 22 Myr. *Terra Nova*, **10**, 27–31, https://doi.org/10.1046/j.1365-3121.1998.00166.x

KODAIRA, S., MJELDE, R., GUNNARSSON, K., SHIOBARA, H. & SHIMAMURA, H. 1998*b*. Structure of the Jan Mayen microcontinent and implications for its evolution. *Geophysical Journal International*, **132**, 383–400, https://doi.org/10.1046/j.1365-246x.1998.00444.x

KORENAGA, J., HOLBROOK, W.S. *ET AL*. 2000. Crustal structure of the southeast Greenland margin from joint refraction and reflection seismic tomography. *Journal of Geophysical Research*, **105**, 21591–21614, https://doi.org/10.1029/2000JB900188

Kvarven, T., Ebbing, J. *et al.* 2014. Crustal structure across the Møre margin, mid-Norway, from wide-angle seismic and gravity data. *Tectonophysics*, **626**, 21–40, https://doi.org/10.1016/j.tecto.2014.03.021

Larsen, H.C. & Marcussen, C. 1992. Sill-intrusion, flood basalt emplacement and deep crustal structure of the Scoresby Sund region, East Greenland. *In*: Storey, B.C., Alabaster, T. & Pankhurst, R.J. (eds) *Magmatism and the Causes of Continental Break-up*. Geological Society, London, Special Publications, **68**, 365–386, https://doi.org/10.1144/GSL.SP.1992.068.01.23

Larsen, H.C. & Saunders, A.D. 1998. Tectonism and volcanism at the southeast Greenland rifted margin: a record of plume impact and later continental rifting. *In*: *Proceedings of the Ocean Drilling Program, Scientific Results, Volume 152*. Ocean Drilling Program, College Station, TX, USA, 503–533, https://doi.org/10.2973/odp.proc.sr.152.240.1998

Larsen, H.C., Saunders, A.D. & Clift, P.D. & The Shipboard Scientific Party 1994. Summary and principal results. *In*: *Proceedings of the Ocean Drilling Program, Scientific Results, Volume 152*. Ocean Drilling Program, College Station, TX, USA, 279–292, https://doi.org/10.2973/odp.proc.ir.152.113.1994

Larsen, H.C., Dahl-Jensen, T. & Hopper, J.R. 1998. Crustal structure along the Leg 152 drilling transect. *In*: *Proceedings of the Ocean Drilling Program, Scientific Results, Volume 152*. Ocean Drilling Program, College Station, TX, USA, 463–476, https://doi.org/10.2973/odp.proc.sr.152.245.1998

Larsen, H.C., Duncan, R.A., Allan, J.F. & Brooks, K. (eds) 1999. *Proceedings of the Ocean Drilling Program, Scientific Results, Volume 163*. Ocean Drilling Program, College Station, TX, USA, https://doi.org/10.2973/odp.proc.sr.163.1999

Libak, A., Mjelde, R., Keers, H., Faleide, J.I. & Murai, Y. 2012. An integrated geophysical study of Vestbakken Volcanic Province, western Barents Sea continental margin, and adjacent oceanic crust. *Marine Geophysical Research*, **33**, 185–207, https://doi.org/10.1007/s11001-012-9155-3

Lundin, E.R. & Doré, A.G. 2011. Hyperextension, serpentinization, and weakening: a new paradigm for rifted margin compressional deformation. *Geology*, **39**, 347–350, https://doi.org/10.1130/G31499.1

Mackenzie, G.D., Shannon, P.M., Jacob, A.W.B., Morewood, N.C., Makris, J., Gaye, M. & Egloff, F. 2002. The velocity structure of the sediments in the southern Rockall Basin: results from new wide-angle seismic modelling. *Marine and Petroleum Geology*, **19**, 989–1003, https://doi.org/10.1016/s0264-8172(02)00133-2

Makris, J., Papoulia, I. & Ziska, H. 2009. Crustal structure of the Shetland-Faeroe Basin from long offset seismic data. *In*: *Faroe Islands Exploration Conference: Proceedings of the 2nd Conference. Annales Societatis Scientiarum Færoensis, Supplementum*, **50**, 30–42.

Merriman, R.J., Taylor, P.N. & Morton, A.C. 1988. Petrochemistry and isotope geochemistry of early Palaeogene basalts forming the dipping reflector sequence SW of Rockall Plateau, NE Atlantic. *In*: Morton, A.C. & Parson, L.M. (eds) *Early Tertiary Volcanism and the Opening of the NE Atlantic*. Geological Society, London, Special Publications, **39**, 123–134, https://doi.org/10.1144/GSL.SP.1988.039.01.12

Miller, D.J. & Christensen, N.I. 1997. Seismic velocities of lower crustal and upper mantle rocks from the slow-spreading Mid-Atlantic Ridge, south of the Kane Transform zone (MARK). *In*: *Proceedings of the Ocean Drilling Program, Scientific Results, Volume 153*. Ocean Drilling Program, College Station, TX, USA, 437–454, https://doi.org/10.2973/odp.proc.sr.153.043.1997

Mjelde, R., Sellevoll, M.A., Shimamura, H., Iwasaki, T. & Kanazawa, T. 1992. A crustal study off Lofoten, N. Norway, by use of 3-component ocean bottom seismographs. *Tectonophysics*, **212**, 269–288, https://doi.org/10.1016/0040-1951(92)90295-H

Mjelde, R., Sellevoll, M.A., Shimamura, H., Iwasaki, T. & Kanazawa, T. 1993. Crustal structure beneath Lofoten, N. Norway, from vertical incidence and wide-angle seismic data. *Geophysical Journal International*, **114**, 116–126, https://doi.org/10.1111/j.1365-246X.1993.tb01471.x

Mjelde, R., Digranes, P. *et al.* 1998. Crustal structure of the northern part of the Vøring Basin, mid-Norway margin, from wide-angle seismic and gravity data. *Tectonophysics*, **293**, 175–205, https://doi.org/10.1016/s0040-1951(98)00090-0

Mjelde, R., Shimamura, H., Kanazawa, T., Kodaira, S., Raum, T. & Shiobara, H. 2003. Crustal lineaments, distribution of lower crustal intrusives and structural evolution of the Vøring Margin, NE Atlantic; new insight from wide-angle seismic models. *Tectonophysics*, **369**, 199–218, https://doi.org/10.1016/s0040-1951(03)00199-9

Mjelde, R., Raum, T., Breivik, A., Shimamura, H., Murai, Y., Takanami, T. & Faleide, J.I. 2005. Crustal structure of the Vøring Margin, NE Atlantic: a review of geological implications based on recent OBS data. *In*: Doré, A.G. & Vining, B.A. (eds) *Petroleum Geology: North-West Europe and Global Perspectives – Proceedings of the 6th Petroleum Geology Conference*. Geological Society, London, 803–813, https://doi.org/10.1144/0060803

Mjelde, R., Raum, T., Breivik, A. & Faleide, J. 2008. Crustal transect across the North Atlantic. *Marine Geophysical Research*, **29**, 73–87, https://doi.org/10.1007/s11001-008-9046-9

Mjelde, R., Faleide, J.I., Breivik, A.J. & Raum, T. 2009. Lower crustal composition and crustal lineaments on the Vøring Margin, NE Atlantic: a review. *Tectonophysics*, **472**, 183–193, https://doi.org/10.1016/j.tecto.2008.04.018

Morgan, R.P.L., Barton, P.J., Warner, M., Morgan, J., Price, C. & Jones, K. 2000. Lithospheric structure north of Scotland – I. P-wave modelling, deep reflection profiles and gravity. *Geophysical Journal International*, **142**, 716–736, https://doi.org/10.1046/j.1365-246x.2000.00151.x

Mosar, J., Eide, E.A., Osmundsen, P.T., Sommaruga, A. & Torsvik, T.H. 2002*a*. Greenland – Norway separation: a geodynamic model for the North Atlantic. *Norwegian Journal of Geology*, **82**, 281–298.

Mosar, J., Torsvik, T.H. & The BAT Team 2002*b*. Opening of the Norwegian and Greenland Seas: plate tectonics in Mid Norway since the Late Permian. *In*: Eide, E.A. (coord.) *BATLAS – Mid Norway Plate Reconstruction Atlas with Global and Atlantic Perspectives*. Geological Survey of Norway, Trondheim, Norway, 48–59.

Nirrengarten, M., Gernigon, L. & Manatschal, G. 2014. Lower crustal bodies in the Møre volcanic rifted margin: geophysical determination and geological implications. *Tectonophysics*, **636**, 143–157, https://doi.org/10.1016/j.tecto.2014.08.004

Olafsson, I., Sundvor, E., Eldholm, O. & Grue, K. 1992. Møre Margin: crustal structure from analysis of expanded spread profiles. *Marine Geophysical Research*, **14**, 137–162, https://doi.org/10.1007/bf01204284

Olesen, O., Ebbing, J. *et al.* 2007. An improved tectonic model for the Eocene opening of the Norwegian–Greenland Sea: use of modern magnetic data. *Marine and Petroleum Geology*, **24**, 53–66, https://doi.org/10.1016/j.marpetgeo.2006.10.008

Olesen, O., Brönner, M. *et al.* 2010. New aeromagnetic and gravity compilations from Norway and adjacent areas: methods and applications. *In*: Vining, B.A. & Pickering, S.C. (eds) *Petroleum Geology: From Mature Basins to New Frontiers – Proceedings of the 7th Petroleum Geology Conference*. Geological Society, London, 559–586, https://doi.org/10.1144/0070559

O'Reilly, B.M., Hauser, F., Jacob, A.W.B. & Shannon, P.M. 1996. The lithosphere below the Rockall Trough: wide-angle seismic evidence for extensive serpentinisation. *Tectonophysics*, **225**, 1–23, https://doi.org/10.1016/0040-1951(95)00149-2

O'Reilly, B.M., Hauser, F., Ravaut, C., Shannon, P.M. & Readman, P.W. 2006. Crustal thinning, mantle exhumation and serpentinization in the Porcupine Basin, offshore Ireland: evidence from wide-angle seismic data. *Journal of the Geological Society, London*, **163**, 775–787, https://doi.org/10.1144/0016-76492005-079

Osmundsen, P.T. & Andersen, T.B. 2001. The middle Devonian basins of western Norway: sedimentary response to large-scale transtensional tectonics? *Tectonophysics*, **332**, 51–68, https://doi.org/10.1016/S0040-1951(00)00249-3

Parkin, C.J. & White, R.S. 2008. Influence of the Iceland mantle plume on oceanic crust generation in the North Atlantic. *Geophysical Journal International*, **173**, 168–188, https://doi.org/10.1111/j.1365-246X.2007.03689.x

Peron-Pinvidic, G., Gernigon, L., Gaina, C. & Ball, P. 2012. Insights from the Jan Mayen system in the Norwegian–Greenland sea – I. Mapping of a microcontinent. *Geophysical Journal International*, **191**, 385–412, https://doi.org/10.1111/j.1365-246X.2012.05639.x

Peron-Pinvidic, G., Manatschal, G. & Osmundsen, P.T. 2013. Structural comparison of archetypal Atlantic rifted margins: a review of observations and concepts. *Marine and Petroleum Geology*, **43**, 21–47, https://doi.org/10.1016/j.marpetgeo.2013.02.002

Pérez-Gussinyé, M. & Reston, T.J. 2001. Rheological evolution during extension at nonvolcanic rifted margins: onset of serpentinization and development of detachments leading to continental breakup. *Journal of Geophysical Research*, **106**, 3961–3975, https://doi.org/10.1029/2000jb900325

Petersen, U.K. & Funck, T. 2016. Review of velocity models in the Faroe–Shetland Channel. *In*: Péron-Pinvidic, G., Hopper, J.R., Stoker, M.S., Gaina, C., Doornenbal, J.C., Funck, T. & Árting, U.E. (eds) *The NE Atlantic Region: A Reappraisal of Crustal Structure, Tectonostratigraphy and Magmatic Evolution*. Geological Society, London, Special Publications, **447**. First published online September 9, 2016, https://doi.org/10.1144/SP447.7

Powell, C.M.R. & Sinha, M.C. 1987. The PUMA experiment west of Lewis, U.K. *Geophysical Journal of the Royal Astronomical Society*, **89**, 259–264, https://doi.org/10.1111/j.1365-246X.1987.tb04417.x

Quirk, D.G., Shakerley, A. & Howe, M.J. 2014. A mechanism for construction of volcanic rifted margins during continental breakup. *Geology*, **42**, 1079–1082, https://doi.org/10.1130/G35974.1

Raum, T. 2000. *Crustal structure and evolution of the Faeroe, Møre and Vøring margins from wide-angle seismic and gravity data*. PhD thesis, University of Bergen, Norway.

Raum, T., Mjelde, R. *et al.* 2002. Crustal structure of the southern part of the Vøring Basin, mid-Norway margin, from wide-angle seismic and gravity data. *Tectonophysics*, **355**, 99–126, https://doi.org/10.1016/S0040-1951(02)00136-1

Raum, T., Mjelde, R. *et al.* 2005. Sub-basalt structures east of the Faroe Islands revealed from wide-angle seismic and gravity data. *Petroleum Geoscience*, **11**, 291–308, https://doi.org/10.1144/1354-079304-627

Raum, T., Mjelde, R. *et al.* 2006. Crustal structure and evolution of the southern Vøring Basin and Vøring Transform Margin, NE Atlantic. *Tectonophysics*, **415**, 167–202, https://doi.org/10.1016/j.tecto.2005.12.008

Ren, S., Faleide, J.I., Eldholm, O., Skogseid, J. & Gradstein, F. 2003. Late Cretaceous–Paleocene tectonic development of the NW Vøring Basin. *Marine and Petroleum Geology*, **20**, 177–206, https://doi.org/10.1016/S0264-8172(03)00005-9

Reynisson, R.F., Ebbing, J., Lundin, E. & Osmundsen, P.T. 2010. Properties and distribution of lower crustal bodies on the mid-Norwegian margin. *In*: Vining, B.A. & Pickering, S.C. (eds) *Petroleum Geology: From Mature Basins to New Frontiers – Proceedings of the 7th Petroleum Geology Conference*. Geological Society, London, 843–854, https://doi.org/10.1144/0070843

Richardson, K.R., Smallwood, J.R., White, R.S., Snyder, D.B. & Maguire, P.K.H. 1998. Crustal structure beneath the Faroe Islands and the Faroe–Iceland Ridge. *Tectonophysics*, **300**, 159–180, https://doi.org/10.1016/s0040-1951(98)00239-x

Ritzmann, O., Jokat, W., Mjelde, R. & Shimamura, H. 2002. Crustal structure between the Knipovich Ridge and the Van Mijenfjorden (Svalbard). *Marine Geophysical Research*, **23**, 379–401, https://doi.org/10.1023/B:MARI.0000018168.89762.a4

RITZMANN, O., MAERCKLIN, N., FALEIDE, J.I., BUNGUM, H., MOONEY, W.D. & DETWEILER, S.T. 2007. A three-dimensional geophysical model of the crust in the Barents Sea region: model construction and basement characterization. *Geophysical Journal International*, **170**, 417–435, https://doi.org/10.1111/j.1365-246X.2007.03337.x

ROBERTS, A.W., WHITE, R.S. & CHRISTIE, P.A.F. 2009. Imaging igneous rocks on the North Atlantic rifted continental margin. *Geophysical Journal International*, **179**, 1024–1038, https://doi.org/10.1111/j.1365-246X.2009.04306.x

ROBERTS, D.G., GINZBERG, A., NUNN, K. & MCQUILLIN, R. 1988. The structure of the Rockall Trough from seismic refraction and wide-angle reflection measurements. *Nature*, **332**, 632–635, https://doi.org/10.1038/332632a0

ROUZO, S., KLINGELHÖFER, F. ET AL. 2006. 2-D and 3-D modelling of wide-angle seismic data: an example from the Vøring volcanic passive margin. *Marine Geophysical Researches*, **27**, 181–199, https://doi.org/10.1007/s11001-006-0001-3

RYSETH, A., AUGUSTSON, J.H. ET AL. 2003. Cenozoic stratigraphy and evolution of the Sørvestsnaget Basin, southwestern Barents Sea. *Norwegian Journal of Geology*, **83**, 107–130.

SAUNDERS, A.D., FITTON, J.G., KERR, A.C., NORRY, M.J. & KENT, R.W. 1997. The North Atlantic Igneous Province. *In*: MAHONEY, J.J. & COFFIN, M.F. (eds) *Large Igneous Provinces: Continental, Oceanic, and Planetary Flood Volcanism*. American Geophysical Union, Geophysical Monograph Series, **100**, 45–93.

SAUNDERS, A.D., LARSEN, H.C. & WISE, S.W., JR. (eds) 1998. *Proceedings of the Ocean Drilling Program, Scientific Results, Volume 152*. Ocean Drilling Program, College Station, TX, USA, https://doi.org/10.2973/odp.proc.sr.152.1998

SCHLINDWEIN, V. & JOKAT, W. 1999. Structure and evolution of the continental crust of northern east Greenland from integrated geophysical studies. *Journal of Geophysical Research*, **104**, 15,227–15,245, https://doi.org/10.1029/1999JB900101

SERANNE, M. & SEGURET, M. 1987. The Devonian basins of western Norway: tectonics and kinematics of an extending crust. *In*: COWARD, M.P., DEWEY, J.F. & HANCOCK, P.L. (eds) *Continental Extensional Tectonics*. Geological Society, London, Special Publications, **28**, 537–548, https://doi.org/10.1144/GSL.SP.1987.028.01.35

SHANNON, P.M., JACOB, A.W.B., O'REILLY, B.M., HAUSER, F., READMAN, P.W. & MAKRIS, J. 1999. Structural setting, geological development and basin modelling in the Rockall Trough. *In*: FLEET, A.J. & BOLDY, S.A.R. (eds) *Petroleum Geology of Northwest Europe: Proceedings of the 5th Conference*. Geological Society, London, 421-431, https://doi.org/10.1144/0050421

STAPLES, R.K., WHITE, R.S., BRANDSDÓTTIR, B., MENKE, W., MAGUIRE, P.K.H. & MCBRIDE, J.H. 1997. Färoe–Iceland Ridge Experiment 1. Crustal structure of northeastern Iceland. *Journal of Geophysical Research*, **102**, 7849–7866, https://doi.org/10.1029/96jb03911

STOREY, M., DUNCAN, R.A. & TEGNER, C. 2007. Timing and duration of volcanism in the North Atlantic Igneous Province: implications for geodynamics and links to the Iceland hotspot. *Chemical Geology*, **241**, 264–281, https://doi.org/10.1016/j.chemgeo.2007.01.016

SURLYK, F. 1990. Timing, style and sedimentary evolution of Late Palaeozoic-Mesozoic extensional basins of East Greenland. *In*: HARDMAN, R.F.P. & BROOKS, J. (eds) *Tectonic Events Responsible for Britain's Oil and Gas Reserves*. Geological Society, London, Special Publications, **55**, 107–125, https://doi.org/10.1144/GSL.SP.1990.055.01.05

SVELLINGEN, W. & PEDERSEN, R.B. 2003. Jan Mayen: a result of ridge–transform–microcontinent interaction. *Geophysical Research Abstracts*, **5**, 12993.

TORSVIK, T.H., AMUNDSEN, H.E.F. ET AL. 2015. Continental crust beneath southeast Iceland. *Proceedings of the National Academy of Sciences*, **112**, E1818–E1827, https://doi.org/10.1073/pnas.1423099112

TSIKALAS, F., FALEIDE, J.I., ELDHOLM, O. & WILSON, J. 2005. Late Mesozoic–Cenozoic structural and stratigraphic correlations between the conjugate mid-Norway and NE Greenland continental margins. *In*: DORÉ, A.G. & VINING, B.A. (eds) *Petroleum Geology: North-West Europe and Global Perspectives – Proceedings of the 6th Petroleum Geology Conference*. Geological Society, London, 785–801, https://doi.org/10.1144/0060785

TUCHOLKE, B.E., SAWYER, D.S. & SIBUET, J.-C. 2007. Breakup of the Newfoundland–Iberia rift. *In*: KARNER, G.D., MANATSCHAL, G. & PINHEIRO, L.M. (eds) *Imaging, Mapping and Modelling Continental Lithosphere Extension and Breakup*. Geological Society, London, Special Publications, **282**, 9–46, https://doi.org/10.1144/sp282.2

VOGT, U., MAKRIS, J., O'REILLY, B.M., HAUSER, F., READMAN, P.W., JACOB, A.W.B. & SHANNON, P.M. 1998. The Hatton Basin and continental margin: crustal structure from wide-angle seismic and gravity data. *Journal of Geophysical Research*, **103**, 12,545–12,566, https://doi.org/10.1029/98jb00604

VOSS, M. & JOKAT, W. 2007. Continent–ocean transition and voluminous magmatic underplating derived from P-wave velocity modelling of the East Greenland continental margin. *Geophysical Journal International*, **170**, 580–604, https://doi.org/10.1111/j.1365-246X.2007.03438.x

VOSS, M. & JOKAT, W. 2009. From Devonian extensional collapse to early Eocene continental break-up: an extended transect of the Kejser Franz Joseph Fjord of the East Greenland margin. *Geophysical Journal International*, **177**, 743–754, https://doi.org/10.1111/j.1365-246X.2008.04076.x

VOSS, M., SCHMIDT-AURSCH, M.C. & JOKAT, W. 2009. Variations in magmatic processes along the East Greenland volcanic margin. *Geophysical Journal International*, **177**, 755–782, https://doi.org/10.1111/j.1365-246X.2009.04077.x

WEIGEL, W., FLÜH, E.R. ET AL. 1995. Investigations of the East Greenland continental margin between 70° and 72° N by deep seismic sounding and gravity studies. *Marine Geophysical Research*, **17**, 167–199, https://doi.org/10.1007/bf01203425

White, R.S. 1997. Rift–plume interaction in the North Atlantic. *Philosophical Transactions of the Royal Society of London, Series A*, **355**, 319–339, https://doi.org/10.1098/rsta.1997.0011

White, R.S. & McKenzie, D. 1995. Mantle plumes and flood basalts. *Journal of Geophysical Research*, **100**, 17543–17585, https://doi.org/10.1029/95jb01585

White, R.S. & Smith, L.K. 2009. Crustal structure of the Hatton and the conjugate east Greenland rifted volcanic continental margins, NE Atlantic. *Journal of Geophysical Research*, **114**, B02305, https://doi.org/10.1029/2008jb005856

White, R.S., McKenzie, D. & O'Nions, R.K. 1992. Oceanic crustal thickness from seismic measurements and rare earth element inversions. *Journal of Geophysical Research*, **97**, 19683–19715, https://doi.org/10.1029/92jb01749

White, R.S., Smith, L.K., Roberts, A.W., Christie, P.A.F. & Kusznir, N.J. 2008. Lower-crustal intrusion on the North Atlantic continental margin. *Nature*, **452**, 460–464, https://doi.org/10.1038/nature06687

Ziegler, P. (ed.) 1988. *Evolution of the Arctic–North Atlantic and the Western Thetys*. American Association of Petroleum Geologists, Memoirs, **43**.

Moho and basement depth in the NE Atlantic Ocean based on seismic refraction data and receiver functions

THOMAS FUNCK[1]*, WOLFRAM H. GEISSLER[2], GEOFFREY S. KIMBELL[3], SOFIE GRADMANN[4], ÖGMUNDUR ERLENDSSON[5], KENNETH McDERMOTT[6] & UNI K. PETERSEN[7]

[1]*Geological Survey of Denmark and Greenland, Øster Voldgade 10, 1350 Copenhagen K, Denmark*

[2]*Alfred Wegener Institute Helmholtz Centre for Polar and Marine Research, Am Alten Hafen 26, 27568 Bremerhaven, Germany*

[3]*British Geological Survey, Keyworth, Nottingham NG12 5GG, UK*

[4]*Geological Survey of Norway, Leiv Eirikssons vei 39, 7040 Trondheim, Norway*

[5]*Iceland Geosurvey, Grensásvegi 9, 108 Reykjavík, Iceland*

[6]*UCD School of Geological Sciences, University College Dublin, Belfield, Dublin 4, Ireland*

[7]*Faroese Earth and Energy Directorate, Brekkutún 1, 110 Tórshavn, Faroe Islands*

**Correspondence: tf@geus.dk*

Abstract: Seismic refraction data and results from receiver functions were used to compile the depth to the basement and Moho in the NE Atlantic Ocean. For interpolation between the unevenly spaced data points, the kriging technique was used. Free-air gravity data were used as constraints in the kriging process for the basement. That way, structures with little or no seismic coverage are still presented on the basement map, in particular the basins off East Greenland. The rift basins off NW Europe are mapped as a continuous zone with basement depths of between 5 and 15 km. Maximum basement depths off NE Greenland are 8 km, but these are probably underestimated. Plate reconstructions for Chron C24 (*c.* 54 Ma) suggest that the poorly known Ammassalik Basin off SE Greenland may correlate with the northern termination of the Hatton Basin at the conjugate margin. The most prominent feature on the Moho map is the Greenland–Iceland–Faroe Ridge, with Moho depths >28 km. Crustal thickness is compiled from the Moho and basement depths. The oceanic crust displays an increased thickness close to the volcanic margins affected by the Iceland plume.

The rifting in the NE Atlantic Ocean region (Fig. 1) occurred in several episodes spanning from the Carboniferous Period to the early Cenozoic break-up (e.g. Ziegler 1988; Doré *et al.* 1999). Seafloor spreading propagated from the Central Atlantic Ocean northwards into the NE Atlantic Ocean (Srivastava & Tapscott 1986), where spreading between Greenland and NW Europe began in the Early Eocene (Storey *et al.* 2007). Large sections of the continental margins fringing the NE Atlantic Ocean are magma-rich margins with extensive flood-basalt volcanism and igneous intrusions (e.g. Eldholm & Grue 1994; Holbrook *et al.* 2001). The amount of break-up-related magmatism depends on the distance to the Iceland plume (Holbrook *et al.* 2001). Furthermore, the plume has profoundly influenced the creation of oceanic crust throughout the NE Atlantic region (Howell *et al.* 2014). The influence of the plume appears to extend further to the south along the Reykjanes Ridge than to the north along the Kolbeinsey Ridge (Hooft *et al.* 2006). The most outstanding feature on the bathymetric map is the Greenland–Iceland–Faroe Ridge (composed of the Greenland–Iceland Ridge, Iceland and the Iceland–Faroe Ridge: Fig. 1), the result of enhanced melting in the Iceland plume (White & McKenzie 1995; Staples *et al.* 1997). Most of the NE Atlantic Ocean has shallower bathymetry and hotter mantle when compared to other oceans (Parkin & White 2008; Artemieva & Thybo 2013).

The complex rifting and spreading history, as well as the interaction with the Iceland plume,

From: Péron-Pinvidic, G., Hopper, J. R., Stoker, M. S., Gaina, C., Doornenbal, J. C., Funck, T. & Árting, U. E. (eds) 2017. *The NE Atlantic Region: A Reappraisal of Crustal Structure, Tectonostratigraphy and Magmatic Evolution*. Geological Society, London, Special Publications, **447**, 207–231.
First published online July 13, 2016, https://doi.org/10.1144/SP447.1

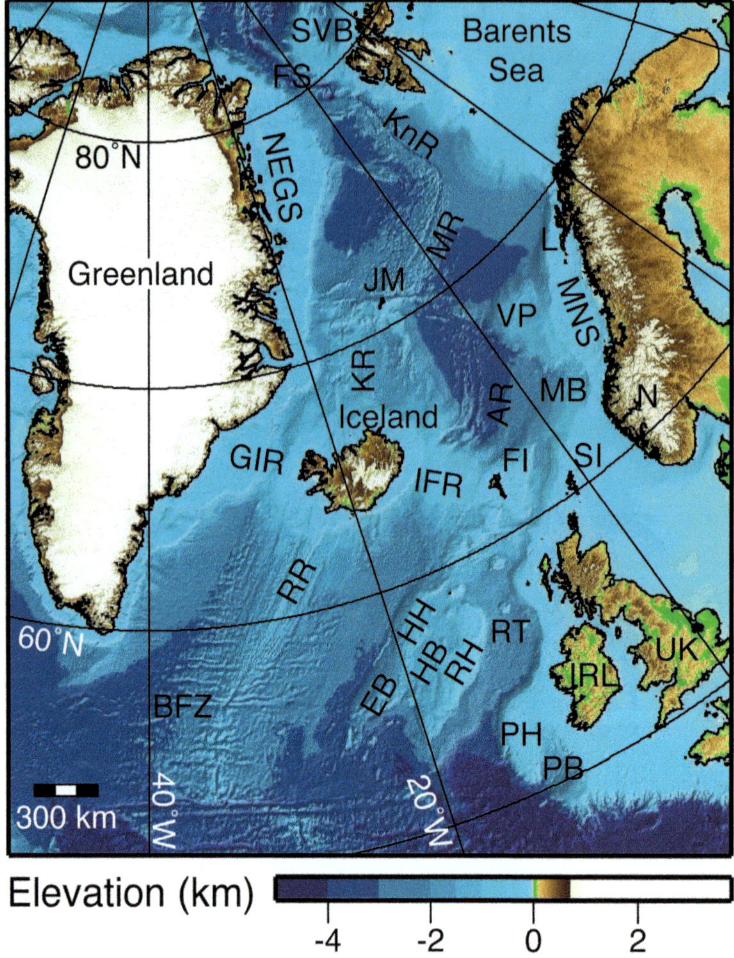

Fig. 1. Physiographical map of the NE Atlantic Ocean using the ETOPO1 Global Relief Model (Amante & Eakins 2009). Abbreviations: BFZ, Bight Fracture Zone; EB, Edoras Bank; FI, Faroe Islands; FS, Fram Strait; GIR, Greenland–Iceland Ridge; HB, Hatton Basin; HH, Hatton High; IFR, Iceland–Faroe Ridge; JM, Jan Mayen; KnR, Knipovich Ridge; KR, Kolbeinsey Ridge; L, Lofoten; MB, Møre Basin; MNS, Mid-Norway Shelf; MR, Mohns Ridge; N, Norway; NEGS, NE Greenland Shelf; PB, Porcupine Basin; PH, Porcupine High; RH, Rockall High; RR, Reykjanes Ridge; RT, Rockall Trough; SI, Shetland Islands; SVB, Svalbard; VP, Vøring Plateau.

have shaped the present configuration of the crust in the NE Atlantic realm. The first-order structure of the crust can be recognized by regional mapping of the Moho and basement depth. A number of maps already exist for the region, such as the global model CRUST1.0 of Laske *et al.* (2013) or the model for the North Atlantic region by Artemieva & Thybo (2013). In addition, large parts of the region are included in compilations that cover the European plate (Grad *et al.* 2009; Molinari & Morelli 2011). In contrast to these existing maps for the NE Atlantic Ocean, the new compilation is exclusively compiled from seismic refraction data supplemented by receiver functions along the bordering land areas. Most seismic refraction lines are acquired along pre-existing multichannel seismic (MCS) data. The interpretation of the MCS data down to the basement is generally used as the starting point for velocity modelling of the seismic refraction lines. Deep crustal seismic reflection data were not considered for two reasons. First, even though the reflection and refraction Moho generally correlate well, there can be deviations (cf. Mooney & Brocher 1987). Second, seismic reflection data require a conversion from time to depth, which is difficult to do when no velocity information is available. To allow for the best possible quality control and internal consistency, existing local compilations were not incorporated into our database. Examples of such compilations are the seismically constrained gravity inversion for Iceland (Kaban *et al.* 2002) or the model for the Barents Sea (Ritzmann *et al.* 2007) that is based on both seismic reflection and refraction data, as well as on gravity modelling.

By incorporating so far unpublished seismic refraction data, the compilation has an unparalleled data density (Fig. 2), even though there are still substantial data gaps. These unpublished datasets comprise lines that have only been presented at conferences or in internal reports. Knowledge of the basement and Moho depth allows the calculation of the crustal thickness from which stretching factors can be estimated (cf. Kimbell *et al.*, this volume,

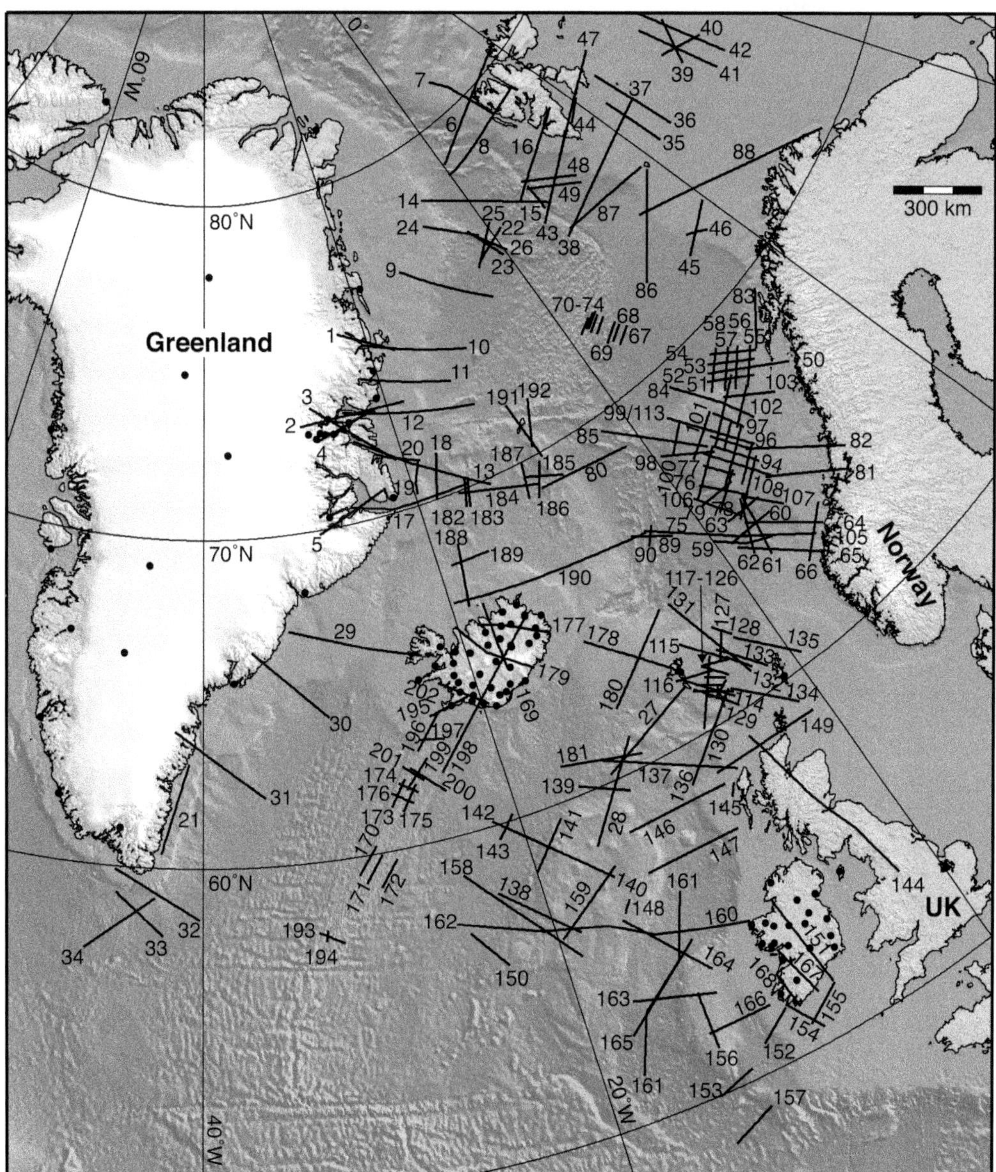

Fig. 2. Data coverage. Solid lines show the location of seismic refraction lines. The labels refer to Table 1 where line names and references are given. Circles mark the positions of receiver functions (see Table 2 for references). The background shows a shaded relief map (ETOPO1: Amante & Eakins 2009).

Table 1. *Seismic refraction studies used in the compilation of data*

Label	Line name	References and comments
1–5	AWI 94300, AWI 94320, AWI 94340, AWI 94360 and AWI 94400	Mandler & Jokat (1998), Schlindwein & Jokat (1999), Schmidt-Aursch & Jokat (2005)
6–8	AWI 99200, AWI 99300 and AWI 99400	Ritzmann & Jokat (2003), Czuba *et al.* (2004, 2005), Ritzmann *et al.* (2004)
9–12	AWI 20030200, AWI 20030300, AWI 20030400 and AWI 20030500	Voss & Jokat (2007, 2009), Voss *et al.* (2009)
13–15	AWI 20090100, AWI 20090200 and AWI 20090250	Jokat (2010), Jokat *et al.* (2012), Hermann & Jokat (2013)
16	AWI 97260 and Barents 98 line 9	Ritzmann *et al.* (2002)
17–20	ARK 1988 lines 3–6	Weigel *et al.* (1995)
21	DLC 94 line 5	Dahl-Jensen *et al.* (1998)
22–24	EAGER 2011 lines 1–3	Gerlings *et al.* (2014), Funck *et al.* (2015)
25–26	GEUS 2002 lines A and B	Døssing *et al.* (2008), Døssing & Funck (2012)
27–28	LOS 2004 lines A and B	Funck *et al.* (2008)
29–32	SIGMA lines 1–4	Korenaga *et al.* (2000), Holbrook *et al.* (2001), Hopper *et al.* (2003), Reiche *et al.* (2011)
33–34	SIGNAL lines 2 and 3	Funck *et al.* (2012)
35–46	Barents 98 lines 1, 2, 3E, 3W, 4–8, 10, A and B	Breivik *et al.* (2002, 2003, 2005), Mjelde *et al.* (2002*b*), Ljones *et al.* (2004)
47	Horsted '05	Czuba *et al.* (2008)
48–49	Knipovich 02 lines 1 and 2	Kandilarov *et al.* (2008, 2010)
50–58	Lofoten 88 lines 1–9	Mjelde (1992), Mjelde *et al.* (1992, 1993, 1995), Mjelde & Sellevoll (1993)
59–63	Møre 99 lines 1–5	Mjelde *et al.* (2009)
64–66	Møre 2009 lines 1–3	Kvarven *et al.* (2014)
67–74	Mohns Ridge 88 lines 2–4 and 6–10	Klingelhöfer *et al.* (2000)
75–80	OBS 2000 lines 1–3, 5, 6 and 8	Breivik *et al.* (2006), Raum *et al.* (2006), Rouzo *et al.* (2006), Mjelde *et al.* (2008)
81–85	OBS 2003 lines 3, 4, 8, 10 and 11	Breivik *et al.* (2008, 2009, 2011) for line 8: Mjelde (pers. comm.)
86–87	OBS 2008 lines 1 and 2	Czuba *et al.* (2011), Libak *et al.* (2012*a*, *b*)
88	PETROBAR 07	Clark *et al.* (2013)
89–90	Valdivia 59–87 lines IV and V	Grevemeyer *et al.* (1997)
91–97	Vøring 92 lines 1–7	Mjelde *et al.* (1997*a*, *b*, 2003), Digranes *et al.* (1998). Note: lines 1, 2, 3 and 5 are not labelled on the map (Fig. 2)
98–112	Vøring 96 lines 1–7, 8A, 8B, 9–14	Mjelde *et al.* (1998, 2001, 2003), Raum (2000), Berndt *et al.* (2001), Raum *et al.* (2002, 2006). Note: lines 7 and 11–14 are not labelled on the map (Fig. 2)
113	Vøring 99 line AB	Mjelde *et al.* (2005, 2007). Note: the line is not labelled on the map (Fig. 2)
114–115	AMG 95 lines 1 and 2	Raum *et al.* (2005)
116	FAST	Richardson *et al.* (1999), Smallwood *et al.* (2001)
117–128	FLARE lines 1–12	Richardson *et al.* (1999), White *et al.* (1999), Fliedner & White (2003), internal reports (Faroese Earth and Energy Directorate)
129–130	Foerbas lines 1 and 2	Internal reports (Faroese Earth and Energy Directorate)
131	iSIMM Faroes	Eccles *et al.* (2007), Roberts *et al.* (2009)
132–133	Mobil lines 1 and 2	Hughes *et al.* (1998), Makris *et al.* (2009)
134–138	AMP lines A, C–E and L	Klingelhöfer *et al.* (2005), Kelly *et al.* (2007), England (pers. comm.)
139	BANS-1	Klingelhöfer *et al.* (2005)
140–143	iSIMM Hatton dip line/strike line/dip line (W)/western strike line	Smith *et al.* (2005), Parkin & White (2008), White & Smith (2009)
144	LISPB	Bamford *et al.* (1977, 1978), Barton (1992)
145	PUMA	Powell & Sinha (1987)

(*Continued*)

Table 1. *Continued*

Label	Line name	References and comments
146–147	BP/Britoil lines 86-002 and 86-005	Roberts *et al.* (1988)
148	Hatton Bank line A	Scrutton (1970, 1972), Bunch (1979)
149	W-reflector profile	Warner *et al.* (1996), Morgan *et al.* (2000), Price & Morgan (2000)
150	CAM 77	Barton & White (1997)
151–155	COOLE lines 1, 3A, 3B, 6 and 7	Makris *et al.* (1988), Lowe & Jacob (1989), O'Reilly *et al.* (1991)
156	Rockall Bank Profile A	Bunch (1979)
157	Goban Spur	Bullock & Minshull (2005), Minshull (pers. comm.)
158–159	HADES combined lines 1 and 2, and line 3	Morewood *et al.* (2005), Ravaut *et al.* (2005), Chabert *et al.* (2006), McDermott (pers. comm.)
160–166	RAPIDS lines 1, 2, 2-1, 32, 33, 34 and 4	Makris *et al.* (1991), Hauser *et al.* (1995), O'Reilly *et al.* (1996), Vogt *et al.* (1998), Shannon *et al.* (1999), Mackenzie *et al.* (2002), Morewood *et al.* (2004, 2005)
167–168	VARNET lines A and B	Masson *et al.* (1998), Landes *et al.* (2000), Hauser *et al.* (2008), O'Reilly *et al.* (2010)
169	B96	Menke *et al.* (1998)
170–172	BK 80 lines X, Y and Z	Bunch & Kennett (1980)
173–176	CAM 71–74	Smallwood *et al.* (1995), Smallwood & White (1998)
177–178	FIRE Land and Offshore	White *et al.* (1996), Staples *et al.* (1997), Richardson *et al.* (1998)
179	ICEMELT	Darbyshire *et al.* (1998, 2000)
180	IFR	Sedov & Makris (2001), Bohnhoff & Makris (2004)
181	IS 2004	Erlendsson & Blischke (2013), Gunnarsson (pers. comm.)
182–187	JMKR-95 lines 1–6	Kodaira *et al.* (1997, 1998*a*, *b*), Mjelde *et al.* (2002*a*)
188–190	KRISE lines 1, 4 and 7	Hooft *et al.* (2006), Furmall (2010), Brandsdóttir *et al.* (2015)
191–192	OBS-JM-2006 lines 1 and 2	Kandilarov *et al.* (2012)
193–194	RAMESSES lines 1 and 2	Navin *et al.* (1998), Sinha *et al.* (1998)
195–197	RISE lines A, B and D	Weir *et al.* (2001)
198–201	RRISP-77 lines 1 and 3–5	Angenheister *et al.* (1980), Gebrande *et al.* (1980), Goldflam *et al.* (1980), Jacoby *et al.* (2007)
202	SIST	Bjarnason *et al.* (1993)

in press). Crustal thickness is also an important parameter for deformable plate reconstructions and basin modelling.

Dataset

The core study area for the mapping of the Moho and basement depth comprises the NE Atlantic Ocean, extending from the Bight Fracture Zone and the southern limit of the Edoras Bank, Rockall and Porcupine highs in the south, to the Fram Strait and western Barents Sea in the north (Fig. 1). While the focus was on the offshore region, Iceland was an integral part of this study.

The primary database behind the compilation consists of velocity–depth models derived from seismic refraction data. Figure 2 shows the location of all lines that were used in this study, and Table 1 provides the line names and references. The 202 lines under consideration were acquired between 1969 and 2011, but only 10 of them prior to 1980. There are numerous other seismic refraction lines in the study area that were not included. The reason for dismissal was mainly associated with age, which frequently was associated with a limited resolution of the velocity models and often the published information would not allow for a proper quality assessment. The majority of lines that were considered in this study were experiments that used ocean-bottom seismometers (either equipped with geophones or hydrophones or both) and seismic land stations as receivers, and airgun arrays or explosives as the source. Occasionally, some additional sonobuoys were used to receive the seismic signals. Only two lines incorporated expanded spread profiles (ESP)

Table 2. *Receiver function studies used in the compilation of data*

Region	References
Greenland	Gregersen *et al.* (1988), Dahl-Jensen *et al.* (2003), Kumar *et al.* (2007), Schiffer *et al.* (2014)
Iceland	Schlindwein (2006), Kumar *et al.* (2007)
Faroe Islands	Harland *et al.* (2009)
Ireland and UK	Tomlinson *et al.* (2006), Licciardi *et al.* (2014)

and 12 lines from the Faeroes Large Aperture Research Experiment (FLARE) (e.g. Fliedner & White 2003) used long-offset seismic lines by towing two multichannel streamers with maximum offsets of 38 km.

The secondary dataset used in this study consisted of receiver functions analysis. This method was based on the observation of teleseismic events at either permanent or temporary deployed stations. Receiver function analysis provides information on the depth of the Moho beneath the station and can fill in some gaps in the primary dataset. Regions of particular interest were Iceland, Greenland and Ireland. Table 2 provides information on the receiver function studies used in the compilation, while the station location is shown in Figure 2.

Compilation

The database consisting of the velocity models from the seismic refraction lines and the Moho depths from the receiver function studies was used to compile both the depth to the basement and the depth to the Moho in the NE Atlantic Ocean. The Moho is generally characterized by an increase in P-wave velocity to values greater than 7.6 km s^{-1} (White *et al.* 1992). This increase is often associated with a prominent wide-angle reflection (PmP). However, serpentinization processes in the mantle rock can reduce the seismic velocity to values as low as 4.8 km s^{-1} (Christensen 2004). In these cases, the Moho depth is measured at the top of the (partially) serpentinized mantle.

For the definition of basement, the top of the igneous crust is used in the oceanic domain. Landwards of the continent–ocean boundary, basement is measured at the top of the crystalline crust. That way, any volcanic rocks located above the continental crystalline crust are considered as part of the sedimentary column. Off NE Greenland, the top of the crystalline crust could not be resolved on some of the lines owing to the presence of consolidated Palaeozoic and Mesozoic sedimentary rocks with velocities exceeding 5 km s^{-1} and a lack of a clear seismic discontinuity at the basement. These lines include AWI 20030200 (Voss *et al.* 2009), AWI 20030300 (Voss *et al.* 2009), AWI 20030400 (Voss & Jokat 2007) and AWI 20030500 (Voss & Jokat 2007, 2009), and here the 5.7 km s^{-1} velocity contour was used as a proxy for the basement. This value differs from the commonly assumed 6.0 km s^{-1} velocity contour as the base of sediments, as the existing data indicate that the crystalline Caledonian basement both in East Greenland and in northern Svalbard can display P-wave velocities down to 5.5 km s^{-1} (Schmidt-Aursch & Jokat 2005; Czuba *et al.* 2005). Therefore, we used the 5.7 km s^{-1} contour as a compromise to avoid an overestimation of the sedimentary thickness.

The original digital velocity models and navigation data were obtained for a majority of the seismic refraction lines, thus avoiding possible inaccuracies that result from digitizing figures from papers. For the compilation of the basement map, only the portions of the lines with seismic control on the basement depth were used. In the next step, the misfit at the cross-points was investigated. In the case of misfits, the original velocity models and the underlying documentation were revisited to check the basement constraints. Data in the vicinity of the cross-point was then excluded along the line with the poorer constraints.

In the case of Iceland, the basement is mostly equivalent to the topography, with exception of areas that are covered by glaciers and thin sediments (mostly beach sands). The existing seismic refraction lines on Iceland often use simplified versions of the topography as the large shot and receiver spacing makes it unnecessary to take the small-scale topographical variations into account. In addition, there are a number of shot and receiver positions that are projected onto the seismic lines, which introduces erroneous elevations. For this reason, the basement points in Iceland and the adjacent coastal zone (see the extent in Fig. 3) were removed from the dataset and replaced with the ETOPO1 Global Relief Model (Amante & Eakins 2009). Regions with glaciers and beach sands were not assigned a basement depth.

All data points were converted to a Lambert conformal projection with 40°W as the central meridian, and 55°N and 75°N as standard parallels. Along the seismic lines, the basement depth was sampled at a spacing of 1 km. Together with the basement points around Iceland, a basement map was compiled employing the universal kriging technique (Stein 1999). Prior to this, a block mean with a point spacing of 10 km was applied to maintain lateral resolution in areas with dense line spacing, such as along the Norwegian margin. Universal kriging within the SAGA GIS tool (Böhner & Antonić

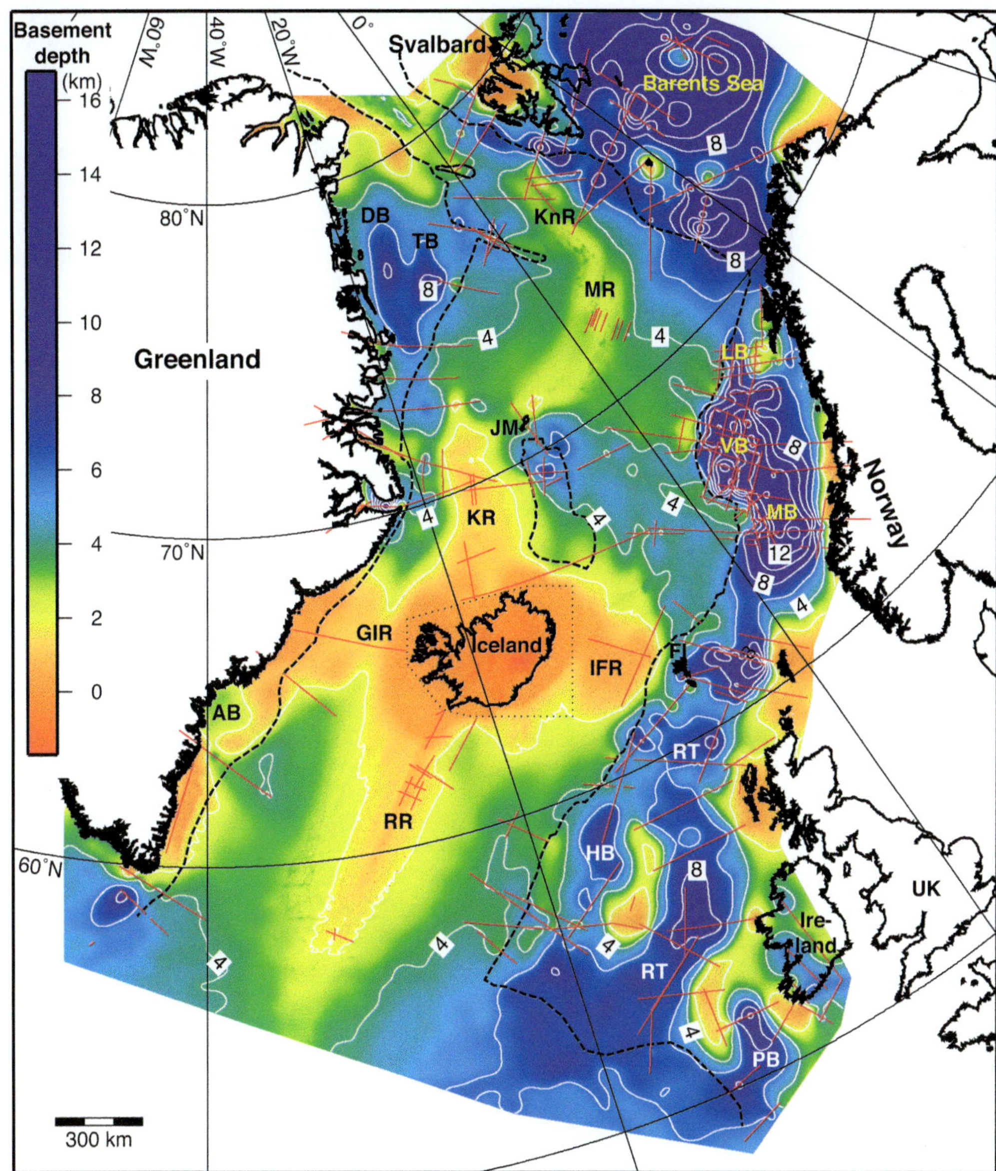

Fig. 3. Basement depth (below sea level). White lines show the contours with a 2 km interval. Seismic constraints are shown in red. Within the dotted line around Iceland, the basement is approximated by the topography. The dashed line indicates the continent–ocean boundary (Funck *et al.* 2014). Abbreviations: AB, Ammassalik Basin; DB, Danmarkshavn Basin; FI, Faroe Islands; GIR, Greenland–Iceland Ridge; HB, Hatton Basin; IFR, Iceland–Faroe Ridge; JM, Jan Mayen; KnR, Knipovich Ridge; KR, Kolbeinsey Ridge; LB, Lofoten Basin; MB, Møre Basin; MR, Mohns Ridge; PB, Porcupine Basin; RR, Reykjanes Ridge; RT, Rockall Trough; TB, Thetis Basin; UK, United Kingdom; VB, Vøring Basin.

2008; Olaya & Conrad 2008) was applied using free-air gravity as a constraint for the interpolation and extrapolation. The gravity data were obtained from the DTU10 grid (Andersen *et al.* 2010; see also fig. 1b in Haase *et al.*, this volume, in review). Owing to the lack of seismic constraints, the onshore areas – with the exception of Iceland, western Ireland, Svalbard, the Faroe Islands and some other smaller islands – were clipped from the resulting basement map shown in Figure 3.

For the compilation of the Moho depth, velocity models along the seismic refraction lines were reviewed to reject portions of the profiles on which the Moho is not constrained by Moho reflections (PmP) or mantle refractions (Pn). Not all publications display the ray coverage to assess the model constraints. In these cases, the outer portions of the models were excluded. The consistency checks at the cross-points of the lines were performed the same way as for the basement depth. Two lines were completely removed from the Moho dataset. The first line is the IFR profile (Bohnhoff & Makris 2004) across the Iceland–Faroe Ridge, which has a substantially lower Moho depth (23 km) than the FIRE offshore line (Richardson *et al.* 1998) along the ridge (>30 km) (Fig. 4). Given that the SIGMA line 1 (Holbrook *et al.* 2001) on the conjugate Greenland–Iceland Ridge displays Moho depths greater than 30 km, the IFR profile was eliminated from the Moho dataset.

The second dataset to be dismissed was DLC 94 line 5 (Dahl-Jensen *et al.* 1998) along the SE coast of Greenland. Along this line, the Moho depth varies between 39 and 52 km, which is substantially more than the maximum Moho depth of 33 km on the two nearby lines 3 (Hopper *et al.* 2003) and 4 (Holbrook *et al.* 2001) of the SIGMA experiment. DLC 94 line 5 has a difficult geometry, with a crooked shot line along the coast and only three receivers onshore. Hence, there might be out-of-plane phases mimicking a Moho. Alternatively, the deep Moho could be an intra-mantle reflection similar to the one observed off West Greenland (Gerlings *et al.* 2009) in an area that was affected by the Iceland plume. The depth at the top of the high-velocity zone (7.5 km s^{-1}) on DLC 94 line 5 varies between 26 and 34 km, and would, in fact, be in reasonable agreement with the interpreted Moho on SIGMA lines 3 and 4.

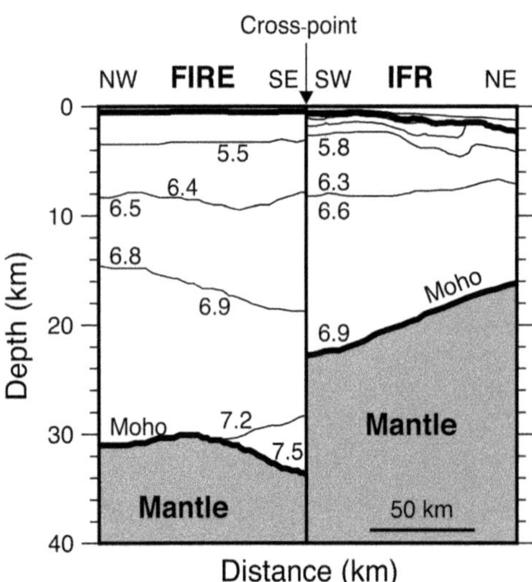

Fig. 4. Comparison of P-wave velocity models of the intersecting seismic refraction lines FIRE Offshore (Richardson *et al.* 1998) and IFR (Bohnhoff & Makris 2004) on the Iceland–Faroe Ridge. Velocities are specified in km s^{-1}.

Moho depths obtained from receiver functions were reviewed critically before they were added to the database. Of particular concern were the measurements on Iceland, as Schlindwein (2006) pointed out that the P–S converted phases there are only weak, which is why the use of the receiver function technique may be limited. For this reason, the 53 available measurements by Kumar *et al.* (2007) were cross-checked with the seismic refraction data and the seismically controlled gravity inversion of Kaban *et al.* (2002). In this process, 14 receiver functions were rejected, as they were not consistent with the other methods. Most of the excluded stations are located close to the coast.

West of the main mapping area, the cleaned dataset of Moho points was supplemented with the global crustal model of Laske *et al.* (2013) at a resolution of 1° (Fig. 5). In the east, the dataset was extended with the European Moho map of Grad *et al.* (2009) at a resolution of 0.1° (Fig. 5). The available computational power required a resampling of the European Moho on a 50 km raster in the Lambert projection described above. To equalize the data distribution, a block average filter was applied to the entire dataset using a 10 km raster. The Moho map (Fig. 6) was then interpolated using the SAGA GIS tool Global Ordinary Kriging. The variance in the logarithmic form was used as a quality measure and the inverse distance was applied as an interpolation method. Finally, the crustal thickness (Fig. 7) was calculated from the difference between the Moho and basement depth.

Results

The basement, Moho and crustal thickness maps compiled from the seismic data are shown in Figures 3, 6 and 7, respectively. A brief description of the main observations is given in this section, while some features of the maps are discussed in more detail in the following section.

On the basement map (Fig. 3), the Greenland–Iceland–Faroe Ridge stands out as a zone of elevated basement. On the Greenland–Iceland Ridge, the maximum basement depth along SIGMA line 1 is 1.6 km below sea level (Holbrook *et al.* 2001). In Iceland, the basement is above sea level and the adjacent mid-oceanic spreading ridges (the

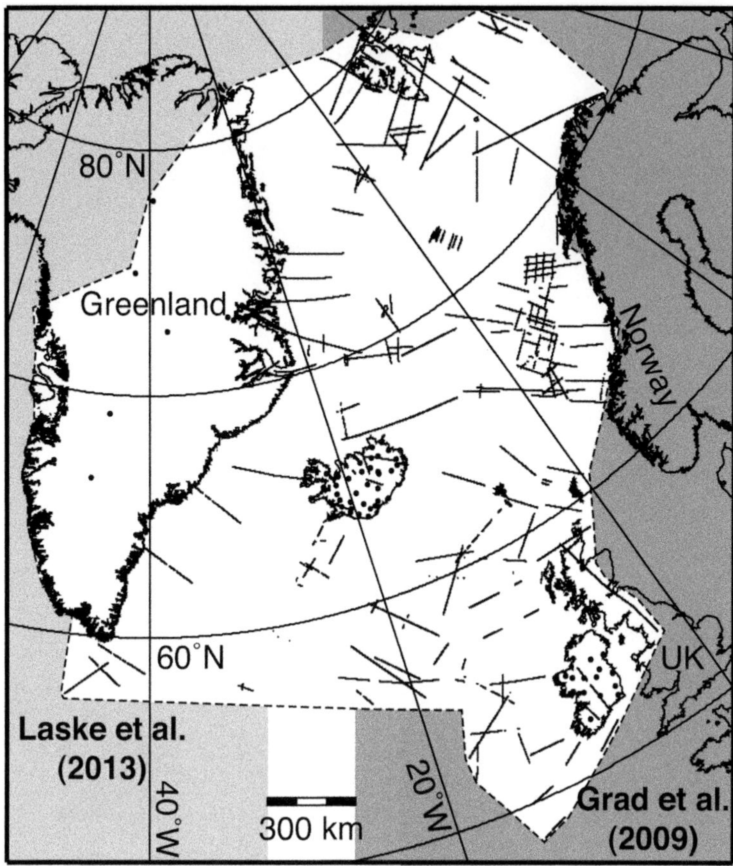

Fig. 5. Datasets used for the compilation of the Moho depth. Circles and lines show the receiver functions and seismic refraction lines, respectively.

Reykjanes and Kolbeinsey ridges) are also characterized by elevated basement. However, the basement depth on these ridges increases with increasing distance from Iceland.

Another prominent feature on the basement map (Fig. 3) is the continuous basin along the NW European margin that extends from the southern Rockall Trough to the Lofoten Basin. The maximum basement depth in this zone is 15 km in the Møre and Vøring basins off mid-Norway. The SW Barents Shelf is also characterized by a deep basement, often exceeding 8 km in depth and up to a maximum depth of 17 km. The sedimentary basins off East Greenland are not well covered by seismic refraction profiles, but there is an indication for basement depths exceeding 8 km off NE Greenland. A smaller basin in the south (the Ammassalik Basin) will be discussed below.

Similar to the basement map, the Greenland–Iceland and Iceland–Faroe ridges are very prominent on the depth to Moho map (Fig. 6). The minimum Moho depth on these ridges is 29 km, while the maximum Moho depth beneath Iceland is 39 km. South and north of Iceland, the Moho is generally shallowest along the spreading ridges (5–10 km) from where the Moho deepens towards the continent–ocean boundary (15–20 km). This increase in Moho depth is related to the cooling of the lithosphere with age and to excess magmatism around the time of break-up, which resulted in anomalously thick oceanic crust close to the continent–ocean boundary (Holbrook *et al.* 2001). The basins along the NW European margins are associated with a relatively shallow Moho depth, best seen in the Rockall Trough (12 km) and the Porcupine Basin (10 km).

Onshore Greenland, the Moho is deepest in the southern part with a depth of 40–45 km, while the NW part displays depths of 35–40 km (Fig. 6). In NW Europe, the deepest Moho in the mapping area is found in Scandinavia (47 km: Grad *et al.* 2009). In Ireland and the UK, the Moho is

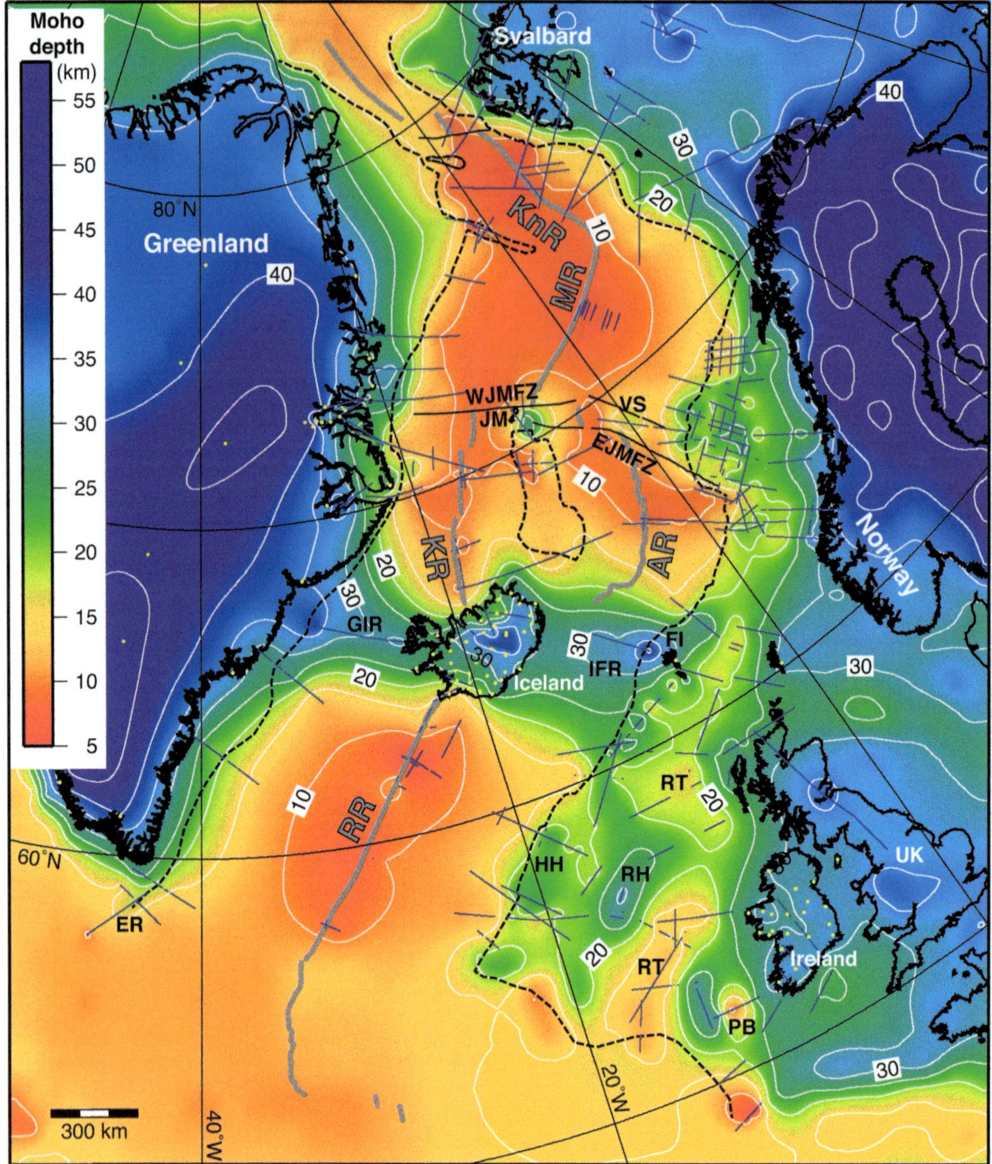

Fig. 6. Moho map (depth below sea level). White lines show the contours with a 5 km interval. Solid lines and yellow circles mark the location of seismic refraction data and receiver functions, respectively. The dashed line indicates the continent–ocean boundary (Funck *et al.* 2014). Grey lines mark active and extinct spreading ridges. Abbreviations: AR, Aegir Ridge; EJMFZ, East Jan Mayen Fracture Zone; ER, Eirik Ridge; FI, Faroe Islands; GIR, Greenland–Iceland Ridge; HH, Hatton High; IFR, Iceland–Faroe Ridge; JM, Jan Mayen; KnR, Knipovich Ridge; KR, Kolbeinsey Ridge; MR, Mohns Ridge; PB, Porcupine Basin; RR, Reykjanes Ridge; RH, Rockall High; RT, Rockall Trough; UK, United Kingdom; VS, Vøring Spur; WJMFZ, West Jan Mayen Fracture Zone.

substantially shallower, varying mainly between 30 and 36 km.

The crustal thickness map (Fig. 7) displays similar characteristics to the Moho map (Fig. 6). The Greenland–Iceland–Faroe Ridge has a minimum thickness of 28 km. South of the ridge, the thickness of the oceanic crust varies mostly between 5 and 8 km, not too dissimilar from the average oceanic crust thickness of 7 km (White *et al.* 1992). North of Iceland, the crust produced along the Kolbeinsey

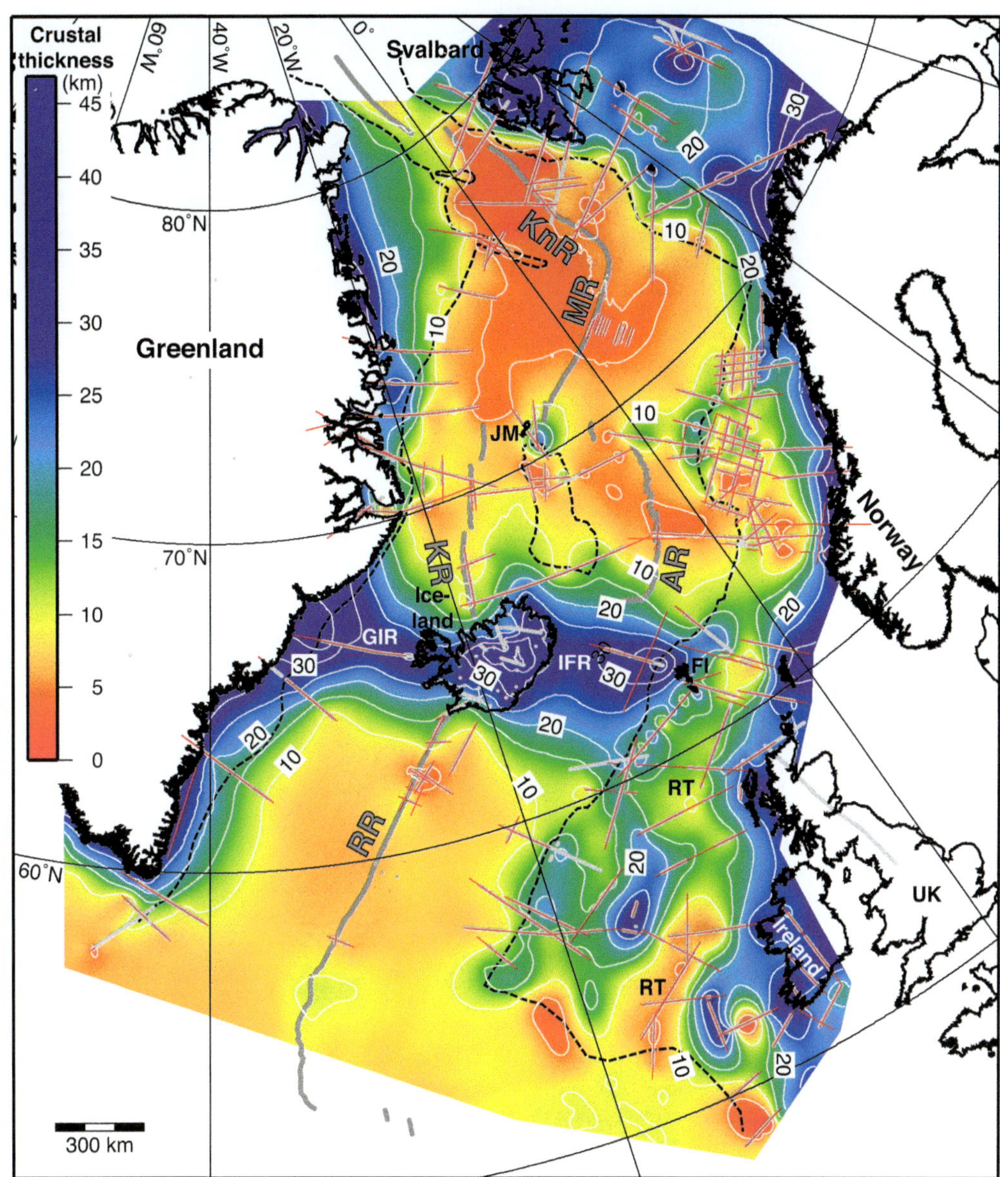

Fig. 7. Crustal thickness map with a contour interval of 5 km (white lines). Data points with Moho and basement constraints are shown in grey and red, respectively. The dashed line indicates the continent–ocean boundary (Funck *et al.* 2014). Grey lines mark active and extinct spreading ridges. Abbreviations: AR, Aegir Ridge; FI, Faroe Islands; GIR, Greenland–Iceland Ridge; IFR, Iceland–Faroe Ridge; JM, Jan Mayen; KnR, Knipovich Ridge; KR, Kolbeinsey Ridge; MR, Mohns Ridge; Reykjanes Ridge; RT, Rockall Trough; UK, United Kingdom.

Ridge is around 8 km thick (e.g. Kodaira *et al.* 1997), while the crust that formed at the now extinct Aegir Ridge is mostly thinner than 6 km (e.g. Breivik *et al.* 2006). Further north along the Mohns Ridge, crustal thickness decreases to 4–5 km (Klingelhöfer *et al.* 2000), and along the Knipovich Ridge segment, the thickness varies between 3 km (Hermann & Jokat 2013) and 7 km (Kandilarov *et al.* 2010).

Within the Rockall Trough, the crust thickens from 5 km in the south to 10 km in the north (Fig. 7). Further to the north, within the Faroe–Shetland Trough, the thickness of the crystalline crust thins again to values as small as 7 km (Makris *et al.* 2009). Likewise, the basins off mid-Norway are characterized by a thin crust with a minimum thickness of 5 km (e.g. Raum 2000).

Discussion

Conjugate margin comparison

The maps displaying basement depth (Fig. 3), Moho depth (Fig. 6) and crustal thickness (Fig. 7) are compiled from seismic refraction data that are unevenly distributed across the study area (Fig. 2). In particular, the NW European margin is much better sampled with seismic data when compared to the East Greenland continental margin. This offers the opportunity to learn from NW Europe when studying the structures found along the East Greenland coast. At the same time, the conjugate margin comparison can reveal potential shortcomings of the compilation that result from the distribution or interpretation of data.

Along the SE Greenland margin, the Ammassalik Basin shows up as a 3 km-deep basement low (Fig. 3). The structure is not sampled by seismic refraction data, but was introduced by the kriging algorithm that used free-air gravity (Andersen *et al.* 2010) as a constraint. Hence, the basin outline correlates with the corresponding gravity low in that area. There are few seismic reflection data available that could independently confirm the shape and depth of the basin. Hopper *et al.* (1998) were the first to notice the presence of a sedimentary basin in this region: they interpreted the structure as a rift system with a sediment infill corresponding to 1 s two-way travel time (TWT). Reprocessing of the seismic data suggests a probable sediment thickness of at least 3 km (Gerlings *et al.*, this volume, in review).

When the basement depth is reconstructed for Chron C24 (Fig. 8) using the rotation poles of Gaina (2014), the Ammassalik Basin lies just to the north of the Anton Dohrn Lineament Complex at the conjugate NW European margin. Kimbell *et al.* (2005) identified three individual lineaments within this complex. The continent–ocean boundary is shifted landwards across the complex (when moving from south to north), while the axis of the Rockall Trough is offset seawards by some 200 km. The Hatton Basin does not continue northwards of the Anton Dohrn Lineament Complex. This is why the Ammassalik Basin could be a NW-shifted continuation of the Hatton Basin that was left on the Greenland side at the time of final break-up. This shift would essentially be similar to the one observed between the northern and southern Rockall Trough. Hence, knowledge of the poorly studied Ammassalik Basin can probably be increased by comparison to the Hatton Basin.

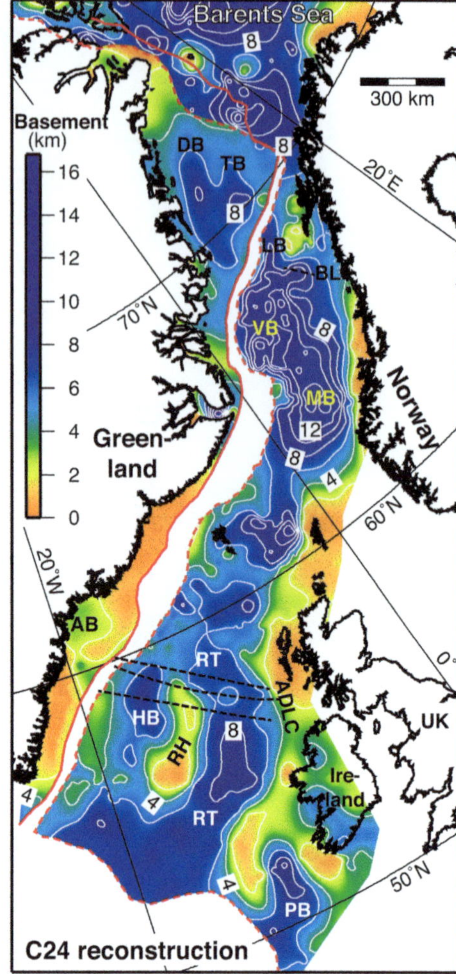

Fig. 8. Reconstruction of the basement depth for Chron C24 using the rotation pole of Gaina (2014). Europe is fixed in this reconstruction and the rotation of Greenland is carried out with GMT (Generic Mapping Tools) software (Wessel *et al.* 2013). White lines show the contours with a 2 km interval. The solid and dashed red lines mark the continent–ocean boundary (COB) of Greenland and NW Europe, respectively (Funck *et al.* 2014). Note that there is an overlap of the COB in the northernmost area. Dashed lines indicate lineaments. Abbreviations: AB, Ammassalik Basin; ADLC, Anton Dohrn Lineament Complex; BL, Bivrost Lineament; DB, Danmarkshavn Basin; HB, Hatton Basin; LB, Lofoten Basin; MB, Møre Basin; PB, Porcupine Basin; RH, Rockall High; RT, Rockall Trough; TB, Thetis Basin; UK, United Kingdom; VB, Vøring Basin.

In the Hatton Basin, the total sediment thickness is more than 6 km in the northern part (Hopper *et al.*, this volume, in prep) and this could be an indication of the possible depth of the Ammassalik Basin. Hence, the kriging technique with gravity as a constraint may underestimate the basement depth if seismic data are lacking.

Such an underestimation of basement depth also occurs on the NE Greenland continental shelf. The basement reconstruction for Chron C24 (Fig. 8) shows that the deep basins off mid-Norway (the Møre, Vøring and Lofoten basins) correlate with the basins off NE Greenland (the Danmarkshavn and Thetis basins). However, the Norwegian basins are up to 15 km thick, while the maximum depth off Greenland seems to be only 8 km. The seismic refraction constraints within the basins off NE Greenland are restricted to the southern edge of the Danmarkshavn Basin (line AWI 20030300) and the eastern edge of the Thetis Basin (line AWI 20030200) (Fig. 3). In addition, the basement on these two lines is only poorly resolved by the seismic data and is therefore approximated by the 5.7 km s^{-1} velocity contour. Newly released seismic reflection data indicate a maximum sediment thickness of 18 km (Hopper *et al.*, this volume, in prep), which is in better agreement with the mid-Norwegian basins and also with the SW Barents Sea.

While the gravity constraints used in the kriging procedure could outline the general structure of the basins not covered by seismic refraction data, this method seems to underestimate the basement depth, as seen in the two examples from Greenland. Initially, gravity constraints were also tested for the kriging of the Moho depth. In particular, filtered versions of the Bouguer gravity anomaly were used for this purpose. However, these resulted in some features, such as crustal roots, that were difficult to explain in some cases. This is why the final kriging of the Moho was performed without gravity constraints. Instead, regional and global datasets (Grad *et al.* 2009; Laske *et al.* 2013) were used to constrain the surrounding areas. Assuming isostasy, the basins on the continental shelves should be associated with a shallowing of the Moho. While this is the case for basins covered with seismic data (e.g. the Rockall Trough), the basins off Greenland (the Ammassalik, Danmarkshavn and Thetis basins) do not show a corresponding expression in the Moho depth. In these areas, the Moho depth obtained from seismically controlled gravity inversion (Haase *et al.*, this volume, in review) provides greater structural detail.

Thickness of oceanic crust

The oceanic crust in the NE Atlantic Ocean displays variations in thickness ranging between 2 and 40 km (Fig. 9) compared to an average of 7 km for normal oceanic crust (White *et al.* 1992). Areas with thick oceanic crust can generally be related to an increased magmatism associated with the Iceland plume. The Greenland–Iceland–Faroe Ridge is characterized by a 28–40 km-thick crust, which White (1997) interpreted as the interaction of a rising mantle plume with a spreading ridge. He suggested that the thickness variations were related to variations in the temperature of the mantle or mantle flow rates, or both.

South of Iceland, the initial oceanic crustal thickness at break-up is generally greater than 15 km (Fig. 9) and decreases with increasing distance from the plume (cf. Holbrook *et al.* 2001). Over time, the thickness of the oceanic crust decreased to values of around 8 km at a distance of 250 km from the continent–ocean boundary. Refraction seismic experiments on the Reykjanes Ridge away from Iceland indicate variations in the crustal thickness that ranged from 4 to 9 km (Bunch & Kennett 1980; Smallwood *et al.* 1995; Navin *et al.* 1998; Smallwood & White 1998; Jacoby *et al.* 2007).

North of Iceland, the initial spreading after break-up was along the Aegir Ridge, which lasted until 30 Ma when spreading there became extinct (Gaina *et al.* 2009). At that time, the Kolbeinsey Ridge started to develop from the south, with final detachment of the Jan Mayen microcontinent from East Greenland occurring at 20 Ma (Chron C6b) (Gaina *et al.* 2009; Peron-Pinvidic *et al.* 2012). The thickness of the oceanic crust that was formed in the northern portions of the Aegir and Kolbeinsey ridges differs significantly (Fig. 9). Between the Jan Mayen microcontinent and mid-Norway, the initial crustal thickness at break-up was around 11 km (Breivik *et al.* 2006), which is less than that observed in SE Greenland at a similar distance from the Iceland plume (19 km: Hopper *et al.* 2003). Close to the extinct Aegir Ridge, the crustal thickness is as little as 4 km (Breivik *et al.* 2006). In contrast, the crust at the Kolbeinsey Ridge has a thickness of between 7 and 10 km (Kodaira *et al.* 1997). Breivik *et al.* (2006) speculated that the thin oceanic crust between the Jan Mayen microcontinent and mid-Norway is caused by interaction with the Iceland plume. They suggested that the construction of the magmatic Greenland–Iceland–Faroe Ridge to the south depleted the mantle. Asthenospheric flow transported this depleted mantle to the Aegir Ridge, giving a lower than normal magma productivity. Howell *et al.* (2014) employed three-dimensional (3D) numerical models that simulated a plume interacting with rifting continents and spreading ridges. Their results support a plume with a relatively low flux (95–128 $m^3\ s^{-1}$) to explain the restriction of the relatively thick crust in the southern part of the Aegir Ridge.

The Kolbeinsey Ridge terminates at the West Jan Mayen Fracture Zone. Northwards of the fracture

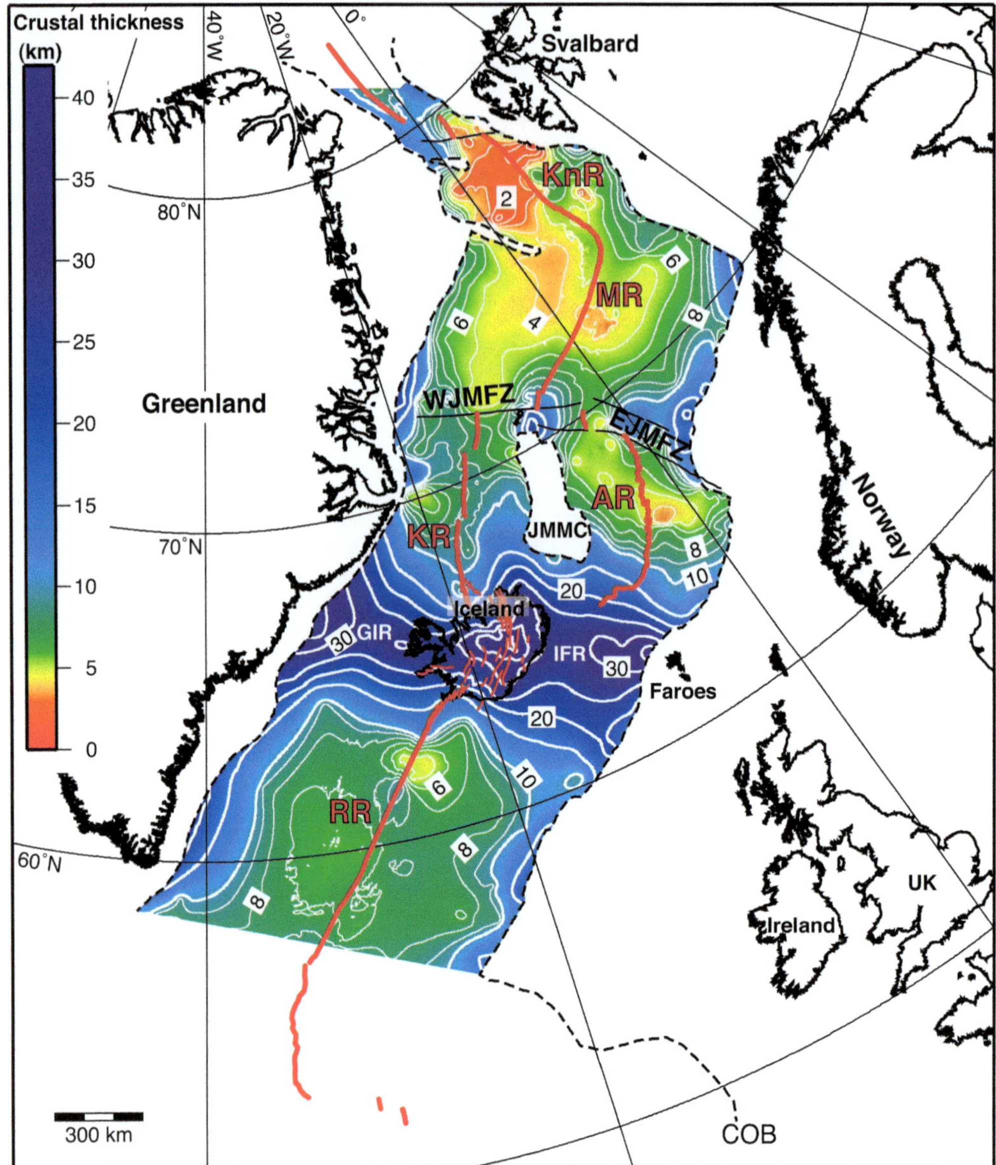

Fig. 9. The thickness of oceanic crust. Contours from 2 to 8 km are shown as thin white lines with an interval of 2 km. Bold white lines are contours >10 km, where a contour interval of 5 km is used. The continent–ocean boundary (COB) is marked by a dashed line. Red lines indicate active and extinct spreading ridges. Abbreviations: AR, Aegir Ridge; EJMFZ, East Jan Mayen Fracture Zone; GIR, Greenland–Iceland Ridge; IFR, Iceland–Faroe Ridge; JMMC, Jan Mayen microcontinent; KnR, Knipovich Ridge; KR, Kolbeinsey Ridge; MR, Mohns Ridge; RR, Reykjanes Ridge; WJMFZ, West Jan Mayen Fracture Zone.

zone, the crustal thickness decreases markedly (Fig. 9). The Mohns Ridge produced thicker than normal oceanic crust just after break-up, but most of the younger crust is only between 4 and 6 km thick. The main exception is the southernmost part of the Mohns Ridge, which is affected by the volcanism on the Jan Mayen islands. Rickers *et al.* (2013) could identify two distinct low-velocity zones in the mantle: one centred beneath Iceland and one beneath the northern Kolbeinsey Ridge. At depth, the latter anomaly covers the whole length of the Kolbeinsey Ridge and extends beyond the

West Jan Mayen Fracture Zone to the southern Mohns Ridge. Just to the north of the West Jan Mayen Fracture Zone, close to Jan Mayen, the crustal thickness is 10 km (Kandilarov *et al.* 2012). Based on the distinct mantle velocity anomalies, Rickers *et al.* (2013) see two seperate hotspots beneath Iceland and Jan Mayen, which can be also supported by isotope studies (Schilling *et al.* 1999).

On a series of seismic refraction lines on the eastern flank of the Mohns Ridge, Klingelhöfer *et al.* (2000) modelled a mean crustal thickness of 4 km that was produced at a full spreading rate of 14 mm a^{-1}. For spreading rates greater than 20 mm a^{-1}, there is little variation in crustal thickness: however, the thickness drops rapidly for lower rates (Dick *et al.* 2003). Crust that was produced at the Knipovich Ridge (a full spreading rate of 14 mm a^{-1}: DeMets *et al.* 1990, 1994) is often just 3 km thick (Hermann & Jokat 2013), in particular to the west of the present location of the ridge. To the east, the crust is generally slightly thicker. Even though there are some asymmetries in crustal accretion at the Knipovich Ridge (Gaina 2014), a systematic difference in crustal thickness on either side of the ridge is difficult to explain. This is why it is worthwhile revisiting the datasets on which the crustal thickness compilation is based.

The main profile that is constraining the crustal thickness west of the Knipovich Ridge is line AWI 20090200 (labelled as number 14 in Fig. 2) (Hermann & Jokat 2013). This line displays a 2–3 km-thick oceanic crust (Fig. 10) that lacks a distinct oceanic layer 3 (Hermann & Jokat 2013). Some 30 km to the south, another profile extends from the Knipovich Ridge towards the NE (Barents 98 line 8, labelled as number 43 in Fig. 2). Close to the ridge, this line exhibits 6–7 km-thick crust (Fig. 10) with distinct oceanic layers 2 and 3 (Ljones *et al.* 2004). A further profile in close proximity is Knipovich 02 line 1 (labelled as numbers 48 in Fig. 2) (Kandilarov *et al.* 2008). Here, the crust is 4 km thick adjacent to the ridge (Fig. 10), which is in better agreement with line AWI 20090200. However, Kandilarov *et al.* (2008) also interpreted the presence of an oceanic layer 3, in contrast to the model of line AWI 20090200 (Hermann & Jokat 2013), on the western side of the Knipovich Ridge. Although crust formed at ultraslow-spreading ridges can display large variations in thickness and velocity structure (Jokat *et al.* 2003; Minshull *et al.* 2006), some differences may also result from different approaches in the velocity modelling and the subsequent interpretation of the model. This is why the three lines will be briefly reviewed here.

The total thickness of oceanic layer 2 is similar on lines AWI 20090200 and Barents 98 line 1 (Fig. 10), even though the level of detail in the models is different. On the AWI line, Hermann & Jokat (2013) differentiated three sublayers (2A, 2B and 2C), while no such subdivision is seen on Barents 98 line 8 (Ljones *et al.* 2004). However, the velocity range observed within layer 2 is not too dissimilar. The main difference here is that the AWI line resolves velocities as low as 2.7 km s^{-1} at the top of the oceanic crust that are not resolved on Barents 98 line 8.

The potentially critical issue on Barents 98 line 8 is the interpreted oceanic layer 3 (Ljones *et al.* 2004) with velocities of 6.4–6.9 and 7.2 km s^{-1} in layers 3A and 3B, respectively (Fig. 10). Ljones *et al.* (2004) stated that there were a total of 107 travel time picks for reflections between layers 2 and 3A, while only 11 Moho reflections (PmP) were picked. With a shot spacing of 200 m, these 11 picks correspond to a 2 km-long segment along which the PmP is observed. With such a limited number of observations, the Moho appears not to have been mapped reliably by reflections. In contrast, the numerous observed reflections between layers 2 and 3 are also unusual, as this is commonly not a reflective boundary. Minshull *et al.* (2006) did not report a single such reflection in their dataset from the ultraslow-spreading SW Indian Ridge, and the same is true for the Mohns Ridge (Klingelhöfer *et al.* 2000). However, both of these other studies show frequent reflections from the Moho. This is why an alternate explanation for the interpreted layer 3 should be looked into. Instead of having a gabbroic composition, a partially serpentinized mantle rock may explain the seismic observations.

The Moho interpretation on line AWI 20090200 is also unclear. Despite a modelled velocity contrast of >1.5 km s^{-1} (Fig. 10), not a single Moho reflection is observed in a 140 km-wide zone adjacent to the Knipovich Ridge (Hermann & Jokat 2013). Hence, there might not be a sharp Moho, as indicated in the model, but rather a gradual velocity transition from layer 2C into the mantle. In this case, the interpreted layer 2C or parts of it could, in fact, be a highly serpentinized mantle. Nevertheless, there are some large misfits between the observed and calculated travel-time curves along the AWI line. For example, ocean-bottom seismometer (OBS) 216 close to the Knipovich Ridge (fig. 2c in Hermann & Jokat 2013) shows that the calculated travel times of the mantle refraction are up to 300 ms too fast, which could mean that the Moho locally could be up to 1.8 km deeper than shown in the velocity model.

When the AWI profile is compared with the Knipovich 02 line 1 (Fig. 10), there is a good match between the combined layers 2A and 2B on line AWI 20090200 (Hermann & Jokat 2013), and oceanic layer 2 on the Knipovich 02 line 1 (Kandilarov *et al.* 2008). This applies to both layer thickness and velocities. The underlying layer is interpreted as

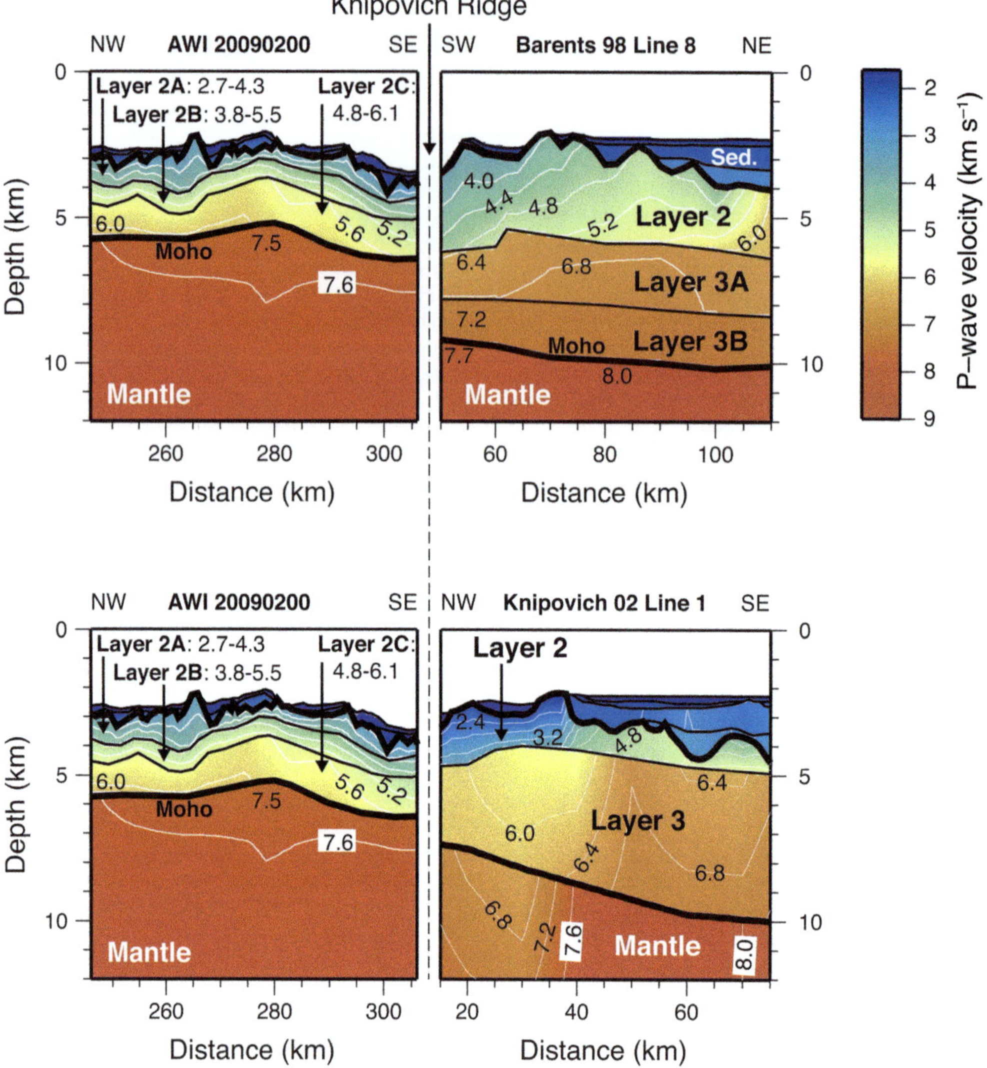

Fig. 10. Comparison of the P-wave velocity model of seismic refraction line AWI 20090200 (Hermann & Jokat 2013) with (top) the Barents 98 line 8 (Ljones *et al.* 2004), and (bottom) the Knipovich 02 line 1 (Kandilarov *et al.* 2008). For all lines, a 60 km-wide portion adjacent to the Knipovich Ridge is shown. Velocities are specified in km s^{-1}. Sed., sediments.

layer 2C on the AWI profile (4.8–6.1 km s^{-1}) and as layer 3 on the Knipovich 02 line 1 (5.6–7.0 km s^{-1}). Hence, based on the modelled velocities, the different interpretations seem to be justified. However, as discussed earlier, there is also the possibility that part of the interpreted layer 2C on the AWI line could be highly serpentinized mantle.

The definition of the Moho along the Knipovich 02 line 1 seems to be based mainly on mantle refractions (Pn). Kandilarov *et al.* (2008) showed only one station that recorded a Moho reflection (PmP) and here the misfit is up to 200 ms. Close to the Knipovich Ridge, the Moho is difficult to model with Pn phases alone, as the model indicates only a small velocity increase across the Moho. With that, there is no distinct change in the phase velocity of the first arrivals between the crustal and mantle arrivals that would allow for an unequivocal phase interpretation. From the previous discussion, it can be seen that the differences in the crustal structure of the lines at the Knipovich Ridge may, in fact, be less than that indicated by the velocity models.

In order to decide whether the differences are real or not, the lines should be reanalysed together.

Conclusions

The maps of the basement and Moho depth are useful tools to use in the discussion of first-order crustal features in the NE Atlantic realm. Although the region is not evenly covered with seismic refraction data and receiver function analyses, there is an adequate coverage of the large-scale tectonic structures. Data used in this compilation were acquired over a timespan of more than 40 years, and are therefore very variable in the shot and receiver spacing, as well as in the modelling technique. In addition, our understanding of rifting processes at continental margins has considerably changed over this period, which, of course, is reflected in the conceptual models on which the interpretation of the velocity models are based. Nevertheless, the basement and Moho depth are generally fairly robust features of velocity models, with the exception of areas with thin crust. Here, the mantle rock can be partially serpentinized and display velocities as low as 4.8 km s^{-1} at a serpentinization rate of 100% (Christensen 2004), and, hence, may erroneously be interpreted as crustal rock. One important measure of quality control for the basement and Moho compilation was the fit at cross-points to check for internal consistency of the datasets. While this helped to eliminate some erroneous data from the compilation, the maps are only as good as the underlying velocity models.

Using seismic refraction data to map the basement depth has two advantages compared to seismic reflection data. One is that no time-to-depth conversion is necessary as the velocity models are developed in the depth domain. The other is that the basement is often better defined on seismic refraction data, in particular in areas with deep sedimentary basins or where basalts are interbedded with sedimentary rocks. Many seismic refraction lines benefit from coincident reflection data by incorporating detailed basement relief from there. The disadvantage is that the data coverage with seismic refraction lines is not nearly as good as that with reflection lines. However, using the kriging method with gravity as a constraint, a rather detailed basement map could be compiled from the seismic refraction data (Fig. 3). That way, even basement lows that are not covered by seismic data are imaged, such as the Ammassalik Basin off SE Greenland. The basin depths of these seismically unconstrained features may be erroneous, but at least allow a qualitative assessment and comparison with the conjugate margin to be made (Fig. 8).

Our results show that the continuous rift basins extending from the southern Rockall Trough to the Lofoten Basin carry on to the Danmarkshavn and Thetis basins of NE Greenland (Fig. 8). Similarly, the Hatton Basin off Ireland seems to continue along with the Ammassalik Basin in SE Greenland. These correlations are most prominent on the basement map (Figs 3 & 8), while details of the Moho geometry beneath the Greenlandic basins are not resolved. This is related to a lack of seismic refraction data in these regions. However, 3D gravity modelling based on the seismic constraints is able to show a continuity of these features on the crustal thickness and Moho maps (Haase *et al.*, this volume, in review).

The crustal thickness that is compiled from the basement and Moho maps can be used to calculate crustal stretching factors and first-order crustal strength profiles, which has done by Kimbell *et al.* (this volume, in press) for the NW European margin. In addition, deformable plate reconstructions depend on good estimates of crustal thickness. Apparent inconsistencies in the crustal thickness map, such as the generally thinner oceanic crust west of the Knipovich Ridge compared to the east, can help to identify datasets that it might be worthwhile remodelling/reanalysing.

This work was part of the NAG-TEC project and was sponsored by BayernGas Norge AS, BP Exploration Operating Company Limited, Bundesanstalt für Geowissenschaften und Rohstoffe (BGR), Chevron East Greenland Exploration A/S, ConocoPhillips Skandinavia AS, DEA Norge AS, Det norske oljeselskap ASA, DONG E&P A/S, E.ON Norge AS, ExxonMobil Exploration and Production Norway AS, Japan Oil, Gas, and Metals National Corporation (JOGMEC), Maersk Oil, Nalcor Energy – Oil and Gas Inc., Nexen Energy ULC, Norwegian Energy Company ASA (Noreco), Repsol Exploration Norge AS, Statoil (U.K.) Limited, and Wintershall Holding GmbH. Reviews from Rolf Mjelde and an anonymous referee helped improve the manuscript. The work of Kenneth McDermott was supported by the Geological Survey of Ireland. Geoffrey S. Kimbell publishes with permission of the Executive Director of the British Geological Survey (Natural Environment Research Council).

References

Amante, C. & Eakins, B.W. 2009. *ETOPO1 1 Arc-Minute Global Relief Model: procedures, Data Sources and Analysis.* NOAA Technical Memorandum NESDIS NGDC-24 National Geophysical Data Center, NOAA, Washington, DC, https://doi.org/10.7289/V5C8276M

Andersen, O.B., Knudsen, P. & Berry, P.A.M. 2010. The DNSC08GRA global marine gravity field from double retracked satellite altimetry. *Journal of Geodesy*, **84**, 191–199, https://doi.org/10.1007/s00190-009-0355-9

Angenheister, G., Bjornsson, S. *et al.* 1980. Reykjanes Ridge Iceland Seismic Experiment (RRISP 77). *Journal of Geophysics*, **47**, 228–238.

Artemieva, I.M. & Thybo, H. 2013. EUNAseis: a seismic model for Moho and crustal structure in Europe, Greenland, and the North Atlantic region. *Tectonophysics*, **609**, 97–153, https://doi.org/10.1016/j.tecto.2013.08.004

Bamford, D., Nunn, K., Prodehl, C. & Jacob, B. 1977. LISPB–III. Upper crustal structure of northern Britain. *Journal of the Geological Society, London*, **133**, 481–488, https://doi.org/10.1144/gsjgs.133.5.0481

Bamford, D., Nunn, K., Prodehl, C. & Jacob, B. 1978. LISPB – IV. Crustal structure of Northern Britain. *Geophysical Journal of the Royal Astronomical Society*, **54**, 43–60, https://doi.org/10.1111/j.1365-246X.1978.tb06755.x

Barton, P.J. 1992. LISPB revisited: a new look under the Caledonides of northern Britain. *Geophysical Journal International*, **110**, 371–391, https://doi.org/10.1111/j.1365-246X.1992.tb00881.x

Barton, A.J. & White, R.S. 1997. Crustal structure of Edoras Bank continental margin and mantle thermal anomalies beneath the North Atlantic. *Journal of Geophysical Research*, **102**, 3109–3129, https://doi.org/10.1029/96jb03387

Berndt, C., Mjelde, R., Planke, S., Shimamura, H. & Faleide, J.I. 2001. Controls on the tectono-magmatic evolution of a volcanic transform margin: the Vøring Transform Margin, NE Atlantic. *Marine Geophysical Research*, **22**, 133–152, https://doi.org/10.1023/a:1012089532282

Bjarnason, I.T., Menke, W., Flóvenz, Ó.G. & Caress, D. 1993. Tomographic image of the Mid-Atlantic plate boundary in southwestern Iceland. *Journal of Geophysical Research*, **98**, 6607–6622, https://doi.org/10.1029/92jb02412

Böhner, J. & Antonić, O. 2008. Land-surface parameters specific to topo-climatology. *In*: Hengl, T. & Reuter, H.I. (eds) *Geomorphometry: Concepts, Software, Applications*. Developments in Soil Science, **33**. Elsevier, Amsterdam, 195–226, https://doi.org/10.1016/S0166-2481(08)00008-1

Bohnhoff, M. & Makris, J. 2004. Crustal structure of the southeastern Iceland-Faeroe Ridge (IFR) from wide aperture seismic data. *Journal of Geodynamics*, **37**, 233–252, https://doi.org/10.1016/j.jog.2004.02.004

Brandsdóttir, B., Hooft, E.E.E., Mjelde, R. & Murai, Y. 2015. Origin and evolution of the Kolbeinsey Ridge and Iceland Plateau, N-Atlantic. *Geochemistry, Geophysics, Geosystems*, **16**, 612–634, https://doi.org/10.1002/2014gc005540

Breivik, A.J., Mjelde, R., Grogan, P., Shimamura, H., Murai, Y., Nishimura, Y. & Kuwano, A. 2002. A possible Caledonide arm through the Barents Sea imaged by OBS data. *Tectonophysics*, **355**, 67–97, https://doi.org/10.1016/S0040-1951(02)00135-X

Breivik, A.J., Mjelde, R., Grogan, P., Shimamura, H., Murai, Y. & Nishimura, Y. 2003. Crustal structure and transform margin development south of Svalbard based on ocean bottom seismometer data. *Tectonophysics*, **369**, 37–70, https://doi.org/10.1016/s0040-1951 (03)00131-8

Breivik, A.J., Mjelde, R., Grogan, P., Shimamura, H., Murai, Y. & Nishimura, Y. 2005. Caledonide development offshore–onshore Svalbard based on ocean bottom seismometer, conventional seismic, and potential field data. *Tectonophysics*, **401**, 79–117, https://doi.org/10.1016/j.tecto.2005.03.009

Breivik, A.J., Mjelde, R., Faleide, J.I. & Murai, Y. 2006. Rates of continental breakup magmatism and seafloor spreading in the Norway Basin-Iceland plume interaction. *Journal of Geophysical Research*, **111**, B07102, https://doi.org/10.1029/2005jb004004

Breivik, A.J., Faleide, J.I. & Mjelde, R. 2008. Neogene magmatism northeast of the Aegir and Kolbeinsey ridges, NE Atlantic: spreading ridge–mantle plume interaction? *Geochemistry, Geophysics, Geosystems*, **9**, Q02004, https://doi.org/10.1029/2007gc001750

Breivik, A.J., Faleide, J.I., Mjelde, R. & Flueh, E.R. 2009. Magma productivity and early seafloor spreading rate correlation on the northern Vøring Margin, Norway – constraints on mantle melting. *Tectonophysics*, **468**, 206–223, https://doi.org/10.1016/j.tecto.2008.09.020

Breivik, A.J., Mjelde, R., Raum, T., Faleide, J.I., Murai, Y. & Flueh, E.R. 2011. Crustal structure beneath the Trøndelag Platform and adjacent areas of the Mid-Norwegian margin, as derived from wide-angle seismic and potential field data. *Norwegian Journal of Geology*, **90**, 141–161.

Bullock, A.D. & Minshull, T.A. 2005. From continental extension to seafloor spreading: crustal structure of the Goban Spur rifted margin, southwest of the UK. *Geophysical Journal International*, **163**, 527–546, https://doi.org/10.1111/j.1365-246X.2005.02726.x

Bunch, A.W.H. 1979. A detailed seismic structure of Rockall Bank (55°N, 15°W) – a synthetic seismogram analysis. *Earth and Planetary Science Letters*, **45**, 453–463, https://doi.org/10.1016/0012-821x(79)90144-4

Bunch, A.W.H. & Kennett, B.L.N. 1980. The crustal structure of the Reykjanes Ridge at 59° 30′N. *Geophysical Journal of the Royal Astronomical Society*, **61**, 141–166, https://doi.org/10.1111/j.1365-246X.1980.tb04310.x

Chabert, A., Ravaut, C., Readman, P.W., O'Reilly, B.M. & Shannon, P.M. 2006. Structure of the Hatton Basin (North Atlantic) from wide-angle and reflection seismic data. American Geophysical Union Fall Meeting, 11–15 December 2006, San Francisco, CA, USA. EOSTrans: AGU, 87 (52), Fall Meeting Supplement, Abstract T53A-1581.

Christensen, N.I. 2004. Serpentinites, peridotites, and seismology. *International Geology Review*, **46**, 795–816, https://doi.org/10.2747/0020-6814.46.9.795

Clark, S.A., Faleide, J.I. *et al.* 2013. Stochastic velocity inversion of seismic reflection/refraction traveltime data for rift structure of the southwest Barents Sea. *Tectonophysics*, **593**, 135–150, https://doi.org/10.1016/j.tecto.2013.02.033

Czuba, W., Ritzmann, O., Nishimura, Y., Grad, M., Mjelde, R., Guterch, A. & Jokat, W. 2004. Crustal structure of the continent–ocean transition zone along two deep seismic transects in north-western Spitsbergen. *Polish Polar Research*, **25**, 205–221.

Czuba, W., Ritzmann, O., Nishimura, Y., Grad, M., Mjelde, R., Guterch, A. & Jokat, W. 2005. Crustal structure of northern Spitsbergen along the deep seismic transect between the Molloy Deep and Nordaustlandet.

Geophysical Journal International, **161**, 347–364, https://doi.org/10.1111/j.1365-246X.2005.02593.x

Czuba, W., Grad, M. *et al.* 2008. Seismic crustal structure along the deep transect Horsted'05, Svalbard. *Polish Polar Research*, **29**, 279–290.

Czuba, W., Grad, M. *et al.* 2011. Continent–ocean-transition across a trans-tensional margin segment: off Bear Island, Barents Sea. *Geophysical Journal International*, **184**, 541–554, https://doi.org/10.1111/j.1365-246X.2010.04873.x

Dahl-Jensen, T., Thybo, H., Hopper, J. & Rosing, M. 1998. Crustal structure at the SE Greenland margin from wide-angle and normal incidence seismic data. *Tectonophysics*, **288**, 191–198, https://doi.org/10.1016/s0040-1951(97)00292-8

Dahl-Jensen, T., Larsen, T.B. *et al.* 2003. Depth to Moho in Greenland: receiver-function analysis suggests two Proterozoic blocks in Greenland. *Earth and Planetary Science Letters*, **205**, 379–393, https://doi.org/10.1016/s0012-821x(02)01080-4

Darbyshire, F.A., Bjarnason, I.T., White, R.S. & Flóvenz, Ó.G. 1998. Crustal structure above the Iceland mantle plume imaged by the ICEMELT refraction profile. *Geophysical Journal International*, **135**, 1131–1149, https://doi.org/10.1046/j.1365-246X.1998.00701.x

Darbyshire, F.A., White, R.S. & Priestley, K.F. 2000. Structure of the crust and uppermost mantle of Iceland from a combined seismic and gravity study. *Earth and Planetary Science Letters*, **181**, 409–428, https://doi.org/10.1016/s0012-821x(00)00206-5

DeMets, C., Gordon, R.G., Argus, D.F. & Stein, S. 1990. Current plate motions. *Geophysical Journal International*, **101**, 425–478, https://doi.org/10.1111/j.1365-246X.1990.tb06579.x

DeMets, C., Gordon, R.G., Argus, D.F. & Stein, S. 1994. Effect of recent revisions to the geomagnetic reversal time scale on estimates of current plate motions. *Geophysical Research Letters*, **21**, 2191–2194, https://doi.org/10.1029/94gl02118

Dick, H.J.B., Lin, J. & Schouten, H. 2003. An ultraslow-spreading class of ocean ridge. *Nature*, **426**, 405–412, https://doi.org/10.1038/nature02128

Digranes, P., Mjelde, R., Kodaira, S., Shimamura, H., Kanazawa, T., Shiobara, H. & Berg, E.W. 1998. A regional shear-wave velocity model in the central Vøring Basin, N. Norway, using three-component Ocean Bottom Seismographs. *Tectonophysics*, **293**, 157–174, https://doi.org/10.1016/s0040-1951(98)00093-6

Doré, A.G., Lundin, E.R., Jensen, L.N., Birkeland, Ø., Eliassen, P.E. & Fichler, C. 1999. Principal tectonic events in the evolution of the northwest European Atlantic margin. *In*: Fleet, A.J. & Boldy, S.A.R. (eds) *Petroleum Geology of Northwest Europe: Proceedings of the 5th Conference*. Geological Society, London, 41–61, https://doi.org/10.1144/0050041

Døssing, A. & Funck, T. 2012. Greenland Fracture Zone–East Greenland Ridge(s) revisited: indications of a C22-change in plate motion? *Journal of Geophysical Research*, **117**, B01103, https://doi.org/10.1029/2011jb008393

Døssing, A., Dahl-Jensen, T., Thybo, H., Mjelde, R. & Nishimura, Y. 2008. East Greenland Ridge in the North Atlantic Ocean: an integrated geophysical study of a continental sliver in a boundary transform fault setting. *Journal of Geophysical Research*, **113**, B10107, https://doi.org/10.1029/2007jb005536

Eccles, J.D., White, R.S., Robert, A.W., Christie, P.A.F. & the iSIMM Team 2007. Wide angle converted shear wave analysis of a North Atlantic volcanic rifted continental margin: constraint on sub-basalt lithology. *First Break*, **25**, 63–70, https://doi.org/10.3997/1365-2397.2007026

Eldholm, O. & Grue, K. 1994. North Atlantic volcanic margins: dimensions and production rates. *Journal of Geophysical Research*, **99**, 2955–2968, https://doi.org/10.1029/93jb02879

Erlendsson, Ö. & Blischke, A. 2013. *Law of the Sea. Seismic Reflection and Refraction Datasets. Status on the Processing and Interpretation Work.* Report. Iceland GeoSurvey (ÍSOR), Reykjavik.

Fliedner, M.M. & White, R.S. 2003. Depth imaging of basalt flows in the Faeroe–Shetland Basin. *Geophysical Journal International*, **152**, 353–371, https://doi.org/10.1046/j.1365-246X.2003.01833.x

Funck, T., Andersen, M.S., Keser Neish, J. & Dahl-Jensen, T. 2008. A refraction seismic transect from the Faroe Islands to the Hatton-Rockall Basin. *Journal of Geophysical Research*, **113**, B12405, https://doi.org/10.1029/2008jb005675

Funck, T., Andrup-Henriksen, G., Dehler, S.A. & Louden, K.E. 2012. The crustal structure of the Eirik Ridge at the southern Greenland continental margin. *In*: *Geological Association of Canada–Mineralogical Association of Canada Joint Annual Meeting*, St John's, Newfoundland & Labrador, 27–29 May 2012, Abstract Volume **35**, p. 47.

Funck, T., Hopper, J.R. *et al.* 2014. Crustal structure. *In*: Hopper, J.R., Funck, T., Stoker, M., Artling, U., Péron-Pindivic, G., Gaina, C. & Doornenbal, H. (eds) *NAG-TEC Atlas: Tectonostratigraphic Atlas of the North-East Atlantic Region.* Geological Survey of Denmark and Greenland (GEUS), Copenhagen, 69–126.

Funck, T., Gerlings, J. *et al.* 2015. The East Greenland Ridge: geophysical mapping and geological sampling reveal a highly segmented continental sliver. *In*: Smelror, M. (ed.) *7th International Conference on Arctic Margins – ICAM 2015*, Trondheim, Norway, 2–5 June 2015, NGU Report 2015.032. Geological Survey of Norway, Trondheim, p. 45.

Furmall, A.V. 2010. *Melt production and ridge geometry over the past 10 Myr on the southern Kolbeinsey Ridge, Iceland.* Master thesis, University of Oregon.

Gaina, C. 2014. Plate reconstructions and regional kinematics. *In*: Hopper, J.R., Funck, T., Stoker, M., Artling, U., Péron-Pindivic, G., Gaina, C. & Doornenbal, H. (eds) *NAG-TEC Atlas: Tectonostratigraphic Atlas of the North-East Atlantic Region.* Geological Survey of Denmark and Greenland (GEUS), Copenhagen, 53–66.

Gaina, C., Gernigon, L. & Ball, P. 2009. Palaeocene–Recent plate boundaries in the NE Atlantic and the formation of the Jan Mayen microcontinent. *Journal of the Geological Society, London*, **166**, 601–616, https://doi.org/10.1144/0016-76492008-112

Gebrande, H., Miller, H. & Einarsson, P. 1980. Seismic structure of Iceland along RRISP-profile I. *Journal of Geophysics*, **47**, 239–249.

Gerlings, J., Funck, T., Jackson, H.R., Louden, K.E. & Klingelhöfer, F. 2009. Seismic evidence for plume-derived volcanism during formation of the continental margin in southern Davis Strait and northern Labrador Sea. *Geophysical Journal International*, **176**, 980–994, https://doi.org/10.1111/j.1365-246X.2008.04021.x

Gerlings, J., Funck, T., Castro, C.F. & Hopper, J.R. 2014. The East Greenland Ridge – a continental sliver along the Greenland Fracture Zone. *Geophysical Research Abstracts*, **16**, EGU2014–EGU2879.

Gerlings, J., Hopper, J.R., Fyhn, M.B.W. & Frandsen, N.K. In review. Mesozoic rift basins on the SE Greenland shelf, near Ammassalik. *In*: Péron-Pinvidic, G., Hopper, J.R., Stoker, M.S., Gaina, C., Doornenbal, J.C., Funck, T. & Árting, U.E. (eds) *The NE Atlantic Region: A Reappraisal of Crustal Structure, Tectonostratigraphy and Magmatic Evolution*. Geological Society, London, Special Publications, **447**.

Goldflam, P., Weigel, W. & Loncarevic, B.D. 1980. Seismic structure along RRISP – profile 1 on the southeast flank of the Reykjanes Ridge. *Journal of Geophysics*, **47**, 250–260.

Grad, M., Tiira, T. & ESC Working Group 2009. The Moho depth map of the European Plate. *Geophysical Journal International*, **176**, 279–292, https://doi.org/10.1111/j.1365-246X.2008.03919.x

Gregersen, S., Clausen, C. & Dahl-Jensen, T. 1988. Crust and upper mantle structure in Greenland. *In*: *Proceedings of the 19th General Assembly of the European Seismological Commission*. Nauka, Moscow, 467–469.

Grevemeyer, I., Weigel, W., Dehghani, G.A., Whitmarsh, R.B. & Avedik, F. 1997. The Aegir Rift: crustal structure of an extinct spreading axis. *Marine Geophysical Research*, **19**, 1–23, https://doi.org/10.1023/a:1004288815129

Haase, C., Ebbing, J. & Funck, T. In review. A 3D regional custal model of the Northeast Atlantic based on seismic and gravity data. *In*: Péron-Pinvidic, G., Hopper, J.R., Stoker, M.S., Gaina, C., Doornenbal, J.C., Funck, T. & Árting, U.E. (eds) *The NE Atlantic Region: A Reappraisal of Crustal Structure, Tectonostratigraphy and Magmatic Evolution*. Geological Society, London, Special Publications, **447**.

Harland, K.E., White, R.S. & Soosalu, H. 2009. Crustal structure beneath the Faroe Islands from teleseismic receiver functions. *Geophysical Journal International*, **177**, 115–124, https://doi.org/10.1111/j.1365-246X.2008.04018.x

Hauser, F., O'Reilly, B.M., Jacob, A.W.B., Shannon, P.M., Makris, J. & Vogt, U. 1995. The crustal structure of the Rockall Trough: differential stretching without underplating. *Journal of Geophysical Research*, **100**, 4097–4116, https://doi.org/10.1029/94jb02879

Hauser, F., O'Reilly, B.M., Readman, P.W., Daly, J.S. & Van den Berg, R. 2008. Constraints on crustal structure and composition within a continental suture zone in the Irish Caledonides from shear wave wide-angle reflection data and lower crustal xenoliths. *Geophysical Journal International*, **175**, 1254–1272, https://doi.org/10.1111/j.1365-246X.2008.03945.x

Hermann, T. & Jokat, W. 2013. Crustal structures of the Boreas Basin and the Knipovich Ridge, North Atlantic. *Geophysical Journal International*, **193**, 1399–1414, https://doi.org/10.1093/gji/ggt048

Holbrook, W.S., Larsen, H.C. *et al.* 2001. Mantle thermal structure and active upwelling during continental breakup in the North Atlantic. *Earth and Planetary Science Letters*, **190**, 251–266, https://doi.org/10.1016/s0012-821x(01)00392-2

Hooft, E.E.E., Brandsdóttir, B., Mjelde, R., Shimamura, H. & Murai, Y. 2006. Asymmetric plume-ridge interaction around Iceland: the Kolbeinsey Ridge Iceland Seismic Experiment. *Geochemistry, Geophysics, Geosystems*, **7**, Q05015, https://doi.org/10.1029/2005gc001123

Hopper, J.R., Lizarralde, D. & Larsen, H.C. 1998. Seismic investigations offshore South-East Greenland. *Geology of Greenland Survey Bulletin*, **180**, 145–151.

Hopper, J.R., Dahl-Jensen, T. *et al.* 2003. Structure of the SE Greenland margin from seismic reflection and refraction data: implications for nascent spreading center subsidence and asymmetric crustal accretion during North Atlantic opening. *Journal of Geophysical Research*, **108**, 2269, https://doi.org/10.1029/2002jb001996

Hopper, J.R. *et al.* In prep. Sediment thickness and residual topography of the North Atlantic: estimating dynamic topography around Iceland. *In*: Péron-Pinvidic, G., Hopper, J.R., Stoker, M.S., Gaina, C., Doornenbal, J.C., Funck, T. & Árting, U.E. (eds) *The NE Atlantic Region: A Reappraisal of Crustal Structure, Tectonostratigraphy and Magmatic Evolution*. Geological Society, London, Special Publications, **447**.

Howell, S.M., Ito, G. *et al.* 2014. The origin of the asymmetry in the Iceland hotspot along the Mid-Atlantic Ridge from continental breakup to present-day. *Earth and Planetary Science Letters*, **392**, 143–153, https://doi.org/10.1016/j.epsl.2014.02.020

Hughes, S., Barton, P.J. & Harrison, D. 1998. Exploration in the Shetland-Faeroe Basin using densely spaced arrays of ocean-bottom seismometers. *Geophysics*, **63**, 490–501, https://doi.org/10.1190/1.1444350

Jacoby, W.R., Weigel, W. & Fedorova, T. 2007. Crustal structure of the Reykjanes Ridge near 62°N, on the basis of seismic refraction and gravity data. *Journal of Geodynamics*, **43**, 55–72, https://doi.org/10.1016/j.jog.2006.10.002

Jokat, W. 2010. *The Expedition of the Research Vessel 'Polarstern' to the Arctic in 2009 (ARK-XXIV/3)*. Reports on Polar and Marine Research, 615. Alfred Wegener Institute for Polar and Marine Research, Bremerhaven, Germany, https://doi.org/10013/epic.35320

Jokat, W., Ritzmann, O., Schmidt-Aursch, M.C., Drachev, S., Gauger, S. & Snow, J. 2003. Geophysical evidence for reduced melt production on the Arctic ultraslow Gakkel mid-ocean ridge. *Nature*, **423**, 962–965, https://doi.org/10.1038/nature01706

Jokat, W., Kollofrath, J., Geissler, W.H. & Jensen, L. 2012. Crustal thickness and earthquake distribution south of the Logachev Seamount, Knipovich Ridge.

Geophysical Research Letters, **39**, L08302, https://doi.org/10.1029/2012gl051199

Kaban, M.K., Flóvenz, Ó.G. & Pálmason, G. 2002. Nature of the crust-mantle transition zone and the thermal state of the upper mantle beneath Iceland from gravity modelling. *Geophysical Journal International*, **149**, 281–299, https://doi.org/10.1046/j.1365-246X.2002.01622.x

Kandilarov, A., Mjelde, R., Okino, K. & Murai, Y. 2008. Crustal structure of the ultra-slow spreading Knipovich Ridge, North Atlantic, along a presumed amagmatic portion of oceanic crustal formation. *Marine Geophysical Research*, **29**, 109–134, https://doi.org/10.1007/s11001-008-9050-0

Kandilarov, A., Landa, H., Mjelde, R., Pedersen, R., Okino, K. & Murai, Y. 2010. Crustal structure of the ultra-slow spreading Knipovich Ridge, North Atlantic, along a presumed ridge segment center. *Marine Geophysical Research*, **31**, 173–195, https://doi.org/10.1007/s11001-010-9095-8

Kandilarov, A., Mjelde, R. *et al.* 2012. The northern boundary of the Jan Mayen microcontinent, North Atlantic determined from ocean bottom seismic, multichannel seismic, and gravity data. *Marine Geophysical Research*, **33**, 55–76, https://doi.org/10.1007/s11001-012-9146-4

Kelly, A., England, R.W. & Maguire, P.K.H. 2007. A crustal seismic velocity model for the UK, Ireland and surrounding seas. *Geophysical Journal International*, **171**, 1172–1184, https://doi.org/10.1111/j.1365-246X.2007.03569.x

Kimbell, G.S., Ritchie, J.D., Johnson, H. & Gatliff, R.W. 2005. Controls on the structure and evolution of the NE Atlantic margin revealed by regional potential field imaging and 3D modelling. *In*: Doré, A.G. & Vining, B.A. (eds) *Petroleum Geology: North-West Europe and Global Perspectives – Proceedings of the 6th Petroleum Geology Conference*. Geological Society, London, 933–945, https://doi.org/10.1144/0060933

Kimbell, G.S., Stewart, M.A. *et al.* In press. Controls on the location of compressional deformation on the NW European margin. *In*: Péron-Pinvidic, G., Hopper, J.R., Stoker, M.S., Gaina, C., Doornenbal, J.C., Funck, T. & Árting, U.E. (eds) *The NE Atlantic Region: A Reappraisal of Crustal Structure, Tectonostratigraphy and Magmatic Evolution*. Geological Society, London, Special Publications, **447**, https://doi.org/10.1144/SP447.3

Klingelhöfer, F., Géli, L., Matias, L., Steinsland, N. & Mohr, J. 2000. Crustal structure of a super-slow spreading centre: a seismic refraction study of Mohns Ridge, 72°N. *Geophysical Journal International*, **141**, 509–526, https://doi.org/10.1046/j.1365-246x.2000.00098.x

Klingelhöfer, F., Edwards, R.A., Hobbs, R.W. & England, R.W. 2005. Crustal structure of the NE Rockall Trough from wide-angle seismic data modeling. *Journal of Geophysical Research*, **110**, B11105, https://doi.org/10.1029/2005jb003763

Kodaira, S., Mjelde, R., Gunnarsson, K., Shiobara, H. & Shimamura, H. 1997. Crustal structure of the Kolbeinsey Ridge, North Atlantic, obtained by use of ocean bottom seismographs. *Journal of Geophysical Research*, **102**, 3131–3151, https://doi.org/10.1029/96JB03487

Kodaira, S., Mjelde, R., Gunnarsson, K., Shiobara, H. & Shimamura, H. 1998*a*. Evolution of oceanic crust on the Kolbeinsey Ridge, north of Iceland, over the past 22 Myr. *Terra Nova*, **10**, 27–31, https://doi.org/10.1046/j.1365-3121.1998.00166.x

Kodaira, S., Mjelde, R., Gunnarsson, K., Shiobara, H. & Shimamura, H. 1998*b*. Structure of the Jan Mayen microcontinent and implications for its evolution. *Geophysical Journal International*, **132**, 383–400, https://doi.org/10.1046/j.1365-246x.1998.00444.x

Korenaga, J., Holbrook, W.S. *et al.* 2000. Crustal structure of the southeast Greenland margin from joint refraction and reflection seismic tomography. *Journal of Geophysical Research*, **105**, 21,591–21,614, https://doi.org/10.1029/2000JB900188

Kumar, P., Kind, R., Priestley, K. & Dahl-Jensen, T. 2007. Crustal structure of Iceland and Greenland from receiver function studies. *Journal of Geophysical Research*, **112**, B03301, https://doi.org/10.1029/2005jb003991

Kvarven, T., Ebbing, J. *et al.* 2014. Crustal structure across the Møre margin, mid-Norway, from wide-angle seismic and gravity data. *Tectonophysics*, **626**, 21–40, https://doi.org/10.1016/j.tecto.2014.03.021

Landes, M., Prodehl, C., Hauser, F., Jacob, A.W.B. & Vermeulen, N.J. 2000. VARNET-96: influence of the Variscan and Caledonian orogenies on crustal structure in SW Ireland. *Geophysical Journal International*, **140**, 660–676, https://doi.org/10.1046/j.1365-246X.2000.00035.x

Laske, G., Masters, G., Ma, Z. & Pasyanos, M. 2013. Update on CRUST1.0 – A 1-degree global model of Earth's crust. *Geophysical Research Abstracts*, **15**, EGU2013–EGU2658.

Libak, A., Eide, C.H., Mjelde, R., Keers, H. & Flüh, E.R. 2012*a*. From pull-apart basins to ultraslow spreading: results from the western Barents Sea Margin. *Tectonophysics*, **514–517**, 44–61, https://doi.org/10.1016/j.tecto.2011.09.020

Libak, A., Mjelde, R., Keers, H., Faleide, J.I. & Murai, Y. 2012*b*. An integrated geophysical study of Vestbakken Volcanic Province, western Barents Sea continental margin, and adjacent oceanic crust. *Marine Geophysical Research*, **33**, 185–207, https://doi.org/10.1007/s11001-012-9155-3

Licciardi, A., Piana Agostinetti, N., Lebedev, S., Schaeffer, A.J., Readman, P.W. & Horan, C. 2014. Moho depth and V_p/V_s in Ireland from teleseismic receiver functions analysis. *Geophysical Journal International*, **199**, 561–579, https://doi.org/10.1093/gji/ggu277

Ljones, F., Kuwano, A., Mjelde, R., Breivik, A., Shimamura, H., Murai, Y. & Nishimura, Y. 2004. Crustal transect from the North Atlantic Knipovich Ridge to the Svalbard Margin west of Hornsund. *Tectonophysics*, **378**, 17–41, https://doi.org/10.1016/j.tecto.2003.10.003

Lowe, C. & Jacob, A.W.B. 1989. A north-south seismic profile across the Caledonian Suture zone in Ireland. *Tectonophysics*, **168**, 297–318, https://doi.org/10.1016/0040-1951(89)90224-2

MACKENZIE, G.D., SHANNON, P.M., JACOB, A.W.B., MOREWOOD, N.C., MAKRIS, J., GAYE, M. & EGLOFF, F. 2002. The velocity structure of the sediments in the southern Rockall Basin: results from new wide-angle seismic modelling. *Marine and Petroleum Geology*, **19**, 989–1003, https://doi.org/10.1016/s0264-8172(02)00133-2

MAKRIS, J., EGLOFF, R., JACOB, A.W.B., MOHR, P., MURPHY, T. & RYAN, P. 1988. Continental crust under the southern Porcupine Seabight west of Ireland. *Earth and Planetary Science Letters*, **89**, 387–397, https://doi.org/10.1016/0012-821X(88)90 125-2

MAKRIS, J., GINZBURG, A., SHANNON, P.M., JACOB, A.W.B., BEAN, C.J. & VOGT, U. 1991. A new look at the Rockall region, offshore Ireland. *Marine and Petroleum Geology*, **8**, 410–416, https://doi.org/10.1016/0264-8172(91)90063-7

MAKRIS, J., PAPOULIA, I. & ZISKA, H. 2009. Crustal structure of the Shetland–Faeroe Basin from long offset seismic data. *In*: *Faroe Islands Exploration Conference: Proceedings of the 2nd Conference. Annales Societatis Scientiarum Færoensis, Supplementum*, **50**, 30–42.

MANDLER, H.A.F. & JOKAT, W. 1998. The crustal structure of Central East Greenland: results from combined land–sea seismic refraction experiments. *Geophysical Journal International*, **135**, 63–76, https://doi.org/10.1046/j.1365-246X.1998.00586.x

MASSON, F., JACOB, A.W.B., PRODEHL, C., READMAN, P.W., SHANNON, P.M., SCHULZE, A. & ENDERLE, U. 1998. A wide-angle seismic traverse through the Variscan of southwest Ireland. *Geophysical Journal International*, **134**, 689–705, https://doi.org/10.1046/j.1365-246x.1998.00572.x

MENKE, W., WEST, M., BRANDSDÓTTIR, B. & SPARKS, D. 1998. Compressional and shear velocity structure of the lithosphere in northern Iceland. *Bulletin of the Seismological Society of America*, **88**, 1561–1571.

MINSHULL, T.A., MULLER, M.R. & WHITE, R.S. 2006. Crustal structure of the Southwest Indian Ridge at 66°E: seismic constraints. *Geophysical Journal International*, **166**, 135–147, https://doi.org/10.1111/j.1365-246X.2006.03001.x

MJELDE, R. 1992. Shear waves from three-component ocean bottom seismographs off Lofoten, Norway, indicative of anisotropy in the lower crust. *Geophysical Journal International*, **110**, 283–296, https://doi.org/10.1111/j.1365-246X.1992.tb00874.x

MJELDE, R. & SELLEVOLL, M.A. 1993. Seismic anisotropy inferred from wide-angle reflections off Lofoten, Norway, indicative of shear-aligned minerals in the upper mantle. *Tectonophysics*, **222**, 21–32, https://doi.org/10.1016/0040-1951(93)90187-O

MJELDE, R., SELLEVOLL, M.A., SHIMAMURA, H., IWASAKI, T. & KANAZAWA, T. 1992. A crustal study off Lofoten, N. Norway, by use of 3-component ocean bottom seismographs. *Tectonophysics*, **212**, 269–288, https://doi.org/10.1016/0040-1951(92)90295-H

MJELDE, R., SELLEVOLL, M.A., SHIMAMURA, H., IWASAKI, T. & KANAZAWA, T. 1993. Crustal structure beneath Lofoten, N. Norway, from vertical incidence and wide-angle seismic data. *Geophysical Journal International*, **114**, 116–126, https://doi.org/10.1111/j.1365-246X.1993.tb01471.x

MJELDE, R., SELLEVOLL, M.A., SHIMAMURA, H., IWASAKI, T. & KANAZAWA, T. 1995. S-wave anisotropy off Lofoten, Norway, indicative of fluids in the lower continental crust? *Geophysical Journal International*, **120**, 87–96, https://doi.org/10.1111/j.1365-246X.1995.tb05912.x

MJELDE, R., KODAIRA, S. *ET AL.* 1997*a*. Comparison between a regional and semi-regional crustal OBS model in the Vøring Basin, Mid-Norway Margin. *Pure and Applied Geophysics*, **149**, 641–665, https://doi.org/10.1007/s000240050045

MJELDE, R., KODAIRA, S., SHIMAMURA, H., KANAZAWA, T., SHIOBARA, H., BERG, E.W. & RIISE, O. 1997*b*. Crustal structure of the central part of the Vøring Basin, mid-Norway margin, from ocean bottom seismographs. *Tectonophysics*, **277**, 235–257, https://doi.org/10.1016/S0040-1951(97)00028-0

MJELDE, R., DIGRANES, P. *ET AL.* 1998. Crustal structure of the northern part of the Vøring Basin, mid-Norway margin, from wide-angle seismic and gravity data. *Tectonophysics*, **293**, 175–205, https://doi.org/10.1016/s0040-1951(98)00090-0

MJELDE, R., DIGRANES, P. *ET AL.* 2001. Crustal structure of the outer Vøring Plateau, offshore Norway, from ocean bottom seismic and gravity data. *Journal of Geophysical Research*, **106**, 6769–6791, https://doi.org/10.1029/2000jb900415

MJELDE, R., AURVÅG, R., KODAIRA, S., SHIMAMURA, H., GUNNARSSON, K., NAKANISHI, A. & SHIOBARA, H. 2002*a*. V_p/V_s-ratios from the central Kolbeinsey Ridge to the Jan Mayen Basin, North Atlantic; implications for lithology, porosity and present-day stress field. *Marine Geophysical Research*, **23**, 123–145, https://doi.org/10.1023/a:1022439707307

MJELDE, R., BREIVIK, A.J. *ET AL.* 2002*b*. Geological development of the Sørvestsnaget Basin, SW Barents Sea, from ocean bottom seismic, surface seismic and potential field data. *Norwegian Journal of Geology*, **82**, 183–202.

MJELDE, R., RAUM, T., DIGRANES, P., SHIMAMURA, H., SHIOBARA, H. & KODAIRA, S. 2003. V_p/V_s ratio along the Vøring Margin, NE Atlantic, derived from OBS data: implications on lithology and stress field. *Tectonophysics*, **369**, 175–197, https://doi.org/10.1016/s0040-1951(03)00198-7

MJELDE, R., RAUM, T. *ET AL.* 2005. Continent-ocean transition on the Vøring Plateau, NE Atlantic, derived from densely sampled ocean bottom seismometer data. *Journal of Geophysical Research*, **110**, B05101, https://doi.org/10.1029/2004jb003026

MJELDE, R., RAUM, T., MURAI, Y. & TAKANAMI, T. 2007. Continent–ocean-transitions: review, and a new tectono-magmatic model of the Vøring Plateau, NE Atlantic. *Journal of Geodynamics*, **43**, 374–392, https://doi.org/10.1016/j.jog.2006.09.013

MJELDE, R., RAUM, T., BREIVIK, A. & FALEIDE, J. 2008. Crustal transect across the North Atlantic. *Marine Geophysical Research*, **29**, 73–87, https://doi.org/10.1007/s11001-008-9046-9

MJELDE, R., RAUM, T., KANDILAROV, A., MURAI, Y. & TAKANAMI, T. 2009. Crustal structure and evolution of the outer Møre Margin, NE Atlantic. *Tectonophysics*, **468**, 224–243, https://doi.org/10.1016/j.tecto.2008.06.003

MOLINARI, I. & MORELLI, A. 2011. EPcrust: a reference crustal model for the European Plate. *Geophysical*

Journal International, **185**, 352–364, https://doi.org/10.1111/j.1365-246X.2011.04940.x

MOONEY, W.D. & BROCHER, T.M. 1987. Coincident seismic reflection/refraction studies of the continental lithosphere: a global review. *Reviews of Geophysics*, **25**, 723–742, https://doi.org/10.1029/RG025i004p00723

MOREWOOD, N.C., SHANNON, P.M. & MACKENZIE, G.D. 2004. Seismic stratigraphy of the southern Rockall Basin: a comparison between wide-angle seismic and normal incidence reflection data. *Marine and Petroleum Geology*, **21**, 1149–1163, https://doi.org/10.1016/j.marpetgeo.2004.07.006

MOREWOOD, N.C., MACKENZIE, G.D., SHANNON, P.M., O'REILLY, B.M., READMAN, P.W. & MAKRIS, J. 2005. The crustal structure and regional development of the Irish Atlantic margin region. *In*: DORÉ, A.G. & VINING, B.A. (eds) *Petroleum Geology: North-West Europe and Global Perspectives – Proceedings of the 6th Petroleum Geology Conference*. Geological Society, London, 1023–1033, https://doi.org/10.1144/0061023

MORGAN, R.P.L., BARTON, P.J., WARNER, M., MORGAN, J., PRICE, C. & JONES, K. 2000. Lithospheric structure north of Scotland – I. P-wave modelling, deep reflection profiles and gravity. *Geophysical Journal International*, **142**, 716–736, https://doi.org/10.1046/j.1365-246x.2000.00151.x

NAVIN, D.A., PEIRCE, C. & SINHA, M.C. 1998. The RAMESSES experiment – II. Evidence for accumulated melt beneath a slow spreading ridge from wide-angle refraction and multichannel reflection seismic profiles. *Geophysical Journal International*, **135**, 746–772, https://doi.org/10.1046/j.1365-246X.1998.00709.x

OLAYA, V. & CONRAD, O. 2008. Geomorphometry in SAGA. *In*: HENGL, T. & REUTER, H.I. (eds) *Geomorphometry: Concepts, Software, Applications*. Developments in Soil Science, **33**. Elsevier, Amsterdam, 293–308, https://doi.org/10.1016/S0166-2481(08)00012-3

O'REILLY, B.M., SHANNON, P.M. & VOGT, U. 1991. Seismic studies in the North Celtic Sea Basin: implications for basin development. *Journal of the Geological Society, London*, **148**, 191–195, https://doi.org/10.1144/gsjgs.148.1.0191

O'REILLY, B.M., HAUSER, F., JACOB, A.W.B. & SHANNON, P.M. 1996. The lithosphere below the Rockall Trough: wide-angle seismic evidence for extensive serpentinisation. *Tectonophysics*, **255**, 1–23, https://doi.org/10.1016/0040-1951(95)00149-2

O'REILLY, B.M., HAUSER, F. & READMAN, P.W. 2010. The fine-scale structure of upper continental lithosphere from seismic waveform methods: insights into Phanerozoic crustal formation processes. *Geophysical Journal International*, **180**, 101–124, https://doi.org/10.1111/j.1365-246X.2009.04420.x

PARKIN, C.J. & WHITE, R.S. 2008. Influence of the Iceland mantle plume on oceanic crust generation in the North Atlantic. *Geophysical Journal International*, **173**, 168–188, https://doi.org/10.1111/j.1365-246X.2007.03689.x

PERON-PINVIDIC, G., GERNIGON, L., GAINA, C. & BALL, P. 2012. Insights from the Jan Mayen system in the Norwegian–Greenland sea – I. Mapping of a microcontinent. *Geophysical Journal International*, **191**, 385–412, https://doi.org/10.1111/j.1365-246X.2012.05639.x

POWELL, C.M.R. & SINHA, M.C. 1987. The PUMA experiment west of Lewis, U.K. *Geophysical Journal of the Royal Astronomical Society*, **89**, 259–264, https://doi.org/10.1111/j.1365-246X.1987.tb04417.x

PRICE, C. & MORGAN, J. 2000. Lithospheric structure north of Scotland – II. Poisson's ratios and waveform modelling. *Geophysical Journal International*, **142**, 737–754, https://doi.org/10.1046/j.1365-246x.2000.00204.x

RAUM, T. 2000. *Crustal structure and evolution of the Faeroe, Møre and Vøring margins from wide-angle seismic and gravity data*. PhD thesis, University of Bergen, Norway.

RAUM, T., MJELDE, R. ET AL. 2002. Crustal structure of the southern part of the Vøring Basin, mid-Norway margin, from wide-angle seismic and gravity data. *Tectonophysics*, **355**, 99–126, https://doi.org/10.1016/S0040-1951(02)00136-1

RAUM, T., MJELDE, R. ET AL. 2005. Sub-basalt structures east of the Faroe Islands revealed from wide-angle seismic and gravity data. *Petroleum Geoscience*, **11**, 291–308, https://doi.org/10.1144/1354-079304-627

RAUM, T., MJELDE, R. ET AL. 2006. Crustal structure and evolution of the southern Vøring Basin and Vøring Transform Margin, NE Atlantic. *Tectonophysics*, **415**, 167–202, https://doi.org/10.1016/j.tecto.2005.12.008

RAVAUT, C., CHABERT, A. ET AL. 2005. Wide-angle seismic imaging of the west Hatton margin (North Atlantic): Results of the HADES experiment. *Geophysical Research Abstracts*, **7**, 05509.

REICHE, S., REID, I. & THYBO, H. 2011. Crustal structure of the Greenland-Iceland Ridge inferred from wide-angle seismic data. *Geophysical Research Abstracts*, **13**, EGU2011–EG11623.

RICHARDSON, K.R., SMALLWOOD, J.R., WHITE, R.S., SNYDER, D.B. & MAGUIRE, P.K.H. 1998. Crustal structure beneath the Faroe Islands and the Faroe–Iceland Ridge. *Tectonophysics*, **300**, 159–180, https://doi.org/10.1016/s0040-1951(98)00239-x

RICHARDSON, K.R., WHITE, R.S., ENGLAND, R.W. & FRUEHN, J. 1999. Crustal structure east of the Faroe Islands: mapping sub-basalt sediments using wide-angle seismic data. *Petroleum Geoscience*, **5**, 161–172, https://doi.org/10.1144/petgeo.5.2.161

RICKERS, F., FICHTNER, A. & TRAMPERT, J. 2013. The Iceland–Jan Mayen plume system and its impact on mantle dynamics in the North Atlantic region: evidence from full-waveform inversion. *Earth and Planetary Science Letters*, **367**, 39–51, https://doi.org/10.1016/j.epsl.2013.02.022

RITZMANN, O. & JOKAT, W. 2003. Crustal structure of northwestern Svalbard and the adjacent Yermak Plateau: evidence for Oligocene detachment tectonics and non-volcanic breakup. *Geophysical Journal International*, **152**, 139–159, https://doi.org/10.1046/j.1365-246X.2003.01836.x

RITZMANN, O., JOKAT, W., MJELDE, R. & SHIMAMURA, H. 2002. Crustal structure between the Knipovich Ridge and the Van Mijenfjorden (Svalbard). *Marine Geophysical Research*, **23**, 379–401, https://doi.org/10.1023/B:MARI.0000018168.89762.a4

Ritzmann, O., Jokat, W., Czuba, W., Guterch, A., Mjelde, R. & Nishimura, Y. 2004. A deep seismic transect from Hovgård Ridge to northwestern Svalbard across the continental-ocean transition: a sheared margin study. *Geophysical Journal International*, **157**, 683–702, https://doi.org/10.1111/j.1365-246X.2004.02204.x

Ritzmann, O., Maercklin, N., Faleide, J.I., Bungum, H., Mooney, W.D. & Detweiler, S.T. 2007. A three-dimensional geophysical model of the crust in the Barents Sea region: model construction and basement characterization. *Geophysical Journal International*, **170**, 417–435, https://doi.org/10.1111/j.1365-246X.2007.03337.x

Roberts, A.W., White, R.S. & Christie, P.A.F. 2009. Imaging igneous rocks on the North Atlantic rifted continental margin. *Geophysical Journal International*, **179**, 1024–1038, https://doi.org/10.1111/j.1365-246X.2009.04306.x

Roberts, D.G., Ginzberg, A., Nunn, K. & McQuillin, R. 1988. The structure of the Rockall Trough from seismic refraction and wide-angle reflection measurements. *Nature*, **332**, 632–635, https://doi.org/10.1038/332632a0

Rouzo, S., Klingelhöfer, F. *et al.* 2006. 2-D and 3-D modelling of wide-angle seismic data: an example from the Vøring volcanic passive margin. *Marine Geophysical Research*, **27**, 181–199, https://doi.org/10.1007/s11001-006-0001-3

Schiffer, C., Balling, N., Jacobsen, B.H., Stephenson, R.A. & Nielsen, S.B. 2014. Seismological evidence for a fossil subduction zone in the East Greenland Caledonides. *Geology*, **42**, 311–314, https://doi.org/10.1130/G35244.1

Schilling, J.G., Kingsley, R., Fontignie, D., Poreda, R. & Xue, S. 1999. Dispersion of the Jan Mayen and Iceland mantle plumes in the Arctic: a He-Pb-Nd-Sr isotope tracer study of basalts from the Kolbeinsey, Mohns, and Knipovich Ridges. *Journal of Geophysical Research*, **104**, 10,543–10,569, https://doi.org/10.1029/1999jb900057

Schlindwein, V. 2006. On the use of teleseismic receiver functions for studying the crustal structure of Iceland. *Geophysical Journal International*, **164**, 551–568, https://doi.org/10.1111/j.1365-246X.2006.02861.x

Schlindwein, V. & Jokat, W. 1999. Structure and evolution of the continental crust of northern east Greenland from integrated geophysical studies. *Journal of Geophysical Research*, **104**, 15,227–15,245, https://doi.org/10.1029/1999JB900101

Schmidt-Aursch, M.C. & Jokat, W. 2005. The crustal structure of central East Greenland – I: from the Caledonian orogen to the Tertiary igneous province. *Geophysical Journal International*, **160**, 736–752, https://doi.org/10.1111/j.1365-246X.2005.02515.x

Scrutton, R.A. 1970. Results of a seismic refraction experiment on Rockall Bank. *Nature*, **227**, 826–827, https://doi.org/10.1038/227826a0

Scrutton, R.A. 1972. The crustal structure of Rockall Plateau microcontinent. *Geophysical Journal of the Royal Astronomical Society*, **27**, 259–275, https://doi.org/10.1111/j.1365-246X.1972.tb06092.x

Sedov, V.V. & Makris, J. 2001. Structure of the Earth's crust of the Iceland-Faeroe Ridge. *Oceanology*, **41**, 900–906.

Shannon, P.M., Jacob, A.W.B., O'Reilly, B.M., Hauser, F., Readman, P.W. & Makris, J. 1999. Structural setting, geological development and basin modelling in the Rockall Trough. *In*: Fleet, A.J. & Boldy, S.A.R. (eds) *Petroleum Geology of Northwest Europe: Proceedings of the 5th Conference*. Geological Society, London, 421–431, https://doi.org/10.1144/0050421

Sinha, M.C., Constable, S.C. *et al.* 1998. Magmatic processes at slow spreading ridges: implications of the RAMESSES experiment at 57° 45′ N on the Mid-Atlantic Ridge. *Geophysical Journal International*, **135**, 731–745, https://doi.org/10.1046/j.1365-246X.1998.00704.x

Smallwood, J.R. & White, R.S. 1998. Crustal accretion at the Reykjanes Ridge, 61–62°N. *Journal of Geophysical Research*, **103**, 5185–5201, https://doi.org/10.1029/97jb03387

Smallwood, J.R., White, R.S. & Minshull, T.A. 1995. Sea-floor spreading in the presence of the Iceland plume: the structure of the Reykjanes Ridge at 61°40′N. *Journal of the Geological Society, London*, **152**, 1023–1029, https://doi.org/10.1144/GSL.JGS.1995.152.01.24

Smallwood, J.R., Towns, M.J. & White, R.S. 2001. The structure of the Faeroe-Shetland Trough from integrated deep seismic and potential field modelling. *Journal of the Geological Society, London*, **158**, 409–412, https://doi.org/10.1144/jgs.158.3.409

Smith, L.K., White, R.S., Kusznir, N.J. & iSIMM Team. 2005. Structure of the Hatton Basin and adjacent continental margin. *In*: Doré, A.G. & Vining, B.A. (eds) *Petroleum Geology: North-West Europe and Global Perspectives – Proceedings of the 6th Petroleum Geology Conference*. Geological Society, London, 947–956, https://doi.org/10.1144/0060947

Srivastava, S.P. & Tapscott, C.R. 1986. Plate kinematics of the North Atlantic. *In*: Vogt, P.R. & Tucholke, B.E. (eds) *The Geology of North America, Volume M, The Western North Atlantic Region*. Geological Society of America, Boulder, CO, 379–404.

Staples, R.K., White, R.S., Brandsdóttir, B., Menke, W., Maguire, P.K.H. & McBride, J.H. 1997. Färoe-Iceland Ridge Experiment 1. Crustal structure of northeastern Iceland. *Journal of Geophysical Research*, **102**, 7849–7866, https://doi.org/10.1029/96jb03911

Stein, M.L. 1999. *Interpolation of Spatial Data – Some Theory for Kriging*. Springer Series in Statistics. Springer, New York, https://doi.org/10.1007/978-1-4612-1494-6

Storey, M., Duncan, R.A. & Swisher, C.C. 2007. Paleocene-Eocene thermal maximum and the opening of the Northeast Atlantic. *Science*, **316**, 587–589, https://doi.org/10.1126/science.1135274

Tomlinson, J.P., Denton, P., Maguire, P.K.H. & Booth, D.C. 2006. Analysis of the crustal velocity structure of the British Isles using teleseismic receiver functions. *Geophysical Journal International*, **167**,

223–237, https://doi.org/10.1111/j.1365-246X.2006.03044.x

Vogt, U., Makris, J., O'Reilly, B.M., Hauser, F., Readman, P.W., Jacob, A.W.B. & Shannon, P.M. 1998. The Hatton Basin and continental margin: crustal structure from wide-angle seismic and gravity data. *Journal of Geophysical Research*, **103**, 12,545–12,566, https://doi.org/10.1029/98jb00604

Voss, M. & Jokat, W. 2007. Continent–ocean transition and voluminous magmatic underplating derived from P-wave velocity modelling of the East Greenland continental margin. *Geophysical Journal International*, **170**, 580–604, https://doi.org/10.1111/j.1365-246X.2007.03438.x

Voss, M. & Jokat, W. 2009. From Devonian extensional collapse to early Eocene continental break-up: an extended transect of the Kejser Franz Joseph Fjord of the East Greenland margin. *Geophysical Journal International*, **177**, 743–754, https://doi.org/10.1111/j.1365-246X.2008.04076.x

Voss, M., Schmidt-Aursch, M.C. & Jokat, W. 2009. Variations in magmatic processes along the East Greenland volcanic margin. *Geophysical Journal International*, **177**, 755–782, https://doi.org/10.1111/j.1365-246X.2009.04077.x

Warner, M., Morgan, J., Barton, P., Morgan, P., Price, C. & Jones, K. 1996. Seismic reflections from the mantle represent relict subduction zones within the continental lithosphere. *Geology*, **24**, 39–42, https://doi.org/10.1130/0091-7613(1996)024<0039:SRFTMR>2.3.CO;2

Weigel, W., Flüh, E.R. *et al.* 1995. Investigations of the East Greenland continental margin between 70° and 72° N by deep seismic sounding and gravity studies. *Marine Geophysical Research*, **17**, 167–199, https://doi.org/10.1007/bf01203425

Weir, N.R.W., White, R.S., Brandsdóttir, B., Einarsson, P., Shimamura, H. & Shiobara, H. 2001. Crustal structure of the northern Reykjanes Ridge and Reykjanes Peninsula, southwest Iceland. *Journal of Geophysical Research*, **106**, 6347–6368, https://doi.org/10.1029/2000jb900358

Wessel, P., Smith, W.H.F., Scharroo, R., Luis, J. & Wobbe, F. 2013. Generic mapping tools: improved version released. *Eos, Transactions of the American Geophysical Union*, **94**, 409–410, https://doi.org/10.1002/2013eo450001

White, R.S. 1997. Rift–plume interaction in the North Atlantic. *Philosophical Transactions of the Royal Society of London, Series A*, **355**, 319–339, https://doi.org/10.1098/rsta.1997.0011

White, R.S. & McKenzie, D. 1995. Mantle plumes and flood basalts. *Journal of Geophysical Research*, **100**, 17,543–17,585, https://doi.org/10.1029/ 95jb01585

White, R.S. & Smith, L.K. 2009. Crustal structure of the Hatton and the conjugate east Greenland rifted volcanic continental margins, NE Atlantic. *Journal of Geophysical Research*, **114**, B02305, https://doi.org/10.1029/2008jb005856

White, R.S., McKenzie, D. & O'Nions, R.K. 1992. Oceanic crustal thickness from seismic measurements and rare earth element inversions. *Journal of Geophysical Research*, **97**, 19,683–19,715, https://doi.org/10.1029/92jb01749

White, R.S., Minshull, T.A. *et al.* 1996. Seismic images of crust beneath Iceland contribute to long-standing debate. *Eos, Transactions of the American Geophysical Union*, **77**, 197–201, https://doi.org/10.1029/96eo00134

White, R.S., Fruehn, J., Richardson, K.R., Cullen, E., Kirk, W., Smallwood, J.R. & Latkiewicz, C. 1999. Faeroes Large Aperture Research Experiment (FLARE): imaging through basalt. *In*: Fleet, A.J. & Boldy, S.A.R. (eds) *Petroleum Geology of Northwest Europe: Proceedings of the 5th Conference*. Geological Society, London, 1243–1252, https://doi.org/10.1144/0051243

Ziegler, P. (ed.) 1988. *Evolution of the Arctic–North Atlantic and the Western Thetys*. American Association of Petroleum Geologists, Memoirs, **43**.

A 3D regional crustal model of the NE Atlantic based on seismic and gravity data

C. HAASE[1]*, J. EBBING[1,2] & T. FUNCK[3]

[1]*Geological Survey of Norway, Leiv Eirikssons vei 39, 7040 Trondheim, Norway*

[2]*Present address: Department of Geosciences, Kiel University, 24118 Kiel, Germany*

[3]*Geological Survey of Denmark and Greenland, Øster Voldgade 10, 1350 Copenhagen K, Denmark*

**Correspondence: claudia.haase@ngu.no*

Abstract: We present a 3D regional crustal model for the North Atlantic, which is based on the integration of seismic constraints and gravity data. The model addresses the crustal thickness geometry, and includes information on sedimentary thickness, the presence of high-velocity zones in the lower crust, and information on the crustal density distribution in the continental and oceanic domains. Using an iterative forward- and inverse-modelling approach, we adhere to the seismic constraints within their uncertainty, but manage to enhance the crustal geometry in areas where seismic data are sparse or absent. A number of basins are resolved with more detail. Recently released seismic reflection data beneath the NE Greenland Shelf allowed for a major improvement of the crustal thickness estimates. Estimated Moho depths beneath the basins there vary between 15 and 25 km, which is compatible with the conjugate Norwegian margin. A major lower-crustal seismic velocity anomaly in the vicinity of the Greenland–Iceland–Faroe Ridge complex is supported by density modelling. We discuss the validity and uncertainties of our model assumptions and discuss the correlation with the main structural elements of the North Atlantic.

Knowledge about the location of the crust–mantle interface (Moho) and the structure of the crust is essential to understand the geological evolution and present-day tectonic setting. The definition of the Moho interface is strongly based on seismic interpretations because it usually shows a strong signal in the data. The same applies to the identification of the top of the basement and different sedimentary sequences. However, seismic data are spatially limited to profiles of individual surveys. In addition, imaging of these interfaces is not always possible because of signal loss (e.g. sub-basalt imaging: Martini & Bean 2002) or because of poor data quality. Data gaps must then be interpolated to facilitate a reconstruction of the geological setting.

Gravity data are often used to validate seismic interpretations as the main density contrast in the Earth is expected between the crust and mantle. High-resolution shipborne data are often acquired simultaneously with seismic surveys, while airborne data are collected on designated surveys. In addition, gravity models generated from satellite data (either satellite altimetry or by direct measurements) cover the entire Earth almost uniformly and are publicly available (e.g. Pail *et al.* 2010; Bouman *et al.* 2013; Andersen *et al.* 2014).

We present results of a 3D regional modelling study of the North Atlantic between Greenland and NW Europe. During the Northeast Atlantic Geoscience Tectonostratigraphic Atlas project (NAG-TEC), a joint project between the geological surveys of NW Europe, an unprecedented amount of geophysical data, including seismic and potential field data, were compiled in one comprehensive atlas (Hopper *et al.* 2014). This source of information was used for the compilation of a novel, seismic-based depth to Moho map as presented by Funck *et al.* (2016), followed by the construction of a lithospheric model in the NE Atlantic region using gravity data. On the scale of the North Atlantic, gravity measurements can be used to study the extent and depth of sedimentary basins, as well as variations in crustal thickness. In order to build the 3D regional model, forward- and inverse-modelling techniques were applied. We used gravity inversion to verify the seismic interpretations of the Moho and to improve the estimations in regions with no seismic coverage. The procedure is described in

From: Péron-Pinvidic, G., Hopper, J. R., Stoker, M. S., Gaina, C., Doornenbal, J. C., Funck, T. & Árting, U. E. (eds) 2017. *The NE Atlantic Region: A Reappraisal of Crustal Structure, Tectonostratigraphy and Magmatic Evolution*. Geological Society, London, Special Publications, **447**, 233–247.
First published online October 12, 2016, updated October 19, 2016, https://doi.org/10.1144/SP447.8

the following, together with a discussion of the resulting implications.

Data

The lithospheric 3D model is based on seismic interpretations, well information and potential field data, as well as on structural information. The study area is outlined by the yellow polygon in the elevation map in Figure 1a.

Seismic data

A wealth of seismic surveys covers the NE Atlantic Ocean and adjacent areas. These data include seismic refraction/wide-angle reflection surveys and receiver functions. During the NAG-TEC project, these data were compiled in one atlas and provided the basis for a new Moho and crustal structure compilation (Funck *et al.* 2014). Supplemented by elevation and gravity data, the seismic interpretations were interpolated and extended into regions without seismic coverage using kriging techniques (Funck *et al.* 2016).

An alternative model based on a similar, although less comprehensive, seismic database has been published by Artemieva & Thybo (2013). This model is, unfortunately, not yet available digitally, but differences to the compilation by Funck *et al.* (2016) are mostly subtle. To some degree, the differences are related to variations in the database but also depend on the inclusion of seismic reflection data, as done by Artemieva & Thybo (2013). The use of reflection data requires a proper velocity function for the time-to-depth conversion, which is not always available.

Gravity data

A variety of shipborne and airborne gravity data exist for the North Atlantic (see Olesen *et al.* 2010 and references therein). Available non-commercial data cover most of the onshore areas, except Greenland where coverage is very sparse. Offshore data are mainly available in the shelf regions in the eastern part of the NE Atlantic but sparser in the central part and lacking in the western part (Fig. 1a). As gravity data are measured from diverse platforms and can be rather old, the data differ in quality and resolution.

Therefore, in order to have a consistent and continuous set of gravity data, we use the satellite-derived DTU10 global gravity field model. A detailed description of the processing is given in Andersen (2010) and Andersen *et al.* (2010). Offshore, the DTU10 model is based on satellite altimetry, onshore it relies on the EGM2008 model (Pavlis *et al.* 2012), which incorporates most of the onshore terrestrial data shown in Figure 1a. In the North Atlantic, satellite altimetry-derived free-air data have an estimated accuracy and resolution of approximately 3 mGal over 10–15 km wavelength (Andersen *et al.* 2010), compared to approximately 1 mGal over 5–10 km wavelength for shipborne data (Dragoi-Stavar & Hall 2009). In the Arctic Ocean, the difference between satellite altimeter data and airborne surveys is around 2–3 mGal on a regional scale (Childers *et al.* 2001). Hence, instead of compiling a new gravity dataset, we use the existing DTU10 model as it provides an acceptable accuracy for our regional modelling study.

The DTU10 model provides free-air gravity (Fig. 1b), which is dominated by topography/bathymetry. Spreading ridges (extinct and active) show up clearly in the data (e.g. Aegir Ridge, Mohns Ridge, Knipovich Ridge, Kolbeinsey Ridge and Reykjanes Ridge), as do major tectonic features such as shelf edges (which may indicate the continent–ocean transition (COT)) and major fracture zones (e.g. Jan Mayen Fracture Zone, Senja Fracture Zone, Greenland Fracture Zone and Bight Fracture Zone).

From the free-air anomaly, the Bouguer anomaly was calculated (Fig. 1c). The Bouguer correction takes into account a standard rock density of 2670 kg m^{-3} for the onshore areas. The Greenland ice sheet thickness was taken from the ETOPO1 Global Relief Model (Amante & Eakins 2009) and is corrected with an ice density of 940 kg m^{-3}. Offshore, the water column is filled with sediments with a density of 2200 kg m^{-3}. This approach eliminates the strong density contrast between water and sediments at the seafloor. The Bouguer anomaly highlights crustal thickness variations across the study area. Positive values correlate with thin but denser oceanic crust, whereas negative values appear in areas of thick, less dense continental crust.

Finally, the tilt derivative (Miller & Singh 1994) of the isostatic anomaly was calculated. The isostatic anomaly is obtained by removing the gravimetric effect of an isostatic model of topographical relief from the Bouguer anomaly. Here, an Airy–Heiskanen root was used, with a normal reference crustal thickness of 32 km and a density contrast of 400 kg m^{-3} at the crust–mantle interface. The gravity signal from the base of the crust has a long wavelength character. Correcting for these effects enhances crustal anomalies and intra-crustal density variations. The tilt derivative is then calculated from the vertical and horizontal gradients of the anomaly, and enhances the edges of these intra-crustal anomalies and highlights linear features (Fig. 1d). It also delineates major structural elements more clearly than the free-air anomaly and

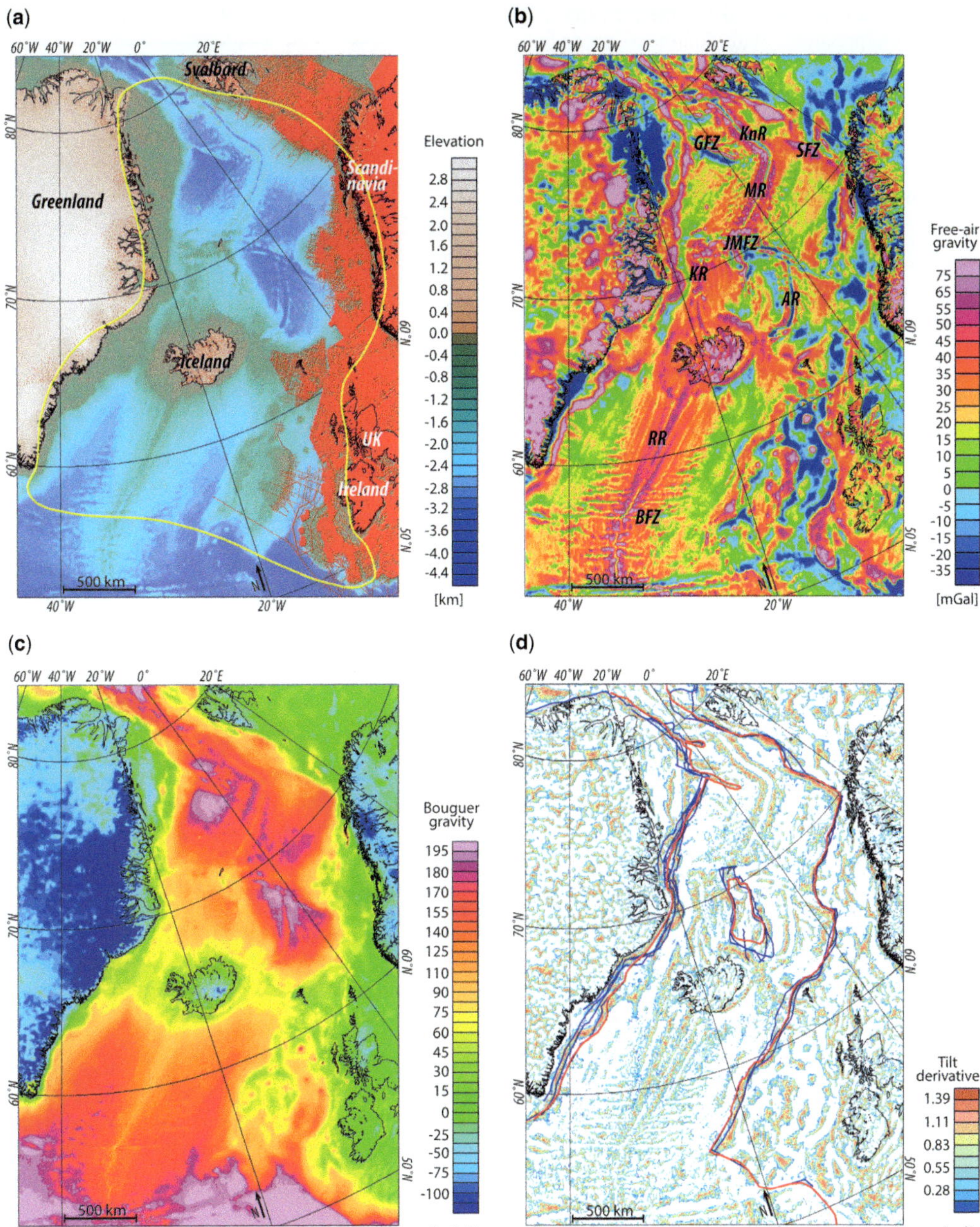

Fig. 1. Elevation and gravity maps based on the DTU10 global gravity field model (Andersen 2010; Andersen *et al.* 2010). (**a**) Bathymetry and coverage with airborne, seaborne and terrestrial gravity data (red dots and lines). The yellow polygon outlines the study area. (**b**) Free-air anomaly. (**c**) Bouguer gravity anomaly, including a correction for ice; (**d**) Tilt derivative of the isostatic anomaly considering an Airy–Heiskanen root, with a reference crustal thickness of 32 km and a density contrast of 400 kg m^{-3} at the crust–mantle interface. Red lines indicate the interpretation of the continent-ocean boundary (COB) after Funck *et al.* (2014); blue lines show COB interpretations after Escher & Pulvertaft (1995), Breivik *et al.* (1999), Scott (2000), Mosar *et al.* (2002*a*, *b*), Sigmond (2002), Hamann *et al.* (2005), Kimbell *et al.* (2005), Tsikalas *et al.* (2005), Gaina *et al.* (2009), Voss *et al.* (2009) and Peron-Pinvidic *et al.* (2012). BFZ, Bight Fracture Zone; AR, Aegir Ridge; GFZ, Greenland Fracture Zone; JMFZ, Jan Mayen Fracture Zone; KnR, Knipovich Ridge; MR, Mohns Ridge; RR, Reykjanes Ridge; SFZ, Senja Fracture Zone.

can, for example, be used to interpret the COT. Recent interpretations of the continent–ocean boundary (COB) are shown by polygons in Figure 1d. The isostatic anomaly was not directly used during the inversion procedure.

Methodology

A 3D density model covering an area of 2600 × 6400 km was built, with a vertical extent of 250 km. The initial model was set up using the GM-SYS 3D modelling software (Geosoft 2014). The model is defined by stacked grids that represent major geological boundaries where density contrasts occur. The resolution of the grids is equal for all horizons and, for this regional study, a grid spacing of 10 km was chosen in order to preserve small-wavelength structures in the upper crust.

The model was later transferred to the forward-modelling software IGMAS+ (Schmidt *et al.* 2010), which facilitates easier access to the model geometry than provided by GM-SYS 3D for the grid-based model. Main components in IGMAS+ models are vertical sections that span the entire model and form a 3D model via triangulation. The model geometry can be modified interactively along these sections. For the structural inversion (Moho depth) and lateral density inversion, the 3D model was transferred back to GM-SYS 3D. Although the two software packages have a different definition of model geometry, the calculated responses are identical, except for some numerical noise.

The model is based mainly on the information described above and some external information for supplementary modelling. Initial key horizons are topography, top basement and Moho. Below, a description of all model layers is given, together with the assigned layer densities (see also Table 1).

Table 1. *Initial density model*

Layer	Density (kg m^{-3})
Water column	2200 (replacement density*)
Sediments	2200–2700 (vertical distribution)
Oceanic crust	2850
Continental crust	
• upper	2750
• lower	2950
Lower crustal bodies	3100
Lithospheric mantle	
• down to 700°C isotherm	3300
• down to 900°C isotherm	3260
• down to base of lithosphere	3240
Asthenospheric mantle	3200

*The replacement density for water refers to a sediment-filled ocean and is in accordance with the Bouguer anomaly correction. Densities for the lithospheric mantle are calculated after Sandwell (2001) with $\rho(T_1) = \rho_m[1 - \alpha(T_1 - T_m)]$, where T_1 is the temperature at the thermal boundary layer, T_m is the mantle temperature, ρ_m is the mantle density (3200 kg m^{-3}) and $\alpha = 3.1 \times 10^{-5}$. C^{-1} is the thermal expansion coefficient.

Initial model

The topography/bathymetry is taken from the DTU10 model and is based on satellite altimeter measurements. Higher-resolution data are available, for example, from the ETOPO1 (Amante & Eakins 2009) or the IBCAO (Jakobsson *et al.* 2012) compilations. However, the DTU10 model is consistent with the gravity model used and provides the necessary resolution. For the modelling, the Bouguer corrected gravity data were used and, therefore, the corresponding density of 2200 kg m^{-3} was assigned to the water column. This is equivalent to replacing the water column with sediments to eliminate the effect of bathymetry and topography.

We did not differentiate between individual sedimentary layers but use the total sediment sequence thickness from Funck *et al.* (2014) and Hopper *et al.* (this volume, in review) to define the depth to the top of the basement. We assigned a density–depth function with an exponential trend that increases from 2200 kg m^{-3} at the seabed to a maximum of 2700 kg m^{-3} at a depth of 8 km below the seafloor and beyond. This accounts for higher densities due to increased compaction with depth. The bottom of the total sediment sequence is defined as the top basement horizon. Interbedded basalts landwards of the COB are generally considered part of the sedimentary column if a deeper basement is imaged. However, given the diversity of data and their quality, it cannot be excluded that top basalt was erroneously interpreted as top basement. In this case, sub-basalt sedimentary rock sequences are missing in the thickness compilation (Funck *et al.* 2014).

The continental crystalline crust is divided by an intra-crustal horizon at a general depth of 20 km, dividing the layer into an upper and lower crust. In the shelf areas, the distribution of upper and lower crust goes over to a constant ratio to simulate a shear model. Densities of 2750 and 2950 kg m^{-3} are assigned to the upper and lower continental crust, respectively, accounting for increasing densities with depth. The oceanic crust differs in composition and is modelled with a density of 2850 kg m^{-3}. The COB follows the interpretation of Funck *et al.* (2014), as shown in Figure 1d, and primarily follows the landwards-most points of seismically defined oceanic crust. In addition, gravity and magnetic data were used for the compilation of the COB, and final adjustments were made using break-up reconstructions (Funck *et al.* 2014).

A new depth to Moho map for the study area, based on seismic refraction data and receiver functions, was compiled by Funck *et al.* (2014) (Fig. 2a). Seismic depth information is only available along the profiles and was then interpolated for regions without data coverage. The interpolation process and the seismic database are described in more detail in Funck *et al.* (2016).

In addition to the crust–mantle interface, regions displaying a high-velocity lower crust (HVLC) were mapped based on the seismic data. These regions occur in the COT zone, and velocities in the lower crust reach values of up to 7800 m s^{-1}. The information on these zones is sparse and limited to the seismic lines. In the 3D density model, these areas are modelled as lower crustal bodies (LCB) with a higher than typical lower crustal density of 3100 kg m^{-3}. The LCB density lies within the typical range of densities for this lower crustal feature (3000–3300 kg m^{-3}: e.g. Mjelde *et al.* 2009; Kvarven *et al.* 2014; Nirrengarten *et al.* 2014).

Gravity data in the oceanic domain contain a thermal effect that is related to the deep lithosphere. The effect is expressed by a long-wavelength gravity field that originates from density changes related to lithospheric cooling with age. In the continental domain, where the lithosphere is thicker, this thermal effect is also present. This gravity field can be modelled and approximated by estimating the depths to isotherms and assigning matching densities. In the oceanic domain, the isotherm calculations are based on the age of the crust as presented in Gaina (2014) and Gaina *et al.* (this volume, in review). Here, the approach from Sandwell (2001 and references therein) was used for the calculation of the isotherms and densities. In the continental domain, the definition of the lithospheric thickness and the depth to the isotherms are based on Artemieva & Mooney (2001) and Artemieva (2006). As the isotherms for the entire study area are based on these two different approaches, a rather sharp transition from the oceanic to the continental domain occurs. This transition has been smoothed by forward modelling of the ultra-long-wavelength gravity field (here we used the IGMAS+ software). The density values for the lithospheric and asthenospheric mantle are relative. This means that the decrease in density with depth is caused by the thermal field, while the effect of pressure and composition, which would lead to an increase of densities with depth, is neglected. The use of absolute

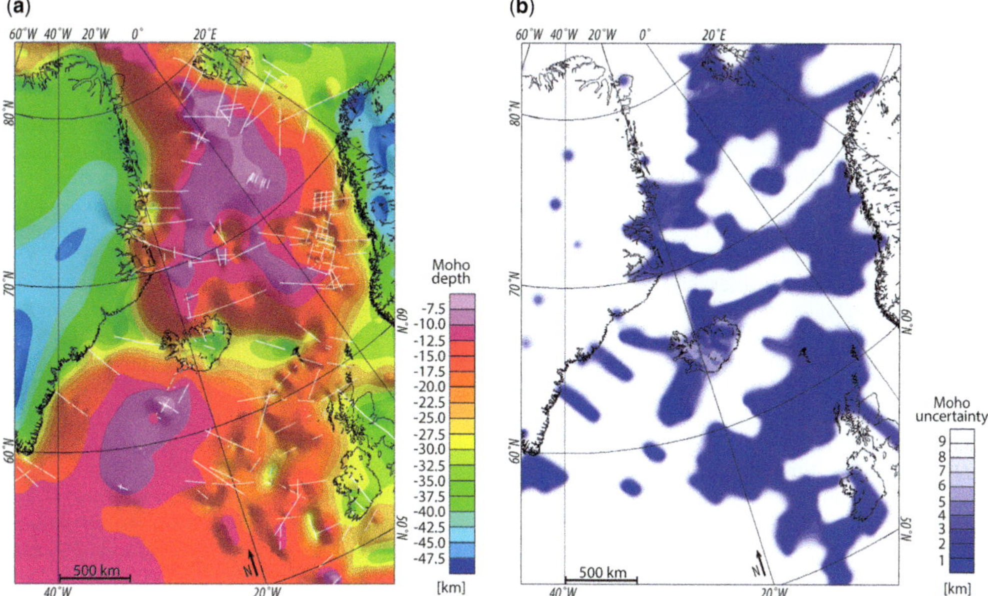

Fig. 2. (**a**) Moho depth map based on seismic interpretations and interpolations (after Funck *et al.* 2014). White lines show seismic profile locations. (**b**) Uncertainty of the Moho depth estimates used as constraints for the gravity inversion. Values along the seismic profiles range from 0.2 to 6.0 km, areas between the profiles have an assigned uncertainty of 10 km.

pressure- and temperature-dependent densities does not affect the gravity field by more than 1%, and would require significantly more complicated computations. Compared to errors in the data and the regional extent of the model, it is acceptable to use the simpler approach and avoid the extensive calculations. The uppermost part of the lithospheric mantle has a density of 3300 kg m^{-3}. The 700°C and 900°C isotherms were then used with densities of 3260 and 3240 kg m^{-3}, respectively. The 1300°C isotherm is regarded as the base of the lithosphere, and a density of 3200 kg m^{-3} was assigned to the asthenospheric mantle underneath.

Inversion

After the initial model set-up and preceding forward modelling, a structural gravity inversion was calculated to verify the Moho depths derived from seismic interpretations and to obtain estimates on the Moho depth in regions with no seismic coverage. The software package GM-SYS 3D, which allows for a structural inversion with the aid of a constraining grid, was used. In order to limit the changes applied to the initial Moho interface, a constraining grid that is based on the confidence in estimates of the Moho depth along seismic refraction profiles (Funck *et al.* 2014) was defined. The uncertainties given for the seismic Moho estimates have a mean of ± 1.5 km, with a maximum of ± 6 km in a few locations. For areas without seismic coverage, an uncertainty of ± 10 km is used. A minimum curvature gridding was used to create the uncertainty grid with a grid-node spacing of 10 km (Fig. 2b).

In addition to the constraining grid, vertical changes of an interface in GM-SYS 3D are restricted by the horizons above and below. Considering that Moho changes are already limited to a maximum of 10 km due to the constraining grid, this is not problematic. However, the grid defining the top of the LCBs is coincident with the Moho grid in all areas with no observed LCB, preventing upwards movement of the Moho. In oceanic areas, the intra-crustal layer boundary is often coincident or very close to the grids defining the Moho and the top of the LCB. Along the mid-Atlantic ridges, downwards movement of the Moho grid is limited by the isotherms that were added to model the thermal effect of the lithospheric mantle. In order to bypass these limitations, the named grids were removed from the model geometry. This simplifies the density structure in a way that the model has too much density in the mantle and too little in the crust. In turn, the calculated gravity will be affected and the measured Bouguer gravity needs to be adjusted to remain comparable. The gravity effects from the residual densities were calculated and the measured gravity corrected accordingly. With this, the inversion can be run with only the main model horizons and the adjusted gravity data to which a 60 km low-pass filter was applied that removes small wavelength signals that are not considered to originate from the Moho.

The inversion reached its convergence criterion within 15 iterations, providing a gravity misfit of 16 mGal standard deviation. Owing to the 10 km resolution of the Moho, in combination with the constraining grid based on the seismic uncertainties, the inverted Moho interface contains some small wavelength features that are considered as artefacts. In order to remove these features, the resulting Moho grid was smoothed with a 100 km-wavelength low-pass Butterworth filter. Finally, the removed layers are put back into the model and the final fit to the measured gravity is calculated.

Results and discussion

The final fit between measured Bouguer anomaly (Fig. 1c) and calculated anomaly from the 3D model has a standard deviation of 22 mGal. The residual gravity map is shown in Figure 3a. The misfit is acceptable for this regional model, where many local intra-crustal sources had to be neglected. An additional source of error and distortion in the model are inaccuracies in the cover sequence, as the compilation includes a significant amount of high-density volcanic rocks (e.g. on the Norwegian margin or around the Faroe Islands) that are not accounted for in the density distribution. Considering the differences between measured and calculated anomalies, the remaining residuals are mainly of short wavelengths, which usually are not associated with the crust–mantle interface. In some areas, large misfits coincide with the tight constraints on the initial data. Here, further adjustment of the model was prevented. This is especially apparent to the south of Iceland along the Reykjanes Ridge, in the Hatton Basin, along the Kolbeinsey Ridge and in the vicinity of the Jan Mayen microcontinent. Also noticeable are the jumps in residuals along the COB: for example, along the Hatton High, the Faroe Platform, the Vøring Basin and the Kangerlussuaq Basin, to name just a few. The density distribution at the COT is more complex than implemented in our model and also contains volcanic sequences, which partly explains the misfit. The correlation of residuals and constraints may suggest that the seismic constraints for Moho and COB need revisiting in some areas.

The residuals in the centre of the study area (Fig. 3a) correlate well with an upper-mantle seismic velocity anomaly that runs from south of Iceland to the Norwegian coast (Bijwaard & Spakman 1999; Weidle & Maupin 2008). The residuals

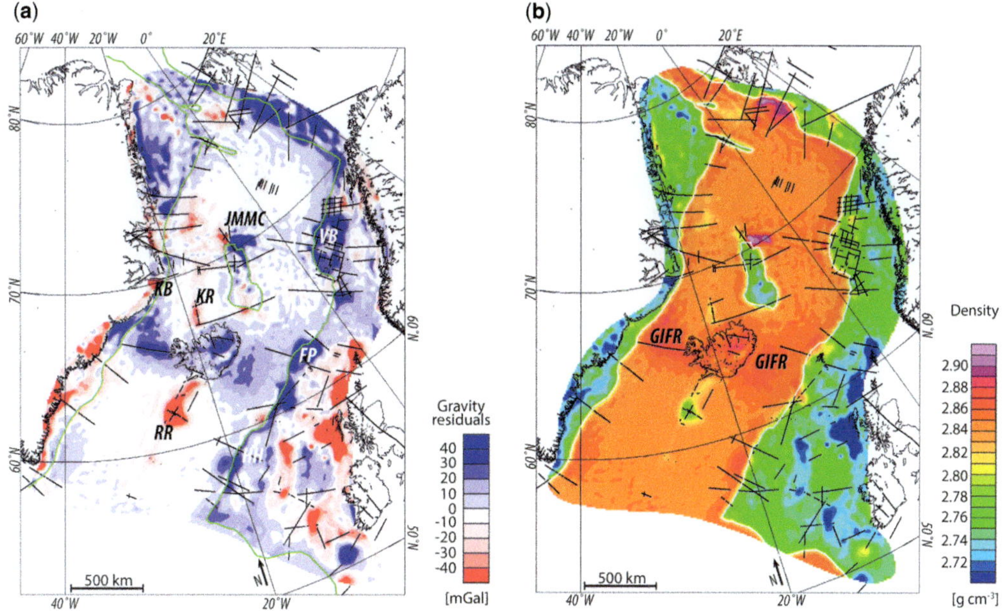

Fig. 3. (**a**) Residual gravity anomaly (measured minus calculated anomaly) after structural inversion. Red colours indicate areas where there is too much mass in the model and blue areas indicate a mass deficit in the model. Black lines show the locations of the seismic lines where gravity inversion was constrained within the confidence level of the seismic data (Fig. 2b). (**b**) Results of density inversions for upper-crustal densities. Initial densities were 2750 kg m^{-3} in the continental domain and 2850 kg m^{-3} in the oceanic domain. The green and white lines show the COB after Funck *et al.* (2014). FP, Faroe Platform; GIFR, Greenland–Iceland–Faroe Ridge; HH, Hatton High; JMMC, Jan Mayen microcontinent; KB, Kangerlussuaq Basin; KR, Kolbeinsey Ridge; RR, Reykjanes Ridge; VB, Vøring Basin.

become smaller towards Norway but this correlation suggests that the definition of the lithospheric mantle might be too simple. The issue could be followed up by combining and improving the model with satellite-derived gravity gradients that are more sensitive to density anomalies at larger depths (Bouman *et al.* 2015) and by focusing more on the mantle composition, as done by Afonso *et al.* (2007). However, the focus of this work was not on a detailed modelling of the mantle but on the crustal structure. The central residuals also correlate with the Greenland–Iceland–Faroe Ridge complex and will be further addressed in the following discussion.

Depth to Moho

The final depth to Moho after gravity inversion is given in Figure 4. Compared to the initial Moho depth (Fig. 2a), the new map provides more detail in areas that lack seismic coverage and previously only showed smooth interpolations. Along the seismic lines, the gravity inversion indicates similar Moho depths as in the initial map. This is related to the fact that changes are limited to the assigned uncertainty in the seismic data (Fig. 2b). However, structural features between the seismic lines become better defined compared to the initial interpolations and a few examples are listed here. The northwards prolongation of the Rockall Basin is clearer, as are the outlines of the Hatton and Møre basins. Off SE Greenland, the Ammassalik Basin (Gerlings *et al.* this volume, in review) is characterized by a shallowing of the Moho that was not seen previously. The NE part of the Norway Basin is now more clearly outlined; the gravity inversion indicates that the shallow Moho extends up to the COB. Very distinct improvements are seen in the Danmarkshavn and Thetis basins. Seismic data in this area are sparse due to the ice cover. However, recently released seismic reflection data allowed for a significantly improved sediment thickness map (Geissler *et al.*, this volume, in review) covering this area. These data are included in our 3D model, and allowed for an improved and more detailed image of the Moho depth (Fig. 4) when compared to the rather smooth Moho map based on extrapolation of the seismic refraction data (Fig. 2a). Moho depths beneath the NE Greenland Shelf vary between 15 and 25 km, and are consistent with the Norwegian margin on the conjugate side.

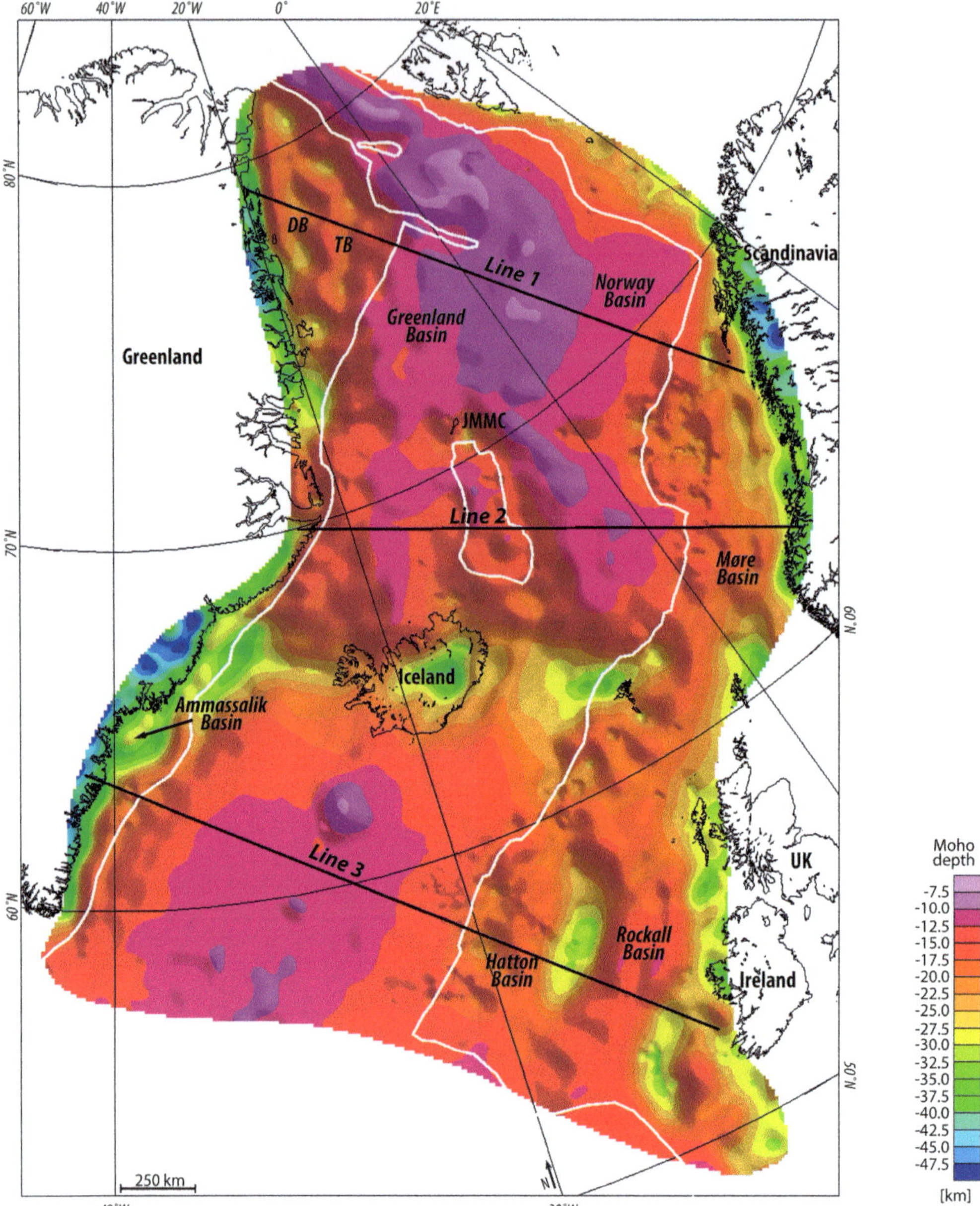

Fig. 4. Depth to Moho based on the structural gravity inversion. The white lines show the interpretation of the COB after Funck *et al.* (2014) and the black lines mark the locations of the 2D cross-sections shown in Figure 6. DB, Danmarkshavn Basin; JMMC, Jan Mayen microcontinent; TB, Thetis Basin.

Crustal thickness

The crustal thickness is defined as the thickness of the crystalline crust from top basement to the Moho (Fig. 5). Underneath the sedimentary basins, crust is usually thin and major basins, such as the Rockall, Hatton, Møre and North Danmarkshavn basins, are characterized by a crustal thickness of around 10 km, but can locally be less than 6 km (Møre and Rockall basins). The igneous complex comprising the Greenland–Iceland–Faroe Ridge is characterized by a crustal thickness of around

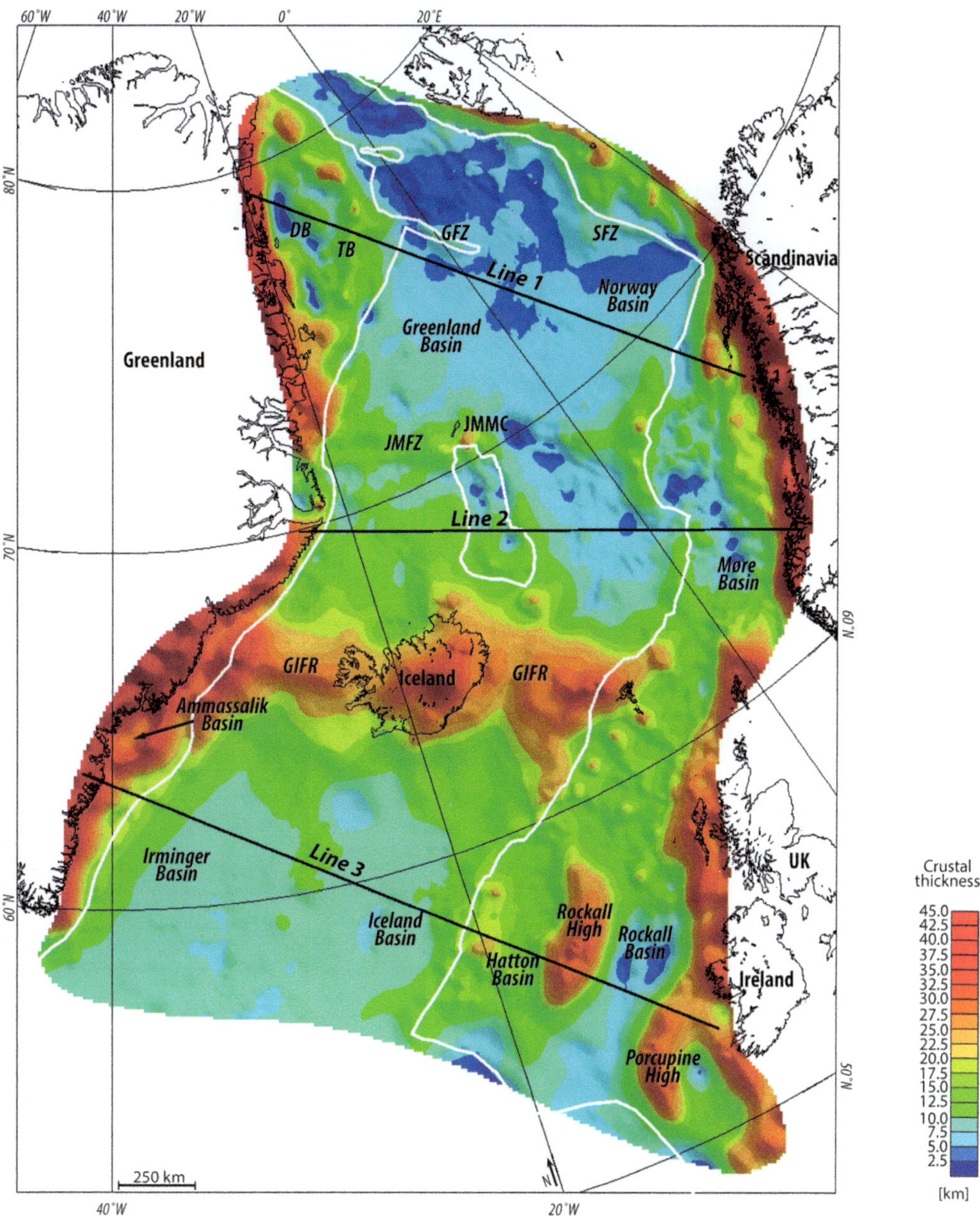

Fig. 5. Crustal thickness based on the structural gravity inversion. The white lines show the interpretation of the COB after Funck *et al.* (2014) and the black lines mark the locations of the 2D cross-sections shown in Figure 6. DB, Danmarkshavn Basin; GIFR, Greenland–Iceland–Faroe Ridge complex; GFZ, Greenland Fracture Zone; JMFZ, Jan Mayen Fracture Zone; SFZ, Senja Fracture Zone; TB, Thetis Basin.

30 km. Similar values are observed beneath the Rockall and Porcupine highs. The central part of the study area overlaps with previous studies incorporating gravity data (e.g. Greenhalgh & Kusznir 2007; Torsvik *et al.* 2015). The crustal thickness of the model here is slightly thicker than in these studies. However, the results still produce gravity residuals here and, therefore, hold a degree of uncertainty. Unfortunately, neither of the previous two studies shows how well the models fit the data.

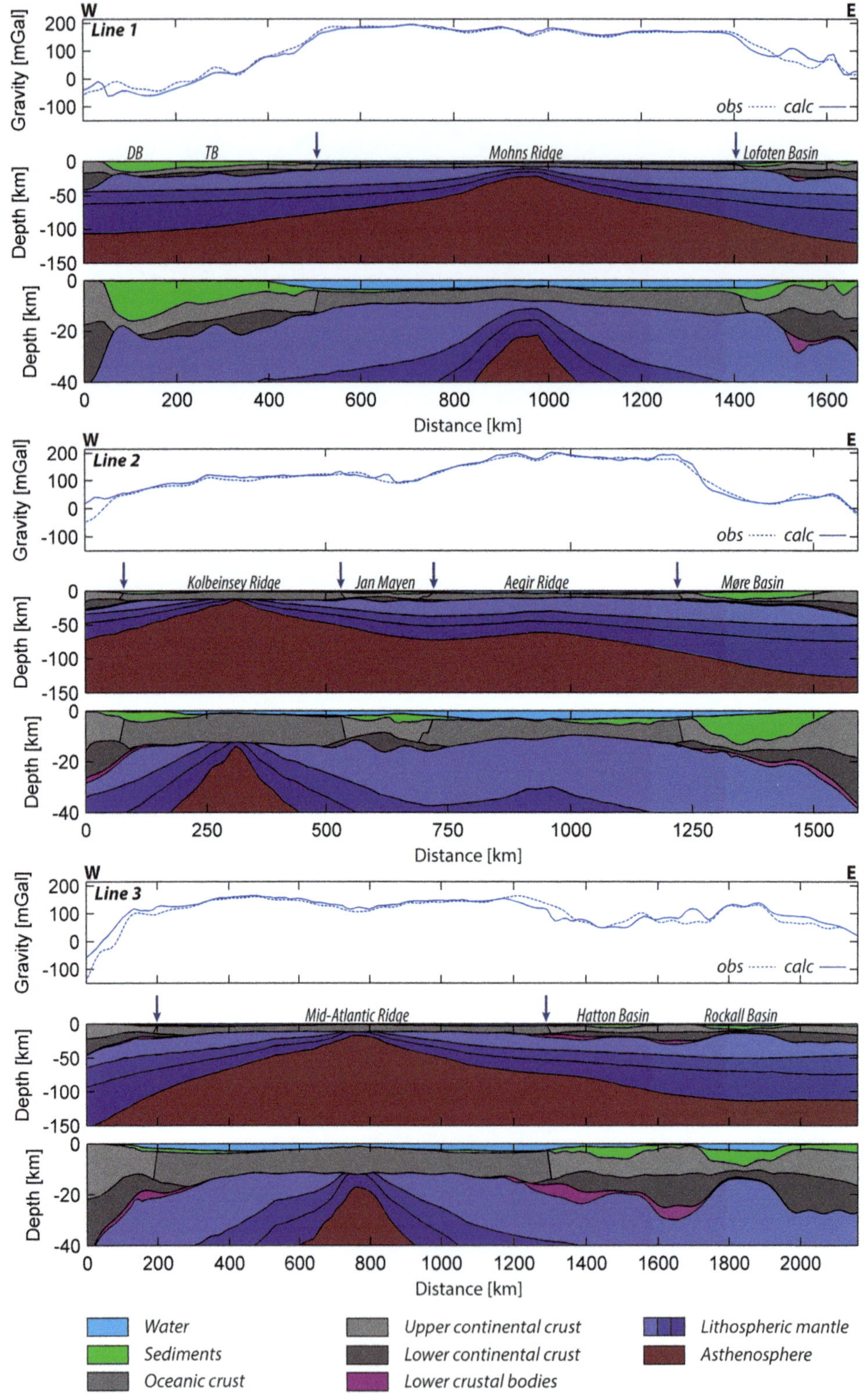

W
E
Line 1
Gravity [mGal]
obs calc
DB
TB
Mohns Ridge
Lofoten Basin
Depth [km]
Distance [km]
Line 2
Kolbeinsey Ridge
Jan Mayen
Aegir Ridge
Møre Basin
Line 3
Mid-Atlantic Ridge
Hatton Basin
Rockall Basin
Water
Sediments
Oceanic crust
Upper continental crust
Lower continental crust
Lower crustal bodies
Lithospheric mantle
Asthenosphere

Larger discrepancies occur at the northern part of the Jan Mayen microcontinent, where the crustal thickness is, in most parts, shallower in comparison with that reported in Torsvik *et al.* (2015). This part, however, is well constrained by seismic data (Fig. 2) and the results here agree, for example, with Kodaira *et al.* (1998). Along the Reykjanes Ridge, the results are strongly influenced by the seismic constraints, as mentioned above (Fig. 3a). Similarly, the seismic constraints are responsible for the thinner crust around the Faroe and British islands compared to that discussed in Torsvik *et al.* (2015).

2D cross-sections

In Figure 6, three vertical Trans-Atlantic cross-sections through the final lithospheric model are shown. The location of the cross-sections is shown in Figures 4 and 5. The profiles show a generally good fit between observed and calculated gravity, but major residuals are present: (1) around most of the COBs; (2) in areas where volcanics are being neglected in the sediment sequence; (3) close to onshore areas; and (4) partly where seismic information is sparse or does not allow for further adjustments within the seismic error bounds. These four cases of increased misfit will now be examined in more detail.

(1) In the density model, the COB of Funck *et al.* (2014) is used to distinguish between oceanic and continental upper crust, as well as oceanic and continental lithospheric mantle. In recent years, alternative interpretations of the COB have been published (see Funck *et al.* 2014 for a summary) differing by up to approximately 80 km off East Greenland (Fig. 1d). A number of the published COBs cross-cut gravity anomalies on the tilt-derivative map (Fig. 1d), which is unusual as the tilt derivatives generally correlate with crustal structures. However, the COB on glaciated margins is complicated by the thick wedge of glacial sediments, the strong lateral density contrasts of which might interfere with the crustal gravity signal. Uncertainties in the location of the COB are reflected in the 2D sections. Even though we implemented a slightly inclined COB (due to the size of the grid cells and minimum-curvature gridding), the changes in crustal densities at this boundary are sharp. In reality, the setting is much more complex and sharp changes are unlikely. Furthermore, the density distribution of the glacial sediments may not be well accounted for. These simplifications are then reflected in the results.

(2) The model uses the total sediment thickness for the estimation of the top basement depth. This sequence, as described earlier, also contains basalts and volcanic rocks, the higher densities of which are now missing in the model. Depending on the volume, this effect can be of the order of tens of mGal. This most probably explains part of the errors in the eastern portions of the lines (e.g. parts of the Lofoten, Møre and Hatton basins). No volcanic rocks were mapped in the Danmarkshavn Basin (Geissler *et al.*, this volume, in review) and the mass deficit here might, instead, be explained by too low sediment densities in this very deep (*c.* 18 km) basin (Line 1).

(3) The onshore parts of Norway and Greenland mark the borders of the study area, and discrepancies here are to be expected. Owing to the lack of seismic information, these areas were only roughly modelled to reduce edge effects. Detailed modelling of the intra-crustal geometry and even of upper-mantle sources (due to varying composition), as in more dedicated modelling studies, was not carried out. Examples for such studies (3D and 2D) can be found for Scandinavia (Ebbing *et al.* 2012; Gradmann *et al.* 2014) and Greenland (Medvedev *et al.* 2013; Schiffer *et al.* 2015).

(4) The western part of Line 1 crosses the COB and the Danmarkshavn and Thetis basins. This is an area where the initial Moho depths were interpolated based on very sparse seismic refraction coverage. Owing to the absence of detailed seismic information and the probability that local, unmodelled sources are present (e.g. high-density LCB as mapped on the conjugate Norwegian side: Funck *et al.* 2014), the gravity does not match perfectly.

Implications from density inversion

Part of the shorter wavelength gravity residuals (Fig. 3a) can be attributed to variations in crustal density. A density inversion, allowing for laterally varying densities in the upper crustal layer, confirms

Fig. 6. Three cross-sections through the final 3D lithospheric model. (Top panels) Observed and calculated gravity. (Middle panels) Model cross-section over the entire model depth. (Bottom panels) Enlargement of the cross-sections showing only the top 40 km. Locations of the sections are indicated by black lines in Figures 4 and 5. The blue arrows indicate the locations of the COB. Densities are listed in Table 1; the upper-crustal density is given in Figure 3b. DB, Danmarkshavn Basin; TB, Thetis Basin.

this (Fig. 3b). The initial densities of 2850 and 2750 kg m^{-3} for oceanic and upper crust, respectively, were allowed to vary between 2700 and 2950 kg m^{-3}.

The resulting density distribution (Fig. 3b) is close to the initial values in most parts of the study area. Larger changes are mainly observed in the oceanic domain, where large misfits after structural inversion occurred due to (1) the seismic constraints and (2) along the Greenland–Iceland–Faroe Ridge complex. Along this ridge, the velocity of the lower oceanic crust deviates from standard values and exceeds 7 km s^{-1} (Richardson *et al.* 1998; Holbrook *et al.* 2001), which consequently implies higher densities than the 2850 kg m^{-3} used here. The increased crustal densities suggested by the inversion can be attributed to the lower-crustal mass deficit. Trials have shown that an increase in lower-crustal densities by up to 100 kg m^{-3} provides a noticeable improvement in the gravity fit. Integration of such density changes in the model requires more detailed modelling of the respective area than done in this study. Furthermore, uncertainties in the COB are reflected in the density results, in particular by increased densities around and south of the Faroe Islands and at the conjugate side offshore SE Greenland. The previously discussed neglect of volcanic rocks in the sediment cover also contributes to density changes in the upper crust where the inversion tries to compensate for the mass deficit.

Conclusions

Seismic interpretations and a global gravity model were used to construct a regional 3D density model for the NE Atlantic. While the initial geometry of the density model is defined by the seismic information, the gravity data helped to verify and improve Moho depth estimations, especially in areas of sparse seismic coverage. The Moho depth model enhances local details based on a reasonable model set-up. Owing to the regional character of the model, intra-crustal sources (LCB) are only modelled to a certain extent and distortions caused by neglected basalts in the sediment cover are not taken into account. Considering the ambiguity of potential field data, the uncertainties and applied simplifications, many possible interpretations for the North Atlantic crustal structure exist. The model here presents one of the possible solutions, the one that fits the input constraints best. It is well suited as a starting point for further, more local and detailed studies.

The crustal thickness model outlines the main structural elements of the North Atlantic region. The crustal model is well constrained and some of the remaining gravity residuals probably originate from deeper sources. The model is therefore an important contribution to identify the reason for observed upper-mantle velocity anomalies (Bijwaard & Spakman 1999; Weidle & Maupin 2008), and to understand the evolution of the North Atlantic topography and its surroundings. Furthermore, the inverse- and forward-modelling procedure can be applied to test the reliability of other seismically derived crustal thickness models (e.g. Artemieva & Thybo 2013).

The modelling results, together with the tilt-derivative gravity map, can be used to better understand the location of the COB. The residuals in the gravity (Fig. 3a) and results from the crustal density inversion (Fig. 3b), as well as a more detailed look along 2D sections (Fig. 5), show the additional value of gravity modelling. However, for a full discussion and better resolution of the boundary, higher-resolution modelling is required. In this context, the integration of magnetic data is essential.

This work formed part of the NAG-TEC project and we acknowledge the support of the industry sponsors (in alphabetical order): Bayerngas Norge AS; BP Exploration Operating Company Ltd, Bundesanstalt für Geowissenschaften und Rohstoffe (BGR); Chevron East Greenland Exploration A/S; ConocoPhillips Skandinavia AS; DEA Norge AS; Det norske oljeselskap ASA; DONG E&P A/S; E.ON Norge AS; ExxonMobil Exploration and Production Norway AS; Japan Oil, Gas and Metals National Corporation (JOGMEC); Maersk Oil; Nalcor Energy – Oil and Gas Inc.; Nexen Energy ULC; Norwegian Energy Company ASA (Noreco); Repsol Exploration Norge AS; Statoil (UK) Ltd; and Wintershall Holding GmbH. We would also like to thank Sofie Gradmann and Odleiv Olesen for fruitful and constructive discussions, as well as two anonymous reviewers.

References

Afonso, J.C., Ranalli, G. & Fernàndez, M. 2007. Density structure and buoyancy of the oceanic lithosphere revisited. *Geophysical Research Letters*, **34**, L10302, https://doi.org/10.1029/2007GL029515

Amante, C. & Eakins, B.W. 2009. *ETOPO1 1 Arc-Minute Global Relief Model: Procedures, Data Sources and Analysis*. NOAA Technical Memorandum NESDIS NGDC-24. National Geophysical Data Center, NOAA, Boulder, CO, USA, https://doi.org/10.7289/V5C8276M

Andersen, O.B. 2010. The DTU10 gravity field and mean sea surface. Paper presented at the *Second International Symposium of the Gravity Field of the Earth (IGFS2)*, 20–22 September 2010, Fairbanks, AK, USA.

Andersen, O.B., Knudsen, P. & Berry, P.A.M. 2010. The DNSC08GRA global marine gravity field from double retracked satellite altimetry. *Journal of Geodesy*, **84**, 191–199, https://doi.org/10.1007/s00190-009-0355-9

ANDERSEN, O.B., KNUDSEN, P., KENYON, S. & HOLMES, S. 2014. Global and arctic marine gravity field from recent satellite altimetry (DTU13). *In*: *76th EAGE Conference and Exhibition 2014, Extended Abstracts, Amsterdam*, https://doi.org/10.3997/2214-4609.20140897

ARTEMIEVA, I.M. 2006. Global 1 × 1° thermal model TC1 for the continental lithosphere: implications for lithosphere secular evolution. *Tectonophysics*, **416**, 245–277, https://doi.org/10.1016/j.tecto.2005.11.022

ARTEMIEVA, I.M. & MOONEY, W.D. 2001. Thermal thickness and evolution of Precambrian lithosphere: a global study. *Journal of Geophysical Research*, **106**, 16,387–16,414, https://doi.org/10.1029/2000JB900439

ARTEMIEVA, I.M. & THYBO, H. 2013. EUNAseis: a seismic model for Moho and crustal structure in Europe, Greenland, and the North Atlantic region. *Tectonophysics*, **609**, 97–153, https://doi.org/10.1016/j.tecto.2013.08.004

BIJWAARD, H. & SPAKMAN, W. 1999. Tomographic evidence for a narrow whole mantle plume below Iceland. *Earth and Planetary Science Letters*, **166**, 121–126, https://doi.org/10.1016/S0012-821X(99)00004-7

BOUMAN, J., FLOBERGHAGEN, R. & RUMMEL, R. 2013. More than 50 years of progress in satellite gravimetry. *Eos, Transactions of the American Geophysical Union*, **94**, 269–270, https://doi.org/10.1002/2013EO310001

BOUMAN, J., EBBING, J. *ET AL*. 2015. GOCE gravity gradient data for lithospheric modeling. *International Journal of Applied Earth Observation and Geoinformation*, **35**, 16–30, https://doi.org/10.1016/j.jag.2013.11.001

BREIVIK, A.J., VERHOEF, J. & FALEIDE, J.I. 1999. Effect of thermal contrasts on gravity modeling at passive margins: results from the western Barents Sea. *Journal of Geophysical Research*, **104**, 15,293–15,311, https://doi.org/10.1029/1998jb900022

CHILDERS, V.A., MCADOO, D.C., BROZENA, J.M. & LAXON, S.W. 2001. New gravity data in the Arctic Ocean – Comparison of airborne and ERS gravity. *Journal of Geophysical Research*, **106**, 8871–8886, https://doi.org/10.1029/2000JB900405

DRAGOI-STAVAR, D. & HALL, S. 2009. Gravity modeling of the ocean–continent transition along the South Atlantic margins. *Journal of Geophysical Research*, **114**, B09401, https://doi.org/10.1029/2008JB006014

EBBING, J., ENGLAND, R.W., KORJA, T., LAURITSEN, T., OLESEN, O., STRATFORD, W. & WEIDLE, C. 2012. Structure of the Scandes lithosphere from surface to depth. *Tectonophysics*, **536**, 1–24, https://doi.org/10.1016/j.bbr.2011.03.031

ESCHER, J.C. & PULVERTAFT, T.C.R. 1995. *Geological Map of Greenland, Scale 1:2 500 000*. Geological Survey of Greenland, Copenhagen, Denmark.

FUNCK, T., HOPPER, J.R. *ET AL*. 2014. Crustal structure. *In*: HOPPER, J.R., FUNCK, T., STOKER, M., ÁRTING, U., PERON-PINDIVIC, G., DOORNENBAL, H. & GAINA, C. (eds) *NAGTEC Atlas: Tectonostratigraphic Atlas of the North-East Atlantic Region*. Geological Survey of Denmark and Greenland (GEUS), Copenhagen, Denmark, 69–126.

FUNCK, T., GEISSLER, W.H., KIMBELL, G.S., GRADMANN, S., ERLENDSSON, Ö., MCDERMOTT, K.G. & PETERSEN, U.K. 2016. Moho and basement depth in the NE Atlantic Ocean based on seismic refraction data and receiver functions. *In*: PÉRON-PINVIDIC, G., HOPPER, J.R., STOKER, M.S., GAINA, C., DOORNENBAL, J.C., FUNCK, T. & ÁRTING, U.E. (eds) *The NE Atlantic Region: A Reappraisal of Crustal Structure, Tectonostratigraphy and Magmatic Evolution*. Geological Society, London, Special Publications, **447**. First published online July 13, 2016, https://doi.org/10.1144/SP447.1

GAINA, C. 2014. Plate reconstructions and regional kinematics. *In*: HOPPER, J.R., FUNCK, T., STOKER, M., ÁRTING, U., PERON-PINDIVIC, G., DOORNENBAL, H. & GAINA, C. (eds) *NAGTEC Atlas: Tectonostratigraphic Atlas of the North-East Atlantic Region*. Geological Survey of Denmark and Greenland, Copenhagen, Denmark, 53–66.

GAINA, C., GERNIGON, L. & BALL, P. 2009. Palaeocene–Recent plate boundaries in the NE Atlantic and the formation of the Jan Mayen microcontinent. *Journal of the Geological Society, London*, **166**, 601–616, https://doi.org/10.1144/0016-76492008-112

GAINA, C., NASUTI, A., KIMBELL, G.S. & BLISCHKE, A. In review. Break-up and seafloor spreading domains in the NE Atlantic. *In*: PÉRON-PINVIDIC, G., HOPPER, J.R., STOKER, M.S., GAINA, C., DOORNENBAL, J.C., FUNCK, T. & ÁRTING, U.E. (eds) *The NE Atlantic Region: A Reappraisal of Crustal Structure, Tectonostratigraphy and Magmatic Evolution*. Geological Society, London, Special Publications, **447**.

GEISSLER, W.H., HOPPER, J.R. *ET AL*. In review. Seismic volcanostratigraphy of the East Greenland continental margin revisited. *In*: PÉRON-PINVIDIC, G., HOPPER, J.R., STOKER, M.S., GAINA, C., DOORNENBAL, J.C., FUNCK, T. & ÁRTING, U.E. (eds) *The NE Atlantic Region: A Reappraisal of Crustal Structure, Tectonostratigraphy and Magmatic Evolution*. Geological Society, London, Special Publications, **447**.

GEOSOFT. 2014. *Geosoft Oasis Montaj Extension – GM-SYS 3D Modelling*. Feature Sheet, Geosoft Inc, Toronto, Canada.

GERLINGS, J., HOPPER, J.R., FYHN, M.B.W. & FRANDSEN, N. In review. Mesozoic and older rift basin on the SE Greenland shelf near Ammassalik. *In*: PÉRON-PINVIDIC, G., HOPPER, J.R., STOKER, M.S., GAINA, C., DOORNENBAL, J.C., FUNCK, T. & ÁRTING, U.E. (eds) *The NE Atlantic Region: A Reappraisal of Crustal Structure, Tectonostratigraphy and Magmatic Evolution*. Geological Society, London, Special Publications, **447**.

GRADMANN, S., EBBING, J. & FULLEA, J. 2014. Integrated geophysical modelling of a lateral transition zone in the lithospheric mantle under Norway and Sweden. *Geophysical Journal International*, **194**, 1358–1373, https://doi.org/10.1093/gji/ggt213

GREENHALGH, E.E. & KUSZNIR, N.J. 2007. Evidence for thin oceanic crust on the extinct Aegir Ridge, Norwegian Basin, NE Atlantic derived from satellite gravity inversion. *Geophysical Research Letters*, **34**, L06305, https://doi.org/10.1029/2007GL029440

HAMANN, N.E., WHITTAKER, R.C. & STEMMERIK, L. 2005. Geological development of the Northeast Greenland Shelf. *In*: DORE, A.G. & VINING, B.A. (eds) *Petroleum*

Geology: Northwest Europe and Global Perspectives – Proceedings of the 6th Petroleum Geology Conference. Geological Society, London, 887–902, https://doi.org/10.1144/0060887

Holbrook, W.S., Larsen, H.C. *et al.* 2001. Mantle thermal structure and active upwelling during continental breakup in the North Atlantic. *Earth and Planetary Science Letters*, **190**, 251–266, https://doi.org/10.1016/s0012-821x(01)00392-2

Hopper, J.R., Funck, T., Stoker, M., Árting, U., Peron-Pinvidic, G., Doornenbal, H. & Gaina, C. 2014. *Tectonostratigraphic Atlas of the North-East Atlantic Region*. Geological Survey of Denmark and Greenland (GEUS), Copenhagen, Denmark.

Hopper, J.R. *et al.* In review. Sediment thickness and residual topography of the North Atlantic: estimating dynamic topography around Iceland. *In*: Péron-Pinvidic, G., Hopper, J.R., Stoker, M.S., Gaina, C., Doornenbal, J.C., Funck, T. & Árting, U.E. (eds) *The NE Atlantic Region: A Reappraisal of Crustal Structure, Tectonostratigraphy and Magmatic Evolution*. Geological Society, London, Special Publications, **447**.

Jakobsson, M., Mayer, L. *et al.* 2012. The International Bathymetric Chart of the Arctic Ocean (IBCAO) Version 3.0. *Geophysical Research Letters*, **39**, L12609, https://doi.org/10.1029/2012gl052219

Kimbell, G.S., Ritchie, J.D., Johnson, H. & Gatliff, R.W. 2005. Controls on the structure and evolution of the NE Atlantic margin revealed by regional potential field imaging and 3D modeling. *In*: Doré, A.G. & Vining, B.A. (eds) *Petroleum Geology: Northwest Europe and Global Perspectives – Proceedings of the 6th Petroleum Geology Conference*. Geological Society, London, 933–945, https://doi.org/10.1144/0060933

Kodaira, S., Mjelde, R., Gunnarsson, K., Shiobara, H. & Shimamura, H. 1998. Structure of the Jan Mayen microcontinent and implications for its evolution. *Geophysical Journal International*, **132**, 383–400, https://doi.org/10.1046/j.1365-246x.1998.00444.x

Kvarven, T., Ebbing, J. *et al.* 2014. Crustal structure across the Møre margin, mid-Norway, from wide-angle seismic and gravity data. *Tectonophysics*, **626**, 21–40, https://doi.org/10.1016/j.tecto.2014.03.021

Martini, F. & Bean, C.J. 2002. Interface scattering v. body scattering in subbasalt imaging and application of prestack wave equation datuming. *Geophysics*, **67**, 1593–1601, https://doi.org/10.1190/1.1512750

Medvedev, S., Souche, A. & Hartz, E.H. 2013. Influence of ice sheet and glacial erosion on passive margins of Greenland. *Geomorphology*, **193**, 36–46, https://doi.org/10.1016/j.geomorph.2013.03.029

Miller, H.G. & Singh, V. 1994. Potential field tilt – a new concept for location of potential field sources. *Journal of Applied Geophysics*, **32**, 213–217, https://doi.org/10.1016/0926-9851(94)90022-1

Mjelde, R., Raum, T., Kandilarov, A., Murai, Y. & Takanami, T. 2009. Crustal structure and evolution of the outer Møre margin, NE Atlantic. *Tectonophysics*, **468**, 224–243, https://doi.org/10.1016/j.tecto.2008.06.003

Mosar, J., Eide, E.A., Osmundsen, P.T., Sommaruga, A. & Torsvik, T.H. 2002*a*. Greenland–Norway separation: a geodynamic model for the North Atlantic. *Norwegian Journal of Geology*, **82**, 281–298.

Mosar, J., Lewis, G. & Torsvik, T.H. 2002*b*. North Atlantic sea-floor spreading rates: implications for the Tertiary development of inversion structures of the Norwegian–Greenland Sea. *Journal of the Geological Society, London*, **159**, 503–515, https://doi.org/10.1144/0016-764901-135

Nirrengarten, M., Gernigon, L. & Manatschal, G. 2014. Lower crustal bodies in the Møre volcanic rifted margin: Geophysical determination and geological implementations. *Tectonophysics*, **636**, 143–157, https://doi.org/10.1016/j.tecto.2014.08.004

Olesen, O., Brönner, M. *et al.* 2010. New aeromagnetic and gravity compilations from Norway and adjacent areas: methods and applications. *In*: Vining, B.A. & Pickering, S.C. (eds) *Petroleum Geology: From Mature Basins to New Frontiers – Proceedings of the 7th Petroleum Geology Conference*. Geological Society, London, 559–586, https://doi.org/10.1144/0070559

Pail, R., Goiginger, H. *et al.* 2010. Combined satellite gravity field model GOCO01S derived from GOCE and GRACE. *Geophysical Research Letters*, **37**, L20314, https://doi.org/10.1029/2010GL044906

Pavlis, N.K., Holmes, S.A., Kenyon, S.C. & Factor, J.K. 2012. The development and evaluation of the Earth Gravitational Model 2008 (EGM2008). *Journal of Geophysical Research: Solid Earth*, **117**, B04406, https://doi.org/10.1029/2011JB008916

Peron-Pinvidic, G., Gernigon, L., Gaina, C. & Ball, P. 2012. Insights from the Jan Mayen system in the Norwegian–Greenland sea – I. Mapping of a microcontinent. *Geophysical Journal International*, **191**, 385–412, https://doi.org/10.1111/j.1365-246X.2012.05639.x

Richardson, K.R., Smallwood, J.R., White, R.S., Snyder, D.B. & Maguire, P.K.H. 1998. Crustal structure beneath the Faroe Islands and the Faroe–Iceland Ridge. *Tectonophysics*, **300**, 159–180, https://doi.org/10.1016/s0040-1951(98)00239-x

Sandwell, D.T. 2001. *Cooling of the Oceanic Lithosphere and Ocean Floor Topography*. University of California, San Diego, CA, USA, http://topex.ucsd.edu/geodynamics/07cooling.pdf

Schiffer, C., Jacobsen, B.H., Balling, N., Ebbing, J. & Nielsen, S.B. 2015. The East Greenland Caledonides – teleseismic signature, gravity and isostasy. *Geophysical Journal International*, **203**, 1400–1418, https://doi.org/10.1093/gji/ggv373

Schmidt, S., Götze, H.-J., Alvers, M. & Fichler, C. 2010. IGMAS+ a new 3D Gravity, FTG and magnetic modeling software. *In*: Zipf, A., Behncke, K., Hillen, F. & Schefermeyer, J. (eds) *Die Welt im Netz (Geoinformatik 2010, Kiel, 17.3. – 19.3.2010)*. Akademische Verlagsgesellschaft AKA GmbH, Osnabrück, 57–63.

Scott, R.A. 2000. Mesozoic–Cenozoic evolution of East Greenland: implications of a reinterpreted continent–ocean boundary location. *Polarforschung*, **68**, 83–91, https://doi.org/10013/epic.29794.d001

Sigmond, E.M.O. 2002. *Geological Map, Land and Sea Area of Northern Europe, Scale 1:4 000 000*. Geological Survey of Norway, Trondheim, Norway.

Torsvik, T.H., Amundsen, H.E.F. *et al.* 2015. Continental crust beneath southeast Iceland. *Proceedings of the National Academy of Sciences*, **112**, E1818–E1827, https://doi.org/10.1073/pnas.1423099112

Tsikalas, F., Faleide, J.I., Eldholm, O. & Wilson, J. 2005. Late Mesozoic–Cenozoic structural and stratigraphic correlations between the conjugate mid-Norway and NE Greenland continental margins. *In*: Doré, A.G. & Vining, B.A. (eds) *Petroleum Geology: Northwest Europe and Global Perspectives – Proceedings of the 6th Petroleum Geology Conference*. Geological Society, London, 785–801, https://doi.org/10.1144/0060785

Voss, M., Schmidt-Aursch, M.C. & Jokat, W. 2009. Variations in magmatic processes along the East Greenland volcanic margin. *Geophysical Journal International*, **177**, 755–782, https://doi.org/10.1111/j.1365-246X.2009.04077.x

Weidle, C. & Maupin, V. 2008. An upper-mantle S-wave velocity model for Northern Europe from Love and Rayleigh group velocities. *Geophysical Journal International*, **175**, 1154–1168, https://doi.org/10.1111/j.1365-246X.2008.03957.x

Controls on the location of compressional deformation on the NW European margin

G. S. KIMBELL[1]*, M. A. STEWART[2], S. GRADMANN[3], P. M. SHANNON[4], T. FUNCK[5], C. HAASE[3], M. S. STOKER[2] & J. R. HOPPER[5]

[1]*British Geological Survey, Keyworth, Nottingham NG12 5GG, UK*

[2]*British Geological Survey, The Lyell Centre, Research Avenue South, Edinburgh EH14 4AP, UK*

[3]*Geological Survey of Norway, Leiv Eirikssons vei 39, 7040 Trondheim, Norway*

[4]*University College Dublin, Belfield, Dublin 4, Ireland*

[5]*Geological Survey of Denmark and Greenland, Øster Voldgade 10, 1350 Copenhagen K, Denmark*

**Correspondence: gsk@bgs.ac.uk*

Abstract: The distribution of Cenozoic compressional structures along the NW European margin has been compared with maps of the thickness of the crystalline crust derived from a compilation of seismic refraction interpretations and gravity modelling, and with the distribution of high-velocity lower crust and/or partially serpentinized upper mantle detected by seismic experiments. Only a subset of the mapped compressional structures coincide with areas susceptible to lithospheric weakening as a result of crustal hyperextension and partial serpentinization of the upper mantle. Notably, partially serpentinized upper mantle is well documented beneath the central part of the southern Rockall Basin, but compressional features are sparse in that area. Where compressional structures have formed but the upper mantle is not serpentinized, simple rheological modelling suggests an alternative weakening mechanism involving ductile lower crust and lithospheric decoupling. The presence of pre-existing weak zones (associated with the properties of the gouge and overpressure in fault zones) and local stress magnitude and orientation are important contributing factors.

The NW European margin records an extensional history spanning from the end of the Caledonian Orogeny to the onset of break-up in the Early Eocene. The Mesozoic part of this evolution progressed from mosaic-like fragmentation of Pangaea in the Permo-Triassic to more systematic east–west extension in the Jurassic, which produced basins elongated in a north–south direction. This was succeeded by NW–SE extension in the Early Cretaceous, and the formation of large basins that cross-cut the older trends (Doré *et al.* 1999). The Cretaceous extension had a major impact on the morphology of the margin, as it involved substantial thinning of the crust beneath a chain of large NE-trending basins (the Lofoten, Vøring, Møre, Faroe–Shetland and Rockall basins: Fig. 1). In the south, similarly high stretching factors beneath the Porcupine Basin include an important contribution from Jurassic extension, whereas the early evolution of the (less stretched) Hatton Basin is poorly understood. From the Eocene onwards, the evolution of parts of the margin was characterized by compressional deformation, with the formation of a series of domes (Fig. 1) (e.g. Doré *et al.* 2008; Ritchie *et al.* 2008; Tuitt *et al.* 2010), although there is also evidence of earlier compression (Doré *et al.* 2008; Tuitt *et al.* 2010; Lundin *et al.* 2013). The aim of this paper is to use results from the NAG-TEC project (Hopper *et al.* 2014) to examine the relationship between the siting of the post-break-up compressional structures and the morphology and rheology of the margin established as a consequence of the preceding extensional history.

A key recent hypothesis that addresses this issue has been provided by Lundin & Doré (2011), who argued that the Cenozoic compressional structures were preferentially sited where the lithosphere had previously been highly stretched. They proposed that the lithosphere in these areas was weakened as a result of crustal hyperextension and consequent partial serpentinization of the upper mantle. Hyperextension is a deformation mode affecting

From: PÉRON-PINVIDIC, G., HOPPER, J. R., STOKER, M. S., GAINA, C., DOORNENBAL, J. C., FUNCK, T. & ÁRTING, U. E. (eds) 2017. *The NE Atlantic Region: A Reappraisal of Crustal Structure, Tectonostratigraphy and Magmatic Evolution.* Geological Society, London, Special Publications, **447**, 249–277.
First published online August 12, 2016, https://doi.org/10.1144/SP447.3

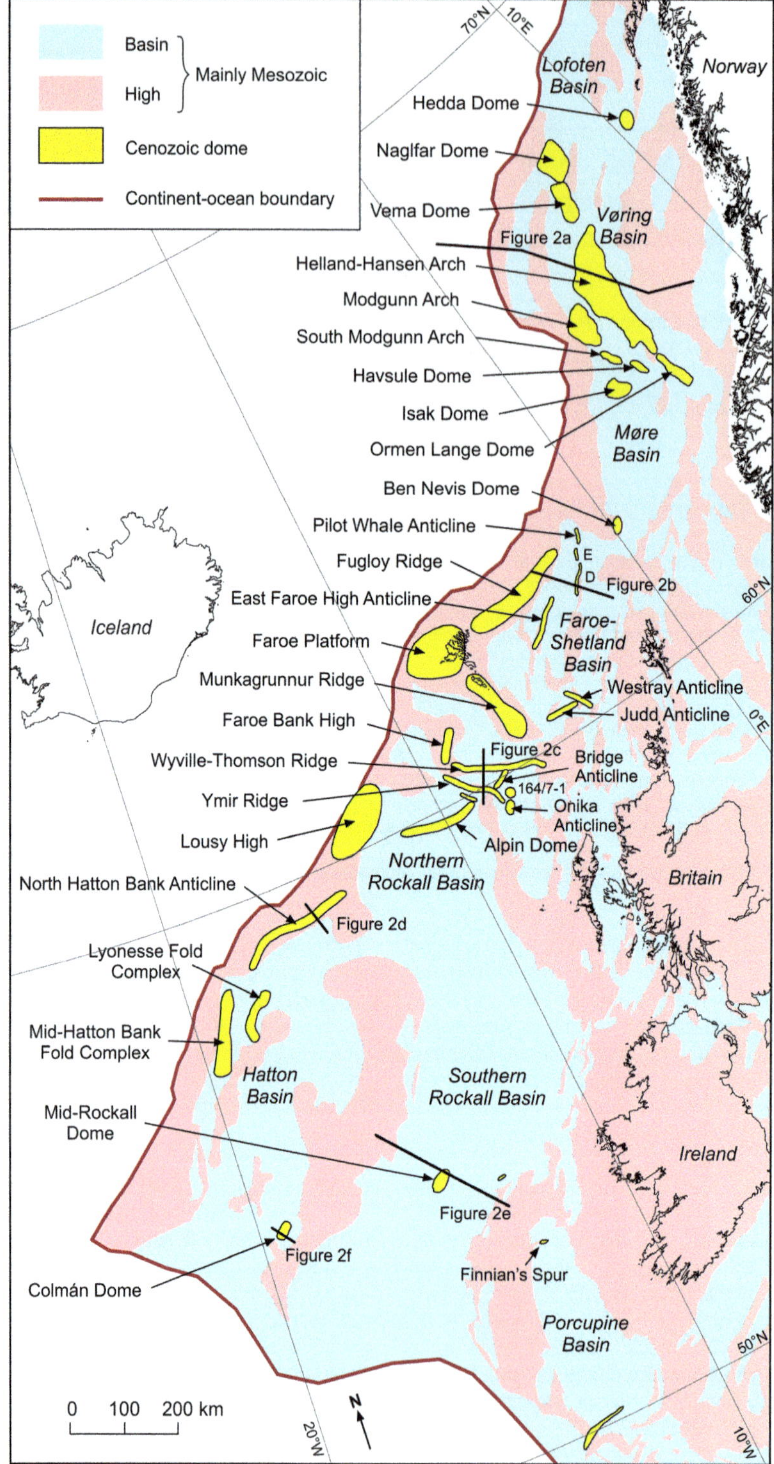

Fig. 1. Location of Cenozoic anticlinal structures on the NW European margin, superimposed on a generalized representation of the structural fabric primarily developed during Mesozoic times.

continental crust that has been thinned to less than 10 km (Unternehr *et al.* 2010; Peron-Pinvidic *et al.* 2013). Serpentinization occurs when water interacts with ultramafic rocks within the stability field of the serpentine minerals (<460°C: Schwartz *et al.* 2013). It is possible for surface water to reach the upper mantle if the entire crust is brittle and is penetrated by faults, but if crustal temperatures are sufficiently high then a ductile lower-crustal layer may form that inhibits such penetration. Although lithospheric extension enhances the geothermal gradient, this applies over a reduced crustal thickness, potentially leading to cooling and embrittlement of the lower crust. The conditions that favour embrittlement are slow extension and low sedimentation rates (a sedimentary layer can have a 'blanketing' effect, increasing temperatures in the underlying crust). Pérez-Gussinyé & Reston (2001) found that stretching factors of between about 3 and 5 were required for embrittlement to occur, depending on the strain rate. The simulations of Rüpke *et al.* (2013) produced similar results when there was no sedimentation, but demonstrated that higher factors were required when the sediment supply was increased.

In the following sections, we: (i) review the compressional structures observed on the NW European margin; (ii) consider their distribution in relation to the characteristics of the underlying crust and the potential for serpentinization of the upper mantle; and (iii) use rheological modelling to explore how the strength of the lithosphere under compression is influenced by its structure, composition (including serpentinization) and thermal state. We use this information to consider how the spatial and temporal variation of multiple factors may have influenced the pattern of deformation observed today.

Cenozoic compressional structures

The following sections briefly review the compressional structures observed on the NW European margin (Fig. 1). These are considered in five broad geographical groupings: Norwegian margin; Faroe–Shetland area; southern Faroe–Shetland to northern Rockall; Hatton margin; and southern Rockall–Hatton–Porcupine.

Norwegian margin

The Helland-Hansen Arch overlies a thick (6–7 km) Cretaceous sedimentary sequence on the eastern side of the Rås Basin, within the Vøring Basin (Figs 1 & 2a) (Blystad *et al.* 1995; Gómez & Vergés 2005). With an axial length of 280 km and maximum amplitude of 1000 m, it is the largest of a number of buried compressional features observed on the Norwegian margin. The structural configuration (Fig. 2a) suggests that at least a component of its Cenozoic growth can be attributed to reverse movements on the Fles Fault Complex (Blystad *et al.* 1995; Brekke 2000). This growth was initiated in the Mid-Eocene–Oligocene, with a second phase occurring in the Miocene (Brekke 2000; Lundin & Doré 2002; Stoker *et al.* 2005*a*). Further growth occurred during the Plio-Pleistocene; Gómez & Vergés (2005) considered this phase to be dominated by differential compaction, and estimated that it may account for up to 70% of the total amplitude of the arch.

Also within the Vøring Basin are the NNE-trending Vema and Naglfar domes (Fig. 1). Both structures underwent initial growth from the Late Eocene to the Early Oligocene, followed by further uplift in the Mid-Miocene continuing to the present day, evidence for which includes active mud diapirs that reach the seafloor (Blystad *et al.* 1995; Gómez & Vergés 2005; Lundin & Doré 2011). East of the Vema Dome, Lundin & Doré (2002) identified the Hedda Dome, with Late Eocene and Early Miocene growth phases. The NW-trending Ormen Lange Dome lies south of the Helland-Hansen Arch and along the same trend (Fig. 1): initial uplift was in the Late Eocene, with another growth phase in the Late Miocene (Blystad *et al.* 1995). West of the Helland-Hansen Arch is the Mogdunn Arch (Fig. 1), and further south are the South Modgunn Arch, the Havsule Dome and the Isak Dome: all of these structures were active in the Miocene (Blystad *et al.* 1995; Lundin & Doré 2002; Doré *et al.* 2008).

Doré *et al.* (2008) identified a set of early 'tectonomagmatic' domes on the Norwegian margin, which are of Late Cretaceous–Paleocene age and related to break-up magmatism. Examples include the Gjallar Ridge (not shown in Fig. 1), and precursors of the Vema and Isak domes.

Faroe–Shetland area

The Fugloy Ridge is a large anticlinal structure that extends ENE from the Faroe Islands. It is asymmetrical, with a steeper SE limb (Fig. 2b), and has a maximum fold amplitude of 3000 m and a bathymetric expression along its length. Boldreel & Andersen (1993, 1995) recognized Paleocene–Eocene and Eocene–Oligocene compressive growth phases, and thinning and onlap in seismic reflection data suggest an enhanced phase of growth in the Mid-Miocene (Fig. 2b), as well as possible further growth during Pliocene to Recent times (Ritchie *et al.* 2003; Johnson *et al.* 2005; Stoker *et al.* 2005*a*). Post-rift thermal subsidence in the flanking Faroe–Shetland Basin has probably enhanced the Fugloy Ridge rather than it being a solely compressional feature (Ritchie *et al.* 2011).

To the east of the Fugloy Ridge lie a number of smaller NE- to NNE-trending compressional

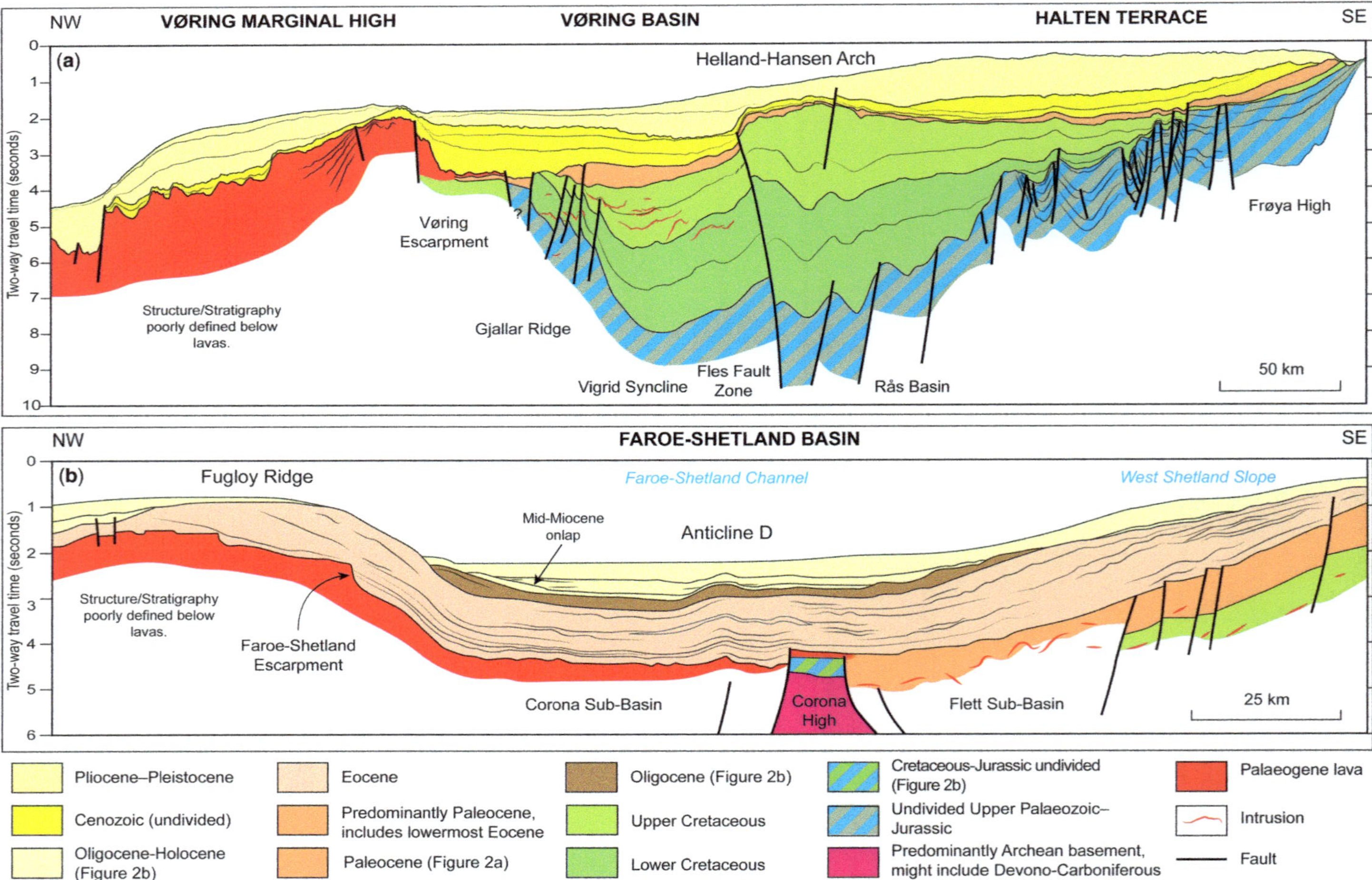

Fig. 2. Geoseismic sections across compressional features, modified from Stoker *et al.* (2014) and references therein. (**a**) Section across the mid-Norwegian margin and the Vøring Basin showing the Helland-Hansen Arch (fig. 7.29c of Stoker *et al.* 2014). (**b**) Section highlighting the Fugloy Ridge and Anticline D compressional structures with Mid-Miocene onlap onto Eocene rocks (fig. 7.21a of Stoker *et al.* 2014).

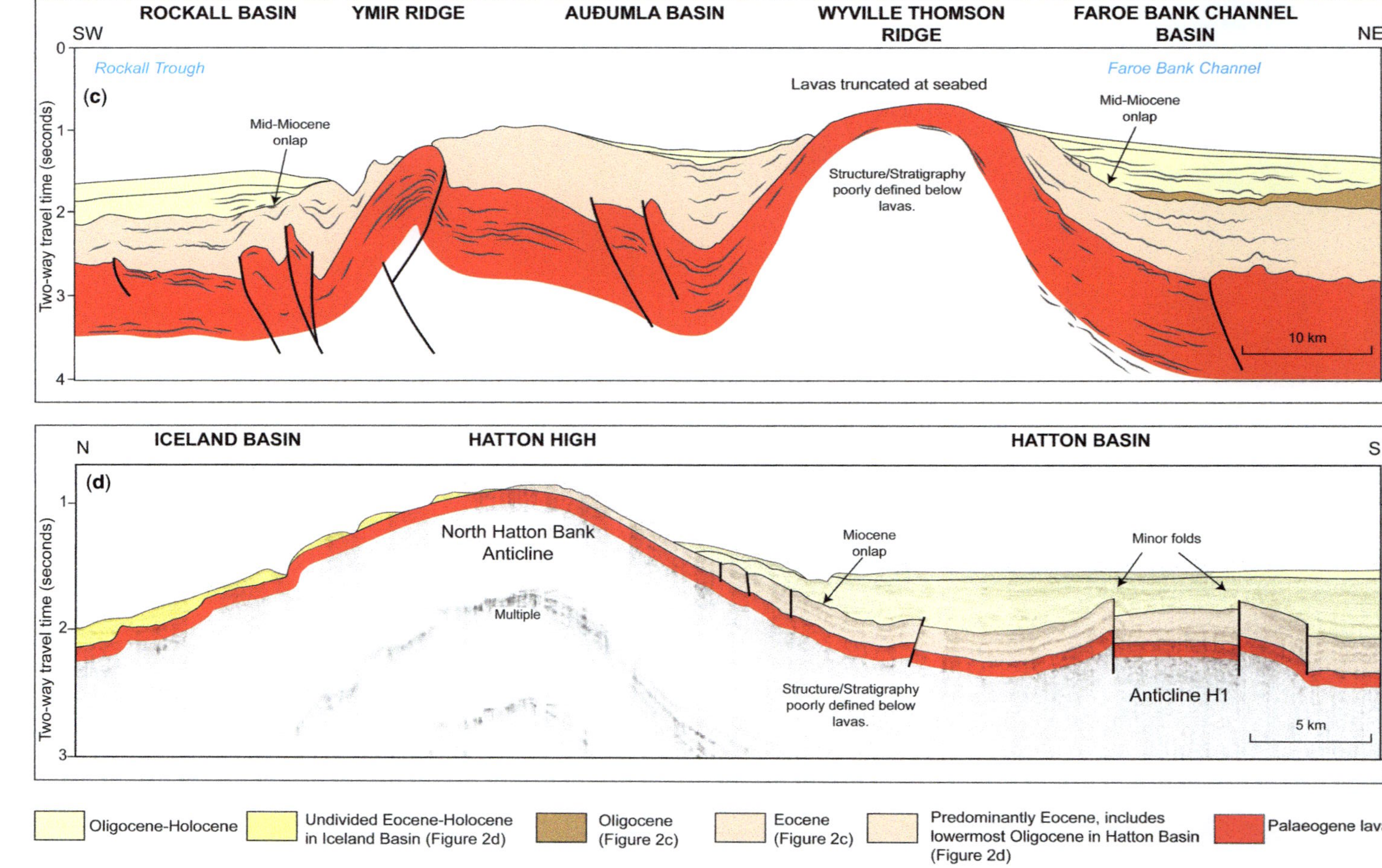

Fig. 2. (**c**) Section showing the Ymir and Wyville Thomson ridges with Mid-Miocene onlap onto the Eocene (fig. 7.21e of Stoker *et al.* 2014). (**d**) Interpreted seismic section (BGS02/02-15) across the Hatton High showing the North Hatton Bank Anticline and Miocene onlap onto the Eocene in the Hatton Basin (fig. 7.25e of Stoker *et al.* 2014).

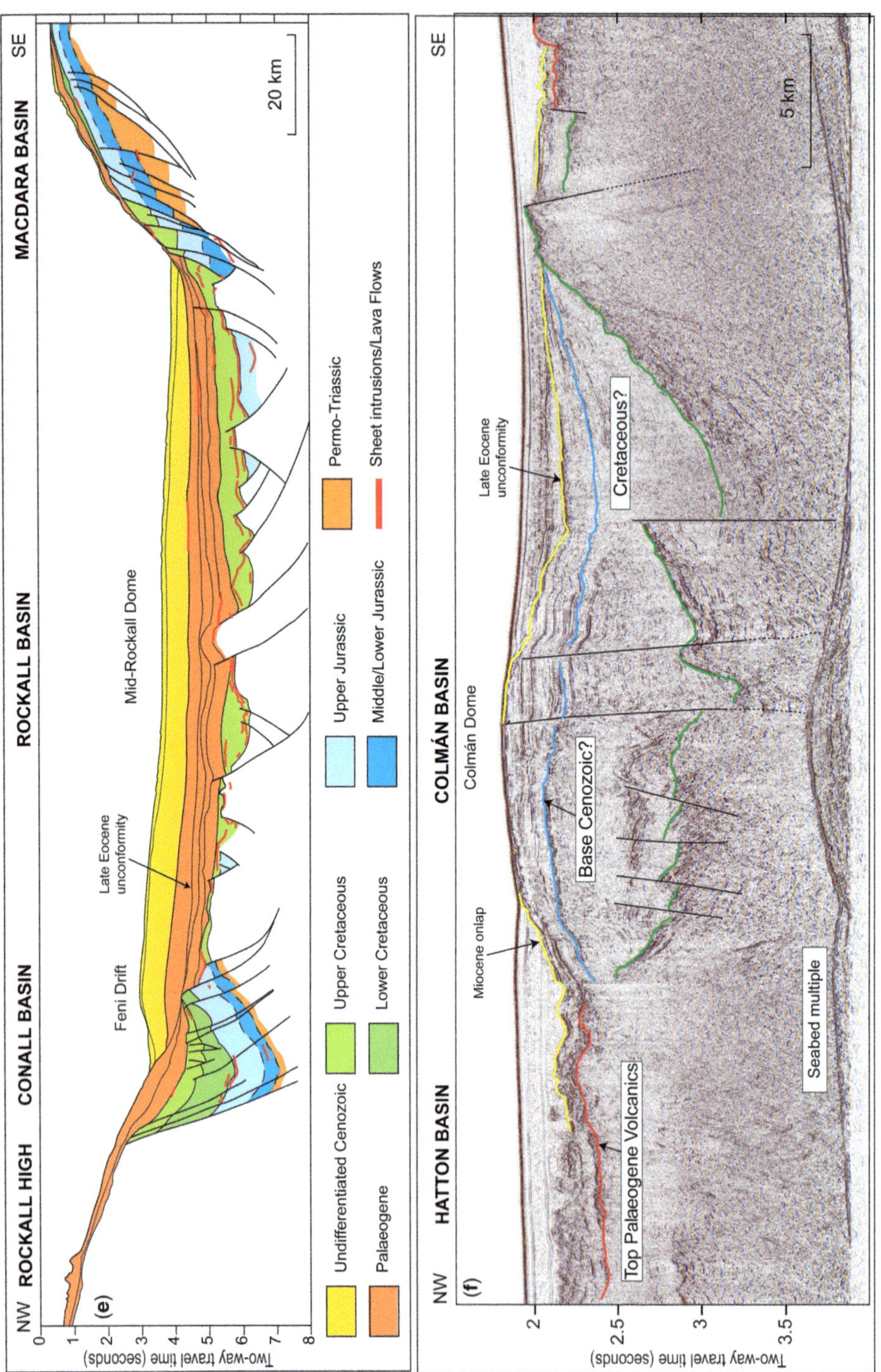

Fig. 2. (**e**) Geoseismic section across the southern Rockall Basin showing the Mid-Rockall Dome and the overlying Late Eocene unconformity (Naylor *et al.* 1999; fig. 7.41e of Stoker *et al.* 2014). (**f**) Seismic section showing the Colmán Dome in the Colmán Basin and the Miocene onlap onto the interpreted Late Eocene unconformity (yellow). Seismic data in (f) was provided courtesy of the Department of Communications, Energy and Natural Resources, Ireland; interpretation based on an unpublished BGS report (D. McInroy pers. comm.).

structures, including the East Faroe High and features termed Anticlines B, D (see Fig. 2b), E and F (the Pilot Whale Anticline) (Davies & Cartwright 2002; Davies *et al.* 2004; Ritchie *et al.* 2003, 2008; Johnson *et al.* 2005). Only selected features are shown in Figure 1: see Ritchie *et al.* (2008) for more detail. These structures were typically initiated in the Mid-Miocene, with further growth in the Early Pliocene; seabed topography suggests that some are still active at the present day (Ritchie *et al.* 2008). Only the younger phase of growth is evident in the case of the Pilot Whale Anticline, which is associated with a number of diapiric structures and mud mounds rising up to 120 m above seabed (Holmes *et al.* 2003). The Ben Nevis Dome is a buried anticline, overlain by basalts, which lies 12 km SE of the large, igneous, Brendan Dome (Fig. 1). Hodges *et al.* (1999) proposed that it formed as a result of the latest Paleocene uplift of a Jurassic fault block driven by igneous intrusion from the Brendan Dome and inversion of the Cretaceous fill of an adjacent half-graben. Rohrman (2007) suggested its formation at around 65–60 Ma (i.e. Paleocene), coincident with the intrusion of an underlying igneous pluton.

In the SW Faroe–Shetland Basin lie the Judd and Westray anticlines, which trend to the west and NW respectively (Fig. 1) (Smallwood & Kirk 2005; Ritchie *et al.* 2008; Stoker *et al.* 2013). The Judd Anticline is 33 km long, and had growth phases in the Early and Mid-Eocene and probably in the Oligocene. The Westray Anticline is 45 km long and intersects the eastern part of the Judd Anticline (Fig. 1). Growth occurred during the Eocene, Late Eocene–Oligocene and the Miocene, with further growth possible up to the present day (Ritchie *et al.* 2008).

To the south of the Faroe Platform, the Munkagrunnur Ridge is a NNW-trending 135 km-long anticline with seabed expression (Boldreel & Andersen 1993; Johnson *et al.* 2005; Ritchie *et al.* 2008, 2011). Boldreel & Andersen (1993) suggested that the ridge is part of a set of ramp-anticlines, along with the Ymir and Wyville Thomson ridges, which formed during the Late Paleocene–Eocene above a northwards-dipping crustal fault in response to seafloor spreading. The Faroe Bank Channel, which separates the Wyville Thomson and Munkagrunnur ridges, may be a synclinal feature formed in response to the growth of those anticlines, especially in the late Palaeogene–early Neogene, and is an important part of the North Atlantic deep-water circulation system (Stoker *et al.* 2005*a*).

Southern Faroe–Shetland Basin to the northern Rockall Basin

The Wyville Thomson Ridge is a 200 km-long WNW-trending basalt anticline with a maximum fold amplitude of 4000 m, a width of around 20 km and clear bathymetric expression (Boldreel & Andersen 1993; Johnson *et al.* 2005; Stoker *et al.* 2005*a*, *b*; Ritchie *et al.* 2011). A well-imaged succession on the NE side of the ridge shows thinning of Paleocene lavas and Early and Mid-Eocene sediments towards its axis, indicative of growth periods (Fig. 2c) (Boldreel & Andersen 1998; Johnson *et al.* 2005). Further thinning and growth is recognized in the early Oligocene (Johnson *et al.* 2005; Tuitt *et al.* 2010), and a series of unconformities indicates a major growth period in the middle Miocene (Fig. 2c) (Boldreel & Andersen 1998; Johnson *et al.* 2005; Stoker *et al.* 2005*a*, *b*).

The Ymir Ridge is a NW-trending asymmetrical anticline up to 100 km in length, with a maximum amplitude of 3500 m and bathymetric expression along most of its length. It is characterized by the presence of reverse faults, especially apparent in the Central Ymir Ridge identified by Ziska & Varming (2008), with growth folding observed in the Eocene–Oligocene succession of the SW flank (Fig. 2c) (Boldreel & Andersen 1993; Johnson *et al.* 2005; Ritchie *et al.* 2008). A number of unconformities also indicate middle Miocene growth.

The Bridge Anticline trends NE between the Wyville Thomson Ridge and the Ymir Ridge (Fig. 1), and has an amplitude of up to 1200 m: it was interpreted by Tuitt *et al.* (2010) as originating in the Mid-Eocene. The Onika Anticline trends NNE with an amplitude of 800 m and is inferred to have grown during the Late Eocene (Tuitt *et al.* 2010). The Alpin Dome is an eastwards-trending anticline up to 150 km in length; onlap and thinning suggest growth during the Late Eocene and Oligocene (Tuitt *et al.* 2010), with middle Miocene compression also suggested by Stoker *et al.* (2005*a*) and Ritchie *et al.* (2008). A domal structure drilled by well 164/7-1 is inferred to have an igneous origin (Archer *et al.* 2005).

The NE-trending Lousy High (Fig. 1) marks the western edge of the northern Rockall Basin and forms an elongated bathymetric feature 125 km long (Tuitt *et al.* 2010; Ritchie *et al.* 2013). A latest Eocene age is suggested by Tuitt *et al.* (2010).

Hatton margin

The North Hatton Bank Anticline (Figs 1 & 2d) (Hitchen 2004; Johnson *et al.* 2005) can be traced for 220 km, and has a wavelength of 40 km and a maximum amplitude of 1900 m (Johnson *et al.* 2005; Tuitt *et al.* 2010). Tuitt *et al.* (2010) recognized three discrete examples of onlap onto the structure within the Eocene in seismic reflection data, with Late Eocene reverse faulting and onlap also observed by Johnson *et al.* (2005), indicative of a major growth phase (Fig. 2d). The NE-trending

Lyonesse Fold Complex has a length of 50 km, a maximum amplitude of 100 m and an age of formation constrained by folding of the Late Eocene unconformity (Johnson *et al.* 2005). Folding in the Mid-Hatton Bank Fold Complex post-dates the Albian rocks recovered from shallow boreholes (Hitchen 2004) and pre-dates the overlying flat-lying Cenozoic strata, which include Early Eocene and Late Eocene unconformities (Johnson *et al.* 2005, fig. 9).

Southern Rockall, Hatton and Porcupine basins

The southern Rockall Basin contains few obvious Cenozoic compressional structures, with the clearest of these lying beneath the central part of the basin (Figs 1 & 2e) (Naylor *et al.* 1999; McDonnell & Shannon 2001; Morewood *et al.* 2004). Resolution is limited by a lack of seismic coverage, but the structure (herein termed the Mid-Rockall Dome) appears to be a NE-trending broad dome underlain by a set of eastwards-dipping late Cretaceous–earliest Cenozoic faults. Growth faulting and onlap of the Paleocene–earliest Eocene strata suggest structural (probably fault) control on its location and, together with truncation by the regional Late Eocene unconformity, a Late Eocene age for its formation.

Another inversion structure (herein termed the Colmán Dome) is apparent in the Colmán Basin on the SE flank of the Hatton Basin, where there is pronounced doming of the Late Eocene unconformity and an expression at the seabed (Fig. 2f). This suggests an Oligocene or younger age of formation. The seismic section shown in Figure 2 was acquired in an area where the Palaeogene volcanic rocks are thin or absent, providing a 'window' through which the underlying structure can be viewed. A deeper sequence is imaged that is likely to be mainly Cretaceous in age (possibly underlain by Jurassic rocks), although the age of individual reflectors is poorly constrained. The structures revealed strongly suggest that the dome was formed by inversion of the underlying basin.

A compressional structure is observed in the south of the Porcupine Basin (southernmost feature in Fig. 1). It is approximately 100 km in length and was described by Masson & Parson (1983) as changing from a complex faulted anticline in the west to a simple monocline in the east. Those authors inferred that it developed in late Eocene time along a pre-existing geological lineament. It has an east–west orientation, and is coincident along part of its northern margin with a large fault (Naylor *et al.* 2002).

Just south of Finnian's Spur in the Porcupine Basin (Fig. 1) is a possible inversion structure identified by Naylor *et al.* (2002) and associated with the Finnian's Spur basement ridge between the Porcupine and North Porcupine basins. The feature is apparent within the Upper Cretaceous section above the southern bounding fault of the spur, and is probably Palaeogene in age.

Geophysical evidence for hyperextension and mantle serpentinization

Figure 3 is a map of the thickness of the crystalline crust along the NW European margin, based on the compilation of seismic refraction data of Funck *et al.* (2014, this volume, in press). Figure 4 shows the crustal thickness derived by gravity modelling (Funck *et al.* 2014; Haase *et al.*, this volume, in review). The display is designed to highlight areas where hyperextension has occurred. The two maps are not fully independent because constraints based on the seismic refraction data were applied during the gravity inversion, and the refraction interpolation used a kriging technique guided by the gravity anomaly. The gravity inversion, however, includes a number of additional constraints, such as sediment thickness based on seismic reflection data and wells (see Hopper *et al.*, this volume, in prep). Stretching factors (original thickness divided by the stretched thickness) are also indicated on these figures, but these are based on a pre-rift crustal thickness that is difficult to define accurately. A reference value of 30 km is estimated around Britain and Ireland on the basis of observed crustal thicknesses in areas where topography lies close to sea level and Mesozoic basins are absent (e.g. Jacob *et al.* 1985; Lowe & Jacob 1989). In the Norwegian sector, larger crustal thicknesses have been detected in the coastal region (e.g. 35–40 km on profile 2 of Kvarven *et al.* 2014) and used as a reference in the construction of gravity models (e.g. 35 km used by Ebbing 2007). Skogseid *et al.* (2000) discussed the issues associated with defining pre-rift thickness on this margin and presented a methodology for inferring it with respect to Late Jurassic–Early Cretaceous rifting. Their models suggest values of about 30 km for the Møre and Vøring basins, and so provide some justification for extending the UK/Ireland thickness assumption into this area.

The degree of stretching shown in Figures 3 and 4 will be underestimated where the mapped crystalline crust includes igneous additions or partially serpentinized upper mantle. Igneous additions are likely to occur adjacent to the continental margin and these or serpentinized upper mantle may be present further inboard on the Norwegian margin, but the stretching is less likely to be underestimated in the Rockall Basin area (see the further discussion below).

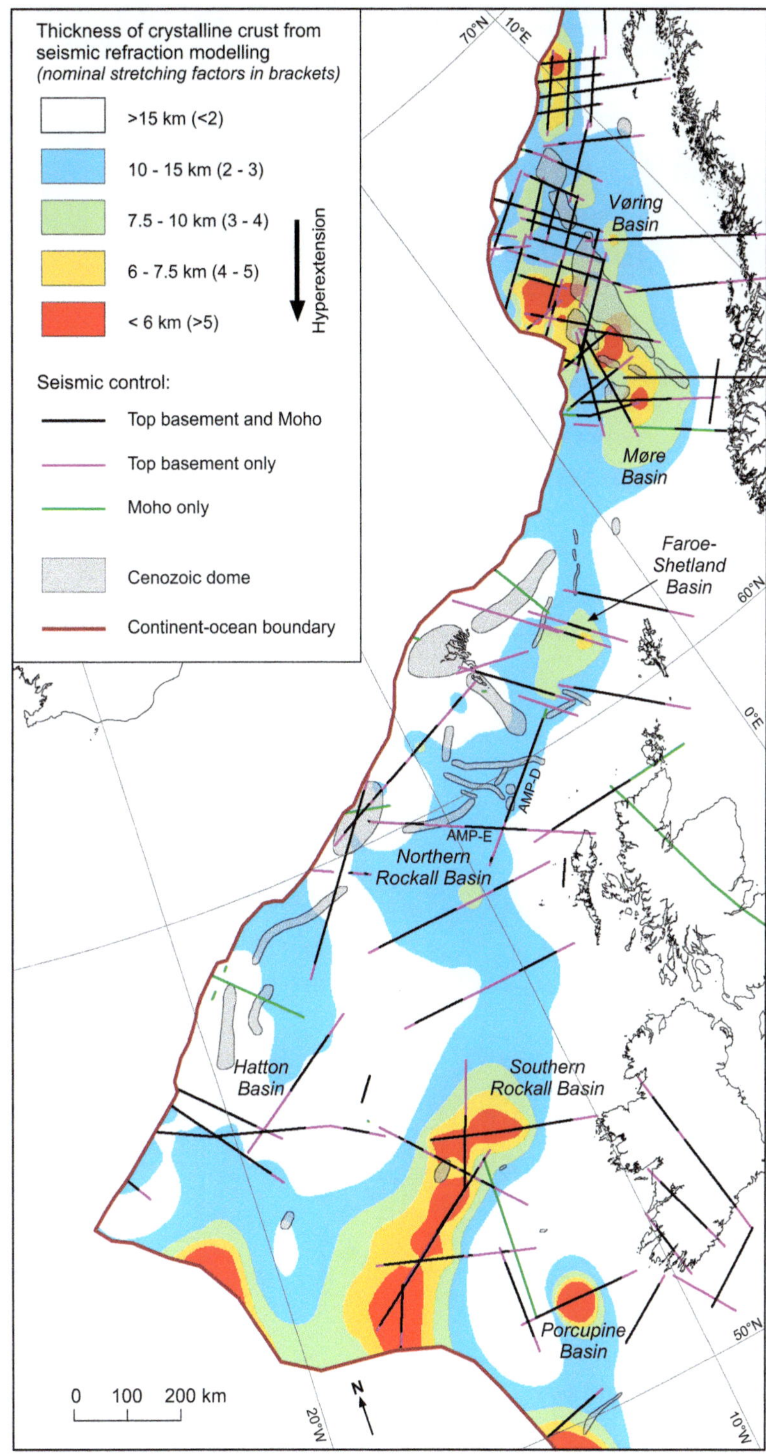

Fig. 3. Thickness of crystalline crust on the NW European margin based on the seismic refraction compilation of Funck *et al.* (2014, this volume, in press). Nominal stretching factors are based on an initial crustal thickness of 30 km (see the text).

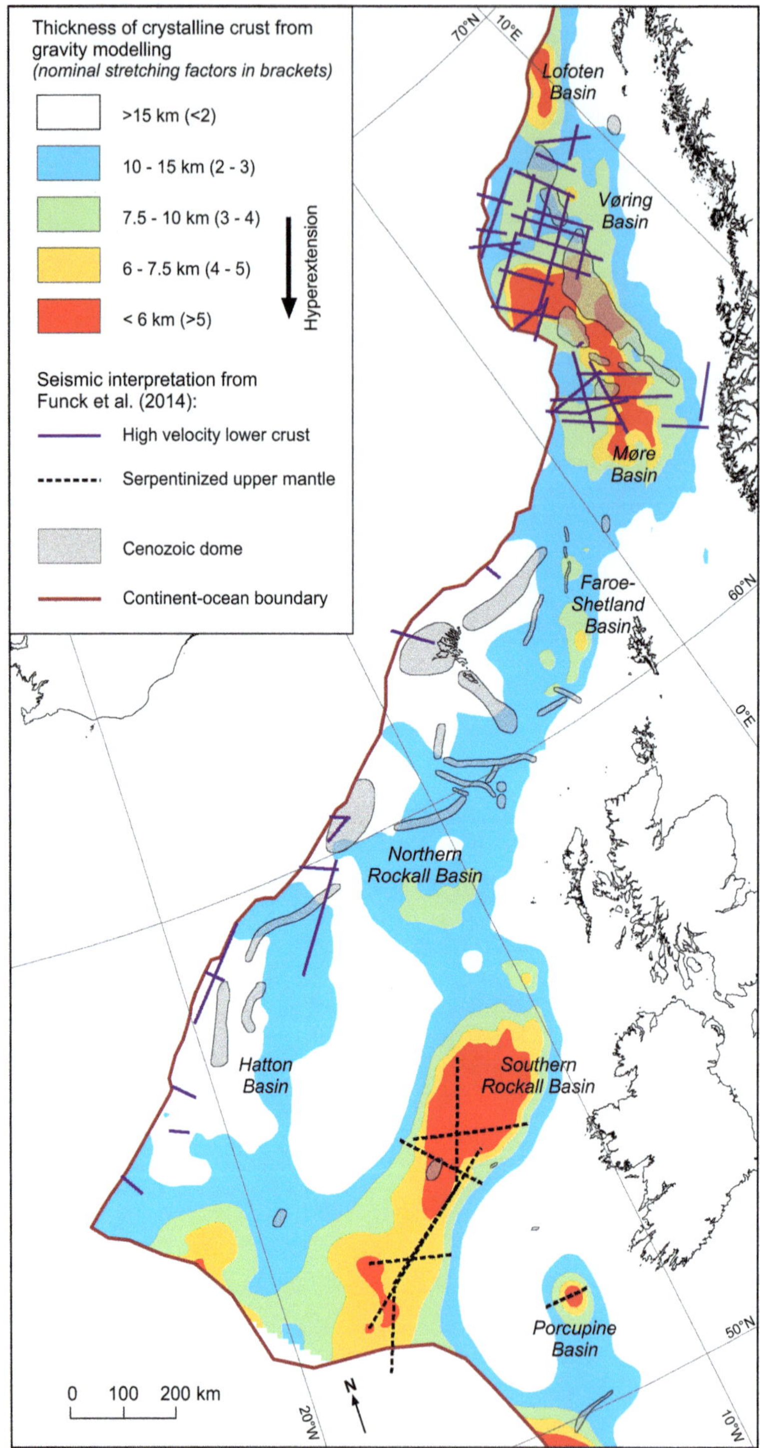

Fig. 4. Thickness of crystalline crust on the NW European margin based on the gravity inversion described by Funck *et al.* (2014) and Haase *et al.* (this volume, in review). Note that alternative interpretations of the high velocity lower crust in the northern part of the area include serpentinized upper mantle (see the text).

The maps indicate several areas where the crust is thin enough to be potentially conducive to serpentinization of the upper mantle (crystalline crustal thickness <10 km: green, orange and red zones in Figs 3 & 4). These cover much of the Norwegian margin (the Lofoten Basin, southern Vøring Basin and the Møre Basin), the southern Rockall Basin and the Porcupine Basin. They are less extensive beneath the Faroe–Shetland Basin and absent beneath most of the northern Rockall Basin. There are differences in detail (e.g. the model derived from gravity inversion extends the hyperextended zone in the southern Rockall Basin further to the north), but these broad observations are compatible with both maps.

Seismic refraction surveys can provide evidence of serpentinization through its influence on the seismic properties of the upper-mantle rocks. There is a linear relationship between the degree of serpentinization and the P-wave velocity, with a reduction from about 8.0 km s^{-1} in unserpentinized peridotite to 4.5–5.0 km s^{-1} when fully serpentinized to lizardite/chrysotile (Christensen 2004). This overlaps with the seismic velocity range of lower-crustal and gabbroic rocks, leading to a potential ambiguity in interpretation. The relatively high Poisson's ratio of the serpentinized rocks (Christensen 2004) can help to reduce the ambiguity, as can the presence of a gradational rather than sharp transition to normal mantle velocities at depth (Minshull 2009). There are, nonetheless, areas where the distinction between high-velocity lower crust (HVLC) and serpentinized upper mantle remains equivocal, as discussed in the following subsections. Figure 4 shows where units falling within these two categories have previously been identified along the NW European margin, based on the categorization of Funck *et al.* (2014).

Norwegian margin

High-velocity lower crust (HVLC) has been mapped over large parts of the Norwegian margin, and its origin has been much debated. Alternative explanations fall into three categories: basic igneous rocks (underplated material); high-grade metamorphic rocks; and partially serpentinized upper mantle (Mjelde *et al.* 2002, 2009*b*).

In the north, the Lofoten margin appears to be devoid of HVLC, despite the fact that extensive volcanic flows and seaward-dipping reflectors (SDRs) occur and have been linked to underplating in other areas. In the central segment of the Vøring margin, HVLC with a thickness of up to 8 km has been mapped by many seismic refraction and reflection experiments, and additionally constrained using gravity data and isostatic modelling (e.g. Gernigon *et al.* 2003, 2004; Ebbing *et al.* 2006; Mjelde *et al.* 2009*a*; Reynisson *et al.* 2010). HVLC has also been detected on the Møre margin, south of the Jan Mayen Lineament, and in that area an additional zone has been detected in the east, close to the shelf edge (Kvarven *et al.* 2014).

The HVLC of the central and northern Vøring margin exhibits velocities of 7.2–7.4 km s^{-1} (locally up to 7.6 km s^{-1}) and ratios of compressional to shear wave velocity (V_p/V_s) of 1.8–1.9, which do not, in themselves, differentiate between the alternative explanations for its origin. Gernigon *et al.* (2003, 2004) showed that there is not a simple correlation between observable crustal extension and the distribution and thickness of high-velocity lower-crustal bodies. On the contrary, on the Vøring margin, such bodies tend to correlate with structural highs (and, thus, with thicker overlying crystalline crust). Furthermore, correlation with the regional fault system suggests that the HVLC was already in place before break-up. Gernigon *et al.* (2003, 2004) interpreted the HVLC to comprise high-pressure metamorphic rocks (granulite/eclogite facies), as observed in the Caledonian nappes onshore. Magmatic underplating associated with the influence of a thermal anomaly prior to ocean opening between Norway and Greenland has been suggested by Skogseid *et al.* (1992) and Mjelde *et al.* (1997). A lower crust containing mafic intrusions emplaced close to the time of break-up is favoured by Mjelde *et al.* (2009*a*). Reynisson *et al.* (2010), Lundin & Doré (2011) and Rüpke *et al.* (2013) advocate partially serpentinized upper mantle. The HVLC of the southern Vøring margin (adjacent to the Jan Mayen Fracture Zone) exhibits P-wave velocities of up to 8.4 km s^{-1}, which suggest an eclogitic rather than peridotitic nature (Mjelde *et al.* 2009*a*).

The HVLC beneath the outer part of the Møre margin has been mapped in a refraction seismic study by Mjelde *et al.* (2009*b*), and corroborated by reflection seismic and gravity data (Nirrengarten *et al.* 2014). Here, the overlying crust is less than 10 km thick, making a serpentinized upper-mantle origin possible, although extensive volcanic sills and SDRs suggest significant magmatic activity. Any of the three main origins for high-velocity lower crust is thus plausible here. The inner zone of the HVLC on the Møre margin is located near the coast along the shelf edge and correlates well with a regional gravity high (Olafsson *et al.* 1992; Kvarven *et al.* 2014; Nirrengarten *et al.* 2014). The HVLC extends beneath crystalline crust that is more than 15 km thick, suggesting that fluid penetration and serpentinization of the uppermost mantle is unlikely. Eclogitic crust from the Western Gneiss Region is widespread on the adjacent mainland, making an eclogitic origin for the HVLC more likely.

HVLC adjacent to the ocean margin

The hypotheses for the HVLC on the Norwegian margin are not mutually exclusive, and proponents of partially serpentinized upper mantle beneath parts of the margin have also embraced a break-up-related magmatic origin where the HVLC lies adjacent to the continent–ocean boundary (Reynisson *et al.* 2010; Lundin & Doré 2011; Rüpke *et al.* 2013). A magmatic origin has also been the preferred interpretation for the HVLC detected adjacent to the ocean margin in a series of seismic refraction experiments further south. Near the Faroe Islands and beneath the northern part of Hatton Bank, the iSIMM experiments revealed a 40–50 km-wide zone of HVLC that was interpreted to be due to the intrusion of basic igneous sills at the time of break-up (White *et al.* 2008; Roberts *et al.* 2009; White & Smith 2009). Funck *et al.* (2008) identified an extension of the HVLC beneath the northern part of the Hatton Basin below crystalline crust about 10 km thick, and interpreted this to be due to magmatic additions to the crust rather than partial serpentinization of the upper mantle. Further south, a zone of HVLC has been detected beneath the Hatton continental margin, overlain by crust in which the lower layer has been completely attenuated (Vogt *et al.* 1998; Shannon *et al.* 1999). The HVLC can be traced southwards beneath the SDRs on Edoras Bank (Barton & White 1997).

Faroe–Shetland Basin

The crustal thickness beneath the Faroe–Shetland Basin is poorly resolved by seismic refraction experiments and there are some contradictory results (Petersen & Funck, this volume, in press). A minimum thickness of 7–8 km has been identified (Makris *et al.* 2009), although gravity modelling suggests that values of 10–12 km are more widespread (Haase *et al.*, this volume, in review). Raum *et al.* (2005) and Makris *et al.* (2009) report upper-mantle velocities of about 8.0 km s^{-1} beneath the basin.

Northern Rockall Basin

A seismic refraction profile across the northern part of the Rockall Basin (AMP-E: Fig. 3) (Klingelhöfer *et al.* 2005) detected crystalline crust with a minimum thickness of 10–12 km. The Moho beneath this profile was primarily identified using PmP reflections, so it is not possible to infer the velocity structure of the underlying mantle. However, an orthogonal line extending from the northern Rockall Basin into the southern Faroe–Shetland Basin (AMP-D: Fig. 3) (Klingelhöfer *et al.* 2005) did detect diving rays into the mantle (Pn), beneath 11–13 km-thick crust in an area affected by compressional deformation, and these indicated an upper-mantle velocity of 8.0–8.2 km s^{-1}. An early (unreversed) refraction experiment also detected clear Pn arrivals indicative of an upper-mantle velocity of 8.2 $\pm$ 0.17 km s^{-1} beneath the northern Rockall Basin (Bott *et al.* 1979). Thus, there is no direct evidence at present for serpentinization beneath this basin.

Southern Rockall Basin

Seismic refraction data indicate that the crystalline crust beneath the southern part of the Rockall Basin is typically 5–7 km thick (Shannon *et al.* 1999; Morewood *et al.* 2005). This is underlain by a 3–10 km-thick zone with P-wave velocities of 7.5–7.8 km s^{-1}, which was interpreted as a partially serpentinized upper mantle by O'Reilly *et al.* (1996). The geometry of the zone, its velocity range and the V_p/V_s ratios (1.80–1.83) were considered to favour this interpretation over one involving igneous underplating. The velocities indicate a degree of serpentinization of 10–15%. Further support for a serpentinization model comes from the subsidence discrepancy along the axis of the basin (O'Reilly *et al.* 1996). The observed subsidence is consistently 400–600 m less than modelled subsidence over the zone of interpreted serpentinization, while there is no discrepancy to the north and south, making it less likely that the abnormal subsidence is due to a thermal anomaly (Joppen & White 1990).

Hauser *et al.* (1995) and O'Reilly *et al.* (1996) observed that the slow stretching rates necessary to explain hyperextension without magmatism were most easily explained by a differential stretching model in which the upper and middle crust were more extended than the lower crust and upper mantle. This interpretation was supported by a comparison between the three-layer velocity model for adjacent, unstretched crust (Jacob *et al.* 1985; Lowe & Jacob 1989) and the velocity structure observed beneath the central part of the basin.

Porcupine Basin

A seismic refraction experiment across the Porcupine Basin resolved severely and asymmetrically stretched (<2 km-thick) continental crust beneath thick sedimentary cover in the centre of the basin (O'Reilly *et al.* 2006). The velocity of the uppermost part of the mantle beneath the thinned crust lies in the range 7.2–7.5 km s^{-1}, increasing to around 8 km s^{-1} over a depth of about 20 km. O'Reilly *et al.* (2006) interpreted this zone to be serpentinized upper mantle, and discounted the alternative explanation of magmatic underplating on the basis of the absence of a characteristic double PmP reflection (typically seen from the top and bottom of such bodies) and the lack of evidence for extensive magmatic activity in this part of the basin at the time of stretching. The velocities encountered suggest a degree of serpentinization of up to 25%.

Rheological modelling

Construction of yield strength envelopes

The yield-strength envelope is used to predict the way in which lithospheric strength varies with depth and to analyse how this is influenced by factors such as the thickness and composition of the crust, the presence of fluids, the temperature regime, and the strain rate. We have constructed a series of such envelopes representing the range of conditions encountered along the NW European margin in order to see whether these can help to explain the distribution of compressional structures. This method is commonly used to explore lithospheric strength profiles in various geological settings (e.g. Jackson 2002; Burov & Watts 2006; Cloetingh *et al.* 2008; Burov 2011 and references therein), but has only rarely been applied to a lithosphere that includes partially serpentinized upper mantle (Escartín *et al.* 1997*a*; Pérez-Gussinyé & Reston 2001).

The envelopes were constructed by combining predictions about the brittle and ductile behaviour of lithospheric rocks. At shallow depth, brittle deformation dominates and is controlled by friction effects that are mainly dependent on the depth of burial. As temperature increases with depth, a point is reached beyond which ductile deformation occurs at lower stresses than brittle failure. In oceanic lithosphere, the result is typically a single strong layer with a thickness that increases with age. In continental lithosphere, more than one strong layer may occur as a result of changes in lithology. For example, the strong upper crust may be decoupled from a strong upper mantle by a ductile lower crust (the 'jelly sandwich' model) (Hopper & Buck 1998).

Yield-strength envelopes are presented in Figure 5 for simple models representing 'normal' (30 km) thickness continental crust (Fig. 5a, b), highly extended crust as encountered, for example, in the northern Rockall Basin (Fig. 5c, d: 5 km of sediments over 12 km of crystalline crust) and hyperextended crust as in the southern Rockall Basin (Fig. 5e, f: 6 km of sediments over 6 km of crystalline crust). The envelopes shown in Figure 5g, j are more representative of the Norwegian margin, where hyperextended crust is overlain by a much thicker sedimentary layer. In Figure 5g, h, 12 km of sediments overlie a crust with a thickness of 8 km. The Moho depth is the same in Figure 5i, j, but in that case a 10 km sedimentary layer overlies 5 km of crystalline crust underlain by 5 km of high-velocity lower crust (HVLC).

In each example, the compressive-strength envelope was calculated on the basis of both 'cool' and 'warm' geotherms (panels on the left- and right-hand side of Fig. 5, respectively). Table 1 lists the thermal parameters assumed in calculating these. The boundary conditions for an initial steady-state model were a surface temperature of 0°C, and alternative mantle heat flows of 25 and 40 mW m^{-2}, respectively. In the stretched models, the small residual thermal effect of the original Mesozoic rifting event was simulated in a simple fashion by the addition of a thermal anomaly calculated using the method and parameters of McKenzie (1978), assuming instantaneous Early Cretaceous (140 Ma) rifting.

The geotherms do not incorporate the details of factors such as the distribution of heat-producing elements in the crust (and the way this is influenced by stretching), and the influence of porosity and lithology on the thermal conductivity of the sedimentary layer. They are not intended to provide a specific simulation, but, rather, a range of temperature profiles that allows an assessment of the sensitivity of lithospheric rheology to thermal conditions. Even with these limitations, it might be hoped that a comparison of measured heat flows with predicted values (indicated towards the bottom of each panel in Fig. 5) would provide a general indication of the present-day strength envelopes across the region. In practice, problems with the availability and accuracy of heat-flow measurements severely limit the confidence with which that calibration can be made. Measured heat flows in basinal settings can be distorted by factors such as recent sedimentation and erosion, refraction into basement highs, and hydrothermal convection. In very broad terms, the representative models for the southern part of the area (Fig. 5a–f) appear to bracket the typical heat flows observed there, whereas the more extensive heat-flow data available for the Norwegian margin and, in particular, the Vøring Basin (Sundvor *et al.* 2000; Ritter *et al.* 2004) suggest a thermal state closer to the warm models than the cool models.

The strength in the brittle parts of the yield-strength envelope was calculated with reference to the Anderson theory of faulting (Anderson 1951; Turcotte & Schubert 2002):

$$\Delta\sigma = \frac{2f_s(\rho gz - \rho_w gz)}{\sqrt{(1+f_s^2)} - f_s}$$

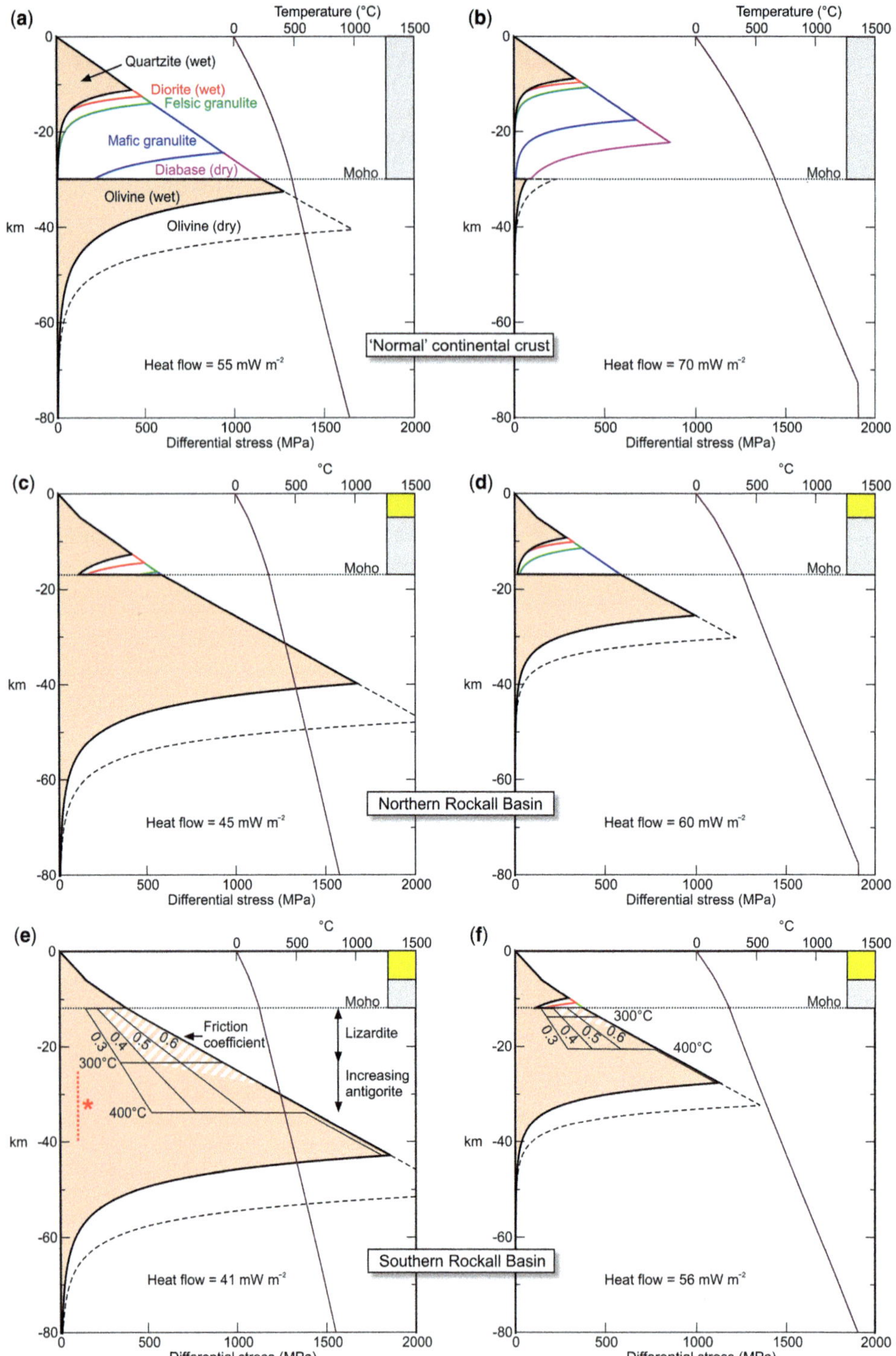

Fig. 5. Theoretical yield-strength envelopes for different simple crustal configurations: (**a**) and (**b**) 30 km-thick crystalline crust; (**c**) and (**d**) 5 km of sediments over 12 km of crystalline crust; (**e**) and (**f**) 6 km of sediments over 6 km of crystalline crust.

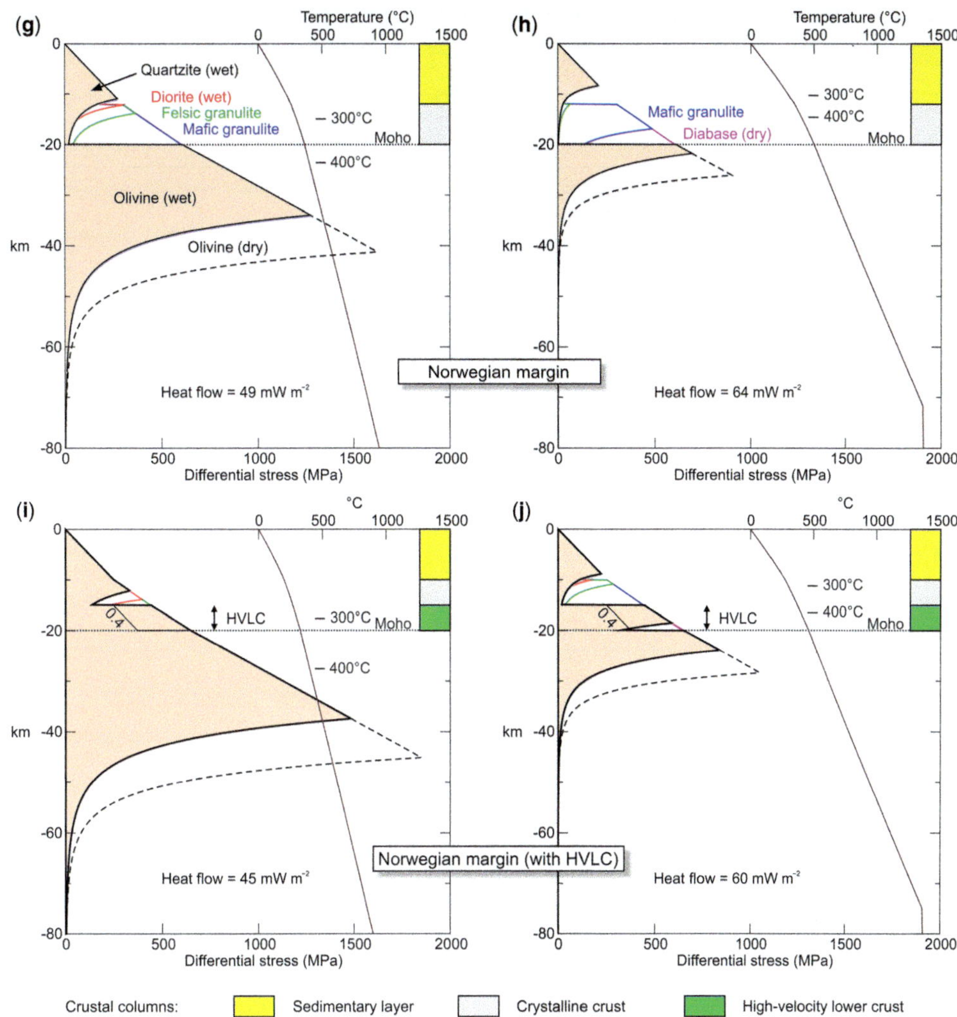

Fig. 5. (**g**) and (**h**) 12 km of sediments over 8 km of crystalline crust; (**i**) and (**j**) 10 km of sediments over 10 km of crystalline crust, of which 5 km is HVLC. The envelopes are for failure under compression (brittle failure occurs at lower stresses under extension). Depths are relative to ground surface/seabed. Each configuration is modelled with alternative 'cool' (a, c, e, g & i) and 'warm' (b, d, f, h & j) conditions (geotherms are shown on the right-hand side of the figures). The additional lines on the example shown in (e) and (f) illustrate the effect of different coefficients of friction within a zone of partially serpentinized upper mantle, and the hachured zone illustrates how the yield-strength envelope might be modified within this zone (see the text). The 0.4 friction coefficient is also shown in the zone of high-velocity lower crust in (i) and (j). The red dashed line marked with an asterix in (e) indicates the depth range within which there is potential for weakening associated with the ductile deformation of antigorite (Hilairet *et al.* 2007).

where $\Delta\sigma$ is the maximum horizontal deviatoric stress that can be withstood under compression, f_s is the coefficient of internal friction (a value of 0.6 was usually assumed, but see further discussion below), z is depth, ρgz is the lithostatic pressure and $\rho_w gz$ is the hydrostatic pressure.

The ductile parts of the yield-strength envelopes were assumed to be dominated by dislocation creep, following a power-law rheology:

$$\Delta\sigma = \left(\frac{\dot{\varepsilon}}{Ae^{-Q/RT}}\right)^{1/n}$$

where $\dot{\varepsilon}$ is the strain rate, A is the power-law stress constant, Q is the activation energy, n is the power-law exponent, R is the universal gas constant (8.3144621 J mol^{-1} K^{-1}) and T is the temperature

Table 1. *Density and thermal parameters assumed in the construction of the yield-strength envelopes*

Density and thermal parameters			
Rock unit	Density ($kg\ m^{-3}$)	Thermal conductivity ($W\ m^{-1}\ K^{-1}$)	Heat production ($\mu W\ m^{-3}$)
Sediments	2200	2.0	1.0
Crystalline crust	2850	2.5	1.0
HVLC	3000	2.5	0.3
Mantle	3300	*	0
Serpentinized mantle	3200	*	0

*Temperature-dependent thermal conductivity using the formulation of Xu *et al.* (2004) to define the lattice component and that of Hofmeister (1999) to define the radiative component (see also McKenzie *et al.* 2005).
HVLC, high-velocity lower crust (mafic).

in Kelvin. The parameters assumed for different lithologies are given in Table 2. A strain rate of $10^{-17}\ s^{-1}$ has been assumed, which is equivalent, for example, to a shortening of 0.5% over a period of 15 Ma. This is a higher rate than suggested by Gómez & Vergés (2005) for the Helland-Hansen Arch and the Vema Dome, but similar to that predicted by applying the shortening inferred by Vågnes *et al.* (1998) for the Ormen Lange Dome and the Helland-Hansen Arch over the timescales for the development of those structures indicated by Doré *et al.* (2008). The ductile behaviour of a range of possible crustal rocks has been simulated, ranging from weak (felsic and wet) to strong (mafic and dry). The filled parts of the strength envelopes shown in Figure 5 are based on a crust with the rheological properties of wet quartzite and a mantle with those of wet olivine, but the supplementary curves enable a range of alternatives to be visualized. For example, the effect of a mid-crustal lithological boundary can be assessed by drawing a horizontal line at the desired depth and adopting the curves for the preferred lithologies above and below that line.

Rheological behaviour of partially serpentinized upper mantle

Escartín *et al.* (2001) reported laboratory investigations into the strength of partially serpentinized peridotites. They concluded that the strength is not a simple function of the degree of serpentinization, but that there is an abrupt weakening at a serpentinite content of 10–15% or less (similar to the degree of serpentinization inferred to exist beneath the hyperextended basins on the NW European margin). The weakening is greater where the dominant serpentinite phase is lizardite rather than antigorite, and this is likely to be the case at temperatures below 300°C. Lizardite is progressively replaced by antigorite at higher temperatures, with the latter mineral becoming ubiquitous at 390°C and secondary olivine crystallization commencing at 460°C (Schwartz *et al.* 2013).

Experimental determinations of the coefficient of friction of lizardite typically lie in the range 0.3–0.5 (e.g. Escartín *et al.* 1997*b*), although lower values (0.15–0.35) have been reported by Reinen *et al.* (1994) and higher values (up to about 0.6)

Table 2. *Ductile deformation parameters assumed in the construction of the yield-strength envelopes*

Ductile deformation parameters (dislocation creep)				
Rock/mineral	A ($MPa^{-n}\ s^{-1}$)	n	Q ($kJ\ mol^{-1}$)	Source
Quartzite (wet)	1.1×10^{-4}	4	223	Burov & Watts (2006)
Diorite (wet)	3.8×10^{-2}	2.4	219	Carter & Tsenn (1987)
Felsic granulite	8×10^{-3}	3.1	243	Wilks & Carter (1990)
Mafic granulite	1.4×10^{4}	4.2	445	Wilks & Carter (1990)
Diabase (dry)	8	4.7	485	Burov & Watts (2006)
Olivine (wet)	417	4.48	498	Burov & Watts (2006)
Olivine (dry)	4.85×10^{4}	3.5	535	Burov & Watts (2006)
Antigorite	2.82×10^{-15}	3.8	8.9	Hilairet *et al.* (2007)

A, the power-law stress constant; n, the power-law exponent; Q, the activation energy.

by Moore *et al.* (1997). Chrysotile has a lower coefficient of friction than lizardite at room temperature, but this increases to similar values at around 200°C (Moore *et al.* 1997). Antigorite samples have higher friction coefficients, similar to those of unserpentinized rocks. There is a transition from localized to distributed (approximating to ductile) deformation in lizardite at around 200 MPa (depth of 9 km; Escartín *et al.* 1997*b*), although laboratory evidence suggests some pressure-dependence beyond this transition that can be simulated by assuming an appropriate coefficient of friction when calculating the yield-strength envelope. For example, the lizardite samples tested by Escartín *et al.* (1997*b*) lie along a trend equivalent to a friction coefficient of 0.3 at pressures of up to 950 MPa (depth of 33 km), and the partially serpentinized samples tested by Escartín *et al.* (2001) lie between the 0.3 and 0.5 coefficients at pressures of 300–450 MPa (depth of 12–17 km). Amiguet *et al.* (2012, 2014) reported weaker behaviour in high-pressure (1–8 GPa; 34–250 km depth) experiments on lizardite samples: they identified a strength under shear stress of about 100 MPa, with little pressure-dependence, associated with the dislocation glide of foliated lizardite along a basal plane. This weakening is dependent on crystal alignment, and a stronger rheology applies when the crystal orientation inhibits such glide (the locked geometry of Amiguet *et al.* 2012). Temperatures at these depths in the rheological models presented in this paper, however, lie outside the stability zone for lizardite (the experiments of Amiguet *et al.* (2012) were designed to simulate conditions in subduction zones).

To illustrate the potential influence of partial serpentinization on the yield-strength envelopes, alternative profiles are shown on the hyperextended example (Fig. 5e, f) for friction coefficients of 0.3, 0.4 and 0.5 in a zone extending from the Moho to the 400°C isotherm. From the above discussion, the most confidently identified zone of weakening is associated with lizardite and lies above the 300°C geotherm (shown in Fig. 5e, f). Below that, an increasing proportion of antigorite causes strengthening between 300 and 400°C, perhaps relatively rapidly if the effect is highly non-linear, as suggested by Escartín *et al.* (2001). A potential zone of lithospheric weakening is shown schematically with hachures in Figure 5e, f.

Hilairet *et al.* (2007) provided laboratory data for the deformation of antigorite, and observed a transition from brittle to ductile behaviour between 0.7 and 1 GPa (25–34 km). Their inferred power-law parameters (Table 2) indicate substantial weakening of antigorite below the brittle–ductile transition. If these parameters are included in a calculation for the partially serpentinized upper mantle using the aggregate flow laws of Tullis *et al.* (1991), a relatively high degree of serpentinization (30–40%) is required for the transition to appear in the envelopes (i.e. greater than indicated by seismic experiments). The problem with this approach is that it assumes that the phases are well dispersed, whereas interconnections of the serpentinite phase are likely to lead to greater weakening (Escartín *et al.* 2001). It follows that there is a zone between about 25 and 40 km (460°C) in the 'cool' hyperextended model in Figure 5e that might be weakened by ductile deformation of antigorite. With higher heat flow (Fig. 5f), the relevant pressures occur below the antigorite stability zone.

The present-day thickness and mineralogy of the zone of partial serpentinization will be a function of the conditions that applied when it was being formed (see Rüpke *et al.* 2013) and subsequent phase transformations. Simple sensitivity trials of the type illustrated here do not simulate the complexity of this process, but the zone of influence suggested for the 'cool' southern Rockall Basin example (Fig. 5e) is broadly compatible with the thickness of partial serpentinization indicated by seismic investigation (up to 10 km: O'Reilly *et al.* 2006). From the seismic evidence, it appears unlikely that the possible deeper zone of weakening associated with the ductile deformation of antigorite occurs below that basin, as normal upper-mantle velocities were encountered at depths of less than 25 km below seabed (O'Reilly *et al.* 2006). As an additional observation: the seismic velocity of antigorite lies between those of lizardite and normal upper mantle (Ji *et al.* 2013), so the lizardite–antigorite phase change may contribute to the velocity gradient seen at the base of the partially serpentinized zones.

Comparison of the yield-strength envelopes

The yield-strength envelopes provide only a general indication of strength variations within the lithosphere. In addition to the temperature uncertainties outlined above, the behaviour under stress is simulated using parameters derived from laboratory measurements on specimens at scales and strain rates that differ by many orders of magnitude from those that apply to the geological processes under consideration. The brittle predictions assume that the rock fails most readily along pre-existing, suitably orientated fractures, but under high-pressure conditions failure may occur more easily along new fractures (Zang *et al.* 2007; Pauselli *et al.* 2010). The transition from brittle to ductile behaviour probably involves semi-brittle failure at lower stresses than indicated by the 'spikes' in the yield-strength envelopes (Chester 1995; Kohlstedt *et al.* 1995). Despite these limitations, the envelopes are useful in identifying the sensitivity of lithospheric strength to lithology and temperature, and the distribution of this strength in different regimes.

The yield-strength envelopes for 30 km-thick continental crust suggest that a 'jelly sandwich' strength profile may apply under cooler conditions (Fig. 5a); coupling between the strong layers will not occur unless there is thick and mafic mid to lower crust. As temperatures increase, the lithosphere is considerably weakened (Fig. 5b) through both the thickening of the ductile lower crust and the weakening of the upper mantle, to the extent that the lithospheric strength may reside solely in a relatively thin crustal layer ('crème brûlée' strength profile: Jackson 2002; Burov & Watts 2006).

Beneath the northern Rockall Basin (Fig. 5c, d), the predicted zone of strong upper mantle is thicker. There is likely to be coupling between crust and upper mantle at lower temperatures (Fig. 5c), but at higher temperatures decoupling may occur if the crust is relatively felsic and wet (Fig. 5d). The strong zone in the upper mantle beneath the southern Rockall Basin (Fig. 5e, f) extends to greater depth, and it appears unlikely that a ductile lower crust will develop under the range of temperatures simulated, regardless of the crustal composition. Although there is some lithospheric weakening as a result of partial serpentinization of the upper mantle, this zone is predicted to retain some strength and potentially maintain the coupling between the crust and mantle, at least under compression.

On the Norwegian margin, where a thick sedimentary layer overlies thin crystalline crust, a weak lower-crustal zone is likely to form if the crust is relatively felsic (Fig. 5g, h). Where high-velocity lower crust is present, and this is assumed to be due to mafic rocks (either high-grade basement or underplating), it will be stronger (Fig. 5i, j). If such a layer is interpreted as partially serpentinized upper mantle, the stress threshold for brittle deformation may be reduced as described above (a coefficient of friction of 0.4 is illustrated in Fig. 5i, j). With higher heat flows, which appear more likely in this area (Sundvor *et al.* 2000; Ritter *et al.* 2004), the predicted temperatures within this layer move out of the stability zone for lizardite (note the depths of the 300 and 400°C isotherms in Fig. 5j), so the likelihood of this being a source of weakening is reduced. The pressure in the layer is lower than the threshold for ductile deformation of antigorite identified by Hilairet *et al.* (2007).

Discussion

The information presented in Figures 3 and 4 suggests that there is not a consistent relationship between the siting of Cenozoic compressional structures on the NW European margin and zones where the crust had previously been hyperextended and the underlying mantle potentially serpentinized. While the Cenozoic domes in the Norwegian sector do lie in the vicinity of a hyperextended zone, the majority of Cenozoic inversion structures in the Faroe–Shetland–northern Rockall–Hatton area are located above crust that is less extended. There are fewer Cenozoic compressional structures in the southern Rockall and Porcupine basins, where hyperextension and serpentinization have been confidently identified.

Simple rheological models reveal the sensitivity of lithospheric strength to the thickness, composition and temperature of the crust. The weakest lithosphere occurs where the crust is felsic and thick, and the heat flow is high. This leads to a ductile lower crust overlain by a relatively thin upper-crustal brittle layer within which short-wavelength compressional structures may form. When stretching has occurred, its influence on the geothermal gradient is countered by the reduced crustal depth over which the gradient applies and a reduction in the influence of crustal heat production. This can lead to relatively strong (coupled) lithosphere where the stretching factor is high and there has been sufficient thermal re-equilibration. Partial serpentinization of the upper mantle weakens the lithosphere but, despite its non-linear nature (Escartín *et al.* 2001), does not necessarily lead to full decoupling between strong crust and strong upper mantle. This could explain the general lack of short-wavelength compressional structures over the southern Rockall and Porcupine basins, and also the lithospheric strength during sediment loading of those basins suggested by gravity anomalies (Kimbell *et al.* 2004).

In the Faroe–Shetland–northern Rockall–Hatton area, where serpentinization of the upper mantle appears less likely, lithospheric weakening and propensity to compressional deformation may be associated with the development of ductile lower crust. This is normally the case when the crust is relatively thick: however, in the more extended areas, relatively warm conditions would be required at the time of deformation. A thermal anomaly associated with the North Atlantic Igneous Province could have facilitated the development of the earlier (Paleocene–Eocene) structures. This is likely to be particularly relevant to the development of domes near the continent–ocean transition, and the presence of sill complexes of similar age in the basins along the margin indicates that the thermal anomaly extended further inboard. Its influence would, however, have reduced at the time of Neogene deformation. The Miocene and younger structures in the northern part of the Faroe–Shetland Basin are therefore more difficult to explain in this way, although there are some indications of a higher-temperature regime at that time in the results of apatite fission track and vitrinite reflectance studies (e.g. Mark *et al.* 2008). Where a particularly strong thermal anomaly occurred, conditions could lie outside the range indicated in Figure 5, although a qualitative assessment of the impact can be made on the basis of extrapolation of the trends seen in that figure.

The criteria for appropriate conditions for the development of Cenozoic compressional structures in the Norwegian sector depend on the interpretation of the high-velocity lower crust (HVLC) identified there. If it is of mafic composition, weakening is most likely to have occurred where the overlying rock thickness (including the sedimentary layer) and geothermal gradient are large enough to allow a ductile zone to form above it. If the HVLC is actually partially serpentinized upper mantle, thin crust and/or a relatively low geothermal gradient may be a prerequisite for this layer to lie within the stability zone for lizardite. Higher temperatures would still weaken the lithosphere, through the thinning of the strong zone in the upper mantle and the potential development of a ductile layer in the overlying crust, but serpentinization could play a reduced role. The spatial relationship between the Cenozoic domes and the HVLC is inconclusive because there is overlap between these in the Vøring Basin but some divergence in the Møre Basin (Figs 4 & 6), although some spatial offset may be admissible if there was reactivation of low-angle detachments.

The contours in Figure 6 show the thickness of potentially weak crust in the Norwegian area, as measured from the seabed to the top of the HVLC, or to the Moho where HVLC was not detected. This version applies to the present day; the layer was up to 1–1.5 km thinner during the Miocene. Experimentation with mapping the temperature at the base of the potentially weak layer, which would be a more direct proxy for the presence of a weak zone, indicated a very similar pattern to that shown on the map, but required more assumptions that could distract from the simple point we wish to make. This is that the domes do not appear to have formed over the thinnest/coolest crust, but in an area where there is a general increase in the thickness of the potentially weak layer towards the east and SE, typically to more than 15 km. The 'Taper Break' of Redfield & Osmundsen (2013, 2014) lies just inboard of the compressional domes, and those authors related it to a concentration of earthquakes in that area. They argued that the earthquakes resulted from loading with Plio-Pleistocene glacial sediments (Byrkjeland *et al.* 2000) in combination with the relative weakening of the lithosphere by previous hyperextension, and used a flexed plate model to explain how locally derived bending stresses could generate the relatively deep (>15 km) compressive failure suggested by focal-plane solutions (Hicks *et al.* 2000). The hyperextended, weak lithosphere that extends oceanwards from this area was seen as 'protecting' it from the influence of far-field stresses. This (present-day) model does not appear to provide a full explanation for the earlier development of the Cenozoic domes, bearing in mind that they are upper-crustal structures with growth that is inferred to be due, at least in part, to far-field stresses. An alternative interpretation for the Norwegian domes is that the lithosphere outboard of them is strong (coupled) and thus aids the transmission of stresses originating from the oceanic area to the edge of a zone where weaker lithosphere facilitates deformation. The yield-strength profiles suggest that such coupling is feasible, given that the examples in Figure 5i, j coincide with the nominal 15 km 'threshold' for the thickness of the weak crustal layer and that this layer is thinner in the area in question. This hypothesis bears similarities to that of Mjelde *et al.* (2003, 2005) in which the distribution of compressional structures was influenced by rigid blocks underlain by HVLC. There is a possible analogue on the Hatton margin, where the compressional structures lie just inboard of the approximately 40 km-wide zone adjacent to the ocean margin affected by mafic intrusion (Johnson *et al.* 2005; White & Smith 2009).

Péron-Pinvidic *et al.* (2008) observed a correlation between a zone underlain by shallow, highly serpentinized exhumed continental mantle and Cenozoic compressional deformation on the Iberian margin, and concluded that this was the weakest zone on that margin. The estimated degree of serpentinization is 60–100% in the top 2–3 km of the upper mantle and 25–45% in the underlying layer, with a gradual transition to normal mantle values at depths of about 10 km below seabed (Chian *et al.* 1999). Landwards of this is a zone of hyperextended continental crust (2–5 km thick) underlain by a layer with a P-wave velocity of 7.3–7.9 km s^{-1}, which was interpreted as weakly serpentinized (<25%) upper mantle (Chian *et al.* 1999). Cenozoic deformational structures are rare in this zone. The geotherms beneath the two areas are unlikely to differ substantially, since the observed heat flow is similar (Louden *et al.* 1997) and the presence of a large proportion of lizardite will reduce the thermal conductivity of the highly serpentinized upper mantle to values closer to that of continental crust (Horai 1971). Brittle deformation is likely to dominate at the depths where the lateral strength contrasts are greatest, so it is the low effective coefficient of friction in the highly serpentinized upper mantle (and particularly in fault zones within this layer?) that is likely to explain the weakening. The degree of serpentinization beneath the

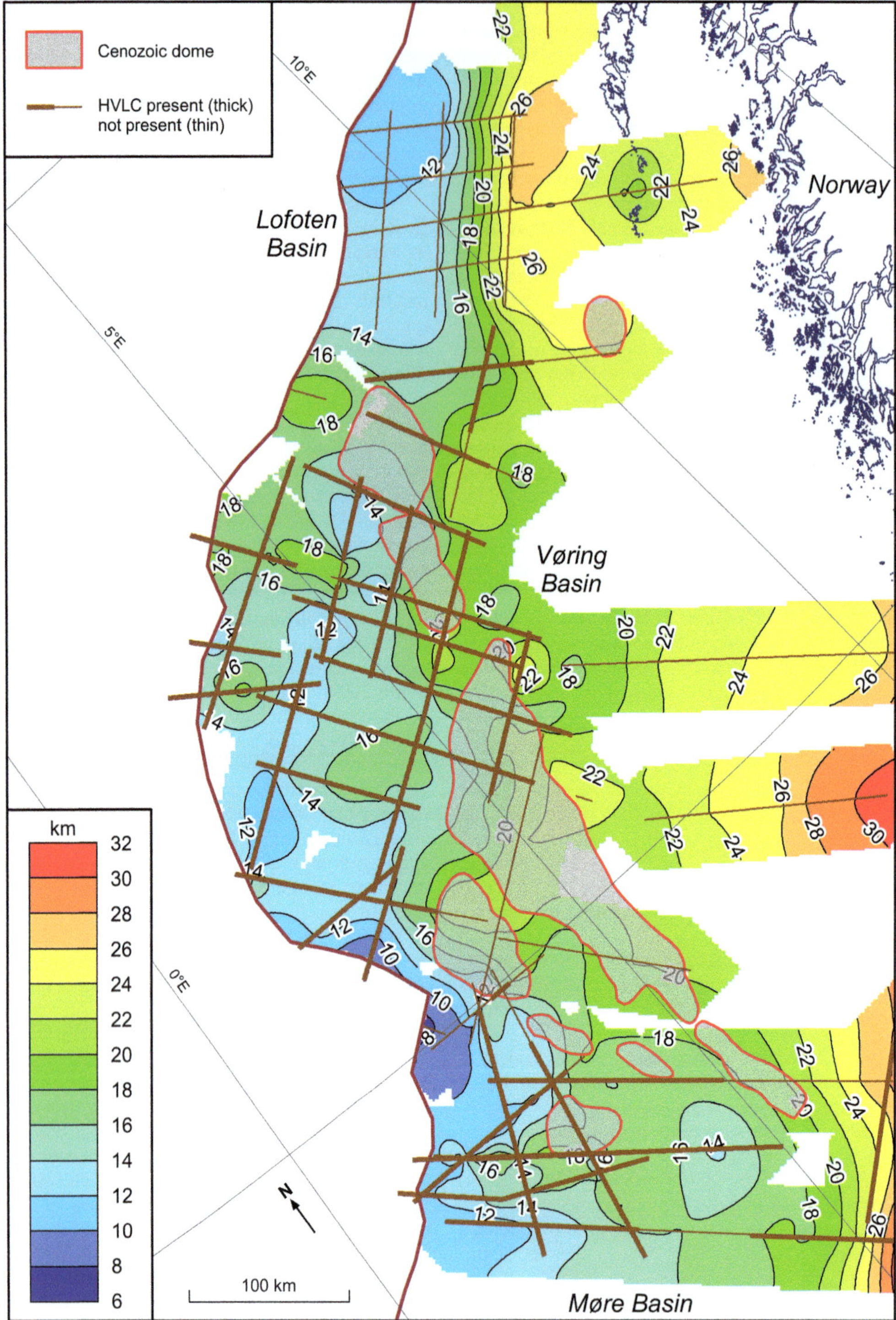

Fig. 6. Locations of Cenozoic domes and high-velocity lower crust (HVLC) on the Norwegian margin superimposed on a contour map of the thickness of potentially weak continental crust (i.e. measured from the seabed to the top of the HVLC). Contours are at 2 km intervals.

hyperextended areas on the NW European margin is closer to that of the stronger zone on the Iberian margin, so the latter margin does not provide an effective analogue for a general relationship between partial serpentinization and compressional deformation further north.

The way in which the lithosphere was stretched during the Mesozoic will have influenced the likelihood of both upper-mantle serpentinization and, consequently, its Cenozoic strength profile. Where the upper part of the crust is extended more than the lower part, its heat production (which is typically concentrated in the upper crust) will be strongly attenuated, leading to a reduction in lower-crustal temperatures and a greater likelihood of embrittlement (and, thus, water ingress and serpentinization). If stretching increases with depth, the cooling effect will be smaller and may not be sufficient for embrittlement to occur. Such a contrast in stretching style may explain the presence of serpentinized upper mantle beneath the southern Rockall and Porcupine basins (Hauser *et al.* 1995; O'Reilly *et al.* 1996, 2006), and its absence beneath the Orphan Basin on the Eastern Canadian conjugate margin, where stretching factors are similar (Welford *et al.* 2012). In the Cenozoic strength profiles, the lower temperatures in the former case reduces the likelihood of ductile lower crust developing above the serpentinized zone (although this may be offset by the thermal blanketing effect of a subsequently deposited, thick sedimentary sequence).

Reactivation of Mesozoic extensional structures provides control over the siting of at least some of the Cenozoic domes observed along the margin. In some cases, this relationship can be seen in seismic reflection data (e.g. the examples shown in Fig. 2a, e, f), but elsewhere it is inferred rather than proven because of difficulties with imaging basin architecture beneath Palaeogene igneous rocks (Doré *et al.* 2008). Reactivation provides a possible explanation for the siting of compressional structures in the areas where the predicted lithospheric strength does not appear optimum: for example, within basins rather than over the thicker crust on their flanks. Van Wees & Beekman (2000) reconstructed the rheological evolution of several basins that have undergone inversion and found evidence for strengthening rather than weakening at the time of inversion, relative to initial values; they concluded that the susceptibility to inversion was explained by pre-existing weak zones. Upper-crustal failure related to fault reactivation indicates that the orientation and frictional resistance of a fault zone can result in it failing more readily than the neighbouring rock mass, with fluid overpressure being an important factor, as well as the lithology of the fault gouge (Sibson 1995; Turner & Williams 2004). At depth, faults may detach into ductile lower crust or partially serpentinized upper mantle, depending on the setting. In the latter case, the weakening will be enhanced if there are appropriately orientated major structures containing foliated lizardite aligned such that dislocation glide is facilitated. Fluid overpressure may be enhanced in such zones as a result of water produced by the dehydration of the serpentine minerals (Raleigh & Paterson 1965).

As well as specific examples of reactivation, there is a more general relationship between groups of compressional structures and the boundaries between the Vøring and Møre basins in the north and the Faroe–Shetland and northern Rockall basins in the south. The implication is that the complex of structures that facilitated the transfer of Mesozoic extension within and between these basins also created preferential sites for compressional reactivation. The Mogdunn Arch, South Mogdunn Arch, Havsule Dome and Ormen Lange Dome are arrayed along the northern of the two transfer zones (Doré *et al.* 2008), and the Wyville Thomson Ridge and Ymir Ridge may correspond to reactivation of individual strands within the southern one (Kimbell *et al.* 2005).

Temporal and spatial variations in the stress field have to be taken into account when considering the spatial distribution of compressional structures. Although the growth histories of the observed folds are complex, in broad terms two dominant phases can be identified: one in the Palaeogene (most commonly the Eocene) and one in the Neogene (dominantly in the Miocene). This is illustrated, for example, by the chronological charts of Doré *et al.* (2008, fig. 4) and Ritchie *et al.* (2008, fig. 3). Although the influence of transient thermal weakening at these times may be a contributing factor, temporal variations in the stress field are a likely primary explanation for this timing. The stress regime is the result of multiple influences, including ridge-push and other plate-boundary forces, mantle drag, and Alpine/Pyrenean/Eurekan orogenic stresses. These will not be discussed in detail here (see the useful review by Doré *et al.* 2008), except to acknowledge that the propensity of a particular area to compressional deformation at a particular time will depend on local stress magnitude and orientation, as well as lithospheric rheology and the availability of suitably orientated pre-existing structures. Compressional stresses generated by differences in gravitational potential energy (GPE), including those associated with the Miocene growth of the Iceland Insular Margin (Doré *et al.* 2008) and the influence of the Southern Scandes (Pascal & Cloetingh 2009), will be greatest in areas of low GPE, possibly contributing to the development of compressional structures in deeper basinal settings. Plate break-up adjacent to the Faroe–Shetland–north Rockall area was complex and protracted, and included the separation and rotation of the Jan Mayen microcontinent; this may have generated stresses that contributed to the concentration of compressional structures in that area (Stoker *et al.* 2013). Borehole breakouts in the Faroe–Shetland Basin indicate that the contemporary maximum horizontal compressive stress is orientated approximately NW–SE in the northern part of the basin, but that more variable stress orientations occur in its central and SW parts, and these have been interpreted in terms of deflections towards weak faults (Holford *et al.* 2016). This is compatible with the basin-parallel trends of compressional structures in the north of the basin, and variable trends of the Judd and Westray anticlines in the south.

An example of susceptibility to compressional deformation changing with time is provided by the two sections in the southern part of the area illustrated in Figure 2e, f. Pre-latest Eocene deformation is observed in the central part of the Rockall Basin (Fig. 2e), but there is little evidence of short-wavelength deformation in the rocks overlying the Late Eocene unconformity here or elsewhere in the southern Rockall Basin. However, the example from the Colmán Basin (Fig. 2f) illustrates that similarly orientated short-wavelength compressional deformation did occur at these later times, at a nearby location where the crust is thicker and upper-mantle serpentinization less likely. The implication is that the hyperextended lithosphere of the southern Rockall Basin has strengthened with time. If such strengthening is due to cooling, it can be correlated with the thickening of strong (coupled) lithosphere, as illustrated in the 'warm' and 'cool' examples in Figure 5f, e, respectively. It is unlikely to relate to a change in the thickness and mineralogy of the partially serpentinized zone, given the relationship between temperature and the stability of the serpentine minerals, but might be assisted by the loss of transient overpressure associated with earlier dehydration.

The emphasis in the discussion so far has been on the development of short-wavelength (<*c.* 100 km) compressional structures. Longer-wavelength (hundreds of km) deformation is evident along the NW European margin, in particular as regional 'sagging' at the end of the Eocene and intra-Pliocene 'tilting' (Praeg *et al.* 2005; Stoker *et al.* 2005*b*). Long-wavelength lithospheric folds can form as a result of in-plane compressive stresses and this can occur coevally with the development of short-wavelength structures in rheologically stratified lithosphere, with the former associated with the deformation of strong upper mantle and the latter with deformation of decoupled upper crust (Cloetingh *et al.* 1999; Cloetingh *et al.* 2008; Cloetingh & Burov 2011). Numerical modelling is required to test whether the observed long-wavelength features are reproducible by in-plane stress, and this topic is not considered in the present paper. Previous modelling has suggested amplitudes of hundreds of metres rather than the observed kilometre-scale 'sagging' and 'tilting' of Stoker *et al.* (2005*b*) and Praeg *et al.* (2005), leading those authors to prefer an alternative mechanism involving shallow mantle convection.

Conclusions

- A comparison between the locations of Cenozoic compressional structures on the NW European margin and regional lithospheric models produced by the NAG-TEC project (Hopper *et al.* 2014) suggests that there is not a simple relationship between the distribution of the structures and the degree of Mesozoic lithospheric stretching. In particular, an association with areas of hyperextension and partial serpentinization of the upper mantle, as proposed by Lundin & Doré (2011), is not consistently observed. The strongest evidence for such an association occurs on the Norwegian margin, but there is little support for it in the southern Rockall Basin where a well-resolved zone of partially serpentinized upper mantle is overlain by largely undeformed Cenozoic strata.
- A range of compressional structures formed in the Faroe–Shetland–northern Rockall area during the Cenozoic. Although the lithosphere in parts of this area (the deepest parts of the Faroe–Shetland and northern Rockall basins) has undergone substantial extension, current evidence suggests that the amount of stretching is insufficient to have resulted in upper-mantle serpentinization. The data are sparse, but there is as yet no indication of reduced upper-mantle velocities that would be indicative of such serpentinization. These conclusions are particularly tentative in the northern Faroe–Shetland Basin, where the deep structure is poorly understood.
- Rheological modelling reveals the sensitivity of the strength of the lithosphere to crustal configuration, lithological variation and temperature profile. The modelling has not been used to address the state of the lithosphere during Mesozoic extension (for this see, e.g., Pérez-Gussinyé & Reston 2001; Rüpke *et al.* 2013), but rather to consider conditions that may have applied later, during Cenozoic compression. Even so, the uncertainties in this modelling are such that it is only used to provide some rheological context for the observed spatial distributions rather than make detailed predictions.
- There are particular uncertainties in modelling the rheology of partially serpentinized upper mantle, but the current results suggest that

conflicting factors affect the degree of lithospheric weakening associated with the presence of such a zone. Lower temperatures promote the stability of the serpentine minerals, but also lead to a thickening of underlying strong upper mantle, which may be coupled to a strong crust, depending on the effective coefficient of friction of the intervening, partially serpentinized zone. Alignment of lizardite crystals and fluid overpressure in fault zones could lead to enhanced weakening, and the overpressure could be a transient phenomenon that only applied during a particular phase of basin evolution.

- In areas where short-wavelength compressional structures occur and serpentinized upper mantle is not present, thermally related lithospheric weakening, involving decoupling by ductile lower crust, may have facilitated their development.
- Although (or, perhaps, because) it is the most thoroughly investigated part of the study area, the Norwegian margin is the most controversial in terms of its deep structure and, in particular, whether the high-velocity lower crust (HVLC) there is of magmatic or metamorphic origin, or is associated with partially serpentinized upper mantle. The weakening implicit in the latter interpretation is compatible with the development of overlying compressional structures, but the spatial relationships admit an alternative interpretation in which the domes formed at the edge of a rheologically weaker zone against a 'buttress' underpinned by mafic (metamorphic or igneous) lower crust. More detailed investigation is required, involving the development of alternative lithological and rheological models along selected deep seismic profiles, flexural/gravity modelling to investigate spatial variations in lithospheric strength indicated by the response to sediment loading, and numerical simulation of the effect of compression on the alternative models.
- Rheological modelling can help in understanding the general factors that influence the propensity of an area to compressional deformation, but these are complemented (and may be outweighed) by the presence of specific weak zones associated with pre-existing structures.

This study has employed the results of the NAG-TEC project, and we acknowledge the support of the industry sponsors of that project (in alphabetical order): Bayerngas Norge AS; BP Exploration Operating Company Limited, Bundesanstalt für Geowissenschaften und Rohstoffe (BGR); Chevron East Greenland Exploration A/S; ConocoPhillips Skandinavia AS; DEA Norge AS; Det norske oljeselskap ASA; DONG E&P A/S; E.ON Norge AS; ExxonMobil Exploration and Production Norway AS; Japan Oil, Gas and Metals National Corporation (JOGMEC); Maersk Oil; Nalcor Energy – Oil and Gas Inc.; Nexen Energy ULC, Norwegian Energy Company ASA (Noreco); Repsol Exploration Norge AS; Statoil (U.K.) Limited and Wintershall Holding GmBH. We are very grateful to Mick Hanrahan at the Irish Petroleum Affairs Division of the Department of Communications, Energy and Natural Resources for his assistance with the permission to publish Figure 2f. The contribution of Kimbell, Stewart and Stoker is made with the permission of the Executive Director of the British Geological Survey (Natural Environment Research Council).

References

AMIGUET, E., REYNARD, B., CARACAS, R., VAN DE MOORTÈLE, B., HILAIRET, N. & WANG, Y. 2012. Creep of phyllosilicates at the onset of plate tectonics. *Earth and Planetary Science Letters*, **345–348**, 142–150, https://doi.org/10.1016/j.epsl.2012.06.033

AMIGUET, E., VAN DE MOORTELE, B., REYNARD, B., CORDIER, P. & HILAIRET, N. 2014. Deformation mechanisms and rheology of serpentines in experiments and in nature. *Journal of Geophysical Research*, **119**, 4640–4655, https://doi.org/10.1002/2013JB010791

ANDERSON, E.M. 1951. *The Dynamics of Faulting and Dyke Formation with Applications to Britain*. Oliver & Boyd, Edinburgh.

ARCHER, S.G., BERGMAN, S.C., ILIFFE, J., MURPHY, C.M. & THORNTON, M. 2005. Palaeogene igneous rocks reveal new insights into the geodynamic evolution and petroleum potential of the Rockall Trough, NE Atlantic Margin. *Basin Research*, **17**, 171–201, https://doi.org/10.1111/j.1365-2117.2005.00260.x

BARTON, A.J. & WHITE, R.S. 1997. Crustal structure of Edoras Bank continental margin and mantle thermal anomalies beneath the North Atlantic. *Journal of Geophysical Research: Solid Earth*, **102**, 3109–3129, https://doi.org/10.1029/96jb03387

BLYSTAD, P., BREKKE, H., FRERSETH, R.B., LARSEN, B.T., SKOGSEID, J. & TØRUDBAKKEN, B. 1995. *Structural Elements of the Norwegian Continental Shelf. Part 2: The Norwegian Sea Region*. Norwegian Petroleum Directorate Bulletin, **8**.

BOLDREEL, L.O. & ANDERSEN, M.S. 1993. Late Paleocene to Miocene compression in the Faroe–Rockall area. *In*: PARKER, J.R. (ed.) *Petroleum Geology of NW Europe: Proceedings of the 4th Conference*. Geological Society, London, 1025–1034, https://doi.org/10.1144/0041025

BOLDREEL, L.O. & ANDERSEN, M.S. 1995. The relationship between the distribution of Tertiary sediments, tectonic processes and deep-water circulation around the Faeroe Islands. *In*: SCRUTTON, R.A., STOKER, M.S., SHIMMIELD, G.B. & TUDHOPE, A.W. (eds) *The Tectonics, Sedimentation and Palaeoceanography of the North Atlantic Region*. Geological Society, London, Special Publications, **90**, 145–158, https://doi.org/10.1144/gsl.sp.1995.090.01.09

BOLDREEL, L.O. & ANDERSEN, M.S. 1998. Tertiary compressional structures on the Faroe–Rockall Plateau in relation to northeast Atlantic ridge-push and Alpine foreland stresses. *Tectonophysics*, **300**, 13–28, https://doi.org/10.1016/s0040-1951(98)00231-5

Bott, M.H.P., Armour, A.R., Himsworth, E.M., Murphy, T. & Wylie, G. 1979. Explosion seismology investigation of the continental-margin west of the Hebrides, Scotland, at 58°N. *Tectonophysics*, **59**, 217–231, https://doi.org/10.1016/0040-1951(79)90046-5

Brekke, H. 2000. The tectonic evolution of the Norwegian Sea Continental Margin with emphasis on the Vøring and Møre Basins. *In*: Nøttvedt, A. (ed.) *Dynamics of the Norwegian Margin*. Geological Society, London, Special Publications, **167**, 327–378, https://doi.org/10.1144/GSL.SP.2000.167.01.13

Burov, E.B. 2011. Rheology and strength of the lithosphere. *Marine and Petroleum Geology*, **28**, 1402–1443, https://doi.org/10.1016/j.marpetgeo.2011.05.008

Burov, E.B. & Watts, A.B. 2006. The long-term strength of continental lithosphere: 'jelly sandwich' or 'crème brûlée'? *GSA Today*, **16**, 4–10.

Byrkjeland, U., Bungum, H. & Eldholm, O. 2000. Seismotectonics of the Norwegian continental margin. *Journal of Geophysical Research*, **105**, 6221–6236, https://doi.org/10.1029/1999JB900275

Carter, N.L. & Tsenn, M.C. 1987. Flow properties of continental lithosphere. *Tectonophysics*, **136**, 27–63, https://doi.org/10.1016/0040-1951(87)90333-7

Chester, F.M. 1995. A rheologic model for wet crust applied to strike-slip faults. *Journal of Geophysical Research*, **100**, 13,033–13,044, https://doi.org/10.1029/95JB00313

Chian, D.P., Louden, K.E., Minshull, T.A. & Whitmarsh, R.B. 1999. Deep structure of the ocean-continent transition in the southern Iberia Abyssal Plain from seismic refraction profiles: Ocean Drilling Program (Legs 149 and 173) transect. *Journal of Geophysical Research*, **104**, 7443–7462, https://doi.org/10.1029/1999JB900004

Christensen, N.I. 2004. Serpentinites, peridotites, and seismology. *International Geology Review*, **46**, 795–816, https://doi.org/10.2747/0020-6814.46.9.795

Cloetingh, S. & Burov, E. 2011. Lithospheric folding and sedimentary basin evolution: a review and analysis of formation mechanisms. *Basin Research*, **23**, 257–290, https://doi.org/10.1111/j.1365-2117.2010.00490.x

Cloetingh, S., Burov, E. & Poliakov, A. 1999. Lithosphere folding: primary response to compression? (from central Asia to Paris basin). *Tectonics*, **18**, 1064–1083, https://doi.org/10.1029/1999TC900040

Cloetingh, S., Beekman, F., Ziegler, P.A., van Wees, J.-D. & Sokoutis, D. 2008. Post-rift compressional reactivation potential of passive margins and extensional basins. *In*: Johnson, H., Doré, A.G., Gatliff, R.W., Holdsworth, R., Lundin, E.R. & Ritchie, J.D. (eds) *The Nature and Origin of Compression in Passive Margins*. Geological Society, London, Special Publications, **306**, 27–70, https://doi.org/10.1144/sp306.2

Davies, R., Cloke, I., Cartwright, J., Robinson, A. & Ferrero, C. 2004. Post-breakup compression of a passive margin and its impact on hydrocarbon prospectivity: an example from the Tertiary of the Faroe–Shetland Basin, United Kingdom. *American Association of Petroleum Geologists Bulletin*, **88**, 1–20.

Davies, R.J. & Cartwright, J. 2002. A fossilized Opal A to Opal C/T transformation on the northeast Atlantic margin: support for a significantly elevated palaeogeothermal gradient during the Neogene? *Basin Research*, **14**, 467–486, https://doi.org/10.1046/j.1365-2117.2002.00184.x

Doré, A.G., Lundin, E.R., Jensen, L.N., Birkeland, O., Eliassen, P.E. & Fichler, C. 1999. Principal tectonic events in the evolution of the northwest European Atlantic margin. *In*: Fleet, A.J. & Boldy, S.A.R. (eds) *Petroleum Geology of Northwest Europe: Proceedings of the 5th Conference*. Geological Society, London, 41–61, https://doi.org/10.1144/0050041

Doré, A.G., Lundin, E.R., Kusznir, N.J. & Pascal, C. 2008. Potential mechanisms for the genesis of Cenozoic domal structures on the NE Atlantic margin: pros, cons and some new ideas. basins. *In*: Johnson, H., Doré, A.G., Gatliff, R.W., Holdsworth, R., Lundin, E.R. & Ritchie, J.D. (eds) *The Nature and Origin of Compression in Passive Margins*. Geological Society, London, Special Publications, **306**, 1–26, https://doi.org/10.1144/sp306.1

Ebbing, J. 2007. Isostatic density modelling explains the missing root of the Scandes. *Norwegian Journal of Geology*, **87**, 13–20.

Ebbing, J., Lundin, E., Olesen, O. & Hansen, E.K. 2006. The mid-Norwegian margin: a discussion of crustal lineaments, mafic intrusions, and remnants of the Caledonian root by 3D density modelling and structural interpretation. *Journal of the Geological Society, London*, **163**, 47–59, https://doi.org/10.1144/0016-764905-029

Escartín, J., Hirth, G. & Evans, B. 1997*a*. Effects of serpentinization on the lithospheric strength and the style of normal faulting at slow-spreading ridges. *Earth and Planetary Science Letters*, **151**, 181–189, https://doi.org/10.1016/S0012-821X(97)81847-X

Escartín, J., Hirth, G. & Evans, B. 1997*b*. Nondilatant brittle deformation of serpentinites: implications for Mohr-Coulomb theory and the strength of faults. *Journal of Geophysical Research*, **102**, 2897–2913, https://doi.org/10.1029/96JB02792

Escartín, J., Hirth, G. & Evans, B. 2001. Strength of slightly serpentinized peridotites: implications for the tectonics of oceanic lithosphere. *Geology*, **29**, 1023–1026, https://doi.org/10.1130/0091-7613(2001)029<1023:SOSSPI>2.0.CO;2

Funck, T., Andersen, M.S., Neish, J.K. & Dahl-Jensen, T. 2008. A refraction seismic transect from the Faroe Islands to the Hatton–Rockall Basin. *Journal of Geophysical Research*, **113**, B12405, https://doi.org/10.1029/2008JB005675

Funck, T., Hopper, J.R. *et al*. 2014. Crustal structure. *In*: Hopper, J.R., Funck, T., Stoker, M., Árting, U., Péron-Pinvidic, G., Doornenbal, H. & Gaina, C. (eds) *NAG-TEC Atlas: Tectonostratigraphic Atlas of the North-East Atlantic Region*. Geological Survey of Denmark and Greenland (GEUS), Copenhagen, 69–126.

Funck, T., Geissler, W.H., Kimbell, G.S., Gradmann, S., Erlendsson, Ö., McDermott, K. & Petersen, U.K. In press. Moho and basement depth in the NE Atlantic Ocean based on seismic refraction data and receiver functions. *In*: Péron-Pinvidic, G., Hopper,

J.R., Stoker, M.S., Gaina, C., Doornenbal, J.C., Funck, T. & Árting, U.E. (eds) *The NE Atlantic Region: A Reappraisal of Crustal Structure, Tectonostratigraphy and Magmatic Evolution*. Geological Society, London, Special Publications, **447**, https://doi.org/10.1144/SP447.1

Gernigon, L., Ringenbach, J.C., Planke, S., Le Gall, B. & Jonquet-Kolsto, H. 2003. Extension, crustal structure and magmatism at the outer Vøring Basin, Norwegian margin. *Journal of the Geological Society, London*, **160**, 197–208, https://doi.org/10.1144/0016-764902-055

Gernigon, L., Ringenbach, J.C., Planke, S. & Le Gall, B. 2004. Deep structures and breakup along volcanic rifted margins: insights from integrated studies along the outer Vøring Basin (Norway). *Marine and Petroleum Geology*, **21**, 363–372, https://doi.org/10.1016/j.marpetgeo.2004.01.005

Gómez, M. & Vergés, J. 2005. Quantifying the contribution of tectonics v. differential compaction in the development of domes along the Mid-Norwegian Atlantic margin. *Basin Research*, **17**, 289–310, https://doi.org/10.1111/j.1365-2117.2005.00264.x

Haase, C., Ebbing, J. & Funck, T. In review. A 3D regional crustal model of the Northeast Atlantic based on seismic and gravity data. *In*: Péron-Pinvidic, G., Hopper, J.R., Stoker, M.S., Gaina, C., Doornenbal, J.C., Funck, T. & Árting, U.E. (eds) *The NE Atlantic Region: A Reappraisal of Crustal Structure, Tectonostratigraphy and Magmatic Evolution*. Geological Society, London, Special Publications, **447**.

Hauser, F., O'Reilly, B.M., Jacob, A.W.B., Shannon, P.M., Makris, J. & Vogt, U. 1995. The crustal structure of the Rockall Trough: differential stretching without underplating. *Journal of Geophysical Research*, **100**, 4097–4116, https://doi.org/10.1029/94jb02879

Hicks, E.C., Bungum, H. & Lindholm, C.D. 2000. Stress inversion of earthquake focal mechanism solutions from onshore and offshore Norway. *Norsk Geologisk Tidsskrift*, **80**, 235–250.

Hilairet, N., Reynard, B., Wang, Y., Daniel, I., Merkel, S., Nishiyama, N. & Petitgirard, S. 2007. High-pressure creep of serpentine, interseismic deformation, and initiation of subduction. *Science*, **318**, 1910–1913, https://doi.org/10.1126/science.1148494

Hitchen, K. 2004. The geology of the UK Hatton-Rockall margin. *Marine and Petroleum Geology*, **21**, 993–1012, https://doi.org/10.1016/j.marpetgeo.2004.05.004

Hodges, S., Line, C. & Evans, R. 1999. The other millennium dome. Paper SPE 56895 presented at the *Society of Petroleum Engineers, Offshore Europe Oil and Gas Exhibition and Conference*, 7–10 September 1999, Aberdeen, UK.

Hofmeister, A.M. 1999. Mantle values of thermal conductivity and the geotherm from phonon lifetimes. *Science*, **283**, 1699–1706, https://doi.org/10.1126/science.283.5408.1699

Holford, S.P., Tassone, D.R., Stoker, M.S. & Hillis, R.R. 2016. Contemporary stress orientations in the Faroe–Shetland region. *Journal of the Geological Society, London*, **173**, 142–152, https://doi.org/10.1144/jgs2015-048

Holmes, R., Hobbs, P.R.N. *et al.* 2003. *DTI Strategic Environmental Assessment Area 4 (SEA4): Geological Evolution Pilot Whale Diapirs and Stability of the Seabed Habitat*. British Geological Survey Commissioned Report, CR/03/082. British Geological Survey, Nottingham, UK.

Hopper, J.R. & Buck, W.R. 1998. Styles of extensional decoupling. *Geology*, **26**, 699–702, https://doi.org/10.1130/0091-7613(1998)026<0699:SOED>2.3.CO;2

Hopper, J.R., Funck, T., Stoker, M., Árting, U., Peron-Pinvidic, G., Doornenbal, H. & Gaina, C. (eds). 2014. *NAG-TEC Atlas: Tectonostratigraphic Atlas of the North-East Atlantic Region*. Geological Survey of Denmark and Greenland (GEUS), Copenhagen.

Hopper, J.R., Fatah, R.A., Gaina, C., Geissler, W., Doornenbal, H., Funck, T. & Kimbell, G.S. In prep. Sediment thickness and residual topography of the North Atlantic: estimating dynamic topography around Iceland. *In*: Péron-Pinvidic, G., Hopper, J.R., Stoker, M.S., Gaina, C., Doornenbal, J.C., Funck, T. & Árting, U.E. (eds) *The NE Atlantic Region: A Reappraisal of Crustal Structure, Tectonostratigraphy and Magmatic Evolution*. Geological Society, London, Special Publications, **447**.

Horai, K. 1971. Thermal conductivity of rock-forming minerals. *Journal of Geophysical Research*, **76**, 1278–1308, https://doi.org/10.1029/JB076i005 p01278

Jackson, J. 2002. Strength of the continental lithosphere: time to abandon the jelly sandwich? *GSA Today*, **12**, 4–9.

Jacob, A.W.B., Kaminski, W., Murphy, T., Phillips, W.E.A. & Prodehl, C. 1985. A crustal model for a northeast-southwest profile through Ireland. *Tectonophysics*, **113**, 75–103, https://doi.org/10.1016/0040-1951(85)90111-8

Ji, S., Li, A., Wang, Q., Long, C., Wang, H., Marcotte, D. & Salisbury, M. 2013. Seismic velocities, anisotropy, and shear-wave splitting of antigorite serpentinites and tectonic implications for subduction zones. *Journal of Geophysical Research*, **118**, 1015–1037, https://doi.org/10.1002/jgrb.50110

Johnson, H., Ritchie, J.D., Hitchen, K., McInroy, D.B. & Kimbell, G.S. 2005. Aspects of the Cenozoic deformational history of the northeast Faroe–Shetland Basin, Wyville–Thomson Ridge and Hatton Bank areas. *In*: Doré, A.G. & Vining, B.A. (eds) *Petroleum Geology: North-West Europe and Global Perspectives – Proceedings of the 6th Petroleum Geology Conference*. Geological Society, London, 993–1007, https://doi.org/10.1144/0060993

Joppen, M. & White, R.S. 1990. The structure and subsidence of Rockall Trough from two-ship seismic experiments. *Journal of Geophysical Research: Solid Earth*, **95**, 19821–19837, https://doi.org/10.1029/JB095iB12p19821

Kimbell, G.S., Gatliff, R.W., Ritchie, J.D., Walker, A.S.D. & Williamson, J.P. 2004. Regional three-dimensional gravity modelling of the NE Atlantic margin. *Basin Research*, **16**, 259–278, https://doi.org/10.1111/j.1365-2117.2004.00232.x

Kimbell, G.S., Ritchie, J.D., Johnson, H. & Gatliff, R.W. 2005. Controls on the structure and evolution of the NE Atlantic margin revealed by regional potential field imaging and 3D modelling. *In*: Doré, A.G. & Vining, B.A. (eds) *Petroleum Geology: North-West Europe and Global Perspectives – Proceedings of the 6th Petroleum Geology Conference*. Geological Society, London, 933–945, https://doi.org/10.1144/0060933

Klingelhöfer, F., Edwards, R.A., Hobbs, R.W. & England, R.W. 2005. Crustal structure of the NE Rockall Trough from wide-angle seismic data modeling. *Journal of Geophysical Research*, **110**, B11105, https://doi.org/10.1029/2005JB003763

Kohlstedt, D.L., Evans, B. & Mackwell, S.J. 1995. Strength of the lithosphere: constraints imposed by laboratory experiments. *Journal of Geophysical Research*, **100**, 17,587–17,602, https://doi.org/10.1029/95JB01460

Kvarven, T., Ebbing, J. *et al.* 2014. Crustal structure across the Møre margin, mid-Norway, from wide-angle seismic and gravity data. *Tectonophysics*, **626**, 21–40, https://doi.org/10.1016/j.tecto.2014.03.021

Louden, K.E., Sibuet, J.C. & Harmegnies, F. 1997. Variations in heat flow across the ocean-continent transition in the Iberia abyssal plain. *Earth and Planetary Science Letters*, **151**, 233–254, https://doi.org/10.1016/S0012-821X(97)81851-1

Lowe, C. & Jacob, A.W.B. 1989. A north-south seismic profile across the Caledonian Suture zone in Ireland. *Tectonophysics*, **168**, 297–318, https://doi.org/10.1016/0040-1951(89)90224-2

Lundin, E.R. & Doré, A.G. 2002. Mid-Cenozoic post-breakup deformation in the 'passive' margins bordering the Norwegian-Greenland Sea. *Marine and Petroleum Geology*, **19**, 79–93, https://doi.org/10.1016/s0264-8172(01)00046-0

Lundin, E.R. & Doré, A.G. 2011. Hyperextension, serpentinization, and weakening: a new paradigm for rifted margin compressional deformation. *Geology*, **39**, 347–350, https://doi.org/10.1130/g31499.1

Lundin, E.R., Doré, A.G., Rønning, K. & Kyrkjebø, R. 2013. Repeated inversion and collapse in the Late Cretaceous–Cenozoic northern Vøring Basin, offshore Norway. *Petroleum Geoscience*, **19**, 329–341, https://doi.org/10.1144/petgeo2012-022

Makris, J., Papoulia, I. & Ziska, H. 2009. Crustal structure of the Shetland–Faeroe Basin from long offset seismic data. *In*: Varming, T. & Ziska, H. (eds) *Faroe Island Exploration Conference: Proceedings of the 2nd Conference*. Faroese Academy of Sciences, Tórshavn, 30–42.

Mark, D.F., Green, P.F., Parnell, J., Kelley, S.P., Lee, M.R. & Sherlock, S.C. 2008. Late Palaeozoic hydrocarbon migration through the Clair field, West of Shetland, UK Atlantic margin. *Geochimica et Cosmochimica Acta*, **72**, 2510–2533, https://doi.org/10.1016/j.gca.2007.11.037

Masson, D.G. & Parson, L.M. 1983. Eocene deformation on the continental margin SW of the British Isles. *Journal of the Geological Society, London*, **140**, 913–920, https://doi.org/10.1144/gsjgs.140.6.0913

McDonnell, A. & Shannon, P.M. 2001. Comparative Tertiary stratigraphic evolution of the Porcupine and Rockall basins. *In*: Shannon, P.M., Haughton, P.D.W. & Corcoran, D.V. (eds) *The Petroleum Exploration of Ireland's Offshore Basins*. Geological Society, London, Special Publications, **188**, 323–344, https://doi.org/10.1144/GSL.SP.2001.188.01.19.

McKenzie, D. 1978. Some remarks on the development of sedimentary basins. *Earth and Planetary Science Letters*, **40**, 25–32, https://doi.org/10.1016/0012-821X(78)90071-7

McKenzie, D., Jackson, J. & Priestley, K. 2005. Thermal structure of oceanic and continental lithosphere. *Earth and Planetary Science Letters*, **233**, 337–349, https://doi.org/10.1016/j.epsl.2005.02.005

Minshull, T.A. 2009. Geophysical characterisation of the ocean–continent transition at magma-poor rifted margins. *Comptes Rendus Geoscience*, **341**, 382–393, https://doi.org/10.1016/j.crte.2008.09.003

Mjelde, R., Kodaira, S., Shimamura, H., Kanazawa, T., Shiobara, H., Berg, E.W. & Riise, O. 1997. Crustal structure of the central part of the Vøring Basin, mid-Norway margin, from ocean bottom seismographs. *Tectonophysics*, **277**, 235–257, https://doi.org/10.1016/S0040-1951(97)00028-0

Mjelde, R., Kasahara, J. *et al.* 2002. Lower crustal seismic velocity-anomalies; magmatic underplating or serpentinized peridotite? Evidence from the Vøring Margin, NE Atlantic. *Marine Geophysical Researches*, **23**, 169–183, https://doi.org/10.1023/a:1022480304527

Mjelde, R., Iwasaki, T., Shimamura, H., Kanazawa, T., Kodaira, S., Raum, T. & Hajime, S. 2003. Spatial relationship between recent compressional structures and older high-velocity crustal structures; examples from the Vøring Margin, NE Atlantic, and Northern Honshu, Japan. *Journal of Geodynamics*, **36**, 537–562, https://doi.org/10.1016/S0264-3707(03)00087-5

Mjelde, R., Raum, T., Breivik, A.J., Shimamura, H., Murai, Y., Takanami, T. & Faleide, J.I. 2005. Crustal structure of the Vøring Margin, NE Atlantic: a review of geological implications based on recent OBS data. *In*: Doré, A.G. & Vining, B.A. (eds) *Petroleum Geology: North-West Europe and Global Perspectives – Proceedings of the 6th Petroleum Geology Conference*. Geological Society, London, 803–813, https://doi.org/10.1144/0060803

Mjelde, R., Faleide, J.I., Breivik, A.J. & Raum, T. 2009*a*. Lower crustal composition and crustal lineaments on the Vøring Margin, NE Atlantic: a review. *Tectonophysics*, **472**, 183–193, https://doi.org/10.1016/j.tecto.2008.04.018

Mjelde, R., Raum, T., Kandilarov, A., Murai, Y. & Takanami, T. 2009*b*. Crustal structure and evolution of the outer Møre margin, NE Atlantic. *Tectonophysics*, **468**, 224–243, https://doi.org/10.1016/j.tecto.2008.06.003

Moore, D.E., Lockner, D.A., Ma, S. & Summers, R. 1997. Strengths of serpentinite gouges at elevated temperatures. *Journal of Geophysical Research*, **102**, 14,787–14,801, https://doi.org/10.1029/97JB00995

Morewood, N.C., Shannon, P.M. & Mackenzie, G.D. 2004. Seismic stratigraphy of the southern Rockall Basin: a comparison between wide-angle seismic and normal incidence reflection data. *Marine and Petroleum Geology*, **21**, 1149–1163, https://doi.org/10.1016/j.marpetgeo.2004.07.006

MOREWOOD, N.C., MACKENZIE, G.D., SHANNON, P.M., O'REILLY, B.M., READMAN, P.W. & MAKRIS, J. 2005. The crustal structure and regional development of the Irish Atlantic margin region. *In*: DORÉ, A.G. & VINING, B.A. (eds) *Petroleum Geology: North-West Europe and Global Perspectives – Proceedings of the 6th Petroleum Geology Conference*. Geological Society, London, 1023–1033, https://doi.org/10.1144/0061023

NAYLOR, D., SHANNON, P.M. & MURPHY, N. 1999. *Irish Rockall Basin Region – A Standard Structural Nomenclature System*. Petroleum Affairs Division, Dublin, Special Publications, **1/99**.

NAYLOR, D., SHANNON, P. & MURPHY, N. 2002. *Porcupine-Goban Region – A Standard Structural Nomenclature System*. Petroleum Affairs Division, Dublin, Special Publications, **1/02**.

NIRRENGARTEN, M., GERNIGON, R.S. & MANATSCHAL, G. 2014. Lower crustal bodies in the Møre volcanic rifted margin: Geophysical determination and geological implications. *Tectonophysics*, **636**, 143–157, https://doi.org/10.1016/j.tecto.2014.08.004

OLAFSSON, I., SUNDVOR, E., ELDHOLM, O. & GRUE, K. 1992. Møre margin: crustal structure from analysis of expanded spread profiles. *Marine Geophysical Researches*, **14**, 137–163, https://doi.org/10.1007/BF01204284

O'REILLY, B.M., HAUSER, F., JACOB, A.W.B. & SHANNON, P.M. 1996. The lithosphere below the Rockall Trough: wide-angle seismic evidence for extensive serpentinisation. *Tectonophysics*, **255**, 1–23, https://doi.org/10.1016/0040-1951(95)00149-2

O'REILLY, B.M., HAUSER, F., RAVAUT, C., SHANNON, P.M. & READMAN, P.W. 2006. Crustal thinning, mantle exhumation and serpentinization in the Porcupine Basin, offshore Ireland: evidence from wide-angle seismic data. *Journal of the Geological Society, London*, **163**, 775–787, https://doi.org/10.1144/0016-76492005-079

PASCAL, C. & CLOETINGH, S.A.P.L. 2009. Gravitational potential stresses and stress field of passive continental margins: insights from the south-Norway shelf. *Earth and Planetary Science Letters*, **277**, 464–473, https://doi.org/10.1016/j.epsl.2008.11.014

PAUSELLI, C., RANALLI, G. & FEDERICO, C. 2010. Rheology of the Northern Apennines: lateral variations of lithospheric strength. *Tectonophysics*, **484**, 27–35, https://doi.org/10.1016/j.tecto.2009.08.029

PÉREZ-GUSSINYÉ, M. & RESTON, T.J. 2001. Rheological evolution during extension at nonvolcanic rifted margins: onset of serpentinization and development of detachments leading to continental breakup. *Journal of Geophysical Research*, **106**, 3961–3975, https://doi.org/10.1029/2000jb900325

PÉRON-PINVIDIC, G., MANATSCHAL, G., DEAN, S.M. & MINSHULL, T.A. 2008. Compressional structures on the West Iberia rifted margin: controls on their distribution. *In*: JOHNSON, H., DORÉ, A.G., GATLIFF, R.W., HOLDSWORTH, R., LUNDIN, E.R. & RITCHIE, J.D. (eds) *The Nature and Origin of Compression in Passive Margins*. Geological Society, London, Special Publications, **306**, 169–183, https://doi.org/10.1144/sp306.8

PERON-PINVIDIC, G., MANATSCHAL, G. & OSMUNDSEN, P.T. 2013. Structural comparison of archetypal Atlantic rifted margins: a review of observations and concepts. *Marine and Petroleum Geology*, **43**, 21–47, https://doi.org/10.1016/j.marpetgeo.2013.02.002

PETERSEN, U.K. & FUNCK, T. In press. Review of seismic refraction modelling in the Faroe–Shetland channel. *In*: PÉRON-PINVIDIC, G., HOPPER, J.R., STOKER, M.S., GAINA, C., DOORNENBAL, J.C., FUNCK, T. & ÁRTING, U.E. (eds) *The NE Atlantic Region: A Reappraisal of Crustal Structure, Tectonostratigraphy and Magmatic Evolution*. Geological Society, London, Special Publications, **447**, https://doi.org/10.1144/SP447.7

PRAEG, D., STOKER, M.S., SHANNON, P.M., CERAMICOLA, S., HJELSTUEN, B., LABERG, J.S. & MATHIESEN, A. 2005. Episodic Cenozoic tectonism and the development of the NW European 'passive' continental margin. *Marine and Petroleum Geology*, **22**, 1007–1030, https://doi.org/10.1016/j.marpetgeo.2005.03.014

RALEIGH, C.B. & PATERSON, M.S. 1965. Experimental deformation of serpentinite and its tectonic implications. *Journal of Geophysical Research*, **70**, 3965–3985, https://doi.org/10.1029/JZ070i016p03965

RAUM, T., MJELDE, R. *ET AL*. 2005. Sub-basalt structures east of the Faroe Islands revealed from wide-angle seismic and gravity data. *Petroleum Geoscience*, **11**, 291–308, https://doi.org/10.1144/1354-079304-627

REDFIELD, T.F. & OSMUNDSEN, P.T. 2013. The long-term topographic response of a continent adjacent to a hyperextended margin: a case study from Scandinavia. *Geological Society of America Bulletin*, **125**, 184–200, https://doi.org/10.1130/B30691.1

REDFIELD, T.F. & OSMUNDSEN, P.T. 2014. Some remarks on the earthquakes of Fennoscandia: a conceptual seismological model drawn from the perspective of hyperextension. *Norwegian Journal of Geology*, **94**, 233–262, https://doi.org/10.17850/njg94-4-01

REINEN, L.A., WEEKS, J.D. & TULLIS, T.E. 1994. Frictional behavior of lizardite and antigorite serpentinites: experiments, constitutive models, and implications for natural faults. *Pure and Applied Geophysics*, **143**, 317–358, https://doi.org/10.1007/BF00874334

REYNISSON, R.F., EBBING, J., LUNDIN, E., OSMUNDSEN, P.T.A. & USOV, S. 2010. Properties and distribution of lower crustal bodies on the mid-Norwegian margin. *In*: VINING, B. & PICKERING, S. (eds) *Petroleum Geology – From Mature Basins to New Frontiers. Proceedings of the 7th Petroleum Geology Conference*. Geological Society, London, UK, 843–854, https://doi.org/10.1144/0070843

RITCHIE, J.D., JOHNSON, H. & KIMBELL, G.S. 2003. The nature and age of Cenozoic contractional deformation within the NE Faroe–Shetland Basin. *Marine and Petroleum Geology*, **20**, 399–409, https://doi.org/10.1016/s0264-8172(03)00075-8

RITCHIE, J.D., JOHNSON, H., QUINN, M.F. & GATLIFF, R.W. 2008. Cenozoic compressional deformation within the Faroe–Shetland Basin and adjacent areas. *In*: JOHNSON, H., DORÉ, A.G., GATLIFF, R.W., HOLDSWORTH, R., LUNDIN, E.R. & RITCHIE, J.D. (eds) *The Nature and Origin of Compression in Passive Margins*. Geological Society, London, Special Publications, **306**, 121–136, https://doi.org/10.1144/SP306.5

RITCHIE, J.D., ZISKA, H., KIMBELL, G.S., QUINN, M. & CHADWICK, A. 2011. Structure. *In*: RITCHIE, J.D.,

ZISKA, H., JOHNSON, H. & EVANS, D. (eds) *The Geology of the Faroe–Shetland Basin, and Adjacent Areas*. British Geological Survey Research Report, RR/11/01, Jarðfeingi Research Report, RR/11/01. British Geological Survey, Nottingham, 9–70.

RITCHIE, J.D., JOHNSON, H., KIMBELL, G.S. & QUINN, M. 2013. Structure. *In*: HITCHEN, K., JOHNSON, H. & GATLIFF, R.W. (eds) *The Geology of the Rockall Basin and Adjacent Areas*. British Geological Survey Research Report, RR/12/03. British Geological Survey, Nottingham, 10–46.

RITTER, U., ZIELINSKI, G.W., WEISS, H.M., ZIELINSKI, R.L.B. & SAETTEM, J. 2004. Heat flow in the Voring Basin, Mid-Norwegian Shelf. *Petroleum Geoscience*, **10**, 353–365, https://doi.org/10.1144/1354-079303-616

ROBERTS, A.W., WHITE, R.S. & CHRISTIE, P.A.F. 2009. Imaging igneous rocks on the North Atlantic rifted continental margin. *Geophysical Journal International*, **179**, 1024–1038, https://doi.org/10.1111/j.1365-246X.2009.04306.x

ROHRMAN, M. 2007. Prospectivity of volcanic basins: trap delineation and acreage de-risking. *American Association of Petroleum Geologists Bulletin*, **91**, 915–939, https://doi.org/10.1306/12150606017

RÜPKE, L.H., SCHMID, D.W., PEREZ-GUSSINYE, M. & HARTZ, E. 2013. Interrelation between rifting, faulting, sedimentation, and mantle serpentinization during continental margin formation – including examples from the Norwegian Sea. *Geochemistry, Geophysics, Geosystems*, **14**, 4351–4369, https://doi.org/10.1002/ggge.20268

SCHWARTZ, S., GUILLOT, S. *ET AL.* 2013. Pressure-temperature estimates of the lizardite/antigorite transition in high pressure serpentinites. *Lithos*, **178**, 197–210, https://doi.org/10.1016/j.lithos.2012.11.023

SHANNON, P.M., JACOB, A.W.B., O'REILLY, B., HAUSER, F., READMAN, P.W. & MAKRIS, J. 1999. Structural setting, geological development and basin modelling in the Rockall Trough. *In*: FLEET, A.J. & BOLDY, S.A.R. (eds) *Petroleum Geology of Northwest Europe: Proceedings of the 5th Conference*. Geological Society, London, 421–431, https://doi.org/10.1144/0050421

SIBSON, R.H. 1995. Selective fault reactivation during basin inversion: potential for fluid redistribution through fault-valve action. *In*: BUCHANAN, J.G. & BUCHANAN, P.G. (eds) *Basin Inversion*. Geological Society, London, Special Publications, **88**, 3–19, https://doi.org/10.1144/gsl.sp.1995.088.01.02

SKOGSEID, J., PEDERSEN, T., ELDHOLM, O. & LARSEN, B.T. 1992. Tectonism and magmatism during NE Atlantic continental break-up: the Vøring Margin. *In*: STOREY, B.C., ALABASTER, T. & PANKHURST, R.J. (eds) *Magmatism and the Causes of Continental Break-up*. Geological Society, London, Special Publications, **68**, 305–320, https://doi.org/10.1144/gsl.sp.1992.068.01.19

SKOGSEID, J., PLANKE, S., FALEIDE, J.I., PEDERSEN, T., ELDHOLM, O. & NEVERDAL, F. 2000. NE Atlantic continental rifting and volcanic margin formation. *In*: NØTTVEDT, A. (ed.) *Dynamics of the Norwegian Margin*. Geological Society, London, Special Publications, **167**, 295–326, https://doi.org/10.1144/GSL.SP.2000.167.01.12

SMALLWOOD, J.R. & KIRK, W.J. 2005. Paleocene exploration in the Faroe–Shetland Channel: disappointments and discoveries. *In*: DORÉ, A.G. & VINING, B.A. (eds) *Petroleum Geology: North-West Europe and Global Perspectives – Proceedings of the 6th Petroleum Geology Conference*. Geological Society, London, 977–997, https://doi.org/10.1144/0060977

STOKER, M.S., HOULT, R.J. *ET AL.* 2005*a*. Sedimentary and oceanographic responses to early Neogene compression on the NW European margin. *Marine and Petroleum Geology*, **22**, 1031–1044, https://doi.org/10.1016/j.marpetgeo.2005.01.009

STOKER, M.S., PRAEG, D. *ET AL.* 2005*b*. Neogene evolution of the Atlantic continental margin of NW Europe (Lofoten Islands to SE Ireland): anything but passive. *In*: DORÉ, A.G. & VINING, B.A. (eds) *Petroleum Geology: North-West Europe and Global Perspectives – Proceedings of the 6th Petroleum Geology Conference*. Geological Society, London, 1057–1076, https://doi.org/10.1144/0061057

STOKER, M.S., LESLIE, A.B. & SMITH, K. 2013. A record of Eocene (Stronsay Group) sedimentation in BGS borehole 99/3, offshore NW Britain: implications for early post-rift development of the Faroe–Shetland Basin. *Scottish Journal of Geology*, **49**, 133–148.

STOKER, M.S., DOORNENBAL, H., HOPPER, J.R. & GAINA, C. 2014. Tectonostratigraphy. *In*: HOPPER, J.R., FUNCK, T., STOKER, M., ÁRTING, U., PERON-PINVIDIC, G., DOORNENBAL, H. & GAINA, C. (eds) *NAG-TEC Atlas: Tectonostratigraphic Atlas of the North-East Atlantic Region*. Geological Survey of Denmark and Greenland (GEUS), Copenhagen, 129–212.

SUNDVOR, E., ELDHOLM, O., GLADCZENKO, T.P. & PLANKE, S. 2000. Norwegian–Greenland Sea thermal field. *In*: NØTTVEDT, A. (ed.) *Dynamics of the Norwegian Margin*. Geological Society, London, Special Publications, **167**, 397–410, https://doi.org/10.1144/GSL.SP.2000.167.01.15

TUITT, A., UNDERHILL, J.R., RITCHIE, J.D., JOHNSON, H. & HITCHEN, K. 2010. Timing, controls and consequences of compression in the Rockall–Faroe area of the NE Atlantic margin. *In*: VINING, B. & PICKERING, S. (eds) *Petroleum Geology: From Mature Basins to New Frontiers. Proceedings of the 7th Petroleum Geology Conference*. Geological Society, London, 963–977, https://doi.org/10.1144/0070963

TULLIS, T.E., HOROWITZ, F.G. & TULLIS, J. 1991. Flow laws of polyphase aggregates from end-member flow laws. *Journal of Geophysical Research*, **96**, 8081–8096, https://doi.org/10.1029/90JB02491

TURCOTTE, D.L. & SCHUBERT, G. 2002. *Geodynamics*. 2nd edn. Cambridge University Press, Cambridge.

TURNER, J.P. & WILLIAMS, G.A. 2004. Sedimentary basin inversion and intra-plate shortening. *Earth-Science Reviews*, **65**, 277–304, https://doi.org/10.1016/j.earscirev.2003.10.002

UNTERNEHR, P., PÉRON-PINVIDIC, G., MANATSCHAL, G. & SUTRA, E. 2010. Hyper-extended crust in the South Atlantic: in search of a model. *Petroleum Geoscience*, **16**, 207–215, https://doi.org/10.1144/1354-079309-904

VÅGNES, E., GABRIELSEN, R.H. & HAREMO, P. 1998. Late Cretaceous–Cenozoic intraplate contractional deformation at the Norwegian continental shelf: timing, magnitude and regional implications. *Tectonophysics*, **300**, 29–46, https://doi.org/10.1016/S0040-1951(98)00232-7

VAN WEES, J.D. & BEEKMAN, F. 2000. Lithosphere rheology during intraplate basin extension and inversion. *Tectonophysics*, **320**, 219–242, https://doi.org/10.1016/S0040-1951(00)00039-1

VOGT, U., MAKRIS, J., O'REILLY, B.M., HAUSER, F., READMAN, P.W., JACOB, A.W.B. & SHANNON, P.M. 1998. The Hatton Basin and continental margin: crustal structure from wide-angle seismic and gravity data. *Journal of Geophysical Research*, **103**, 12,545–12,566, https://doi.org/10.1029/98jb00604

WELFORD, J.K., SHANNON, P.M., O'REILLY, B.M. & HALL, J. 2012. Comparison of lithosphere structure across the Orphan Basin–Flemish Cap and Irish Atlantic conjugate continental margins from constrained 3D gravity inversions. *Journal of the Geological Society, London*, **169**, 405–420, https://doi.org/10.1144/0016-76492011-114

WHITE, R.S. & SMITH, L.K. 2009. Crustal structure of the Hatton and the conjugate east Greenland rifted volcanic continental margins, NE Atlantic. *Journal of Geophysical Research: Solid Earth*, **114**, B02305, https://doi.org/10.1144/0016-76492011-114

WHITE, R.S., SMITH, L.K., ROBERTS, A.W., CHRISTIE, P.A.F., KUSZNIR, N.J. & TEAM, I. 2008. Lower-crustal intrusion on the North Atlantic continental margin. *Nature*, **452**, 460–464, https://doi.org/10.1038/nature06687

WILKS, K.R. & CARTER, N.L. 1990. Rheology of some continental lower crustal rocks. *Tectonophysics*, **182**, 57–77, https://doi.org/10.1016/0040-1951(90)90342-6

XU, Y., SHANKLAND, T.J., LINHARDT, S., RUBIE, D.C., LANGENHORST, F. & KLASINSKI, K. 2004. Thermal diffusivity and conductivity of olivine, wadsleyite and ringwoodite to 20 GPa and 1373 K. *Physics of the Earth and Planetary Interiors*, **143**, 321–336, https://doi.org/10.1016/j.pepi.2004.03.005

ZANG, S.X., WEI, R.Q. & NING, J.Y. 2007. Effect of brittle fracture on the rheological structure of the lithosphere and its application in the Ordos. *Tectonophysics*, **429**, 267–285, https://doi.org/10.1016/j.tecto.2006.10.006

ZISKA, H. & VARMING, T. 2008. Palaeogene evolution of the Ymir and Wvyille Thomson ridges, European North Atlantic Margin. *In*: JOHNSON, H., DORÉ, A.G., GATLIFF, R.W., HOLDSWORTH, R., LUNDIN, E.R. & RITCHIE, J.D. (eds) *The Nature and Origin of Compression in Passive Margins*. Geological Society, London, Special Publications, **306**, 153–168, https://doi.org/10.1144/SP306.7

Isostasy as a tool to validate interpretations of regional geophysical datasets – application to the mid-Norwegian continental margin

SOFIE GRADMANN[1]*, CLAUDIA HAASE[1] & JÖRG EBBING[1,2]

[1]*Geological Survey of Norway, Leiv Eirikssons vei 39, 7040 Trondheim, Norway*

[2]*Present address: Department of Geosciences, Kiel University, 24118 Kiel, Germany*

**Correspondence: sofie.gradmann@ngu.no*

Abstract: Isostasy is a well understood concept, yet rarely applied to its full capacity in regional interpretations of crustal structures. In this study, we utilize a recent density model for the entire NE Atlantic, based on refraction seismic data and gravity inversion, to calculate isostatically balanced bathymetry along the mid-Norwegian margin. Since gravity and isostatically balanced elevation are independent observables but both depend on the underlying density model, consistencies and discrepancies point towards model deficits, erroneously interpreted or poorly understood areas.

Four areas of large isostatic residuals are identified. Along the outer Vøring Margin, a mass deficit points to more extensive high-density bodies or a shallower Moho than currently mapped. Farther seaward, along the Vøring Marginal High, a mass excess indicates inaccurate mapping of the continent–ocean boundary and surrounding structures. A number of eclogitic bodies along the proximal mid-Norwegian margin have been described in recent publications and their presence is now also confirmed by isostatic calculations. Major elevation and gravity residuals along the transition between the Vøring and Møre margins signify that the structure of this region is poorly understood and modifications to the mapped continent–ocean transition may be required.

The concept of isostasy has been applied to several regional studies worldwide, helping to estimate lithospheric strength and explain the shape and origin of topography (e.g. Watts 2001; Kaban *et al.* 2003). Local studies, however, which investigate the crustal structure in more detail by employing refraction and deep reflection seismic data as well as gravity modelling, often ignore the additional insight that could be gained from isostatic considerations. Positive exceptions to this are, for example, Watts *et al.* (1985), Zeyen & Fernàndez (1994), Maţenco *et al.* (1997), Hirsch *et al.* (2009) and Fullea *et al.* (2010), where isostasy has entered the model building and interpretation. Methods for joint analysis and inversion of gravity and isostasy have, for example, been developed by Zeyen & Fernàndez (1994), Stewart *et al.* (2000), Braitenberg *et al.* (2004) and Salem *et al.* (2014).

Only a few of the studies involving isostasy cover the mid-Norwegian margin (Kimbell *et al.* 2004; Ebbing 2007) and we show in this paper that simple isostatic considerations can also improve the understanding of the crustal structure.

Continental margins may be especially well suited for isostatic studies since the crust is thinned and its deeper structure and Moho are more easily imaged by seismic methods than, for example, those of cratonic lithosphere. Seismic mapping is especially extensive (and accessible) along hydrocarbon-bearing continental margins such as the mid-Norwegian one. Furthermore, thinned crust and lithosphere are generally weaker, implying that lateral density variations and Moho undulations will have a larger expression in the bathymetry or basin shape. Thermal weakening during rifting and breakup as well as rigidification during subsequent cooling additionally affect the flexure of the margin.

The enormous amount of geophysical and geological data that exists for the region of the NE Atlantic has been gathered, analysed and compiled in recent years into a comprehensive tectonostratigraphic atlas (the NAG-TEC Atlas, Hopper *et al.* 2014). Based on these data, a new seismic-derived Moho grid has been compiled (Funck *et al.* 2016) and subsequently further improved using gravity inversion (Haase *et al.* 2016). The latter study also

From: Péron-Pinvidic, G., Hopper, J. R., Stoker, M. S., Gaina, C., Doornenbal, J. C., Funck, T. & Árting, U. E. (eds) 2017. *The NE Atlantic Region: A Reappraisal of Crustal Structure, Tectonostratigraphy and Magmatic Evolution.* Geological Society, London, Special Publications, **447**, 279–297.
First published online February 23, 2017, https://doi.org/10.1144/SP447.13

constructed a first-order density model over the entire NE-Atlantic, which we utilize here to forward calculate an isostatically consistent bathymetry that reflects all internal loading of the model. The resulting differences from the true bathymetry enable us to detect consistencies and discrepancies between the datasets mapping the deep subsurface structure. Since gravity and isostatically balanced elevation are independent observables but both depend on the underlying density model, consistencies and discrepancies can point towards poorly understood or erroneously interpreted areas. The obtained elevation residuals are then discussed with respect to existing seismic profiles and structural interpretations along the margin. By considering residuals of shorter wavelengths only, we focus on the crustal structure, which is much better mapped and discussed than the underlying lithospheric mantle. Previous studies that investigated regional isostatic and gravity modelling in the NE Atlantic region (Kimbell *et al.* 2004; Reynisson *et al.* 2010) had less complete datasets at hand and focussed on somewhat different questions. This study merely tests an existing subsurface model and does not attempt to improve it.

Geological history of the mid-Norwegian margin

The mid-Norwegian margin, like any passive rifted margin, consists of numerous rifted blocks and interjacent sedimentary basins and sub-basins (Fig. 1). The margin is subdivided into the three major segments of the Møre, Vøring and Lofoten margins. The large shelf areas of the North Sea and Barents Sea extend to the south and north, respectively. The structure of the mid-Norwegian margin reflects a long period of extension following the orogenic collapse of the Caledonian orogeny. Earliest extension during the post-orogenic collapse initiated in Devonian–Carboniferous times, forming large detachment faults and basins that are now partly exposed in onshore areas (e.g. Seranne & Seguret 1987; Osmundsen & Andersen 2001, and references therein) and possibly buried by younger sediments in the offshore domain. Carboniferous to early Mesozoic extension created the early, proximal rift basins (e.g. the Trøndelag Platform; Brekke 2000). Distal rift basins formed during Middle Jurassic to Early/middle-Cretaceous (e.g. the Lofoten, Vøring and Møre basins; Brekke 2000). Late Cretaceous to Paleocene extension shaped the outer ridges and sub-basin system defining the pre-breakup rift complex (Gernigon *et al.* 2003; Ren *et al.* 2003). Volcanic activity affected the distal margins in the Late Paleocene but was most significant later, during the breakup and volcanic margin formation in the early Eocene (*c.* 54 Ma, Talwani & Eldholm 1977; Skogseid & Eldholm 1987). This volcanic activity was part of the North Atlantic Igneous Province (Eldholm & Grue 1994; Meyer *et al.* 2007) and is reflected along the present-day mid-Norwegian outer margin by extensive regions with seaward-dipping reflectors and massive volcanic flows. Magmatic activity also led to intrusions into the thinned continental crust and underplating in the continent–ocean transition zone. Oceanic spreading along the Aegir and Mohn's ridges led to the complete separation of the Greenland and Eurasian plates. Continued, contemporaneous Cenozoic rifting in East Greenland and associated breakup in late Oligocene (*c.* 25–24 Ma) led to the formation of a second spreading axis and to the separation of the Jan Mayen microcontinent from Greenland (Talwani & Eldholm 1977; Vogt *et al.* 1980; Nunns 1982; Gudlaugsson *et al.* 1988; Gaina *et al.* 2009; Gernigon *et al.* 2015). Today, the Kolbeinsey Ridge is the active spreading centre linking the Reykjanes Ridge in the south to the Mohn's Ridge in the north.

A large number of seismic refraction profiles cover the mid-Norwegian margin but become more sparse to the north (Fig. 1). Some profiles are several decades old and large receiver spacing (10–20 km) often prevents the detailed mapping of smaller-scale features. Nevertheless, this extensive grid predominantly shaped our present-day understanding of the deep margin structures.

The 300 km wide Møre Margin is characterized by a narrow platform and a rapid thinning of the continental crust. The main area of crustal thinning is confined to a narrow zone with a width of only a few tens of kilometres (Olafsson *et al.* 1992; Raum 2000; Mjelde *et al.* 2008; Osmundsen & Ebbing 2008; Kvarven *et al.* 2014; Nirrengarten *et al.* 2014). West of this major necking zone lies the nearly 200 km wide and long Cretaceous Møre Basin, which reaches depths of more than 15 km. This basin is underlain by highly thinned crust (in places less than 10 km thick) and high-velocity, high-density lower crustal bodies (LCBs) have been mapped both outbound and inbound of the necking zone but not directly underneath it (Olafsson *et al.* 1992; Kvarven *et al.* 2014; Nirrengarten *et al.* 2014). The outer limit of the margin comprises the Møre Marginal High which is overlain by a series of volcanic flows. Whereas some authors interpret the basement of the Møre Marginal High as thick continental crust (e.g. Reynisson *et al.* 2010), others suggest a sediment succession of several kilometres thickness beneath the flood basalts (Nirrengarten *et al.* 2014).

The Jan Mayen corridor, which continues as the Jan Mayen Fracture Zone in the oceanic domain, offsets the much wider Vøring Margin (450 km)

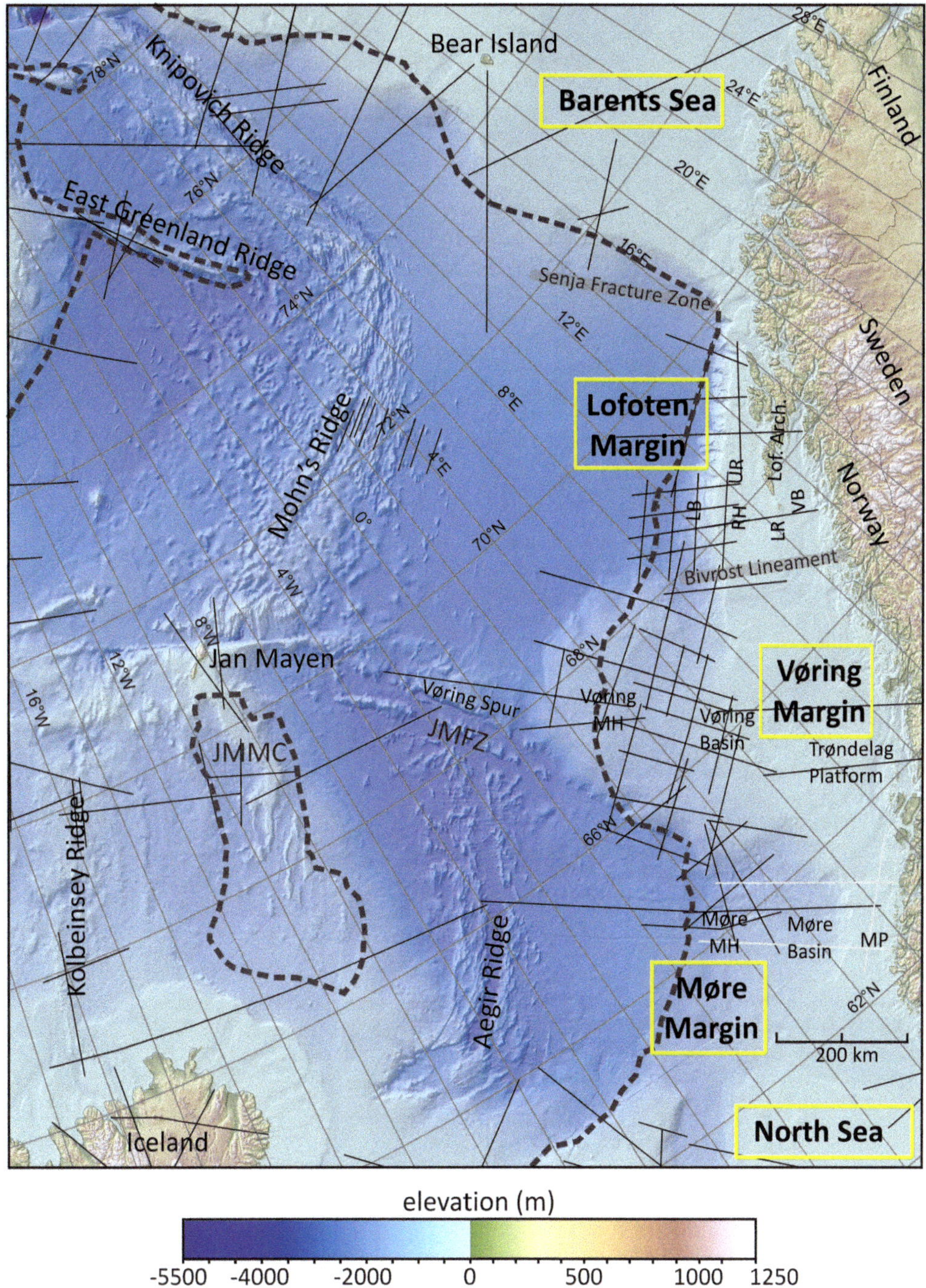

Fig. 1. Bathymetry of the mid-Norwegian margin and adjacent Atlantic Ocean. The locations of refraction seismic profiles that entered the NAG–TEC compilation are marked as black lines, and newer profiles as light grey lines. The grey dashed line marks the COB compiled by Funck *et al.* (2014). Bathymetric data is taken from IBCAO dataset and local compilations (Olesen *et al.* 2010; Jakobsson *et al.* 2012); onshore elevation is from ETOPO1 (Amante & Eakins 2009). Lof. Arch., Lofoten Archipelago; LB, Lofoten Basin; RH, Røst High; UR, Utrøst Ridge; LR, Lofoten Ridge; VB, Vesterålen Basin; MH, Marginal High; MP, Møre Platform; JMFZ, Jan Mayen Fracture Zone; JMFZ, Jan Mayen microcontinent.

from the Møre Margin. The Vøring segment comprises the wide, proximal Trøndelag Platform, a domain of terraces and fault complexes, the 200 km wide Vøring Basin and the outer Vøring Marginal High (Brekke 2000). The crust in the Vøring segment thins gradually over a distance of more than 200 km from over 30 km under the platform domain in the east to a minimum of 8 km under the distal sag basin. Extensive LCBs underlie most of the outer margin and often constitute nearly 50% of the total crustal thickness (Mjelde *et al.* 2009*a*). P-wave velocities are generally between 7.2 and 7.5 km s^{-1} within these bodies, but supposedly increase to significantly more than 8.0 km s^{-1} in an area at the SE edge of the margin (Raum *et al.* 2006), although such values are not confirmed by other studies (Rouzo *et al.* 2006). Densities of these bodies are estimated to range between 3000 and 3300 kg m^{-3} (Mjelde *et al.* 2009*a*; Kvarven *et al.* 2014; Nirrengarten *et al.* 2014). Extensive lava flows and sills characterize the outer Vøring Basin and adjacent Vøring Marginal High, making it diffcult to accurately determine depth to both basement and Moho. Commonly used interpretations of the COB (continent–ocean boundary, here used as the outer limit of the continent–ocean transition zone) place it along the outer Vøring Margin, where thickened, probably underplated, crust is mapped; yet other interpretations exist (cf. Gernigon *et al.* 2004; Peron-Pinvidic *et al.* 2013; Funck *et al.* 2014; Eagles *et al.* 2015). Seaward of the COB of the outer Vøring Margin, a topographic high underlain by up to 15 km thick crust with velocities less than 7.2 km s^{-1} constitutes the Vøring Spur (Breivik *et al.* 2008; Gernigon *et al.* 2009).

The Bivrost lineament marks the northern end of the Vøring Margin (Planke *et al.* 1991; Blystad *et al.* 1995; Olesen *et al.* 2002; Mjelde *et al.* 2003; Mjelde *et al.* 2005*a*). This lineament constitutes a distinct change in petrophysical properties (density, seismic velocity, susceptibility) in the offshore continental domain (Ebbing *et al.* 2006) but has no equivalent expression (e.g. a fracture zone) in the oceanic domain (Olesen *et al.* 2007). The northward located Lofoten Margin narrows from more than 250 km in the south to *c.* 50 km at the Senja Fracture Zone in the north. The Lofoten–Utrøst ridge system separates the inner Vestfjorden Basin from the outer Lofoten Basin. Crustal thickness undulates between 15 and 25 km below the Lofoten Archipelago, Lofoten Ridge and Utrøst Ridge and rapidly thins to less than 10 km under the Lofoten Basin (Avedik *et al.* 1984; Goldschmidt-Rokita *et al.* 1988; Mjelde *et al.* 1993). A pronounced shallowing of the Moho has been mapped under the Lofoten Ridge (Mjelde *et al.* 1993). In contrast, the Røst High to the west is associated with crustal thickening. The transition to the oceanic domain is relatively abrupt but seismic imaging is obstructed by extensive volcanic flows. No LCBs have been mapped here.

Data and methods

Geophysical data and density model

Numerous geophysical surveys have been conducted over the past decades along the mid-Norwegian margin. A comprehensive tectonostratigraphic atlas for the entire NE Atlantic region has recently been compiled based on reflection and refraction seismic experiments, magnetic and gravity surveys, as well as borehole and heatflow measurements (the NAG-TEC atlas; Hopper *et al.* 2014). Datasets that did not enter the compilation include the vast amounts of proprietory data (2D/3D reflection seismic, borehole data) as well as a number of still unpublished aeromagnetic datasets (e.g. Gernigon *et al.* 2012*b*; Brönner *et al.* 2014).

The datasets relevant for the isostatic study presented here are the bathymetry, sedimentary and crustal horizons, as well as densities from a recent gravity inversion model (Haase *et al.* 2016). Their model covers the entire NE Atlantic and aims to refine the Moho horizon where it is not or only poorly covered by seismic profiles. It also includes the sub-Moho lithospheric thermal structure to first order. All datasets are described in detail in Haase *et al.* (2016) and are briefly summarized here for completion. The layers and respective density values of the model are listed in Table 1. The entire model has a lateral resolution of 10 × 10 km and extends to 250 km depth. A subset of this model is used for the isostatic calculations of the mid-Norwegian margin, covering an area of 1200 × 1550 km.

Layer geometry of subsurface model

The newest, most detailed bathymetric grid consists of the IBCAO dataset (Jakobsson *et al.* 2012) and

Table 1. *Density values of model layers*

Layer	Density (kg m^{-3})
Water	1030
Sediments	2200–2700
Upper continental crust	2750
Lower continental crust	2950
Lower crustal bodies	3100
Oceanic crust	2850
Lithospheric mantle	
Moho to 700°C isotherm	3300
700–900°C isotherm	3260
900°C isotherm to base lithosphere (LAB)	3240
Asthenospheric mantle	3200

a more local compilation along the Norwegian margin (Olesen *et al.* 2010). For the gravity inversion and the current isostatic study, less resolution is required and the DTU10 elevation model is used, which is based on satellite altimetry (Andersen *et al.* 2010; Andersen 2010).

The top basement is derived from the total sediment sequence thickness, compiled from existing datasets and seismic refraction data compiled in the NAG-TEC atlas (Funck *et al.* 2014; Hopper this volume, in prep). Individual sedimentary layers are not differentiated, yet a vertical division into 2 km thick sublayers is introduced to be able to represent a density structure. The total sedimentary sequence also contains volcanic flows and in places, the top basement might merge with the top of the basalts rather than the top of the crystalline crust, especially around the continent–ocean transition. This should be kept in mind when interpreting isostatic compensation.

The Moho along the mid-Norwegian margin is constrained by numerous refraction seismic profiles (locations shown in Fig. 1). A refraction seismic-based grid for the Moho depth was constructed for the entire NE Atlantic by Funck *et al.* (2016) using kriging techniques supplemented by bathymetric and gravity data. This Moho horizon was further refined by gravity inversion (Haase *et al.* 2016), adhering to the seismic uncertainties along the seismic profiles. In the density model of the gravity inversion, the crystalline crust is divided into two layers, consistent with many seismic observations from thinned continental domains. Whereas the intra-crustal boundary is at constant 20 km depth in the onshore domains, it separates the offshore continental crust into equally thick upper and lower crustal layers (prior to inversion and adjustments of the Moho). This is a large simplification compared with the highly variable depth of this intra-crustal boundary as seen in seismic profiles, but such variations could not be included consistently across the entire study area owing to insuffcient seismic coverage. The high-velocity, high-density LCBs are also included according to the record of refraction seismic data (e.g. Mjelde *et al.* 2009*a*). A division into continental and oceanic domain is required, and the COB follows the interpretation of Funck *et al.* (2014), using magnetic lineations, tectonic reconstructions and the landward-most points of seismically defined oceanic crust. The lithosphere-asthenosphere boundary (LAB) is included as the 1300°C isotherm in the oceanic domain, based on the age of the crust by Gaina *et al.* (2016) and isotherm calculations after Sandwell (2001). In the continental domain, the LAB is also modelled as an isotherm, based on the definitions by Artemieva & Mooney (2001) and Artemieva (2006). The transition between the two regions has been smoothed to avoid long-wavelength effects (Haase *et al.* 2016). The LAB here represents the boundary between conductive and convective heat transfer, such that temperatures (and here also densities) are constant below the LAB. The sub-Moho lithosphere is subdivided along the 700 and 900°C isotherms with respective depths calculated according to Sandwell (2001, and references therein).

Densities of subsurface model

Sediment densities increase stepwise with burial depth mimicking an exponential trend. The respective densities for the 2 km thick intervals are 2200, 2270, 2340, 2650 and 2700 kg m^{-3}, where the last value represents any sediments buried more than 8 km. These density values are typical for clastic sediments but do not consider the denser layers of basalts and volcanics that are present in parts of the study area. Crustal densities are 2750 and 2950 kg m^{-3} for the continental upper and lower crust, respectively. The density in the oceanic domain of 2850 kg m^{-3} is consistent with previous gravity models of the mid-Norwegian margin (e.g. Ebbing *et al.* 2006). The lower continental crust extends slightly under the oceanic crust and tapers over a distance of *c.* 40 km to ensure a gradual density change across the COB (representing a zone of mafically intruded crust). This is a crude approximation of the actual continent–ocean transition zone, of which the velocity structure has been seismically imaged in many places along the margin. The lower crustal bodies are assigned a constant density of 3100 kg m^{-3}, although a range of densities has been suggested by previous gravity models (3000–3300 kg m^{-3}, e.g. Mjelde *et al.* 2009*a*; Kvarven *et al.* 2014; Nirrengarten *et al.* 2014). Densities below the Moho consider the thermal regime to first order and decrease stepwise at the 700, 900 and 1300°C isotherm from 3300 kg m^{-3} at the Moho down to 3260, 3240 and 3200 kg m^{-3}, respectively. The LAB of this model is thus only a thermal but not compositional boundary. The densities do not include pressure-related density increases through compaction or phase changes. Furthermore any lateral density changes are omitted which originate, for example, from the effects of the Iceland plume or from compositional variations.

Evaluation of the density model

We use the density model exactly as it has been presented by Haase *et al.* (2016). Because the density model was developed to match the regional gravity data of the entire NE Atlantic, but our study covers the smaller area of the mid-Norwegian margin, certain limitations and oversimplifications are expected and have to be kept in mind for the interpretation of the results. By applying isostatic considerations, we can point out areas where the density model does not fit. These might represent additional lateral density variation which had not been included in the subsurface model, located in sediments, crust or uppermost mantle. We do not attempt to improve the density model in this study but suggest a future joint inversion using gravity and isostatic calculations. Isostatic tests alone may not allow for new insights into the crustal structure of the margin but when evaluated in context of additional data – here for example regional seismic profiles – it can give clues as to the previously undetected variations in density, composition or thermal regime.

The gravity misfit for the regional density model is 22 mGal standard deviaton with the largest residuals remaining around the Greenland–Iceland–Faroe corridor, some oceanic ridges, as well as around the continent–ocean transitions. The long-wavelength residuals can be attributed to large-scale lateral variations in temperature and composition introduced by the Iceland plume and imaged, for example, by teleseismic tomography (Bijwaard & Spakman 1999; Weidle & Maupin 2008). Many of the smaller-wavelength residuals must be attributed to the crustal structure, which is only coarsely represented in this regional model. Deviations are thus to be expected around the very coarsely represented structures of the LCBs, the COB, the intra-crustal boundary and compositional density variation in the lithospheric mantle.

Long-wavelength residuals as introduced by an incomplete thermal model, the neglect of sublithospheric density variations or the forced smoothing of the LAB across the continent–ocean transition will affect the calculated long-wavelength bathymetric features but not the smaller-scale ones, which are subject of this study. We thereby acknowledge that the thermal regime assumed for the density model is only a first-order approximation. Additional modifications would lead to long-wavelength changes, which are not considered relevant for the smaller-scale structures discussed in this study.

Isostatic calculations

In order to check the density model for isostatic equilibrium, we need to calculate the buoyancy of each vertical column of the model (each 10×10 km) and how it could be balanced by vertical movement, i.e. a modified bathymetry. We calculate the pressure at a compensation depth z_{comp} and its deviation from a reference pressure P_{ref}. We then calculate the flexural response of this load distribution with a given flexural strength of the lithosphere (effective elastic thickness) using the open source software *tisc* by Garcia-Castellanos (2002) and Garcia-Castellanos *et al.* (2003). The resulting deflection d_i thus represents the difference between the isostatically compensated bathymetry of the model and the true bathymetry, in the following sections referred to as elevation residual.

In the end-member case where lithosphere has no flexural strength and the system is locally isostatically balanced, each rock column is considered to be balanced independently of its neighbours and

$$d_i = \frac{P_{\text{ref}} - P_i}{\Delta \rho g}, \qquad (1)$$

where the index i refers to an individual rock column, P_i is the pressure at compensation depth, P_{ref} is a reference pressure, g is gravitational acceleration and $\Delta\rho = \rho_{bottom} - \rho_{top}$ is the density difference between material below the column (mantle) and above (water or air). The pressure at compensation depth is calculated by integrating the densities of each rock column from compensation depth to the surface

$$P_i = \int_{z_{\text{comp}}}^{h_i} \rho(z) g \, dz,$$

where h_i is the elevation. The deflection d_i is positive if $P_i < P_{ref}$ and negative if $P_i > P_{ref}$. The new (forward calculated) isostatically balanced elevation h^{bal} is then given by

$$h_i^{\text{bal}} = h_i^0 + d_i. \qquad (2)$$

If isostatic adjustment raises a column from below sea-level ($h^0 < 0$) to above sea-level ($h^{bal} > 0$), the equations need to be modified:

$$h_i^{\text{bal}} = \frac{P_{\text{ref}} - P_i}{\rho_a g} - \frac{|h_i^0|(\rho_a - \rho_w)}{\rho_a g}, \qquad (3)$$

where ρ_a and ρ_w are the densities of the asthenosphere (sublithospheric mantle) and water, respectively. Air density is considered to be negligibly small.

If flexural isostasy is assumed, the load ΔP that causes a deflection of the lithosphere, must be convoluted with a flexural filter. This filter effectively suppresses short wavelengths and depends on the flexural rigidity D (Turcotte & Schubert 1982; Watts 2001). In the spectral domain, it can be written as

$$Y(k) = \frac{L(k)}{g(\rho_a - \rho_w)} * \Phi_e(k) \qquad (4)$$

where $Y(k)$ and $L(k)$ are the Fourier transforms of the deflection and the load, respectively. In our case, the load is given by the pressure distribution P_i at the base of the compensation depth.

$\Phi_e(k)$ represents the flexural response function:

$$\Phi_e(k) = \left(\frac{Dk^4}{\rho_a} + 1\right)^{-1} \quad (5)$$

with D being the flexural rigidity of an elastic plate. This parameter can also be expressed through the effective elastic thickness T_{eff}, the Young's modulus E and the Poisson ratio ν:

$$D = \frac{ET_{eff}^3}{12(1-\nu^2)} \quad (6)$$

Values of T_{eff} range from approximately 5–100 km (e.g. Tesauro *et al.* 2013). These values are determined from natural loading experiments (removal of ice loads, large lakes, seamounts) or spectral correlation of gravity and topographic data. For oceanic domains, the effective elastic thickness roughly corresponds to the depth of the 450°C isotherm (Watts 2001). For continental domains T_{eff} is usually much smaller than the thickness of the lithosphere measured with seismic methods (or even the crust in some places). These different values of seismic (or thermal) and effective elastic thickness of the lithosphere arise because the lithosphere is not a homogenous elastic layer. Instead, it consists of layers of multiple lithologies with a net response that resembles that of a single, yet thinner, elastic layer. We test a range of effective elastic thicknesses from 5 to 20 km.

The compensation depth z_{comp} is chosen to be 250 km, which is far below the lithosphere–asthenosphere boundary (usually less than 100 km in oceanic and offshore domains) and therefore well within the region of viscoplastic deformation.

The reference pressure is commonly calculated from a reference column. This is often chosen to be a column of oceanic lithosphere with well-known density distribution or a pure asthenospheric column representative of a mid-oceanic ridge. In this study, the density structure of the lithospheric mantle is highly simplified and thereby not directly comparable with the absolute values. We thus use the average pressure at compensation depth across the study area as the reference pressure. This reference pressure is representative for large parts of the oceanic domains and the resulting elevation adjustment will here accordingly be zero. This approach also leads to fairly equally distributed positive and negative elevation adjustments along the margin and is therefore well suited to investigate the local density structures in the study area. A different reference pressure would lead to an overall shift in the calculated bathymetry and thus in the residuals.

Results of isostatic calculations

When comparing a forward calculated dataset with an observed one, it is important to consider data content of similar wavelength. The existing bathymetric data has thus been regridded to a 10 × 10 km grid in order to be comparable with the forward calculated bathymetry (Fig. 2a).

Figure 2b and c show the calculated bathymetry with $T_{eff} = 5$ km and the corresponding residuals, respectively. The overall elevation is well reproduced, yet residuals of several hundreds of metres remain in the continent–ocean transition area around the Jan Mayen microcontinent, the SW Barents Sea and the mid-Norwegian margin. Calculations with higher effective elastic thickness of $T_{eff} = 20$ km (Fig. 2d) highlight the more regional residuals of longer wavelength than in the case with $T_{eff} = 5$ km. Calculations based on the pre-inversion density model with a purely seismically derived Moho (not shown here) display the same patterns of residuals, yet much higher magnitudes. This shows that, although isostasy was not considered in the gravity inversion, the modifications are in line with isostatic balancing. Both gravity and elevation are thus sensitive to the same density structure.

Applying a constant effective elastic thickness for the entire study area is a strong simplification. Lateral variations must exist for oceanic and continental domains, for highly stretched and very thick crust. Furthermore, a temporal change of the rigidity (weak when young and hot, strong when old and cold) is to be expected. This would lead to different compensation of the crystalline crustal structure and the subsequent basin fill, as explained, for example, by Kimbell *et al.* (2004). In this study, where only the regional trends are analysed, a constant effective elastic thickness is appropriate. But further, local investigations need to take potential spatial and temporal variations into account.

Choice of compensation depth

The compensation depth z_{comp} to be used in isostatic modelling is not a well-defined value. Here, we take the view that the convective asthenospheric mantle provides no isostatic support and thus choose a compensation depth below the LAB at 250 km. Since no density variations are expected below this depth, an even larger depth for z_{comp} would not change the result. Results of calculations with shallower compensation depths are shown in Figure 3.

A compensation depth of 100 km (Fig. 3a) yields a similar distribution of the local elevation residuals along the mid-Norwegian margin. However, the long-wavelength residuals, in particular around the mid-oceanic ridges, are much higher. This reflects

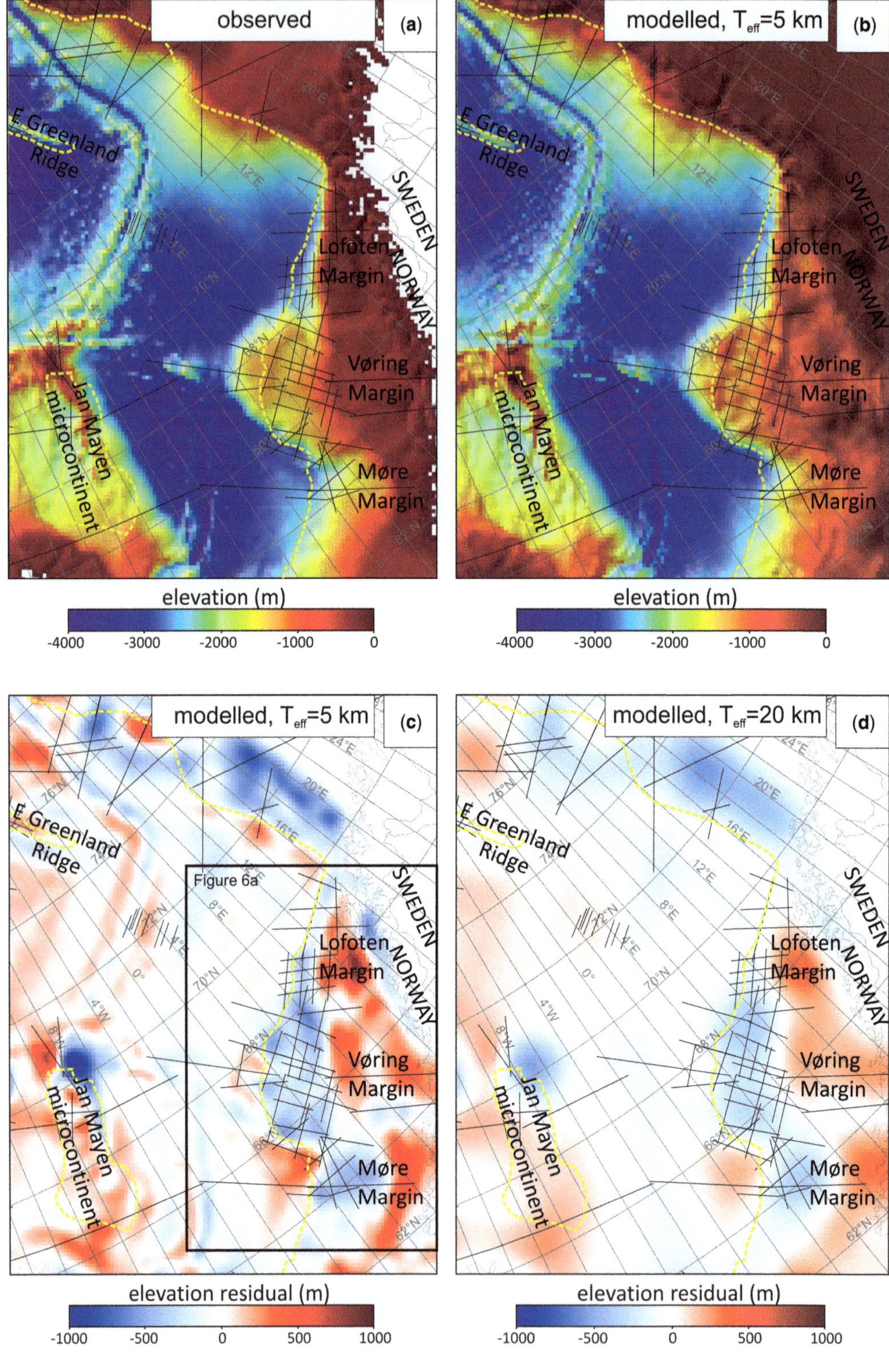
observed
(a)
modelled, T_{eff}=5 km
(b)
modelled, T_{eff}=5 km
(c)
modelled, T_{eff}=20 km
(d)
E Greenland Ridge
Jan Mayen microcontinent
Lofoten Margin
Vøring Margin
Møre Margin
SWEDEN
NORWAY
Figure 6a
elevation (m)
-4000 -3000 -2000 -1000 0
elevation residual (m)
-1000 -500 0 500 1000

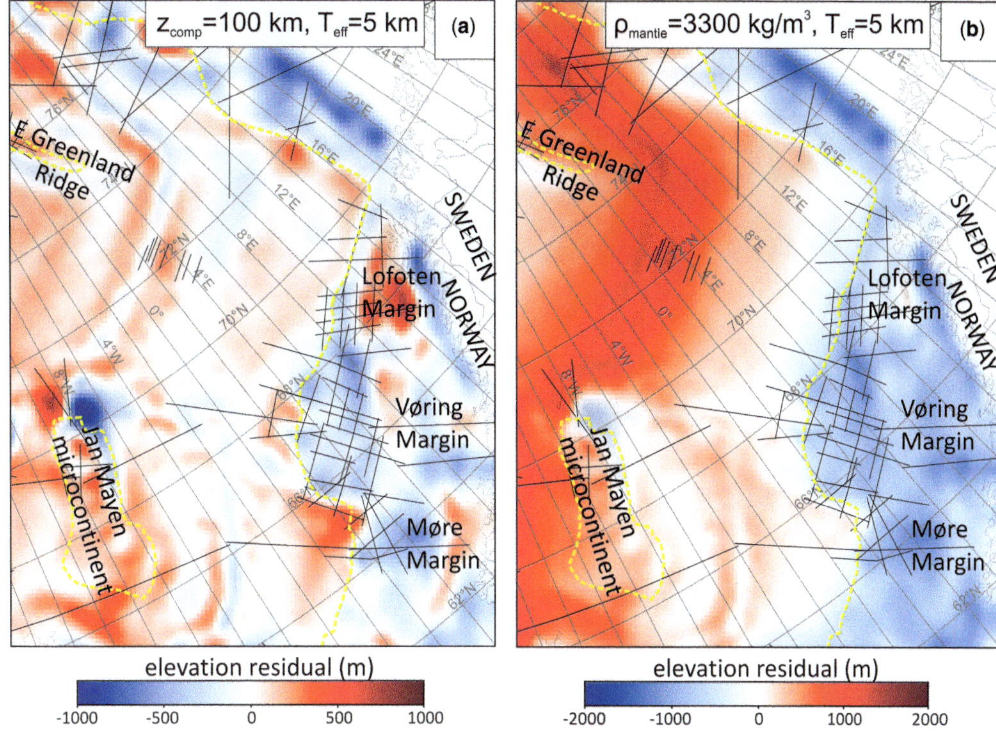

Fig. 3. (**a**) Elevation residual from isostatic calculations using the inversion model of Haase *et al.* (2016) and a compensation depth of 100 km ($E_{res} = E_{obs} - E_{calc}$). A long-wavelength trend from the ridges to the continent can be seen, reflecting the missing thermal effects of the lithospheric mantle below the compensation depth of 100 km. (**b**) Elevation residual from isostatic calculations using the inversion model of Haase *et al.* (2016) but assigning constant densities of 3300 kg m^{-3} to the lithospheric mantle. This can be considered equivalent to assuming isostatic compensation at Moho depth. The prominent long-wavelength residuals that are present between the mid-oceanic ridges and the continent again show the missing thermal effects of the lithosphere. Dashed yellow lines mark the COB, and black lines the location of refraction seismic profiles.

that the thermal structure of the oceanic domain and the passive margin comprises major density variations at 100 km depth around the spreading ridges, but relatively uniform densities deep under the continental margin. Kimbell *et al.* (2004) suggest that a compensation depth of 90 km is suffcient, yet they employ a slightly different thermal structure with maximum temperatures of 1100°C and no significant lateral density variations at greater depths.

Figure 3b shows the elevation residual if isostatic compensation is assumed to occur at Moho depth. This is here implemented by a constant density of the underlying mantle of 3300 kg m^{-3}. The residuals clearly show the influence of the thermal structure on elevation. Omitting the lithospheric mantle in isostatic studies can thus give undesired long-wavelength residuals. This is even more the case if compositional density changes are present within the lithospheric mantle. These are also not included in the density model used here. The calculations furthermore do not include the dynamic pressure exerted by flowing mantle (strict definition

Fig. 2. (**a**) Bathymetry of the mid-Norwegian margin and the adjacent Atlantic Ocean from the DTU10 elevation model (Andersen 2010; Andersen *et al.* 2010) resampled to a 10 × 10 km grid. In this and the following panels, the yellow dashed lines mark the location of the COB as interpreted and brought forward by Funck *et al.* (2014). Black lines mark the refraction seismic profiles that constrained the geometry of the density model. (**b**) Forward calculated bathymetry using the density model of Haase *et al.* (2016) and an effective elastic thickness of $T_{eff} = 5$ km. (**c**) Corresponding elevation residual ($E_{res} = E_{obs} - E_{calc}$). Black frame marks area shown in Figure 4. (**d**) Elevation residual from isostatic bathymetry calculations using the density model of Haase *et al.* (2016) and an effective elastic thickness of $T_{eff} = 20$ km.

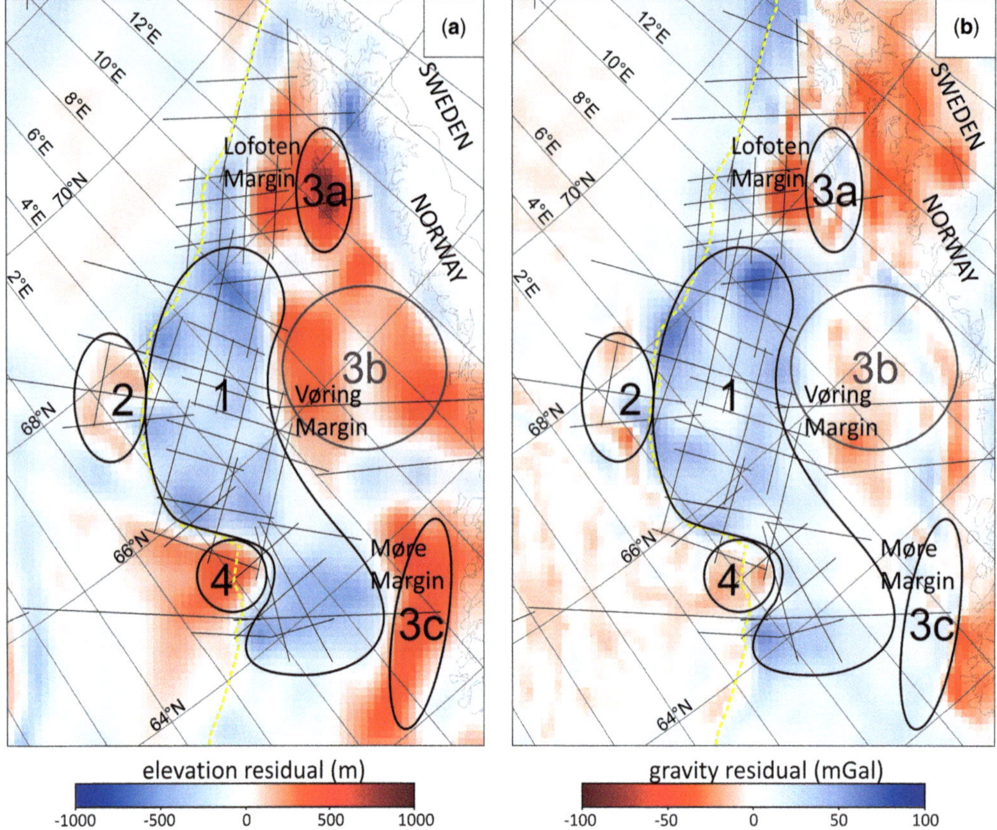

Fig. 4. (**a**) Close-up of the elevation residual along the mid-Norwegian margin ($E_{res} = E_{obs} - E_{calc}$). Four areas with high residuals are marked. A negative (blue) residual points to a mass deficit in the model. (**b**) Close-up of the gravity residual along the mid-Norwegian margin based on same density model by Haase *et al.* (2016) ($G_{res} = G_{obs} - G_{calc}$). A positive (blue) residual points to a mass deficit in the model. Dashed yellow line marks the COB, and black lines the location of refraction seismic profiles.

of dynamic topography), merely the thermally induced density differences in the lithosphere and asthenosphere.

We now focus on the short-wavelength residuals ($T_{eff} = 5$ km) displayed on the mid-Norwegian margin where a multitude of refraction seismic profiles are available that constrain the subsurface model and allow us to interpret the elevation residuals. Four areas of high residuals are highlighted (Fig. 4a) and will be discussed in the following. These 'unbalanced' areas are also marked on the map of gravity residuals (Fig. 4b), which result from the inversion by Haase *et al.* (2016).

Area 1

Area 1 shows a large mass deficit in an area where seismic data coverage is high and subsurface geometries should therefore be well determined. The gravity response of the subsurface model equally indicates a mass deficit in the density model here (Fig. 4b). As Haase *et al.* (2016) point out, the densities assumed for the sedimentary layers are representative for clastic sediments but do not consider the contribution of sills and lava flows. These flood basalts and sills are massive along the mid-Norwegian margin (location shown as green transparent area on Fig. 5) and additionally hamper seismic imaging on the outer margin. Thus, seismic penetration is poor and the structures below the volcanics – in particular the thickness of the basalts themselves and of the underlying sediments as well as the Moho depth – are poorly resolved. These uncertainties could justify significant modifications to the subsurface model, which could yield a better fit to gravity data and isostatically compensated bathymetry. The current mass deficit in the model could, for example, be compensated by thinner sediments (shallower basement) or a shallower Moho.

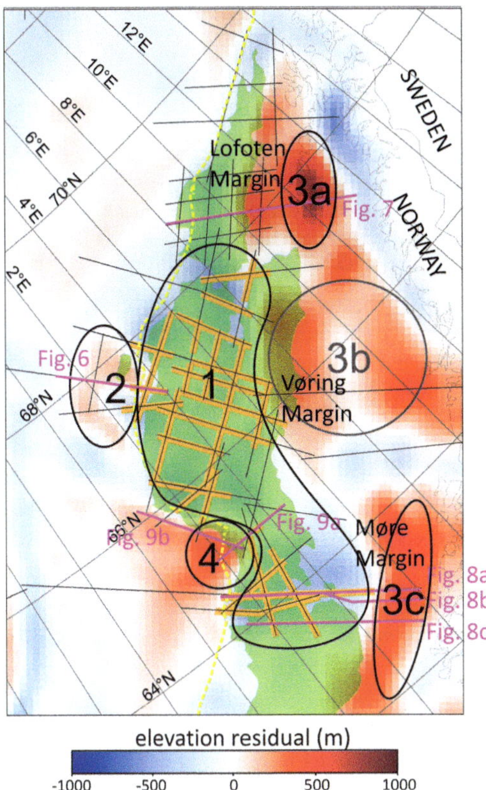

Fig. 5. Elevation residual along the mid-Norwegian margin with extent of the shallow volcanics (green shading) and the seismically imaged lower crustal bodies (orange lines). Location of refraction seismic profiles shown in Figures 6–9 are marked. Dashed yellow line marks the COB, and black lines the location of refraction seismic profiles.

In addition to the volcanics near the surface, magmatic activity has probably affected the lower crust as well. The seismic velocities around Area 1 are increased with respect to the more landward, stretched continental crust (Funck *et al.* 2014). These high velocities probably go hand in hand with higher densities, but regions of transitional crust are not represented in the current density model. An inversion of the gravity residual for crustal densities also indicates a need for higher densities in this area (Haase *et al.* 2016).

An alternative explanation of the regional residuals can be found in the LCBs (high-velocity, high-density lower crustal bodies), which in large parts of the mid-Norwegian margin correlate with the extent of the volcanic flows (Reynisson *et al.* 2010). Figure 5 illustrates that their extent fairly well matches the extent of Area 1, where residuals for both bathymetry and gravity are high. Calculations with an LCB density of 3150 kg m^{-3} instead of 3100 kg m^{-3} show only a slight improvement (10–20% of the gravity and elevation residual). The LCB density increase alone is thus not suffcient to explain the residuals. Higher densities for the sedimentary section in this area strongly improves the gravity residuals but has little effect on the elevation residuals.

Area 2

Seaward of Area 1 exists a small area of positive elevation and gravity residuals, which is crossed by several refraction seismic profiles (Mjelde *et al.* 2001, 2005*b*; Mjelde *et al.* 2007; Breivik *et al.* 2008). An exemplary refraction seismic profile is shown in Figure 6 together with the crustal structure of the density model. The elevation and gravity residuals correlate with the positive bathymetry outboard of the mapped COB (Figs 1 & 2a). This indicates that the density structure across the COB may not be as simple as included in the present subsurface model and perhaps even that a further seaward located COB might be more appropriate (Gernigon *et al.* 2012*a*; Peron-Pinvidic *et al.* 2013; Eagles *et al.* 2015).

Area 3

An overall mass excess is seen in an elongated, coast-parallel region, which is crossed by a few refraction seismic profiles (Fig. 5). The gravity residual indicates a slight mass deficit for Areas 3a and 3c and a generally good fit for the central Area 3b. Since elevation and gravity residuals are derived from the same density model, these seemingly contradicting results must be explained by the different methods. Isostatic adjustment should be less sensitive to small wavelength features as their respective loads can be supported by a sufficiently rigid lithospheric plate. Yet, even with a stronger lithosphere ($T_{\mathrm{eff}} = 20$ km), the calculated elevation is too low (Fig. 2d) and a significant residual remains. This indicates that the excess mass in the density model is a more regional feature. Because the gravity signal is more sensitive to shallower depths and the isostatic compensation does not depend on the vertical distribution of mass within a model column, the mass excess revealed in Areas 3a and 3c probably stems from greater depth.

The velocity structure of the refraction seismic profile crossing Area 3a is shown in Figure 7 together with the crustal structure of the density model. The continental shelf is underlain by thinned continental crust that forms several basement highs, neighboured by sedimentary basins. Whereas the Røst High and Marmæle Spur are underlain

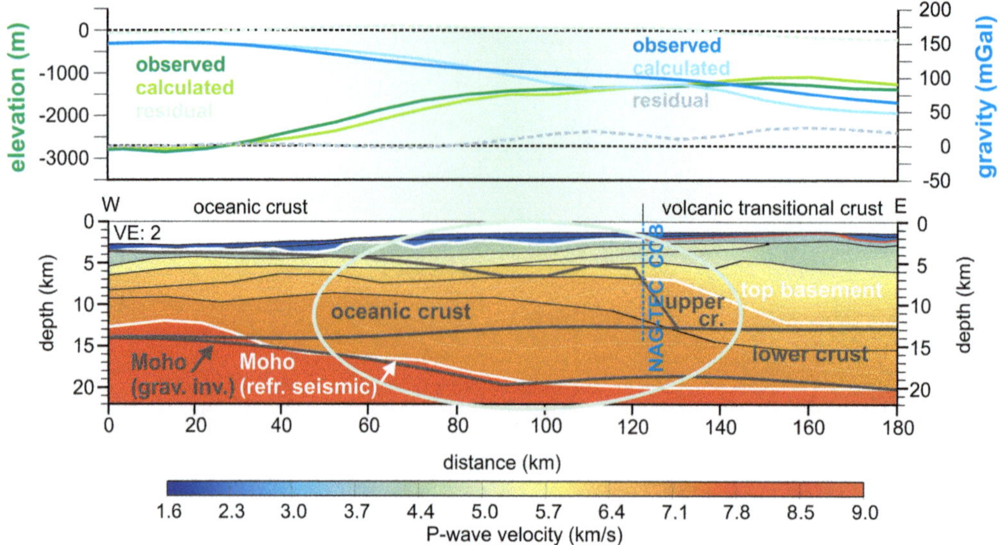

Fig. 6. Refraction seismic profile across Area 2 (pale green ellipse) imaging a relatively thick crust in what has been considered the oceanic domain (eastern part of profile 11-03, published in full extent by Breivik *et al.* 2008). Crustal structure of density model used in this study is shown in dark grey (Haase *et al.* 2016). Top panel shows observed and calculated elevation and gravity signals as well as the respective residuals. Location shown in Figure 5.

by thickened crust, the Lofoten Ridge is located above a Moho high. This scenario is in disagreement with local isostasy. Similar to these results, Reynisson *et al.* (2010) also calculated that for achieving complete isostatic balancing, the respective Moho needed to be several kilometres deeper than the seismically imaged Moho. A deeper Moho however, would not be compatible with the gravity data but could be compensated, for example, by a different composition of the overlying basement high (Reynisson *et al.* 2010). It was originally proposed that the Lofoten Ridge is a large-scale horst structure (Mjelde *et al.* 1993) and that the isostatic imbalance is too small to be locally compensated. Recent studies (yet still based on older seismic data) interpret a high-velocity body in this area (Mjelde *et al.* 2013). The reflector originally interpreted as the Moho would then be the top of a high-density body with mantle-type velocities. While the refraction seismic data did not detect lateral variations of subcrustal velocities, a deep mantle reflector *c.* 10 km below the detected Moho suggests that some deeper structures are still present below the Lofoten Ridge (Mjelde *et al.* 1993). Reflection seismic data show yet a different Moho horizon with an up to 4 km shallower Moho east of the Lofoten Ridge (Fig. 7, Mjelde *et al.* 1993). This shows that seismic methods do not unambiguously map the deep structures here, and differences in determining seismic velocities may play a role.

A similar scenario can be brought forward for Area 3c (Fig. 8). The refraction seismic line of Mjelde *et al.* (2008) shows again that a relative basement high is underlain by a Moho high, but the gravity inversion modified this Moho high to a generally landward-dipping horizon. Further outboard, the crust is drastically thinned and comprises a high-velocity lower layer, which pinches out towards the shelf. Seismic investigations from the last few years (Kvarven *et al.* 2014; Nirrengarten *et al.* 2014) and older, expanding spread profiles and sonobuoy data from neighbouring areas (Olafsson *et al.* 1992) describe an additional high-velocity lens in the area showing the mass excess (Fig. 8b, c). Slightly higher subcrustal velocities were also detected in the refraction seismic profile entering the current subsurface model (Fig. 8a, Mjelde *et al.* 2008), but only much more recently interpreted by the same authors as lower crustal eclogitic bodies (Mjelde *et al.* 2013). This high-velocity, high-density lower crustal lens is thought to be a remnant of old eclogized crust which was once part of the Caledonian orogen but broke up and transformed during rifting (Nirrengarten *et al.* 2014). While these new interpretations did not enter the subsurface starting model for the gravity inversion, the respective results nevertheless show a deeper Moho in this area (Fig. 8a).

For the central Area 3b, no indications for LCBs have been brought forward (Breivik *et al.* 2011).

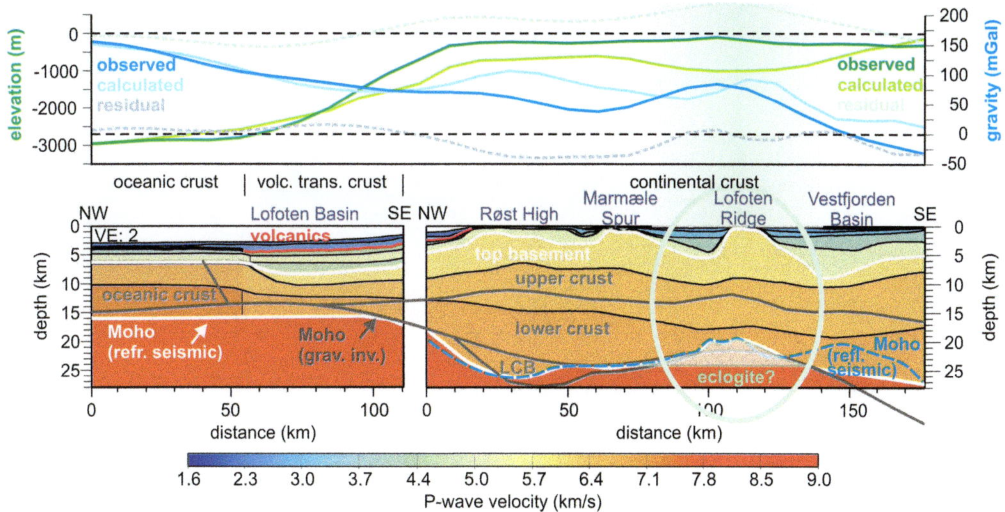

Fig. 7. Refraction seismic profile across Area 3a (pale green ellipse) imaging a basement high above a Moho high (after Mjelde *et al.* 1993). Crustal structure of density model used in this study is shown in dark grey (Haase *et al.* 2016). Moho depth of reflection seismic survey (Mjelde *et al.* 1993) and recently interpreted eclogites (Mjelde *et al.* 2013) are additionally marked. Top panel shows observed and calculated elevation and gravity signals as well as the respective residuals. Location shown in Figure 5.

Here, the Trøndelag Platform is underlain by relatively thick continental crust (more than 20 km) and the simple structure assumed in the gravity model is probably not adequate to capture the effects of internal layering and inhomogeneities on the gravity and isostasy signal. A better mapping of the margin structure and more detailed modelling is needed here.

While the suggested Moho modifications for Area 3 (additional LCB, lower Moho) would be sufficient to reduce the gravity residual, a positive elevation residual still remains that points to excess mass (missing buoyancy). A local, further lowering of the Moho could remove this mismatch, yet the densities of this lower crustal structure remain unresolved. Whereas isostatic considerations require low crustal densities, gravity modelling by Kvarven *et al.* (2014) suggests densities as high as 3300 kg m^{-3}. On the other hand, Nirrengarten *et al.* (2014) suggest a density of 3100 kg m^{-3}, closer to the values used in our study. A separate explanation could be lateral, compositional density variations in the uppermost mantle. Indications for such variations are seen in the refraction seismic data (see Fig. 8a, b) and discussed, for example, by Kimbell *et al.* (2004) for an area NW of Scotland.

Area 4

Area 4 exhibits a strong mass excess, which is seen in both the elevation and the gravity residuals (Fig. 4a, b). This region is situated in the corner between the outer Møre and Vøring margins in the outer part of the Jan Mayen corridor. Reynisson *et al.* (2010) mapped an equal mass excess (isostatic Moho shallower than refraction seismic Moho) slightly landward, on the outer Møre Margin. The area under investigation is considered to extend across the COB, covering both oceanic and transitional crust. It is also highly affected by the shearing and differential opening along the Jan Mayen Fracture Zone (Talwani & Eldholm 1977; Skogseid & Eldholm 1987; Gernigon *et al.* 2015). Refraction seismic profiles that partially cover this area are shown in Figure 9. The crust is generally thin (*c.* 7 km) and no LCB has been mapped. For the area under discussion here, it is also plausible that the crustal densities are not accurately represented in the model, namely that the actual densities are lower than those assumed. This could be explained by a different, farther seaward located course of the COB or a different, more abrupt, density transition across the COB.

Discussion and conclusions

Isostatic calculations are an easy way to provide additional support for seismic interpretation and gravity modelling. Especially where new concepts meet older data interpretations, the consideration of isostatic elevation residuals may provide additional insight. This is, for example, the case for the

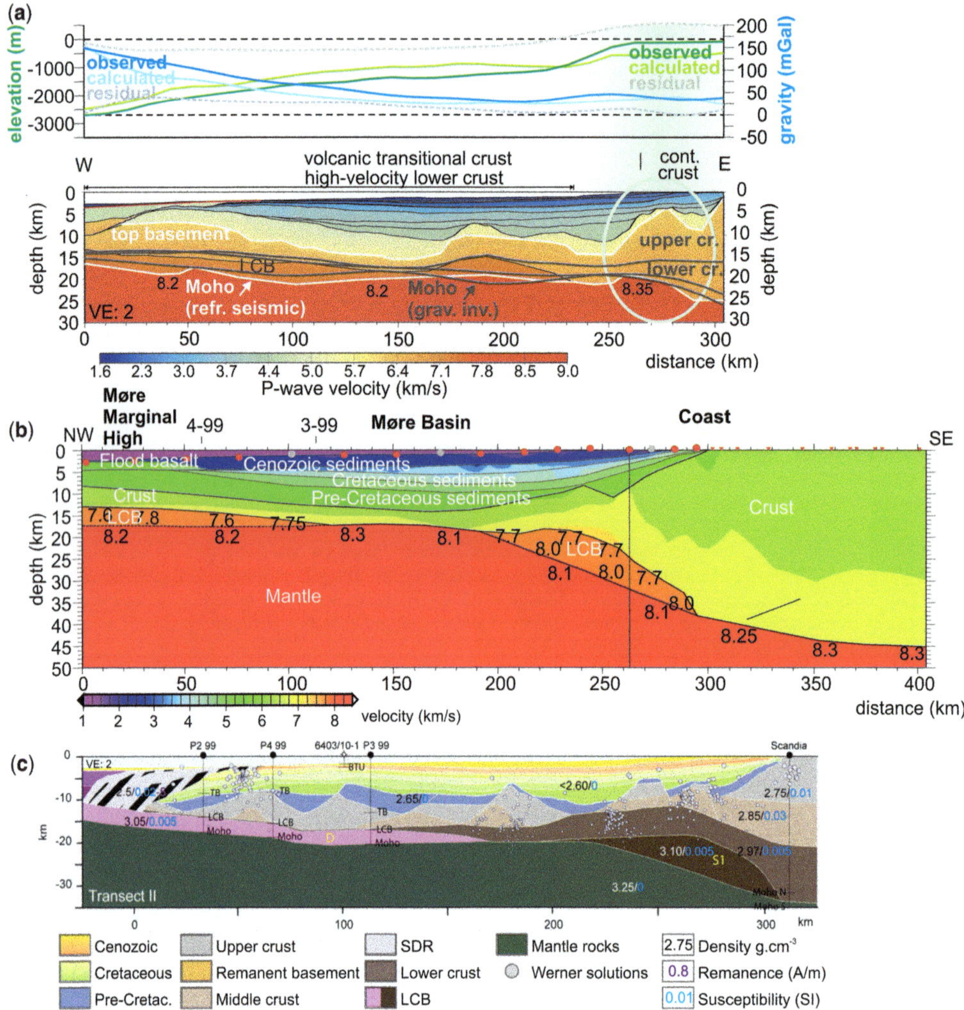

Fig. 8. Refraction and reflection seismic profiles across Area 3c (pale green ellipse) imaging a basement high above a Moho high or above a high-density lower crustal body. Locations are shown in Figure 5. (**a**) Profile 96-8a from (Mjelde *et al.* 2008) together with observed and calculated elevation and gravity signals. The original interpretation does not include a high-density high-velocity body below the Moho but this has been suggested in a later publication (Mjelde *et al.* 2013). Sub-crustal seismic velocities derived from refraction seismic data are indicated. Crustal structure of density model used in this study is shown in dark grey (Haase *et al.* 2016). (**b**) Profile 2 from Kvarven *et al.* (2014) showing a high-velocity lower crustal body in Area 3c. The corresponding density was modelled to be 3300 kg m^{-3}, equal to that of the underlying mantle. (**c**) Reflection seismic profile II from Nirrengarten *et al.* (2014) showing a high-velocity, high-density lower crustal body in Area 3c. The geometry of the latter two profiles did not enter the starting model of the gravity inversion.

eclogitic bodies along the proximal mid-Norwegian margin that have been discussed in recent publications and whose presence is now also confirmed by isostatic calculations. Even where no other interpretations are available, elevation residuals may point out areas that require renewed investigation through more data or renewed data analysis. This is, for example, the case for the outer Vøring Marginal High (Area 2) or the complex area SW of the Vøring Margin (Area 4).

Limitations

Elevation residuals can stem from inadequate isostatic calculations (e.g. inappropriate effective elastic thickness T_{eff} or compensation depth) or an

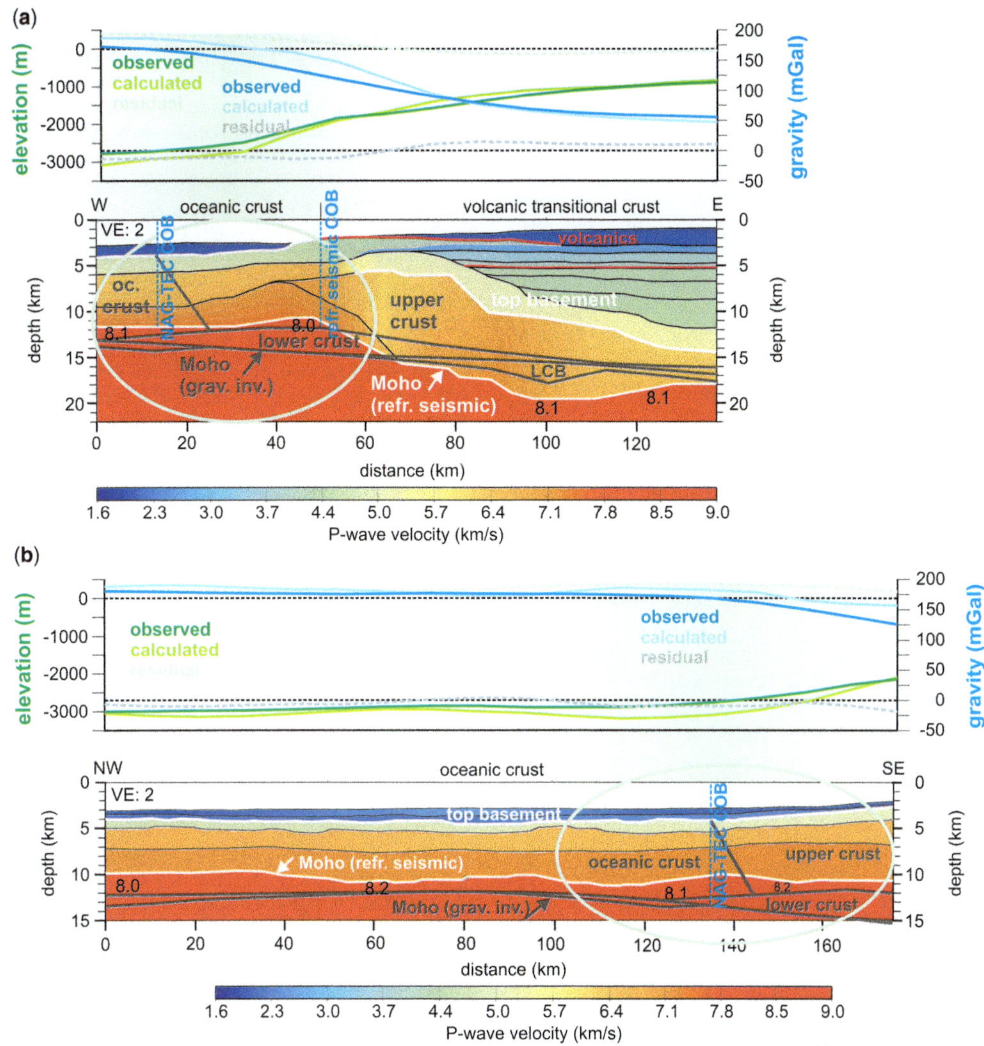

Fig. 9. Refraction seismic profiles across Area 4 (pale green ellipse). Locations are shown in Figure 5. Crustal structure of density model used in this study is shown in dark grey (Haase *et al.* 2016). (**a**) Line 99–5 from Mjelde *et al.* (2009*b*) shows a thin but high-velocity (high-density) crust, overlain by thick volcanics. The COB location based on magnetic data lies *c.* 25 km west of the seismically derived one. Top panel shows observed and calculated elevation and gravity signals as well as the respective residuals. (**b**) Profile 96-8b from Raum *et al.* (2006) runs mainly parallel to the COB to the NE. Refraction seismic data suggest that it extends across oceanic crust only, whereas plate reconstructions depict continental crust in the southeastern part of the profile. Observed and calculated elevation and gravity signals are shown.

inadequate subsurface density model. If the erroneous density structure is relatively shallow, it will also affect the gravity residuals (see e.g. Area 1). If it is deep, it will probably remain undetected by the gravity calculations. An erroneous subsurface model can result from wrongly assigned densities, wrong interpretation of refraction seismic data and poor quality or even lack of data, which in turn may lead to erroneous data interpretations. The latter seems to be the case for the deep LCBs along the mid-Norwegian margin (Areas 3a and c).

The deep structure, in particular the mantle densities, and the choice of compensation depth have a large effect on the long-wavelength elevation residual. The choice of the reference column, however, determines only the absolute, not the relative, values. In areas where thermal anomalies create lateral density changes, these need to be taken into account. Isostatic compensation is usually not reached at the Moho level.

The mantle densities of the current subsurface model decrease with depth, following the thermal trend. They do not include pressure-dependent density increase with depth as is the case in the natural system. This pressure dependency is a nearly uniform density increase without significant lateral variations. It therefore does not affect the gravity residuals nor the relative elevation residuals. Lateral density variations in the uppermost mantle, however, which could originate from different degrees of melting and depletion, can have a major effect.

Isostatic calculations comprise a simple, additional tool to validate interpretations of deep seismic data and gravity models. They can either be applied separately, as in this study, or included in a joint data analysis or joint inversion with gravity. Isostasy is only sensitive to regional structures, but, different from other methods, is equally sensitive to shallow and deep structures.

With this simple study we hope to have demonstrated that interpretation of a single refraction seismic profile can give misleading results, where high uncertainties arise from resolving deep structures (e.g. Moho, LCB and mantle reflections) or sub-basalt features. Different seismic methods (here expanding spread profiles, reflection seismic) and additional geophysical datasets (gravity, bathymetry) should always be considered for the interpretation.

This work formed part of the NAG-TEC project and we acknowledge the support of the industry sponsors (in alphabetical order): Bayerngas Norge AS; BP Exploration Operating Company Limited, Bundesanstalt für Geowissenschaften und Rohstoffe (BGR); Chevron East Greenland Exploration A/S; ConocoPhillips Skandinavia AS; DEA Norge AS; Det norske oljeselskap ASA; DONG E&P A/S; E.ON Norge AS; ExxonMobil Exploration and Production Norway AS; Japan Oil, Gas and Metals National Corporation (JOGMEC); Maersk Oil; Nalcor Energy Oil and Gas Inc.; Nexen Energy ULC, Norwegian Energy Company ASA (Noreco); Repsol Exploration Norge AS; Statoil (UK) Limited, and Wintershall Holding GmbH. We furthermore would like to thank the reviewers for their critical evaluation of the manuscript and Laurent Gernigon for further improvements of the text.

References

AMANTE, C. & EAKINS, B. 2009. *ETOPO1 1 Arc-Minute Global Relief Model: Procedures, Data Sources and Analysis*. Technical report, National Geophysical Data Center, National Oceanographic and Atmospheric Administration https://doi.org/10.7289/V5C8276M. NOAA Technical Memorandum NESDIS NGDC-24.

ANDERSEN, O.B. 2010. The DTU10 gravity field and mean sea surface. *In*: *Second International Symposium of the Gravity Feld of the Earth (IGFS2)*, Fairbanks, AK.

ANDERSEN, O.B., KNUDSEN, P. & BERRY, P.A.M. 2010. The DNSC08GRA global marine gravity field from double retracked satellite altimetry. *Journal of Geodesy*, **84**, 191–199, https://doi.org/10.1007/s00190-009-0355-9

ARTEMIEVA, I.M. 2006. Global 1 × 1 thermal model TC1 for the continental lithosphere: implications for lithosphere secular evolution. *Tectonophysics*, **416**, 245–277, https://doi.org/10.1016/j.tecto.2005.11.022

ARTEMIEVA, I.M. & MOONEY, W.D. 2001. Thermal thickness and evolution of Precambrian lithosphere: a global study. *Journal of Geophysical Research: Solid Earth*, **106**, 16 387–16 414, https://doi.org/10.1029/2000JB900439

AVEDIK, F., BERENDSEN, D. *ET AL.* 1984. Seismic investigations along the Scandinavian 'Blue Norma' profile. *Annales Geophysicae*, **2**, 571–577.

BIJWAARD, H. & SPAKMAN, W. 1999. Tomographic evidence for a narrow whole mantle plume below Iceland. *Earth and Planetary Science Letters*, **166**, 121–126, https://doi.org/10.1016/S0012-821X(99)00004-7

BLYSTAD, P., BREKKE, H., FAERSETH, R., LARSEN, B., SKOGSEID, J. & TØRUDBAKKEN, B. 1995. Structural elements of the Norwegian continental shelf. Part II: the Norwegian Sea Region. *Norwegian Petroleum Directorate Bulletin*, **8**, 1–45.

BRAITENBERG, C., PAGOT, E., WANG, Y. & FANG, J. 2004. Bathymetry and crustal thickness variations from gravity inversion and flexural isostasy. *In*: HWANG, C., SHUM, C.K. & LI, J. (eds) *Satellite Altimetry for Geodesy, Geophysics and Oceanography: Proceedings of the International Workshop on Satellite Altimetry, a joint workshop of IAG Section III Special Study Group SSG3.186 and IAG Section II*, 8–13 September 2002, Wuhan, China. Springer, Berlin, 143–149.

BREIVIK, A.J., FALEIDE, J.I. & MJELDE, R. 2008. Neogene magmatism northeast of the Aegir and Kolbeinsey ridges, NE Atlantic: spreading ridge–mantle plume interaction? *Geochemistry, Geophysics, Geosystems*, **9**, Q02004, https://doi.org/10.1029/2007GC001750

BREIVIK, A.J., MJELDE, R., RAUM, T., FALEIDE, J.I., MURAI, Y. & FLUEH, E.R. 2011. Crustal structure beneath the Trøndelag Platform and adjacent areas of the mid-Norwegian margin, as derived from wide-angle seismic and potential field data. *Norwegian Journal of Geology*, **90**, 141–161.

BREKKE, H. 2000. The tectonic evolution of the Norwegian Sea continental margin with emphasis on the Vøring and Møre basins. *In*: NØTTVEDT, A. (ed.) *Dynamics of the Norwegian Margin*. Geological Society, London, Special Publications, **167**, 327–378, https://doi.org/10.1144/GSL.SP.2000.167.01.13

BRÖNNER, M., GERNIGON, L. & NASUTI, A. 2014. *Lofoten-Vesterålen aeromagnetic survey 2011 – LOVAS-11 – acquisition, processing and interpretation report.* NGU Report **2013.060**.

EAGLES, G., PÉREZ-DÍAZ, L. & SCARSELLI, N. 2015. Getting over continent ocean boundaries. *Earth-Science*

Reviews, **151**, 244–265, https://doi.org/10.1016/j.earscirev.2015.10.009

EBBING, J. 2007. Isostatic density modelling explains the missing root of the Scandes. *Norwegian Journal of Geology*, **87**, 13–20.

EBBING, J., LUNDIN, E., OLESEN, O. & HANSEN, E.K. 2006. The mid-Norwegian margin: a discussion of crustal lineaments, mafic intrusions, and remnants of the Caledonian root by 3D density modelling and structural interpretation. *Journal of the Geological Society*, **163**, 47–59, https://doi.org/10.1144/0016-764905-029

ELDHOLM, O. & GRUE, K. 1994. North-Atlantic volcanic margins – dimensions and production rates. *Journal of Geophysical Research – Solid Earth*, **99**, 2955–2968, https://doi.org/10.1029/93JB02879

FULLEA, J., FERNÀNDEZ, M., AFONSO, J., VERGÉS, J. & ZEYEN, H. 2010. The structure and evolution of the lithosphere–asthenosphere boundary beneath the Atlantic–Mediterranean Transition Region. *Lithos*, **120**, 74–95, https://doi.org/10.1016/j.lithos.2010.03.003

FUNCK, T., HOPPER, J.R. *ET AL.* 2014. *Tectonostratigraphic Atlas of the North-East Atlantic Region*. Geological Survey of Denmark and Greenland (GEUS), Copenhagen, chapter 6, Crustal Structure.

FUNCK, T., GEISSLER, W., KIMBELL, G., GRADMANN, S., ERLENDSSON, Ø., MCDERMOTT, K. & PETERSEN, U. 2016. Moho and basement depth in the NE Atlantic Ocean based on seismic refraction data and receiver functions. *In*: PERON-PINVIDIC, G., HOPPER, J., STOKER, M., GAINA, C., DOORNENBAL, J., FUNCK, T. & ARTING, U. (eds) *The NE Atlantic Region: A Reappraisal of Crustal Structure, Tectonostratigraphy and Magmatic Evolution*. Geological Society, London, Special Publications, **447**. https://doi.org/10.1144/SP447.1

GAINA, C., GERNIGON, L. & BALL, P. 2009. Palaeocene–Recent plate boundaries in the NE Atlantic and the formation of the Jan Mayen microcontinent. *Journal of the Geological Society*, **166**, 601–616, https://doi.org/10.1144/0016-76492008-112

GAINA, C., KIMBELL, G. & BLISCHKE, A. 2016. Seafloor spreading domains in the NE Atlantic. *In*: PERON-PINVIDIC, G., HOPPER, J., STOKER, M., GAINA, C., DOORNENBAL, J., FUNCK, T. & ARTING, U. (eds) *The NE Atlantic Region: A Reappraisal of Crustal Structure, Tectonostratigraphy and Magmatic Evolution*. Geological Society, London, Special Publications, **447**, https://doi.org/10.1144/SP447.6

GARCIA-CASTELLANOS, D. 2002. Interplay between lithospheric flexure and river transport in foreland basins. *Basin Research*, **14**, 89–104, https://doi.org/10.1046/j.1365-2117.2002.00174.x

GARCIA-CASTELLANOS, D., VERGS, J., GASPAR-ESCRIBANO, J. & CLOETINGH, S. 2003. Interplay between tectonics, climate, and fluvial transport during the Cenozoic evolution of the Ebro Basin (NE Iberia). *Journal of Geophysical Research: Solid Earth*, **108**, https://doi.org/10.1029/2002JB002073

GERNIGON, L., RINGENBACH, J.C., PLANKE, S., LE GALL, B. & JONQUET-KOLSTO, H. 2003. Extension, crustal structure and magmatism at the outer Vøring Basin, Norwegian margin. *Journal of the Geological Society*, **160**, 197–208, https://doi.org/10.1144/0016-764902-055

GERNIGON, L., RINGENBACH, J.C., PLANKE, S. & LE GALL, B. 2004. Deep structures and breakup along volcanic rifted margins: insights from integrated studies along the outer Vøring Basin (Norway). *Marine and Petroleum Geology*, **21**, 363–372, https://doi.org/10.1016/j.marpetgeo.2004.01.005

GERNIGON, L., OLESEN, O. *ET AL.* 2009. Geophysical insights and early spreading history in the vicinity of the Jan Mayen Fracture Zone, Norwegian–Greenland Sea. *Tectonophysics*, **468**, 185–205, https://doi.org/10.1016/j.tecto.2008.04.025

GERNIGON, L., GAINA, C., OLESEN, O., BALL, P., PERON-PINVIDIC, G. & YAMASAKI, T. 2012*a*. The Norway Basin revisited: from continental breakup to spreading ridge extinction. *Marine and Petroleum Geology*, **35**, 1–19, https://doi.org/10.1016/j.marpetgeo.2012.02.015

GERNIGON, L., KOZIEL, J. & NASUTI, A. 2012*b*. *Jan Mayen aeromagnetic survey JAS-12 – part A: acquisition, processing*. NGU Report **2012.069**.

GERNIGON, L., BLISCHKE, A., NASUTI, A. & SAND, M. 2015. Conjugate volcanic rifted margins, seafloor spreading, and microcontinent: insights from new high-resolution aeromagnetic surveys in the Norway Basin. *Tectonics*, **34**, 907–933, https://doi.org/10.1002/2014TC003717

GOLDSCHMIDT-ROKITA, A., SELLEVOLL, M.A., HIRSCHLEBER, H.B. & AVEDIK, F. 1988. Results of two seismic refraction profiles off Lofoten, Northern Norway. *Geological Survey of Norway, Trondheim*, **3**, 49–57.

GUDLAUGSSON, S.T., GUNNARSSON, K., SAND, M. & SKOGSEID, J. 1988. Tectonic and volcanic events at the Jan Mayen Ridge microcontinent. *In*: MORTON, A.C. & PARSON, L.M. (eds) *Early Tertiary Volcanism and the Opening of the NE Atlantic*. Geological Society, London, Special Publications, **39**, 85–93, https://doi.org/10.1144/GSL.SP.1988.039.01.09

HAASE, C., EBBING, J. & FUNCK, T. 2016. A 3D regional crustal model of the Northeast Atlantic based on seismic and gravity data. *In*: PERON-PINVIDIC, G., HOPPER, J., STOKER, M., GAINA, C., DOORNENBAL, J., FUNCK, T. & ARTING, U. (eds) *The NE Atlantic Region: A Reappraisal of Crustal Structure, Tectonostratigraphy and Magmatic Evolution*. Geological Society, London, Special Publications, **447**, https://doi.org/10.1144/SP447.8

HIRSCH, K.K., BAUER, K. & SCHECK-WENDEROTH, M. 2009. Deep structure of the western South African passive margin Results of a combined approach of seismic, gravity and isostatic investigations. *Tectonophysics*, **470**, 57–70, https://doi.org/10.1016/j.tecto.2008.04.028

HOPPER, J. In prep. Sediment thickness and residual topography of the North Atlantic: estimating dynamic topography around Iceland. *In*: PERON-PINVIDIC, G., HOPPER, J., STOKER, M., GAINA, C., DOORNENBAL, J., FUNCK, T. & ARTING, U. (eds) *The NE Atlantic Region: A Reappraisal of Crustal Structure, Tectonostratigraphy and Magmatic Evolution*. Geological Society of London, Special Publications, **447**.

HOPPER, J.H., FUNCK, T., STOKER, M., ARTING, U., PERON-PINVIDIC, G., DOORNENBAL, H. & GAINA, C. (eds) 2014. *Tectonostratigraphic Atlas of the*

North-East Atlantic Region. Geological Survey of Denmark and Greenland (GEUS), Copenhagen.

JAKOBSSON, M., MAYER, L. ET AL. 2012. The International Bathymetric Chart of the Arctic Ocean (IBCAO) version 3.0. *Geophysical Research Letters*, **39**, https://doi.org/10.1029/2012GL052219

KABAN, M.K., SCHWINTZER, P., ARTEMIEVA, I.M. & MOONEY, W.D. 2003. Density of the continental roots: compositional and thermal contributions. *Earth and Planetary Science Letters*, **209**, 53–69, https://doi.org/10.1016/S0012-821X(03)00072-4

KIMBELL, G.S., GATLIFF, R.W., RITCHIE, J.D., WALKER, A.S.D. & WILLIAMSON, J.P. 2004. Regional three-dimensional gravity modelling of the NE Atlantic margin. *Basin Research*, **16**, 259–278, https://doi.org/10.1111/j.1365-2117.2004.00232.x

KVARVEN, T., EBBING, J. ET AL. 2014. Crustal structure across the Møre margin, mid-Norway, from wide-angle seismic and gravity data. *Tectonophysics*, **626**, 21–40, https://doi.org/10.1016/j.tecto.2014.03.021

MAŢENCO, L., ZOETEMEIJER, R., CLOETINGH, S. & DINU, C. 1997. Lateral variations in mechanical properties of the Romanian external Carpathians: inferences of flexure and gravity modelling. *Tectonophysics*, **282**, 147–166, https://doi.org/10.1016/S0040-1951(97)00217-5

MEYER, R., VAN WIJK, J. & GERNIGON, L. 2007. The North Atlantic Igneous Province: a review of models for its formation. *Geological Society of America, Special Papers*, **430**, 525–552, http://specialpapers.gsapubs.org/content/430/525.abstract, https://doi.org/10.1130/2007.2430(26)

MJELDE, R., SELLEVOLL, M.A., SHIMAMURA, H., IWASAKI, T. & KANAZAWA, T. 1993. Crustal structure beneath Lofoten, N Norway, from vertical incidence and wide-angle seismic data. *Geophysical Journal International*, **114**, 116–126, https://doi.org/10.1111/j.1365-246X.1993.tb01471.x

MJELDE, R., DIGRANES, P. ET AL. 2001. Crustal structure of the outer Vøring Plateau, offshore Norway, from ocean bottom seismic and gravity data. *Journal of Geophysical Research – Solid Earth*, **106**, 6769–6791, https://doi.org/10.1029/2000JB900415

MJELDE, R., SHIMAMURA, H., KANAZAWA, T., KODAIRA, S., RAUM, T. & SHIOBARA, H. 2003. Crustal lineaments, distribution of lower crustal intrusives and structural evolution of the Vøring Margin, NE Atlantic; new insight from wide-angle seismic models. *Tectonophysics*, **369**, 199–218, https://doi.org/10.1016/S0040-1951(03)00199-9

MJELDE, R., RAUM, T., BREIVIK, A., SHIMAMURA, H., MURAI, Y., TAKANAMI, T. & FALEIDE, J.I. 2005*a*. Crustal structure of the Vøring Margin, NE Atlantic: a review of geological implications based on recent OBS data. *In*: DORÉ, A.G. & VINING, B.A. (eds) *Petroleum Geology: North-West Europe and Global Perspectives – Proceedings of the 6th Petroleum Geology Conference*. Geological Society, London, Petroleum Geology Conference Series, **6**, 803–813, https://doi.org/10.1144/0060803

MJELDE, R., RAUM, T. ET AL. 2005*b*. Continent–ocean transition on the Vøring Plateau, NE Atlantic, derived from densely sampled ocean bottom seismometer data. *Journal of Geophysical Research – Solid Earth*, **110**, https://doi.org/10.1029/2004JB003026

MJELDE, R., RAUM, T., MURAI, Y. & TAKANAMI, T. 2007. Continent–ocean-transitions: review, and a new tectono-magmatic model of the Vøring Plateau, NE Atlantic. *Journal of Geodynamics*, **43**, 374–392, https://doi.org/10.1016/j.jog.2006.09.013

MJELDE, R., RAUM, T., BREIVIK, A.J. & FALEIDE, J.I. 2008. Crustal transect across the North Atlantic. *Marine Geophysical Research*, **29**, 73–87, https://doi.org/10.1007/s11001-008-9046-9

MJELDE, R., FALEIDE, J.I., BREIVIK, A.J. & RAUM, T. 2009*a*. Lower crustal composition and crustal lineaments on the Vøring Margin, NE Atlantic: a review. *Tectonophysics*, **472**, 183–193, https://doi.org/10.1016/j.tecto.2008.04.018

MJELDE, R., RAUM, T., KANDILAROV, A., MURAI, Y. & TAKANAMI, T. 2009*b*. Crustal structure and evolution of the outer Møre Margin, NE Atlantic. *Tectonophysics*, **468**, 224–243, https://doi.org/10.1016/j.tecto.2008.06.003

MJELDE, R., GONCHAROV, A. & MÜLLER, R.D. 2013. The moho: boundary above upper mantle peridotites or lower crustal eclogites? A global review and new interpretations for passive margins. *Tectonophysics*, **609**, 636–650, https://doi.org/10.1016/j.tecto.2012.03.001

NIRRENGARTEN, M., GERNIGON, L. & MANATSCHAL, G. 2014. Lower crustal bodies in the Møre volcanic rifted margin: geophysical determination and geological implications. *Tectonophysics*, **636**, 143–157, https://doi.org/10.1016/j.tecto.2014.08.004

NUNNS, A. 1982. The structure and evolution of the Jan Mayen Ridge and surrounding regions. *In*: WATKINS, J. & DRAKE, C. (eds) *Studies in Continental Margin Geology*. AAPG, Memoirs, Tulsa, OK, USA, **34**, 193–208.

OLAFSSON, I., SUNDVOR, E., ELDHOLM, O. & GRUE, K. 1992. Møre margin – crustal structure from analysis of expanded spread profiles. *Marine Geophysical Researches*, **14**, 137–162, https://doi.org/10.1007/BF01204284

OLESEN, O., LUNDIN, E. ET AL. 2002. Bridging the gap between the onshore and offshore geology in Nordland, northern Norway. *Norwegian Journal of Geology*, **82**, 243–262.

OLESEN, O., EBBING, J. ET AL. 2007. An improved tectonic model for the Eocene opening of the Norwegian–Greenland Sea: use of modern magnetic data. *Marine and Petroleum Geology*, **24**, 53–66, https://doi.org/10.1016/j.marpetgeo.2006.10.008

OLESEN, O., BRÖNNER, M. ET AL. 2010. New aeromagnetic and gravity compilations from norway and adjacent areas: methods and applications. *In*: VINING, B.A. & PICKERING, S.C. (eds) *Petroleum Geology: From Mature Basins to New Frontiers – Proceedings of the 7th Petroleum Geology Conference*. Geological Society, London, **7**, 559–586, https://doi.org/10.1144/0070559

OSMUNDSEN, P. & ANDERSEN, T. 2001. The middle Devonian basins of western Norway: sedimentary response to large-scale transtensional tectonics? *Tectonophysics*, **332**, 51–68, https://doi.org/10.1016/S0040-1951(00)00249-3

Osmundsen, P.T. & Ebbing, J. 2008. Styles of extension offshore mid-Norway and implications for mechanisms of crustal thinning at passive margins. *Tectonics*, **27**, https://doi.org/10.1029/2007TC002242

Peron-Pinvidic, G., Manatschal, G. & Osmundsen, P.T. 2013. Structural comparison of archetypal Atlantic rifted margins: a review of observations and concepts. *Marine and Petroleum Geology*, **43**, 21–47, https://doi.org/10.1016/j.marpetgeo.2013.02.002

Planke, S., Skogseid, J. & Eldholm, O. 1991. Crustal structure off norway, 62-degrees to 70-degrees north. *Tectonophysics*, **189**, 91–107, https://doi.org/10.1016/0040-1951(91)90489-F

Raum, T. 2000. *Crustal structure and evolution of the Faeroe, Møre and Vøring margins from wide-angle seismic and gravity data*. PhD thesis, University of Bergen, Norway.

Raum, T., Mjelde, R. *et al.* 2006. Crustal structure and evolution of the southern Vøring basin and Vøring transform margin, NE Atlantic. *Tectonophysics*, **415**, 167–202, https://doi.org/10.1016/j.tecto.2005.12.008

Ren, S.C., Faleide, J.I., Eldholm, O., Skogseid, J. & Gradstein, F. 2003. Late Cretaceous-Paleocene tectonic development of the NW Vøring Basin. *Marine and Petroleum Geology*, **20**, 177–206, https://doi.org/10.1016/S0264-8172(03)00005-9

Reynisson, R.F., Ebbing, J., Lundin, E. & Osmundsen, P.T. 2010. Properties and distribution of lower crustal bodies on the mid-Norwegian margin. *In*: Vining, B.A. & Pickering, S.C. (eds) *Petroleum Geology: From Mature Basins to New Frontiers – Proceedings of the 7th Petroleum Geology Conference*. Geological Society, London, **7**, 843–854, https://doi.org/10.1144/0070843

Rouzo, S., Klingelhofer, F. *et al.* 2006. 2-D and 3-D modelling of wide-angle seismic data: an example from the Vøring volcanic passive margin. *Marine Geophysical Researches*, **27**, 181–199, https://doi.org/10.1007/s11001-006-0001-3

Salem, A., Green, C., Stewart, M. & De Lerma, D. 2014. Inversion of gravity data with isostatic constraints. *Geophysics*, **79**, A45–A50, https://doi.org/10.1190/geo2014-0220.1

Sandwell, D. 2001. *Cooling of the Oceanic Lithosphere and Ocean Floor Topography*. University of California, San Diego, CA, http://topex.ucsd.edu/geodynamics/07cooling.pdf

Seranne, M. & Seguret, M. 1987. The Devonian Basins of western Norway: tectonics and kinematics of an extending crust. *In*: Coward, M.P., Dewey, J.F. & Hancock, P.L. (eds) *Continental Extensional Tectonics*. Geological Society, London, Special Publications, **28**, 537–548, https://doi.org/10.1144/GSL.SP.1987.028.01.35

Skogseid, J. & Eldholm, O. 1987. Early Cenozoic crust at the Norwegian continental margin and the conjugate Jan-Mayen Ridge. *Journal of Geophysical Research – Solid Earth and Planets*, **92**, 11471–11491, https://doi.org/10.1029/JB092iB11p11471

Stewart, J., Watts, A.B. & Bagguley, J.G. 2000. Three-dimensional subsidence analysis and gravity modelling of the continental margin offshore Namibia. *Geophysical Journal International*, **141**, 724–746, https://doi.org/10.1046/j.1365-246x.2000.00124.x

Talwani, M. & Eldholm, O. 1977. Evolution of the Norwegian–Greenland Sea. *Geological Society of America Bulletin*, **88**, 969–999, https://doi.org/10.1130/0016-7606(1977)88¡969:EOTNS¿2.0.CO;2

Tesauro, M., Kaban, M.K. & Cloetingh, S.A. 2013. Global model for the lithospheric strength and effective elastic thickness. *Tectonophysics*, **602**, 78–86, https://doi.org/10.1016/j.tecto.2013.01.006

Turcotte, D. & Schubert, G. 1982. *Geodynamics: Applications of Continuum Physics to Geological Problems*. John Wiley and Sons, New York.

Vogt, P., Johnson, G. & Kristjansson, L. 1980. Morphology and magnetic isochrons north of Iceland. *Journal of Geophysics*, **47**, 67–80.

Watts, A. 2001. *Isostasy and Fexure of the Lithosphere*. Cambridge University Press, Cambridge.

Watts, A.B., ten Brink, U.S., Buhl, P. & Brocher, T.M. 1985. A multichannel seismic study of lithospheric flexure across the Hawaiian–Emperor seamount chain. *Nature*, **315**, 105–111, https://doi.org/10.1038/315105a0

Weidle, C. & Maupin, V. 2008. An upper-mantle S-wave velocity model for Northern Europe from Love and Rayleigh group velocities. *Geophysical Journal International*, **175**, 1154–1168, https://doi.org/10.1111/j.1365-246X.2008.03957.x

Zeyen, H. & Fernàndez, M. 1994. Integrated lithospheric modeling combining thermal, gravity, and local isostasy analysis: Application to the NE Spanish Geotransect. *Journal of Geophysical Research: Solid Earth*, **99**, 18089–18102, https://doi.org/10.1029/94JB00898

The Jan Mayen microcontinent: an update of its architecture, structural development and role during the transition from the Ægir Ridge to the mid-oceanic Kolbeinsey Ridge

A. BLISCHKE[1]*, C. GAINA[2], J. R. HOPPER[3], G. PÉRON-PINVIDIC[4], B. BRANDSDÓTTIR[5], P. GUARNIERI[3], Ö. ERLENDSSON[6] & K. GUNNARSSON[6]

[1]*Iceland GeoSurvey, Branch at Akureyri, Rangárvöllum, 602 Akureyri, Iceland*

[2]*Centre for Earth Evolution and Dynamics, University of Oslo, Sem Sælands vei 24, PO Box 1048, Blindern, NO-0316 Oslo, Norway*

[3]*Geological Survey of Denmark and Greenland, Øster Voldgade 10, DK 1350 Copenhagen, Denmark*

[4]*Geological Survey of Norway, Postboks 6315 Sluppen, Trondheim 7491, Norway*

[5]*Institute of Earth Science, Science Institute, University of Iceland, Askja, Sturlugata 7, 101 Reykjavík, Iceland*

[6]*Iceland GeoSurvey, Grensásvegi 9, 108 Reykjavík, Iceland*

**Correspondence: Anett.Blischke@isor.is*

Abstract: We present a revised tectonostratigraphy of the Jan Mayen microcontinent (JMMC) and its southern extent, with the focus on its relationship to the Greenland–Iceland–Faroe Ridge area and the Faroe–Iceland Fracture Zone. The microcontinent's Cenozoic evolution consists of six main phases corresponding to regional stratigraphic unconformities. Emplacement of Early Eocene plateau basalts at pre-break-up time (56–55 Ma), preceded the continental break-up (55 Ma) and the formation of seawards-dipping reflectors (SDRs) along the eastern and SE flanks of the JMMC. Simultaneously with SDR formation, orthogonal seafloor spreading initiated along the Ægir Ridge (Norway Basin) during the Early Eocene (C24n2r, 53.36 Ma to C22n, 49.3 Ma). Changes in plate motions at C21n (47.33 Ma) led to oblique seafloor spreading offset by transform faults and uplift along the microcontinent's southern flank. At C13n (33.2 Ma), spreading rates along the Ægir Ridge started to decrease, first south and then in the north. This was probably complemented by intra-continental extension within the JMMC, as indicated by the opening of the Jan Mayen Basin – a series of small pull-apart basins along the microcontinent's NW flank. JMMC was completely isolated when the mid-oceanic Kolbeinsey Ridge became fully established and the Ægir Ridge was abandoned between C7 and C6b (24–21.56 Ma).

The Jan Mayen microcontinent (JMMC) is a structural entity encompassing the Jan Mayen Ridge and the surrounding area, including the Jan Mayen Basin, the Jan Mayen Basin South, the Jan Mayen Trough and the Southern Ridge Complex (SRC) (Fig. 1; Table 1). The JMMC is bordered to the north by the east and west segments of the Jan Mayen Fracture Zone and the volcanic complex of Jan Mayen Island (Svellingen & Pedersen 2003). To the south, it is bordered by the NE coastal shelf of Iceland, to the east by the Norway Basin and to the west by the Kolbeinsey Ridge. Early descriptions of the JMMC considered only the Jan Mayen Ridge (Vogt *et al.* 1970; Talwani *et al.* 1976*a*), a steep-flanked bathymetric horst structure with water depths varying between 200 and 2500 m that extends south from Jan Mayen Island. However, based on modern datasets, it is now accepted that the microcontinent is much larger than this and encompasses a number distinct, structurally controlled tectonic features that were formed by a succession of tectonic and volcanic events (e.g. Scott *et al.* 2005; Gaina *et al.* 2009; Peron-Pinvidic *et al.* 2012*a*, *b*; Gernigon *et al.* 2012). In total, the JMMC is 400–450 km long, and varies in width from 100 km in the north to 310 km in the south.

From: PÉRON-PINVIDIC, G., HOPPER, J. R., STOKER, M. S., GAINA, C., DOORNENBAL, J. C., FUNCK, T. & ÁRTING, U. E. (eds) 2017. *The NE Atlantic Region: A Reappraisal of Crustal Structure, Tectonostratigraphy and Magmatic Evolution*. Geological Society, London, Special Publications, **447**, 299–337.
First published online September 8, 2016, https://doi.org/10.1144/SP447.5

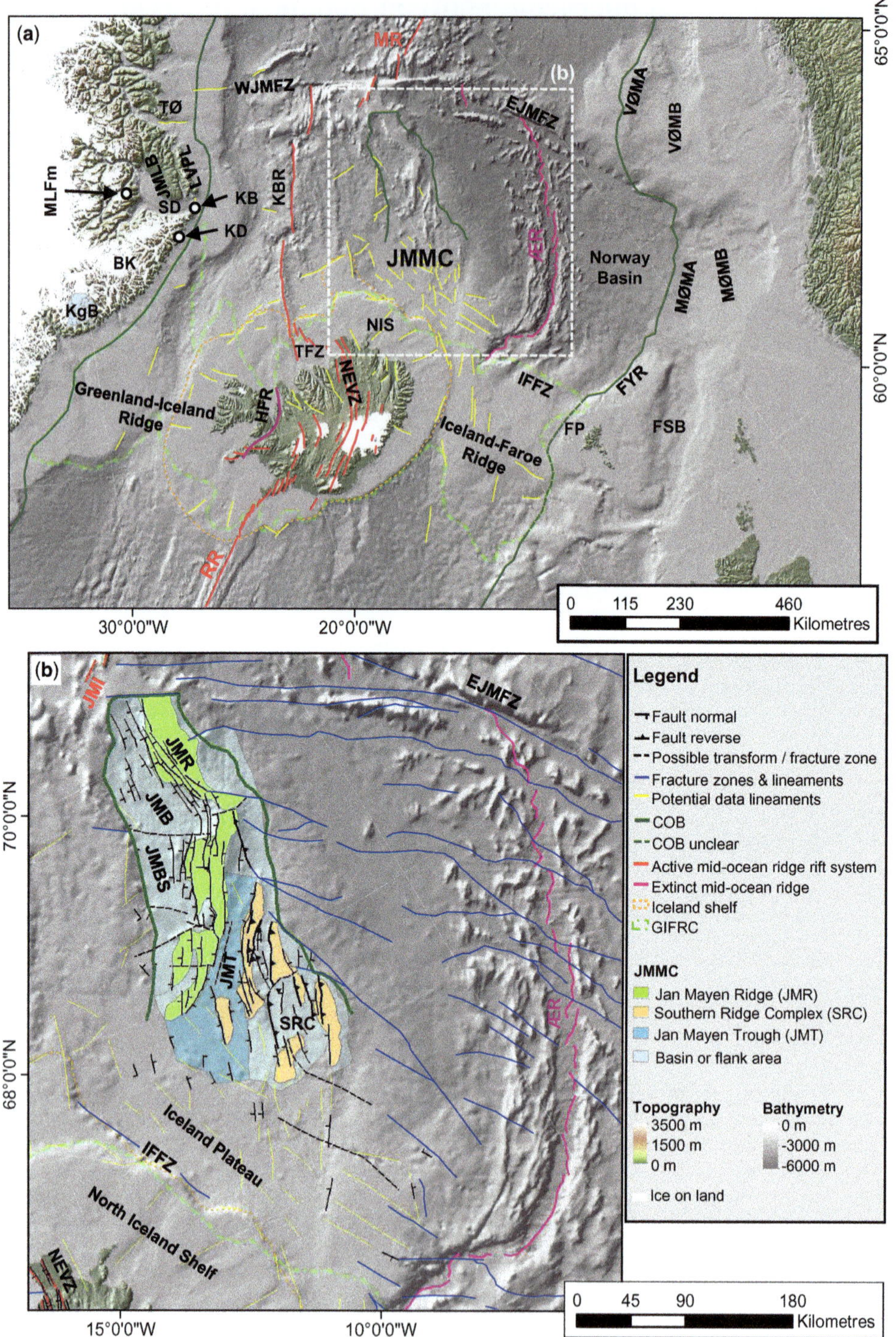

Fig. 1. Overview map (**a**) of the study area with the location of structural elements identified on potential field data. Structural elements map (**b**) for the JMMC study with mapped faults, fractures zones and lineaments based on this study and modified after Peron-Pinvidic *et al.* (2012*a*) and Gernigon *et al.* (2015) (for label keys, see Table 1). The background image is shaded bathymetry (IBCAO 3.0: Jakobsson *et al.* 2012; Amante & Eakins 2009).

Table 1. *Explanation of structural element abbreviations and label key, modified after Gunnarsson* et al. *(1989), Jóhannesson (2011), Hjartarson & Sæmundsson (2014), Hopper* et al. *(2014) and Magnúsdóttir* et al. *(2015)*

Abbreviation and label key					
General features:		**Central NE Atlantic:**			
COB	Continent–ocean boundary	JMMC	Jan Mayen microcontinent	IB	Iceland Basin
SDR	Seawards-dipping reflector	BR	Buðli Ridge		
Mid-oceanic ridges:		EFBN	Jan Mayen East Flank Basins North	GIFRC	Faroe–Iceland–Greenland Ridge Complex
ÆR	Ægir mid-oceanic ridge	EFBS	Jan Mayen East Flank Basins South	GIR	Greenland–Iceland Ridge
MR	Mohn's mid-oceanic ridge	HC/JMWIP	Hakarenna Channel/ Jan Mayen West Igneous Province South	HFR	Húnaflóa Rift
KBR	Kolbeinsey mid-oceanic ridge	HR	Högni Ridge	ICE	Iceland onshore
RR	Reykjanes mid-oceanic ridge	JMT/HT	Jan Mayen Trough/ Hléssund Trough	IFR	Iceland–Faroe Island Ridge
JMI	Jan Mayen Island System	JMB	Jan Mayen Basin	NEVZ	Northeast Volcanic Zone
Transfer systems and fracture zones:		JMRN	Jan Mayen Ridge North		
EJMFZ	East Jan Mayen Fracture Zone	LYR	Lyngvi Ridge	**Central Norway Margin:**	
IFFZ	Iceland–Faroe Fracture Zone	SFB/JMBS	Sörlahryggur Flank Basin/Jan Mayen Basin South	MØMA	Møre Marginal High
MIRFTS	Mid-Iceland Rift Transfer System	SHR	Sörlahryggur Ridge	MØMB	Møre Basin
SISZ	South Iceland Seismic Zone	WIPN	Jan Mayen West Igneous Province North	VØMA	Vøring Marginal High
TFZ	Tjörnes Fracture Zone			VØMB	Vøring Basin
WJMFZ	West Jan Mayen Fracture Zone	SRCCC	Jan Mayen microcontinent–Southern Ridge Complex (SRC) – continental crust		
Central East Greenland Margin:		FR	Fáfnir Ridge	**Faroe Islands Atlantic Margin:**	
BK	Blosseville Kyst	OR	Otur Ridge	FYR	Fugloy Ridge
JMLB	Jameson Land Basin			FP	Faroe Platform
KD	Kap Dalton outcrop site	SRCTC	Jan Mayen microcontinent–Southern Ridge Complex – transitional crust	FSB	Faroe–Shetland Basin
KB	Kap Brewster outcrop site	DR	Dreki Ridge		
KgB	Kangerlussuaq Basin	LR	Langabrún Ridge		
LVPL	Liverpool Land High	ORS	Otur Ridge southern spur		
ML Fm	Milne Land Formation outcrop site	TR	Treitel Ridge		
SD	Scoresby Sund				
TØ	Trail Ø	NIS	North Iceland Shelf		
		IP	Iceland Plateau		
		IPR	Iceland Plateau Rift		

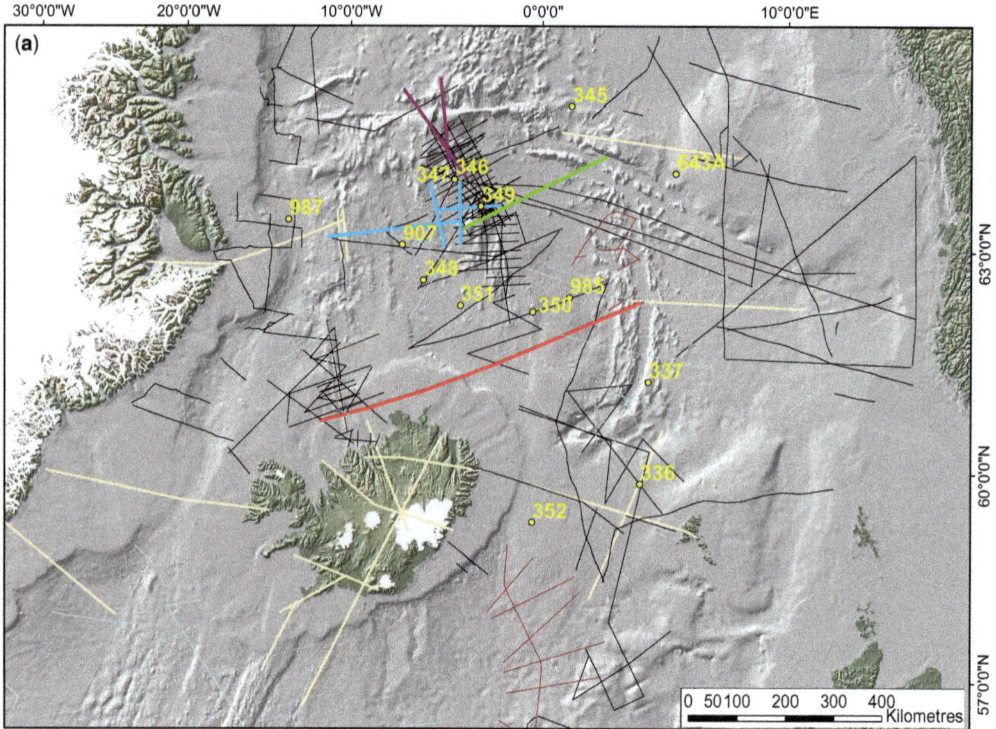

Fig. 2. Regional map showing shaded bathymetry (Amante & Eakins 2009; Jakobsson *et al.* 2012) and (**a**) refraction and reflection seismic lines and boreholes. Legend see Figure 2b.

The microcontinent is bounded on all sides by oceanic crust, although its southern limit remains poorly constrained. This is due, in part, to sparse data coverage south of 68° N (Fig. 2), but also to the occurrence of numerous intrusive and extrusive volcanic rocks that limit seismic imaging of the underlying features. Previous interpretations of the continent–ocean transition (COT) along the JMMC margins were mainly based on magnetic and/or gravity data (e.g. Vogt *et al.* 1970; Talwani & Eldholm 1977; Åkermoen 1989; Doré *et al.* 1999; Lundin & Doré 2002; Rey *et al.* 2003; Gaina *et al.* 2009; Gernigon *et al.* 2012) and seismic reflection data (Gunnarsson *et al.* 1989; Scott *et al.* 2005; Peron-Pinvidic *et al.* 2012*a*, *b*). Breivik *et al.* (2012) considered crustal velocity information from wide-angle data with potential field data to derive the location of the COT.

The purpose of this paper is to establish a detailed tectonic and stratigraphic framework for the JMMC based on a new regional database of geological and geophysical data. The analysis includes interpretation of new seismic reflection data, as well as recent geological findings, from on- and offshore central East Greenland (e.g. Larsen *et al.* 2013; Guarnieri 2015). This study has been facilitated by the interpretation of recently acquired commercial seismic reflection data that were made available for the project, together with older seismic reflection and refraction data collected offshore Iceland since the early 1970s (Fig. 2). Revised $^{39}Ar–^{40}Ar$ dates of East Greenland basalt samples (e.g. Tegner *et al.* 2008; Larsen *et al.* 2013), and an improved coverage of magnetic data and interpretations (CAMP-GM: Gaina *et al.* 2011; Gernigon *et al.* 2015), are also considered. Pre- and post-break-up sedimentary strata and igneous complexes, together with volcanostratigraphic seismic characterization, have been revisited, together with a reassessment of the seawards-dipping reflector sequences (SDRs), igneous complexes, sill and dyke intrusions, and hydrothermal vent complexes.

The JMMC margins are compared to the conjugate margins: the central East Greenland margin and the Møre margin off Norway (Blystad *et al.* 1995). The western JMMC margin is linked to central East Greenland, where the Palaeozoic–Mesozoic Jameson Land Basin is located (JMLB: Henriksen 2008) (Fig. 1). The segmentation and extent of the southern area of the JMMC and its link to the oblique opening of the Norway Basin are also considered. Finally, the question of how the igneous

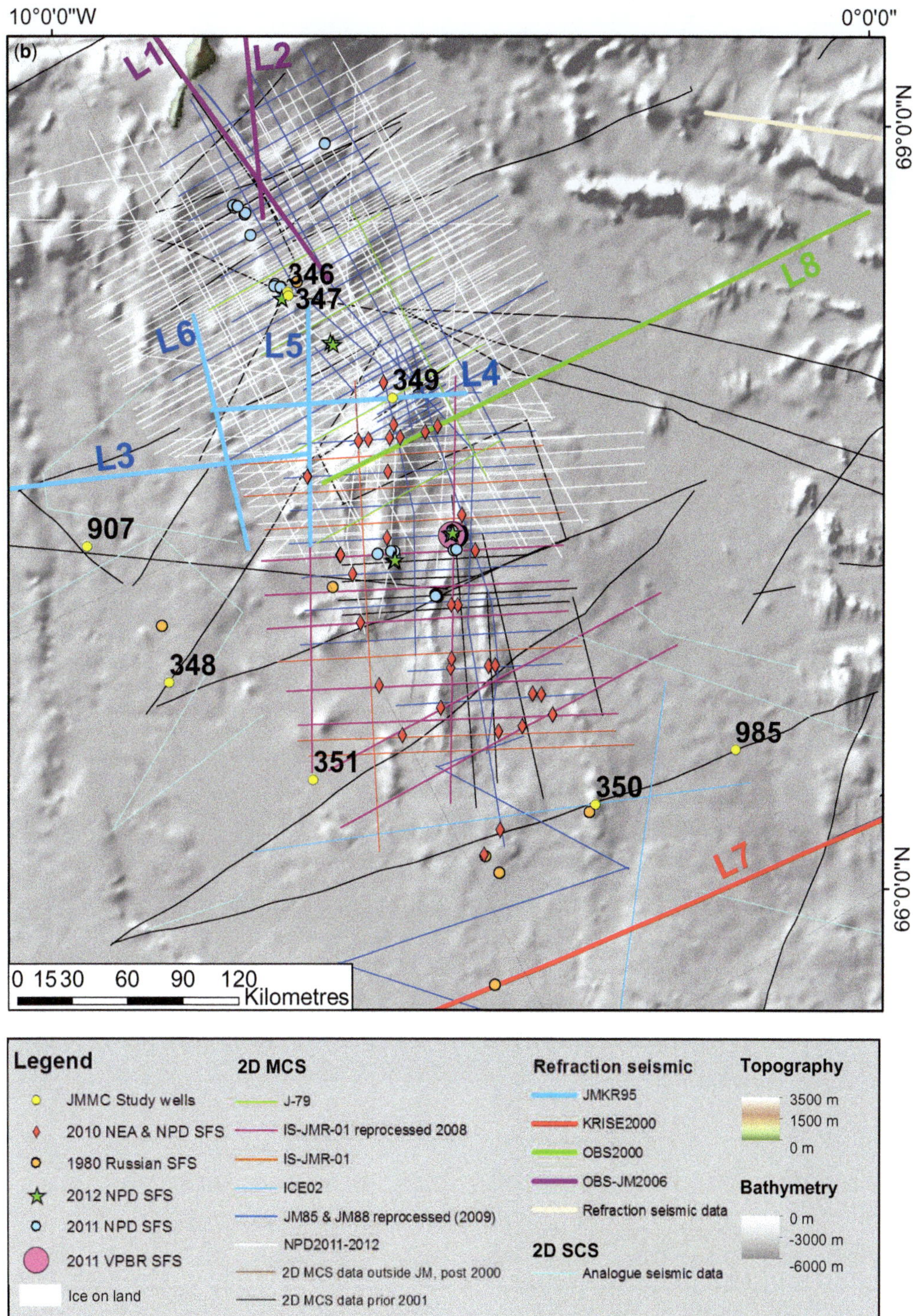

Fig. 2. (**b**) seabed sampling sites (NEA, National Energy Authority, Iceland; NPD, Norwegian Petroleum Directorate (2013); Spectrum ASA; TGS; SFS seafloor samples; VBPR, Volcanic Basin Petroleum Research AS).

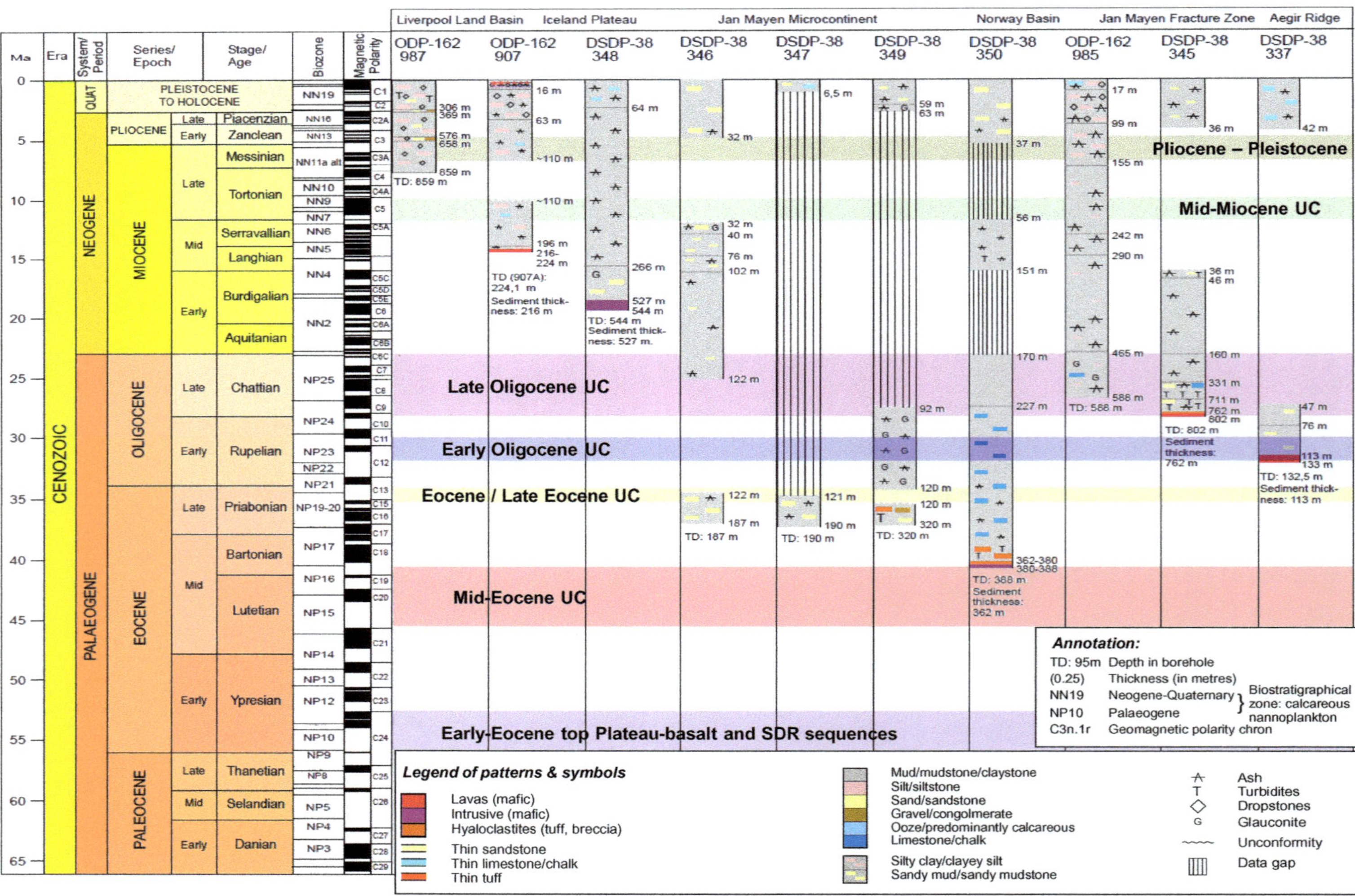

Liverpool Land Basin
Iceland Plateau
Jan Mayen Microcontinent
Norway Basin
Jan Mayen Fracture Zone
Aegir Ridge
ODP-162 987
ODP-162 907
DSDP-38 348
DSDP-38 346
DSDP-38 347
DSDP-38 349
DSDP-38 350
ODP-162 985
DSDP-38 345
DSDP-38 337
Pliocene – Pleistocene
Mid-Miocene UC
Late Oligocene UC
Early Oligocene UC
Eocene / Late Eocene UC
Mid-Eocene UC
Early-Eocene top Plateau-basalt and SDR sequences
TD: 659 m
TD (907A): 224,1 m
Sediment thickness: 216 m
TD: 544 m
Sediment thickness: 527 m.
TD: 187 m
TD: 190 m
TD: 320 m
TD: 388 m
Sediment thickness: 362 m
TD: 588 m
TD: 802 m
Sediment thickness: 762 m
TD: 132,5 m
Sediment thickness: 113 m
Ma
Era
System/Period
Series/Epoch
Stage/Age
Biozone
Magnetic Polarity
CENOZOIC
QUAT
NEOGENE
PALAEOGENE
PLEISTOCENE TO HOLOCENE
PLIOCENE
MIOCENE
OLIGOCENE
EOCENE
PALEOCENE
Piacenzian
Zanclean
Messinian
Tortonian
Serravallian
Langhian
Burdigalian
Aquitanian
Chattian
Rupelian
Priabonian
Bartonian
Lutetian
Ypresian
Thanetian
Selandian
Danian
Annotation:
TD: 95m Depth in borehole
(0.25) Thickness (in metres)
NN19 Neogene-Quaternary
NP10 Palaeogene
Biostratigraphical zone: calcareous nannoplankton
C3n.1r Geomagnetic polarity chron
Legend of patterns & symbols
Lavas (mafic)
Intrusive (mafic)
Hyaloclastites (tuff, breccia)
Thin sandstone
Thin limestone/chalk
Thin tuff
Mud/mudstone/claystone
Silt/siltstone
Sand/sandstone
Gravel/congolmerate
Ooze/predominantly calcareous
Limestone/chalk
Silty clay/clayey silt
Sandy mud/sandy mudstone
Ash
Turbidites
Dropstones
Glauconite
Unconformity
Data gap

events along the southern half of the JMMC are related to the Blosseville Kyst, the Iceland–Faroe Fracture Zone system that forms the NE limit of the Greenland–Iceland–Faroe Ridge Complex (GIFRC) (Árting 2014), and the Iceland Plateau (Fig. 1; Table 1) is addressed. The new interpretation has been used to model the detailed kinematics of the JMMC from pre-break-up time to the present day.

Geological setting of the central NE Atlantic

Several distinct rifting episodes and the subsequent break-up of the supercontinent Pangea led to the formation of a series of segmented rifted margins along the North Atlantic Ocean (Ziegler 1988). Extensional episodes are recognized from Devonian and Carboniferous times, initiated by the collapse of the Caledonian mountain belt (e.g. Andersen & Jamtveit 1990). Devonian onshore rift basins along East Greenland (Henriksen 2008) and SW Norway (Osmundsen & Andersen 1994, 2001; Osmundsen *et al.* 2002) are well documented, including their complex relationship to large-scale transtensional tectonics (Osmundsen & Andersen 2001). These basins are interpreted to extend in the central and northern part of the NE Atlantic during the Carboniferous, and were not affected by the Variscan Orogeny (Hopper *et al.* 2014), which occurred at the same time and influenced the NE and SE regions of the NE Atlantic, the North Sea and northern Europe (Pharaoh *et al.* 2010). During the Permian and Triassic periods, the entire NE Atlantic region was subjected to extension (e.g. Doré *et al.* 1999; Brekke 2000). At that time, a first rifting phase led to minor rotational block faulting and westwards tilted half-graben along East Greenland (Seidler 2000), forming terrestrial to shallow marine basins that discordantly covered the old Devonian–Carboniferous basin (Stemmerik 2000). The entire NE Atlantic system went through two major rifting phases during the Late Jurassic and a major Cretaceous rifting phase from the late Early Cretaceous (Aptian–Albian) to Late Cretaceous (Lundin & Doré 1997, 2011; Stoker *et al.* 2016), leading to significant crustal thinning in the central parts of the corridor and forming deep basins. The Cretaceous rifting phase may have included a hyper-extension (Peron-Pinvidic *et al.* 2013), resulting in exhumation of deep crust and possibly mantle, as suggested by Osmundsen *et al.* (2002) or Osmundsen & Ebbing (2008).

During Late Paleocene and pre-break-up time, early volcanism associated with the North Atlantic igneous province occurred. Regionally extensive landwards flows consisting of subaerial and submarine lava flows onto adjacent elevated margins were emplaced during this time (Horni *et al.* 2016). Infilling of pre-existing basin areas formed escarpments and hyaloclastite deltas (Planke *et al.* 2000; Horni *et al.* 2016). Intense magmatism occurred at this time just SW of the JMMC, close to the Kangerlussuaq Basin and the southern extent of the Blosseville Kyst (e.g. Tegner *et al.* 2008; Brooks 2011). Magma-rich margins formed during the Early Eocene (56–55 Ma), in association with final rupture of the lithosphere and the onset of seafloor spreading of the NE Atlantic (e.g. Talwani & Eldholm 1977). The resulting North Atlantic continental margins contain SDR sequences observed on seismic reflection data (Hinz 1981). The break-up process was also accompanied by the emplacement of sill and dyke complexes into the margin flank areas. Oceanic crust was first formed in the Norway Basin at the end of chron C25 or the beginning of C24r (*c.* 55 Ma) forming the Ægir mid-oceanic ridge (e.g. Talwani & Eldholm 1977; Gaina *et al.* 2009).

The JMMC structure and stratigraphy observed between its eastern and western margins is profoundly segmented. A first-order boundary within the microcontinent is between the Jan Mayen Ridge and the SRC. Updated datasets suggest that the JMMC internal segmentation is probably related to the complex multistage seafloor spreading processes on both sides of the microcontinent.

Published plate tectonic reconstructions indicate a westwards migration of the plate boundary from the Norway Basin towards the Kolbeinsey mid-oceanic ridge (Nunns 1983*a*, *b*; Nunns *et al.* 1983; Lundin & Doré 2005; Doré *et al.* 2008; Gaina *et al.* 2009), suggesting a gradual separation of the microcontinent from East Greenland during the Early Miocene (Talwani & Eldholm 1977; Gunnarsson *et al.* 1989). Larsen *et al.* (2013) suggested that early rifting between the JMMC and East Greenland coast may have occurred from 49 to 44 Ma, with a direction semi-parallel to the Ægir mid-oceanic ridge system. This event generated increased igneous activity and structural deformation along the NE extent of the Blosseville Kyst (Fig. 1).

Fig. 3. The JMMC stratigraphic summary chart, partly based on DSDP and ODP boreholes (Talwani *et al.* 1976*a*, *b*; Manum & Schrader 1976; Manum *et al.* 1976*a*, *b*; Raschka *et al.* 1976; Nilsen *et al.* 1978; Thiede *et al.* 1995; Jansen *et al.* 1996; Channell *et al.* 1999*a*, *b*; Butt *et al.* 2001). This is used to tie the known shallow Cenozoic stratigraphy and unconformities to the seismic reflection data (see the type section in Fig. 4). The Pliocene–Pleistocene correlation marker is based on sedimentary core records (Talwani *et al.* 1976*a*, *b*).

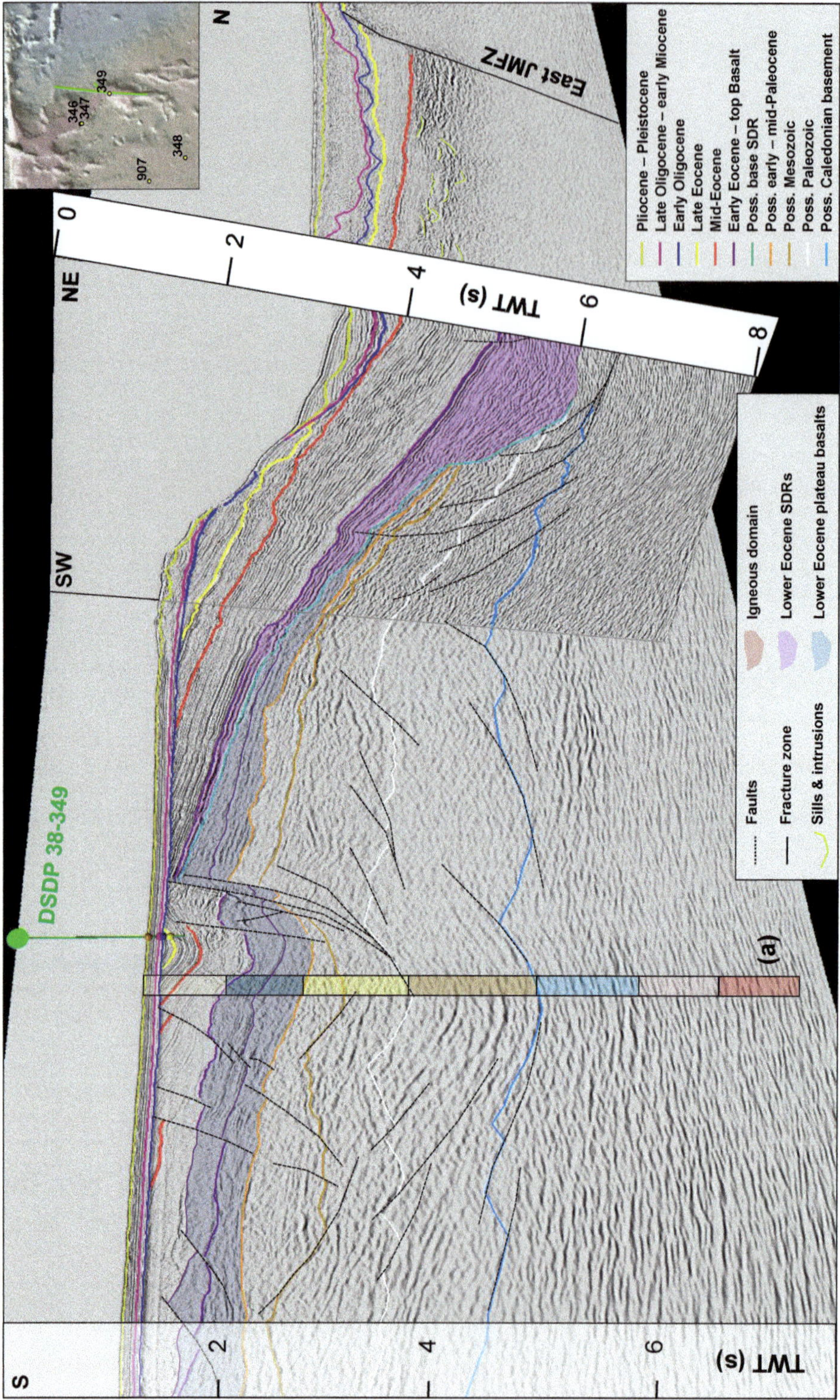

Fig. 4. Type section showing sedimentary basins below the Cenozoic succession and the seismic refraction velocity interval interpretation intersection (**a**) (Table 3). The SDR sequence (purple) is interpreted to overlie a thick basalt sequence (blue) that is likely to be equivalent to the plateau basalts exposed on East Greenland. The east–west line is courtesy of Spectrum ASA, and the north–south line is courtesy of NPD.

Data overview and methods

Geophysical datasets consist of magnetic and gravity anomaly compilations (Haase & Ebbing 2014; Nasuti & Olesen 2014) (Fig. 2), 2D multichannel seismic reflection data (2D MCS data) and seismic refraction data (Johansen *et al.* 1988; Olafsson & Gunnarsson 1989; Kodaira *et al.* 1998; Brandsdóttir *et al.* 2015). Eight shallow Ocean Drilling Program (ODP) and Deep Sea Drilling Program (DSDP) boreholes (legs 38, 151 and 162) (Eldholm & Windish 1974; Talwani & Udintsev 1976; Eldholm *et al.* 1987, 1989) provide some of the few samples from the JMMC (Figs 2–5). Results from seafloor sampling campaigns carried out in 1973 by Geodekyan *et al.* (1980), in 2010 by the National Energy Authority of Iceland (OS) and the Norwegian Petroleum Directorate (NPD), in 2012 by the NPD (Sandstå *et al.* 2012), and in 2012 by the Volcanic Basin Petroleum Research (VBPR) and TGS (Polteau *et al.* 2012) were also taken into account. Finally, recently revised ^{40}Ar–^{39}Ar dating of East Greenland coastal basalts and onshore unconformities within the igneous successions of the Blosseville Kyst area are considered (Larsen *et al.* 2013).

Seismic reflection data includes only a few 2D multichannel surveys from before 2001, of which the JM-85-88 results were reprocessed in 2009. More recent surveys include IS-JMR-01 (2001), ICE-02 (2002), WI-JMR-08 (2008), NPD-11 (2011) and NPD-12 (2012) (Fig. 2; Table 2). The reprocessed dataset was used for detailed volcanostratigraphic seismic characterization, which facilitated mapping and identification of structural elements, sedimentary sequences, SDR sequences, and sill and dyke complexes. Multibeam bathymetry data were used to map structural trends and features at the seafloor. This high-resolution bathymetry data in combination with seismic reflection data enabled us to differentiate strike-slip from normal fault systems and slump faulting along the steep escarpments of the microcontinent's ridges.

There are no deep drill holes on the JMMC. For this reason, the older history and stratigraphic correlations are inferred by comparison to better-known analogue areas along the conjugate margins, in particular the Jameson Land Basin (Surlyk *et al.* 1973; Surlyk & Noe-Nygaard 2001; Surlyk 1977, 1978, 1990, 1991, 2003; Henriksen 2008), and the mid-Norway Møre and Vøring margins (e.g. Brekke *et al.* 1999; Osmundsen *et al.* 2002; Faleide *et al.* 2010) (Fig. 1a; Table 1).

Correlation of stratigraphic information to seismic reflection data

Information from DSDP Leg 38 (sites 346, 348, 349 and 350) and from seafloor samples retrieved by the NPD in 2011, 2012 and 2013 (Fig. 2b; Table 2) provide key constraints for tying it to the seismic reflection data along the central part of the JMMC (Fig. 3). This permits mapping of unconformities and post-basalt stratigraphy across the JMMC (Figs 4, 5 & 6). Some uncertainties in local stratigraphic correlations still exist, notably along the collapsed western flank of the JMMC and between the dislocated southern ridges. The sub-basalt sequences are only visible in the central area of the JMMC and around DSDP Leg 38 site 349. Seismic reflection data in vicinity of this drill site has been interpreted as possible Mesozoic and Palaeozoic strata based on comparisons to the Jameson Land Basin (Blischke *et al.* 2014*a*).

In addition to the well ties, onshore and offshore stratigraphic relationships along the western conjugate margin (Blosseville Kyst in East Greenland) provide information on the basalt stratigraphy that can be used for interpreting volcanic horizons on the seismic reflection dataset. This will be discussed in detail in the following section on 'Stratigraphic setting'.

Basement tie on reflection data using velocity interpretations

Significant uncertainty surrounds the full extent of the JMMC, primarily to the south but also to the east and west. Of particular focus here is the southern extent of the JMMC towards the Icelandic Shelf (Fig. 1). Talwani & Eldholm (1977) and Brandsdóttir *et al.* (2015) suggested that the JMMC terminates south of the SRC (Fig. 5). However, other studies propose severely stretched and fragmented continental crust and/or exhumed altered mantle for the southernmost part of the JMMC (Gaina *et al.* 2009; Breivik *et al.* 2012; Peron-Pinvidic *et al.* 2012*a*; Gernigon *et al.* 2015; Torsvik *et al.* 2015). Sparse data in combination with the inherent non-uniqueness of geophysical modelling and interpretation makes this particularly challenging. To better constrain this region, the 2D seismic reflection dataset was analysed in combination with the available seismic refraction data and crustal velocity models, as well as available well control along the ridges. This enabled an interpretation of the nature of acoustic basement and different crustal type domains (Figs 4 & 5).

Seismic refraction velocity model. Three ocean-bottom seismometer (OBS) experiments have been carried out for the larger JMMC area (Fig. 2; Table 2). The JMKR-95 survey included east–west- and SW–NE- orientated profiles (Kodaira *et al.* 1998), with profile JMKR95-L4 crossing the Jan Mayen Ridge just south of DSDP borehole

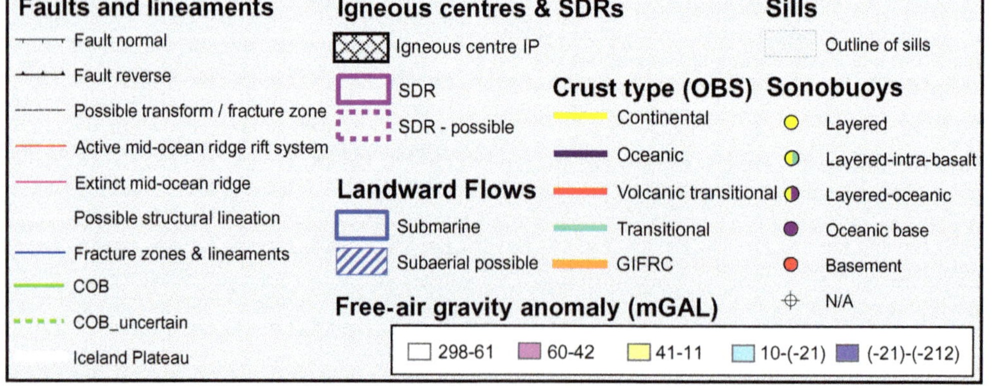
0 30 60 120
Kilometres
WJMFZ
MR
JMI
EJMFZ
345
346
347
349
907
348
351
350
985
337
C6b
C24n2r
ÆR
KBR
IPR
IP
NIS
TFZ
IFFZ
68°0'0"N
66°0'0"N
64°0'0"N
20°0'0"W
15°0'0"W
10°0'0"W
Faults and lineaments
Fault normal
Fault reverse
Possible transform / fracture zone
Active mid-ocean ridge rift system
Extinct mid-ocean ridge
Possible structural lineation
Fracture zones & lineaments
COB
COB_uncertain
Iceland Plateau
Igneous centres & SDRs
Igneous centre IP
SDR
SDR - possible
Landward Flows
Submarine
Subaerial possible
Crust type (OBS)
Continental
Oceanic
Volcanic transitional
Transitional
GIFRC
Sills
Outline of sills
Sonobuoys
Layered
Layered-intra-basalt
Layered-oceanic
Oceanic base
Basement
N/A
Free-air gravity anomaly (mGAL)
298-61
60-42
41-11
10-(-21)
(-21)-(-212)

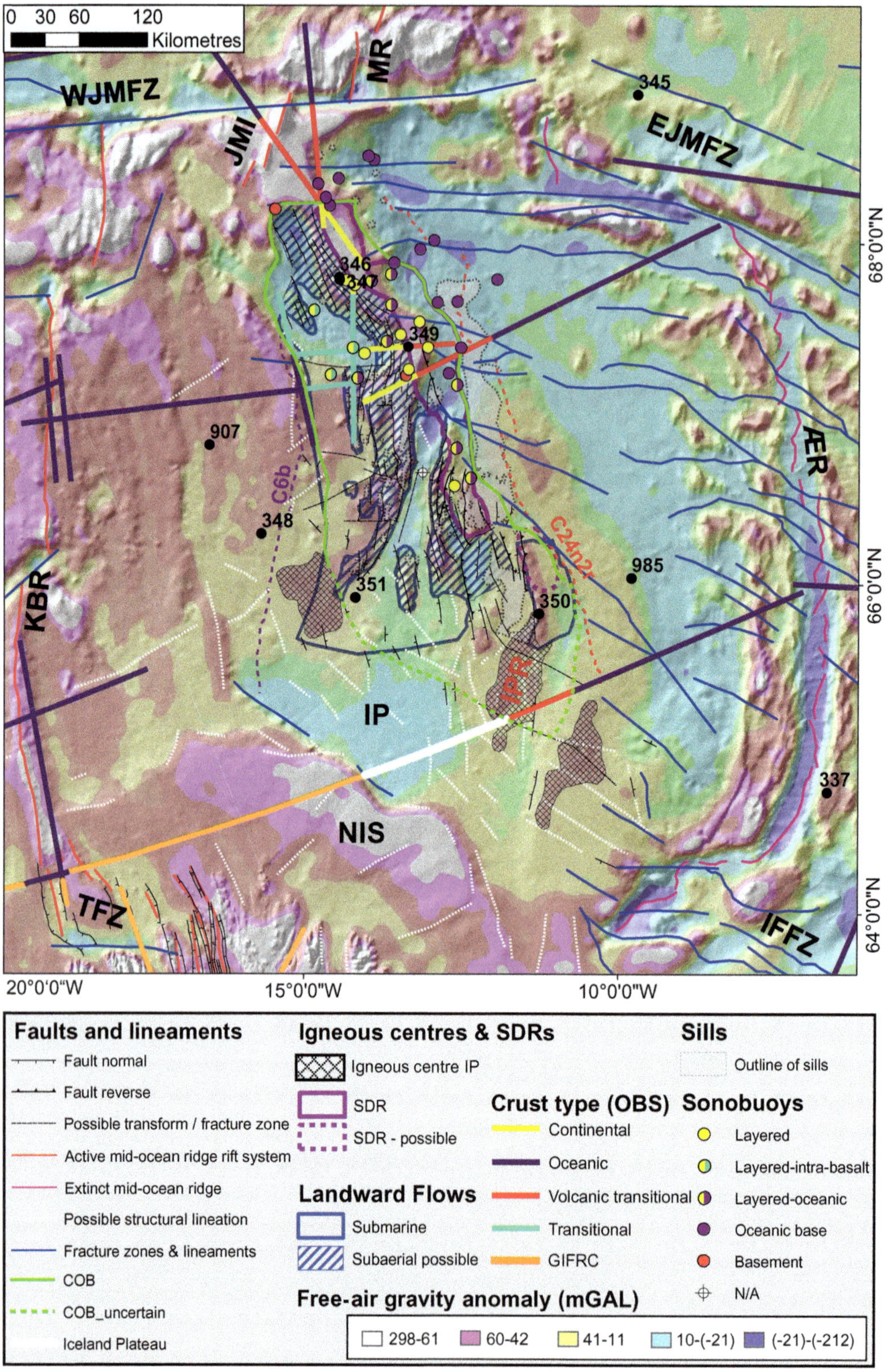

Leg 38 site 349. Line 8 of the OBS2000 survey lies 30 km south of that same borehole, crossing the SW end of the JMMC and the NW end of the EJMFZ (Mjelde *et al.* 2002, 2007; Breivik *et al.* 2012). The southern extent of the JMMC onto the Iceland Plateau was investigated as part of the KRISE survey in 2000 (Brandsdóttir *et al.* 2015) (L7 on Fig. 2b). A third survey in 2006 focused on the northern region of the JMMC, consisting of a NW–SE (L1) and a north–south profile (L2) across Jan Mayen Island (Kandilarov *et al.* 2012). Refraction data from all these surveys were used to constrain the southern extent of the JMMC.

Sonobuoys deployed during a 1985 seismic reflection survey provide velocity information for the upper layers of the microcontinent (Olafsson & Gunnarsson 1989). Based on these, we were able to better constrain the igneous crust of the JMMC, especially the area within the SDR sequences. Each sonobuoy location was assigned to a velocity-profile domain and incorporated into the volcanic facies map (Table 2; Fig. 5). Distinct velocity-profile domains are defined: layered, layered-intra-basalt, layered-oceanic, oceanic basement and basement of the microcontinent. The layered domain corresponds to velocity layers within the range 1.7–3.2 km s^{-1}, which are interpreted as post-break-up sediments; velocity layers between 3.9 and 5.5 km s^{-1} across the crest area of the JMMC, and distinct velocity interval breaks, are, however, most likely to correspond to pre-break-up sedimentary sections that correlate directly to seismic refraction data. The layered-intra-basalt domain corresponds to 270–470 m-thick basaltic layers (4–5 km s^{-1}) within the post-break-up sedimentary section (1.8–2.5 km s^{-1}) of the Jan Mayen Basin. A distinct velocity domain within the SDR area of the eastern flank is termed the layered-oceanic domain. The oceanic basement domain is characterized by thin low-velocity sediment layers (<2.5 km s^{-1}) on top of a high-velocity layer (4–5 km s^{-1}) that gradually and smoothly increases towards the base (5–6 km s^{-1}). These oceanic basement velocity domains were also compared to seismic refraction data interpretations (Breivik *et al.* 2012). One velocity profile at the crest of the JMMC is inferred to represent continental crust, as an abrupt velocity layer increase to 5.5 km s^{-1} was recorded below the thin post-break-up sediment cover (1.9–2.2 km s^{-1}).

Seismic velocities derived from wide-angle data were used as a basis for the depth and stratigraphic thickness estimations across the JMMC (Figs 4, 5, 6 & 7; Table 3). Relatively high-velocity values (4.4–5.6 km s^{-1}) have been assigned to the deeper layers above the acoustic basement where reflectivity is observed and interpreted as older pre-Cenozoic sedimentary sequences. This is similar to what is observed along the conjugate Norwegian Shelf, where the Mesozoic–Palaeozoic sections are usually interpreted to range between 4 and 5.5 km s^{-1} (Mjelde *et al.* 2008, 2009).

Stratigraphic setting

The following subsections summarize the interpretations of the Palaeozoic–Cenozoic succession over the JMMC. The total thickness of interpreted sediments is variable across the area and may reach up to 18 km along the eastern flank of the JMMC. The microcontinent contains several major unconformities and related structures that are linked to the complex tectonomagmatic processes on both sides.

A type section was constructed to provide a framework for mapping unconformity horizons and stratigraphic geometries along the JMMC (Figs 4 & 8). The section is based on bathymetric, borehole and seismic refraction data, combined with a dense grid of seismic reflection data. The section is orientated north–south along the strike of the Lyngvi Ridge, a central and stable block of the JMMC (Fig. 4, LYR in Fig. 8; Table 1).

The presence of Palaeogene volcanic rocks on the JMMC makes it difficult to interpret older strata below on seismic sections. Some local uncertainties in stratigraphic correlations still exist, notably along the microcontinent's collapsed western flank and between the dislocated southern ridges. Sub-basalt sequences are only visible within the central area of the microcontinent in the vicinity of DSDP Leg 38 site 349, where seismic reflection and refraction data have been compared to the Mesozoic and Palaeozoic strata of the Jameson Land Basin area of the East Greenland margin.

Fig. 5. Volcanic facies map based on the interpretation of seismic reflection and refraction data and information from wells, in addition to free-air gravity anomaly data (DTU2010: Andersen 2010). Refraction information includes velocity profile interpretations of wide-angle data and crustal-type interpretations (modified after Funck *et al.* 2014), as well as sonobuoy velocity profile interpretations (Olafsson & Gunnarsson 1989). Magnetic anomalies C6b and C24n2r are from Gernigon *et al.* (2015), showing the onset of oceanic seafloor spreading east and west of the JMMC. The extent of landwards flows labelled 'subaerial possible' refers to the pre-break-up plateau basalt extent over the area. The areas labelled 'submarine' areas are interpreted primarily by mapping the F-reflector (Gunnarsson *et al.* 1989) and are inferred as being related to the second break-up phase during Late Oligocene.

Table 2. *JMMC database and results that have been reviewed. Data and studies that have been used in this study are marked in column 'A'*

A	Year	Survey ID	Survey lead	Country	Platform name	Data repository	Data types
	1957		NAVO	USA			Aeromagnetic
X	1961–1971	V2304/V2703/V2803	L-DGO	USA	Vema/Conrad	NGDC	Bathymetry; magnetics; gravity; 2D multichannel reflection seismic (2D MCS)
X	1973	V3010	L-DEO	Norway	Vema/Conrad		Bathymetry; magnetics; gravity; 2D MCS
X	1974	DSDP Leg 38	DSDP		Glomar Challenger		Boreholes
	1975	CEPAN-75	CNEXO	France	Jean Charcot	Ifremer	Bathymetry; magnetics; gravity; 2D MCS
	1975	CEPAN-75	CNEXO	France	Jean Charcot	Ifremer	Bathymetry; magnetics; gravity; 2D MCS
	1975	CEPAN-75	CNEXO	France	Jean Charcot	Ifremer	Bathymetry; magnetics; gravity; 2D MCS
X	1975	BGR-75	BGR	Germany	Longva	BGR	2D MCS
X	1976	BGR-76	BGR	Germany	Explora	BGR	2D MCS
	1976	CGG-76	NPD/CGG	Norway			Aeromagnetic
	1977	IOS-77	UD/IOS	England	Shackleton	NGDC	2D MCS
X	1978	RC2114	L-DGO	USA	Robert Conrad	MGDS	Bathymetry; magnetics; gravity; 2D MCS2D MCS
X	1978	WGC-78		USA	Karen Bravo	Western-Geco	2D MCS
X	1979	J-79	NPD	Norway	GECO alpha	NPD	Bathymetry; magnetics; gravity; 2D MCS
X	1980		PAH/SGC	USSR	Akademic Kurchatov		Seafloor sampling
	1983	NGT83/RC2412	L-DGO/BGR	USA/Germany	Prospekta/Conrad		2D MCS, ESP, WA, CDP
	1983	RC2412	L-DEO	Norway	Robert D. Conrad		2D MCS and single-channel reflection seismic (2D SCS), gravimeter, magnetometer, sonar-echosounder
	1984	Arktis II/5	UHH	Germany	Polarstern		Refraction seismic
X	1985	JM-85	NPD/NEA	Norway	Malene Østervold	NPD	Bathymetry; magnetics; free air gravity; Bouguer gravity; magnetic; 2D MCS
X	1985	ODP Leg 104	ODP		JOIDES Resolution		Boreholes
X	1986	UiO-86	UiO	Norway	Håkon Mosby	NPD	2D MCS
X	1987	ESP	IFP	France			ESP; velocity; gravity
X	1988	JM-88	NPD/NEA	Norway	Håkon Mosby	NPD	Bathymetry; magnetics; gravity; 2D MCS; sonobuoy
X	2000	KRISE 2000	UiB	Norway	Håkon Mosby	UiB	2D MCS
X	2001	IS-JMR-01	InSeis	Norway	Polar Princess	CGGVeritas	2D MCS
X	2002	ICE-02	TGS-Nopec	Iceland	Zephyr 1	TGS-NOPEC	2D MCS; gravity
	2003	EW0307	L-DEO	USA	Maurice Ewing	MGDS	2D MCS; gravity; bathymetry cores
X	2005	JAS-05	NGU/NPD	Norway	Piper Navajo	NGU	Aeromagnetic
X	2006	OBS JM-06	UiB/Geomar	Norway/Germany	G. O. SARS	UiB	2D MCS; gravity, magnetics
X	2008	WI-JMR-08	Wavefield InSeis	Norway	Malene Østervold	Spectrum	2D MCS
X	2008	A8-2008	HAFRO/NEA	Iceland	Arni Fridriksson	HAFRO/NEA	Multibeam
X	2009	JM-85-88	Spectrum	Norway	Re-processing	Spectrum	2D MCS
X	2009	SAR-ICE-2009	NEA	Norway	ENVISAT satellite	Fugro NPA	Satellite Synthetic Aperture Radar (SAR)
X	2010	A11-2010	HAFRO/NEA/NPD	Iceland	Arni Fridriksson	HAFRO/NEA/ NPD/Fugro Geolab	Multibeam; seafloor sampling
	2010	B11-2008	HAFRO/NEA	Iceland	Arni Fridriksson	HAFRO/NEA	Bentic survey
X	2011	NPD-11	NPD/UiB	Norway	Harrier Explorer	NPD/PGS	2D MCS; seafloor sampling
X	2011	JMRS11	VPBR/TGS	Norway	TGS	VPBR/TGS	Seafloor sampling
X	2012	NPD-12	NPD/UiB	Norway	Nordic Explorer	NPD/PGS	2D MCS; seafloor sampling
X	2012	JAS-12	NGU/NPD/NEA	Norway	Piper Chieftain	NGU/NPD/NEA	Aeromagnetic

The Palaeozoic

Sub-basalt structures and inferred velocities along the JMMC (Fig. 4) are comparable to the Upper Palaeozoic–Lower Mesozoic rocks of the Jameson Land Basin and Traill Ø (Fig. 1a) areas onshore East Greenland, as well as of the Møre and Vøring basins offshore Norway (e.g. Surlyk *et al.* 1973; Surlyk & Noe-Nygaard 2001; Brekke *et al.* 1999; Osmundsen *et al.* 2002; Surlyk 2003; Henriksen 2008; Faleide *et al.* 2010). The inferred Palaeozoic section on the JMMC is thinner and more condensed in comparison to the East Greenland and Møre–mid-Norway areas. It can be inferred that the JMMC was at that time in a structurally higher position, corresponding to a shallow platform domain between the adjacent Jameson Land and Møre basins. Small sub-basin structures below the Cenozoic section are potentially linked to the region south of the Jameson Land Basin (Fig. 4). It remains uncertain whether older Palaeozoic, in particular Devonian and/or Carboniferous rocks similar to those that crop out along the NW edge of the Jameson Land Basin, underlie the northernmost part of the JMMC.

The Mesozoic

The interpreted Mesozoic–Paleocene interval of the JMMC has a velocity range of between 3.9 and 5.0 km s^{-1} (Table 3) (Kodaira *et al.* 1998; Kandilarov *et al.* 2012). These are similar to velocities interpreted for Mesozoic sequences in the Møre Basin, where pre-Cretaceous- Lower Cretaceous sections show a range of values between 3.85 and 5.35 km s^{-1} (Mjelde *et al.* 2008, 2009). Since subunits within the Mesozoic layers cannot be identified, an average value of 4.4 km s^{-1} is used for a time-depth conversion.

Although controversial, a seafloor sample has been interpreted to contain evidence of a Jurassic oil seep (Polteau *et al.* 2012) (NPD 2012) (Fig. 2b). Waxy bitumen samples from Cenozoic basalts on the Faroe Islands and the Isle of Skye in Scotland have also been associated with mature source rocks (Laier *et al.* 1997; Laier & Nytoft 2004).

The Jameson Land Basin contains deltaic and lacustrine facies of Early Jurassic age that became increasingly influenced by marine processes during Mid–Late Jurassic, including deposits of a black organic-rich mudstone of the Hareelv Formation (Surlyk 2003), which is consistent with the regional setting of the NE Atlantic (Stoker *et al.* 2016). Thus, a trend towards a Mesozoic marine setting in the JMMC area can be inferred from regional structural observations. That would suggest a phase of southwards and eastwards crustal thinning during the Jurassic, which, in turn, may have resulted in a general subsidence of the region and the development of marine conditions (Peron-Pinvidic *et al.* 2012*b*).

The presence of a thin, Lower Cretaceous sedimentary succession across the JMMC seems probable according to results from a structural and stratigraphic comparison with the conjugate margins and the interpretation of the local seismic refraction data (Figs 4–6). During the Cretaceous, regional extension occurred throughout much of the NE Atlantic region (Stoker *et al.* 2016). Thick, deep-marine, sag-type basins formed during this process, including the Danmarkshavn and Thetis basins (Lundin & Doré 1997; Doré *et al.* 1999; Lundin & Doré 2011), the Traill Ø–Hold with Hope area of NE Greenland, and the Norwegian Vøring and Møre basins (Brekke 2000; Osmundsen *et al.* 2002; Faleide *et al.* 2010, 2010; Peron-Pinvidic *et al.* 2012*b*). However, all of the above-mentioned basins had their main extensional to hyperextensional phase during the Mesozoic, whereas the Jameson Land Basin had a main opening phase and faulting during the Palaeozoic (Henriksen 2008). Such hyperextension cannot be seen along central Eastern Greenland or the JMMC area, and possibly formed the western shelf margin of the Vøring basin.

The Cenozoic

Cenozoic sedimentary succession. The pre-break-up Cenozoic sedimentary succession was inferred by comparing onshore geological data from East Greenland to offshore areas of Scoresby Sund and Blosseville Kyst and the seismic reflection data of the JMMC. The post-break-up successions are derived from borehole data and seismic reflection data across the microcontinent, tied to the main unconformity horizons of the post-break-up succession (Figs 3, 4, 6 & 7). Mapping thickness intervals of the Cenozoic sequence along the ridge flanks and in-between the ridge segments enabled us to indicate areas of sediment deposition due to subsidence. Depositional settings for the Cenozoic strata are constrained by borehole data interpretations and onshore analogue comparisons.

Pre-break-up Cenozoic strata. In situ Paleocene dinoflagellate cyst assemblages were found in the uppermost pre-volcanic/pre-break-up sequences at the Blosseville Kyst area (Nøhr-Hansen & Piasecki 2002) (Figs 4, 6 & 7), which is the closest conjugate segment to the JMMC. This sequence corresponds to the upper marine sections of the Ryberg Formation in the Kangerlussuaq Basin (KgB on Fig. 1a) (Soper *et al.* 1976; Nøhr-Hansen *et al.* 2002), representing a regional marker that most likely also covered the JMMC area in the Paleocene. Dark mudstones of the Kap Brewster site (KB on Fig. 1a)

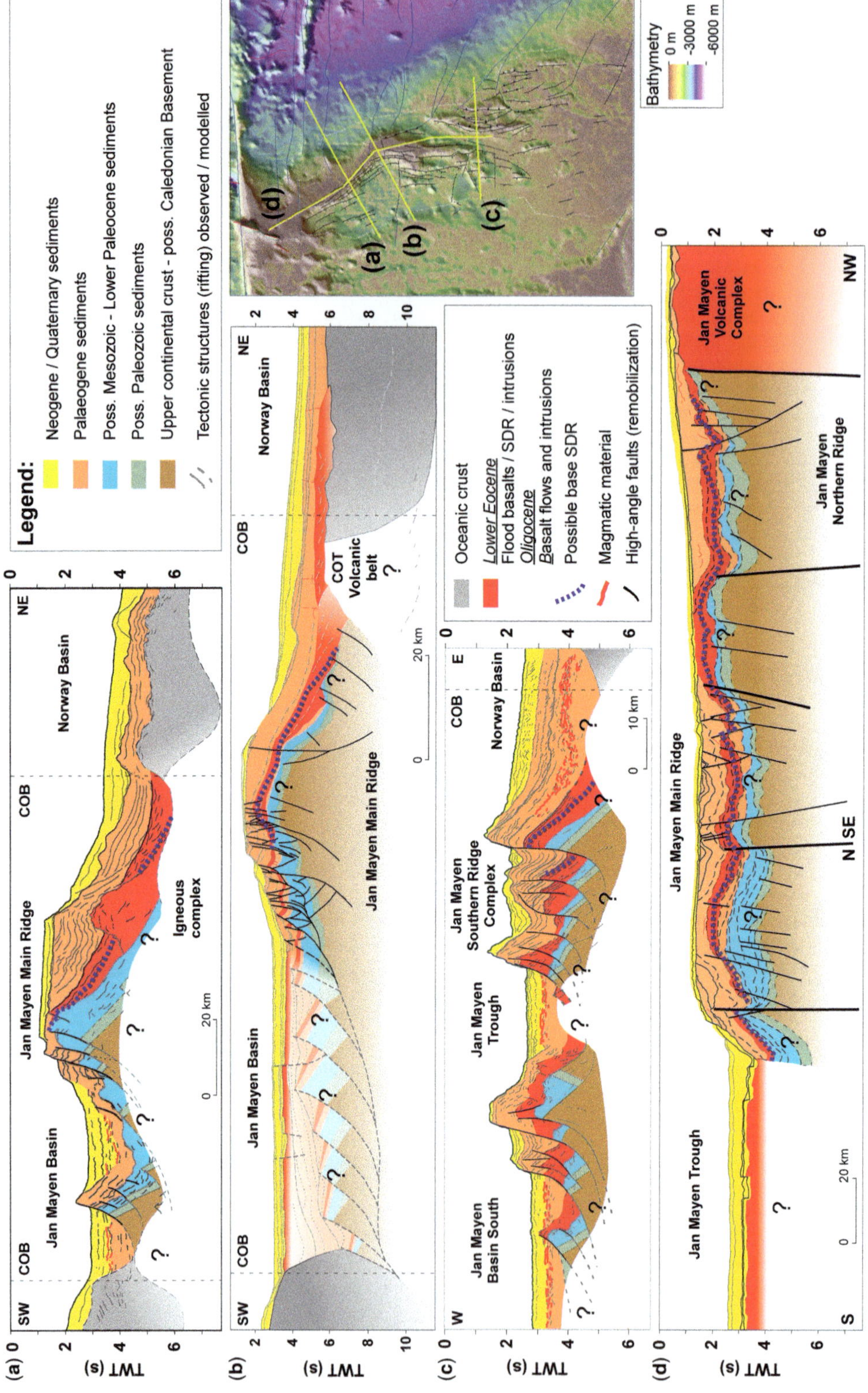

Legend:
Neogene / Quaternary sediments
Palaeogene sediments
Poss. Mesozoic - Lower Paleocene sediments
Poss. Paleozoic sediments
Upper continental crust - poss. Caledonian Basement
Tectonic structures (rifting) observed / modelled
Oceanic crust
Lower Eocene
Flood basalts / SDR / intrusions
Oligocene
Basalt flows and intrusions
Possible base SDR
Magmatic material
High-angle faults (remobilization)
Bathymetry
0 m
-3000 m
-6000 m
(a)
(b)
(c)
(d)
TWT (s)
SW
NE
COB
Jan Mayen Basin
Jan Mayen Main Ridge
Norway Basin
Igneous complex
20 km
COT Volcanic belt
W
E
Jan Mayen Basin South
Jan Mayen Trough
Jan Mayen Southern Ridge Complex
10 km
S
NW
N SE
Jan Mayen Northern Ridge
Jan Mayen Volcanic Complex

contain reworked Cretaceous dinoflagellate cysts, indicating an age range between the late Danian to early Selandian (Nøhr-Hansen *et al.* 2002). Analogue areas for Lower Paleocene deposits include the Hold with Hope and Wollaston Foreland formations in Northeast Greenland (Larsen *et al.* 1999; Nøhr-Hansen 2003, 2012), and the Vøring and Møre basins offshore Norway (Brekke 2000; Faleide *et al.* 2010).

Post-break-up Cenozoic strata. The post-break-up Cenozoic sedimentary section is thickest along the microcontinent's eastern flank (Figs 5, 6 & 8), but has been eroded to a large extent across the highest sections of the ridges (e.g. the Lyngvi Ridge). Based on data from DSDP Leg 38 boreholes located on the northern Jan Mayen Ridge (Talwani *et al.* 1976*b*; Talwani & Eldholm 1977), the Cenozoic succession has been subdivided into a Lower Paleocene–Lower Oligocene unit, unconformably overlain by an Upper Oligocene–Quaternary unit (Figs 6 & 7).

The stratigraphic thickness of post-break-up sediments varies from 0 to 4200 m along the eastern flank of the JMMC (Fig. 8). The Cenozoic units consist predominantly of mudstone, whereas the Lower Paleocene–Lower Oligocene unit includes thin sand and muddy sand beds, which might have been deposited by turbidity currents on the JMMC shelf edge (Figs 3 & 7). The Upper Oligocene–Miocene units probably represent erosional sediments from the JMMC highs, redeposited into the surrounding lows. This can be seen around the highs of the SRC and its small sub-basins and in the borehole records (Talwani *et al.* 1976*b*; Talwani & Eldholm 1977). Above the Mid-Upper Miocene hiatus, Pliocene–Pleistocene deep-marine sediments are present across the microcontinent (Fig. 3). These youngest sediments are cut by deep-sea current features, causing localized erosion along the ridge segments, and were affected by gravitational slumping and faulting from the steep ridge flanks.

During the Pleistocene, several glacial events removed about 1 km of the Iceland Plateau basalts (Walker 1964) and much of the JMMC. The eroded sediments were likely to have been deposited into newly formed basins within the Iceland Plateau (Figs 5 & 9b), between the Iceland shelf and the SRC.

The Cenozoic igneous sequence

The Lower Palaeogene volcanic succession probably includes two major units: the pre-break-up plateau basalt sequence; and the break-up SDR sequence along the eastern flank of the JMMC (Planke *et al.* 2000) (Figs 4, 6 & 7). The southern and central part of the JMMC appear to be covered by Early Eocene plateau basalts, with a layered character of distinct lava-flow events interpreted as landwards flows (Planke *et al.* 2000). Apparent erosional effects at the top of these formations are probably filled with Lower Eocene sediments.

The pre-break-up igneous formations. The plateau basalt equivalent section on the JMMC is subdivided into two major units and appears to increase in thickness from north to south towards the SRC (Fig. 7c). A possible base of the plateau basalts was tied to the velocity model from refraction data (Kodaira *et al.* 1998). From this, an igneous stratigraphic thickness of approximately 1100 m is estimated across the crest of the JMMC. There are indications that the plateau basalt sequence continues to be downfaulted towards the south (Fig. 4), implying the formation of a topographical low south of the SRC (Figs 10 & 11a). This would possibly correlate with the very thick basalt sections in the Kangerlussuaq Basin, at the southernmost extent of the Blosseville Kyst, and the NW area of the Faroe Islands Platform (FP on Fig. 1a). In these regions, the main pre-break-up to break-up phase plateau basalts are well exposed onshore. The succession is estimated to be more than 6 km thick towards the southern extent of the Blosseville Kyst and the Kangerlussuaq area (Brooks 2011), but is progressively younger and dramatically thinner to the north in Scoresby Sund and Jameson Land Basin (Larsen *et al.* 1999, 2014). Based on age dating of exposed dykes that have similar ages to the plateau basalts (Larsen *et al.* 2014), it is inferred that the plateau basalts covered the Jameson Land Basin but were subsequently eroded away. The thickness of the eroded section is estimated at 2–3 km (Mathiesen *et al.* 2000).

The break-up igneous formations. During break-up (54–55 Ma), a magma-rich margin formed along the eastern flank of the JMMC with thick sequences of onlapping lava flows forming SDRs (Figs 4, 5 & 7).

Fig. 6. JMMC tectonostratigraphic type sections that are based on seismic reflection and refraction data interpretations, tied to shallow borehole data only. Possible Palaeozoic–Mesozoic formations of the JMMC are inferred from the structural and stratigraphic setting in comparison to the East Greenland analogue areas (Hamann *et al.* 2005). Seismic velocity models from refraction data are consistent with interpretation. Modified from Peron-Pinvidic *et al.* (2012*a*, *b*) and Blischke *et al.* (2014*b*. The volcanic margin is clear (see the sections in **a–c**) along the eastern flank of the microcontinent. The magmatic anomaly poor western margin that was formed during the second break-up (**d**) appears as a sharp boundary along the western margin of the microcontinent.

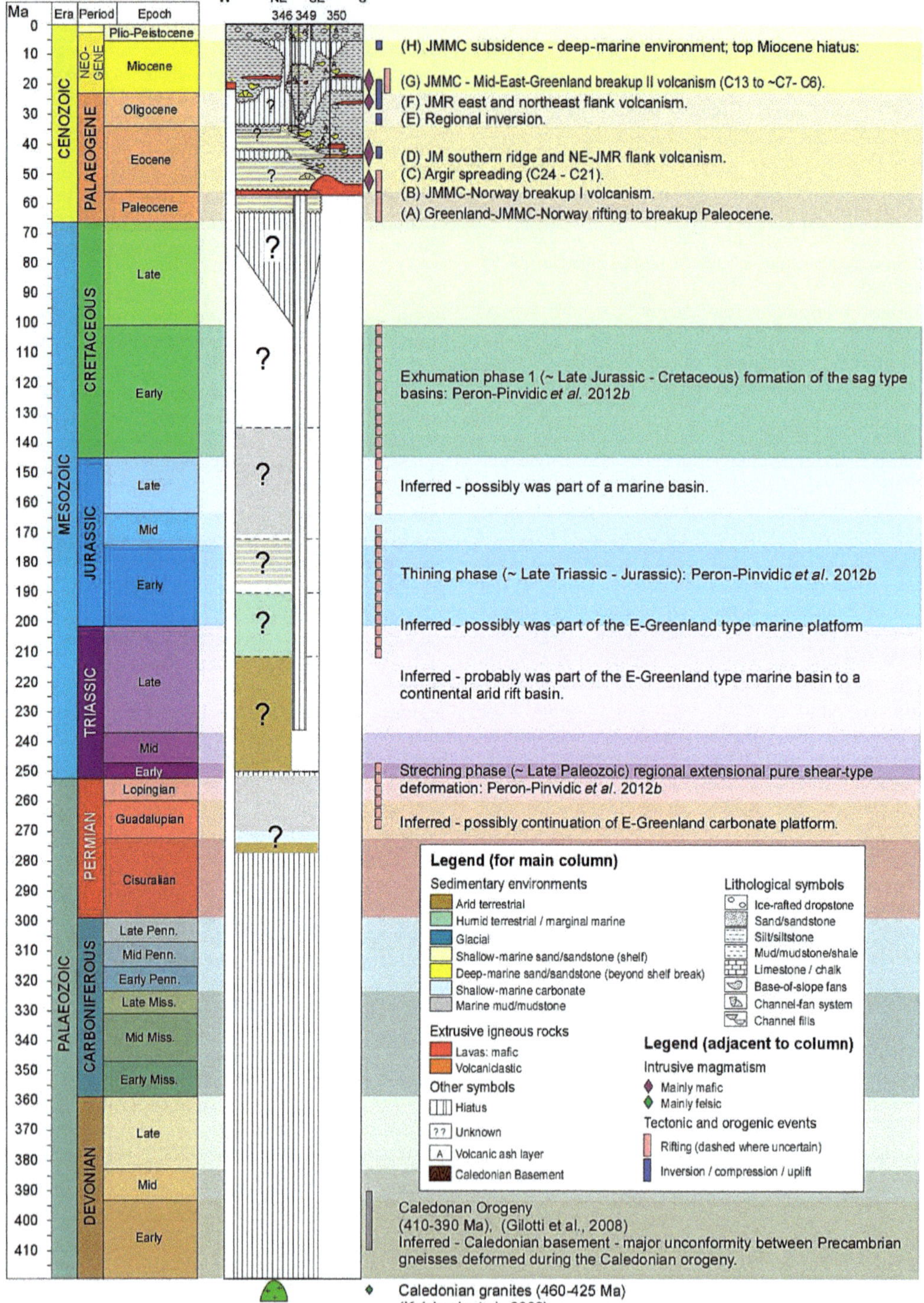

Fig. 7. Tectonostratigraphic chart of the JMMC region based on borehole data, seismic interpretation and analogue studies (Talwani & Eldholm 1977; Åkermoen 1989; Gunnarsson *et al.* 1989; Rey *et al.* 2003; Harðarsson *et al.* 2008;

These SDRs are considered to be subaerial lava flows onlapping higher terranes, stacking onto previous flows as rifting and seafloor spreading initiated along the margins (Hinz 1981; Mutter *et al.* 1982; Planke *et al.* 2000; Berndt *et al.* 2001). The SDRs along the eastern flank onlap westwards onto the crest of the main ridge and the SRC: they are up to 4–6 km thick at the NE end of the JMMC, thinning from north to south.

The Mid-Eocene–Neogene volcanostratigraphy

The Eocene sedimentary succession is intruded by many sills and dykes, especially along the eastern and SE flanks, coinciding with the Early–Middle Eocene time (49–44 Ma). During this period, igneous activity affected the entire southern extent of the microcontinent. The occurrence of a series of ridge jumps from east to west across the Iceland Plateau have been suggested by several authors (e.g. Gaina *et al.* 2009; Brandsdóttir *et al.* 2015) (Figs 5 & 9). The youngest substantial igneous event on the JMMC is expressed as a flat-lying, opaque reflection in seismic data, the so called 'F-Reflector' (Gunnarsson *et al.* 1989), which covers most of the northern Jan Mayen Basin, the western margin of the JMMC (Fig. 6b, c) and much of the Jan Mayen Trough (Figs 5 & 6d). This reflection is believed to correspond to regionally extensive composite sheets of flat-lying lava flows and intrusive rocks that covered the underlying structures in very shallow and unconsolidated wet sediment possibly during the Late Oligocene (28–22 Ma) (Gunnarsson *et al.* 1989). This corresponds to the time of plate boundary relocation from the Ægir mid-oceanic ridge to the Kolbeinsey mid-oceanic ridge (Gaina *et al.* 2009). No SDR type formations are observed along the JMMC western margin. Since the complete separation from the East Greenland margin, only the northern extent of the JMMC has been affected by volcanic activity, which is related to the present-day Jan Mayen Island volcanic system.

Kinematic reconstruction of the central NE Atlantic region

A series of detailed kinematic reconstructions for the JMMC tectonic blocks and surrounding areas is presented in Figures 10 and 11. Plate reconstruction parameters for the relative motion of the JMMC and conjugate margins were calculated using an interactive fitting method using GPlates (http://www.gplates.org: Boyden *et al.* 2011; see also Gaina *et al.* this volume, in review). Rotation parameters for Greenland relative to Eurasia are based on Gaina *et al.* (this volume, in review). The geographical extent of the individual JMMC tectonic blocks was guided by the interpretation of Peron-Pinvidic *et al.* (2012*a*) and Gernigon *et al.* (2015).

The model includes six stages: (1) the pre-break-up stage ending at 56–55 Ma; (2) the break-up stage at chron C24n2r (53.36 Ma) equivalent to chron C24B of Gunnarsson *et al.* (1989), associated with the formation of a wide volcanic margin; (3) an early intra-JMMC rifting phase around C22n (49.3 Ma); (4) a fully established intra-JMMC rift phase and the beginning of rift transfer from east to west around C21n (47.33 Ma); (5) westwards rift transfer and the initial western JMMC margin break-up phase around C13n (33.1 Ma); and (6) the complete isolation of JMMC by establishing the Kolbeinsey mid-oceanic ridge around C6b (21.56 Ma).

Structural elements included in plate tectonic reconstructions

We combined magnetic and gravity anomaly data interpretations with other structural, geological and geophysical data for defining several distinct structural elements that constitute independent

Fig. 7. (*Continued*) Gilotti *et al.* 2008; Kalsbeek *et al.* 2008; Gaina *et al.* 2009; Erlendsson 2010; Gernigon *et al.* 2012, 2015; Peron-Pinvidic *et al.* 2012*a*, *b*; Stoker *et al.* 2016). The chronostratigraphic scheme is based on Gradstein *et al.* 2012. The pre-break-up stratigraphic section is inferred from seismic data and analogue comparisons, primarily the Jameson Land Basin, East Greenland. (**a**) Greenland–JMMC–Norway rifting to break-up. (**b**) JMMC–Norway break-up I volcanism: emplacement of the plateau basalts, SDR, dyke complexes. (**c**) Aegir spreading (C24–C21): extension along the JMMC– mid-East Greenland rift. (**d**) JM southern ridge and NE–JMR flank volcanism. (**e**) Regional inversion causing an erosional hiatus across the main ridges, localized erosion into surrounding lows as marine fan–turbidite deposits, and minor reverse faulting. (**f**) JMR east and NE flank volcanism. (**g**) JMMC–mid-East Greenland break-up II volcanism (C13; *c.* C7–C6): the formation of composite sheets of flat-lying, shallow intrusions and lavas into shallow soft sediment (the F-reflector), and fault-connected dykes and sill intrusions simultaneously with the establishment of oceanic crust from the Kolbeinsey mid-oceanic ridge and formation of the Jan Mayen Basin. At this time, erosion along the Southern Ridge Complex occurred. (**h**) JMMC subsidence and establishment of a deep-marine environment. The Kolbeinsey mid-oceanic ridge continues to establish itself as the main spreading centre. Sedimentary sequences include ash layers. The top Miocene hiatus was possibly caused by regional uplift in conjunction with west to east migration of the main rift axis on Iceland.

Table 3. *Reflection seismic interval velocity estimates used for building a time-depth conversion model based on refraction data interpretations*

Seismic intervals	OBS model (Kodaira *et al.* 1998)		OBS model (Kandilarov *et al.* 2012)	Velocity data used for time-depth conversion	Estimated stratigraphic thickness across the JMMC	Estimated stratigraphic thickness OBS model Figure 4 intersection
	P-wave model layer	V_P (km s^{-1})		V_P (km s^{-1})	Thickness ranges (km)	TST estimate (km)
Seabed	Water depth		–	1.49	0–3.2	0.95
Plio-Pleistocene	Cenozoic sediments	2.0–3.5	1.7–2.2	1.8	0.1–1.1	0.06
Late Oligocene–Early-Miocene				2.0	0–1.05	0.09
Early Oligocene			2.2–3.2	2.3	0–2.05	0.29
Late Eocene						
Mid-Eocene				3.0	0–2.84	0.65
Basalts–Early Eocene	Basalt (SDR)	4.0–5.0	3.2–4.1	4.0	0–6	1.48
Early–Mid Paleocene	Possible Lower	3.9–4.7	3.9–5.0	4.4	0–4.5	2.25
Possible Mesozoic	Paleocene–Mesozoic					
Possible Palaeozoic	Possible Palaeozoic	5.0–5.3	5.0–5.5	5.2	0–4	3.82
Estimated stratigrahic and igneous section thickness above Caledonian basement:						8.64
Possible Caledonian Basement	Continental upper crust	5.5–6.7	5.5–6.5	5.6	0–15	2.79
Continental lower crust – sub-basalt	Continental lower crust	6.7–6.8	6.5–7.2	6.8	0–10	2.96
Estimated stratigrahic, igneous section, and crustal stratigraphic thickness:						14.4

Displayed are the possible stratigraphic thickness ranges across the highly variable mapped JMMC, and one specific stratigraphic thickness profile for the reflection seismic data intersection with the OBS model by Kodaira *et al.* (1998).

kinematic model blocks (isochrons and rotation model, see Gaina *et al.* this volume, in review). Interpreted fault and transfer systems are linked to stratigraphic thickness changes, mapped unconformities, age data control from borehole data and other geological information. Reconstructions include the present-day coastline for better reference (Figs 10 & 11).

We have compiled a number of structural lineaments in the JMMC area, which are based on published work, bathymetry, free-air gravity anomaly and derivatives, magnetic anomaly (Hopper *et al.* 2014), and seismic reflection data. These proposed structural lineaments (Fig. 1a) are based on features that can be inferred from at least two, and preferably three, different potential datasets.

We consider several observations that may give information on how the JMMC evolved during the multiphased break-up. Along the eastern margin, Palaeogene rocks dip steeply towards the Norway Basin (Fig. 6) and exhibit normal faulting associated with rapid subsidence within the northern and southern eastern-flank basin (EFBN and EFBS in Fig. 8; Table 1), in association with SDR emplacement and the early establishment of the Ægir mid-oceanic ridge system. The western margin, along the Jan Mayen Ridge North, the Sörlahryggur Flank Basin and the Sörlahryggur Ridge (JMRN, SFB, SHR in Fig. 8; Table 1), displays a more gentle west-facing listric normal fault system (Fig. 6), where rotated crustal blocks are downfaulted towards the Jan Mayen Basin (JMB in Fig. 8; Table 1) along major detachment faults. This indicates a distinct phase of extension and basin formation before the final break-up of the JMMC to the west. In addition, some minor reverse faulting occurred along the SE segments within the SRC, including the Fáfnir Ridge, the Otur Ridge, the Otur Ridge southern spur, the Langabrún Ridge and the Dreki Ridge (FR, OR, ORS, LR and DR in Figs 6 & 8; Table 1), as a result of regional inversion during the Late Eocene–Early Miocene.

Stratigraphic unconformities represent major tectonostratigraphic markers. Several major and small-scale unconformities in the JMMC stratigraphy have been described (Figs 3, 4 & 7). Three unconformities are present within Eocene sections: (1) an Early Eocene main break-up unconformity at C24 (56–53 Ma); (2) a main Middle Eocene unconformity at C19–C20 (47–41 Ma) associated with the initiation of ridge transition from the Ægir mid-oceanic ridge to extension concentrated further west; and (3) an unconformity of the Late Eocene age at chron C15 (*c.* 35 Ma). Following these Eocene events, two major erosional events affected the microcontinent during the Oligocene. An unconformity that marks a major truncation surface at 33 Ma (C12–C11) can be observed across all ridge areas of the microcontinent and correlates to a change in the seafloor spreading direction along the Ægir mid-oceanic ridge axis before the cessation of the mid-oceanic ridge (Gernigon *et al.* 2015) (Fig. 11d). This transtensional phase resulted in small-scale reverse faulting across the SRC. A second unconformity marks a major hiatus in the Late Oligocene, which corresponds to the complete cessation of the Ægir mid-oceanic ridge. The microcontinent was isolated completely, as the Kolbeinsey mid-oceanic ridge became fully established along the western margin at chron C6b (22–21 Ma) (Fig. 11e).

Pre-break-up stage ending around 56–55 Ma

Major structural elements and subdivisions of the JMMC at pre-break-up time align fairly well with published regional trends and lineaments interpreted on the NE Atlantic continental margins (Hamann *et al.* 2005; Tsikalas *et al.* 2005, 2008; Vogt & Jung 2009; Gaina *et al.* 2009, 2013; Gernigon *et al.* 2012, 2015) (Fig. 10). Three main trends are observed (Fig. 10): (1) a north–south trend similar to the strike-slip fault systems of the Shetland Islands, which is also aligned with the Jameson Land Basin axis and the main boundary fault of the Liverpool Land High; (2) an east–west trend parallel to the strike-slip fault system proposed by Guarnieri (2015), forming the northern limit of the Faroe–Shetland region; and (3) a SE–NW trend that separates the JMMC from the Vøring margin to the north and the Faroe Islands region to the south. The latter two subdivisions also form the boundaries of several gaps in the reconstructions shown by Gaina *et al.* (this volume, in review).

At pre-break-up time, the JMMC was most probably a 40–100 km-wide crustal fragment, with stratigraphic and crustal geometries corresponding to the conjugate central East Greenland margin (Gaina *et al.* 2009; Gernigon *et al.* 2015) (Figs 1a, 10 & 11). The JMMC was bounded to the north by the proto-Jan Mayen Fracture Zone (proto-JMFZ in Fig. 10). The northern segment of the microcontinent near the proto-JMFZ that could be mapped shows stratigraphic and structural similarities with the Scoresby Sund and Blosseville Kyst areas.

The NNW–SSE-orientated axis of the Jameson Land Basin terminates abruptly at the Blosseville Kyst (Engkilde & Surlyk 2003), with no major east–west-striking fault structures marking its southern boundary in the Scoresby Sund area. This indicates that a deep basin may continue south underneath the Blosseville Kyst. The kinematic models here suggest that the central and southern JMMC were attached to that part of East Greenland prior to break-up, with the NNW–SSE-striking Liverpool Land high lining up with

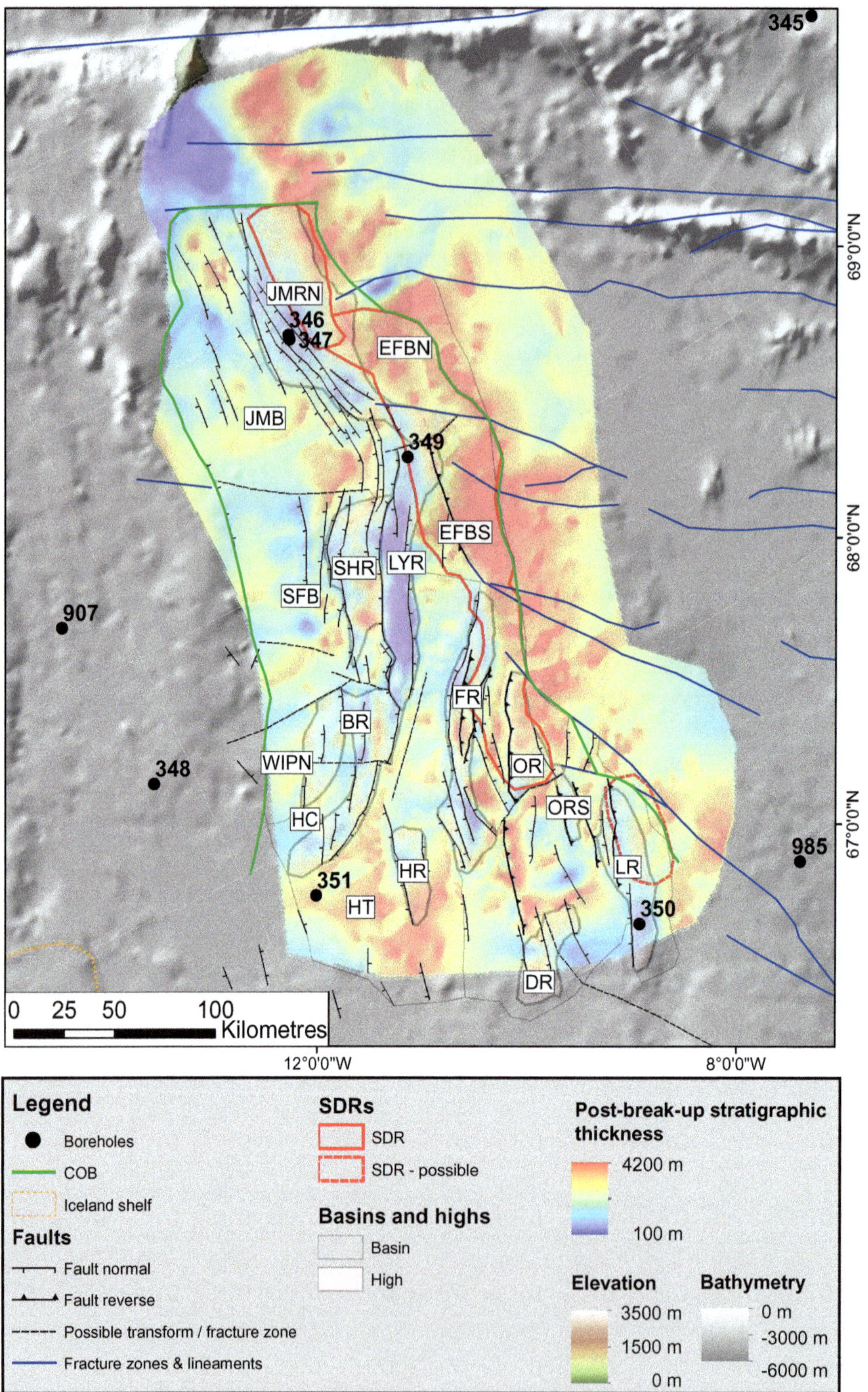
345
69°0'0"N
JMRN
346
347
EFBN
JMB
349
EFBS
68°0'0"N
SHR
LYR
SFB
907
FR
BR
WIPN
OR
348
ORS
67°0'0"N
HC
985
HR
LR
351
HT
350
DR
0 25 50 100
Kilometres
12°0'0"W
8°0'0"W
Legend
Boreholes
COB
Iceland shelf
Faults
Fault normal
Fault reverse
Possible transform / fracture zone
Fracture zones & lineaments
SDRs
SDR
SDR - possible
Basins and highs
Basin
High
Post-break-up stratigraphic thickness
4200 m
100 m
Elevation
3500 m
1500 m
0 m
Bathymetry
0 m
-3000 m
-6000 m

the northern Lyngvi Ridge, the central high of the JMMC (Fig. 8).

The southern boundary is less clear, but was probably influenced by the large-scale transform system proposed by Guarnieri (2015). This, in turn, was linked to the development of the Greenland–Iceland–Faroe Ridge Complex, a subdomain of the North Atlantic Igneous Province. The GIFRC forms a complex WNW–ESE-striking ridge structure that includes the Greenland–Iceland Ridge, the entire Iceland shelf and the Iceland–Faeroe Ridge.

The southernmost part of the microcontinent, where East Greenland links to the Faroe Platform and the Hatton Bank, remains far more uncertain owing to a lack of data constraints (Breivik *et al.* 2012; Brandsdóttir *et al.* 2015; Gernigon *et al.* 2015; Torsvik *et al.* 2015). Still, the stratigraphic mapping of the JMMC (event B in Fig. 7) suggests a potential link between the southern extent of the JMMC and the pre-break-up/break-up successions along the Blosseville Kyst of central East Greenland margin, the NW margin of the Faroe Platform and the northern edge of the Iceland Faroe Ridge (Figs 10 & 11a). The interpretation of seismic reflection data across the central part of the JMMC indicates two possible Early Eocene plateau basalt-equivalent sections that appear to increase in thickness from north to south (Fig. 4).

A gap in our pre-break-up reconstruction situated to the south of the JMMC is assumed to have been filled by either stretched continental crust (Torsvik *et al.* 2015) and/or pre-break-up formations of Palaeozoic–Early Paleocene age, similar to those known from onshore East Greenland and the Norwegian shelf margin (Brekke 2000).

Break-up stage around C24n2r (53.36 Ma)

The break-up stage (Fig. 11a and event (C) in Fig. 7) is marked by large-scale extrusive volcanism leading to the formation of the plateau basalts onshore East Greenland (Storey *et al.* 2007*a*, *b*). The plateau basalts mapped extend from East Greenland and across the Faroe Islands area. They have an estimated thickness of more than 6 km in East Greenland (Brooks 2011 and over 7 km at the Faroe Islands (Árting 2014).

The main structural elements of the JMMC are parallel to the overall trends of the basins and highs of the surrounding regions, except for the east Jan Mayen Fracture Zone, which appears to be linked to an initial rift centre just at the NW edge of the JMMC, as proposed by Gaina *et al.* (2009). This coincides with the formation of SDR sequences along the eastern margin, which reach a stratigraphic thickness of 4–6 km at the northeasternmost flank (Figs 4 & 5).

Early Eocene (53.36 Ma) volcanism marks the establishment of the Ægir mid-ocean ridge system at C24n2r (e.g. Gernigon *et al.* 2015), separating the mid-Norwegian Vøring and Møre basins from the Central East Greenland margin. Large igneous complexes were mapped on seismic reflection data along the eastern flank of the JMMC located close to fracture/fault zones (Figs 7a, b & 9). These complexes most probably formed after the initial emplacement of SDR sequences, cutting through that sequence and the initial oceanic crust, forming a 40–80 km-wide volcanic margin (Fig. 5) as Eocene sediments onlap and fill in those features.

Initial intra-JMMC rifting phase around C22n (49.3 Ma)

The initial intra-JMMC rifting phase at chron C22n (49.3 Ma; Fig. 11b) is marked by the southwards propagation of the Ægir mid-oceanic ridge and the establishment of a continuous spreading system in the Norway Basin. The EJMFZ extends from the Mohn's mid-oceanic ridge, which was established to the NE of the JMMC at C24 and separated the Vøring margin from the NE Greenland margin. A distinct rift segment can be seen along the SE margin of the JMMC, forming the eastern extent of the Iceland Plateau, here referred to as Iceland Plateau rift I (IPR-I: Fig. 11b, c). During the Lower Eocene, thick sediment wedges formed along the central eastern and NE flank of the microcontinent (Figs 7 & 8).

An uneven north to south change in spreading rate from intermediate to slow (Gernigon *et al.* 2015) led to oblique spreading within the Norway Basin, initiating a V-shaped mid-oceanic ridge structure and a counterclockwise rotation of the JMMC. This resulted in extension of the entire southern half of the microcontinent and the formation of small ridge segments. This extension widened the area across the southern half of the microcontinent from the original 100 km to approximately 150 km. This also explains the crustal

Fig. 8. Post-break-up stratigraphic thickness map includes estimates from the Lower Eocene (top SDR/flood basalts) to present day. The main ridges are heavily eroded and have only a very thin sediment cover. The low areas along the east flank show a very thick stratigraphic section. The thicker stratigraphic sections along the west flank of the ridge are related to the Jan Mayen Basin, as well as to several lows between the main structural highs. The stratigraphic thinning towards the Jan Mayen Island volcanic complex to the north and towards the south, and the area of borehole 350, where the Iceland Plateau Rift is proposed. For an explanation of abbreviations, see Table 1.

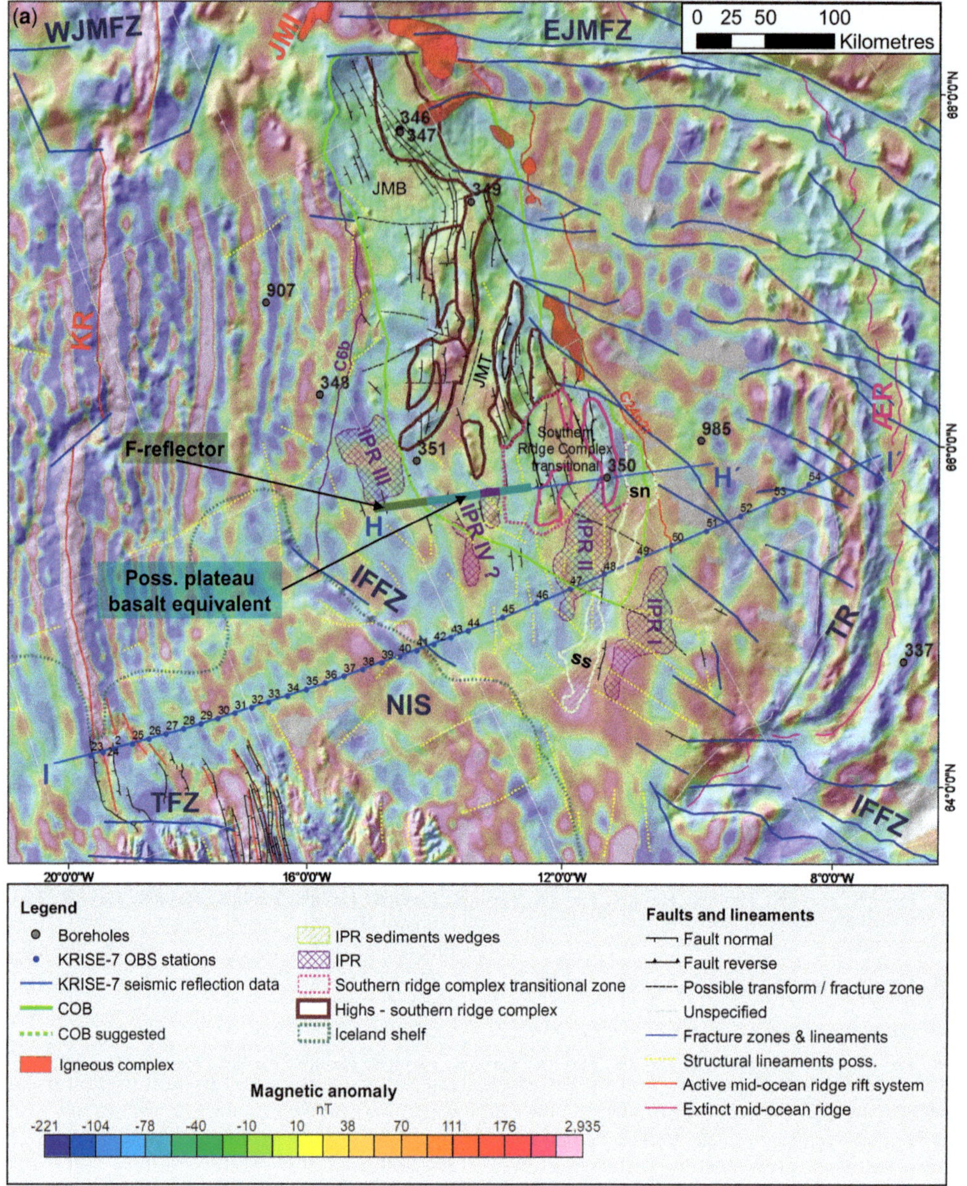

Fig. 9. Structure and volcanic elements map (**a**) of the JMMC and the Iceland Plateau. The proposed Iceland Plateau rifts I, II and III (IPR I, IPR II and IPR III) are interpreted on seismic reflection data. Gravity and magnetic anomalies are linked to segments of older microcontinent transitional crust 'Southern Ridge Complex transition', pre-break-up plateau basalts segments and youngest second break-up volcanics 'F-marker'. The Iceland Plateau rift II was described by Brandsdóttir *et al.* (2015) based on interpretation of seismic refraction data. The sediment wedges (sn and ss) correspond to observations on seismic reflection data (in b) and correlate to negative gravity anomalies west of the COB. Interpreted profiles.

thinning trend from north to south that has been observed on refraction data in the JMMC area (Kandilarov *et al.* 2012).

The Ægir mid-oceanic ridge system did not directly link up with the Reykjanes mid-oceanic ridge system to the south. Instead, the link to the Reykjanes mid-oceanic ridge was marked by a complex system of transforms and off-ridge volcanic systems that formed an Iceland-type oceanic crust within the proto-GIFRC (Árting 2014). In order to

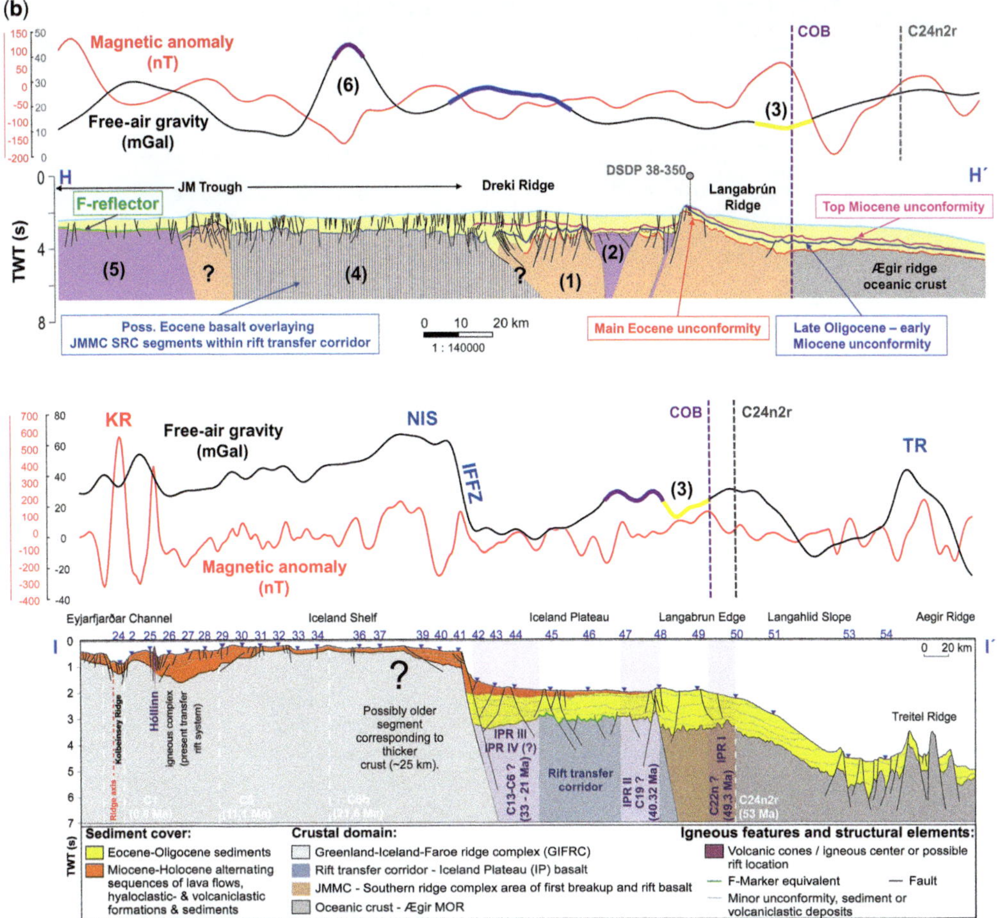

Fig. 9. (**b**) I–I′ from Brandsdóttir *et al.* (2015) and H–H′ across the SRC. The H–H′ profile based on TGS seismic reflection data crosses the southernmost area where the SRC is clearly observed and ties to DSDP site 350. Section H–H′ has several areas marked to explain the subdivision of the mapped structures: (1) The Dreki Ridge transfer system (possibly active from the Mid-Eocene to Oligocene with Miocene infill). (2) A graben structure that separated the Eocene ridges of the SRC. The strike of the graben is parallel to the Iceland Plateau rift II structure. (3) A sediment wedge that is thickest close to the COB. This is observed along the entire eastern edge of the JMMC and helps to define the COB location. (4) Areas where deep faulting appears absent and structures are distinctly different from elsewhere. A basalt cover is indicated, possibly equivalent to the plateau basalts. (5) A typical section where the F-reflector is observed and is connected to IPR III. (6) A positive gravity anomaly that might be related to another possible rift complex, but not confirmed on seismic reflection data. A clear signature on the magnetic anomaly data (Fig. 9a) south of intersection H–H′ is observed, however. Section I–I′ shows the Iceland Plateau rift between the Langabrun Edge and the Iceland Shelf. It includes a segment of thick crust between the C6 magnetic chron just west of the youngest rift system on NE Iceland (the Eyjarfjarðar Channel/Kolbeinsey mid-oceanic ridge).

maintain a spatial balance within the reconstruction, the areas of the Iceland Plateau and East Iceland must have been involved in this early stage of break-up, forming new basaltic crust and possibly SDR overlying older crust segments that remain as segments in-between off-mid-oceanic ridge volcanic segments (Erlendsson & Blischke 2013; Blischke *et al.* 2014*b*).

Fully established intra-JMMC rift phase and the beginning of rift migration around C21n: 47.33 Ma

The oblique seafloor spreading direction recorded by the Norway Basin oceanic crust caused large strike-slip/transfer fault systems that affected the eastern flank of the microcontinent (event D on

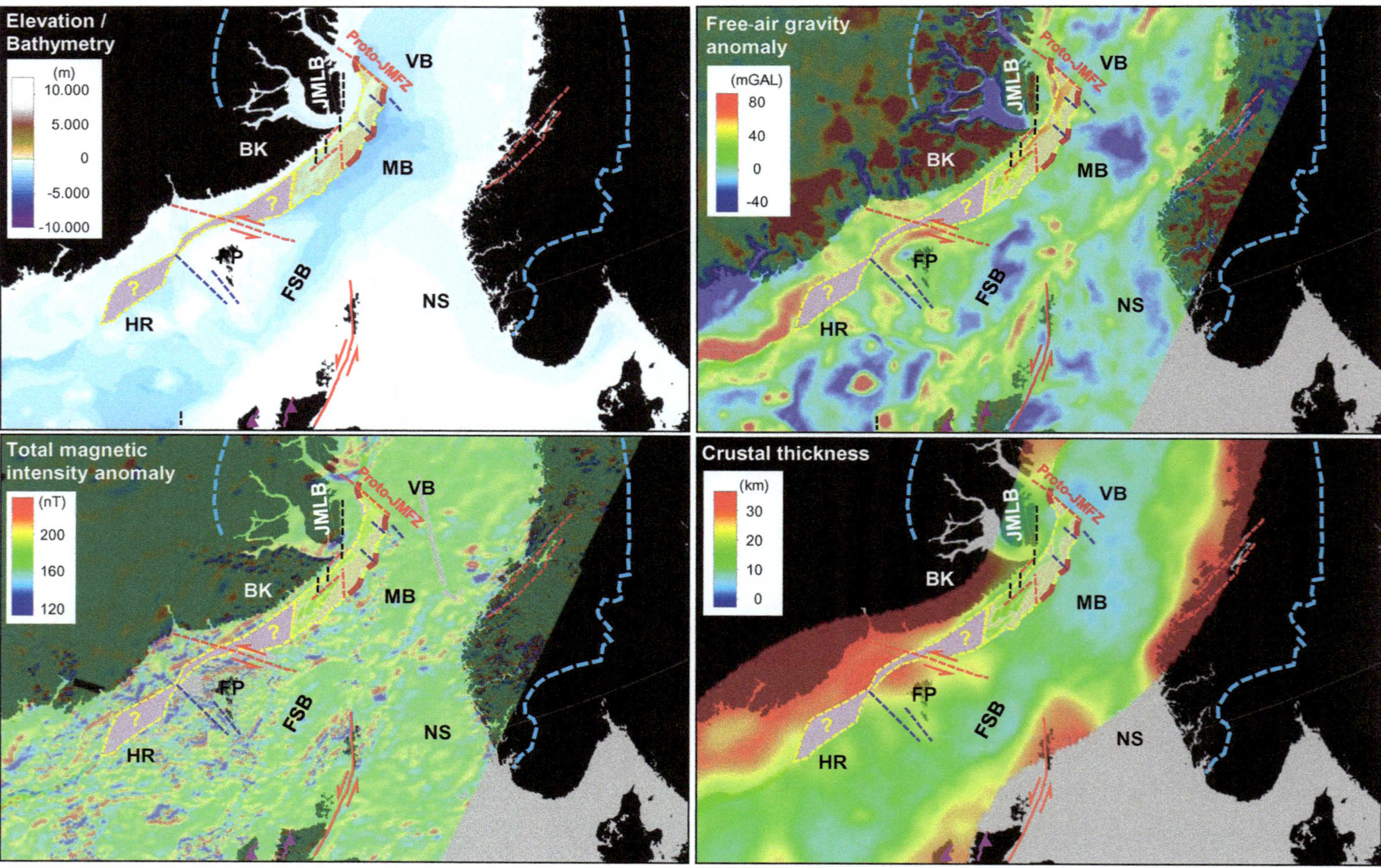

Fig. 10. *Continued*

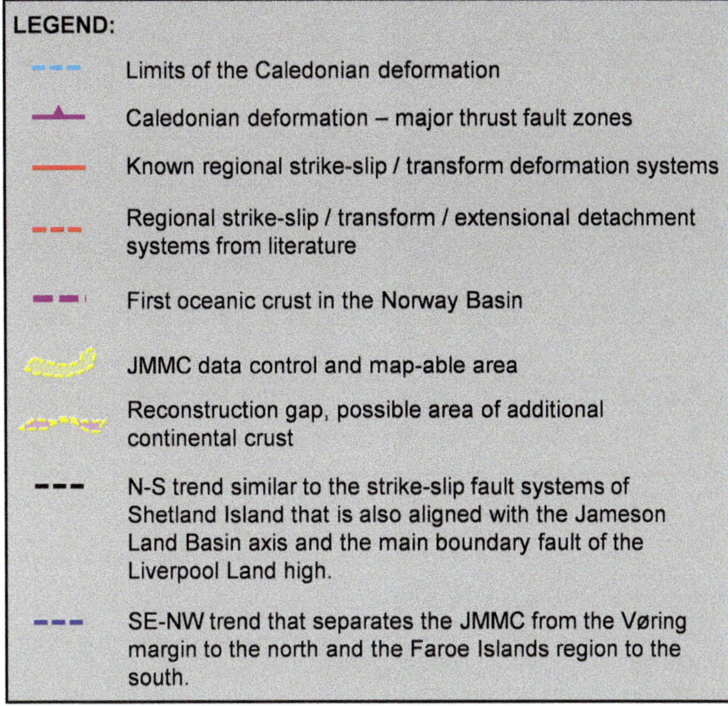

Fig. 10. Pre-break-up setting of the central NE Atlantic region, showing reconstructed present-day bathymetry, magnetic, gravity and crustal thickness data. Crustal thickness is based on the gravity inversion (Funck *et al.* 2014). The reconstruction is at pre-break-up stage ending at 56–55 Ma and is fixed to the European Plate. Features displayed are modified from data, and interpretations by Osmundsen & Andersen (2001), Torsvik *et al.* (2001), Foulger *et al.* (2005), Henriksen (2008), Gaina *et al.* (2009), Boyden *et al.* (2011), Peron-Pinvidic *et al.* (2012*a*, 2013), Gasser (2014), Hopper *et al.* (2014), Gernigon *et al.* (2015), Guarnieri (2015) and Torsvik *et al.* (2001, 2015). Regions marked are: BK, Blosseville Kyst; FP, Faroe Plateau; FSB, Faroe–Shetland Basin; HR, Hatton–Rockall margin and basin; JLB, Jameson Land Basin; MB, Møre Basin; NS, North Sea; VB, Vøring Basin.

Fig. 7) and subdivided the JMMC into the northern Jan Mayen Ridge and the SRC. The Ægir mid-oceanic ridge appears to terminate at the GIFRC (Fig. 11c).

A series of en echelon magnetic anomalies across the Iceland Plateau, referred to as the Iceland Plateau Rift System (Fig. 11c), is interpreted as marking the onset of a propagating rift system towards the Ægir mid-oceanic ridge (Figs 5 & 9), A NW–SE-striking fault system that links up to the Iceland–Faroe Fracture Zone and terminates the north–south fault trend of the SRC, marks the southern extent of the microcontinent. The NW–SE-striking fault system is in direct alignment with the volcanically active area of the Blosseville Kyst (F on Fig. 11c), where a major coast-parallel dyke swarm belonging to the Igtertivâ Formation magmatism is thought to have been caused by a regional extensional event at 49–44 Ma (Larsen *et al.* 2013). This event is also marked by a distinct unconformity, where the base consists of sediments that are dated to 49.09 ± 0.48 Ma and the top is formed by sediments intercalated with lava flows of the Bopladsdalen Formation, which are dated to 43.77 ± 1.08 Ma (Larsen *et al.* 2013). This Late–Middle Eocene time interval (49–44 Ma) coincides with an increase in observed sills and intrusions within the sediment stratigraphy (Fig. 8) of the JMMC area.

Westwards rift transfer and the initial western JMMC margin break-up phase around C13n: 33.1 Ma

By chron C13n (33.2 Ma), the Norway Basin seafloor spreading had changed from slow to ultraslow spreading (Gernigon *et al.* 2015), while, on

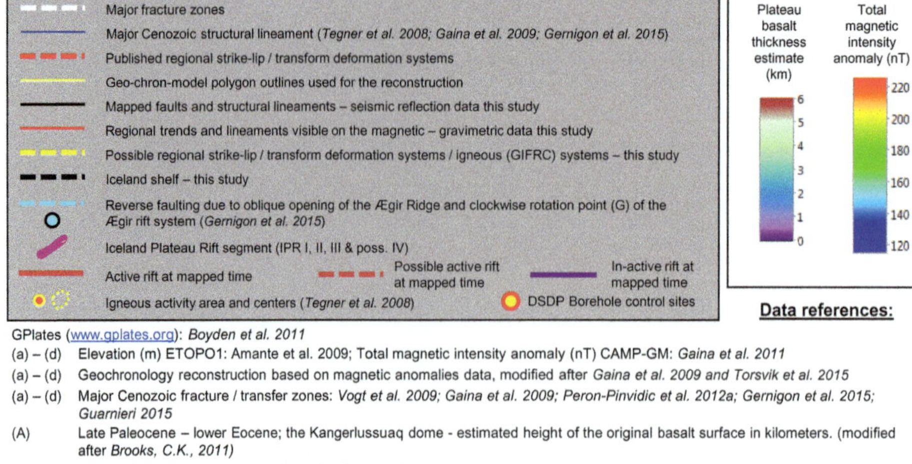

Data references:

GPlates (www.gplates.org): *Boyden et al. 2011*

(a) – (d) Elevation (m) ETOPO1: Amante et al. 2009; Total magnetic intensity anomaly (nT) CAMP-GM: *Gaina et al. 2011*

(a) – (d) Geochronology reconstruction based on magnetic anomalies data, modified after *Gaina et al. 2009 and Torsvik et al. 2015*

(a) – (d) Major Cenozoic fracture / transfer zones: *Vogt et al. 2009; Gaina et al. 2009; Peron-Pinvidic et al. 2012a; Gernigon et al. 2015; Guarnieri 2015*

(A) Late Paleocene – lower Eocene; the Kangerlussuaq dome - estimated height of the original basalt surface in kilometers. (modified after *Brooks, C.K., 2011)*

(B) Iceland Plateau rift location modified after *Brandsdóttir et al. 2015*

(C) Initial Paleogene rifting north and along the NE JMMC: *Gaina et al. 2009*

(D) Seaward dipping reflectors (SDR): *Thordarson, T. & Larsen, G. 2007; Berndt et al. 2001*

(E) Regional strike – slip fault / transfer system: *Gaina et al. 2009; Guarnieri 2015*

(F) Intrusive rift activity within the Blosseville plateau basalt formations at Kap Dalton (49 – 44 Ma) and alkaline sill intrusions (36 – 34 Ma): *Larsen et al. 2013*

(G) Possible initial Paleogene rifting location and structural lineaments: *Tegner et al. 2008*

(H) Rift location at ca. 25 Ma: *Harðarsson et al. 2008*

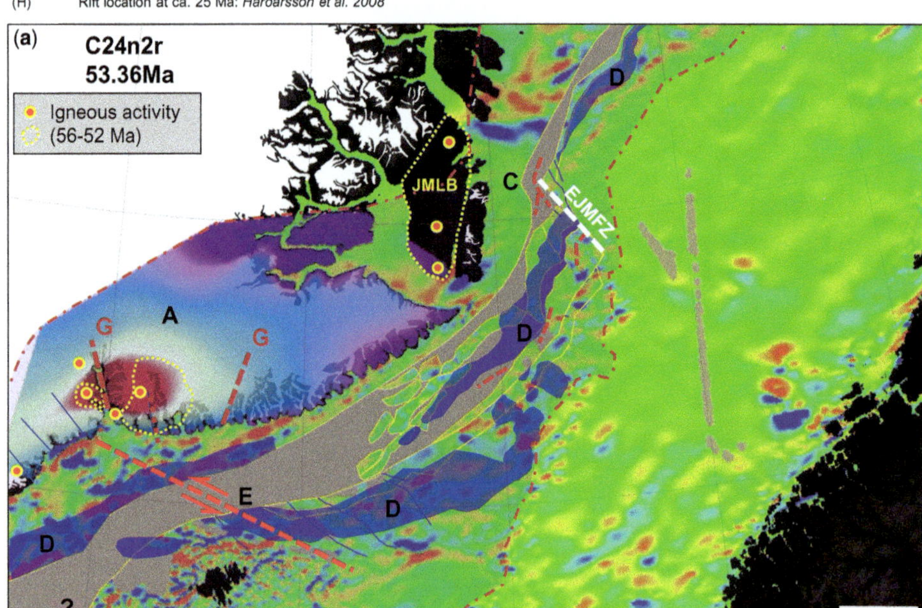

Fig. 11. Central NE Atlantic plate reconstructions relative to the European Plate: (**a**) break-up stage around chron C24n2r (53.36 Ma).

the western part of the JMMC, the proto-Kolbeinsey mid-oceanic ridge was forming. Two igneous complexes were identified on seismic reflection and gravity data, and are here referred to as the Iceland Plateau rift II and III systems (Figs 9 & 11d). Iceland Plateau rift II is located parallel to magnetic anomaly C19n (40.32 Ma) and forms the eastern shelf limit of Iceland (Brandsdóttir *et al.* 2015). Iceland Plateau rift III is observed on seismic reflection, gravity and magnetic anomaly data, and appears to have affected the nearby Hakarenna Channel (HC in Fig. 9). This area of the

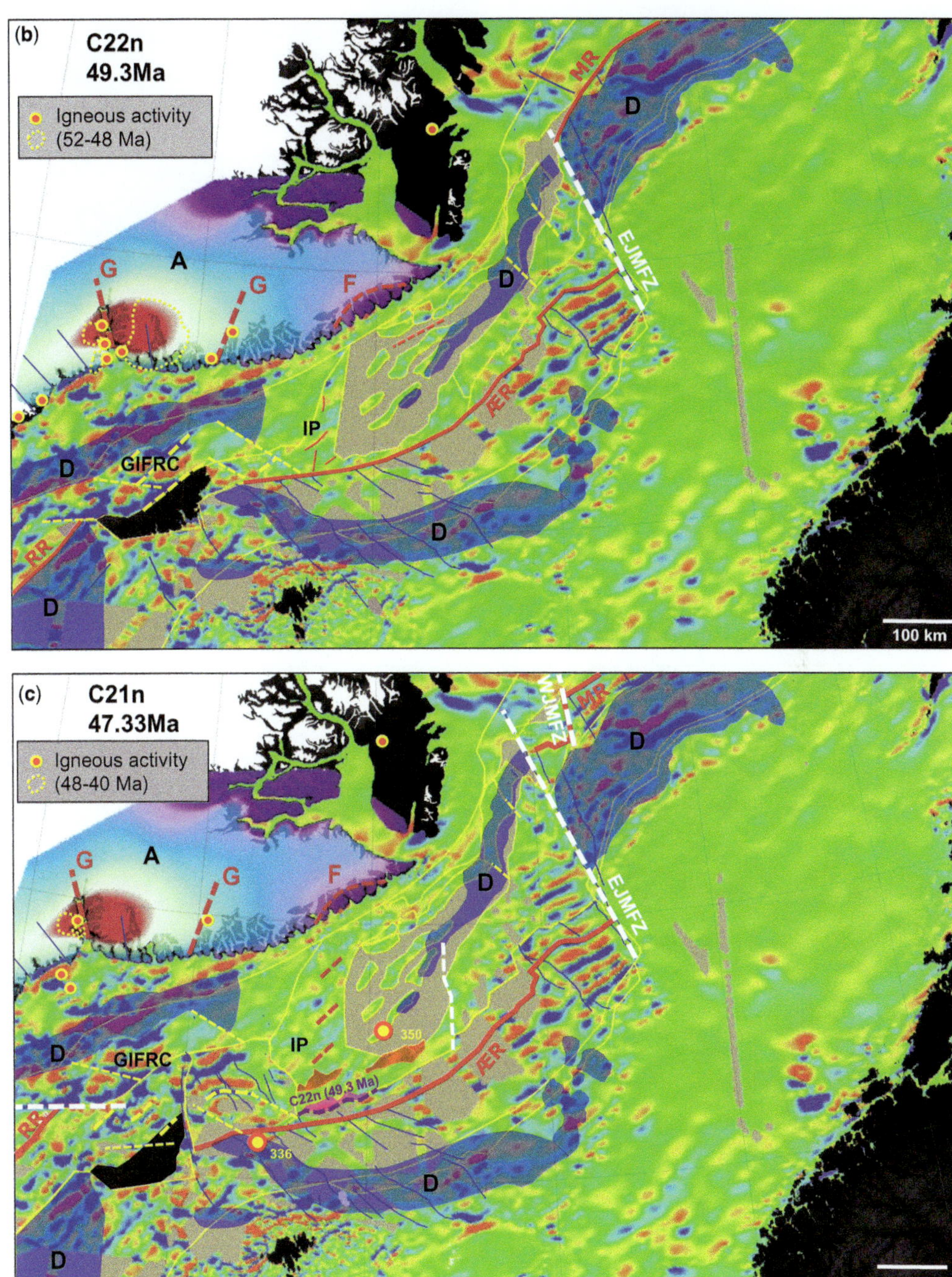

Fig. 11. (**b**) initial intra-JMMC rifting phase around chron C22n (49.3 Ma); (**c**) fully established seafloor spreading east of JMMC rift phase and beginning of intra-JMMC rift migration around chron C21n (47.33 Ma).

microcontinent was located parallel to the offshore Blosseville Kyst region at chron C13n, and lines up with observed offshore igneous centres (Árting 2014). This is interpreted as clear evidence that the rift transition had reached the SW corner of the JMMC by chron C13n.

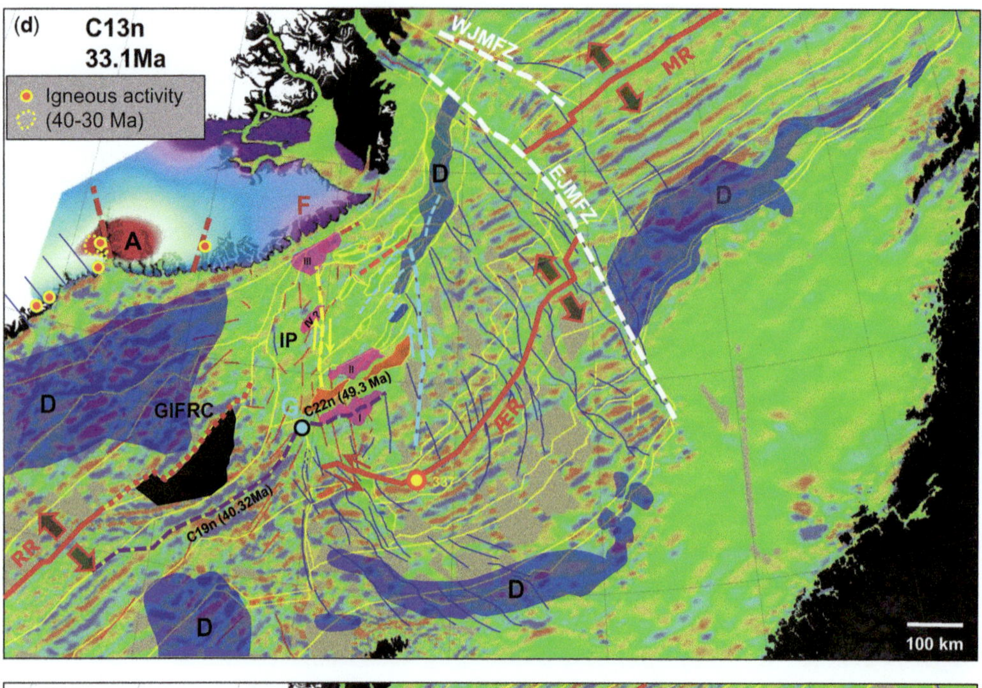

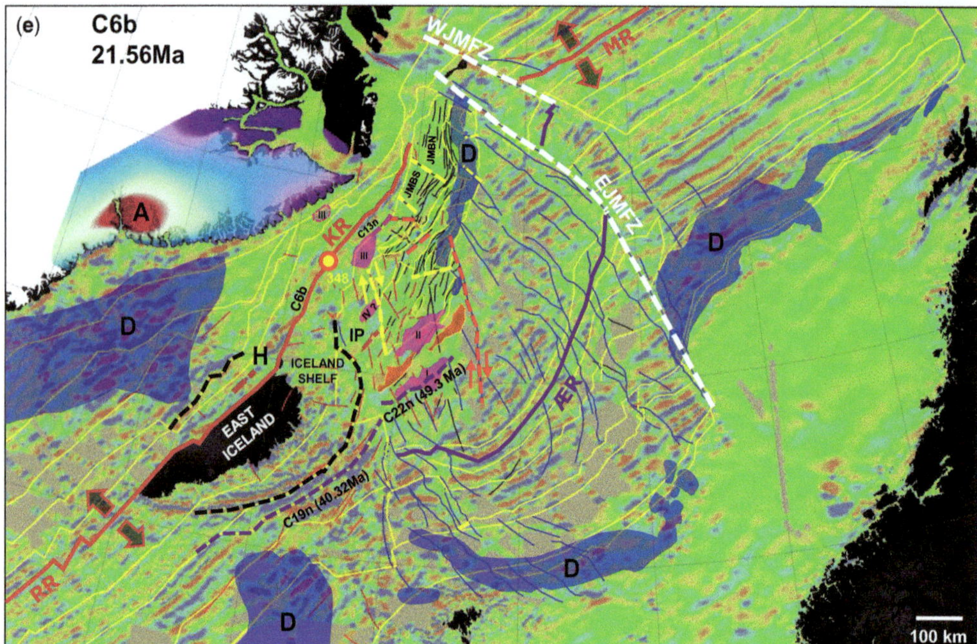

Fig. 11. (**d**) further westwards rift transfer and initial western JMMC margin break-up phase around chron C13n (33.1 Ma); and (**e**) complete isolation of the JMMC through the establishment of the Kolbeinsey mid-oceanic ridge around chron C6b (21.56 Ma). The kinematic reconstruction considers the JMMC as several independent tectonic blocks. Plate reconstruction parameters for the relative motion of the JMMC and conjugate margins were calculated using an interactive fitting method in GPlates (http://www.gplates.org: Boyden *et al.* 2011; see also Gaina *et al.* this volume, in review). The reconstructions include data interpretations for fault and transfer systems linked to stratigraphic thickness changes, mapped unconformities, age data from magnetic anomalies and borehole data. Present-day topography is included in the displays for a better reference to today's coastline.

During this phase, the Jan Mayen Basin and Jan Mayen Trough (HT in Fig. 8; Table 1) both developed as a consequence of Middle Eocene and Late Oligocene–Early Miocene extension prior to the establishment of the Kolbeinsey mid-oceanic ridge (Peron-Pinvidic *et al.* 2012*b*). The Jan Mayen Basin was initiated as a series of small pull-apart basins along the NW flank of the JMMC (Fig. 11d, e). These small basins spatially compensated for the rapid extension occurring during this second break-up phase (Fig. 6a, b). Rotation and compression along the SE JMMC, initiated in the previous phase, probably continued until the cessation of seafloor spreading in the Norway Basin. This resulted in inversion structures that are also observed in the stratigraphic record in the form of the Lower Oligocene unconformity across the microcontinent's ridges (Figs 3 & 6, event E in Fig. 7).

Extension of the SRC widened the area to about 310 km. The nature of the basement at depth still remains unconstrained, but can be interpreted in terms of highly stretched and probably very thin continental crust. This thin crust is covered by thick basalt formations emplaced since break-up. During rift transfer, the area was intersected by the Iceland Plateau rift systems that separated the stretched composite crust into segments, forming a transitional crust between continental crust and the oceanic crust of the Iceland Plateau. These basalt formations are probably Late Oligocene–Early Miocene and filled the topographical lows between the main structures along the Jan Mayen Ridge. They are indicated by the F-Marker on seismic reflection data (Figs 5 & 6).

Complete isolation of the JMMC by establishing the Kolbeinsey mid-oceanic ridge around C6b: 21.56 Ma

The final separation of the JMMC occurred around C6b at 21.56 Ma (e.g. Gernigon *et al.* 2015) (Fig. 11e), when the Kolbeinsey mid-oceanic ridge reached the JMFZ and the plate boundary in the Greenland Sea. This phase coincides with the initiation of Iceland as an insular province and the formation of the GIFRC by increased igneous activity (Harðarsson *et al.* 2008). The offshore area NW of Iceland, and the onshore and offshore areas of East Iceland, began to form the Icelandic Plateau basalts continuously throughout the Late Miocene (Walker 1964; Sæmundsson 1979; Thordarson & Larsen 2007).

Rift propagations have been shown to occur in a NW direction within the Iceland Plateau–GIFRC area up to the Late Miocene (7.2–5.3 Ma). At around 7 Ma, a northeastwards oceanic rift relocation occurred (Jóhannesson & Sæmundsson 2009) from the Húnaflóa rift centre (HFR on Fig. 1) in the NW of Iceland to the NE volcanic zone. As a consequence of substantially increased volcanic activity, new land formed and the GIFRC has been documented to have been located above or close to sea level (Denk *et al.* 2011). Well data from the highest ridges of the JMMC show a Mid–Late Miocene hiatus with marine sedimentation continuing only from the Pliocene onwards (Fig. 3).

During the Quaternary, the JMMC has only been affected by occasional volcanic activity of the Jan Mayen Island volcanic system, located at the northern edge of the JMMC, and gravitational erosion from the escarpment areas of the steep ridge flanks. In the Pleistocene, several glacial events removed about 1 km of the Iceland Plateau basalt (Walker 1964). The opposite occurred for the area between the Iceland Shelf and the SRC, indicating that the JMMC was subsiding during this last phase, presumably due to lithospheric cooling.

Discussion

The central focus of this contribution is the detailed development of the Jan Mayen microcontinent and its relationship to the surrounding areas to enable a better understanding of the full extent of the continental crust and its age and history. Central goals include:

(1) Establishing a detailed tectonic and stratigraphic framework for the JMMC:
   (a) the pre-break-up section and the JMMC relationship to the surrounding areas, in particular the East Greenland and Norway margins;
   (b) the stratigraphic and igneous record during first break-up;
   (c) the stratigraphic and igneous record during mid-oceanic ridge transfer.
(2) To develop a detailed kinematic model of the JMMC from pre-break-up time to the present day, in particular with respect to the second break-up phase and the formation of a microcontinent.
(3) To assess the Iceland–Faroe Fracture Zone (IFFZ) and the southern extent of the JMMC, in particular the connection to the Greenland–Iceland–Faroe Ridge Complex along the IFFZ, and the Iceland Plateau.

(1a) Stratigraphic records of the pre-break-up section

Where the JMMC type section intersects the seismic refraction profile line 4 of Kodaira *et al.* (1998) (Fig. 4), the pre-break-up sedimentary section, defined as the interval between the basalt and acoustic basement, is approximately 6 km thick. The velocity model indicates that this interval is characterized by velocities of 3.9–5.3 km s^{-1} (Kodaira *et al.* 1998). Palaeozoic–Mesozoic sequences in the Jameson Land Basin along East Greenland also show seismic velocities in the range between 3.5 and 5.5 km s^{-1} (Fechner & Jokat 1996), assuming similar velocities along the central eastern flank of the microcontinent, where the pre-break sections appear thickest and may reach up to 9 km thick.

Overall, the Lyngvi Ridge seismic profile is similar in structural character to the Jameson Land Basin, which is 3–5 s deep at its centre and contains up to 12–16 km of pre-break-up sedimentary sequences (Henriksen 2008) with multiple unconformities, complex faulting patterns and deep intrusive events (Blischke *et al.* 2014*b*).

(1b) Stratigraphic and igneous records during first break-up

The Early Eocene flood basalts, SDRs and igneous centres along the eastern flank of the JMMC are located within an interpreted volcanic-transitional crust consisting of stretched continental crust modified by significant volcanism. The earliest clear oceanic crust produced by the Ægir mid-oceanic ridge spreading centre is indicated by the onset of regular magnetic anomalies (chron C24n2r in Fig. 5) (Gernigon *et al.* 2015). The average thickness of oceanic crust in the Norway Basin is about 5.3 km ± 1 km (Breivik & Mjelde 2003), in contrast to the much thicker 6–10 km (Kandilarov *et al.* 2012) oceanic crust close to the Jan Mayen Island volcanic complex (JMI) and the west Jan Mayen Fracture Zone (WJMFZ in Fig. 5). The crustal thickness of the central area of the microcontinent (Fig. 6b) is 15–18 km based on refraction data (JMKR-95 in Figs 2 & 5), whereas the continental crust under the main Jan Mayen Ridge, close to wells 346 and 347, reaches up to 20 km (Kandilarov *et al.* 2012).

The break-up-related basaltic layers interpreted here along the JMMC are assumed to be correlative to the central East Greenland margin. Nøhr-Hansen (2003) and Larsen *et al.* (2013) reported that the landwards flows (Planke *et al.* 2000) of the plateau basalts overlie Lower Paleocene sediments at Kap Brewster and Kap Dalton (KB and KD on Fig. 1a), and mark an erosional horizon interpreted here as equivalent to the break-up conformity of the central JMMC (Figs 3 & 4).

Further to the west, the Milne Land Formation (56.36 ± 0.25 Ma) of the main plateau basalts discordantly overlies Precambrian gneiss (Storey *et al.* 2007*b*), marking the break-up unconformity for the north Blosseville Kyst and Scoresby Sund region. At break-up time (*c.* 55 Ma), this region was located approximately 400 km to the NE of the Faroe Island Plateau basalts.

The Faroe Islands, which have been covered by more than 7 km-thick landward basalt flows (Passey & Jolley 2009; Passey & Hitchen 2011; Árting 2014), is conjugate to the Kangerlussuaq Basin, located at the southernmost extent of the Blosseville Kyst. This plateau basalt succession has often been assumed to have had a similar stratigraphic thickness across the entire area further to the north, which would have included the proto-Jan Mayen microcontinent area. Seismic reflection data (Fig. 4; Table 3), however, indicate a thinner basaltic section of approximately 1.1–1.5 km over the central part of the JMMC. As noted earlier, the Jameson Land Basin may have been covered by thick flood basalts and 2–3 km of basalt may have been removed (Mathiesen *et al.* 2000). The same may have occurred here. By inference, the flood basalts associated with SDR formation probably covered the entire JMMC and may have been continuous with the flood basalts of the Blosseville Kyst area.

(1c) Stratigraphic and igneous records during mid-oceanic ridge transfer

Along the Blosseville Kyst area, the lower Igtertivâ Formation (C22: *c.* 49 Ma) coincides with the beginning of rift transfer away from the Ægir mid-oceanic ridge (F in Fig. 11b), which is followed by a hiatus between the lower (C22–C21: 49–47 Ma) and the upper Igtertivâ Formation (C20: *c.* 44 Ma) (Larsen *et al.* 2014). As only the Mid-Eocene (chron C20) unconformity (Fig. 3) is observed along the eastern flank of the microcontinent, it is assumed that the area was again above sea level from 49 to 44 Ma, eroding all pre-chron C22 deposits. The rift transfer processes may have contributed to some thermal uplift along the SW and southern flanks of the microcontinent, accompanied by emplacement of igneous complexes and sill intrusions primarily into the Lower Eocene strata along the NE and SE flanks of the JMMC. This is also seen in borehole data (DSDP 38-350: Fig. 3), and on seismic reflection and refraction data (Figs 4–6).

Along the southern flank of the microcontinent, increased magmatism most probably coincided with volcanism within the Greenland–Iceland–Faroe Ridge Complex region. The Eocene sediment succession shows many intrusive sills and dykes, especially along the eastern and SE flanks. These intrusions are primarily located within the Lower Eocene sediment sequences, possibly coinciding emplacement within the Mid-Eocene time interval (49–44 Ma). This time interval correlates well with the increased igneous activity observed along the East Greenland coast (Larsen *et al.* 2014) and may indicate a regional event, and, furthermore, may explain the major unconformity that has been observed along the JMMC in the Mid-Eocene (Figs 3, 4 & 7).

(2) The second break-up phase and the formation of a microcontinent

This second break-up phase between the western edge of the JMMC and the central East Greenland margin is most probably a magma-starved break-up due to the lack of SDR sequences and large-scale magmatic activity. A gradual rift propagation is observed beginning at chron C21 (Fig. 11c) accompanied by large-scale extension of the SRC, crustal thinning across the Iceland Plateau and a listric normal faulting along the western flank (Fig. 5). Extension rates were probably very small, consistent with a reduced magma supply. Lundin *et al.* (2014) suggested that this is likely in areas near the tip of a propagating rift and eventual normal oceanic-crust accretion.

The youngest regionally extensive igneous event indicated on seismic reflection data on the JMMC (Gunnarsson *et al.* 1989) is referred to as the 'F-Reflector', and covers an area of approximately 18 400 km^2 along the western and the SW to southern flanks of the Jan Mayen Ridge and within the Jan Mayen Trough (JMT in Fig. 5). This igneous formation is interpreted as shallow-marine landwards flows emplaced during chrons C13–C6b (33–21.56 Ma), possibly sourced from fissure-type volcanic complexes south and west of the microcontinent. Small lava deltas located on the SW extent of the JMMC on seismic reflection data indicate south–north to SW–NE flow directions.

During the second break-up event, the central East Greenland region was most probably the main sediment source, along with the Jan Mayen ridges and highs. The Jan Mayen Basin, including local low areas along the SW flank of the microcontinent, was filled with sediments sourced from the west as the microcontinent separated from the East Greenland margin (Fig. 8; Table 1). The southern area of the microcontinent towards the Iceland Plateau must have been elevated from the Mid-Oligocene, as the overall sediment stratigraphic thickness decreased from north to south and the Late Oligocene unconformity (Fig. 3) is observed on highs of the SRC (Fig. 6c).

After the Kolbeinsey mid-oceanic ridge had completely separated the microcontinent from East Greenland, the sediment supply was greatly reduced. Intra-Neogene unconformities within the oceanic sediments are observed away from the JMMC, and the occurrence of mounded onlapping sediment packages are observed on the flanks of the Jan Mayen Ridge. These attest to processes of erosion and deposition associated with deep-water bottom currents, which were common around the NE Atlantic region from the Mid-Miocene onwards (e.g. Bohrmann *et al.* 1990; Howe *et al.* 1994; Davies *et al.* 2001; Stoker *et al.* 2005). Borehole information (Talwani *et al.* 1976*b*; Talwani & Eldholm 1977) and interpretation of seismic reflection data show that these sediment sequences are very thin, deep marine, and form thick contourite deposits. Beginning in the Mid-Miocene, the sediment supply direction was from the North Iceland Shelf (NIS) area to the south. Sediment was supplied into the Ægir to Kolbeinsey rift transfer corridor of the Iceland Plateau (Fig. 5) and into the Hléssund Trough (HS in Fig. 8; Table 1), the southernmost extension of the Jan Mayen Trough.

(3) The Iceland–Faroe Fracture Zone and the southern extent of the JMMC

To understand the development of the southernmost part of the JMMC, the present-day development of the onshore areas of Iceland are considered, where complex transfer zones link the Reykjanes and Kolbeinsey mid-oceanic ridges to the main spreading axis (Sæmundsson 1974; Magnúsdóttir *et al.* 2015). These transfer systems result in en echelon orientated volcanic ridge segments, here referred to as flank systems (e.g. the Snæfellsnes and Öræfajökull volcanic zones: e.g. Hards *et al.* 1995; Prestvik *et al.* 2001; Einarsson 2008; Jakobsson *et al.* 2008). The present-day spreading axis is apparently migrating east via ridge jumps (Árting 2014). These observations serve as an analogue to understand how the left-lateral Greenland–Faroe Transfer System described by Guarnieri (2015) might have developed in time (Figs 10 & 11).

This transfer system between the Norway Basin and Kolbeinsey mid-oceanic ridge system that passes south of the JMMC and north of the Iceland–Faroe Ridge has been described previously (e.g. Vogt & Jung 2009; Gernigon *et al.* 2015). The interpretations of the wide-angle data along KRISE Line 7 (Brandsdóttir *et al.* 2015) (Figs 2 & 5), which extends from the Kolbeinsey mid-oceanic ridge to the Aegir mid-oceanic ridge, are important in this context. The western part of the profile crosses the Kolbeinsey mid-oceanic ridge and the NIS, which is part of the GIFRC. The crustal thickness ranges from 12 to 14 km near the Kolbeinsey Ridge, increasing gradually across the Iceland Shelf up to 25 km. Crustal thickness decreases abruptly down to around 8 km across the Iceland Plateau corridor and across the NIS shelf break, with a major fault escarpment dipping NE. Within the volcanic transitional area of the Iceland Plateau Rift (IPR), the crustal thickness again increases to 12 km. The oceanic crust towards the Aegir mid-oceanic ridge is relatively thin, at only 4–5 km thick. A domain characterized by velocity variations in lower-crustal structures across the Iceland Plateau is interpreted as an extinct spreading centre that is part of the Iceland Plateau Rift, which was active at the same time as the Aegir Ridge prior to the initiation of the Kolbeinsey Ridge. The spreading rate during that time decreased along the Aegir Ridge, as more and more of the extension was being taken up further west.

From the Iceland Plateau to the JMMC, a clear change in fault and lineament trends occurs based on the bathymetry and potential field datasets. These trends range from a north–south direction on the JMMC to a NW–SE trend on the Iceland Plateau. The latter trend is in alignment with the structural trend of the Iceland–Faroe Fracture Zone (IFFZ in Figs 1a & 5) and both trends correlate with magnetic anomalies. The junction of those two trends is suggested to mark the most likely southern boundary of the microcontinent as a structural entity. The boundary is probably a volcanic transitional-type crust that incorporates slivers of continental crust along with formed new volcanic crustal accretion. Gernigon *et al.* (2015) proposed a major SE–NW regional dextral strike-slip system from the Ægir mid-oceanic ridge to the centre of the microcontinent at the northern limit of the SRC. This system lies parallel to the Iceland Plateau corridor and the Iceland–Faroe Fracture Zone. In the Dreki Ridge area (Fig. 9a, b), where the segmentation of the SRC is clearly visible, a subdivision of the lineaments can be made. Here, distinct segments are likely to relate to pre-break-up segments of continental crust. Segments of possible Lower Eocene plateau basalts are intersected by possible oceanic crust. If the Iceland Plateau corridor represents a broad dextral SE–NW strike-slip fault zone, then the minimum horizontal stress lies approximately east–west, allowing faults to open and propagate in a north–south direction, which is parallel to the magnetic lineation in that area. These transtensional oblique rift systems were volcanically active and may be similar to the oblique rift segments observed today where the Reykjanes Ridge connects across Iceland to the Eastern Volcanic Zone (Clifton & Schlische 2003; Clifton & Kattenhorn 2005), and within the Northeastern Volcanic Zone (Khodayar 2014) on Iceland (NEVZ on Fig. 1a).

Conclusions

The objective of this study was to construct a detailed tectonostratigraphic history of the Jan Mayen microcontinent with a focus on the southernmost area. This was then integrated into kinematic reconstructions of the central NE Atlantic to better understand the Cenozoic development and the implications for the pre-Cenozoic development of regional rift basins, remnants of which probably underlie the JMMC. Complex structural patterns are observed along the microcontinent's margins, as well as the conjugate East Greenland and Norwegian margins on either side. The new model includes a description of how the southern JMMC structural elements were linked to tectonic features on the Iceland Plateau and Greenland–Iceland–Faroe Ridge Complex.

Mapping the pre- to post-break-up sedimentary strata and igneous complexes together with volcanostratigraphic seismic characterization has facilitated a reassessment and clearer definition of the igneous v. sedimentary domains of the JMMC area throughout its break-up history. The main results include:

- Interpretation of new and vintage geophysical data suggests that a significant pre-Palaeogene stratigraphic history is preserved. However, without deep borehole data, the age of possible sedimentary successions are speculative. Nevertheless, the conjugate Jameson Land and Møre basins are considered to be direct analogue areas for the JMMC. These basins are well constrained, and contain a sedimentary succession that includes Devonian continental sediments, Permo-Triassic continental and marine sequences, Jurassic and Cretaceous shallow- to deep-marine sequences, and Lower Paleocene alluvial to shallow-marine sediments.
- The break-up and post-break-up igneous sequences were separated into plateau basalts of probable Paleocene–Early Eocene age, seawards-dipping reflector sequences, igneous complexes, and sill and dyke intrusions along the flanks of the JMMC.
- A consistent kinematic model for the Cenozoic evolution of the JMMC and surrounding oceanic crust that consists of six main phases is proposed. The boundaries between these phases correlate to major unconformities and related structures. Important events include:
  (1) A pre-break-up stage ending at 56–55 Ma and the emplacement of Lower Eocene plateau basalts across the microcontinent and the Blosseville Kyst region, with an apparent thickening of the basalt sequences to the south, possibly continuing into the Faroe–Iceland–East Greenland corridor. The structures of the JMMC are consistently orientated with major structural lineaments of the surrounding regions prior to break-up. The main trends are aligned with the Jameson Land Basin and Liverpool Land high. The JMMC probably forms the southern extension of the Jameson Land Basin.

(2) A first break-up phase that began at 55 Ma (e.g. Gaina *et al.* 2009) and was associated with the formation of SDR along the east flank of the JMMC, followed by the initiation of seafloor spreading in the Norway Basin along the Ægir Ridge in the Early Eocene (chron C24n2r 53.36 Ma) (Gaina *et al.* 2009; Gernigon *et al.* 2015).
(3) An initial intra-JMMC rifting phase around chron C22n (49.3 Ma) and the establishment of a continuous spreading system in the Norway Basin and forming the eastern extent of the Iceland Plateau, here referred to as Iceland Plateau rift I (IPR-I: Fig. 11b, c). Initial extension of the entire southern half of the microcontinent occurred, widening it from originally 100 km to approximately 150 km. This eventually lead to early stage break-up in the Iceland Plateau area, forming new Lower Eocene volcanic formations (volcanic breccia, intrusions and SDR sequences) along the SE flank of the Southern Ridge Complex (SRC).
(4) The initiation of the southern JMMC rift transition at chron C21n (47.33 Ma) contemporaneous with oblique seafloor spreading east of the JMMC, causing the formation of transform systems and uplift along the southern flank of the JMMC. Volcanic activity occurred along the NE margin of the Blosseville Kyst (Larsen *et al.* 2014).
(5) A westwards rift transfer and initial break-up along the western JMMC around chron C13n–33.1 Ma. Oblique mid-oceanic ridge relocation via a SE–NW en echelon rift system occurred from the southern extent of the microcontinent during the Early Oligocene. Significant volcanism affected the SW area of the JMMC, referred to here as the Iceland Plateau rift III, which can be linked to the Blosseville Kyst margin. Oblique extension occurred along the NW flank of the JMMC, resulting in the opening of the Jan Mayen Basin and a series of small pull-apart basins and igneous intrusions with little to no evidence of SDR formation.
(6) A second break-up phase at chron C6b (21.56 Ma) with complete cessation of seafloor spreading in the Norway Basin (Gernigon *et al.* 2015) and the establishment of the Kolbeinsey mid-oceanic ridge as the main mid-ocean spreading centre.

- The extension of the southern half of the JMMC has been quantified from the original 40–100 km width up to a width of 310 km during the Early Eocene. The SRC is overprinted by volcanic extrusive complexes that consist primarily of Early Eocene basalt flows, which are interpreted as being similar to the plateau basalts exposed along the Blosseville Kyst of Greenland. These are onlapped by clear and well-developed SDRs associated with the final opening of the Norway Basin. Multiple phases of intrusive events appear to have affected the eastern flank of the microcontinent during the Eocene and the southern part of the JMMC during the Late Eocene–Early Oligocene.
- The Iceland–Faroe Fracture Zone across the Iceland Plateau has been mapped as an en echelon transfer system from the Ægir Ridge to the Kolbeinsey Ridge. Detailed mapping of the southern extent of the JMMC supports the Gaina *et al.* (2009) model in which mid-oceanic ridge propagation occurred directly south of the microcontinent, beginning in the latest Early Eocene (49.3 Ma) and continuing throughout the Eocene. This formed at least three rift-flank systems on the Iceland Plateau. These rifts flanks are referred to as Iceland Plateau rift IPR-I, IPR-II and IPR -III (Brandsdóttir *et al.* 2015).

This work is part of a PhD project at the University of Iceland, the National Energy Authority of Iceland (Orkustofnun) and the Iceland GeoSurvey, with data permissions provided by: Spectrum ASA, TGS; the University of Oslo (UiO); the Norwegian Petroleum Directorate (NPD); the Bundesanstalt für Geowissenschaften und Rohstoffe (BGR); and GEUS (Geological Survey of Denmark and Greenland). This project benefitted from the NAG-TEC project. Support from the following industry sponsors of NAG-TEC is gratefully acknowledged (in alphabetical order): Bayerngas Norge AS; BP Exploration Operating Company Ltd; Bundesanstalt für Geowissenschaften und Rohstoffe (BGR); Chevron East Greenland Exploration A/S; ConocoPhillips Skandinavia AS; DEA Norge AS; Det norske oljeselskap ASA; DONG E&P A/S; E.ON Norge AS; ExxonMobil Exploration and Production Norway AS; Japan Oil, Gas and Metals National Corporation (JOGMEC); Maersk Oil; Nalcor Energy – Oil and Gas Inc.; Nexen Energy ULC, Norwegian Energy Company ASA (Noreco); Repsol Exploration Norge AS; Statoil (UK) Ltd; and Wintershall Holding GmBH. C.G. acknowledges support from the Research Council of Norway through its Centres of Excellence funding scheme, project number 223272. In addition, the authors would like to thank Dr Nina Lebedeva-Ivanova and Dr Asbjørn Breivik (CEED, UiO), and Dr Laurent Gernigon (NGU) for all their help, advice and patience throughout this project. Reviews from Dieter Franke, Erik Lundin and the editor, Martyn Stoker, greatly improved this manuscript.

References

Åkermoen, T. 1989. *Jan Mayen-ryggen: et seismisk stratigrafisk og strukturelt studium* (The Jan Mayen Ridge: a seismic stratigraphic and structural study.). Cand. scient. thesis, University of Oslo.

Amante, C. & Eakins, B.W. 2009. *ETOPO1 1 Arc-Minute Global Relief Model: Procedures, Data Sources and Analysis*. NOAA Technical Memorandum NESDIS NGDC-24.

Andersen, O.B. 2010. The DTU10 gravity field and mean sea surface. Paper presented at the *Second International Symposium of the Gravity Field of the Earth (IGFS2)*, 20–22 September 2010, Fairbanks, Alaska, USA.

Andersen, T.B. & Jamtveit, B. 1990. Uplift of deep crust during orogenic extensional collapse: a model based on field studies in the Sogn–Sunnfjord area of Western Norway. *Tectonics*, **9**, 1097–1111.

Árting, U. 2014. Regional volcanism. *In*: Hopper, J.R., Funck, T., Stoker, M., Árting, U., Peron-Pinvidic, G., Doornenbal, H. & Gaina, C. (eds) *Tectonostratigraphic Atlas of the North-East Atlantic Region*. GEUS, Copenhagen, Denmark, 223–290.

Berndt, C., Mjelde, R., Planke, S., Shimamura, H. & Faleide, J.I. 2001. Controls on the tectono-magmatic evolution of a volcanic transform margin: the Vøring Transform Margin, NE Atlantic. *Marine Geophysical Researches*, **22**, 133–152.

Blischke, A., Erlendsson, Ö. & Árnadóttir, S. 2014*a*. *CRUSMID-3D – NORDMIN Status Report 2014 – Crustal Structure and Mineral Deposit Systems: 3D-modelling of Base Metal Mineralization in Jameson Land and Nickel Mineralization in Disko-Nuussuaq*. ÍSOR-2014/056. Closed Report for the Geological Survey of Denmark and Greenland and the National Energy Authority of Iceland, 184–194.

Blischke, A., Hjartarson, A., Erlendsson, Ö., Árnadóttir, S. & Peron-Pinvidic, G. 2014*b*. The Iceland margin, Jan Mayen microcontinent, and adjacent oceanic areas. *In*: Hopper, J.R., Funck, T., Stoker, M., Árting, U., Peron-Pinvidic, G., Doornenbal, H. & Gaina, C. (eds) *Tectonostratigraphic Atlas of the North-East Atlantic Region*. GEUS, Copenhagen, Denmark.

Blystad, P., Brekke, H., Færseth, R.B., Larsen, B.T., Skogseid, J. & Tørudbakken, B. 1995. *Structural Elements of the Norwegian Continental Shelf, Part II. The Norwegian Sea Region*. Norwegian Petroleum Directorate Bulletin, **8**.

Bohrmann, G., Henrich, R. & Thiede, J. 1990. Miocene to Quaternary paleoceanography in the Northern North Atlantic: variability in carbonate and biogenic opal accumulation. *In*: Bleil, U. & Thiede, J. (eds) *Geological History of the Polar Oceans: Arctic Versus Antarctic*. NATO Science Series C, **308**, 647–675.

Boyden, J.A., Müller, R.D. *et al.* 2011. Next-generation plate-tectonic reconstructions using Gplates. *In*: Keller, G. & Baru, C. (eds) *Geoinformatics: Cyberinfrastructure for the Solid Earth Sciences*. Cambridge University Press, Cambridge, 95–116.

Brandsdóttir, B., Hooft, E., Mjelde, R. & Murai, Y. 2015. Origin and evolution of the Kolbeinsey Ridge and Iceland Plateau, N-Atlantic. *Geochemistry, Geophysics, Geosystems*, **16**, 612–634.

Breivik, A. & Mjelde, R. 2003. *Modelling of Profile 8 Across the Jan Mayen Ridge*. Report of the Institute of Solid Earth Physics, University of Bergen, Bergen, Norway.

Breivik, A.J., Mjelde, R., Faleide, J.I. & Murai, Y. 2012. The eastern Jan Mayen microcontinent volcanic margin. *Geophysical Journal International*, **188**, 798–818.

Brekke, H. 2000. The tectonic evolution of the Norwegian Sea continental margin with emphasis on the Vøring and Møre basins. *In*: Nøttvedt, A. (ed.) *Dynamics of the Norwegian Margin*. Geological Society, London, Special Publications, **167**, 327–378, https://doi.org/10.1144/GSL.SP.2000.167.01.13

Brekke, H., Dahlgren, S., Nyland, B. & Magnus, C. 1999. The prospectivity of the Vøring and Møre basins on the Norwegian Sea continental margins. *In*: Fleet, A.J. & Boldy, S.A.R. (eds) *Petroleum Geology of Northwest Europe: Proceedings of the 5th Conference*. Geological Society, London, 261–274, https://doi.org/10.1144/0050261

Brooks, C.K. 2011. *The East Greenland Rifted Volcanic Margin*. Geological Survey of Den-mark and Greenland Bulletin, **24**.

Butt, F.A., Elverhoi, A., Forsberg, C.F. & Solheim, A. 2001. An evolution of the Scoresby Sund Fan, central East Greenland – evidence from ODP Site 987. *Norsk Geologisk Tidsskrift*, **81**, 3–15.

Channell, J.E.T., Amigo, A.E. *et al.* (eds). 1999*a*. Magnetic stratigraphy at Sites 907 and 985 in the Norwegian–Greenland Sea and a revision of the Site 907 composite section. *In*: Raymo, M.E., Jansen, E., Blum, P. & Herbert, T.D. (eds) *Proceedings of the Ocean Drilling Program, Scientific Results*, Volume **162**. Ocean Drilling Program, College Station, TX, 131–148.

Channell, J.E.T., Smelror, M., Jansen, E., Higgins, S.M., Lehman, B., Eidvin, T. & Solheim, A. 1999*b*. Age models for glacial fan deposits off East Greenland and Svalbard (sites 986 and 987). Paper presented at *Proceedings of the Ocean Drilling Program, Scientific Results*, College Station TX (Integrated Ocean Drilling Program Management International, Inc.), 162.

Clifton, A.E. & Kattenhorn, S.A. 2005. Structural architecture of a highly oblique divergent plate boundary segment. *Tectonophysics*, **419**, 27–40.

Clifton, A.E. & Schlische, R.W. 2003. Fracture populations on the Reykjanes Peninsula, Iceland: Comparison with experimental clay models of oblique rifting. *Journal of Geophysical Research*, **108**, 2074.

Davies, R., Cartwright, J., Pike, J. & Line, L. 2001. Early Oligocene initiation of North Atlantic Deep Water formation. *Nature*, **410**, 917–920, https://doi.org/10.1038/35073551

Denk, T., Grímsson, F. & Kvacek, Z. 2011. The Miocene floras of Iceland and their significance for Late Cainozoic North Atlantic biogeography. *Botanical Journal of the Linnean Society*, **149**, 369–417.

Doré, A.G., Lundin, E.R., Jensen, L.N., Birkeland, Ø., Eliassen, P.E. & Fichler, C. 1999. Principal tectonic events in the evolution of the northwest European Atlantic margin. *In*: Fleet, A.J. & Boldy, S.A.R. (eds) *Petroleum Geology of Northwest Europe: Proceedings of the 5th Conference*. Geological Society, London, 41–61, https://doi.org/10.1144/0050041

Doré, A.G., Lundin, E.R., Kusznir, N.J. & Pascal, C. 2008. Potential mechanisms for the genesis of

Cenozoic domal structures on the NE Atlantic margin: pros, cons and some new ideas. *In*: Johnson, H., Doré, T.G., Gatliff, R.W., Holdsworth, R.W., Lundin, E. & Ritchie, J.D. (eds) *The Nature and Origin of Compression in Passive Margins*. Geological Society, London, Special Publications, **306**, 1–26, https://doi.org/10.1144/SP306.1

Einarsson, P. 2008. Plate boundaries, rifts and transforms in Iceland. *Jökull*, **58**, 35–58.

Eldholm, O. & Windish, C.C. 1974. Sediment distribution in the Norwegian-Greenland Sea. *Geological Society of America Bulletin*, **85**, 1661–1676.

Eldholm, O., Thiede, J. *et al.* 1987. *Proceedings of the ODP Initial Reports*, Volume **104**. Ocean Drilling Program, College Station, TX.

Eldholm, O., Thiede, J. & Taylor, E. 1989. The Norwegian continental margin: tectonic, volcanic, and paleo-environmental framework. *In*: Eldholm, O., Thiede, J. *et al.* *Proceedings of the Ocean Drilling Program, Scientific Results*, Volume **104**. Ocean Drilling Program, College Station, TX, 5–26.

Engkilde, M. & Surlyk, F. 2003. Shallow marine syn-rift sedimentation: Middle Jurassic Pelion Formation, Jameson Land, East Greenland. *In*: Ineson, J.R. & Surlyk, F. (eds) *The Jurassic of Denmark and Greenland*. Geological Survey of Denmark and Greenland Bulletin, **1**, 813–863.

Erlendsson, Ö. 2010. *Seismic Investigation of the Jan Mayen Ridge – with a Close Study of Sill Intrusions*. Aarhus University, Aarhus, Denmark.

Erlendsson, Ö. & Blischke, A. 2013. *Haffréttarmál: hljóðendurvarps- og bylgjubrotsmælin-gar: skýrsla um stöðu mála á úrvinnslu og túlkun gagna* (Law of the sea: seismic reflection and refraction database and interpretation status report.). Report ÍSOR-2013/067. ÍSOR (Iceland GeoSurvey), Reykjavík, Iceland.

Faleide, J.I., Bjørlykke, K. & Gabrielsen, R.H. 2010. Geology of the Norwegian continental shelf. *In*: Bjørlykke, K. (ed.) *Petroleum Geoscience: From Sedimentary Environments to Rock Physics*. Springer, New York, 467–499.

Fechner, N. & Jokat, W. 1996. Seismic refraction investigation on the crustal structure of the western Jamson Land Basin, East Greenland. *Journal of Geophysical Research*, **101**, 15,867–15,881.

Foulger, G.R., Natland, J.H. & Anderson, D.L. 2005. A source for Icelandic magmas in re-melted Iapetus crust. *Journal of Volcanology and Geothermal Research*, **141**, 23–44.

Funck, T., Hopper, J.R. *et al.* 2014. Crustal structure. *In*: Hopper, J.R., Funck, T., Stoker, M., Árting, U., Peron-Pinvidic, G., Doornenbal, H. & Gaina, C. (eds) *Tectonostratigraphic Atlas of the North-East Atlantic Region*. Geological Survey of Denmark and Greenland (GEUS), Copenhagen, Denmark.

Gaina, C., Gernigon, L. & Ball, P. 2009. Palaeocene–Recent plate boundaries in the NE Atlantic and the formation of the Jan Mayen microcontinent. *Journal of the Geological Society, London*, **166**, 601–616, https://doi.org/10.1144/0016-76492008-112

Gaina, C., Werner, S.C., Saltus, R., Maus, S. & the CAMP-GM Group 2011. Circum-arctic mapping project: new magnetic and gravity anomaly maps of the arctic. *In*: Spencer, A.M., Embry, A.F., Gautier, D.L., Stoupakova, A. & Sørensen, K. (eds) *Arctic Petroleum Geology*. Geological Society, London, Memoirs, **35**, 39–48, https://doi.org/10.1144/M35.3

Gaina, C., Torsvik, T.H., van Hinsbergen, D., Medvedev, S., Werner, S.C. & Labails, C. 2013. The African Plate: A history of oceanic crust accretion and subduction since the Jurassic. *Tectonophysics*, ISSN 0040-1951.604, s4–25. doi:10.1016/j.tecto.2013.05.037

Gaina, C., Nasuti, A., Kimbell, G.S. & Blischke, A. In review. Break-up and seafloor spreading domains in the NE Atlantic. *In*: Péron-Pinvidic, G., Hopper, J.R., Stoker, T., Gaina, C., Doornebal, H., Funck, T. & Árting, U. (eds) *The NE Atlantic Region: A Reappraisal of Crustal Structure, Tectonostratigraphy and Magmatic Evolution*. Geological Society of London, London, Special Publications, **447**.

Gasser, D. 2014. The Caledonides of Greenland, Svalbard and other Arctic areas: status of research and open questions. *In*: Corfu, F., Gasser, D. & Chew, D.M. (eds) *New Perspectives on the Caledonides of Scandinavia and Related Areas*. Geological Society of London, London, Special Publications, **390**, 93–129, https://doi.org/10.1144/SP390.17

Geodekyan, A.A., Verkhovskaya, Z.I., Sudyin, A.V. & Trotsiuk, V.Ya. 1980. Gases in seawater and bottom sediments. *In*: Udintsev, C.B. (ed.) *Iceland and Mid-Oceanic Ridge: Structure of the Ocean-Floor*. National Research Council, Reykjavik, Iceland, 19–36.

Gernigon, L., Gaina, C., Olesen, O., Ball, P., Peron-Pinvidic, G. & Yamasaki, T. 2012. The Norway Basin revisited: from continental breakup to spreading ridge extinction. *Marine and Petroleum Geology*, **35**, 1–19.

Gernigon, L., Blischke, A., Nasuti, A. & Sand, M. 2015. Conjugate volcanic rifted margins, seafloor spreading, and microcontinent: insights from new high-resolution aeromagnetic surveys in the Norway Basin. *Tectonics*, **34**, 907–933.

Gilotti, J.A., Jones, K.A. & Elvevold, S. 2008. Caledonian metamorphic patterns in Greenland. *In*: Higgins, A.K., Gilotti, J.A. & Smith, M.P. (eds) *The Greenland Caledonides: Evolution of the Northeast Margin of Laurentia*. Geological Society of America, Memoirs, **202**, 201–225.

Gradstein, F.M., Ogg, J.G., Schmitz, M.D. & Ogg, G.M. 2012. *The Geologic Time Scale 2012*. Elsevier, Amsterdam.

Guarnieri, P. 2015. Pre-breakup palaeo-stress state along the East Greenland margin. *Journal of the Geological Society, London*, **172**, 727–739, https://doi.org/10.1144/jgs2015-053

Gunnarsson, K., Sand, M. & Gudlaugsson, S.T. 1989. *Geology and Hydrocarbon Potential of the Jan Mayen Ridge*. Orkustofnun, Reykjavík Report OS-98014 & Norwegian Petroleum Directorate, Stavanger, OD-89-91.

Haase, C. & Ebbing, J. 2014. Gravity data. *In*: Hopper, J.R., Funck, T., Stoker, M.S., Árting, U., Peron-Pinvidic, G., Doornenbal, H. & Gaina, C. (eds) *Tectonostratigraphic Atlas of the North-East Atlantic Region*. Geological Survey of Denmark and Greenland (GEUS), Copenhagen, Denmark.

Hamann, N.E., Wittaker, R.C. & Stemmerik, L. 2005. Geological development of the Northeast Greenland shelf. *In*: Doré, A.G. & Vining, B.A. (eds) *Petroleum Geology: North-West Europe and Global Perspectives – Proceedings of the 6th Petroleum Geology Conference*. Geological Society, London, 887–902, https://doi.org/10.1144/0060887

Hards, V.L., Kempton, P.D. & Thompson, R.N. 1995. The heterogeneous Iceland plume: new insights from the alkaline basalts of the Snaefell volcanic centre. *Journal of the Geological Society, London*, **152**, 1003–1009, https://doi.org/10.1144/GSL.JGS.1995.152.01.21

Harðarsson, B., Fitton, J. & Hjartarson, Á. 2008. Tertiary volcanism in Iceland. *Jökull*, **58**, 161–178.

Henriksen, N. 2008. *Geological History of Greenland – Four Billion Years of Earth Evolution*. Geological Survey of Denmark and Greenland (GEUS), Copenhagen, Denmark.

Hinz, K. 1981. A hypothesis on terrestrial catastrophes wedges of very thick ocean-ward dip-ping layers beneath passive continental margins; their origin and palaeo-environmental significance. *Geologisches Jahrbuch*, **E2**, 3–28.

Hjartarson, A. & Sæmundsson, K. 2014. *Geological Map of Iceland – Bedrock Map, 1:600 000, Datum: ISN93*. Iceland GeoSurvey, Reykjavík.

Hopper, J.R., Funck, T., Stoker, M.S., Árting, U., Peron-Pinvidic, G., Doornenbal, H. & Gaina, C. (eds). 2014. *Tectonostratigraphic Atlas of the North-East Atlantic Region*. Geological Survey of Denmark and Greenland (GEUS), Copenhagen.

Horni, J., Hopper, J.R. *et al.* 2016. Regional distribution of volcanism within the North Atlantic Igneous Province. *In*: Péron-Pinvidic, G., Hopper, J., Stoker, M.S., Gaina, C., Doornenbal, H., Funck, T. & Árting, U. (eds) *The North-East Atlantic Region: A Reappraisal of Crustal Structure, Tectono-Stratigraphy and Magmatic Evolution*. Geological Society, London, Special Publications, **447**, xx–xx, https://doi.org/10.1144/SP447.?

Howe, J.A., Stoker, M.S. & Stow, D.A.V. 1994. Late Cenozoic sediment drift complex, northeast Rockall Trough, North Atlantic. *Palaeoceanography*, **9**, 989–999.

Jansen, E., Raymo, M.E. *et al.* 1996. *Proceedings of the Ocean Drilling Program. Initial Reports, Volume 162*. Ocean Drilling Program, College Station, TX.

Jakobsson, M., Mayer, L. *et al.* 2012. The International Bathymetric Chart of the Arctic Ocean (IBCAO) Version 3.0. *Geophysical Research Letters*, **39**, https://doi.org/10.1029/2012GL052219

Jakobsson, S.P., Jónasson, K. & Sigurdsson, I.A. 2008. The three igneous rock series of Iceland. *Jökull*, **58**, 117–138.

Jóhannesson, H. 2011. *Örnefnagjöf á landgrunninu og utan þess. III. Drekasvæðið, Rep. JHJ–2011–012*, 6 pp., National Energy Authority of Iceland, Reykjavík, Iceland.

Jóhannesson, H. & Sæmundsson, K. 2009. *Geological Map of Iceland. 1:600,000. Tectonics*. 1st edn. Icelandic Institute of Natural History, Reykjavik.

Johansen, B., Eldholm, O., Talwani, M., Stoffa, P.L. & Buhl, P. 1988. Expanding spread profile at the northern Jan Mayen Ridge. *Polar Research*, **6**, 95–104.

Kalsbeek, F., Higgins, A.K., Jepsen, H.F., Frei, R. & Nutman, A.P. 2008. Granites and granites in the East Greenland Caledonides. *In*: Higgins, A.K., Gilotti, J.A. & Smith, M.P. (eds) *The Greenland Caledonides: Evolution of the Northeast Margin of Laurentia*. Geological Society of America, Memoirs, **202**, 227–249.

Kandilarov, A., Mjelde, R. *et al.* 2012. The northern boundary of the Jan Mayen microcontinent, North Atlantic determined from ocean bottom seismic, multichannel seismic, and gravity data. *Marine Geophysical Research*, **33**, 55–76.

Khodayar, M. 2014. *Shift of Þeistareykir Fissure Swarm in Tjörnes Fracture Zone: Case of Pull-apart on Strike-slip?* Technical report, ISOR-2014/074, LV-2014-144, Iceland GeoSurvey for Landsvirkjun (National Energy Company of Iceland), Reykjavik, Iceland.

Kodaira, S., Mjelde, R., Gunnarsson, K., Shiobara, H. & Shimamura, H. 1998. Structure of the Jan Mayen microcontinent and implications for its evolution. *Geophysical Journal International*, **132**, 383–400.

Laier, T. & Nytoft, H.P. 2004. Waxy bitumen in basalts – Skye and Faroe Islands, North Atlantic. *Geochimica et Cosmochimica Acta*, **68**, (11), Supplement 234.

Laier, T., Nytoft, H.P., Jørgensen, O. & Isaksen, G.H. 1997. Hydrocarbon traces in the Tertiary basalts of the Faeroe Islands. *Marine and Petroleum Geology*, **14**, 257–266.

Larsen, L.M., Waagstein, R., Pedersen, A.K. & Storey, M. 1999. Trans-Atlantic correlation of the Palaeogene volcanic successions in the Faeroe Islands and East Greenland. *Journal of the Geophysical Society*, **156**, 1081–1095.

Larsen, L.M., Pedersen, A.K., Sørensen, E.V., Watt, W.S. & Duncan, R.A. 2013. Stratigraphy and age of the Eocene Igtertivâ Formation basalts, alkaline pebbles and sediments of the Kap Dalton Group in the graben at Kap Dalton. East Greenland. *Bulletin of the Geological Society of Denmark*, **61**, 1–18.

Larsen, L.M., Pedersen, A.K., Tegner, T. & Duncan, R.A. 2014. Eocene to Miocene igneous activity in NE Greenland: northward younging of magmatism along the East Greenland margin. *Journal of the Geological Society, London*, **171**, 539–553, https://doi.org/10.1144/jgs2013-118

Lundin, E.R. & Doré, A.G. 1997. A tectonic model for the Norwegian passive margin with implications for the NE Atlantic: early Cretaceous to break-up. *Journal of the Geological Society, London*, **154**, 545–550, https://doi.org/10.1144/gsjgs.154.3.0545

Lundin, E.R. & Doré, A.G. 2002. Mid-Cenozoic post-breakup deformation in the 'passive' margins bordering the Norwegian-Greenland Sea. *Marine and Petroleum Geology*, **19**, 79–93.

Lundin, E.R. & Doré, A.G. 2005. NE Atlantic breakup: a re-examination of the Iceland mantle plume model and the Atlantic–Arctic linkage. *In*: Doré, A.G. & Vining, B.A. (eds) *Petroleum Geology: North-West Europe and Global Perspectives – Proceedings of the 6th Petroleum Geology Conference*. Geological Society, London, 739–754, https://doi.org/10.1144/0060739

LUNDIN, E.R. & DORÉ, A.G. 2011. Hyperextension, serpentinization, and weakening: a new paradigm for rifted margin compressional deformation. *Geology*, **39**, 347–350.

LUNDIN, E.R., REDFIELD, T.F. & PERON-PINDIVIC, G. 2014. Rifted continental margins: geometric influence on crustal architecture and melting. *In*: PINDELL, J., HORN, B. *ET AL.* (eds) *33rd Annual GCSSEPM Foundation Bob F. Perkins Conference. Sedimentary Basins: Origin, Depositional Histories, and Petroleum Systems*. Gulf Coast Section SEPM, Houston, TX, 18–53.

MAGNÚSDÓTTIR, S., BRANDSDÓTTIR, B., DRISCOLL, N. & DETRICK, R. 2015. Postglacial tectonic activity within the Skjálfandadjúp Basin, Tjörnes Fracture Zone, offshore Northern Iceland, based on high resolution seismic stratigraphy. *Marine Geology*, **367**, 159–170. https://doi.org/10.1016/j.margeo.2015.06.004

MANUM, S.B. & SCHRADER, H.J. 1976. Sites 346, 347, and 349. *In*: TALWANI, M. & UDINTSEV, G. (eds) *Initial Reports of the Deep Sea Drilling Project*, Volume **38**. United States Government Printing Office, Washington, DC, 521–594.

MANUM, S.B., RASCHKA, H. & ECKHARDT, F.J. 1976*a*. Site 350. *In*: TALWANI, M. & UDINTSEV, G. (eds) *Initial Reports of the Deep Sea Drilling Project*, Volume **38**. United States Government Printing Office, Washington, DC.

MANUM, S.B., RASCHKA, H., ECKHARDT, F.J., SCHRADER, H., TALWANI, M. & UDINTSEV, G. 1976*b*. Site 337 initial reports of the Deep Sea Drilling Project. *In*: TALWANI, M., UDINTSEV, G. *ET AL.* (eds) *Initial Reports of the Deep Sea Drilling Project*, Volume **38**. United States Government Printing Office, Washington, DC, 117–150,

MATHIESEN, A., BIDSTRUP, T. & CHRISTINSEN, F.G. 2000. Denudation and uplift history of the Jameson Land Basin, East Greenland-constrained from maturity and apatite fission track data. *Global and Planetary Change*, **24**, 275–301.

MJELDE, R., AURVÅG, R., KODAIRA, S., SHIMAMURA, H., GUNNARSSON, K., NAKANISHI, A. & SHIOBARA, H. 2002. Vp/Vs-ratios from the central Kolbeinsey Ridge to the Jan Mayen Basin, North Atlantic; implications for lithology, porosity and present-day stress field. *Marine Geophysical Researches*, **23**, 123–145.

MJELDE, R., ECKHOFF, I. *ET AL.* 2007. Gravity and S-wave modelling across the Jan Mayen Ridge, North Atlantic; implications for crustal lithology. *Marine Geophysical Researches*, **28**, 27–41.

MJELDE, R., BREIVIK, A.J., RAUM, T., MITTELSTAEDT, E., ITO, G. & FALEIDE, J.I. 2008. Magmatic and tectonic evolution of the North Atlantic. *Journal of the Geological Society, London*, **165**, 31–42, https://doi.org/10.1144/0016-76492007-018

MJELDE, R., RAUM, T., KANDILAROV, A., MURAI, Y. & TAKANAMI, T. 2009. Crustal structure and evolution of the outer Møre Margin, NE Atlantic. *Tectonophysics*, **468**, 224–243.

MUTTER, J.C., TALWANI, M. & STOFFA, P.L. 1982. Origin of seaward-dipping reflectors in oceanic crust off the Norwegian margin by 'subaerial sea-floor spreading'. *Geology*, **10**, 3–12.

NASUTI, A. & OLESEN, O. 2014. Magnetic data. *In*: HOPPER, J.R., FUNCK, T., STOKER, M.S., ÁRTING, U., PERON-PINVIDIC, G., DOORNENBAL, H. & GAINA, C. (eds) *Tectonostratigraphic Atlas of the North-East Atlantic Region*. Geological Survey of Denmark and Greenland (GEUS), Copenhagen, Denmark, 41–51.

NILSEN, T.H., KERR, D.R., TALWANI, M. & UDINTSEV, G. 1978. Turbidites, redbeds, sedimentary structures, and trace fossils observed in DSDP Leg 38 cores and the sedimentary history of the Norwegian–Greenland Sea. *In*: *Initial Reports of the Deep Sea Drilling Project. Supplement to Volume 38*. United States Government Printing Office, Washington, DC, 259–288.

NØHR-HANSEN, H. 2003. Dinoflagellate cyst stratigraphy of the Palaeogene strata from the wells Hellefisk-1, Ikermiut-1, Kangâmiut-1, Nukik-1, Nukik-2 and Qulleq-1, offshore West Green-land. *Marine and Petroleum Geology*, **20**, 987–1016.

NØHR-HANSEN, H. 2012. Palynostratigraphy of the Cretaceous-Lower Palaeogene sedimentary succession in the Kangerlussuaq Basin, southern East Greenland. *Review of Palaeobotany and Palynology*, **178**, 59–90.

NØHR-HANSEN, H. & PIASECKI, S. 2002. Paleocene subbasaltic sediments on Savoia Halvø, East Greenland. *Geology of Greenland Survey Bulletin*, **191**, 111–116.

NORWEGIAN PETROLEUM DIRECTORATE 2012. *Submarine Fieldwork on the Jan Mayen Ridge: Integrated Seismic and ROV-Sampling*. Norwegian Petroleum Directorate, Stavanger, Norway, http://www.npd.no/en/Publications/Presentations/Submarine-fieldwork-on-the-Jan-Mayen-Ridge/

NORWEGIAN PETROLEUM DIRECTORATE 2013. *The Petroleum Resources on the Norwegian Continental Shelf*. Norwegian Petroleum Directorate, Stavanger, Norway, http://www.npd.no/en/Publications/Resource-Reports/2013/

NUNNS, A. 1983*a*. The structure and evolution of the Jan Mayen Ridge and surroundings regions. *In*: WATKINS, J.S. & DRAKE, C.L. (eds) *Studies in Continental Margin Geology*. American Association of Petroleum Geologists, Memoirs, **34**, 193–208.

NUNNS, A.G. 1983*b*. Plate tectonic evolution of the Greenland-Scotland Ridge and surrounding regions. *In*: BOTT, M.H.P., SAXOV, S., TALWANI, M. & THIEDE, J. (eds) *Structure and Development of the Greenland–Scotland Ridge*. Plenum, New York, 11–30.

NUNNS, A.G., TALWANI, M. *ET AL.* 1983. Magnetic anomalies over Iceland and surrounding seas. *In*: BOTT, M.H.P., SAXOV, S., TALWANI, M. & THIEDE, J. (eds) *Structure and Development of the Greenland-Scotland Ridge*. Plenum, New York, 661–678.

OLAFSSON, I. & GUNNARSSON, K. 1989. *The Jan Mayen Ridge: Velocity Structure from Analysis of Sonobuoy Data*. OS-89030/JHD-04. Orkustofnun, Reykjavík.

OSMUNDSEN, P.T. & ANDERSEN, T.B. 1994. Caledonian compressional and late-orogenic extensional deformation in the Staveneset area, Sunnfjord, Western Norway. *Journal of Structural Geology*, **16**, 1385–1401.

OSMUNDSEN, P.T. & ANDERSEN, T.B. 2001. The middle Devonian basins of western Norway: sedimentary response to large-scale transtensional tectonics. *Tectonophysics*, **332**, 51–68, https://doi.org/10.1016/s0040-1951(00)00249-3

Osmundsen, P.T. & Ebbing, J. 2008. Styles of extension offshore mid-Norway and implications for mechanisms of crustal thinning at passive margins. *Tectonics*, **27**, TC6016.

Osmundsen, P.T., Sommaruga, A., Skilbrei, J.R. & Olesen, O. 2002. Deep structure of the Mid Norway rifted margin. *Norwegian Journal of Geology*, **82**, 205–224.

Passey, S. & Hitchen, K. 2011. Cenozoic (igneous). *In*: Ritchie, J.D., Ziska, H., Johnson, H. & Evans, D. (eds) *Geology of the Faroe-Shetland Basin and Adjacent Areas*. British Geological Survey and Jarðfeingi Research Report RR/11/01 British Geological Survey, Keyworth, Nottingham, 209–228.

Passey, S.R. & Jolley, D.W. 2009. A revised lithostratigraphic nomenclature for the Palaeogene Faroe Islands basalt Group, NE Atlantic Ocean. *Earth and Environmental Science, Transactions of the Royal Society of Edinburgh*, **99**, 127–158.

Peron-Pinvidic, G., Gernigon, L., Gaina, C. & Ball, P. 2012*a*. Insights from the Jan Mayen system in the Norwegian-Greenland Sea – I: mapping of a microcontinent. *Geophysical Journal International*, **191**, 385–412.

Peron-Pinvidic, G., Gernigon, L., Gaina, C. & Ball, P. 2012*b*. Insights from the Jan Mayen system in the Norwegian-Greenland Sea – II: architecture of a microcontinent. *Geophysical Journal International*, **191**, 413–435.

Peron-Pinvidic, G., Manatschal, G. & Osmundsen, P.T. 2013. Structural comparison of archetypal Atlantic rifted margins: a review of observations and concepts. *Marine and Petroleum Geology*, **43**, 21–47.

Pharaoh, T.C., Dusar, M. *et al.* 2010. Tectonic evolution. *In*: Doornenbal, J.C. & Stevenson, A.G. (eds) *Petroleum Geological Atlas of the Southern Permian Basin Area*. European Association of Geoscientists & Engineers, Houten, The Netherlands, 25–57.

Planke, S., Symonds, P.A., Alvestad, E. & Skogseid, J. 2000. Seismic volcanostratigraphy of large-volume basaltic extrusive complexes on rifted margins. *Journal of Geophysical Research*, **105**, 19335–19351.

Polteau, S., Mazzini, A., Trulsvik, M. & Planke, S. 2012. *JMRS11 – Jan Mayen Ridge Sampling Survey 2011*. VBPR-TGS, Commercial Report, February 2012.

Prestvik, T., Goldberg, S., Karlsson, H. & Grönvold, K. 2001. Anomalous strontium and lead isotope signatures in the off-rift Öræfajökull central volcano in southeast Iceland: evidence for enriched endmember(s) of the Iceland mantle plume? *Earth and Planetary Science Letters*, **190**, 211–220.

Raschka, H., Eckhardt, F.J. & Manum, S.B. 1976. Site 348. *In*: Talwani, M. & Udintsev, G. (eds) *Initial Reports of the Deep Sea Drilling Project*, Volume **38**. United States Government Printing Office, Washington, DC, 595–654.

Rey, S.S., Eldholm, O. & Planke, S. 2003. Formation of the Jan Mayen Microcontinent, the Norwegian Sea. *Eos, Transactions of the American Geophysical Union*, **84**, Abstract T31D-0872.

Sandstå, N.R., Pedersen, R.B., Williams, R.D., Bering, D., Magnus, C., Sand, M. & Brekke, H. 2012. *Submarine fieldwork on the Jan Mayen Ridge; integrated seismic and ROV-sampling*. Norwegian Petroleum Directorate, Stavanger, Norway, http://www.npd.no/

Sæmundsson, K. 1974. Evolution of the Axial Rifting Zone in Northern Iceland and the Tjörnes Fracture Zone. *Geological Society of America Bulletin*, **85**, 495–504.

Sæmundsson, K. 1979. Outline of the geology of Iceland. *Jökull*, **29**, 7–28.

Scott, R.A., Ramsey, L.A., Jones, S.M., Sinclair, S. & Pickles, C.S. 2005. Development of the Jan Mayen microcontinent by linked propagation and retreat of spreading ridges. *In*: Wandås, B.T.G., Nystuen, J.P., Eide, E. & Gradstein, F. (eds) *Onshore–Offshore Relationships on the North Atlantic Margin*. Norwegian Petroleum Society, Oslo, 69–82.

Seidler, L. 2000. Incised submarine canyons governing new evidence of Early Triassic rifting in East Greenland. *Palaeogeography, Palaeoclimatology, Palaeoecology*, **161**, 267–293.

Soper, N.J., Higgins, A.C., Downie, C., Matthews, D.W. & Brown, P.E. 1976. Late Cretaceous-early Tertiary stratigraphy of the Kangerdlugssuaq area, east Greenland, and the age of opening of the northeast Atlantic. *Journal of the Geological Society, London*, **132**, 85–104, https://doi.org/10.1144/gsjgs.132.1.0085

Stemmerik, L. 2000. Late Palaeozoic evolution of the North Atlantic margin of Pangea. Palaeo-geography Palaeo-climatology. *Palaeoecology*, **161**, 95–126.

Stoker, M.S., Praeg, D. *et al.* 2005. Neogene evolution of the Atlantic continental margin of NW Europe (Lofoten Islands to SE Ireland): anything but passive. *In*: Doré, A.G. & Vining, B.A. (eds) *Petroleum Geology: North-West Europe and Global Perspectives – Proceedings of the 6th Petroleum Geology Conference*. Geological Society, London, **6**, 1057–1076, https://doi.org/10.1144/0061057

Stoker, M.S., Stewart, M. *et al.* 2016. An overview of the Upper Paleozoic–Mesozoic stratigraphy of the NE Atlantic region. *In*: Péron-Pinvidic, G., Hopper, J., Stoker, M.S., Gaina, C., Doornenbal, H., Funck, T. & Árting, U. (eds) *The North-East Atlantic Region: A Reappraisal of Crustal Structure, Tectono-Stratigraphy and Magmatic Evolution*. Geological Society, London, Special Publications, **447**, first published online August 11, 2016, https://doi.org/10.1144/SP447.2

Storey, M., Duncan, R.A. & Swisher, C.C., III. 2007*a*. Paleocene-Eocene thermal maximum and opening of the Northeast Atlantic. *Science*, **316**, 587–589.

Storey, M., Duncan, R.A. & Tegner, C. 2007*b*. Timing and duration of volcanism in the North Atlantic Igneous Province: implications for geodynamics and links to the Iceland hotspot. *Chemical Geology*, **241**, 264–281.

Surlyk, F. 1977. *Stratigraphy, Tectonics and Palaeogeography of the Jurassic Sediments of the Areas North of Kong Oscars Fjord, East Greenland*. Bulletin Grønlands Geologiske Undersøgelse, **123**.

Surlyk, F. 1978. *Submarine Fan Sedimentation Along Fault Scarps On Tilted Fault Blocks*

(Jurassic–Cretaceous Boundary, East Greenland). Bulletin Grønlands Geologiske Undersøgelse, **128**.

SURLYK, F. 1990. Timing, style and sedimentary evolution of Late Paleozoic–Mesozoic extensional basins of East Greenland. *In*: HARDMAN, R.F.P. & BROOKS, J. (eds) *Tectonic Events Responsible for Britain's Oil and Gas Reserves*. Geological Society, London, Special Publications, **55**, 107–125, https://doi.org/10.1144/GSL.SP.1990.055.01.05

SURLYK, F. 1991. Tectonostratigraphy of North Greenland. *In*: PEEL, J.S. & SØNDERHOLM, M. (eds) *Sedimentary Basins of North Greenland*. Bulletin Grønlands Geologiske Undersøgelse, **160**, 25–47.

SURLYK, F. 2003. The Jurassic of East Greenland: a sedimentary record of thermal subsidence, onset and culmination of rifting. *In*: INESON, J.R. & SURLYK, F. (eds) *The Jurassic of Denmark and Greenland*. Geological Survey of Denmark and Greenland Bulletin, **1**, 659–722.

SURLYK, F. & NOE-NYGAARD, N. 2001. Cretaceous faulting and associated coarse-grained marine gravity flow sedimentation, Traill Ø, East Greenland. *In*: MARTINSEN, O.J. & DREYER, T. (eds) *Sedimentary Environments Offshore Norway – Paleozoic to Recent*. Norwegian Petroleum Society, Special Publications, **10**, 293–319.

SURLYK, F., CALLOMON, J.H., BROMLEY, R.G. & BIRKELUND, T. 1973. *Stratigraphy of the Lower Jurassic–Lower Cretaceous Sediments of Jameson Land and Scoresby Land, East Greenland*. Bulletin Grønlands Geologiske Undersøgelse, **105**.

SVELLINGEN, W. & PEDERSEN, R. 2003. Jan Mayen: a result of ridge-transform-micro-continent interaction. *Geophysical Research Abstracts*, **5**, 12993.

TALWANI, M. & ELDHOLM, O. 1977. Evolution of the Norwegian-Greenland Sea. *Geological Society of America Bulletin*, **88**, 969–999.

TALWANI, M. & UDINTSEV, G. (eds). 1976. *Initial Reports of the Deep Sea Drilling Project*, Volume **38**. United States Government Printing Office, Washington, DC.

TALWANI, M., UDINTSEV, G. & SHIRSHOV, P.R. 1976*a*. Tectonic synthesis. *In*: TALWANI, M. & UDINTSEV, G. (eds) *Initial Reports of the Deep Sea Drilling Project*, Volume **38**. United States Government Printing Office, Washington, DC, 1213–1242.

TALWANI, M., UDINTSEV, G. & WHITE, S.M. 1976*b*. Introduction and explanatory notes, Leg 38, Deep Sea Drilling Project. *In*: TALWANI, M. & UDINTSEV, G. (eds) *Initial Reports of the Deep Sea Drilling Project*, Volume **38**. United States Government Printing Office, Washington, DC, 3–19.

TEGNER, C., BROOKS, C.K., DUNCAN, R.A., HEISTER, L.E. & BERNSTEIN, S. 2008. ^{40}Ar–^{39}Ar ages of intrusions in East Greenland: rift-to-drift transition over the Iceland hotspot. *Lithos*, **101**, 480–500.

THIEDE, J., FIRTH, J.V. *ET AL*. 1995. Site 907. *In*: MYHRE, A.M., THIEDE, J. *ET AL*. (eds) *Proceedings of the Ocean Drilling Program, Initial Reports*, Volume **151**. Ocean Drilling Program, College Station, TX, 57–111.

THORDARSON, T. & LARSEN, G. 2007. Volcanism in Iceland in historical time: volcano types, eruption styles and eruptive history. *Journal of Geodynamics*, **43**, 118–152.

TORSVIK, T.H., MOSAR, J. & EIDE, E.A. 2001. Cretaceous-Tertiary geodynamics: a North Atlantic exercise. *Geophysical Journal International*, **146**, 850–866.

TORSVIK, T.H., AMUNDSEN, H.E.F. *ET AL*. 2015. Continental crust beneath southeast Iceland. *Proceedings of the National Academy of Sciences of the United States of America*, **112**, 1818–1827.

TSIKALAS, F., ELDHOLM, O. & FALEIDE, J.I. 2005. Crustal structure of the Lofoten-Vesterålen continental margin, off Norway. *Tectonophysics*, **404**, 151–174.

TSIKALAS, F., FALEIDE, J.I. & KUSZNIR, N.J. 2008.Along-strike variations in rifted margin crustal architecture and lithosphere thinning between northern Vøring and Lofoten margin segments off mid-Norway. *In*: ROSENBAUM, G., WEINBERG, R.F. & REGENAUER-LIEB, K. (eds) *Geodynamics of Lithospheric Extension. Tectonophysics*. **458**, 68–81.

VOGT, P.R. & JUNG, W.Y. 2009. Treitel ridge: a unique inside corner hogback on the west flank of extinct Aegir spreading ridge, Norway basin. *Marine Geology*, **267**, 86–100.

VOGT, P.R., ANDERSON, C.N., BRACEY, D.R. & SCHNEIDER, E.M. 1970. North Atlantic Magnetic Smooth Zones. *Journal of Geophysical Research*, **75**, 3955–3968.

WALKER, G.P.I. 1964. Geological investigations in eastern Iceland. *Bulletin of Volcanology*, **27**, 351–363.

ZIEGLER, P.A. 1988. *Evolution of the Arctic-North Atlantic and the Western Tethys*. American Association of Petroleum Geologists, Memoirs, **43**.

The stratigraphy and structure of the Faroese continental margin

JANA ÓLAVSDÓTTIR[1]*, ÓLUVA R. EIDESGAARD[1] & MARTYN S. STOKER[2]

[1]*Jarðfeingi (Faroe Earth and Energy Directorate) Brekkutún 1, Postbox 3059, FO-110 Tórshavn, Faroe Islands*

[2]*British Geological Survey, The Lyell Centre, Research Avenue South, Edinburgh EH14 4AP, UK*

**Correspondence: jana.olavsdottir@jardfeingi.fo*

Abstract: This paper presents a summary of the stratigraphy and structure of the Faroese region. As the Faroese area is mostly covered by volcanic material, the nature of the pre-volcanic geology remains largely unproven. Seismic refraction data provide some indications of the distribution of crystalline basement, which probably comprises Archaean rocks, with the overlying cover composed predominantly of Upper Mesozoic (Cretaceous?) and Cenozoic strata. The Cenozoic succession is dominated by the syn-break-up Faroe Islands Basalt Group, which crops out on the Faroe Islands (where it is up to 6.6 km thick) and shelf areas; post-break-up sediments are preserved in the adjacent deep-water basins, including the Faroe–Shetland Basin. Seismic interpretation of the post-volcanic strata shows that almost every sub-basin in the Faroe–Shetland Basin has been affected by structural inversion, particularly during the Miocene. These effects are also observed on the Faroe Platform, the Munkagrunnur Ridge and the Fugloy Ridge, where interpretation of low-gravity anomalies suggests a large-scale fold pattern. The structure of the Iceland–Faroe Ridge, which borders the NW part of the Faroe area, remains ambiguous. The generally thick crust, together with the absence of well-defined seawards-dipping reflectors, may indicate that much of it is underlain by continental material.

The Faroese continental margin is located in the North Atlantic Ocean on the outer part of the NW European continental margin situated approximately in the central part of the North Atlantic Igneous Province (NAIP) (Fig. 1). Rocks associated with the NAIP cover approximately 99% of the Faroese area. This widespread coverage of the volcanic sequence inhibits our understanding of the region owing to what lies below the volcanic sequence. The situation is further complicated by the fact that the upper part of the NAIP is related to the SW–NE seafloor-spreading trend (Saunders *et al.* 1997), while the lower part of the NAIP is related to a NW–SE ridge-trend volcanism. The latter, in particular, complicates our understanding of the opening of the North Atlantic Ocean because the magnetic chrons are not visible across the Greenland–Iceland–Faroe Ridge (GIFR) (Gernigon *et al.* 2012). Moreover, the crust beneath the Iceland–Faroe Ridge (IFR) is in places relatively thick, even though it is assumed to be oceanic (Richardson *et al.* 1998; Smallwood *et al.* 1999).

The interval between approximately 63 and 56 Ma witnessed the contemporaneous eruption of volcanic material along a 2000 km NW–SE disc-shaped ridge area that extended between West Greenland and Great Britain, and Archer *et al.* (2005), Lundin & Doré (2005) and Ziska & Varming 2008 have all suggested that this pre-break-up volcanism was associated with the development of NW–SE-trending fissures that fed shield volcanoes intermittently located along the line of the fissures. By way of contrast, other researchers believe that the pre-break-up plateau basalts are a consequence of hotspot or plume processes (White 1988; Smallwood & White 2002). The oldest volcanic material in this disc-shaped structure is seen in Great Britain and West Greenland, and is dated to approximately 63 Ma (Storey *et al.* 2007; Ganerød *et al.* 2010). Onshore the Faroe Islands, the oldest drilled volcanic material has an age of approximately 62 Ma (Storey *et al.* 2007), but the volcanic sequence has not been fully penetrated by a well.

The Faroese area is juxtaposed with the IFR towards the NW. The former seafloor-spreading ridge – the Aegir Ridge – is located just north of the area. To the east and south of the Faroe Islands, a complex arrangement of Late Palaeozoic–Mesozoic–Early Cenozoic basins comprise the Faroe–Shetland and north Rockall regions (Figs 1 & 2).

From: Péron-Pinvidic, G., Hopper, J. R., Stoker, M. S., Gaina, C., Doornenbal, J. C., Funck, T. & Árting, U. E. (eds) 2017. *The NE Atlantic Region: A Reappraisal of Crustal Structure, Tectonostratigraphy and Magmatic Evolution.* Geological Society, London, Special Publications, **447**, 339–356.
First published online July 22, 2016, https://doi.org/10.1144/SP447.4

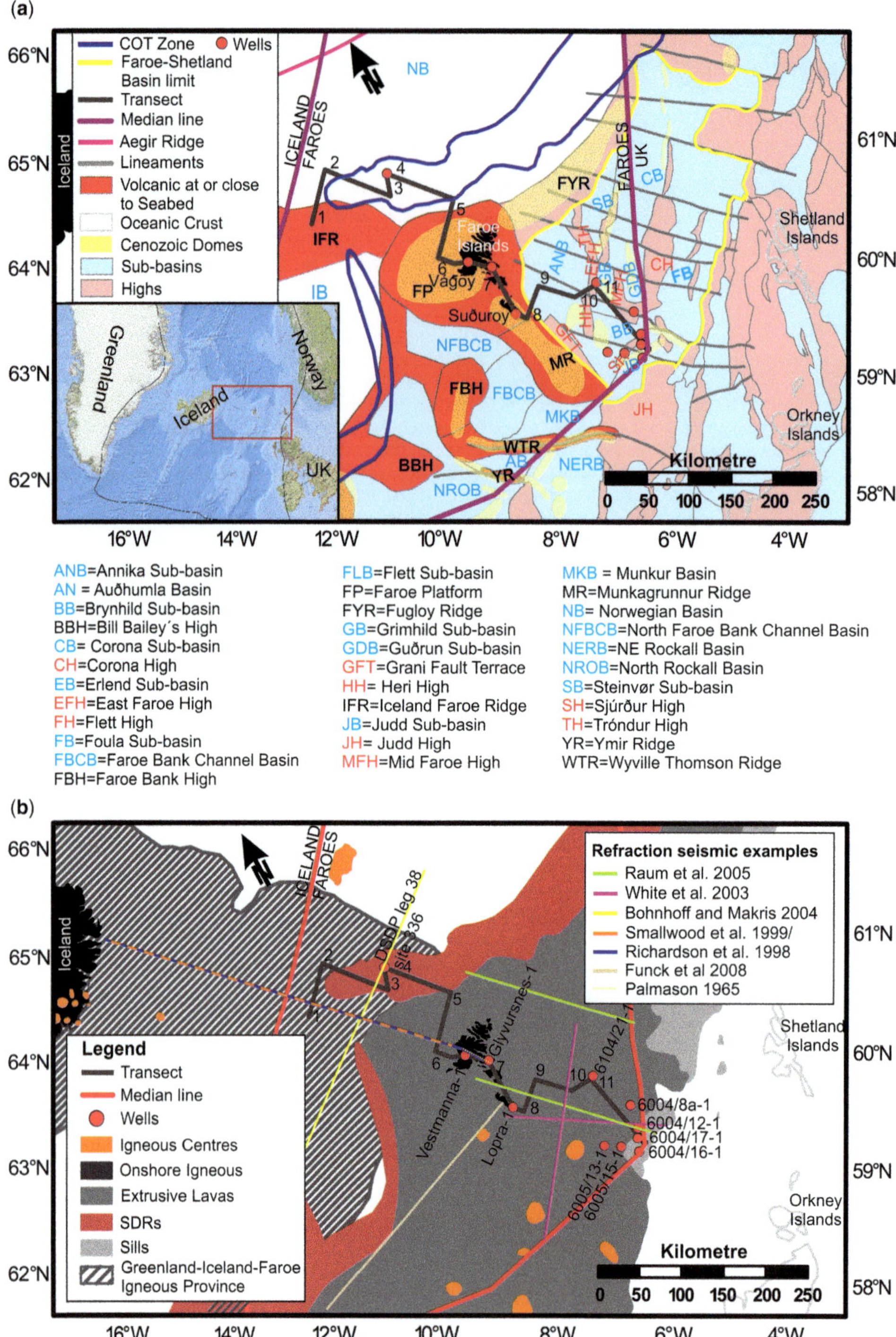

Fig. 1. Maps showing (**a**) the structural framework and (**b**) the volcanic domains of the Faroese continental margin, modified from Ritchie *et al.* (2011) and Haase & Ebbing (2014). The maps also show the location of the shallow crustal transect shown in Figure 3, the boreholes and wells both onshore and offshore the Faroe Islands, and the various refraction seismic lines referred to in the text.

(a)

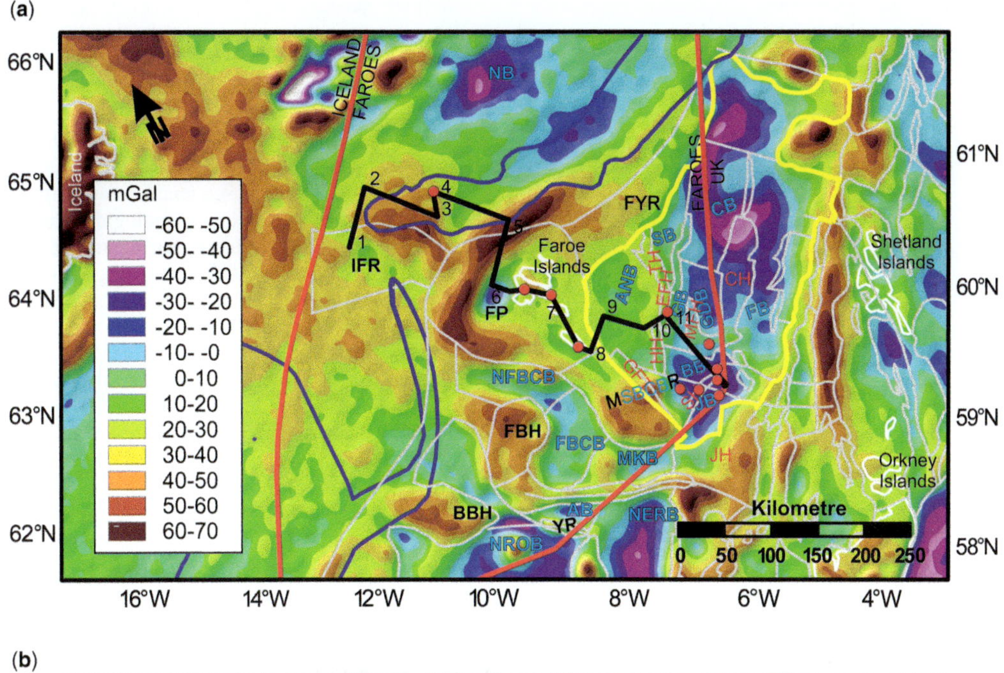

(b)

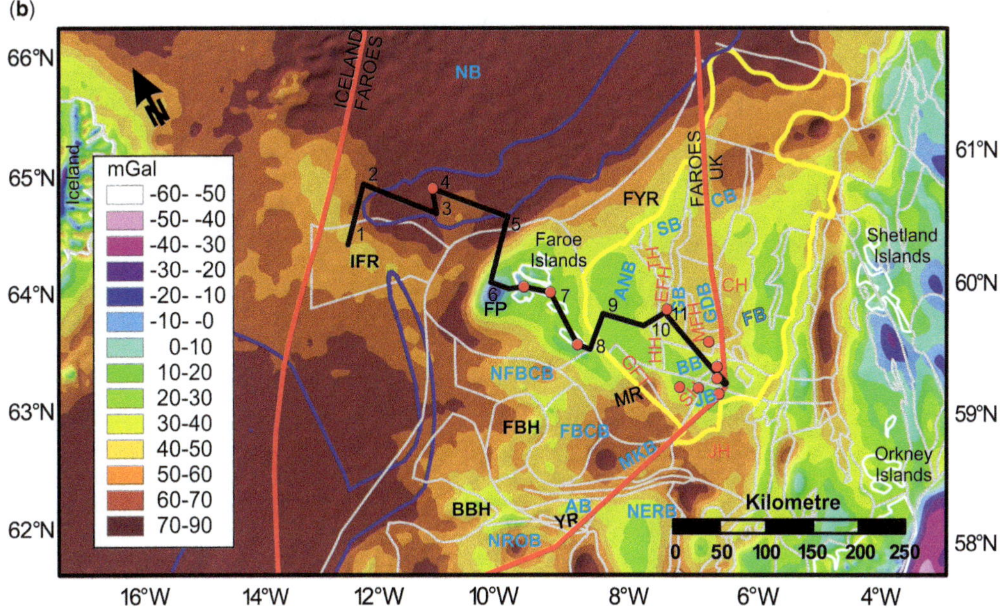

Fig. 2. (**a**) Free-air gravity anomaly map and (**b**) Bouguer gravity anomaly map covering the Faroese continental margin. The maps also retain – for reference – an outline of the main structural elements (detailed in Fig. 1), together with boreholes and wells, and the shallow crustal transect. See Figure 1a for the key to the abbreviations. Both maps are from Haase & Ebbing (2014).

Thus, the geological structure of the Faroese margin is a legacy of a prolonged history of extension and rifting that is related to the fragmentation of Pangaea, which ultimately led to continental break-up to the north and west of the Faroe Islands in the earliest Eocene (Doré *et al.* 1999; Roberts *et al.* 1999; Passey & Hitchen 2011; Ritchie *et al.* 2011; Ólavsdóttir *et al.* 2013; Stoker *et al.*, this volume, in press).

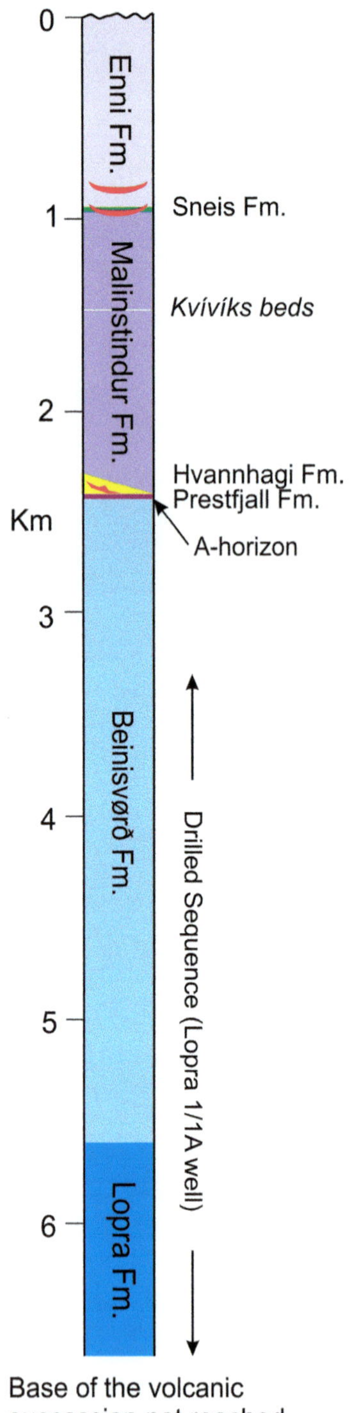

Fig. 3. Stratigraphic subdivision of the onshore Faroe Islands Basalt Group (FIBG), including the drilled sequence from the Lopra 1/1A well. The figure is modified from Passey & Jolley (2009).

A thick succession of Paleocene–Eocene pre- and syn-break-up volcanic rocks – the Faroe Islands Basalt Group (FIBG) (Fig. 3) that is part of the NAIP – covers almost the entire Faroese region: extending eastwards from the continental–ocean transition (COT) into the Faroe–Shetland Basin (Passey & Jolley 2009; Passey & Hitchen 2011) (Figs 1b & 4). As a consequence, very little is known about the pre-Cenozoic geological framework of the continental margin. On seismic reflection profiles, the quality of the data imaged below the top of the volcanic sequence is of poor resolution, which has hindered interpretation of the pre-volcanic strata. Mesozoic and Upper Palaeozoic rocks as old as Devonian in age (Smith & Ziska 2011) have been proved in wells in the UK sector of the Faroe–Shetland Basin. Although a number of sub-basins and highs have been interpreted in the western half (Faroese sector) of the Faroe–Shetland Basin, it is not possible at present – owing to the limited subsurface imaging of the basalt and the lack of well penetrations – to get a clear picture of the underlying pre-volcanic succession. Refraction seismic data have been used to suggest that Mesozoic and Palaeozoic sedimentary rocks might occur between the basement and the volcanic strata (Richardson *et al.* 1998, 1999; Smallwood & White 2002; White *et al.* 2003, 2008; Bohnhoff & Makris 2004; Raum *et al.* 2005), but this remains unsubstantiated as no well in the Faroese sector has penetrated strata older than Paleocene. Thus, the presence of Upper Palaeozoic and Mesozoic rocks in the Faroese sector of the Faroe–Shetland Basin remains largely inferred. Similarly, the structure and fill of the adjacent basins, including the Munkur, Faroe Bank Channel and north Rockall basins, remains unclear (Keser Neish & Ziska 2005; Stoker *et al.*, this volume, in press).

Consequently, the structural framework of the Faroe–Shetland–north Rockall region remains a 'work in progress'. The general pattern of basins and highs shown in Figure 1a has been established by an iterative process of exploration and interpretation over the last few decades, and the present distribution of structural elements combines the most recent compilations (Keser Neish 2004; Ritchie *et al.* 2011; Funck *et al.* 2014). One of the biggest uncertainties concerns the structure of the major present-day bathymetric highs, such as the Fugloy and Munkagrunnur ridges, as well as the Faroe Platform: in particular, whether these are basement-cored structures or inverted basins. In order to address these issues, this paper presents a regional appraisal of the stratigraphy and structure of the Faroese continental margin. The main objective of the study is to establish what is proven fact and what is inference concerning the structure of the continental margin. In turn,

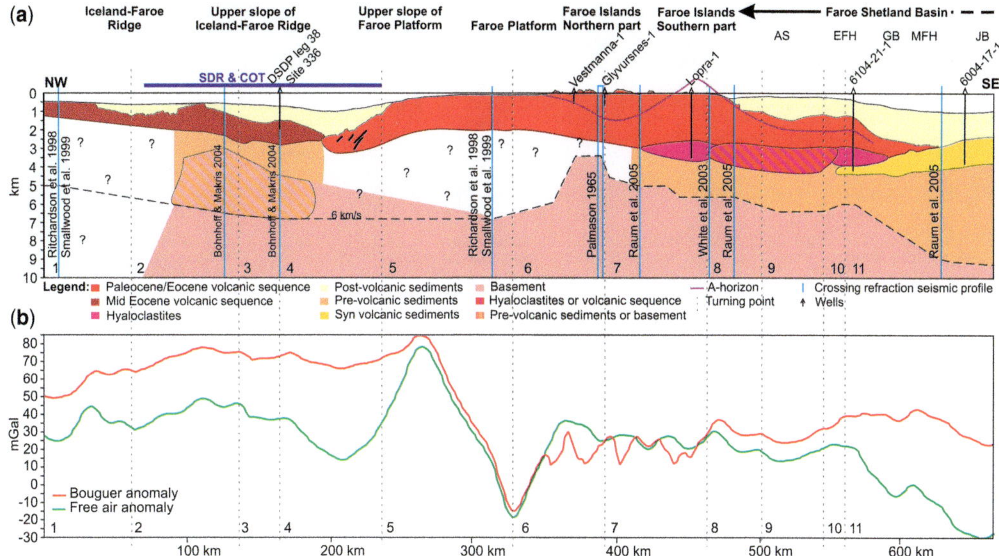

Fig. 4. (**a**) A profile across the Faroese continental margin summarizing the shallow crustal structure as compiled from various refraction seismic studies. The crustal transect is divided into 11 sections that represent specific controlled segments of the profile. The boundaries between the sections are marked with black vertical dashed lines. The vertical purple lines indicate the crossing refraction seismic lines from the various studies utilized in this study. Based on different seismic refraction studies, the top of the crust is marked on the transect at a given location: in addition, the thickness of the volcanic and pre-volcanic sequence can be seen. (**b**) shows the free-air gravity anomaly (green) and Bouguer gravity anomaly (red) values along the transect. Abbreviations: COT, continent–ocean transition; SDRs, seawards-dipping reflectors. See Figure 1a for the key to the abbreviations in the Faroe–Shetland Basin. The location of the profile is shown in Figures 1 and 2.

we hope that this will provide both context and constraints for ongoing discussions of the processes and events that have helped to shape this complex tectonic region.

Geological setting

The Faroe–Shetland Basin dominates the eastern half of the Faroese continental margin (Fig. 1). This NE-trending basin is up to 400 km long and 250 km wide, and consists of a complex amalgam of sub-basins generally separated from one another by north- to NE-trending structural highs (Ritchie *et al.* 2011). The Faroese sector of the Faroe–Shetland Basin comprises the Annika, Brynhild, Grimhild, Guðrun and Steinvør sub-basins, as well as the NW part of the Judd Sub-basin. Whereas their counterparts in the UK sector of the basin, including the Foula, Flett and Judd sub-basins, contain a proven Triassic–Cretaceous fill (Stoker *et al.*, this volume, in press), the known record of sedimentation in the western half of the Faroe–Shetland Basin is, to date, limited to the Cenozoic (Waagstein & Heilmann-Clausen 1995; Andersen *et al.* 2000, 2002; Sørensen 2003; Ólavsdóttir *et al.* 2010, 2013) (Fig. 5). Many of the intervening structural highs consist of, or are underlain by, Archaean (Lewisian affinity) or Proterozoic basement, which are locally capped by Upper Palaeozoic rocks, including Devono-Carboniferous (Ritchie *et al.* 2011; Smith & Ziska 2011). The continuity of the structural highs is interpreted as being disrupted by NW-trending faults and transfer zones or lineaments (Ritchie *et al.* 2011), although the existence and significance of some of these features is debated (Moy & Imber 2009).

According to Ritchie *et al.* (2011), the Fugloy and Munkagrunnur ridges mark the northern and western boundaries, respectively, of the Faroe–Shetland Basin, with the Fugloy Ridge separating the basin from the COT (Fig. 1). Both of these ridges are interpreted as consisting of crystalline basement blocks capped by Mesozoic and/or early Cenozoic rocks (Smallwood *et al.* 2001; Raum *et al.* 2005; Ritchie *et al.* 2011; Funck *et al.* 2014). The present antiformal geometry of the ridges is inferred to have developed in response to later, post-break-up, contractional deformation and/or the effects of differential thermal subsidence, which affected a large part of the continental margin, including the Wyville Thomson Ridge to the SW, particularly

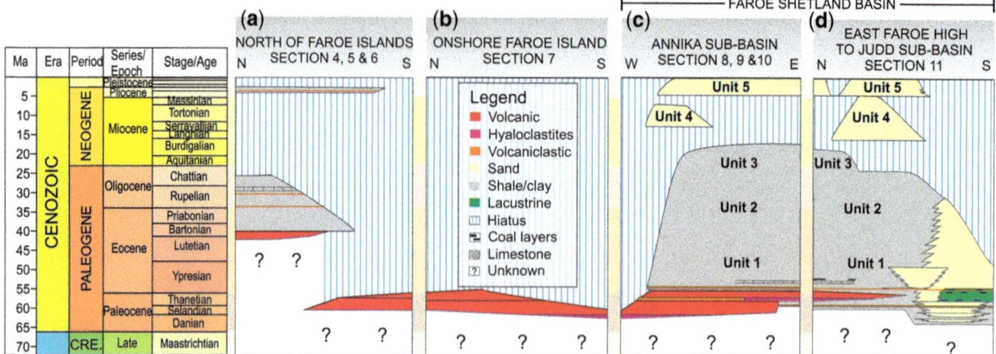

Fig. 5. A regional stratigraphic correlation chart showing the established Cenozoic stratigraphy of the Faroese continental margin. The timescale is based on Gradstein *et al.* (2012).

during the Eocene–Miocene interval (Johnson *et al.* 2005; Ritchie *et al.* 2008). The transition of both the Fugloy Ridge and the Munkagrunnur Ridge with the Faroe Platform is poorly understood.

The post-break-up tectonic movements enhanced the Fugloy and Munkagrunnur ridges as structural highs, and thus helped to create the contemporary bathymetry, whereby the Faroe and West Shetland shelves are separated by the deeper-water Faroe–Shetland Channel, which exceeds a water depth of 1500 m. The latter represents the present-day expression of the Faroe–Shetland Basin, albeit narrower as a consequence of the infilling of the wider Mesozoic basin by episodic shelf-margin progradation of both the East Faroe and West Shetland margins throughout the Cenozoic (Stoker *et al.* 2005, 2013; Ólavsdóttir *et al.* 2010, 2013; Stoker & Varming 2011).

To the NW of the Faroese continental margin, the IFR is a distinct aseismic area of anomalously thick oceanic crust (*c.* 30 km) and shallow ocean floor (á Horni *et al.* 2014) (Fig. 1). This ridge is part of the more extensive GIFR that extends further to the NW beyond Iceland to Greenland, and as such forms a bridge between Greenland and NW Europe. Although it is clear that the GIFR has formed as a result of anomalous melt production, the cause is less certain, and has been variously attributed to the activity of elevated temperatures associated with a mantle plume (White & McKenzie 1989; Smallwood *et al.* 1999) and melting of a fertile upper mantle during plate break-up (Foulger 2002; Lundin & Doré 2005). On the basis of a regional assessment of volcanic seismic facies, á Horni *et al.* (2014) indicated that the boundary between the IFR and the Faroe Platform is marked by a discontinuous cover of volcanic rocks interpreted to be part of the seawards- dipping reflector (SDR) sequence that characterizes the COT (Fig. 1b). However, a key complication along the IFR (and, indeed, the length of the GIFR) is that the volcanic productivity of the region has remained uniformly high from break-up until present. Thus, vertical layering of basaltic flows from different magnetic polarity chrons prevents simple seafloor-spreading anomalies from developing (Richardson *et al.* 1998; Doré *et al.* 1999, 2008; Ritchie *et al.* 2011; Funck *et al.* 2014). In Figures 1–3, the location of the COT is taken from the seafloor-dipping reflector zone of á Horni *et al.* (2014).

Data and methods

In this study, we have summarized all available information from the Faroe continental margin: this includes reflection and refraction seismic data, gravity data, released well data, field observations, and published material. To date, no well has drilled into the rocks that lie below the volcanic sequence, except for some thin siliciclastic layers at the bottom of the Anne Marie well (6004/8a-1: Fig. 1b): thus, there is no physical proof of the composition of the sub-volcanic succession. However, seismic refraction data provide some indications of the nature and character of the sub-volcanic succession, and information has also been compiled from various studies (e.g. Pálmason 1965; Richardson *et al.* 1998; Smallwood & White 2002; White *et al.* 2003; Bohnhoff & Makris 2004; Raum *et al.* 2005) that intersect the chosen transect of this study (Figs 1–3) where important information such as e.g. the top of the crust has been plotted. The transect runs from the SE part of the IFR, across the Faroe Platform and ends in the Faroe–Shetland Basin (the Judd Sub-basin) (Fig. 1). The transect was chosen in an attempt to illustrate a zone that links the better known Faroe–Shetland Basin with the less known Faroe Platform and Iceland–Faroe Ridge areas with respect to the adjacent areas, such as Munkagrunnur and Fugloy ridges. The line of transect was positioned in order to utilize available seismic profiles, where the location of the seismic lines were chosen so that they cut through the wells. Onshore of the Faroe Islands, no reflection seismic data were available and therefore the profile from Waagstein (1988) was used. The section was divided into 11 segments, as shown in Figures 1 and 2, where section 1 is the northernmost part of the transect and section 11 is located in the central Faroe–Shetland Basin.

It is commonly assumed that gravity data cannot be used for distinguishing between crystalline basement rocks in volcanic terrains due to their comparable relatively high density. For example, the density of the Scottish Lewisian Gneiss, which also is expected to underlie the Faroe area, is 2.76–2.84 g cm^{-3} (Watts 1971), whereas investigations into the density of basalt onshore the Faroe Islands give an average density of 2.86 g cm^{-3} (Saxov & Abrahamsen 1964). By way of contrast, the density of volcaniclastic sediments and tuffs is 2.66 g cm^{-3} (Ólavsdóttir *et al.* 2015) and 2.17 g cm^{-3} (Saxov & Abrahamsen 1964), respectively.

We have chosen to calculate the density of the volcanic interval in some of the Faroese wells. The common thought is that the Faroses volcanics contain mostly basalt and should, therefore, have a high density. However, the volcanic interval does not only consist of basalt but also of different volcanic material, such as, for example, hyaloclastites, volcanic sediments, tuffs, vesicular basalt, dolorite, in addition to basalt, which gives the volcanic interval a varying density, as pointed out above. To show this, the average compensated bulk density in three volcanic-rich wells in the Faroese area was found. The mean density of the volcanic interval from 0 to 3137 m in the Lopra-1 well was found to be 2.8 g cm^{-3}, while the density of the volcanic interval from 3100 to 3860 m in the Brugdan-1 well (6104/21-1) was 2.47 g cm^{-3}, and, in a larger interval from 2200 to 4050 m in the Brugdan-II well, the density was found to be 2.67 g cm^{-3} (Fig. 1b). If the volcanic sequence had only consisted of basalt, then the density of the intervals would most likely have been higher, but as it does not, the density is lower than expected. If the volcanic interval had had a higher density than the expected basement at the given position, it would most likely have affected the gravity data, so that areas with thick volcanic intervals – such as sub-basins filled with volcanic material – would act as high gravity anomaly areas.

We note that gravity data are used together with other types of geophysical data to help to resolve the sub-basalt geology in volcanic basins in other parts of the world, such as the Deccan Volcanic Province in India, the Sicily Channel in Italy and the Jungar Basin in China (Diljith *et al.* 2008; Yang *et al.* 2009; Ray *et al.* 2010; Rao *et al.* 2013) where sub-basins have low gravity anomaly values. Thus, we have utilized gravity data in our study as an aid to structural interpretation.

For ease of description, we have presented the results of our study in terms of three main structural domains: the Iceland–Faroe Ridge (IFR); the Faroe Platform (including the Fugloy and Munkagrunnur ridges); and the Faroe–Shetland Basin. We describe each of these domains in terms of what is known about their stratigraphical and structural character, with the latter incorporating seismic and gravity data where available. For the shallow crustal transect, the IFR includes transect sections 1–4; the Faroe Platform includes transect sections 5–7; and the Faroe–Shetland Basin includes transect sections 8–11 (Fig. 4).

Iceland–Faroe Ridge

Stratigraphy

Post-volcanic. On the SE part of the IFR, the post-volcanic sediment cover is thickest adjacent to the Faroe Platform where up to 1 s TWTT (two-way travel time) of Eocene and younger rocks are preserved (Fig. 4). The age of this sediment cover has been proved by DSDP site 336, which was drilled in a water depth of 811 m and had a total penetration below seabed of 515 m. The borehole drilled 168.5 m of Pliocene–Pleistocene marine and glacimarine mud and sporadic sand that unconformably overlies a 295 m-thick sequence of Middle–Upper Eocene–Oligocene marine mudstone, which, in turn, overlies a Middle Eocene volcanic section (see below) in which the borehole terminates (Talwani *et al.* 1976; Stoker & Varming 2011) (Fig. 5a). This sequence forms the eroded featheredge of a much thicker succession (up to 2 km) of post-volcanic sedimentary rocks that are preserved on the upper to mid-slope of the IFR as part of a shelf-margin succession that thickens and progrades northwards into the oceanic Norwegian Basin (Nielsen & van Weering 1998). This succession is also preserved on the northern slope of the Faroe Platform, as well as on the adjacent Fugloy Ridge. At its thickest development, the Eocene–Oligocene succession is up to 1 km thick. Although the Miocene succession is absent in DSDP site 336, a distinctive shelf-margin wedge, up to 550–600 m thick, of probable Miocene–earliest Pliocene age is preserved on the mid-slope region of the IFR. The Pliocene–Pleistocene sequence reaches a maximum thickness of about 200 m in this area of the slope (Nielsen & van Weering 1998).

Volcanic. DSDP site 336 cored 30.5 m of Middle Eocene basalts that have been radiometrically dated by K–Ar to be 41.5 ± 2.5 Ma (Talwani *et al.* 1976; Ganerød *et al.* 2014; Wilkinson *et al.*, this volume, in review), and which form the upper part of the

IFR volcanic section (Figs 1a, 4 & 5a). The basalt is overlain by an 8 m-thick unit of volcanic conglomerate (basaltic rubble), which is overlain by a 13 m-thick unit of red claystone, interpreted as a ferruginous lateritic palaeosol formed by *in situ* weathering of the basaltic ridge (Nilsen 1978; Nilsen & Kerr 1978).

Pre-volcanic. In terms of the pre-volcanic stratigraphy, some information is derived from a NE–SW refraction seismic line orientated approximately perpendicular to the trend of the IFR (Fig. 1b), and which crosses the shallow crustal transect in the southern part of section 2 and at the boundary between sections 3 and 4, where DSDP site 336 is located (Bohnhoff & Makris 2004). At these two intersecting points, a low-velocity layer beneath the volcanic sequence has been proposed, which has been suggested to represent Mesozoic sedimentary rocks, with a thickness of 0.5 and 1.5 km, respectively (Figs 1b & 4).

A pre-volcanic low-velocity layer up to 1 km thick has also been modelled from a refraction seismic line located approximately 30 km in a easterly direction from the junction of transect sections 4 and 5 (Raum *et al.* 2005; Roberts *et al.* 2009) (Fig. 1b). By way of contrast, a refraction seismic line – the FIRE line – orientated along the axis of the IFR did not reveal a low-velocity layer anywhere along the line (Richardson *et al.* 1998; Smallwood *et al.* 1999; Smallwood & White 2002) (Fig. 1b).

Structure. The nature of the crustal structure of the IFR remains ambiguous. Whereas in much of the NE Atlantic region it is relatively easy to define the COT on the basis of magnetic chrons and the SDRs from seismic reflection data, this is not the case for the IFR where neither of these identifiers is definitive. The trend of the SDRs east (Norwegian Basin) and west (Iceland Basin) of the IFR changes direction, so the COT trends – at least in part – along the ridge instead of crossing it (Fig. 1), whereas the magnetic chrons stop when they reach the ridge. In transect section 1, on the SE part of the IFR, Bohnhoff & Makris (2004) interpreted the crust to be about 15 km thick and of a stretched continental character, with the top at a depth of 3 km (Fig. 4). Transect sections 2–4 are located on the COT (Fig. 1) where SDRs are clearly observed.

In terms of the deeper crustal structure, it is interesting to note that the refraction seismic line of Bohnhoff & Makris (2004) is crossed by the FIRE line reported by both Richardson *et al.* (1998) and Smallwood *et al.* (1999) 42 km west of the southern part of transect section 2 (Fig. 1b). Where these two lines intersect, the authors interpret the Moho to be at different depths: Bohnhoff & Makris (2004) suggested a depth of 23 km, whereas Richardson *et al.* (1998) and Smallwood *et al.* (1999) inferred 40 and 35 km, respectively. This represents a variation in the interpreted depth to the Moho of 17 km at the same position, which raises some doubt as to the reliability of the data.

A high free-air anomaly triple junction structure is seen north of the Faroe Islands. One leg crosses eastwards along the western part of the Fugloy Ridge, the second leg is crosses the northern Faroe Platform area and the third leg runs along the SE edge of the IFR (Fig. 2a). The values of the high free-air anomaly triple junction area range from 60 to 70 mGal (Fig. 2a). On the central IFR, just west of the third leg of the triple junction structure, a NW–SE-trending low free-air anomaly area with values down to 10 mGal can be seen.

The triple junction structure is not as clear on the Bouguer anomaly map (Fig. 2b), most probably due to the high values of the Norwegian Basin (up to 90 mGal), but the NW–SE-trending low gravity anomaly values on the central southern IFR can be seen, with values down to 40 mGal (Fig. 2).

Faroe Platform (including the Fugloy and Munkagrunnur ridges)

On the Faroe Platform, volcanic material crops out onshore of the Faroe Islands, as well as at the seabed on the adjacent continental shelf, including the Fugloy and Munkagrunnur ridges. Offshore, there is a general lack of seismic reflection data, and what is available is of poor quality (a consequence of the low signal-to-noise ratio caused by the volcanic sequence): thus, a detailed interpretation of the subvolcanic geology is not possible. By way of contrast, there exists a detailed description of the onshore volcanic succession, which is up to 6.6 km thick, and has been described from a combination of outcrop and well data (e.g. Lopra-1, Glyvursnes-1 and Vestmanna-1) (Figs 1–4).

Stratigraphy

Post-volcanic. On the Faroe Platform, the postvolcanic stratigraphy wholly comprises Quaternary glacial and post-glacial deposits (Jørgensen & Rasmussen 1986).

Volcanic. The onshore volcanic sequence is assigned to the Faroe Islands Basalt Group (FIBG), which is divided into seven formations, comprising (in ascending order): the Lopra, Beinisvørð, Prestfjall, Hvannhagi, Malinstindur, Sneis and Enni formations (Fig. 3) (Passey & Jolley 2009). The Lopra Formation is dominated by volcaniclastic and hyaloclastic rocks containing microfossils indicative of a marine environment, whereas the Beinisvørð, Malinstindur and Enni formations are dominated by basaltic lava flows indicative of an overall terrestrial environment (Rasmussen & Noe-Nygaard 1969, 1970; Berthelsen *et al.* 1984; Passey & Bell 2007; Passey 2009). The basaltic formations are separated by three sedimentary units: the Prestfjall and Hvannhagi formations, which lie between the Beinisvørð Formation and the Malinstindur Formation, and the Sneis Formation, which separates the Malinstindur and Enni formations. These sedimentary units were deposited during significant breaks in the eruption of the lava flows (Passey & Bell 2007). The Prestfjall Formation is a coal-bearing sedimentary sequence (Ellis *et al.* 2002; Passey & Bell 2007); the Hvannhagi Formation comprises interbedded basaltic tuffs and volcaniclastic sedimentary lithologies (Passey & Bell 2007); and the Sneis Formation consists of epi-volcaniclastic mass-flow deposits (Passey 2009). On the basis of radiometric dating and biostratigraphical data, the FIBG is of Paleocene–earliest Eocene age (*c.* 61.0–54.5 Ma) (Waagstein 1988; Waagstein *et al.* 2002; Storey *et al.* 2007; Passey & Jolley 2009; Passey & Hitchen 2011; Mudge 2015) (Figs 3 & 5).

Pre-volcanic. The refraction seismic lines of White *et al.* (2003) and Raum *et al.* (2005) cross the crustal transect in the Faroe Platform area in the central part of transect section 7 and in the northern part of transect section 8, and a pre-volcanic sedimentary layer, up to 2 km thick, has been inferred. In contrast, in the southern part of transect section 6, Richardson *et al.* (1998) and Pálmason (1965) – who shot onshore refraction lines – did not interpret a pre-volcanic sedimentary layer, though their studies were more focused on defining the top of the basement.

Structure. Aspects of the structure of the Faroe Platform can be discerned at several levels: (1) a broad pattern of folding is evident from the gravity data, together with the structural disposition of some of the volcanic formations in the FIBG; (2) fault analysis on the Faroe Islands provides an insight into extensional stresses that have affected the platform; and (3) seismic refraction data provide information on the depths to basement and the Moho.

The gravity anomalies in the area show variations in the free-air and Bouguer anomalies across the Faroe Platform that could suggest a pattern of domes and troughs (Figs 2 & 4). Both free-air and Bouguer anomaly gravity data show low values in the NW and southern part of the islands (Saxov & Abrahamsen 1964; Saxov 1966). The free-air and the Bouguer gravity values in the NW part of the Faroe Islands are −20 and −15 mGal, respectively, while the values in the southern part of the islands are 10 and 20 mGal, respectively. The gravity values in the area between these two low gravity areas (in the central part of the Faroe Islands) range up to 30 mGal for both free-air and Bouguer (Figs 2 & 4). From the onshore stratigraphy, the doming structure in the southern part of the Faroe Islands (Waagstein 1988; Andersen *et al.* 2002; Ólavsdóttir *et al.* 2013) can be observed from the structural disposition of the A-horizon (the unconformable boundary between the Prestfjall and Beinisvørð formations: Figs 3 & 4a) where section 7 and 8 intersect, whereas part of the doming structure in the NW part of the Faroe Islands can be seen in section 6 (Fig. 4a). These domal structures are coincident with the low gravity areas.

In terms of shallow brittle deformation of the Faroe Platform, a preferred orientation and evolution of faults, fractures and dykes onshore the Faroe Islands has been noted by various authors (e.g. Rasmussen & Noe-Nygaard 1969; Geoffroy *et al.* 1994; Walker *et al.* 2011; Ziska 2012). In summary, the fracture pattern reveals a progressive anti-clockwise rotation of extension vectors from NE–SW/east–west to NW–SE immediately prior to and following early Palaeogene continental break-up in this area. The significance of this evolutionary trend is that immediately prior to break-up there was a distinct period of margin-parallel extension, which might have implications for NW–SE-trending sub-volcanic basins/rifts on the platform (see the Discussion later in this paper).

In terms of the deeper structure, a number of refraction seismic lines cross section 5 and the offshore part of section 7 (Richardson *et al.* 1998; Smallwood & White 2002; White *et al.* 2002; Raum *et al.* 2005) (Figs 1b & 4). Of particular note is the interpretation of the FIRE line (Richardson *et al.* 1998; Smallwood & White 2002), which crosses the western part of section 5 of the transect. At this position, the depth to crust is interpreted to be 6 km, while the depth to Moho is interpreted to be 43 km in both of these studies cases. Onshore seismic refraction data suggests that depth to basement is between 3 and 6 km (Pálmason 1965) (Figs 1b & 4).

The extension of the Faroe Platform area to the east and south incorporates two major antiformal structures: the Fugloy Ridge and the Munkagrunnur Ridge. On both of these ridges, the volcanic sequence crops out at the seabed, which results in poor seismic resolution. On the southern part of the Munkagrunnur Ridge, the free-air and the Bouguer gravity data show values up to 60 and 70 mGal, respectively, while the northern part of the ridge has lower values of 20 and 30 mGal, respectively (Fig. 2). This low gravity area is an extension of the low gravity area in the southern part of Faroe Platform, where the island of Suðuroy is situated, and has an overall NW–SE trend. Seismic refraction lines across the Munkagrunnur Ridge indicate a pre-volcanic sediment thickness of up to 4.5 km beneath the western part of the ridge, whereas the volcanic sequence has a thickness of approximately 500–2000 m (White *et al.* 2003). Funck *et al.* (2014) have suggested that the remnants of two pre-break-up volcanoes are preserved on both the eastern and western flanks of the central part of Munkagrunnur Ridge. This interpretation is supported by the high free-air and Bouguer gravity anomalies (Hitchen & Ritchie 1987) of the southern part of the ridge, which range from 50 to 70 mGal (Hitchen & Ritchie 1987).

On the Fugloy Ridge, the western part of the ridge has free-air values of up to 50 mGal, whereas the eastern end displays lower values down to 10 mGal (Fig. 2). On the basis of refraction seismic lines crossing the Fugloy Ridge close to its junction with the Faroe Platform, it has been suggested that Cretaceous and pre-Cretaceous sediments are present below the outcropping volcanic sequence (Raum *et al.* 2005). According to various authors, the Fugloy Ridge has been shaped, at least in part, by contractional deformation, especially during the Miocene (Boldreel & Andersen 1993, 1998; Ólavsdóttir *et al.* 2013), and particularly along the eastern part of the ridge where low gravity values are indicated. However, its deeper interpretation remains ambiguous. The possibility that the ridge is entirely basement-cored has been proposed by Keser Neish (2004) and Ritchie *et al.* (2011).

Faroe–Shetland Basin

Stratigraphy

Post-volcanic. Reactivation of older, Palaeozoic and Mesozoic, structural elements such as the lineaments seem to control the sediment pathways and restrict the depositional areas because the basement highs act as obstacles (Ólavsdóttir *et al.* 2013).

The post-volcanic deposits in the Faroese sector of the Faroe–Shetland Basin have been divided into the following five regional formations/units (1–5) (Ólavsdóttir *et al.* 2013): (1) Lower–Middle Eocene; (2) Upper Eocene–Oligocene; (3) Early Miocene; (4) Middle Miocene–Lower Pliocene; and (5) Lower Pliocene–Holocene (Fig. 5c, d). It is clear that the sediment thickness is significant in the sub-basins, such as Annika, Corona, Guðrun and Steinvør, where the thickness ranges up to 1500, 3500, 2000 and 2000 m, respectively, while the post-volcanic sediments thin out against the ridges (e.g. the East Faroe, Heri, Sjúrður and Tróndur highs), where the thickness ranges up to approximately 1000 m.

Unit 1 is found across the entire Faroe Shetland area, but is thicker in the central part of the Faroe–Shetland Basin east of East Faroe High (Fig. 5c & d). The maximum thickness of Unit 1 is in the Corona Sub-basin, where it has a thickness of approximately 2000 m. Unit 2 appears across almost the entire area, except in the Judd Sub-basin. The maximum thickness of Unit 2 is in the Guðrun Sub-basin, where it has a thickness of 1200 m (Ólavsdóttir *et al.* 2013) (Fig. 5c, d).

Unit 3 is mostly concentrated in the NE part of the Faroese section of the Faroe–Shetland Basin (Fig. 5c, d), with a maximum thickness of 800 m in the Steinvør Sub-basin (Ólavsdóttir *et al.* 2013). Unit 4 has been encountered more sporadically across the Faroe–Shetland Basin (Fig. 5c, d) (Ólavsdóttir *et al.* 2013). The lack of preservation may be controlled by the sub-basin inversion affecting the areas during Miocene time (e.g. Annika, Brinhild, Grimhild and Guðrun sub-basins).

Unit 5 is highly concentrated in the Annika Sub-basin (Fig. 5c, d), with a thickness of 500 m, and forms part of a regional sediment wedge that progrades eastwards from the Faroe Platform, and downlaps and thins into the central part of Faroe–Shetland Basin (Ólavsdóttir *et al.* 2013).

Volcanic. Volcanic rocks underlie most of the Faroe–Shetland Basin, with the exception of the Judd Sub-basin, although, even in this region, the rocks are heavily intruded by sills. There are no wells in the Annika Sub-basin, therefore it is not possible to say much about the composition or thickness of the volcanic sequence in this area. However, Raum *et al.* (2005) have suggested a thickness of up to 3000 m on the basis of refraction seismic data (Fig. 4a). On the East Faroe High, well 6104/21-1 drilled approximately 3000 m of volcanic rocks without reaching the base of the succession (Fig. 4a). On the Mid Faroe High, well 6004/8a-1 drilled a thick pile of volcanic material with a thickness of 1300 m. Information from these two wells further indicates that only approximately 20–25% of the volcanic material is sub-areal basalt: most of the material is volcaniclastic, of which hyaloclastite deposits form an important component, and is especially concentrated towards the base of the succession. A significant volcaniclastic component is also characteristic of the volcanic stratigraphy proved in the Lopra-1 well (Fig. 1), where the lowermost 1000 m of the drilled succession consists of hyaloclastites.

Pre-volcanic. Refraction seismic data from the Faroese part of the Faroe–Shetland Basin suggest that there is a low-velocity layer under the volcanic sequence (White *et al.* 2003; Raum *et al.* 2005). This sequence is further interpreted to be heavily intruded by sills, with an approximate thickness of 2000 m in the Annika Sub-basin, but possibly with an thickness of between 6000 and 7000 m further towards the SE (Raum *et al.* 2005) (Fig. 4a). However, it should be noted that none of the wells so far drilled in the Faroese part of the Faroe–Shetland Basin have penetrated deeper than early Paleocene strata.

Structure. The arrangement of sub-basins and highs that form part of the Faroe–Shetland Basin (Fig. 1) was established by Keser Neish (2004) and Ritchie *et al.* (2011). However, this structural pattern remains to be improved, as the seismic data available for the current interpretation were of poor quality. Potential field data have provided some additional insights into the structural configuration. The gravity anomaly pattern in the Faroe–Shetland Basin is similar to the domes and troughs described from the Faroe Platform. In the Faroe–Shetland Basin, the existence of such a fold pattern is confirmed by seismic data throughout the basin (including the UK sector) where numerous antiforms and synforms have been reported (Sørensen 2003; Johnson *et al.* 2005; Ólavsdóttir *et al.* 2013). In the Faroe–Shetland Basin, some of these antiforms are situated in low gravity areas displaying a NE–SW trend, such as in the Annika, Brinhild, Corona, Grimhild, Guðrun, Judd and Steinvør sub-basins (Figs 1, 2 & 3b), and the age of these dome structures ranges from the Eocene to the Miocene. Other dome structures are situated on high gravity areas such as the Corona, East Faroe, Mid Faroe and Tróndur highs (Figs 1, 2 & 3b). Ritchie *et al.* (2011) interpreted these areas as basement structural high areas. These basement structural high areas are displaying an overall NE–SW trend.

In terms of the deeper structure, Raum *et al.* (2005) suggested that depth to crystalline basement in the northern part of the Annika Sub-basin is 8 km (Figs 1b & 4a): in the central part of the sub-basin, however, another refraction seismic line from the same authors suggests a depth to basement of approximately 5.5 km. In the Judd Sub-basin, the depth to top basement has been estimated at approximately 9 km (Fig. 4a) (Raum *et al.* 2005). To date, no well in the Faroese sector of the Faroe–Shetland Basin has drilled into crystalline basement.

Discussion

Whereas the syn- to post-break-up development of the Faroese continental margin, NE Atlantic Ocean, is fairly well documented, the widespread and commonly thick cover of the Paleocene–Early Eocene FIBG hinders our understanding of its pre-break-up structural and geological framework. In terms of the pre-break-up setting, the following discussion focuses on the key observations and interpretations regarding the nature and character of the sub-basalt (FIBG) crustal structure and stratigraphical framework; we will also address the issue of the ambiguity of the definition of the COT on the IFR.

Sub-basalt (FIBG) crustal structure and stratigraphical framework

The compilation – based on seismic refraction data – of the shallow crustal structure underlying the Faroese continental margin presented in Figure 3 provides a variable interpretation of the sub-basalt geology (Richardson *et al.* 1998, 1999; Smallwood & White 2002; White *et al.* 2003; Bohnhoff & Makris 2004; Raum *et al.* 2005; Funck *et al.* 2008). The compilation has focused on the boundary between the 'basement' and an overlying low-velocity layer, which immediately underlies the FIBG and its lateral correlatives in the Faroe–Shetland Basin and on the IFR. This interpretation is most robust in the area between the central Faroe Platform (including the Faroe Islands) and the Faroe–Shetland Basin. This is consistent with the basement depth model shown in Funck *et al.* (2014) based on several refraction seismic lines in the Faroe–Shetland Basin: however, their model is based on no data coverage in the Faroe Platform area. In contrast, the presence of a low-velocity layer below the NW part of the Faroe Platform and, indeed, the very nature of the shallow crustal layer below the SE part of the IFR remain equivocal. Indeed, this ambiguity even extends to the deeper crustal levels beneath the IFR, where a difference of up to 17 km has been reported for the depth to the Moho (Richardson *et al.* 1998; Smallwood *et al.* 1999; Bohnhoff & Makris 2004). This highlights a variable degree of reliability attached to the interpretation of refraction seismic data, whereby contrasting and conflicting interpretations of the different surveys (and even between interpretations of the same dataset obtained by various researchers) are common. The 370 km-long ISFA profile illustrated by Bohnhoff & Makris (2004) has 43 ocean-bottom seismometer (OBS) stations located perpendicular to the Iceland–Faroe Ridge over a 370 km-long distance where they shoot a 60 l sleeve gun every 250 m along the profile. The FIRE profile illustrated in the work of Richardson *et al.* (1998) and Smallwood *et al.* (1999) has seven OBH stations placed along the Iceland–Faroe Ridge, along which they shot a 9324 in^3 airgun every 75 m. The depth to the Moho in the two studies of the FIRE line are more equal (5 km difference) than with the ISFA profile, where the difference is 12 and 17 km, respectively. The variation in the depth to the Moho in these two profiles could be caused by the difference in line orientation, the variation in the number of OBH/OBS and/or the difference in the size of the airguns/sleeve guns that were used. In addition, the perceived starting model for each study might be different, which might impact on the way in which the study concludes. Such a variation in the results of these refraction data highlights the continuing ambiguity of these types of study.

The basement shown in Figure 3 is generally defined as 'crystalline basement' (Raum *et al.* 2005). The probable composition of this upper-crustal layer is informed by well and outcrop data from the adjacent UK West Shetland area and the East Greenland Kangerlussuaq area, between which the Faroe Islands is commonly placed in pre-break-up reconstructions (e.g. Larsen & Whitham 2005). In the UK sector of the Faroe–Shetland Basin, crystalline basement has been proven in a number of wells to the west and north of Shetland. West of Shetland, the crystalline basement consists predominantly of Archaean rocks (Lewisian Gneiss), which have been assigned the term 'Faroe–Shetland Block' as an indicator of the crustal domain (Skogseid *et al.* 2000). In eastern Greenland, exposed gneissic basement of Archean age is also described from the Kangerlussuaq region (Larsen *et al.* 2006). This forms part of the Archaean–Proterozoic Central Greenland Craton, and has been suggested as a correlative of the Faroe–Shetland Block (Skogseid *et al.* 2000). A separate crustal domain to the north of Shetland, termed the Erlend–East Shetland Block, comprises a mixed Late Archaean–Proterozoic assemblage, with a predominance of Proterozoic ages. The wider correlation of this crustal block is more enigmatic, and could suggest a correlation with the Central Greenland Craton or, possibly, the East Greenland Terrane, located to the north (Doré *et al.* 1999; Skogseid *et al.* 2000).

As noted earlier, the low-velocity layer beneath the volcanic sequence and the top of basement is interpreted fairly consistently between the central part of the Faroe Platform and the Faroe–Shetland Basin (Fig. 4). Raum *et al.* (2005) estimated that this layer ranges from being 2 km thick below the Faroe Platform to a maximum of 8 km in thickness in the Faroe–Shetland Basin. In comparison with the UK sector of the Faroe–Shetland Basin, the composition of the low-velocity layer has been attributed to the occurrence of Upper Palaeozoic–Mesozoic sedimentary rocks. Cretaceous rocks up to 5 km thick have been proved in the UK sector, together with varying thicknesses of Jurassic, Permo-Triassic and Devono-Carboniferous rocks (Ritchie *et al.* 2011; Ellis & Stoker 2014). The continuation of the low-velocity layer beneath the NW part of the Faroese continental margin (and, possibly, on to the SE part of the IFR?) remains uncertain, although the occurrence of Mesozoic basins on the outermost part of the Rockall–Hatton margin, to the SW of the study area (Ellis & Stoker 2014), indicates that there is no reason why a similar arrangement of basins might not exist beneath the NW Faroe Platform, or even the Fugloy Ridge (cf. Raum *et al.* 2005). One important implication of this scenario is the definition of the western margin of the Faroe–Shetland Basin. The current structural framework (i.e. Ritchie *et al.* 2011) locates the boundary of the basin along the margins of the Fugloy and Munkagrunnur ridges (Figs 1 & 2). However, as described previously, these are late-stage (post-break-up) antiformal structures (Johnson *et al.* 2005; Ritchie *et al.* 2008). If basins do exist to the west of the currently accepted boundary, this might warrant a reappraisal of the definition of the Faroe–Shetland Basin.

Support for sub-volcanic basins beneath the Faroe Platform, the Munkagrunnur Ridge and the IFR is provided by the regional gravity maps, which show that there are indications of two NW–SE-trending low gravity areas that could be indicative of sub-basins. Inspection of Figure 2b shows one of these NW–SE-trending gravity lows extending between the area NW of Mykines towards Suðuroy in the SE to the central Munkagrunnur Ridge area; and a second gravity low running between the Annika Sub-basin across the Heri High, the Brynhild Sub-basin and the Sjúrður High into the Judd Sub-basin. These two low gravity areas are sub-parallel and separated by a higher gravity area situated east of the Faroe Islands. A third trending gravity low area can be seen on the SE area of the IFR. One could speculate that the trend of these gravity lows – if they represent sub-volcanic basins – reflects the pre-break-up phase of margin-parallel extension proposed by Geoffroy *et al.* (1994), Lundin & Doré (2005), Walker *et al.* (2011) and Ziska (2012).

Evidence in favour of the southern part of the Faroe Islands/Platform being a former sub-basin can be found in the bottom of the Lopra-1 well, which penetrated through 1000 m of marine hyaloclastic deposits (Ellis *et al.* 2002). Overlying the hyaloclastites is subareal volcanic material dated to an age of approximately 61 Ma (Storey *et al.* 2007). This suggests that such a basin could have acted as a depocentre for pre- and syn-break-up material (Cretaceous–early Paleocene) prior to burial beneath the widespread subareal volcanic rocks. It is interesting to note that traces of hydrocarbons were recorded in samples from the Beinisvørð Formation in Vágoy and Suðuroy (White 1989; Smallwood *et al.* 1999). Smallwood *et al.* (1999) suggested that these hydrocarbons were probably generated from a marine shale source rock, perhaps with a slight marly character, at early to peak oil-window maturity where the age of the source rock is suggested to be Jurassic or younger.

Seismic reflection interpretation throughout the Faroe–Shetland Basin indicates that all of the sub-basins – whether pre- or post-volcanic – in the Faroese part of the Faroe–Shetland Basin were inverted during the Eocene–Holocene (Ólavsdóttir *et al.* 2013). In terms of the possible NW-trending pre-volcanic basins described previously, the area west of the island of Mykines, and in the area where Suðuroy is situated, have been uplifted more than the surrounding areas of the Faroe Platform (Fig. 4) during Miocene–Pliocene time, and have since been affected by erosion of approximately 2000 and 700 m, respectively (Jørgensen 1984, 2006; Andersen *et al.* 2002). Further east, the Annika and Brynhild sub-basins were inverted during the Miocene, and the Heri High was formed at this time. This raises the possibility that this 'structural high', as depicted by Ritchie *et al.* (2011), might be more accurately described as a domal structure formed during the inversion process.

Continent–ocean transition on the Iceland–Faroe Ridge

As described earlier, the vertical layering of basaltic flows from different magnetic polarity chrons has prevented the delineation of a simple pattern of seafloor-spreading anomalies on the IFR: thus, the definition of the COT on the NW margin of the Faroe Islands remains equivocal. The refraction seismic models of Smallwood & White (2002) and Bohnhoff & Makris (2004) suggest crustal thicknesses in excess of 20–30 km in the area north of the currently interpreted COT. On the basis of this modelling, it could be speculated that continental crust extends almost to transect section 1 (Fig. 4). The implication of this is that the southern part of the IFR is an extension of continental crust rather than oceanic crust (Bohnhoff & Makris 2004). In the area SE of Iceland, Torsvik *et al.* (2015) suggested that this part of the IFR is old continental crust with a thickness in excess of 30 km, although the central part of the IFR might be thinner (20 km) (Smallwood *et al.* 1999; Smallwood & White 2002).

The potential admixture or juxtaposition of continental and oceanic (e.g. the SDR sequences) crustal elements on the IFR invites speculation on the general origin of the GIFR. The GIFR is commonly referred to as a Paleocene–Holocene plume track of the Iceland 'hotspot' (e.g. (White 1989; Skogseid *et al.* 2000), despite the fact that it is not time-transgressive in one direction as the Iceland 'hotspot' has never been positioned below the IFR side of the GIFR (Lundin & Doré 2002, 2005; Foulger 2010), except if the hotspot is situated around the spreading ridge all the time, which is probably untenable. As an alternative, Ellis & Stoker (2014) have suggested that the GIFR was formed by purely plate tectonic mechanisms, with lithospheric thinning and variable decompressive upper-mantle melting being facilitated by transtension along a NW-trending rift that extended between Kangerlussuaq (East Greenland) and the Faroe region. The inclusion of continental fragments within this zone of transtension might have been linked to the complex break-up between East Greenland, Jan Mayen and NW Europe.

Conclusions

Key conclusions from this appraisal of regional geological and geophysical data from the Faroes continental margin include the following:

- Seismic refraction studies suggest that the pre-break-up structure of the Faroese continental margin comprises a low-velocity layer overlying 'basement'. The latter most probably consists of Archaean crystalline rocks, as in pre-break-up reconstructions the Faroe region would have been part of the Central Greenland Craton–Faroe–Shetland Block. The low-velocity layer is inferred to comprise Upper Palaeozoic–Mesozoic strata, although, in common with the adjacent UK sector, this succession is probably dominated by Cretaceous rocks.
- The regional gravity maps indicate NW–SE-trending gravity low areas on and adjacent to the Faroe Platform, as well as the IFR, which might represent pre-break-up sub-volcanic basins. These basins might have acted as depocentres for pre- and syn-break-up Cretaceous–Early Paleocene material, including hyaloclastite deposition that preceded the widespread extrusion of subareal basalts.
- The Faroe Islands Basalt Group (FIBG) represents the syn-break-up phase that led to seafloor spreading to the NW of the Faroe Islands at about 54.5 Ma. The combination of radiometric and biostratigraphic data suggests that this occurred in the interval between approximately 61 and 54.5 Ma (i.e. Mid-Paleocene–earliest Eocene).
- The post-break-up development of the Faroese continental margin has been dominated by contractional deformation, which has contributed to the creation of the present-day large-scale bathymetry of the Faroe–Shetland region, including the Fugloy and Munkagrunnur ridges. The preserved pattern of sedimentation is a response to these vertical movements. The combination of gravity and refraction seismic data suggests that present-day highs, such as the Fugloy and Heri highs, might represent inverted basins.
- The identification of the continent–ocean transition (COT) on the IFR remains equivocal. The generally thick crust below the IFR, together

with the absence of well-defined SDRs, might be an indication that much of the IFR is underlain by continental material.

We would like to acknowledge the support of the industry sponsors (in alphabetical order): Bayerngas Norge AS; BP Exploration Operating Company Ltd; Bundesanstalt für Geowissenschaften und Rohstoffe (BGR); Chevron East Greenland Exploration A/S; ConocoPhillips Skandinavia AS; DEA Norge AS; Det norske oljeselskap ASA; DONG E&P A/S; E.ON Norge AS; ExxonMobil Exploration and Production Norway AS; Japan Oil, Gas and Metals National Corporation (JOGMEC); Maersk Oil; Nalcor Energy – Oil and Gas Inc.; Nexen Energy ULC, Norwegian Energy Company ASA (Noreco); Repsol Exploration Norge AS; Statoil (UK) Ltd; and Wintershall Holding GmBH. The contribution of MSS is with the permission of the Executive Director of the British Geological Survey (NERC).

Reference

Á HORNI, J., GEISSLER, W.H. *ET AL.* 2014. Offshore volcanic facies. *In*: HOPPER, J.R., FUNCK, T., STOKER, M.S., ÁRTING, U., PERON-PINVIDIC, G., DOORNENBAL, H. & GAINA, C. (eds) *Tectonostratigraphic Atlas of the North-East Atlantic Region.* Geological Survey of Denmark and Greenland (GEUS), Copenhagen, 235–253.

ANDERSEN, M.S., NIELSEN, T., SØRENSEN, A.B., BOLDREEL, L.O. & KUIJPERS, A. 2000. Cenozoic sediment distribution and tectonic movements in the Faroe region. *Global and Planetary Change*, **24**, 239–259, https://doi.org/10.1016/s0921-8181(00)00011-4

ANDERSEN, M.S., SØRENSEN, A.B., BOLDREEL, L.O. & NIELSEN, T. 2002. Cenozoic evolution of the Faroe Platform: comparing denudation and deposition. *In*: DORÉ, A.G., CARTWRIGHT, J.A., STOKER, M.S., TURNER, J.P. & WHITE, N. (eds) *Exhumation of the North Atlantic Margin: Timing, Mechanisms and Implications for Petroleum Exploration.* Geological Society, London, Special Publications, **196**, 291–311, https://doi.org/10.1144/GSL.SP.2002.196.01.16

ARCHER, S.G., BERGMAN, S.C., ILIFFE, J., MURPHY, C.M. & THORNTON, M. 2005. Palaeogene igneous rocks reveal new insights into the geodynamic evolution and petroleum potential of the Rockall Trough, NE Atlantic Margin. *Basin Research*, **17**, 171–201, https://doi.org/10.1111/j.1365-2117.2005.00260.x

BERTHELSEN, O., NOE-NYGAARD, A. & RASMUSSEN, J. 1984. The Deep Drilling Project 1980–1981 in the Faeroe Islands. *Annales Societatis Scientiarum Færoensis, Supplementum*, **9**.

BOHNHOFF, M. & MAKRIS, J. 2004. Crustal structure of the southeastern Iceland-Faeroe Ridge (IFR) from wide aperture seismic data. *Journal of Geodynamics*, **37**, 233–252.

BOLDREEL, L.O. & ANDERSEN, M.S. 1993. Late Paleocene to Miocene compression in the Faeroe–Rockall area. *In*: PARKER, J.R. (ed.) *Petroleum Geology of Northwest Europe: Proceedings of the 4th Conference.* Geological Society, London, 1025–1034.

BOLDREEL, L.O. & ANDERSEN, M.S. 1998. Tertiary compressional structures on the Faroe–Rockall Plateau in relation to northeast Atlantic ridge-push and Alpine foreland stresses. *Tectonophysics*, **300**, 13–28, https://doi.org/10.1016/s0040-1951(98)00231-5

DILJITH, D.T., KUMAR, N. & SINGH, B. 2008. Imaging of sub-basalt geology in the Deccan Volcanic Province of cental India from gravity studies. *In*: *Proc. SPG 7th International Conference & Symposium on Petroleum, Geophysics*, 14–16 January 2008, Hyderabad, India, 346–351.

DORÉ, A.G., LUNDIN, E.R., JENSEN, L.N., BIRKELAND, Ø., ELIASSEN, P.E. & FICHLER, C. 1999. Principal tectonic events in the evolution of the northwest European Atlantic margin. *In*: FLEET, A.J. & BOLDY, S.A.R. (eds) *Petroleum Geology of Northwest Europe – Proceedings of the 5th Conference.* Geological Society, London, 41–61.

DORÉ, A.G., LUNDIN, E., KUSZNIR, N.J. & PASCAL, C. 2008. Potential mechanisms for the genesis of Cenozoic domal structures on the NE Atlantic margin: pros, cons and some new ideas. *In*: JOHNSON, H., DORÉ, A.G., GATLIFF, R.W., HOLDSWORTH, R., LUNDIN, E. & RITCHIE, J.D. (eds) *The Nature and Origin of Compression in Passive Margins.* Geological Society, London, Special Publications, **306**, 1–26, https://doi.org/10.1144/SP306.1

ELLIS, D. & STOKER, M.S. 2014. The Faroe–Shetland Basin: a regional perspective from the Paleocene to the present day and its relationship to the opening of the North Atlantic Ocean. *In*: CANNON, S.J.C. & ELLIS, D. (eds) *Hydrocarbon Exploration to Exploitation West of Shetlands.* Geological Society, London, Special Publications, **397**, 11–31, https://doi.org/10.1144/SP397.1

ELLIS, D., BELL, B.R., JOLLEY, D.W. & O'CALLAGHAN, M. 2002. The stratigraphy, environment of eruption and age of the Faroes Lava Group, NE Atlantic Ocean. *In*: JOLLEY, D.W. & BELL, B.R. (eds) *The North Atlantic Igneous Province: Stratigraphy, Tectonic, Volcanic and Magmatic Processes.* Geologial Society, London, Special Publications, **197**, 253–269, https://doi.org/10.1144/GSL.SP.2002.197.01.10

FOULGER, G. 2002. Plumes, or plate tectonic processes? *Astronomy and Geophysics*, **43**, 19–23.

FOULGER, G. 2010. *Plates vs Plumes: A Geological Controversy.* Wiley-Blackwell, Oxford.

FUNCK, T., ANDERSEN, M.S., NEISH, J.C.K. & DAHL-JENSEN, T. 2008. A refraction seismic transect from the Faroe Islands to the Hatton–Rockall Basin. *Journal of Geophysical Research*, **113**, B12405, https://doi.org/10.1029/2008JB005675

FUNCK, T., HOPPER, J.R. *ET AL.* 2014. Crustal structure. *In*: HOPPER, J.R., FUNCK, T., STOKER, M.S., ÁRTING, U., PERON-PINVIDIC, G., DOORNENBAL, H. & GAINA, C. (eds) *Tectonostratigraphic Atlas of the North-East Atlantic Region.* Geological Survey of Denmark and Greenland (GEUS), Copenhagen, 69–126.

GANERØD, M., SMETHURST, M.A. *ET AL.* 2010. The North Atlantic Igneous Province reconstructed and its relation to the Plume Generation Zone: the Antrim Lava Group revisited. *Geophysical Journal International*, **182**, 183–202, https://doi.org/10.1111/j.1365-246X.2010.04620.x

GANERØD, M., WILKINSON, C.M. & HENDRIKS, H. 2014. Geochronology. *In*: HOPPER, J., FUNCK, T., STOKER, M.S., ÁRTING, U., PÉRON-PINVIDIC, G., DOORNENBAL, H. & GAINA, C. (eds) *NAG-TEC Atlas: Tectonostratigraphic Atlas of the North-East Atlantic Region*. Geological Survey of Denmark and Greenland (GEUS), Copenhagen, 253–261.

GEOFFROY, L., BERGERAT, F. & ANGELIER, J. 1994. Tectonic evolution of the greenland-Scotland ridge during the Paleogene: new constraints. *Geology*, **22**, 653–656.

GERNIGON, L., GAINA, C., OLESON, O., BALL, P.J., PÉRON-PINVIDIC, G. & YAMASAKI, T. 2012. The Norway Basin revisited: from continental breakup to spreading ridge extinction. *Marine and Petroleum Geology*, **35**, 1–19, https://doi.org/10.1016/j.marpetgeo.2012.02.015

GRADSTEIN, F.M., OGG, J.G., SCHMITZ, M.D. & OGG, G.M. 2012. *The Geologic Time Scale*. Elsevier, Amsterdam.

HAASE, C. & EBBING, J. 2014. Gravity data. *In*: HOPPER, J.R., FUNCK, T., STOKER, M.S., ÁRTING, U., PÉRON-PINVIDIC, G., DOORNENBAL, H. & GAINA, C. (eds) *NAG-TEC Atlas: Tectonostratigraphic Atlas of the North-East Atlantic Region*. Geological Survey of Denmark and Greenland (GEUS), Copenhagen, 29–39.

HITCHEN, K. & RITCHIE, J.D. 1987. Geological review of the West Shetland area. *In*: BROOKS, J. & GLENNIE, K.W. (eds) *Petrouleum Geology of Northwest Europe, Proceedings of the 3rd Conference*. Graham & Trotman, London, 737–749.

JOHNSON, H., RITCHIE, J.D., HITCHEN, K., MCINROY, D.B. & KIMBELL, G.S. 2005. Aspects of the Cenozoic deformational history of the Northeast Faroe–Shetland Basin, Wyville–Thomson Ridge and Hatton Bank areas. *In*: DORÉ, A.G. & VINING, B.A. (eds) *Petroleum Geology: North-West Europe and Global Perspectives – Proceedings of the 6th Petroleum Geology Conference*. Geological Society, London, 993–1007.

JØRGENSEN, G. & RASMUSSEN, J. 1986. Glacial striae, roches moutonnées and ice movements in the Faroe Islands. *Danmarks Geologiske Undersøgelse*, DGU Series C, No. 7.

JØRGENSEN, O. 1984. Zeolite zones in the basaltic lavas of the Faeroe Islands. *In*: BERTHELSEN, O., NOE-NYGAARD, A. & RASMUSSEN, J. (eds) *The Deep Drilling Project 1980–1981 in the Faeroe Islands*. Føroya Fróðskaparfelag (Faroese Scientific Society), Tórshavn, Faroe Islands, 71–91.

JØRGENSEN, O. 2006. The regional distribution of zeolites in the basalts of the Faroe Islands and the significance of zeolites as palaeotemperature indicators. *In*: CHALMERS, J.A. & WAAGSTEIN, R. (eds) *Scientific Results from the Deepened Lopra-1 Borehole, Faroe Islands*. Geological Survey of Denmark and Greenland (GEUS), Copenhagen, 123–156.

KESER NEISH, J. 2004. *A Standard Structural Nomenclature System*. Faroese Geological Survey, Tórshavn.

KESER NEISH, J. & ZISKA, H. 2005. Structure of the Faroe Bank Channel Basin, offshore Faroe Islands. *In*: DORÉ, A.G. & VINING, B.A. (eds) *Petroleum Geology: North-West Europe and Global Perspectives – Proceedings of the 6th Petroleum Geology Conference*. Geological Society, London, 873–885.

LARSEN, M. & WHITHAM, A. 2005. Evidence for a major sediment input point into the Faroe–Shetland Basin from Kangerlussuaq region of southern East Greenland. *In*: DORÉ, A.G. & VINING, B.A. (eds) *Petroleum Geology: North-West Europe and Global Perspectives – Proceedings of the 6th Petroleum Geology Conference*. Geological Society, London, 913–922.

LARSEN, M., KNUDSEN, C., FREI, D., FREI, M., RASMUSSEN, T. & WITHHAM, A.G. 2006. East Greenland and Faroe–Shetland sediment provenance and Palaeogene sand dispersal systems. *Geological Survey of Denmark and Greenland Bulletin*, **10**, 29–32.

LUNDIN, E. & DORÉ, A.G. 2002. Mid-Cenozoic post-breakup deformation in the 'passive' margins bordering the Norwegian–Greenland Sea. *Marine and Petroleum Geology*, **19**, 79–93, https://doi.org/10.1016/s0264-8172(01)00046-0

LUNDIN, E.R. & DORÉ, A.G. 2005. NE Atlantic break-up: a re-examination of the Iceland mantle plume model and the Atlantic–Arctic linkage. *In*: DORÉ, A.G. & VINING, B.A. (eds) *Petroleum Geology: North-West Europe and Global Perspectives – Proceedings of the 6th Petroleum Geology Conference*. Geological Society, London, 739–754.

MOY, D.J. & IMBER, J. 2009. A critical analysis of the structure and tectonic significance of rift-oblique lineaments ('transfer zones') in the Mesozoic–Cenozoic succession of the Faroe-Shetland Basin, NE Atlantic margin. *Journal of the Geological Society, London*, **166**, 831–844, https://doi.org/10.1144/0016-76492009-010

MUDGE, D.C. 2015. Regional controls on Lower Tertiary sandstone distribution in the North Sea and NE Atlantic margin basins. *In*: MCKIE, T., ROSE, P.T.S., HARTLEY, A.J., JONES, D.W. & ARMSTRONG, T.L. (eds) *Tertiary Deep-Marine Reservoirs of the North Sea Region*. Geological Society, London, Special Publications, **403**, 17–42, https://doi.org/10.1144/SP403.5

NIELSEN, T. & VAN WEERING, T.C.E. 1998. Seismic stratigraphy and sedimentary processes at the norwegian Sea margin northeast of the Faeroe Islands. *Marine Geology*, **152**, 141–157.

NILSEN, T.H. 1978. Lower Tertiary laterite on the Iceland-Faeroe Ridge and the Thulean land bridge. *Nature*, **274**, 786–788.

NILSEN, T.H. & KERR, D.R. 1978. Turbidites, redbeds, sedimentary structures and trace fossils observed in DSDP leg 38 cores and the sedimentary history of the Norwegian-Greenland Sea. *In*: *Deep Sea Drilling Project: Supplement to Volumes XXXVIII, XXXIX, XL and XLI*. Government Printing Office, Washington, DC, 259–288.

ÓLAVSDÓTTIR, J., BOLDREEL, L.O. & ANDERSEN, M.S. 2010. Development of a shelf margin delta due to uplift of Munkagrunnur Ridge at the margin of Faroe-Shetland Basin: a seismic sequence stratigraphic study. *Petroleum Geoscience*, **16**, 91–103, https://doi.org/10.1144/1354-079309-014

ÓLAVSDÓTTIR, J., ANDERSEN, M.S. & BOLDREEL, L.O. 2013. Seismic stratigraphic analysis of the Cenozoic sediments in the NW Faroe Shetland Basin – Implications for inherited structural control of sediment distribution. *Marine and Petroleum Geology*, **46**, 19–35, https://doi.org/10.1016/j.marpetgeo.2013.05.012

ÓLAVSDÓTTIR, J., ANDERSEN, M.S. & BOLDREEL, L.O. 2015. Reservoir quality of intra-basalt volcaniclastic units onshore Faroe Islands, North Atlantic Igneous Province, NE Atlantic. *American Association of Petroleum Geologists Bulletin*, **99**, 467–497, https://doi.org/10.1306/08061412084

PÁLMASON, G. 1965. Seismic refraction measurements of the basalt lavas of the Faeroe Islands. *Tectonophysics*, **2**, 475–482.

PASSEY, S.R. 2009. *Recognition of a Faulted Basalt Lava Flow Sequence Through the Correlation of Stratigraphic Marker Units, Skopunarfjørður, Faroe Islands*. Annales Societatis Scientiarum Færoensis, Tórshavn.

PASSEY, S.R. & BELL, B.R. 2007. Morphologies and emplacement mechanisms of the lava flows of the faroe islands basalt group, faroe islands, NE atlantic ocean. *Bulletin of Volcanology*, **70**, 139–156, https://doi.org/10.1007/s00445-007-0125-6

PASSEY, S.R. & HITCHEN, K. 2011. Cenozoic (igneous). *In*: RITCHIE, J.D., ZISKA, H., JOHNSON, H. & EVANS, D. (eds) *Geology of the Faroe–Shetland Basin and Adjacent Areas. BGS Research Report, RR/11/01*. British Geological Survey, Keyworth, Nottingham, 209–228.

PASSEY, S.R. & JOLLEY, D.W. 2009. A revised lithostratigraphic nomenclature for the palaeogene Faroe Islands basalt Group, North Atlantic. *Earth and Environmental Science Transaction of the Royal Society of Edinburgh*, **99**, 127–158.

RAO, B.N., SINGH, B., SINGH, A.P. & TIWARI, V.M. 2013. Resolving sub-basalt Geology from Joint Analysis of Gravity and Magnetic Data over the Deccan Trap of Central India. *Geohorizons*, **18**, 57–63.

RASMUSSEN, J. & NOE-NYGAARD, A. 1969. *Beskrivelse til Geologisk Kort over Færøerne i målestok 1:50.000*. C. A. Reitzels Forlag (Jørgen Sandal), Copenhagen.

RASMUSSEN, J. & NOE-NYGAARD, A. 1970. *Geology of the Faroe Islands (Pre-Quaternary)*. (HENDERSON, G., trans.) Geological Survey of Denmark Series 1, **25**.

RAUM, T., MJELDE, R. *ET AL*. 2005. Sub-basalt structures east of the Faroe Islands revealed from wide-angle seismic and gravity data. *Petroleum Geoscience*, **11**, 291–308, https://doi.org/10.1144/1354-079304-627

RAY, A., WITTE, S. & MCALLISTER, E. 2010. Sicily Channel – integrated interpretation of gravity and magnetic data. *Paper presented at the EGM 2010 International Workshop*, 11–14 April 2010, Capri, Italy.

RICHARDSON, K.R., SMALLWOOD, J.R., WHITE, R.S., SNYDER, D.B. & MAGUIRE, P.K.H. 1998. Crustal structure beneath the Faroe Islands and the Faroe–Iceland Ridge. *Tectonophysics*, **300**, 159–180, https://doi.org/10.1016/s0040-1951(98)00239-x

RICHARDSON, K.R., WHITE, R.S., ENGLAND, R.W. & FRUEHN, J. 1999. Crustal structure east of the Faroe Islands: mapping sub-basalt sediments using wide-angle seismic data. *Petroleum Geoscience*, **5**, 161–172, https://doi.org/10.1144/petgeo.5.2.161

RITCHIE, J.D., JOHNSON, H., QUINN, M.F. & GATLIFF, R.W. 2008. Cenozoic compressional deformation within the Faroe–Shetland Basin and adjacent areas. *In*: JOHNSON, H., DORÉ, A.G., HOLDSWORTH, R.E., GATLIFF, R.W., LUNDIN, E.R. & RITCHIE, J.D. (eds) *The Nature and Origin of Compression in Passive Margins*. Geological Society, London, Special Publications, **306**, 121–136, https://doi.org/10.1144/SP306.5

RITCHIE, J.D., ZISKA, H., KIMBELL, G., QUINN, M. & CHADWICK, A. 2011. Structure. *In*: RITCHIE, J.D., ZISKA, H., JOHNSON, H. & EVANS, D. (eds) *Geology of the Faroe–Shetland Basin and Adjacent Areas. BGS Research Report, RR/11/01*. British Geological Survey, Keyworth, Nottingham, 9–70.

ROBERTS, A.W., WHITE, R.S. & CHRISTIE, P.A.F. 2009. Imaging igneous rocks on the North Atlantic rifted continental margin. *Geophysical Journal International*, **179**, 1024–1038, https://doi.org/10.1111/j.1365-246X.2009.04306.x

ROBERTS, D.G., THOMPSON, M., MITCHENER, B., HOSSACK, J., CARMICHAEL, S. & BJØRNSETH, H.-M. 1999. Palaeozoic to Tertiary Rift and Basin Dynamics: mid-Norway to the Bay of Biscay – a new context for hydrocarbon prospectivity in the deep water frontier. *In*: FLEET, A.J. & BOLDY, S.A.R. (eds) *Petroleum Geology of Northwest Europe: Proceedings of the 5th Conference*. Geological Society, London, 7–40.

SAUNDERS, A.D., FITTON, J.G., KERR, A.C., NORRY, M.J. & KENT, R.W. 1997. The North Atlantic Igneous Province. *In*: MAHONEY, J.J. & COFFIN, M.L. (eds) *Large Igneous Provinnce, Continental Oceanic, and Planetary Flood Volcanism*. American Geophysical Union, Geophysical Monographs, **100**, 45–93.

SAXOV, S. & ABRAHAMSEN, N. 1964. A Note on some gravity and density measurings in the Faroe Islands. *Bollettino di Geofisica Teorica ed Applicata*, **VI**, 249–262.

SAXOV, S. & ABRAHAMSEN, N. 1966. Some geophysical investigations in the Faroe Islands. A preliminary report. *Zeitschrift fur Geophysik*, **32**, 455–471.

SKOGSEID, J., PLAMKE, S., FALEIDE, J.I., PEDERSEN, T., ELDHOLM, O. & NEVERDAL, F. 2000. NE Atlantic continental rifting and volcanic margin formation. *In*: NØTTVEDT, A. (ed.) *Dynamics of the Norwegian Margin*. Geological Society, London, Special Publications, **167**, 295–326, https://doi.org/10.1144/GSL.SP.2000.167.01.12

SMALLWOOD, J.R. & WHITE, R.S. 2002. Ridge-plume interaction in the North Atlantic and its influence on continental breakup and seafloor spreading. *In*: JOLLEY, D.W. & BELL, B.R. (eds) *The North Atlantic Igneous Province: Stratigraphy, Tectonic, Volcanic and Magmatic Processes*. Geological Society, London, Special Publications, **197**, 15–37, https://doi.org/10.1144/GSL.SP.2002.197.01.02

SMALLWOOD, J.R., STAPLES, R.K., RICHARDSON, K.R. & WHITE, R.S. 1999. Crust generated above the Iceland mantle plume: from continental rift to oceanic spreading center. *Journal of Geophysical Research*, **104**, 22,885–822,902.

SMALLWOOD, J.R., TOWNS, M.J. & WHITE, R.S. 2001. The structure of the Faeroe-Shetland Trough from integrated deep seismic and potential field modelling. *Journal of the Geological Society, London*, **158**, 409–412, https://doi.org/10.1144/jgs.158.3.409

SMITH, K. & ZISKA, H. 2011. Devonian and carboniferous. *In*: RITCHIE, J.D., ZISKA, H., JOHNSON, H. & EVANS, D. (eds) *Geology of the Faroe–Shetland Basin and*

Adjacent Areas. BGS Research Report, RR/11/01. British Geological Survey, Keyworth, Nottingham, 79–91.

SØRENSEN, A.B. 2003. Cenozoic basin development and stratigraphy of the Faroes area. *Petroleum Geoscience*, **9**, 189–207, https://doi.org/10.1144/1354-079302-508

STOKER, M.S. & VARMING, T. 2011. Cenozoic (sedimentary). *In*: RITCHIE, J.D., ZISKA, H., JOHNSON, H. & EVANS, D. (eds) *Geology of the Faroe-Shetland Basin and Adjacent Areas. BGS Research Report, RR/11/01*. British Geological Survey, Keyworth, Nottingham, 151–208.

STOKER, M.S., HOULT, R.J. *ET AL.* 2005. Sedimentary and oceanographic responses to early Neogene compression on the NW European margin. *Marine and Petroleum Geology*, **22**, 1031–1044, https://doi.org/10.1016/j.marpetgeo.2005.01.009

STOKER, M.S., LESLIE, A.B. & SMITH, K. 2013. A record of Eocene (Stronsay Group) sedimentation in 1 BGS borehole 99/3, offshore NW Britain: implications for early post-breakup development of the Faroe-Shetland Basin. *Scottish Journal of Geology*, **49**, 1–17, https://doi.org/10.1144/sjg2013-001

STOKER, M.S., STEWART, M.A. *ET AL.* In press. An overview of the Upper Palaeozoic–Mesozoic stratigraphy of the NE Atlantic region. *In*: PÉRON-PINVIDIC, G., HOPPER, J., GAINA, C., STOKER, M.S., DOORNENBAL, H., FUNCK, T. & ÁRTING, U. (eds) *The NE Atlantic Region: A Reappraisal of Crustal Structure, Tectonostratigraphy and Magmatic Evolution*. Geological Society, London, Special Publications, **447**, https://doi.org/10.1144/SP447.2

STOREY, M., DUNCAN, R.A. & TEGNER, C. 2007. Timing and duration of volcanism in the North Atlantic Igneous Province: implications for geodynamics and links to the Iceland hotspot. *Chemical Geology*, **241**, 264–281.

TALWANI, M., UDINTSEV, G. *ET AL.* 1976. Sites 336 and 352. *In*: TALWANI, M., UDINTSEV, G. *ET AL.* (eds) *Initial Reports of the Deep Sea Drilling Project*. Government Printing Office, Washington, DC, 23–116.

TORSVIK, T.H., AMUNDSEN, H.E.F. *ET AL.* 2015. Continental crust beneath southeast Iceland. *Proceedings of the National Academy of Sciences*, **112**, E1818–E1827, https://doi.org/10.1073/pnas.1423099112

WAAGSTEIN, R. 1988. Structure, composition and age of the Faeroe basalt plateau. *In*: MORTON, A.C. & PARSON, L.M. (eds) *Early Tertiary Volcanism and the Opening of the NE Atlantic*. Geological Society, London, Special Publications, **39**, 225–238, https://doi.org/10.1144/GSL.SP.1988.039.01.21

WAAGSTEIN, R. & HEILMANN-CLAUSEN, C. 1995. Petrography and biostratigraphy of Palaeogene volcanclastic sediments dredged from the Faeroes shelf. *In*: SCRUTTON, R.A., STOKER, M.S., SHIMMIELD, G.B. & TUDHOPE, A.W. (eds) *The Tectonics, Sedimentation and Palaeoceanography of the North Atlantic Region*. Geological Society, London, Special Publications, **90**, 179–197, https://doi.org/10.1144/GSL.SP.1995.090.01.11

WAAGSTEIN, R., GUISE, P. & REX, D. 2002. K/Ar and ^{39}Ar/^{40}Ar whole-rock dating of zeolite facies metamorphosed flood basalts: the upper Paleocene basalts of the Faroe Islands, NE Atlantic. *In*: JOLLEY, D.W. & BELL, B.R. (eds) *The North Atlantic Igneous Province: Stratigraphy, Tectonic, Volcanic and Magmatic Processes*. Geological Society, London, Special Publications, **197**, 219–251, https://doi.org/10.1144/GSL.SP.2002.197.01.09

WALKER, R.J., HOLDSWORTH, R.E., IMBER, J. & ELLIS, D. 2011. The development of cavities and clastic infills along fault-related fractures in Tertiary basalts on the NE Atlantic margin. *Journal of Structural Geology*, **33**, 92–106, https://doi.org/10.1016/j.jsg.2010.12.001

WATTS, A.B. 1971. Geophysical investigations on the continental shelf and slope north of Scotland. *Scottish Journal of Geolocy*, **7**, 189–218.

WHITE, R.S. 1988. A hot-spot model for early Tertiary volcanism in the N Atlantic. *In*: MORTON, A.C. & PARSON, L.M. (eds) *Early Tertiary Volcanism and the Opening of the NE Atlantic*. Geological Society, London, Special Publications, **39**, 3–13, https://doi.org/10.1144/GSL.SP.1988.039.01.02

WHITE, R.S. 1989. Initiation of the Iceland plume and opening of the North Atlantic. *In*: TANKARD, A.J. & BALKWILL, H.R. (eds) *Extensional Tectonics and Stratigraphy of the North Atlantic Margins*. American Association of Petroleum Geologists, Memoirs, **46**, 149–154.

WHITE, R.S. & MCKENZIE, D. 1989. Magmatism at rift zones: the generation of volcanic continental margins and flood basalts. *Journal of Geophysical Research*, **94**, 7685–7729.

WHITE, R.S., CHRISTIE, P.A.F. *ET AL.* 2002. iSIMM pusher frontiers of marine seism. *First Break*, **20**, 782–786.

WHITE, R.S., SMALLWOOD, J.R., FLIEDNER, M.M., BOSLAUGH, B., MARESH, J. & FRUEHN, J. 2003. Imaging and regional distribution of basalt flows in the Faeroe-Shetland Basin. *Geophysical Prospecting*, **51**, 215–231, https://doi.org/10.1046/j.1365-2478.2003.00364.x

WHITE, R.S., SMITH, L.K., ROBERTS, A.W., CHRISTIE, P.A.F., KUSZNIR, N.J. & THE iSIMM TEAM. 2008. Lower-crustal intrusion on the North Atlantic continental margin. *Nature*, **452**, 460–464.

WILKINSON, C.M., GANERØD, M., HENDRIKS, B.W.H. & EIDE, E.A. In review. Compilation and appraisal of geochronological data from the North Atlantic Igneous Province (NAIP). *In*: PÉRON-PINVIDIC, G., HOPPER, J., GAINA, C., STOKER, M.S., DOORNENBAL, H., FUNCK, T. & ÁRTING, U. (eds) *The NE Atlantic Region: A Reappraisal of Crustal Structure, Tectonostratigraphy and Magmatic Evolution*. Geological Society, London, Special Publications, **447**.

YANG, H., WEN, B. *ET AL.* 2009. Distribution of hydrocarbon traps in volcanic rocks and optimization for selecting exploration prospects and targets in Junggar Basin: case study in Ludong-Wucaiwan area, NW China. *Petroleum Exploration and Development*, **36**, 419–427.

ZISKA, H. 2012. Fracture orientations onshore Faroe Islands (North Atlantic); evidence for dual rifting episodes in the Palaeogene? *In*: VARMING, T. & ZISKA, H. (eds) *Proceedings of the 3rd Faroe Islands Exploration Conference. Annales Societatis Scientarum Færoensis Supplementum*, **56**, 40–58.

Ziska, H. & Varming, T. 2008. Palaeogene evalution of the Ymir and Wyville Thomson Ridge, European North Atlantic Margin. *In*: Johnson, H., Dore, A.G., Gatliff,. R.W., Holdsworth, R., Lundin, E. & Ritchie, J.D. (eds) *The Nature and Origin of Compression in Passive Margins*. Geological Society, London, Special Publications, **306**, 153–168, https://doi.org/10.1144/SP306.7

Review of velocity models in the Faroe–Shetland Channel

UNI K. PETERSEN[1]* & THOMAS FUNCK[2]

[1]*Faroese Earth and Energy Directorate, Brekkutún 1, FO 110 Tórshavn, Faroe Islands*

[2]*Geological Survey of Denmark and Greenland, Øster Voldgade 10, 1350 Copenhagen K, Denmark*

**Correspondence: up@jf.fo*

Abstract: Over the last few decades, a number of wide-angle seismic experiments have been conducted in the Faroe–Shetland Channel area with the objective of mapping the crustal structure. However, the volcanic rocks covering most of the area present a challenge for the imaging of sub-basalt structures. The results of the seismic studies are consistent in describing the Faroe–Shetland Channel as thinned continental crust and in establishing the presence of sub-basalt sediments. However, the various datasets often show differences in depth to crystalline basement and to the Moho. This paper presents a review of the velocity models in the Faroe–Shetland Channel and analyses the differences at line intersections. Down to top basalt the models are fairly consistent, while there are deviations of up to 1 km s^{-1} in basalt velocities and sub-basalt sediment velocities, 2 km in basalt thickness, 3.2 km in depth to crystalline basement, and 11.7 km in depth to the Moho.

The Faroe–Shetland Channel (Fig. 1) is characterized by Palaeogene volcanic rocks associated with the North Atlantic Igneous Province and the break-up of the North Atlantic (Larsen *et al.* 1999). The volcanic rocks cover large areas around the Faroe Islands and make the imaging of deeper sections challenging. Over the last few decades, a number of wide-angle seismic experiments have been conducted in the Faroe–Shetland Channel area with the objective of mapping the crustal velocity structure.

The different surveys (Hughes *et al.* 1998; Richardson *et al.* 1999; White *et al.* 1999; Fliedner & White 2003; Raum *et al.* 2005; Eccles *et al.* 2007; Makris *et al.* 2009; Roberts *et al.* 2009) describe the crust beneath the Faroe–Shetland Channel as thinned continental crust. Further, they all indicate the presence of sub-basalt sediments. However, a comparison of intersecting profiles shows that there are a number of differences between crossing and adjacent lines (Funck *et al.* 2014). In this paper, we review all velocity models in the Faroe–Shetland Channel that are based on seismic refraction data, and some of them also on gravity data, acquired since 1994. The objective is to quantify the differences at line intersections and to discuss the implications.

The variable seismic properties within the basalt column with a succession of high-impedance contrasts leads to attenuation and scattering of seismic waves during the propagation through the basalts (e.g. Maresh *et al.* 2006). The adverse effect that basalt has on the imaging of sub-basalt structures is very well illustrated by Neish (2004, p. 142, fig. 15). She showed that there is a large area on the Faroese continental shelf where base basalt and sub-basalt structures are difficult to interpret on seismic reflection data.

One of the main objectives for most refraction experiments around the Faroes is the improvement of the sub-basalt imaging (e.g. Fliedner & White 2003), which is difficult to achieve with seismic reflection data alone. Seismic refraction data can hold information on sub-basalt properties in these areas by providing observations of seismic phases at large offsets.

Comparison of velocity models

Table 1 summarizes the acquisition parameters of the seismic lines reviewed in this study. To ease the comparison of the lines, digital versions of the velocity models were produced. For line AMP-D, a digital model was available, while the remaining models were digitized based on publications and existing reports. The digital models were obtained by converting the colour values according to the colour-scale bar in figures (FLARE and Mobil surveys), by reading velocity contours (iSIMM line) or by converting annotated velocity models (AMG95 and FAST surveys). The location of the models is based on digital navigation (FAST, FLARE and iSIMM surveys) or on digitized location maps

From: Péron-Pinvidic, G., Hopper, J. R., Stoker, M. S., Gaina, C., Doornenbal, J. C., Funck, T. & Árting, U. E. (eds) 2017. *The NE Atlantic Region: A Reappraisal of Crustal Structure, Tectonostratigraphy and Magmatic Evolution*. Geological Society, London, Special Publications, **447**, 357–374.
First published online September 9, 2016, https://doi.org/10.1144/SP447.7

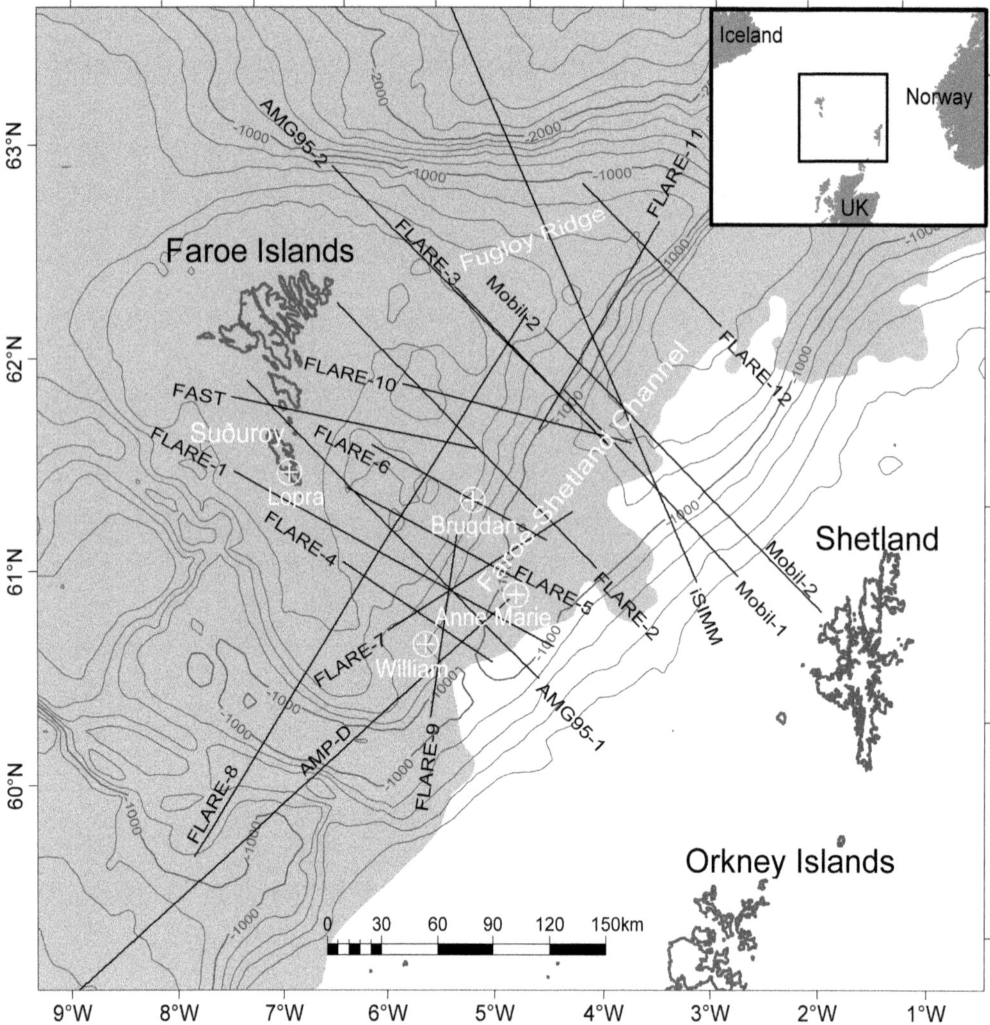

Fig. 1. Location of the reviewed seismic refraction lines (see Table 1 for more details). Bathymetry contours (after Hopper & Gaina 2014) are at 200 m intervals. White crosses show the locations of the Lopra, Brugdan, William and Anne Marie wells. The shaded area shows the basalt cover (after á Horni *et al.* 2014). The inset shows the location of the survey area.

taken from publications (AMG95 and Mobil surveys). Figure 1 shows the location of all wide-angle seismic lines used in this study.

The modelling of travel times is based on ray theory. All velocity models are developed using a combination of forward and inverse modeling, with the exception of the models regarding the Mobil survey that are based on forward modelling only. In addition, the FAST, FLARE and iSIMM surveys use conventional semblance analysis techniques on reflection seismic data for modelling the post-basalt sediments and the top basalt interface. The FLARE, FAST, AMG95 and AMP-D surveys integrate gravity data in the modelling. The modelling procedures for each survey are summarized below. For details on the modeling, we refer readers to the respective publications.

At line intersections, key model features are compared. This includes the depth to crystalline basement and the Moho, as well as the thickness of the basalt sequence and the underlying sediments. In addition, basalt and sub-basalt velocities are compared.

With the exception of FLARE lines 2–12, all velocity models infer Moho for the full length of the profiles, even though large sections of the

Table 1. *Reviewed seismic refraction lines*

Line name	Acquisition year	Acquisition parameters	Publications used for the digitization of the model
AMG95-1 and AMG95-2	1995	**Source**: 79 l airgun array, 150 m shot interval **Receivers**: Ocean-bottom seismographs. 42 stations along AMG95-1 and 40 stations along AMG95-2. 4 km receiver spacing	Raum *et al.* (2005)
AMP-D	1996	**Source**: 120 l airgun array. 120 m shot spacing **Receivers**: Ocean-bottom seismographs, 45 stations with 2– 4 km spacing	Digital model was available (Klingelhöfer *et al.* 2005)
FAST	1994	**Source**: 153 l airgun array. 50 m shot interval **Receivers**: Four land stations with vertical component geophones located on Streymoy Faroe Islands and 6 km streamer	Richardson *et al.* (1999)
FLARE-1	1996	Two-ship acquisition. **Source**: One 85.1 l and one 49.2 l airgun array. 50 m shot interval for each ship. Composite gathers have 100 m shot intervals **Receivers**: One 6 km and one 4.8 km streamer. Offset gap between cables 5.6, 15.2 and 24.8 km. Maximum offset for marine data is 38.4 km. Six additional land seismometers were placed on Suðuroy (Faroe Islands)	White *et al.* (1999)
FLARE-2 and FLARE-3	1996	Two-ship acquisition **Source**: One 85.1 l and one 49.2 l airgun array. 50 m shot interval for each ship. Composite gathers have 100 m shot intervals **Receivers**: One 6-km and one 4.8 km streamer. Offset gap between cables 5.6, 15.2 and 24.8 km. Maximum offset is 38 400 m	Latkiewicz & Kirk (1999)
FLARE-4– FLARE-12	1998	Two-ship acquisition **Source**: One 58.0 l and one 62.3 l airgun array. 50 m shot interval for each ship. Composite gathers have 100 m shot intervals **Receivers**: Two 6 km streamers. Offset gap between cables 6 km. Maximum offset of 18 km	Latkiewicz & Kirk (1999)
iSIMM	2002	**Source**: 103 litre airgun array. 100 m shot interval **Receivers**: Ocean-bottom seismographs, 85 stations with 2–6 km spacing	Roberts *et al.* (2009)
Mobil-1 and Mobil-2	1996	**Source**: 120 litre airgun array. 120 m shot interval **Receivers**: Ocean-bottom seismographs. 24 stations with 3.5 km spacing along Mobil-1 and 42 stations with 3.8 km spacing along Mobil-2	Makris *et al.* (2009)

Moho are not constrained by PmP (reflections from the Moho) and Pn (mantle refractions) arrivals. Where not constrained by seismic data, the Moho is defined by extrapolation, *a priori* information for the area or gravity data. In fact, at no intersection presented here do both profiles have seismic constraints on the depth of the Moho.

Comparison of lines Mobil-1 and AMG95-2

Lines Mobil-1 and AMG95-2 (Fig. 2) are offset by 3.8 km and run parallel to each other in a 35 km-wide zone (Fig. 1). The velocity model for line AMG95-2 (Raum *et al.* 2005) was obtained from forward and inverse modelling of travel times. In addition, information from coincident seismic reflection data was incorporated into the model. No PmP or Pn phases were observed that could determine the Moho depth. For this reason, the Moho was inferred from gravity modelling. Sub-basalt sediment structures and velocities are based on discontinuous horizons interpreted on seismic reflection data in the area.

The initial velocity model for line Mobil-1 (Hughes *et al.* 1998) is based on forward modelling. Similar to line AMG95-2, the Moho depth is

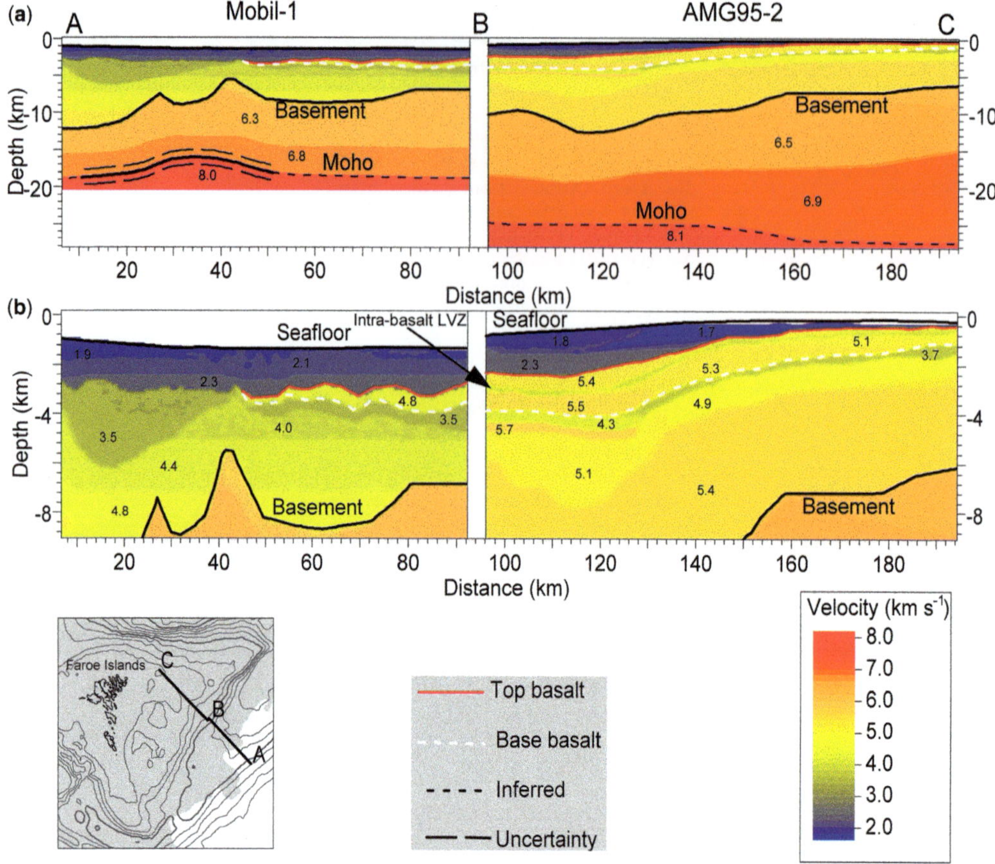

Fig. 2. Comparison of P-wave velocity models of lines Mobil-1 and AMG95-2. (**a**) & (**b**) show different depth scales. The two lines do not intersect but run parallel to each other at a distance of 3.8 km. Annotated velocities are given in km s^{-1}. The map in the lower left-hand corner shows the location of the composite profile. The shaded area shows the basalt cover.

constrained by gravity data as there are no clear PmP observations. Only one ocean-bottom seismometer (OBS) recorded a phase that could be a PmP. However, it is unclear whether the reflection is, indeed, from the Moho and not from intruded or underplated magmatic rocks. In the gravity modelling, the Moho depth was set to 30 km beneath the Shetland Platform and to 40–45 km beneath the Faroe Islands. In the Faroe–Shetland Channel, the Moho was then adjusted to maintain isostatic equilibrium. However, the model shown in Figure 2 is based on the later modelling by Makris *et al.* (2009) without the use of gravity data. They incorporated additional wide-angle seismic data consisting of an OBS array (5 × 8 OBS with a 5 km grid spacing) and a north–south line with 10 OBS crossing this array. This improved the constraints on the Moho along the central part of line Mobil-1 based on PmP reflections. The extent of the line with a seismically constrained Moho in Figure 2 is based on the location of the OBS array, and the uncertainty in Moho depth is estimated from the velocity uncertainties given by Makris *et al.* (2009). The inferred Moho outside the area with seismic constraint appears to be based on extrapolations only. The sub-basalt velocities appear to be based on the extrapolation of velocities from areas not covered by basalts.

While the basalts have a thickness of 2 km on line AMG95-2, they are only 1 km thick on line Mobil-1. It is not clear how Makris *et al.* (2009) determined the thickness of the basalts on line Mobil-1: however, the earlier model by Hughes *et al.* (1998) displays a similar basalt thickness that is constrained by the lateral extent of the refraction in the basalt layer using a velocity gradient obtained from synthetic amplitude modelling. The basalt thickness on line Mobil-1 corresponds

Table 2. *Comparison of features of the velocity models at the intersection of lines Mobil-1 and AMG95-2*

	Mobil-1	AMG95-2
Moho	18.7 km	25 km
Crystalline basement	6.8 km	10 km
Basalt thickness	1 km	2 km
Basalt velocity	4.8 km s^{-1}	5.4 km s^{-1}
Sub-basalt sediment thickness	3.3 km	6 km
Sub-basalt sediment velocities	3.5–4.4 km s^{-1}	4.4–5.5 km s^{-1}

roughly to the upper part of the basalt sequence on line AMG95-2, defined by an intra-basalt low-velocity zone.

A comparison of the key features in the velocity models is given in Table 2. The depth to crystalline basement differs by 3.2 km. Velocities in the sub-basalt sediments are 1.0 km s^{-1} higher on line Mobil-1 than on line AMG95-2. Differences in the sub-basalt velocities can explain some of the misfit in the depth to crystalline basement but are not sufficient to account for the full 3.2 km. The difference in the sub-basalt sediment velocities is difficult to assess, as neither Makris *et al.* (2009) nor Raum *et al.* (2005) provide details on how these velocities are constrained. The difference in depth to Moho is 6.3 km, but at the intersection there is no constraint from seismic data.

Comparison of lines Mobil-2 and iSIMM

Lines Mobil-2 and iSIMM intersect in the central part of the Faroe–Shetland Channel (Fig. 3). The model for line Mobil-2 is taken from Makris *et al.* (2009). Similar to line Mobil-1, an earlier model is presented in Hughes *et al.* (1998). That model shows the Moho to be 6 km deeper, based on gravity modelling. The modelling of line Mobil-2 followed the procedure of line Mobil-1 described above and uses the same OBS array and north–south line dataset to constrain the Moho depth at the centre of the profile. Along the remainder of the profile, the depth to Moho is apparently extrapolated.

The velocity model for line iSIMM (Roberts *et al.* 2009) is based on combined forward and inverse modelling of travel times. In addition, conventional semblance analysis techniques on reflection seismic data were used for the modelling of sediment velocities. Special attention was given to model the step back indicative of the low-velocity zone (LVZ) below the basalts. The velocity in the LVZ was determined by forward modelling for the thickness of the LVZ at various fixed velocities. The best fit was obtained using a velocity of 4.5 km s^{-1}. Crystalline basement was not interpreted. However, the coincident seismic reflection data show a strong reflection across the Fugloy Ridge at about 1.25 s two-way travel time (TWT) below the base of the LVZ, which could represent the crystalline basement (Roberts *et al.* 2009). The Moho is constrained by PmP phases throughout the profile, and the uncertainty in the Moho depth is given by Roberts *et al.* (2009).

The velocity model shown in the paper of Roberts *et al.* (2009) does not specify the sediment velocities above the basalts. However, for a portion of the line, Lau *et al.* (2010) provide such information using a pre-stack depth-migration workflow. They obtain post-basalt sediment velocities of 1.6–2.6 km s^{-1}. This velocity range is similar to the intersecting lines FLARE-11 and Mobil-2.

The comparison of the velocity models of lines Mobil-2 and iSIMM is summarized in Table 3 and reveals a 0.7 km difference in basalt thickness. In fact, the basalt layer on line Mobil-2 peters out at the cross-point with line iSIMM. The velocities of the LVZ deviate by 0.9 km s^{-1} at the top but are similar at the base. The depth to the Moho deviates by 1.9 km. Both lines have constraints on the Moho depth, although not at the actual intersection. The modelling of the lines was carried out without incorporation of gravity data and no *a priori* information on Moho depth was used.

Comparison of lines FLARE-1, AMG95-1 and FAST

The composite transect of lines FLARE-1, AMG95-1 and FAST (Fig. 4) describes a zigzag path running from the southernmost island of the Faroes to the centre of the Faroe–Shetland Channel, back to the Faroe Islands and from there back into the channel. The velocity modelling along line AMG95-1 (Raum *et al.* 2005) was carried out in a similar way as described above for line AMG95-2. However, the PmP and Pn phases constrain the Moho depth on line AMG95-1. Raum *et al.* (2005) provided estimates on the depth uncertainty of interfaces.

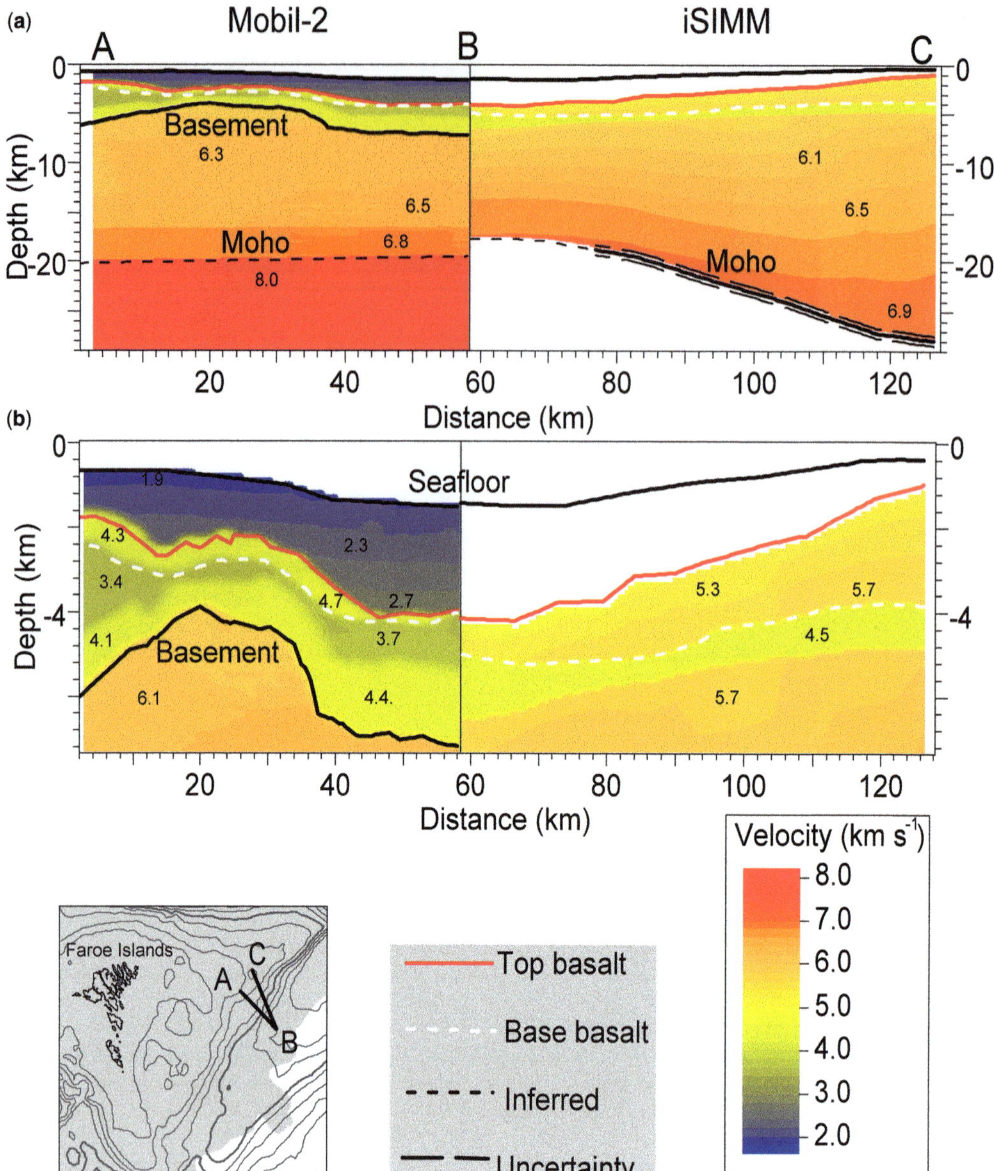

Fig. 3. Comparison of P-wave velocity models of lines Mobil-2 and iSIMM. (**a**) & (**b**) show the different depth scales. Annotated velocities are given in km s^{-1}. The map in the lower left-hand corner shows the location of the composite profile. The shaded area shows the basalt cover. The uncertainty of the Moho depth is set to ± 0.5 km, being representative of the more detailed uncertainties given in Roberts *et al.* (2009).

The shallow velocity structure of line FLARE-1 (Richardson *et al.* 1999; White *et al.* 1999) was determined using conventional semblance analysis techniques and the top basalt was derived from the interpretation of the seismic reflection data. Forward and inverse modelling was used for the velocity analysis in the deeper sections. Constraints on basalt velocities were obtained from vertical seismic profiles (VSP) recorded as part of the FIRE experiment (Richardson *et al.* 1998), the VSP at the Lopra well (Kiørboe & Petersen 1995) and from older seismic refraction data (Palmason (1965). For the lower crust, fixed velocities of 6.4 and 6.8 km s^{-1} at the top and base, respectively, were

Table 3. *Comparison of features of the velocity models at the intersection of lines Mobil-1 and iSIMM*

	Mobil-2	iSIMM
Moho	19.5 km	17.6 km
Crystalline basement	7.1 km	
Basalt thickness	0 km	0.7 km
Basalt velocity	4.7 km s^{-1}	4.8–5.3 km s^{-1}
Sub-basalt sediment thickness	3 km	
Sub-basalt sediment velocities	3.6–4.5 km s^{-1}	4.5 km s^{-1}

used owing to a lack of seismic constraints. These velocities are similar to the ones observed in nearby continental crust in the UK (Richardson *et al.* 1999). Strong PmP reflections recorded on one land station at the NW end of the profile provided some seismic constraints on the Moho depth that helped in the gravity modelling of the Moho along the remainder of the line. Neither Richardson *et al.* (1999) nor White *et al.* (1999) provided estimates on the uncertainty of the Moho depth, which is why a rather large estimate of 2 km is used in Figure 4.

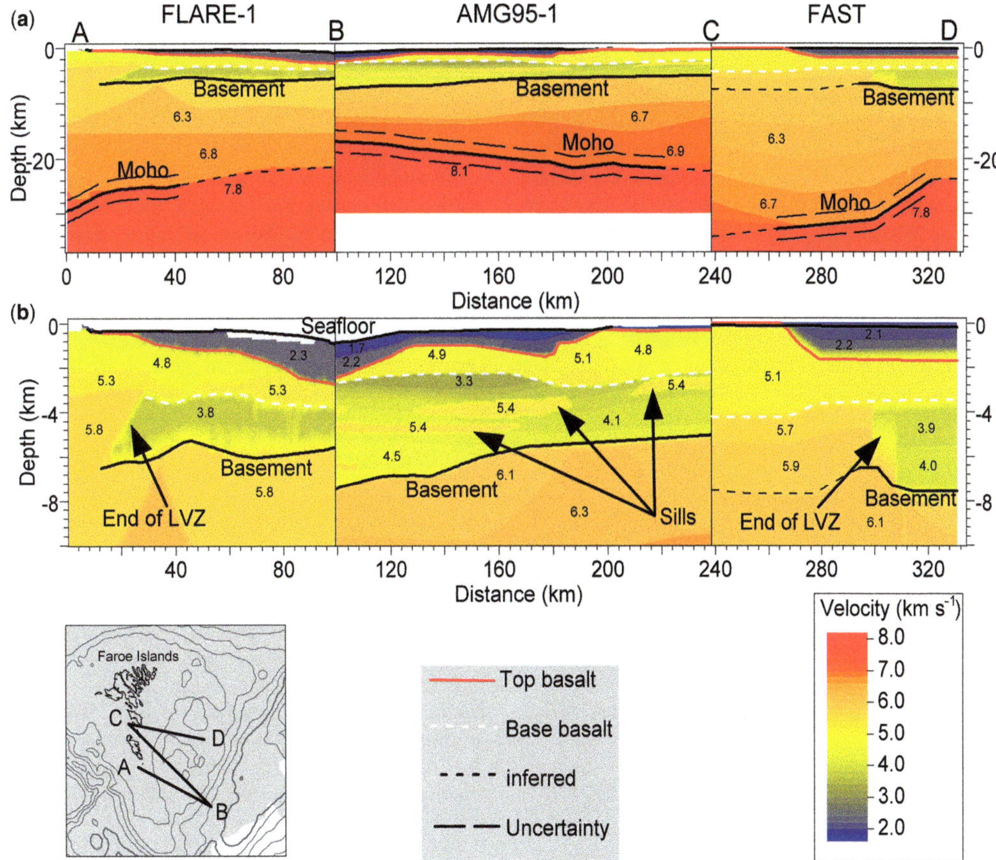

Fig. 4. Comparison of P-wave velocity models of lines FLARE-1, AMG95-1 and FAST. (**a**) & (**b**) show different depth scales. Annotated velocities are given in km s^{-1}. The map in the lower left-hand corner shows the location of the composite profile. The shaded area shows the basalt cover.

Table 4. *Comparison of features of the velocity models at the intersection of lines FLARE-1 and AMG95-1*

	FLARE-1	AMG95-1
Moho	21.5 km	17 km
Crystalline basement	5.6 km	7.4 km
Basalt thickness	1 km	0 km
Basalt velocity	5.25 km s^{-1}	5 km s^{-1}
Sub-basalt sediment thickness	2 km	5 km
Sub-basalt sediment velocities	3.8 km s^{-1}	3.3–4.6 km s^{-1*}

*Not including velocities of sills in the sub-basalt section.

Similar to line FLARE-1, the shallow velocities of line FAST (Richardson *et al.* 1999) were determined using conventional semblance analysis techniques and the top basalt was taken from the interpretation of the stacked profile. The crustal velocity structure was obtained from forward and inverse modelling of travel times from data recorded on the four land seismometers. These seismometers were not deployed along the extension of the shot line but were offset by 27–50 km to the north. High-amplitude wide-angle reflections observed on one of the land stations determined the Moho depth at the NW end. In all other areas, gravity modelling was employed to obtain the depth to the Moho. The uncertainty on the Moho depth was not provided by Richardson *et al.* (1999), and Figure 4 uses therefore an estimate of 2 km similar to line FLARE-1.

Tables 4 and 5 compare the velocity models at their intersections. At the crossing of lines FLARE-1 and AMG95-1, the deviation in basalt thickness of 1 km is most likely to be related to differences in the interpretation of the data. The sub-basalt velocities are around 3.8 km s^{-1} on line FLARE-1, but vary between 3.3 and 4.6 km s^{-1} on line AMG95-1. In addition, the velocity model for line AMG95-1 shows several thick intrusions within the sub-basalt sediments with velocities of 5.15–5.50 km s^{-1}. On lines FLARE-1 and FAST, the LVZ terminates close to the Faroe Islands. This termination is based on observations from the land seismometers. Richardson *et al.* (1999) suggested that the higher velocities beneath the basalts on the Faroe Islands either relate to intrusions or could represent a crystalline basement high. In contrast, the model of line AMG95-1 shows a continuation of the LVZ beneath the Faroe Islands.

Table 5. *Comparison of features of the velocity models at the intersection of lines AMG95-1 and FAST*

	AMG95-1	FAST
Moho	22.3 km	34 km
Crystalline basement	5 km	7.5 km
Basalt thickness	2 km	4 km
Basalt velocity	4.75–5.00 km s^{-1}	5 km s^{-1}

At the intersection of lines AMG95-1 and FAST, the Moho depth is 22.3 and 34 km, respectively. The unreversed ray coverage and the significant offset of the seismic stations relative to the shot line are likely to be the main reasons for the differences. Although the intersection is at an unconstrained location for both lines, this is a quite significant difference, especially when considering that both lines incorporate gravity data in the velocity modelling.

It should be noted that the publications on the FLARE survey use the term 'basement' in the meaning of 'seismic basement' corresponding to a strong sub-basalt reflection and not crystalline basement (e.g. Fliedner & White 2001; Fruehn *et al.* 2001). In Fliedner & White (2003, p. 356, fig. 11), the depth to crystalline basement along line FLARE-1 is interpreted at a depth of 10–12 km, whereas the modelled seismic basement is only 5–7 km deep (Fig. 4).

Comparison of lines AMP-D and AMG95-1

Line AMP-D is a strike line in the Faroe–Shetland Channel and is perpendicular to line AMG95-1 that extends to the Faroe Islands (Fig. 1). The velocity model (Fig. 5) of line AMP-D (Klingelhöfer *et al.* 2005) is based on forward and inverse modelling of travel times. In addition, ray-synthetic seismograms were calculated. Large portions of the Moho are constrained by PmP and Pn phases. In addition, gravity modelling was invoked to obtain the Moho depth in the seismically unconstrained segments. The model for line AMP-D has two distinct LVZs beneath basalt layers. Velocities in these LVZs are not constrained, and are set to 3.95–4.15 km s^{-1} for the upper LVZ and 4.2–4.7 km s^{-1} for the lower LVZ. The modelling procedures for line AMG95-1 are described earlier in this paper.

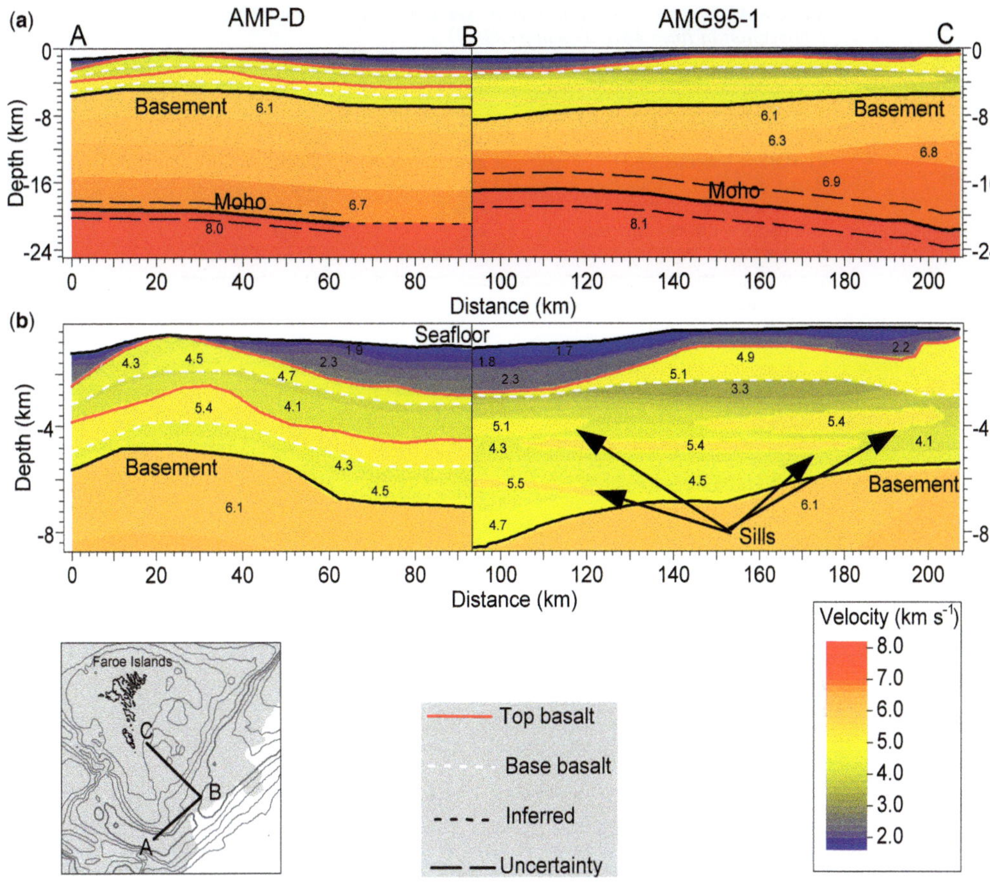

Fig. 5. Comparison of P-wave velocity models of lines AMP-D and AMG95-1. (**a**) & (**b**) show different depth scales. Annotated velocities are given in km s^{-1}. The map in the lower left-hand corner shows the location of the composite profile. The shaded area shows the basalt cover.

A comparison of the model features of the two lines is given in Table 6. The velocity models indicate a less than 0.3 km thin basalt layer at the intersection. Despite this agreement, there are distinct differences in the underlying units. While line AMG95-1 displays several intrusions into the sub-basalt sediments, line AMP-D shows a second continuous basalt layer within these sediments. However, both models use similar velocities for the sub-basalt sediments of 3.95–4.70 km s^{-1} on line AMP-D and 3.7–4.7 km s^{-1} on line AMG95-1. There is a 1.5 km misfit in the depth to basement.

Table 6. *Comparison of features of the velocity models at the intersection of lines AMP-D and AMG95-1*

	AMP-D	AMG95-1
Moho	20.8 km	16.8 km
Crystalline basement	7 km	8.5 km
Basalt thickness (upper)	0.3 km	0.1 km
Basalt velocity	4.6–5.3 km s^{-1}	5–5.5 km s^{-1}
Sub-basalt sediment thickness	3.9 km	5.9 km
Sub-basalt sediment velocities	3.95–4.70 km s^{-1}	3.7–4.7 km s^{-1}*

*Not including velocities of sills in the sub-basalt section.

Table 7. *Comparison of features of the velocity models at the intersection of lines AMP-D and FLARE-1*

	AMP-D	FLARE-1
Moho	21 km	21 km
Crystalline basement	7.1 km	6.0 km*
Basalt thickness	0.3 km	1 km
Basalt velocity	4.70 km s^{-1}	5.25 km s^{-1}
Sub-basalt sediment thickness	4.0 km	3.3 km
Sub-basalt sediment velocities	4.0–4.6 km s^{-1}	3.75–4.00 km s^{-1}

*Seismic basement.

At the Moho level, the deviation between the two lines is 4 km, a misfit that could be reconciled with a Moho uncertainty of 2 km on either line.

Comparison of lines AMP-D and FLARE-1

Line AMP-D in the Faroe–Shetland Channel also crosses line FLARE-1, which extends towards Suðuroy (Fig. 1). The modelling procedures for lines AMP-D and FLARE-1 were described earlier.

At the intersection of lines AMD-D and FLARE-1, basalts of line AMP-D are 0.3 km thick, while they are 1 km on line FLARE-1 (Table 7; Fig. 6). The basement of line FLARE-1 is 1.1 km shallower than on line AMP-D. The discrepancy is, again, related to the fact that line FLARE-1 shows the

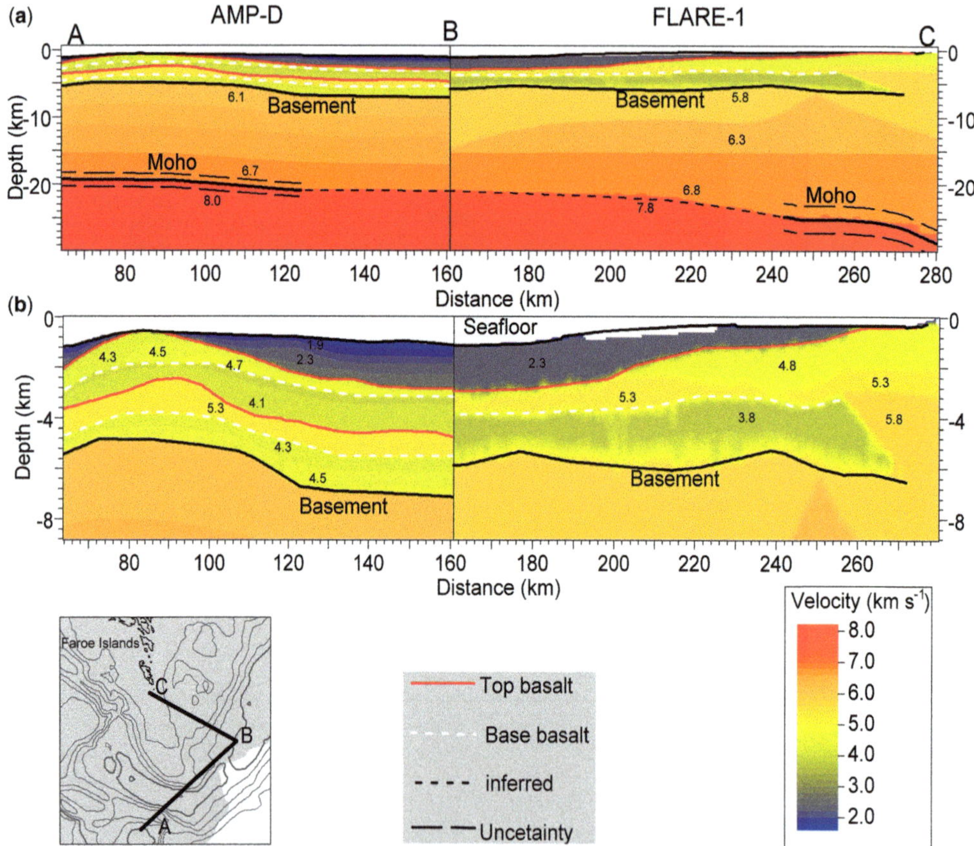

Fig. 6. Comparison of P-wave velocity models of lines FOERBAS and AMP-D–FLARE-1. (**a**) & (**b**) show different depth scales. Annotated velocities are given in km s^{-1}. The map in the lower left-hand corner shows the location of the composite profile. The shaded area shows the basalt cover.

depth to seismic basement and not to crystalline basement, as is the case for line AMP-D. Seismic basement on line FLARE-1 has approximately the same depth as the base of the lower basalt layer on line AMP-D (Fig. 5b), but it is unclear whether this correlation has some relevance or if it is arbitrary. The sub-basalt sediments of line FLARE-1 are modelled as a single-layer, whilst line AMP-D has two distinct LVZs beneath basalt layers. This can also be seen in relation to line AMG95-1 in the same area, for which several thick intrusions are modelled into the sub-basalt sediments (Fig. 5).

Comparison of lines FLARE-1 to 12

The FLARE survey consists of 12 lines spanning the NW part of the Faroe–Shetland Channel (Fig. 1). From these lines, a composite north–south profile (Fig. 7) was constructed to evaluate the internal consistency of the velocity models.

The velocity modelling for lines FLARE-2–FLARE-12 was performed in the same way as described earlier for line FLARE-1 (Latkiewicz & Kirk 1999; White *et al.* 1999, 2003; Fliedner & White 2003). The shallow velocity structure was determined using conventional semblance analysis techniques, while stacked seismic records were used to interpret the top of the basalt. Forward and inverse modelling were employed to determine the velocities of deeper structures. The top of the seismic basement interface was mapped by wide-angle reflections, which also provided constraints on the sub-basalt velocities. Basement refractions were rarely observed owing to the limited shot-receiver distance of no more than 38 km. For this reason, basement velocities are generally unconstrained (Latkiewicz & Kirk 1999; White *et al.* 1999, 2003; Fliedner & White 2003).

The composite line shows that the depth to top basalt and base basalt is consistent at all intersections (Fig. 7). Velocities in the basalts are consistent in most cases, while there are generally larger differences in the sub-basalt velocities – with the largest difference of 1.2 km s^{-1} occurring at the intersection of lines FLARE-8 and FLARE-1.

Comparison of the FLARE survey with lines AMG95-1 and Mobil-2

As there is a rather dense grid of seismic refraction lines in the Faroe–Shetland Channel, there are a few more line intersections to investigate. Here, the intersections of the FLARE survey with the two lines AMG95-1 (Fig. 8) and Mobil-2 (Fig. 9) are presented. A description of the modelling procedures was given earlier.

The base basalt on line AMG95-1 is up to 1.5 km shallower than on lines FLARE-7 and FLARE-8 (Table 8; Fig. 8). Sub-basalt velocities on line FLARE-7 (3.8 km s^{-1}) fall within the velocity range observed on line AMG95-1 (3.2–4.7 km s^{-1}). At the intersection of lines FLARE-8 and AMG95-1, the sub-basalt velocities do not match (Table 9; Fig. 8). The model for line FLARE-8 indicates a rather high velocity of 4.9 km s^{-1}, which contrasts with the velocity range of 3.2–4.5 km s^{-1} on line AMG95-1. Another striking difference is the presence of sills in the sub-basalt section of line AMG95-1, while no such features are resolved in the models of lines FLARE-1 and FLARE-8.

Lines Mobil-2 and FLARE-11 have similar velocities of 4.3 km s^{-1} at the top of the basalt sequence (Table 10; Fig. 9). However, further below there are substantial differences in the interpretation. While the basalts on line FLARE-11 are 1.7 km thick, their thickness is only 0.6 km on line Mobil-2. Maximum velocities in the basalts are also higher on line FLARE-11 (5.3 km s^{-1}) than on line Mobil-2 (4.3 km s^{-1}). Interestingly, the depth to crystalline basement on line Mobil-2 correlates with the base basalt interpretation on line FLARE-11. Owing to the differences in the velocity distribution, it is safe to say that this match is coincidental as the corresponding reflectors do have different travel times on the seismic record.

Discussion

The comparisons given above show that the velocity models at the intersection of seismic refraction lines match reasonably well down to and into the uppermost basalts. The inconsistency at deeper levels is largely related to difficulties in determining the base of the basalts and the velocities in the sub-basalt sediments that, in most cases, represent a LVZ. While seismic refraction modelling is often presented as an alternative to seismic reflection imaging of sub-basalts (Hughes *et al.* 1998; Richardson *et al.* 1999; Fliedner & White 2003; White *et al.* 2003; Klingelhöfer *et al.* 2005; Raum *et al.* 2005; Makris *et al.* 2009), the method clearly has problems of its own regarding this matter. These difficulties relate to the imaging and interpretation of thick basalts, and to the lack of refractions within the sub-basalt sediments due to their character as a LVZ. Lau *et al.* (2010) stated that the discrimination between the basalt sequence and the underlying geology remains the most critical seismic imaging problem in the Faroe–Shetland Channel.

In some models, the offset range of the refracted basalt phase is used as a measure for the basalt thickness (e.g. line Mobil-1: Hughes *et al.* 1998). Sometimes a base basalt reflection can be identified

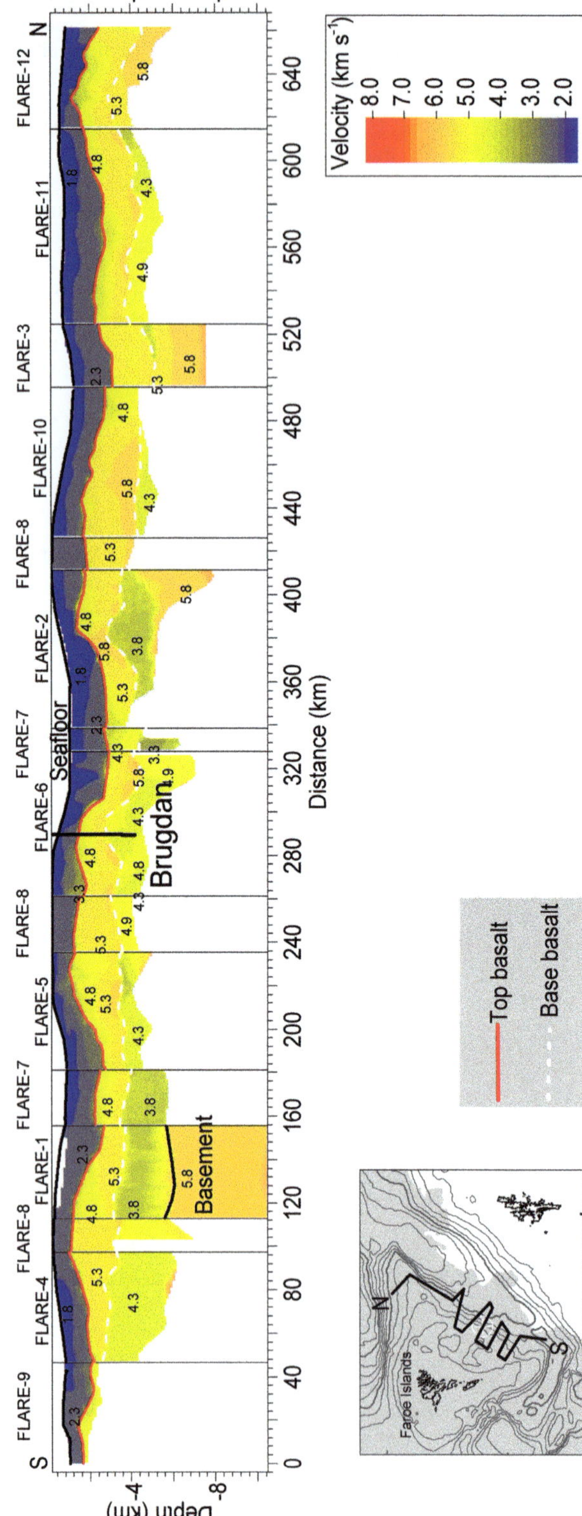

Fig. 7. The composite FLARE P-wave velocity model consists of sections from all 12 of the profiles, with the location and depth of the Brugdan well annotated. The white marker on the map shows the location of the Brugdan well. Annotated velocities are given in km s^{-1}. The map in the lower left-hand corner shows the location of the composite profile. The shaded area shows the basalt cover.

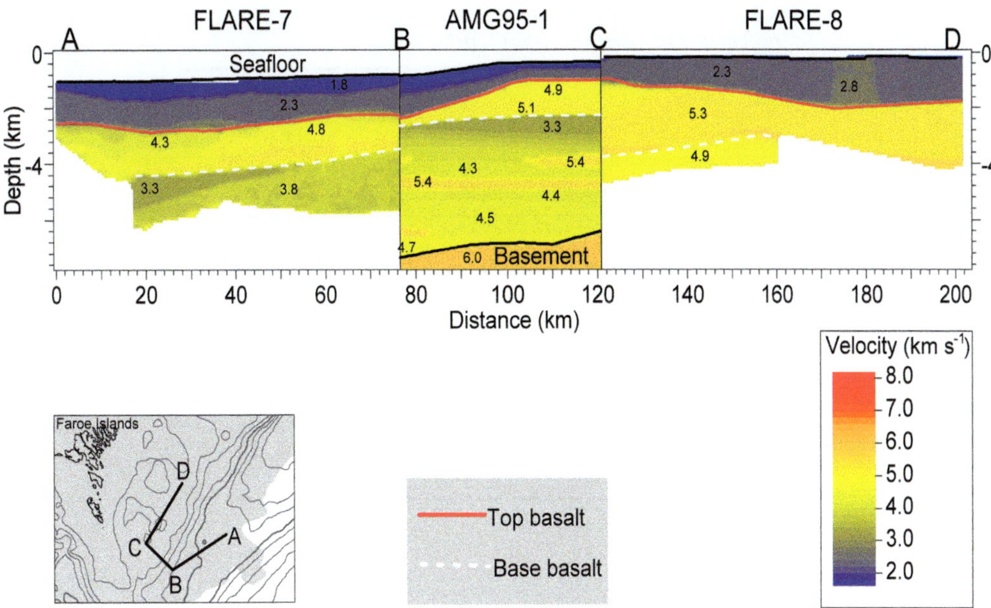

Fig. 8. Comparison of P-wave velocity models of lines FLARE-7, AMG95-1 and FLARE-8. Annotated velocities are given in km s^{-1}. The map in the lower left-hand corner shows the location of the composite profile. The shaded area shows the basalt cover.

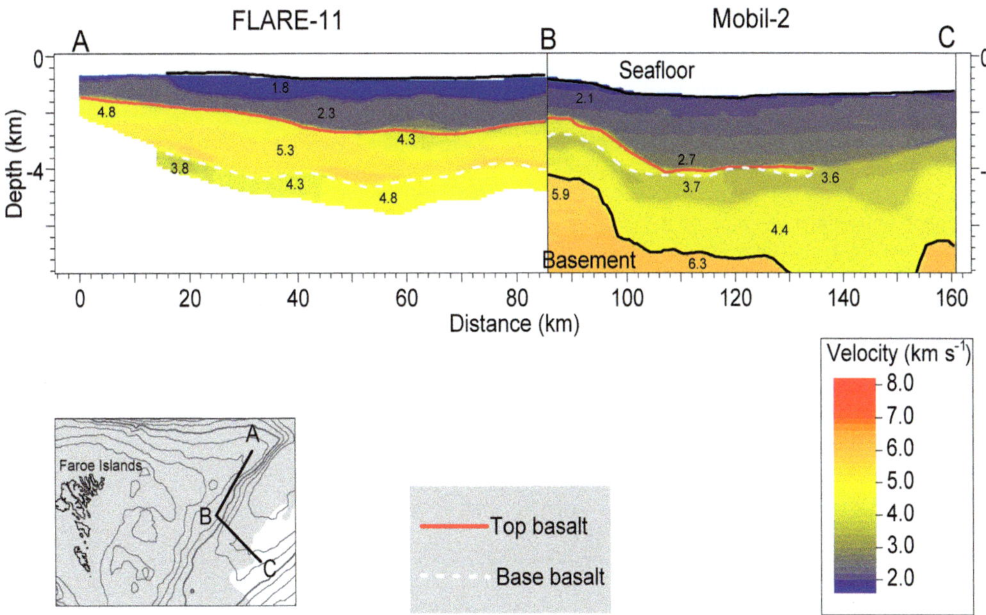

Fig. 9. Comparison of P-wave velocity models of lines FLARE-11 and Mobil-2. Annotated velocities are given in km s^{-1}. The map in the lower left-hand corner shows the location of the composite profile. The shaded area shows the basalt cover.

Table 8. *Comparison of features of the velocity models at the intersection of lines AMG95-1 and FLARE-7*

	Flare-7	AMG95-1
Basalt thickness	1.4 km	0.3 km
Basalt velocity	4.25–5.20 km s^{-1}	5.2 km s^{-1}
Sub-basalt sediment velocities	3.8 km s^{-1}	3.2–4.7 km s^{-1}

Table 9. *Comparison of features of the velocity models at the intersection of lines FLARE-8 and AMG95-1*

	AMG95-1	FLARE-8
Basalt thickness	1.3 km	2.8 km
Basalt velocity	4.8–5.5 km s^{-1}	5.00–5.25 km s^{-1}
Sub-basalt sediment velocities	3.2–4.5 km s^{-1}	4.9 km s^{-1}

(e.g. line FLARE-1: Richardson *et al.* 1999) or interpretations on stacked reflection seismic data are used to infer the basalt thickness (e.g. the FLARE survey: Fliedner & White 2003). However, a strong intra-basalt reflector (cf. Fig. 2) resulting from, for example, an interbedded sedimentary layer poses a problem for all of these methods. First, the intra-basalt reflector can affect the maximum range of the refraction in the basalts. This, in turn, can result in wrong estimates of the total basalt thickness. Second, the reflection from the intra-basalt reflector can be misinterpreted as a reflection from base-basalt. Third, a well-defined reflection from an intra-basalt reflector, in combination with a change in seismic facies, can potentially be misinterpreted as the base of the basalt on stacked seismic reflection data (Petersen 2014).

Such seismic facies changes are observed in a number of settings at volcanic margins. Differences in seismic facies for seawards-dipping reflectors (SDRs), hyaloclastites and flow-foot breccia sequences are shown in Spitzer *et al.* (2008). Planke *et al.* (2000) presented an extensive study of seismic volcanostratigraphy, in which they distinguish various facies such as landward flows, lava delta, inner flows, inner SDRs, outer high and outer SDRs. The Lopra-1 well (Christie *et al.* 2006) is a case where, at the transition from subaerial flows to hyaloclastites, the seismic response changes significantly towards lower amplitudes. Even subaerial basalt formations can display significant differences in seismic facies (Petersen *et al.* 2006, 2015).

The problem of determining base basalt in the absence of a base basalt reflector was illustrated by Varming *et al.* (2012), who showed a seismic reflection profile that connects two wells (6005/15-1 and 6005/13-1) displaced about 20 km from each other. While the one well encountered 30 m of basaltic lava flows and 55 m of volcaniclastic sandstone–siltstone, the other well terminated within the volcanic section after encountering a 1475 m-thick series of basaltic lavas and hyaloclastites. However, the seismic profile connecting the two wells did not show a traceable seismic horizon that could represent this significant change in basalt thickness.

One example of where the total basalt thickness could not be determined from seismic data is the Brugdan well in the Faroe–Shetland Channel (Fig. 1). Here base basalt was drilled at 3745 m, while prior to drilling the base of the basalt was predicted at a depth of 2280 m (Øregaard *et al.* 2007).

Table 10. *Comparison of features of the velocity models at the intersection of lines Mobil-2 and FLARE-11*

	Mobil-2	FLARE-11
Basalt thickness	0.6 km	1.7 km
Basalt velocity	4.3 km s^{-1}	4.3–5.3 km s^{-1}
Base basalt–crystalline basement velocity	3.6–4.4 km s^{-1}	4.8 km s^{-1}

It is not clear on what this prognosis was based, but it was most likely to have been on the interpretation of seismic reflection data. Line FLARE-6 (Fliedner & White 2003) intersects the Brugdan well (Fig. 7). At this location, the modelled depth of base basalt is about 2800 m, which is 945 m less than what was found in the well. The reason for this mismatch may relate to effects from hyaloclastites. From 2542 m down to the base of the volcanic succession, the lithology in the Brugdan well is dominated by hyaloclastites interbedded with basalt and volcaniclastic sediments (Øregaard *et al.* 2007).

The comparison of interval velocities from VSP and velocity log shows consistency with the velocity model of line FLARE-6 (Fig. 10), although post-basalt sediment velocities are too high and, subsequently, the top basalt is about 200 m too deep. At 2800 m, line FLARE-6 models base basalt with a velocity inversion that reaches its minimum of 4.25 km s^{-1} at a depth of 2880 m. If correcting the depth of the modelled velocities according to the discrepancy of the top basalt, the depth of the velocity inversion on line FLARE-6 matches the depth of a velocity inversion actually seen at the top of a 200 m-thick hyaloclastics section in the velocity log at 2686 m. Figure 10 also shows that there is a correlation between the sub-basalt section of the model and the hyaloclastites in the well.

None of the intersecting lines show consistency between different surveys, while intersecting lines within the same survey are fairly consistent, as seen on the FLARE composite profile (Fig. 7). In relation to this, it should be mentioned that the modelling of line iSIMM utilizes the intersecting line FLARE-11 as a constraint (Spitzer & White 2005). This is why the iSIMM profile is consistent with the results of line FLARE-11. All other intersecting lines with regard to different surveys show differences in basalt thickness, depth to crystalline basement, and in velocities and internal structure of the sub-basalt sediments. The Moho, however, is consistent at the intersection of lines Mobil-2 and iSIMM (Fig. 3), as well as at the intersection of lines AMP-D and FLARE-1 (Fig. 6).

The velocity models of the sub-basalt sediments SE of the Faroes display large lateral variations. The models for lines FAST and FLARE-1 show a transition from low sub-basalt velocities (<4.6 km s^{-1}) to higher sub-basalt velocities (>5.7 km s^{-1}) when approaching the Faroes (Fig. 4). The change in velocity is interpreted as relating to the landwards termination of the sub-basalt sediments against a basement high or may represent a zone with intrusions (Richardson *et al.* 1999). In contrast, the sub-basalt sediments on line AMG95-1 continue as a LVZ all the way to the Faroes (Fig. 4).

There also seems to be some correlation between the basalt thickness and whether or not the seismic

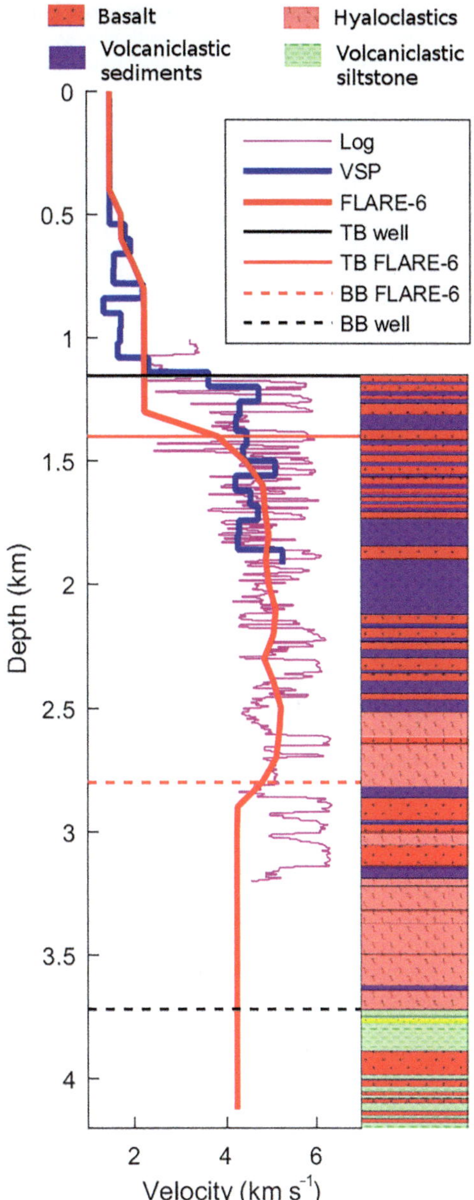

Fig. 10. Comparison of velocities from the Brugdan well with a modelled velocities extract from line FLARE-6 at the location of Brugdan well (Fig. 7). Stratigraphy from the well completion report (Øregaard *et al.* 2007) is inserted to the right. Log, interval velocities based on travel times from velocity logs taken at 5 m intervals; VSP, interval velocities based on VSP travel times taken at 60 m intervals; FLARE-6, velocities extract from line FLARE-6 at the location of the Brugdan well; TB well, top basalt from composite well logs; TB FLARE-6, top basalt from line FLARE-6 at the well location; BB FLARE-6 base basalt from line FLARE-6 at the well location; BB well, base basalt from composite well logs.

line extends beyond the basalt cover in the Faroe–Shetland Channel. In general, the thickness of the basalts tends to be less on lines that reach into the eastern part of the channel with no basalts (lines Mobil-1, Mobil-2 and AMG95-1) than on profiles that are restricted to the basalt-covered area (line AMG95-2 and most of the lines from the FLARE survey).

The largest difference of Moho depth is at the AMG95-1–FAST intersection, with a difference of 11.7 km. Although the intersection is at a location where neither line has constraints from PmP reflections, both lines have a constraint within 20 km of the intersection. Notice, however, that for line FAST the seismic refraction data are from unreversed ray coverage, recorded on land stations with a significant offset to the shot line.

The profile with the best seismic constraints on the Moho depth is the iSIMM line. This is primarily due to the line location across the Fugloy Ridge with significant offsets to either side of the ridge. To the NW, the line extends into oceanic crust, while the thinned continental crust of the Faroe–Shetland Channel is encountered in the SE. This line geometry, together with the use of large airgun sources tuned to produce low-frequency energy, resulted in a good coverage with PmP reflections along the entire length of the profile.

Only for line AMG95-1 are upper-mantle velocities (8.0–8.3 km s^{-1}) constrained by Pn observations. For Mobil-1 and Mobil-2, the upper-mantle velocity (8.0 km s^{-1}) appears to be from Pn observations, although this is not clearly stated, while for lines FLARE-1 and FAST it is assigned at 7.8 km s^{-1} and for line AMG95-2 at 8.1 km s^{-1} without constraint from Pn observations.

All models presented here consider uncertainty bounds to some degree. Even though uncertainties are often related to the accuracy of the observed travel times, it is not at all a trivial task to quantify the velocity and depth uncertainties of a multilayer model. Looking at some of the models from the Faroe–Shetland Channel, Raum *et al.* (2005) and Makris *et al.* (2009) gave a general estimate on the uncertainties for the entire model, while Richardson *et al.* (1999) presented uncertainty estimates only down to the base of the LVZ. Similarly, Roberts *et al.* (2009) concentrated their uncertainty analysis on the LVZ and the depth to the Moho. Klingelhöfer *et al.* (2005) provided uncertainty estimates for a number of interfaces, including the Moho.

Conclusion

The inconsistencies of the velocity models in the Faroe–Shetland Channel call for a remodelling of the seismic data guided and constrained by the latest geological models for the area (e.g. Ritchie *et al.* 2011; Hopper *et al.* 2014; Funck *et al.* 2016). Knowledge from available wells that drilled basalts in the Faroe–Shetland Channel should be integrated. In particular, the Brugdan (6104/21-1), William (6005/13-1) and Anne-Marie (6004/8a-1) wells (Fig. 1) are of importance as they were drilled in areas with significant basalt cover. The Brugdan well intersects the FLARE-6 profile and can, as such, be directly fed into the modelling of the seismic refraction data, while the William and Anne-Marie wells are not located on seismic refraction lines. However, the two latter wells can help to improve the understanding of the seismic facies in different types of basalts, and thereby aid the interpretation of basalt sequences on seismic reflection lines in the vicinity of the Faroe Islands. A reinterpretation of the basalts and other regional structures would put significant constraints on a remodelling of all seismic refraction datasets, in addition to the required consistency at all line intersections.

We acknowledge the support of the industry sponsors (in alphabetical order): Bayerngas Norge AS; BP Exploration Operating Company Ltd; Bundesanstalt für Geowissenschaften und Rohstoffe (BGR); Chevron East Greenland Exploration A/S; ConocoPhillips Skandinavia AS; DEA Norge AS; Det norske oljeselskap ASA; DONG E&P A/S; E.ON Norge AS; ExxonMobil Exploration and Production Norway AS; Japan Oil, Gas and Metals National Corporation (JOGMEC); Maersk Oil; Nalcor Energy – Oil and Gas Inc.; Nexen Energy ULC; Norwegian Energy Company ASA (Noreco); Repsol Exploration Norge AS; Statoil (UK) Ltd; and Wintershall Holding GmBH.

References

Á Horni, J., Geissler, W. *et al.* 2014. Offshore volcanic facies. *In*: Hopper, J.R., Funck, T., Stoker, M., Árting, U., Peron-Pinvidic, G., Doornenbal, H. & Gaina, C. (eds) *Tectonostraticgraphic Atlas of the North-East Atlantic Region*. Geological Survey of Denmark and Greenland (GEUS), Copenhagen, Denmark, 235–253.

Christie, P.A.F., Gollifer, I.D. & Cowper, D. 2006. Borehole seismic studies of a volcanic succesion from the Lopra-1/1A borehole in the Faroe Islands, northern North Atlantic. *In*: Chalmers, J.A. & Waagstein, R. (eds) *Scientific results from the deepened Lopra-1 borehole, Faroe Islands*. Geological Survey of Denmark and Greenland Bulletin, **9**, 23–40.

Eccles, J.D., White, R.S., Roberts, A.W., Christie, P.A.F. & the iSIMM team. 2007. Wide angle converted shear wave analysis of a North Atlantic volcanic rifted continental margin: constraint on sub-basalt lithology. *First Break*, **25**, 63–70, https://doi.org/10.3997/1365-2397.2007026

Fliedner, M.M. & White, R.S. 2001. Sub-basalt imaging in the Faeroe–Shetland Basin with large-offset data. *First Break*, **19**, 247–252, https://doi.org/10.1046/j.0263-5046.2001.00156.x

FLIEDNER, M.M. & WHITE, R.S. 2003. Depth imaging of basalt flows in the Faeroe–Shetland Basin. *Geophysical Journal International*, **152**, 353–371, https://doi.org/10.1046/j.1365-246X.2003.01833.x

FRUEHN, J., FLIEDNER, M.M. & WHITE, R.S. 2001. Integrated wide-angle and near-vertical subbasalt study using large-aperture seismic data from the Faeroe–Shetland region. *Geophysics*, **66**, 1340–1348, https://doi.org/10.1190/1.1487079

FUNCK, T., HOPPER, J.R. *ET AL*. 2014. Crustal structure. *In*: HOPPER, J.R., FUNCK, T., STOKER, M., ÁRTING, U., PERON-PINVIDIC, G., DOORNENBAL, H. & GAINA, C. (eds) *Tectonostratigraphic Atlas of the North-East Atlantic Region*. Geological Survey of Denmark and Greenland (GEUS), Copenhagen, Denmark, 69–126.

FUNCK, T., GEISSLER, W.H., KIMBELL, G.S., GRADMANN, S., ERLENDSSON, Ö., MCDERMOTT, K. & PETERSEN, U.K. 2016. Moho and basement depth in the NE Atlantic Ocean based on seismic refraction data and receiver functions. *In*: PÉRON-PINVIDIC, G., HOPPER, J.R., STOKER, M.S., GAIN, C., DOORNENBAL, J.C., FUNCK, T. & ÁRTING, U.E. (eds) *The NE Atlantic Region: A Reappraisal of Crustal Structure, Tectonostratigraphy and Magmatic Evolution*. Geological Society, London, Special Publications, **447**. First published online July 13, 2016, https://doi.org/10.1144/SP447.1.

HOPPER, J.R. & GAINA, C. 2014. Bathymetry and elevation. *In*: HOPPER, J.R., FUNCK, T., STOKER, M., ÁRTING, U., PERON-PINVIDIC, G., DOORNENBAL, H. & GAINA, C. (eds) *Tectonostraticgraphic Atlas of the North-East Atlantic Region*. Geological Survey of Denmark and Greenland (GEUS), Copehagen, Denmark, 23–28.

HOPPER, J.R., FUNCK, T., STOKER, M.S., ÁRTING, U., PERON-PINVIDIC, G., DOORNENBAL, H. & GAIN, C. (eds) 2014. *Tectonostratigraphic Atlas of the North-East Atlantic Region*. Geological Survey of Denmark and Greenland, Copehagen, Denmark.

HUGHES, S., BARTON, P.J. & HARRISON, D. 1998. Exploration in the Shetland–Faeroe Basin using densely spaced arrays of ocean-bottom seismometers. *Geophysics*, **63**, 490–501, https://doi.org/10.1190/1.1444350

KIØRBOE, L. & PETERSEN, S.A. 1995. Seismic investigation of the Faeroe basalts and their substratum. *In*: SCRUTTON, R.A., SHIMMIELD, G.B. & TUDHOPE, A.W. (eds) *The Tectonics, Sedimentation and Palaeoceanography of the North Atlantic Region*. Geological Society, London, Special Publications, **90**, 111–122, https://doi.org/10.1144/GSL.SP.1995.090.01.06

KLINGELHÖFER, F., EDWARDS, R.A. & HOBBS, R.W. 2005. Crustal structure of the NE Rockall Trough from wide-angle seismic data modeling. *Journal of Geophysical Research*, **110**, B11105, https://doi.org/10.1029/2005JB003763

LARSEN, L.M., WAAGSTEIN, R., PEDERSEN, A.K. & STOREY, M. 1999. Trans-Atlantic correlation of the Palaeogene volcanic successions in the Faeroe Islands and East Greenland. *Journal of the Geological Society, London*, **156**, 1081–1095, https://doi.org/10.1144/gsjgs.156.6.1081

LATKIEWICZ, C.B. & KIRK, W.J. 1999. *Report: F.L.A.R.E., Faroes Large Aperture Research Experiment*. Amerada Hess Ltd, Aberdeen.

LAU, K.W.H., WHITE, R.S. & CHRISTIE, P.A.F. 2010. Integrating streamer and ocean-bottom seismic data for sub-basalt imaging on the Atlantic Margin. *Petroleum Geoscience*, **16**, 349–366, https://doi.org/10.1144/1354-0793/10-023

MAKRIS, J., PAPOULIA, I. & ZISKA, H. 2009. Crustal structure of the Shetland–Faeroe Basin from long offset seismic data. *In*: VARMING, T. & ZISKA, H. (eds) *Faroe Island Exploration Conference: Proceedings of the 2nd Conference*. Annales Societatis Scientiarum Færoensis, Tórshavn, Faroe Islands, Supplementum, **50**, 30–42.

MARESH, J., WHITE, R.S., HOBBS, R.W. & SMALLWOOD, J.R. 2006. Seismic attenuation of Atlantic margin basalts: observations and modeling. *Geophysics*, **71**, B211–B221, https://doi.org/10.1190/1.2335875

NEISH, J.K. 2004. Faroese area: structural interpretation of seismic data in a basalt environment. *In*: VARMING, T. & ZISKA, H. (eds) *Faroe Islands Exploration Conference: Proceedings of the 1st Conference*. Annales Societatis Scientiarum Færoensis, Tórshavn, Faroe Islands, Supplementum, **43**, 131–145.

ØREGAARD, J., ROKSVAAG, P., KIRKEMO, E.G., LEFDAL, F., HUNNES, O., FLATEBØ, T. & HAUGEN, J.E. 2007. *Final Well Report. Well FO 6104/21-1*. Licence No. 006. Brugdan: Statoil.

PALMASON, G. 1965. Seismic refraction measurements of basalt Lavas of Faeroe Islands. *Tectonophysics*, **2**, 475–482, https://doi.org/10.1016/0040-1951(65)90002-8

PETERSEN, U.K. 2014. *Propagation and scattering of reflection seismic waves in a basalt succession*. PhD thesis, University of the Faroe Islands, Tórshavn, https://doi.org/10.13140/RG.2.1.1133.1927

PETERSEN, U.K., ANDERSEN, M.S. & WHITE, R.S. 2006. Seismic imaging of basalts at Glyvursnes, Faroe Islands: hunting for future exploration methods in basalt covered areas. *First Break*, **24**, 45–52, https://doi.org/10.3997/1365-2397.2006006

PETERSEN, U.K., ANDERSEN, M.S. & BROWN, R.J. 2015. *Geophysical aspects of basalt geology and identification of intrabasaltic horizons*. *In*: EIDESGAARD, Ó. & ZISKA, H. (eds) Faroe Islands Exploration Conference: Proceedings of the 4th Conference. Annales Societatis Scientiarum Færoensis, Tórshavn, Faroe Islands, Supplementum, **64**, 76–93, https://doi.org/10.13140/RG.2.1.2387.9765

PLANKE, S., SYMONDS, P.A., ALVESTAD, E. & SKOGSEID, J. 2000. Seismic volcanostratigraphy of large-volume basaltic extrusive complexes on rifted margins. *Journal of Geophysical Research*, **105**, 19,335–19,351.

RAUM, T., MJELDE, A.M. *ET AL*. 2005. Sub-basalt structures east of the Faroe Islands revealed from wide-angle seismic and gravity data. *Petroleum Geoscience*, **11**, 291–308, https://doi.org/10.1144/1354-079304-627

RICHARDSON, K.R., SMALLWOOD, J.R., WHITE, R.S., SNYDER, D.B. & MAGUIRE, P.K.H. 1998. Crustal structure beneath the Faroe Islands and the Faroe–Iceland Ridge. *Tectonophysics*, **300**, 159–180, https://doi.org/10.1016/S0040-1951(98)00239-X

Richardson, K.R., White, R.S., England, R.W. & Fruehn, J. 1999. Crustal structure east of the Faroe Islands: mapping sub-basalt sediments using wide-angle seismic data. *Petroleum Geoscience*, **5**, 161–172, https://doi.org/10.1144/petgeo.5.2.161

Ritchie, J.D., Ziska, H., Johnson, H. & Evans, D. (eds) 2011. *Geology of the Faroe–Shetland Basin and Adjacent Areas*. British Geological Survey, Nothingham.

Roberts, A.W., White, R.S. & Christie, P.A.F. 2009. Imaging igneous rocks on the North Atlantic rifted continental margin. *Geophysical Journal International*, **179**, 1024–1038, https://doi.org/10.1111/j.1365-246X.2009.04306.x

Spitzer, R. & White, R.S. 2005. Advances in seismic imaging through basalts: a case study from the Faroe-Shetland Basin. *Petroleum Geoscience*, **11**, 147–156, https://doi.org/10.1144/1354-079304-639

Spitzer, R., White, R.S. & Christie, P.A.F. 2008. Seismic characterization of basalt flows from the Faroes margin and the Faroe–Shetland basin. *Geophysical Prospecting*, **56**, 21–31, https://doi.org/10.1111/j.1365-2478.2007.00666.x

Varming, T., Ziska, H. & Ólavsdóttir, J. 2012. Exploring for hydrocarbons in a volcanic province – a review of exploration on the Faroese Continental Shelf. *In*: Varming, T. & Ziska, H. (eds) *Faroe Island Exploration Conference: Proceedings of the 3rd Conference*. Annales Societatis Scientiarum Færoensis, Tórshavn, Faroe Islands, Supplementum, **56**, 84–106.

White, R.S., Fruehn, J., Richardson, K.R., Cullen, E., Kirk, W., Smallwood, J.R. & Latkiewicz, C. 1999. Faeroes Large Aperture Research Experiment (FLARE): imaging through basalt. *In*: Fleet, A.J. & Boldy, S.A.R. (eds) *Petroleum Geology of Northwest Europe: Proceedings of the 5th Conference*. Geological Society, London, 1243–1252, https://doi.org/10.1144/0051243

White, R.S., Smallwood, J.R., Fliedner, M.M., Boslaugh, B., Maresh, J. & Fruehn, J. 2003. Imaging and regional distribution of basalt flows in the Faeroe–Shetland Basin. *Geophysical Prospecting*, **51**, 215–231, https://doi.org/10.1046/j.1365-2478.2003.00364.x

Mesozoic and older rift basins on the SE Greenland Shelf offshore Ammassalik

JOANNA GERLINGS[1,2], JOHN R. HOPPER[1]*, MICHAEL B. W. FYHN[1] & NICOLAS FRANDSEN[3]

[1]*Geological Survey of Denmark and Greenland (GEUS), Øster Voldgade 10, DK-1350 Copenhagen K, Denmark*

[2]*Present address: Danish Hydrographic Office – Arctic, Lindholm Brygge 31, DK-9400 Nørresundby, Denmark*

[3]*Niels Bohr Institute (NBI), University of Copenhagen, Blegdamsvej 17, DK-2100 Copenhagen, Denmark*

**Correspondence: jrh@geus.dk*

Abstract: Seismic reflection data and shallow cores from the SE Greenland margin show that rift basins formed by the mid- to Late Cretaceous in the offshore area near Ammassalik. Here termed the Ammassalik Basin, this contribution documents the area using reprocessed older shallow seismic reflection data together with a more recent, commercial deep seismic reflection profile. The data show that the basin is at least 4 km deep and may be regionally quite extensive. Interpretation of gravity anomaly data indicate that the basin potentially covers an area of nearly 100 000 km^2. The sediments in the basins are at least of Cretaceous age, as indicated by a sample from just below the basalt cover that was dated as Albian. Dipping sediment layers in the basins indicate that older sediments are present. Comparison of the data to the conjugate Hatton margin where older basins are exposed beneath the volcanic cover shows similar stratigraphy of similar ages. Reconstructions of the position of the basin during the Permian–Triassic and Jurassic suggest that older sedimentary strata could also be possible. In contrast to the conjugate Hatton margin, possible older strata subcrop out below the seafloor along the shallow margin, providing a future opportunity to sample some of the oldest sediments to determine the onset of rifting between SE Greenland and the Hatton margin.

The SE Greenland margin formed in response to rifting and break-up between Greenland and Europe during the Late Cretaceous–early Eocene (e.g. Larsen & Saunders 1998) (Fig. 1). Based on single-channel seismic data collected in the 1970s, B. Larsen (1980) and H.C. Larsen (1980) hypothesized the presence of Cretaceous sedimentary strata in the offshore areas near Ammassalik. However, the presence or absence of offshore sedimentary basins remained mostly speculative because of the massive cover of volcanic material that erupted just prior to and during final break-up (see Horní *et al.* this volume, in review, and references therein). The problems inherent with seismic imaging below basalt, is that it mostly obscures possible basins below the lava flows. Nevertheless, shallow, high-resolution seismic reflection data collected in 1997 along the shelf near Ammassalik confirmed the presence of apparent sedimentary layering stratigraphically below the basaltic cover (Hopper *et al.* 1998). During a subsequent shallow coring campaign, sedimentary rocks of Albian age were recovered from below the basalt, demonstrating for the first time the presence of older, Mesozoic sedimentary rocks in the offshore area (Thy *et al.* 2007).

In this contribution, the sedimentary basin offshore SE Greenland, here termed the Ammassalik Basin, is investigated further using older seismic reflection data in conjunction with a recent commercial, deep seismic reflection profile and regional gravity anomaly data. Three key seismic lines from the 1997 survey were reprocessed to better image possible basins in the area. The 1997 lines run perpendicular to the coast and the commercial line runs along the margin, tying together the older data. The reprocessing is described briefly, along with an interpretation and assessment of the probable thickness of the basin sediments. The approximate extent and dimensions are interpreted based on gravity data, indicating a substantial basin covering approximately 100 000 km^2. Plate reconstructions and comparison with the conjugate

From: Péron-Pinvidic, G., Hopper, J. R., Stoker, M. S., Gaina, C., Doornenbal, J. C., Funck, T. & Árting, U. E. (eds) 2017. *The NE Atlantic Region: A Reappraisal of Crustal Structure, Tectonostratigraphy and Magmatic Evolution*. Geological Society, London, Special Publications, **447**, 375–392.
First published online April 13, 2017, https://doi.org/10.1144/SP447.15

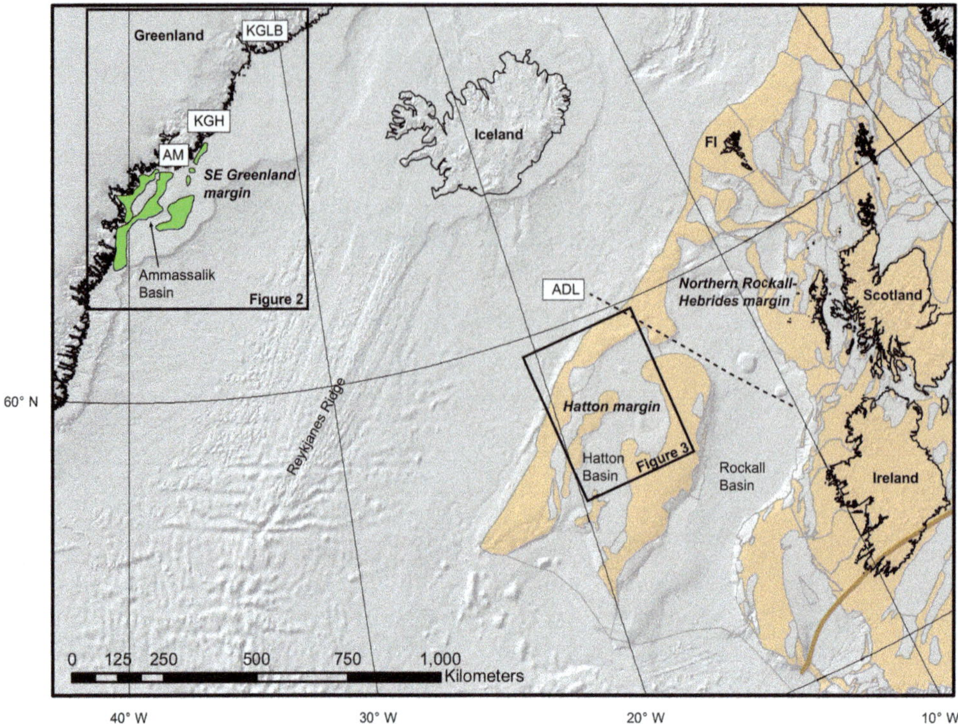

Fig. 1. Regional conjugate setting of the SE Greenland and Hatton margins with shaded relief bathymetry/topography as background. The eastern margins include the structural elements map (basins and highs) compiled in Hopper *et al.* (2014). The orange areas are highs and the grey areas are basins. The Ammassalik Basin, the main focus of this paper, is highlighted in green. *Abbreviations*: ADL, Anton Dohrn Lineament; AM, Ammassalik; FI, Faroe Islands; KGH, Kap Gustav Holm; KGLB, Kangerlussuaq Basin.

margin and regional data suggest that the Ammassalik Basin was part of a well-developed Mesozoic rift system that is also observed along the Hatton margin. Along the northern Rockall and Hebrides margins, NE of the Hatton and Rockall basins, significant basin formation and development occurred throughout the late Palaeozoic and Mesozoic (e.g. Stoker *et al.* 2014, 2016). Several studies have suggested that significant pre-Cretaceous extension and basin formation must have also occurred in the Hatton and Rockall basins (Cole & Peachey 1999), although this is currently unproven and controversial (Stoker *et al.* 2016). The possibility of older sedimentary successions within the Ammassalik Basin is therefore considered here in light of reconstructed stratigraphic distribution maps back to Permian–Triassic times.

Regional setting

Onshore geology

The onshore area along SE Greenland is dominated by the Palaeoproterozoic Nagssugtoqidian Orogen (Fig. 2) (see Kolb 2014 for a recent summary). South of Kap Gustav Holm, no rocks younger than Proterozoic are known to crop out. Large onshore basaltic dykes, which can also be interpreted on the magnetic anomaly map, are Proterozoic in age (Riisager & Rasmussen 2014). The onshore region is thus devoid of any known Palaeozoic or Mesozoic rocks, and the geological history of the region over this time interval is mostly unknown.

North of Ammassalik, the nearest known sedimentary outcrops are found on Kap Gustav Holm (Figs 1 & 2). These are described as Upper Cretaceous–Cenozoic sandstones (Wager 1934), but little subsequent work has been carried out on these outcrops. They were described by Myers *et al.* (1993) as a 150 m-thick sandstone unit containing marine fossils near the top, and metamorphosed by a coastal dyke swarm and a nearby gabbroic intrusion, which is Eocene in age (Lenoir *et al.* 2003). The succession must therefore be older than this.

Further north, Cretaceous–Paleocene sedimentary rocks of the Kangerlussuaq Basin (Fig. 2) have been well studied and are described in Larsen *et al.* (1999). The Kangerlussuaq Basin consists of

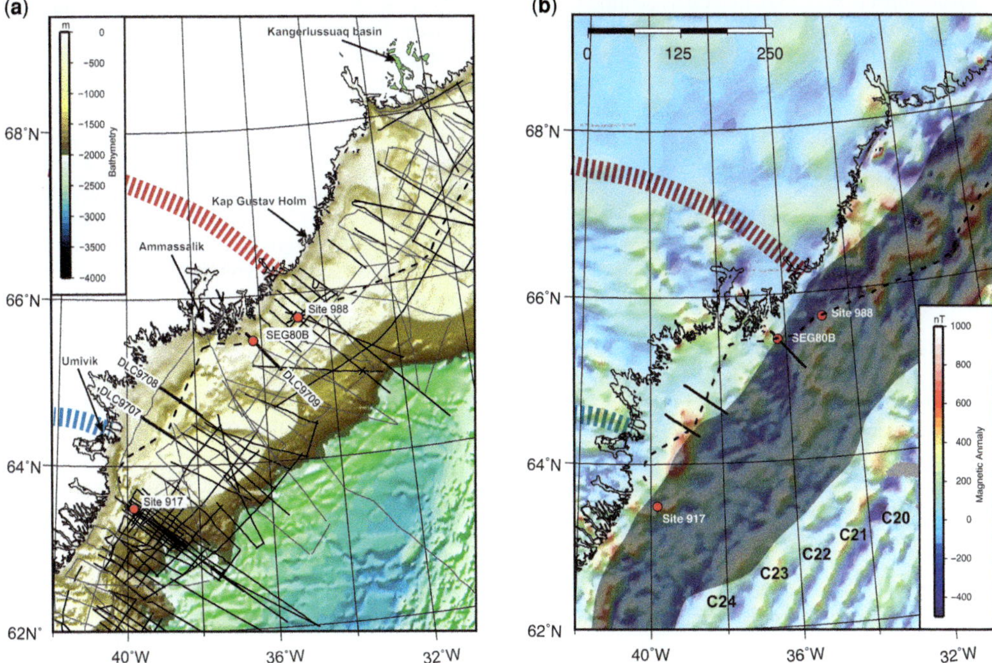

Fig. 2. (**a**) Bathymetric map of the SE Greenland margin near Ammassalik showing the available seismic data along with ODP legs 152 and 163 drill sites 917 and 988, and ODP Leg 163X drill site SEG80B. Thin grey lines indicate older single-channel seismic data from the 1970s; black lines are multichannel seismic data from the 1980s to present. The three DLC97 seismic lines (thick black) discussed in this paper are marked on the map, as is the TGS2012 profile (dashed line). Mid-Cretaceous sediments are exposed in the Kangerlussuaq Basin north of the Ammassalik Basin (e.g. Larsen & Saunders 1998). In addition, possible Cretaceous–Paleocene sediments are found near Kap Gustav Holm (Myers *et al.* 1993). The Nagssugtoqidian Orogen boundaries are marked as dark blue dashed (south) and red dashed (north) curves. (**b**) Map of the magnetic anomaly onshore and offshore of SE Greenland (Nasuti & Olesen 2014). The dark shaded areas offshore represent the area where seaward-dipping reflectors are observed. These define the main volcanic cover along the margin (see Horní *et al.* this volume, in review). Magnetic chrons C20–C24 are marked.

an approximately 1 km-thick Cretaceous–Palaeogene sedimentary succession (Larsen *et al.* 1999; Henriksen *et al.* 2009). The sediments onlap crystalline basement to the east and north: however, the base of the succession is not observed elsewhere (Larsen *et al.* 1999; Henriksen *et al.* 2009). The sedimentary rocks belong to the Kangerdlugssuaq and Blosseville groups (Soper *et al.* 1976; Nielsen *et al.* 1981). The oldest exposed deposits are Upper Aptian–Lower Albian alluvial and shallow-marine deposits. Upper Cretaceous marine mudstones interbedded with thin turbiditic sandstones overlie these successions. In the early Paleocene, submarine fan sandstones were deposited along the northern basin margin, whereas mudstone deposition continued within the basin. Fluvial sheet sandstones and conglomerates of latest Paleocene age overlie unconformably the offshore marine succession (Larsen *et al.* 1999, 2001, 2006). In the mid-Cretaceous, the basin underwent transgression that led to a Late Cretaceous–early Paleocene highstand, which was followed by extensive uplift and basin-wide erosion in the mid-Paleocene. Subsequently, renewed subsidence and extensive volcanism initiated immediately prior to and during break-up.

Offshore geology

The offshore area is dominated by the Palaeogene volcanic province associated with the opening of the North Atlantic (e.g. Larsen & Saunders 1998). Because of its importance for understanding volcanism associated with the development of the North Atlantic Igneous Province, most work along the margin has focused on the basaltic cover. In terms of the volcanic evolution, the margin is well studied, including deep seismic programmes to constrain crustal thickness, composition and volcanic productivity (e.g. Korenaga *et al.* 2000; Holbrook *et al.*

2001; Hopper *et al.* 2003), as well as drilling to sample the volcanic rocks for geochronology and geochemistry (Larsen & Saunders 1998; Tegner *et al.* 1998; Storey *et al.* 2007). The focus here, however, is on the older history and the sedimentary basins stratigraphically underlying the volcanic cover.

Offshore, sedimentary rocks were sampled at two drilling locations. At Ocean Drilling Program (ODP) Site SEG80B, a grey micaceous sandstone with coal flasers was recovered. These sediments appear to be stratigraphically below the basaltic lavas sampled in other cores further seaward (Thy *et al.* 2007). Palynological analyses at GEUS suggested an upper Middle Albian–lower Upper Albian age (Thy *et al.* 2007), correlating to the palynozonation constructed from other parts of East Greenland (Nøhr-Hansen 1993).

At ODP Site 917, metamorphosed sandstones were recovered that are stratigraphically overlain by the oldest Paleocene basalts found in East Greenland (62–61 Ma: Sinton & Duncan 1998). The sedimentary rocks were devoid of nanofossils and could not be dated (Wei 1998), but were inferred to be Upper Cretaceous–lower Paleocene based on stratigraphic relationships and by analogy to the Kangerlussuaq Basin further north (Vallier *et al.* 1998). Metamorphism of the strata was considered to be linked with Paleocene magmatism (Vallier *et al.* 1998), an interpretation that was subsequently questioned by Fyhn *et al.* (2012). A thin layer of unmetamorphosed gravel is sandwiched between the basalts and the metamorphic rocks. The metamorphic character of the sandstones and the non-metamorphic character of the gravels are difficult to reconcile with Paleocene contact metamorphism, and could indicate a much older age to the metasandstones. Fyhn *et al.* (2012) speculated that the metasedimentary rocks could be as old as late Proterozoic or Palaeozoic and could have been metamorphosed during the Caledonian Orogeny.

Tectonic setting

Although many details are still matters for debate, the overall tectonic setting of the region during the Cenozoic is well established (e.g. White 1997; Holbrook *et al.* 2001; Gaina *et al.* 2009; Ellis & Stoker 2014). Northward propagation of seafloor spreading from the Central North Atlantic during the Late Jurassic–Early Cretaceous separated North America from northern Africa and Europe. Seafloor spreading then propagated into the Labrador Sea no later than 62 Ma (Chalmers & Laursen 1995; Srivastava & Roest 1999). A second rift branch propagated to the NE at 56 Ma (magnetic Chron C24), opening the main North Atlantic oceanic basins and separating SE Greenland from the eastern margins along Ireland, the UK and Norway. Continuous seafloor spreading between SE Greenland and northern Europe has occurred since then (e.g. White 1997).

While the early rift branch into the Labrador Sea was largely magma-starved (Chian & Louden 1994; Chian *et al.* 1995), the second rift branch was accompanied by extreme volcanism with a spike in volcanic productivity at break-up time (Holbrook *et al.* 2001; Storey *et al.* 2007). This volcanic event buried the main marginal basins. Consequently, the pre-Cenozoic tectonic evolution and basin history is poorly established, especially off the SE Greenland margin. Along the conjugate margin, basin evolution is linked to better investigated areas further east, where extension and basin formation began in the Palaeozoic as the Caledonian and Variscan orogenic belts collapsed (e.g. Ziegler 1988). In the Rockall Basin, the main rifting occurred in the Early Cretaceous, but reconstructions suggest that older rifting with β stretching factors of up to 1.6 must have occurred (Cole & Peachey 1999). In the Hatton Basin, the main rift phase is generally assumed to be mid-Cretaceous (McInroy & Hitchen 2008). Of particular importance along the Hatton margin are several basalt-free windows where older basins are observed (Fig. 3). Albian sedimentary rocks were sampled in shallow cores taken by the British Geological Survey in 1999 from these basalt-free windows (Hitchen 2004). Analogous to the Rockall Basin, older rifting is thought to be likely (Cole & Peachey 1999). Nevertheless, the pre-Cretaceous history of the outer basins along the Hatton and Rockall margins is presently unconstrained by direct sampling.

Data and methods

The description and dimensions of the Ammassalik Basin are based primarily on seismic reflection data supplemented with gravity anomaly data. Most seismic reflection data collected along the SE Greenland margin was in support of ODP campaigns to sample the basaltic sequences (Larsen & Saunders 1998; Holbrook *et al.* 2001). Thus, most profiles are located further offshore, straddling the continent–ocean transition (Fig. 2a). Less attention was paid to the areas closer to the coast and possible underlying sedimentary sequences.

In this study, data from two seismic surveys were used. In 1997, a high-resolution survey was collected by the Danish Lithosphere Centre in support of a drilling campaign that used a 5 m rock-drill system to drill multiple holes along a stratigraphic transect. Seismic profiles were acquired as close to the coast as conditions would allow in order to locate the eastward edge of the basalt sequences for sampling of the oldest flows. The data were acquired with a high-resolution seismic system consisting

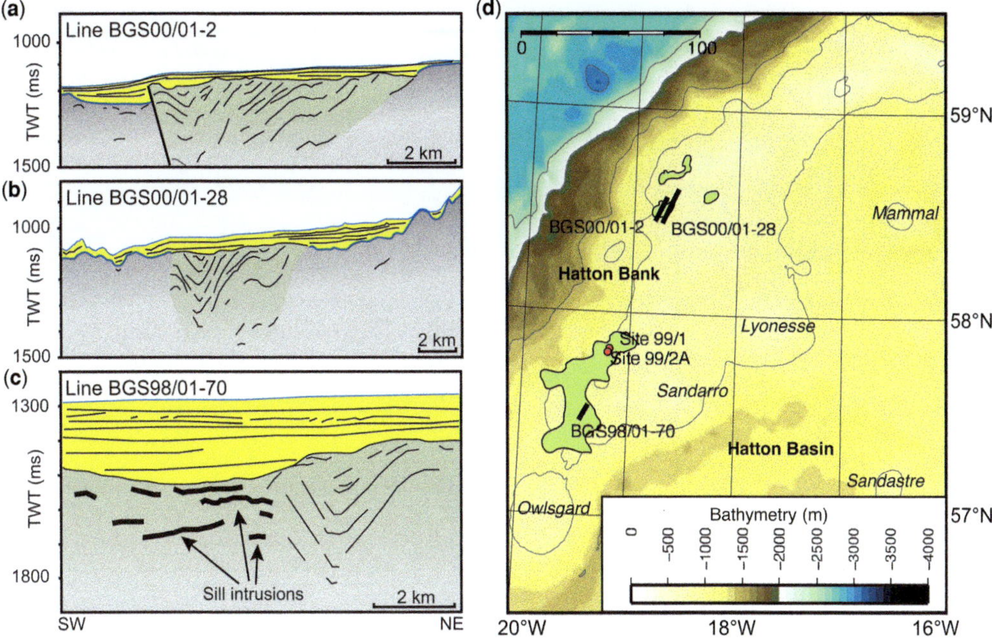

Fig. 3. (**a**)–(**c**) Interpretations of the seismic lines BGS00/01-2, BGS00/01-28 and BGS98/01-70 situated on the Hatton margin that cross basalt-free windows (modified from Hitchen 2004). Green indicates areas where Cretaceous sedimentary rocks are proven or inferred; the dark grey shaded area is basalt cover; light grey may be either basement or sediment; and thick black lines are strong reflections, possibly representing sill intrusions. (**d**) Map of the Hatton Bank showing the location of the three BGS seismic lines, together with the BGS drill sites 99/1 and 99/2A. Green areas indicate the basalt-free windows on the Hatton Bank where Cretaceous sedimentary rocks are proven or inferred. Names in italic represent central igneous complexes. Less than 50 km NE of Mammal, just outside the map, oil slicks are observed that are likely to be the result of natural oil seepage (Hitchen 2004).

of 4 × 40 cubic inch sleeve guns chained together in a small cluster spaced 50 cm apart. Data were recorded on a 96-channel, 594 m streamer with a 6.25 m channel interval. In 2012, a commercial seismic profile was shot by TGS using a 6 km streamer and 3350 cubic inch tuned array. Three of the profiles from the earlier survey, DLC97-07, DLC97-08 and DLC97-09, showed reflectivity suggestive of older sedimentary basins stratigraphically below the basalts (Hopper *et al.* 1998). The commercial seismic line tied these three profiles together and confirmed the presence of significant sedimentary sequences.

Processing of the commercial seismic line followed current industry standards, and included multiple attenuation and pre-stack time migration. The earlier data, however, had only limited processing initially since the focus of the early studies was on the basalt cover, rather than on the underlying sequences. Only brute stacks of DLC97-08 and DLC97-09 have been published previously (Hopper *et al.* 1998; Thy *et al.* 2007, respectively). Therefore, the three DLC97 profiles were reprocessed to better image the sedimentary stratigraphy.

The reprocessing flow included SEG-D to SEG-Y conversion with geometry assignment, a source-signature deconvolution where the source wavelet is derived from averaging the direct wave, normal move-out correction and stack, post-stack Kirchhoff migration, coherence enhancement (dip scan stack: ±2 ms/trace), amplitude balancing, and bandpass filtering (10–20 Hz low-pass cutoff, and either 200–250 Hz or 50–100 Hz high pass-cutoff depending on target area). Because of the short streamer, multiple attenuation was not attempted and it is generally difficult to detect any signal below the first multiple. An exception, described further in the following 'Interpretation of seismic and gravity data'section, is for the case of line DLC97-09, where signals from some deep reflections are strong enough to penetrate through the multiples.

Because of the limited seismic coverage, only four profiles, it is difficult to establish the dimensions of the basin. Therefore, regional gravity data was used to estimate the potential lateral extent of the basin and to help define the main structural trends. The gravity data are from Haase

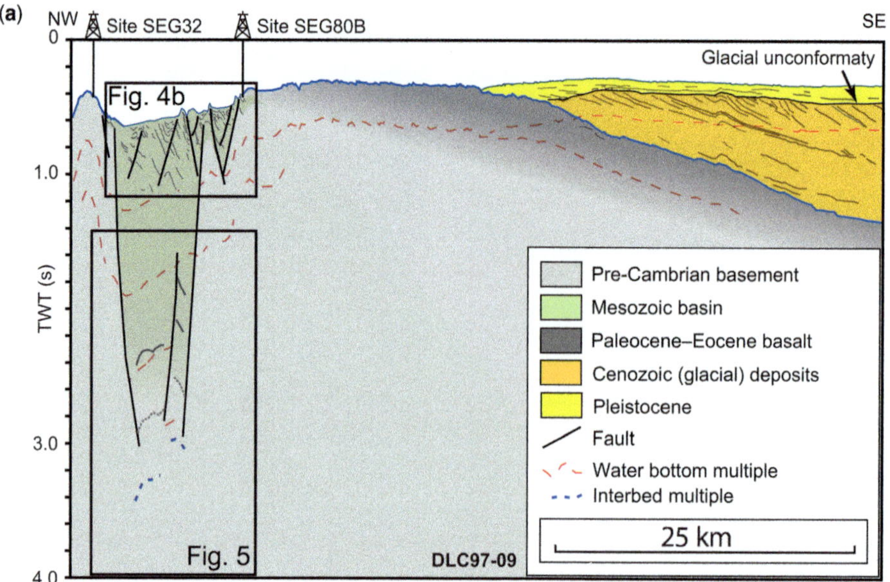
(a)
NW
Site SEG32
Site SEG80B
SE
Glacial unconformaty
Fig. 4b
Fig. 5
DLC97-09
TWT (s)
0
1.0
3.0
4.0
Pre-Cambrian basement
Mesozoic basin
Paleocene–Eocene basalt
Cenozoic (glacial) deposits
Pleistocene
Fault
Water bottom multiple
Interbed multiple
25 km

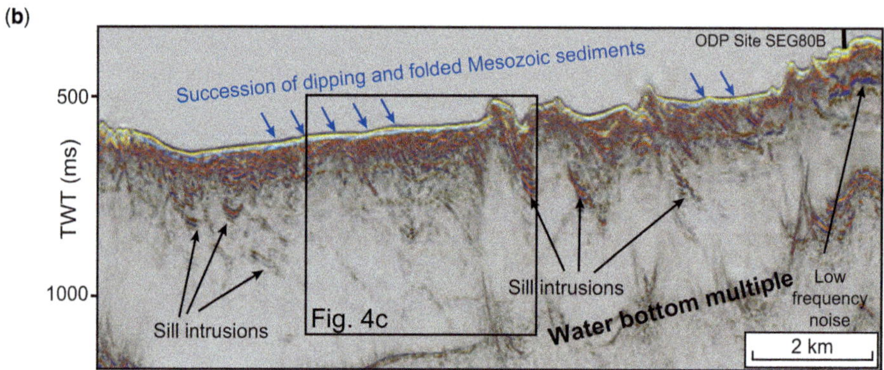
(b)
Succession of dipping and folded Mesozoic sediments
ODP Site SEG80B
500
1000
TWT (ms)
Sill intrusions
Fig. 4c
Sill intrusions
Water bottom multiple
Low frequency noise
2 km

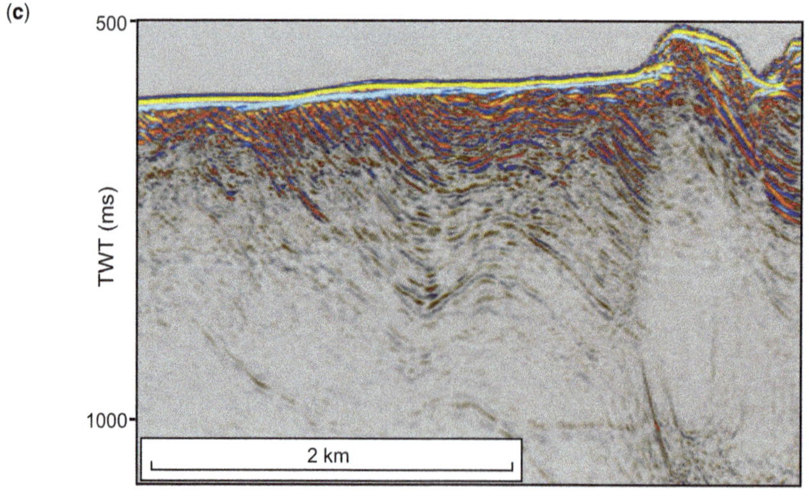
(c)
500
1000
TWT (ms)
2 km

& Ebbing (2014) and Haase *et al.* (2016) and is based on the DTU10 satellite gravity compilation (Andersen 2010).

Interpretation of seismic and gravity data

Profile DLC97-09

DLC97-09 is the northernmost seismic profile with clear indications of older sedimentary stratigraphy. It is approximately 68 km long and runs across the SE Greenland Shelf just north of Ammassalik (Fig. 2). A basement high is imaged (Fig. 4a) on the landward-most part of the line that was sampled during the ODP Leg 163X at Site SEG32B. The samples show that the high is a Proterozoic orthogneiss (Thy *et al.* 2007). Seaward of this, a 12–13 km-wide section of dipping sedimentary strata is imaged, indicating the presence of a significant sedimentary basin. The basin is interpreted to be downfaulted relative to the Proterozoic basement. At the seaward-most edge of the basin close to the uppermost part of the stratigraphic succession, an Albian sandstone was recovered at Site SEG80B (Thy *et al.* 2007). Stratigraphically, these are the probably the youngest of the pre-Cenozoic strata imaged along the profile. Just seaward of this site, a step in the seafloor morphology is interpreted as the overlying basalt cover. This is in accordance with samples taken at all sites seaward of SEG80B. Dating of the basalt samples show they were extruded immediately prior to and during continental break-up (61–56 Ma). Both the sediments and the basalts dip seaward, and the samples consistently become younger seaward. Further seaward, a wedge of post-basalt Cenozoic sediments has been deposited. This sedimentary wedge is clearly imaged in the seismic profile (illustrated in the interpreted seismic section: Fig. 4a), but is not the focus of this paper.

The sedimentary basin imaged between the Precambrian basement and Palaeogene basalts has a thin (*c.* 20 ms TWT (two-way time)) cover of Quaternary sediments overlying an up to 2.2 s dipping sequence of strata. In places, the sedimentary sequences are folded (Fig. 4c) and the section is reminiscent of the sections from Hatton Bank (Fig. 3) (Hitchen 2004). The folding appears to be deformational and some faults may be reversed, although this cannot be demonstrated unambiguously. It is suggested that the folding may be related to post-extensional inversion. Nevertheless, the overall seaward dip of the sedimentary successions suggests that the oldest sedimentary rocks subcrop out to the west. The uppermost and youngest sediments are to the east, where Albian-aged deposits were recovered just below the basalt (Thy *et al.* 2007). Thus, despite the possibility of later inversion, an overall significant stratigraphic thickness is indicated. Based on the substantial stratigraphic thickness indicated below the Albian deposits, we infer that basin development initiated well before the mid-Cretaceous, implying a significant Early Cretaceous rift phase, comparable to that inferred on the Hatton Bank (Cole & Peachey 1999; Hitchen 2004). Given the significant stratigraphic thickness, the presence of even older Mesozoic–Palaeozoic deposits cannot be ruled out.

Strong reflections within the shallow section are likely to be sill intrusions, as would be expected given the strongly volcanic nature of final rifting and break-up. Deeper strong reflections are also observed, however. Calculations of possible multiple energy paths show that these cannot be multiples from any of the main shallow reflections and thus some must be primary reflectivity (Fig. 5). Reflections at roughly 2.3, 2.8 and 3.2 s cross-cut the third and fourth water bottom multiples (Fig. 5). Modelling possible interbed multiples shows that the 3.2 s may be an interbed multiple of the overlying 2.3 and 2.8 s reflections (see Fig. 5, lower-right corner). The 2.8 s reflection may thus represent the base of the basin. Alternatively, the 2.3 s event is the basement reflection. Nevertheless, the basin has a significant depth of 1.7–2.2 s and is bounded by steep normal faults. Assuming P-wave

Fig. 4. (**a**) An interpretation of seismic line DLC97-09. At the most NW part of the line ODP drill site SEG32 is located, where a Proterozoic gneiss was sampled in shallow cores (Thy *et al.* 2007). To the east, the seismic data clearly show a succession of dipping and folded sediments (see also Fig. 3b). The base of this basin is at approximately 2.8 s. At the SE limit, Albian sedimentary rocks were recovered at ODP drill site SEG80B (Thy *et al.* 2007). Immediately east of this site, the basaltic cover is observed along the rest of the line. The approximately 30 km-long most NW part of the line is covered by a wedge of Cenozoic sediments. The two boxes indicate a close-up of the seismic sections in (b) and Figure 5. (**b**) Close-up of the top approximately 600 ms of the sedimentary basin observed on line DLC97-09. The location of the ODP drill site SEG80B is indicated in the most SE part of this seismic image. Note the succession of dipping and folded sediments. The dip direction of the sedimentary layers shows that the subcropping sediments become older towards land (NW). Some strong reflections are probably from sill intrusions. The box indicates a close-up of the seismic section in (c). (**c**) A close-up of folded and dipping strata along line DLC97-09. Notice how thin (*c.* 20 ms) the Cenozoic cover is. The seismic image is strikingly similar to that from the conjugate Hatton Bank (Hitchen 2004).

velocities of 3–4 km s^{-1}, a basin of up to 4 km thick is indicated. This is consistent with regional gravity inversion, which shows that the depth to basement along the shelf here is 4–5 km deep (Haase *et al.* 2016).

Profile DLC97-08

DLC97-08 is an approximately 63 km-long line that runs across the SE Greenland Shelf south of Ammassalik and terminates seaward at the limit of the basalt cover (Fig. 6). The landward-most approximately 6.5 km of the line images the Proterozoic basement with a thin layer of Quaternary sediments on top. The next approximately 15 km of the profile shows a distinct change in seafloor morphology (Fig. 6b). The seafloor shows prominent relief, probably carved out by Neogene glacial erosion. Seismic imaging below the seafloor in the bathymetric low is characterized by chaotic reflectivity, in contrast to the more acoustically transparent continental basement further landward. The section with chaotic reflectivity may indicate a softer substrate than the Proterozoic basement, and thus the reflectivity could indicate the presence of sediments or meta-sediments (see also the following discussion of profile DLC97-07).

Seaward of the segment with chaotic sub-seafloor reflections, a sedimentary basin is clearly imaged and can be traced for another 35 km along the profile (Fig. 6c). The basin is interpreted to be downfaulted relative to the Proterozoic basement. Extensional faults offset and deform the strata, compartmentalizing the basin with a main and deepest section situated in the middle of the profile (Figs 6c & 7a) and second thinner section located at the seaward end of the basin. Smaller, minor pockets of older sediments separated by fault blocks appear throughout the profile (Fig. 7b).

In the main part of the basin (Figs 6c & 7a), a thin layer of horizontally deposited sediment is interpreted as Quaternary cover. It is up to 80 ms thick, although some of this may be ringing of the source and the cover may be even thinner. Seaward-dipping sedimentary strata are reasonably clear immediately below the cover, although the imaging quality quickly degrades with depth. Strong, low-frequency reflections are interpreted as sill intrusions that follow the basin stratigraphy (Fig. 6c) and are even more clear in the first multiple (Fig. 7a). Reflectivity appears to continue to at least the first water bottom multiple at 2.1 s, indicating that the base must lie deeper than the multiple. Although not well imaged, some faulting is also indicated that, together with the steep dip, suggests tectonic activity after deposition. The sediments seem to be more horizontal in the seaward section of the basin, although the imaging here is significantly poorer (see Figs 6a & 7b).

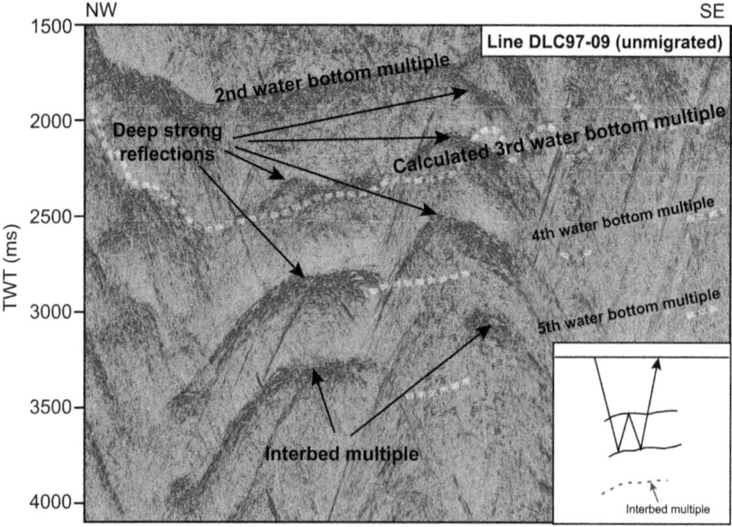

Fig. 5. A close-up of the deeper part (1500–4000 ms) of the sedimentary basin imaged on line DLC97-09; see Figure 4a for the location. Note the strong reflection that cross-cuts the third multiple at 2000–2500 ms, as well as the strong reflections at approximately 3000 and 3300 ms. Calculations of multiple energy show that the lower strong reflection could be an interbed multiple of the two reflections above (see the line drawing in the bottom right-hand corner). This suggests that the middle of strong reflections may be the base of the basin and that the basin is >2.5 s deep.

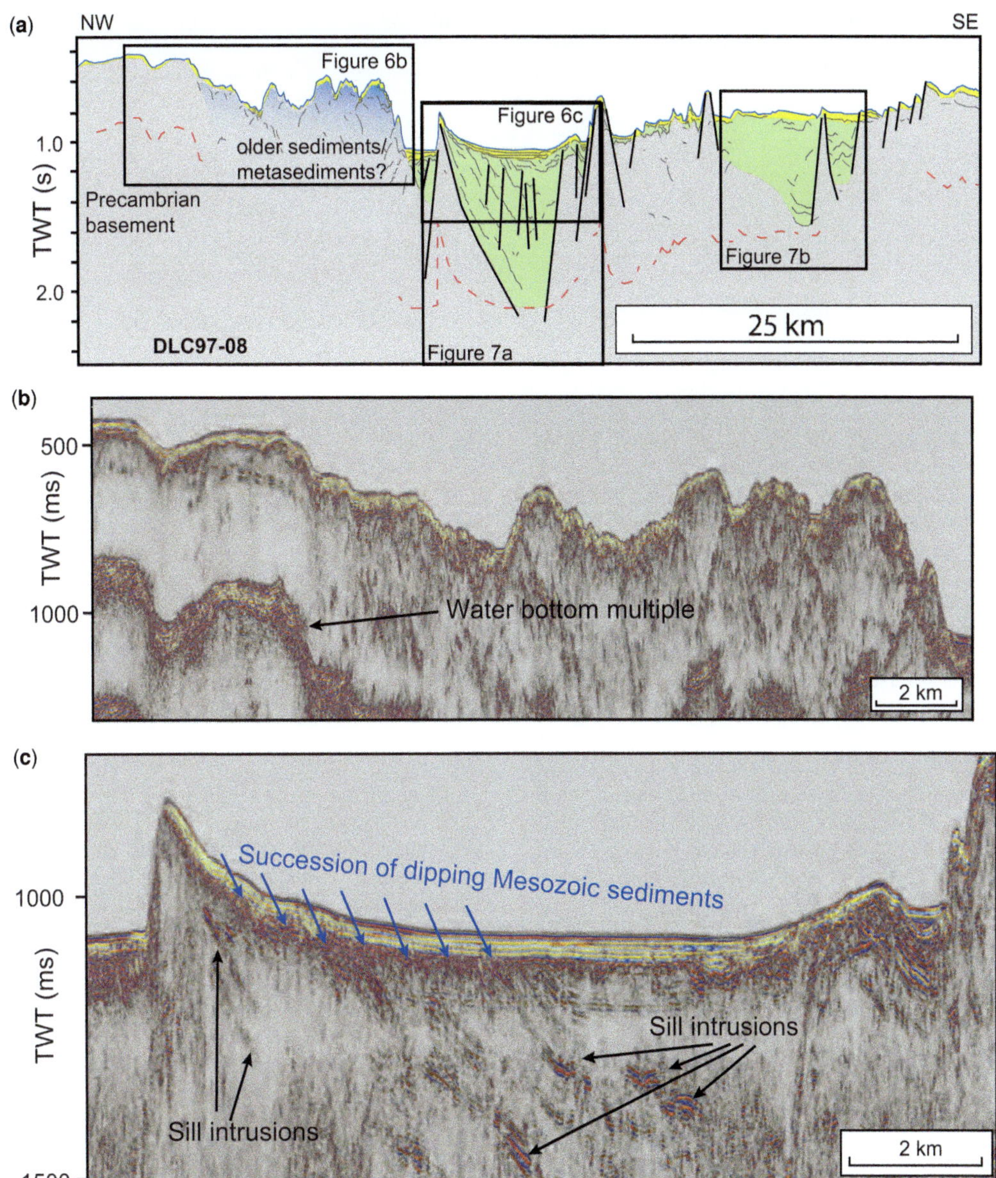

Fig. 6. (**a**) An interpretation of the seismic line DLC97-08. The boxes indicate close-ups of the seismic sections in (b) and (c), and in Figure 7. The legend is shown in Figure 4a. (**b**) Close-up of the landward-most flank of the sedimentary basin. Note the change from a smooth and flat seafloor to a seafloor with high relief from NW to SE. Below the smooth seafloor there is very little reflectivity, except for some low-frequency noise approximately 100 ms below the seafloor. The most NW part of the line is interpreted as Precambrian basement with a thin Cenozoic cover. In contrast, chaotic reflectivity is observed beneath the seafloor with high relief, suggesting a change in the underlying geology (see the discussion in the text). (**c**) Close-up of the top approximately 600 ms of the sedimentary basin observed in the centre of line DLC97-08. Like profile DLC97-09 described earlier, the seismic data show a succession of dipping strata with volcanic sill intrusions and a thin cover (up to *c.* 80 ms) of Quaternary sediments. The dipping strata are presumed to be Cretaceous and possible older sedimentary rocks. Rotated fault blocks are observed throughout the basin, which on this line is approximately 30 km wide. It is not possible to determine the basement depth along this line since it is concealed beneath the strong water bottom multiple. The most SE part of the line may just intersect the basalt cover, as indicated in Figure 2a. However, it is not possible to determine this from the DLC97 data.

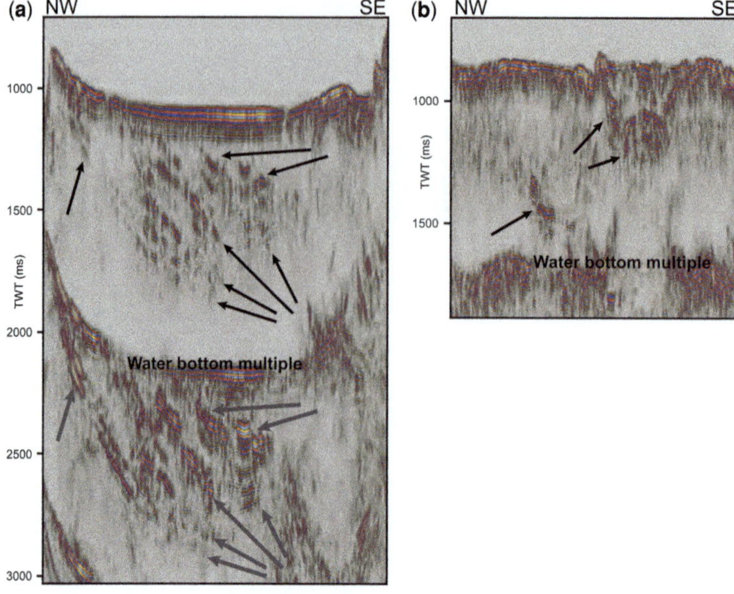

Fig. 7. (**a**) Low-frequency band pass of the seismic data from the central part of line DLC97-08. The strong dipping reflections (sills; marked with black arrows) are more pronounced in the low-frequency range and even more distinguishable in the first multiple (grey arrows). See Figure 6a for the locations of the close-ups. (**b**) Strong reflections are also imaged in the SE part of the line.

The last 6.5 km and seaward-most part of the line images a series of normal faulted blocks with a thin cover of Quaternary sediments.

Profile DLC97-07

DLC97-07 is approximately 50 km long and runs parallel to DLC97-08 to the south. The 18 km landward-most part of the line displays a relatively smooth, low-relief seafloor morphology similar to the landward-most part of line DLC97-08 (Fig. 8). Hence, this part is also interpreted as Proterozoic basement. The seaward-most part of the line shows a seafloor morphology with low relief and evidence of shallow seaward-dipping reflectors interpreted as basalt flows, thus marking the feather edge of the break-up volcanism. In-between the basalt flows and the interpreted Proterozoic basement is a section with high seafloor relief with chaotic reflectivity, very similar to that observed along DLC97-08 (see seaward part of Fig. 6b). Small pockets of Quaternary sediments are deposited in the deepest part of this bathymetric depression. The regions of high relief are probably the result of erosion by continental glaciations. The stronger, more resistant areas were less affected by erosion and show more subdued relief, whereas areas with less resistant bedrock eroded more easily, resulting in the distinct seafloor morphology. Sedimentary and metasedimentary rocks are likely to be more prone to erosion than basement, and we suggest that these bathymetric troughs are part of the Ammassalik Basin. Clear sedimentary stratigraphy like that imaged on profiles DLC97-08 and DLC97-09, however, are not apparent on this profile.

TGS seismic line

A commercial seismic line from 2012 ties the previous three profiles together (Fig. 9). Precise location information is confidential, so tie points along the southern part of the profile are not shown and the horizontal scale is unspecified for the northern portion where it crosses the SEG80B shallow core location. The profile covers two main basinal depressions. The northernmost one crossing the shallow core site is bounded on both sides by apparent basement. Although Proterozoic basement can be difficult to distinguish from a Palaeogene basaltic cover, the basaltic cover often shows a step-like morphology and is characterized by intra-basaltic reflectivity from the stacked lava flows. In contrast, the Precambrian basement appears acoustically transparent. Thus, the area to the north of the northernmost basin is interpreted to include a basaltic cover and the basin may continue to the north beneath it. The Proterozoic crystalline basement

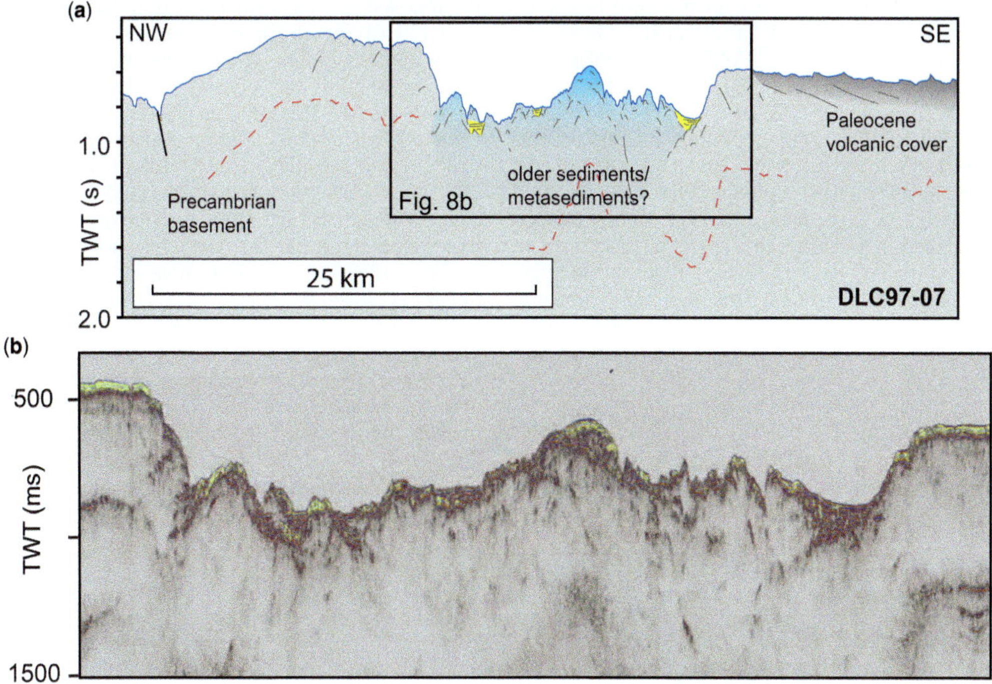

Fig. 8. (**a**) An interpretation of the seismic line DLC97-07. The box indicates a close-up of the seismic data in (b). The most NW part of the line is interpreted to be Precambrian basement with a thin cover of Cenozoic sediments. The most SE part of the line has a more step-like seafloor morphology, possibly indicating a basaltic cover. The legend is shown in Figure 4a. (**b**) Close-up of the central part of line DLC97-07 that crosses a significant bathymetric depression characterized by a high seafloor relief. The seismic image shows the same kind of high-relief seafloor and chaotic reflectivity as DLC97-08 (blue-shaded area: Fig. 5b). This area is tentatively interpreted to be underlain by a sedimentary basin.

separates the northern basin from the main basin further south, however.

The main part of the basin is very well imaged (Fig. 9). At least 2 s (3–4 km, assuming typical P-wave velocities) of sedimentary layering can be interpreted, showing that the sediment thickness here is consistent with that interpreted along DLC97-09. As noted earlier, this is also consistent with a depth to basement of 4–5 km from the regional gravity inversion (Haase *et al.* 2016).

The basins appear to be deformed and a number of intra-basinal extensional structures, graben and half-graben, are interpreted within the southern sub-basin. The Ammassalik Basin is therefore

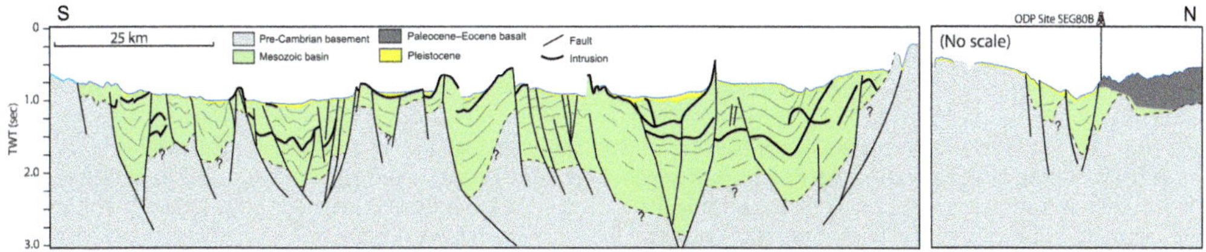

Fig. 9. (**a**) An interpretation of a commercial seismic line acquired by TGS in 2012. The line runs parallel to the coast and crosses the three DLC97 seismic lines. The seismic line confirms the existence of a large sedimentary basin south of Ammassalik, possibly more than 100 km long and approximately 2 s deep. A succession of dipping and folded sediments is observed, similar to those of the DLC97 lines. Steep faults and graben-like structures are also imaged. The older sediments are intruded by sills and covered by a thin layer of Quatnernary sediments. The line crosses a large basement block that separates the larger basin south of Ammassalik from the smaller basin to the north along DLC97-09.

interpreted to have formed in response to rifting. Subsequent deformation has affected the basin and led to its present outline. Folding of the strata also resembles that imaged on DLC97-09 and is similar to the inversion of Cretaceous basins observed on the Hatton Bank (Fig. 3) (Hitchen 2004). Compressional inversion may thus have affected the Ammassalik Basin, although direct evidence of compression in the form of unambiguously imaged compressional faults remains to be documented. Part of the deformation is also interpreted to relate to magmatic intrusion of the basin. Furthermore, fairly steep faults have been interpreted that may be related to a component of strike-slip deformation.

Gravity anomalies

Gravity data were used to estimate the potential size and structural trends of the basin. Details regarding the regional gravity anomalies are given in Haase *et al.* (2016). Figure 10a shows the tilt derivative of the isostatic anomaly. The tilt derivative enhances edges of features that generate anomalies (Miller & Singh 1994). Here, the analysis is restricted to the shelf, where the anomalies reflect the shallow structures. Anomalies to the east are associated with the continent–ocean transition, and shelf-slope edge anomalies are avoided as they are less straightforward to interpret. On land, the anomalies are likely to reflect the topography beneath the ice sheet rather the crustal structure.

Where the seismic line DLC97-08 and the TGS seismic line intersect regions of positive tilt, Proterozoic basement sub-crops at the seabed and regions of positive tilt are interpreted to represent basement highs along the shelf. This is particularly clear on DLC97-08 and suggests that a pronounced basement ridge bounds the main part of the Ammassalik

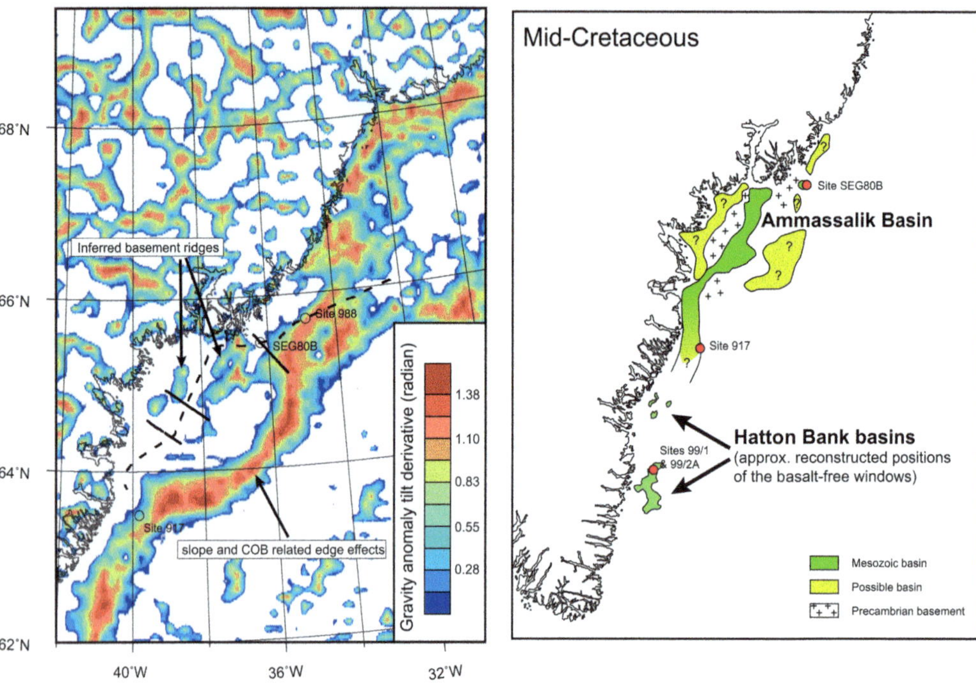

Fig. 10. (**a**) Map of the tilt-derivative of the isostatic gravity anomaly onshore and offshore of SE Greenland. The tilt is positive over interpreted basement highs, and negative over interpreted sedimentary basins. (**b**) Possible extent of the Ammassalik Basin based on available seismic, bathymetric, gravimetric and magnetic data, together with the pre-break-up reconstructed position of the Hatton margin basalt-free windows containing Cretaceous sediments. The main section of the Ammassalik Basin runs from Ammassalik southward towards ODP drill site 917. This main part of the basin varies from a few kilometres to approximately 30 km wide, up to 4 km deep and potentially more than 200 km long. Landward of the main part of the basin there is a basement high. Adjacent to the basement high and along the coast there may be another sedimentary basin intruded by dykes. North of Ammassalik, line DLC97-09 crosses a smaller basin without a basalt cover. The gravity data suggest a potential sedimentary basin along the coast landward of drill site 988 and north of line DLC97-09. The position of the Hatton Bank basins suggests that the Ammassalik was part of a continuous rift system in the Early Cretaceous. The total area covered by the basin areas interpreted on the SE Greenland margin is approximately 100 000 km^2.

Basin to the west. Between this basement high and the coast, negative tilt is again observed, and it is suggested that additional sub-basins could be preserved further west than those documented here (Fig. 10b).

The region of negative tilt that the TGS seismic line follows is interpreted to represent the main part of the basin. To the east, where there is no seismic coverage, a region of negative tilt is separated from the main basin by a linear region of positive tilt. This is interpreted as a basement high separating the main basin from an additional sub-basin to the east, as shown in Figure 10b. Thus, it is suggested that the basin continues beneath the basalt cover here.

The two westernmost areas of negative tilt separated by the basement high merge to the south and a low anomaly continues towards ODP Site 917, where sedimentary and meta-sedimentary rocks just below Paleocene basalt were recovered (Vallier *et al.* 1998). Interpretation of both the seismic and gravity data becomes complicated in this region. The possible southern continuation of the basins along the margin is uncertain, but seems unlikely. Reconstructing the positions of the Hatton Bank basins shown in Figure 3 to their position along Greenland at 80 Ma shows that they are along-strike and just to the south of the southern limit of the Ammassalik Basin, as interpreted here (Fig. 10b). The reconstructed positions are based on the poles of rotation of Gaina (2014).

Discussion

The seismic reflection data presented here show unambiguous sedimentary basins offshore SE Greenland. The main part of the basin appears to be just to the south of Ammassalik, and is documented by lines TGS2012-10, DLC97-07 and DLC97-08. A smaller basin is also clearly imaged just to the east of Ammassalik along the DLC97-09 profile, as well as in the TGS data. Although apparently smaller, the latter basin is significant because it was sampled by a shallow coring campaign, providing key evidence for Albian sandstones below the Paleocene basalts (Thy *et al.* 2007). The significant stratigraphic thickness indicated along the seismic profiles and the observed structural pattern suggest that major Cretaceous rifting affected the region. While the sparse seismic coverage makes establishing dimensions of the basin and interpreting the orientations of key structures and basin confining faults difficult, the interpretation shown Figure 10 indicates a sizable basin of approximately 100 000 km^2. In addition, while Cretaceous sediments are proven by core samples, the existence of pre-Cretaceous strata should also be considered. Possible ages for the basin are discussed based on plate reconstructions back to Permian–Triassic times.

Figure 10b shows reconstructed Campanian (80 Ma) positions of the basalt-free windows where Cretaceous sediments sub-crop out along the Hatton margin. These align well with the Ammassalik Basin trend outlined here. The southward continuation of the Ammassalik Basin is therefore suggested to be located on the Hatton margin. This implies a major transform boundary just south of the ODP legs 152 and 163 drilling transect. The position is approximately conjugate to the Anton Dohrn Lineament (Fig. 1), which is a prominent but complex region where transfer offset between the northern and southern Rockall basins occurs (Kimbell *et al.* 2005). It coincides with offsets in the continent–ocean boundary and is interpreted by Kimbell *et al.* (2005) to be a precursor of the oceanic transform faults that formed during early seafloor spreading (Featherstone *et al.* 1977; Smallwood & White 2002). Kimbell *et al.* (2005) showed that the main offset must predate seafloor spreading. If the interpretation here of the Ammassalik Basin and its southern continuation onto the Hatton margin is correct, then the main offset is likely to have occurred in the Late Cretaceous (Campanian–Maastrichtian) and after the deposition of most of the Cretaceous fill, which must be older than the Albian strata immediately below the basalts.

A central unresolved issue for understanding the regional Mesozoic development is at what point a through-going rift system developed between SE Greenland and NW Britain and Ireland (see Stoker *et al.* 2016 for a full summary). While eventual break-up in the Paleocene shows that by Late Cretaceous time this must have occurred, the older history remains enigmatic. Historically, most reconstructions show that this may already have occurred by the Jurassic (see Stoker *et al.* 2016 and references therein), which has obvious implications for understanding the development of regional petroleum systems.

Cole & Peachey (1999) proposed that pre-Cretaceous rifting and stretching are required in both the Rockall and Hatton basins to eliminate overlap between a small Rockall plate and the main European Plate. They estimated a pre-Cretaceous β stretching factor of 1.6, which would probably have resulted in a substantial and deep basin in these outer-margin areas.

Figure 11 shows reconstructed sediment distribution maps compiled during the NAG-TEC project (Stoker *et al.* 2014). It should be noted that, in these reconstructions, rigid plates are assumed and so margin deformation is not accounted for, and will lead to some distortion in the present-day size of basins v. their pre-stretched lateral extents (Gaina 2014). The Cretaceous fits at 100 and 80 Ma show

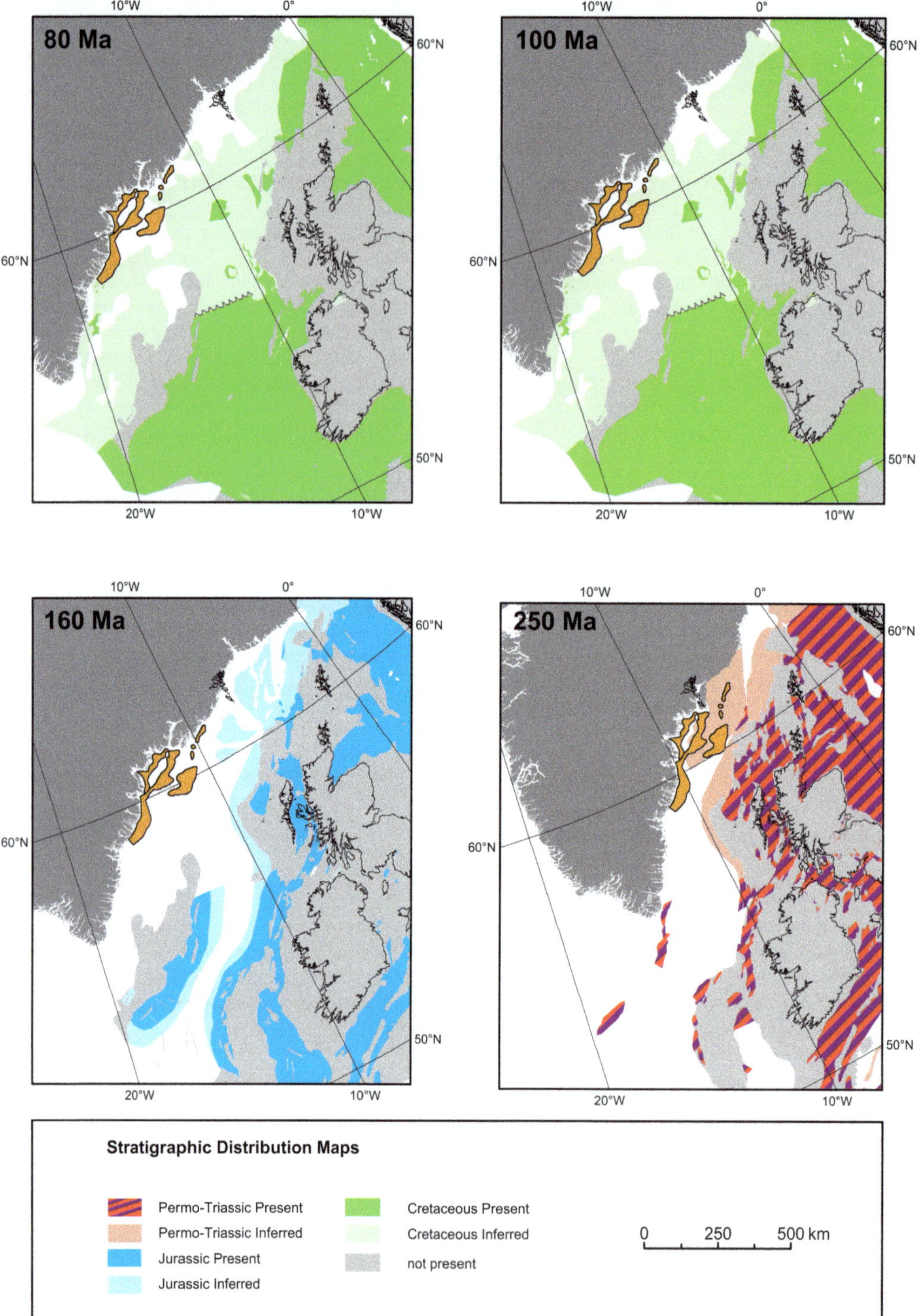

Fig. 11. Plate reconstructions of stratigraphic distribution maps from SE Greenland and the European conjugate margins from the Permian–Triassic (250 Ma: red/purple), Jurassic (160 Ma: blue) and Cretaceous (100 and 80 Ma: green). The Ammassalik Basin is outlined in orange.

that the Ammassalik Basin overlaps well with areas inferred to contain Cretaceous basins along the conjugate margins. The Jurassic (160 Ma) and Permian–Triassic (250 Ma) fits are similarly intriguing, although the uncertainties of these fits are large (see Gaina 2014), and the ridge plate assumptions remain problematic. In particular, since the Cretaceous extension has not been accounted for, the position of the Ammassalik Basin could be further SE towards Ireland and Britain.

On the conjugate British and Irish flank of the North Atlantic, proven Jurassic sequences occur along the inner margins and in the Rockall Basin, but only limited Jurassic sedimentary sequences have been found on the Faroe–Shetland platform, and there is little information on the region between Rockall Basin and Faroe–Shetland region. Thus, it remains equivocal to what extent a significant Jurassic system had developed through the region as opposed to more isolated and disconnected basins (Stoker *et al.* 2016). As shown in the reconstruction, a possible Jurassic rift system would be likely to include parts of the SE Greenland margin, near to where the data presented here show the presence of a significant basin. Considering the Permian–Triassic reconstructions at approximately 250 Ma, the region of the Ammassalik Basin overlaps with areas of inferred Permian–Triassic basins along the NW British margin and the Faroe–Shetland platform. Here it suggested that some of the thick sedimentary accumulations in the Ammassalik Basin imaged in this study could be early Mesozoic and possibly even Palaeozoic in age, although this can ultimately only be proven by sampling the sequences.

Conclusions

The Ammassalik Basin offshore SE Greenland has been investigated using a combination of reprocessed shallow seismic reflection profiles, a recent commercial deep seismic reflection profile and gravity data. The results show the present-day vertical thickness of the basin is at least 2 s, or 4 km. This is consistent with gravity inversion results showing a depth to basement of 4–5 km along the shelf off Ammassalik (Haase *et al.* 2016). In several areas, steeply dipping strata are imaged over several kilometres of profile, indicating that the stratigraphic thickness of the basin fill could also be significant. Apparent folding and possible fault reversal indicate that later inversion and compression also affected the basin. Owing to the sparse seismic data, however, inversion cannot be demonstrated unambiguously.

The main basin fill is interpreted to be a result of rifting and extension between SE Greenland and the Hatton margin prior to eventual break-up and seafloor spreading, which initiated in the latest Paleocene–earliest Eocene. Samples recovered from the stratigraphic top of the succession along one profile are Albian in age (Thy *et al.* 2007), indicating significant Early Cretaceous rifting and basin formation. Reconstructions to the conjugate margin show that the main part of the Ammassalik Basin is along-strike from areas of the Hatton margin where mid-Cretaceous sediments were recovered. This suggests that a well-developed regional rift basin system existed between SE Greenland and the Hatton/Rockall regions by this time. The boundary between the Ammassalik Basin and the basins along the Hatton margin coincides with the Anton Dohrn Lineament. It is suggested that strike-slip and transform motion developed in the Late Cretaceous, with the northern part of the basin system left on the Greenland side and the southern part of the system left on Hatton side after final break-up.

The older history of the basin is unconstrained by the available data. Reconstructions going back to the Jurassic and Permian–Triassic, however, show intriguing possibilities that a through-going rift system separating SE Greenland from the Hatton Margin could be older than Cretaceous. A sampling campaign targeting the oldest sediments where the data here show that they onlap Precambrian basement would provide important information regarding the complete history of rifting of the North Atlantic between SE Greenland and Europe.

We thank TGS for permission to include an interpretation of their seismic profile along the margin in Figure 9. This paper includes some results of the NAG-TEC project and we gratefully acknowledge the NAG-TEC sponsors for permission to include the reconstruction shown in Figure 11. Sponsors in alphabetical order: Bayerngas Norge AS; BP Exploration Operating Company Ltd; Bundesanstalt für Geowissenschaften und Rohstoffe (BGR); Chevron East Greenland Exploration A/S; ConocoPhillips Skandinavia AS; DEA Norge AS; Det norske oljeselskap ASA; DONG E&P A/S; E.ON Norge AS; ExxonMobil Exploration and Production Norway AS; Japan Oil, Gas and Metals National Corporation (JOGMEC); Maersk Oil; Nalcor Energy – Oil and Gas Inc.; Nexen Energy ULC; Norwegian Energy Company ASA (Noreco); Repsol Exploration Norge AS; Statoil (UK) Ltd; and Wintershall Holding GmBH.

References

Andersen, O.B. 2010. The DTU10 gravity field and mean sea surface – improvements in the Arctic Ocean. Paper presented at *Second International Symposium of the Gravity Field of the Earth (IGFS2)*, 20–22 September 2010, University of Alaska Fairbanks, Fairbanks, Alaska.

Chalmers, J.A. & Laursen, K.H. 1995. Labrador Sea: the extent of continental and oceanic crust and the timing of the onset of seafloor spreading. *Marine and Petroleum Geology*, **12**, 205–217, https://doi.org/10.1016/0264-8172(95)92840-s

Chian, D. & Louden, K.E. 1994. The continent–ocean crustal transition across the southwest Greenland margin. *Journal of Geophysical Research*, **99**, 9117–9135, https://doi.org/10.1029/93jb03404

Chian, D.P., Keen, C., Reid, I. & Louden, K.E. 1995. Evolution of nonvolcanic rifted margins: new results from the conjugate margins of the Labrador Sea. *Geology*, **23**, 589–592.

Cole, J.E. & Peachey, J. 1999. Evidence for pre-Cretaceous rifting in the Rockall Trough: an analysis using quantitative plate tectonic modelling. *In*: Fleet, A.J. & Boldy, S.A.R. (eds) *Petroleum Geology of Northwest Europe: Proceedings of the 5th Conference*. Geological Society, London, Petroleum Geology Conference Series, **5**, 359–370, https://doi.org/10.1144/0050359

Ellis, D. & Stoker, M.S. 2014. The Faroe–Shetland Basin: a regional perspective from the Paleocene to the present day and its relationship to the opening of the North Atlantic Ocean. *In*: Cannon, S.J.C. & Ellis, D. (eds) *Hydrocarbon Exploration to Exploitation West of Shetlands*. Geological Society, London, Special Publications, **397**, 11–31, https://doi.org/10.1144/SP397.1

Featherstone, P.S., Bott, M.H.P. & Peacock, J.H. 1977. Structure of the continental margin of southeastern Greenland. *Geophysical Journal International*, **48**, 15–27, https://doi.org/10.1111/j.1365-246X.1977.tb01282.x

Fyhn, M.B.W., Rasmussen, T.M., Dahl-Jensen, T., Weng, W.L., Bojesen-Koefoed, J.A. & Nielsen, T. 2012. Geological assessment of the East Greenland margin. *Geological Survey of Denmark and Greenland Bulletin*, **26**, 61–64.

Gaina, C. 2014. Plate reconstructions and regional kinematics. *In*: Hopper, J.R., Funck, T., Stoker, M.S., Árting, U., Péron-Pinvidic, G., Doornenbal, H. & Gaina, C. (eds) *Tectonostratigraphic Atlas of the North-East Atlantic Region*. Geological Survey of Denmark and Greenland, Copenhagen, Denmark, 53–68.

Gaina, C., Gernigon, L. & Ball, P. 2009. Palaeocene–Recent plate boundaries in the NE Atlantic and the formation of the Jan Mayen microcontinent. *Journal of the Geological Society, London*, **166**, 601–616, https://doi.org/10.1144/0016-76492008-112

Haase, C. & Ebbing, J. 2014. Gravity data. *In*: Hopper, J.R., Funck, T., Stoker, M., Árting, U., Peron-Pindivic, G., Gaina, C. & Doornenbal, H. (eds) *Tectonostratigraphic Atlas of the North-East Atlantic Region*. Geological Survey of Denmark and Greenland (GEUS), Copenhagen, 29–39.

Haase, C., Ebbing, J. & Funck, T. 2016. A 3D regional crustal model of the NE Atlantic based on seismic and gravity data. *In*: Péron-Pinvidic, G., Hopper, J.R., Stoker, M.S., Gaina, C., Doornenbal, J.C., Funck, T. & Árting, U.E. (eds) *The NE Atlantic Region: A Reappraisal of Crustal Structure, Tectonostratigraphy and Magmatic Evolution*. Geological Society, London, Special Publications, **447**. First published online October 12, 2016, updated October 19, 2016, https://doi.org/10.1144/SP447.8

Henriksen, N., Higgins, A.K., Kalsbeek, F. & Pulvertaft, T.C.R. 2009. *Greenland from Archaean to Quaternary*. Geological Survey of Denmark and Greenland, Copenhagen, Denmark.

Hitchen, K. 2004. The geology of the UK Hatton–Rockall margin. *Marine and Petroleum Geology*, **21**, 993–1012, https://doi.org/10.1016/j.marpetgeo.2004.05.004

Holbrook, W.S., Larsen, H.C. *et al.* 2001. Mantle thermal structure and active upwelling during continental breakup in the North Atlantic. *Earth and Planetary Science Letters*, **190**, 251–266, https://doi.org/10.1016/s0012-821x(01)00392-2

Hopper, J.R., Lizarralde, D. & Larsen, H.C. 1998. Seismic investigations offshore SE Greenland. *Geology of Greenland Survey Bulletin*, **180**, 145–151.

Hopper, J.R., Dahl-Jensen, T. *et al.* 2003. Structure of the SE Greenland margin from seismic reflection and refraction data: implications for nascent spreading center subsidence and asymmetric crustal accretion during North Atlantic opening. *Journal of Geophysical Research*, **108**, 2269, https://doi.org/10.1029/2002jb001996

Hopper, J.R., Funck, T., Stoker, M., Árting, U., Peron-Pindivic, G., Gaina, C. & Doornenbal, H. (eds) 2014. *Tectonostratigraphic Atlas of the North-East Atlantic Region*. Geological Survey of Denmark and Greenland (GEUS), Copenhagen.

Horní, J., Hopper, J.R., In review *et al.* Regional distribution of volcanism within the North Atlantic igneous province. *In*: Péron-Pinvidic, G., Hopper, J.R., Stoker, M.S., Gaina, C., Doornenbal, J.C., Funck, T. & Árting, U.E. (eds) *The NE Atlantic Region: A Reappraisal of Crustal Structure, Tectonostratigraphy and Magmatic Evolution*. Geological Society, London, Special Publications, **447**.

Kimbell, G.S., Ritchie, J.D., Johnson, H. & Gatliff, R.W. 2005. Controls on the structure and evolution of the NE Atlantic margin revealed by regional potential field imaging and 3D modelling. *In*: Doré, A.G. & Vining, B.A. (eds) *Petroleum Geology: North-West Europe and Global Perspectives – Proceedings of the 6th Petroleum Geology Conference*. Geological Society, London, Petroleum Geology Conference Series, **6**, 933–945, https://doi.org/10.1144/0060933

Kolb, J. 2014. Structure of the Palaeoproterozoic Nagssugtoqidian Orogen, SE Greenland: model for the tectonic evolution. *Precambrian Research*, **225**, 809–822, https://doi.org/10.1016/j.precamres.2013.12.015

Korenaga, J., Holbrook, W.S. *et al.* 2000. Crustal structure of the southeast Greenland margin from joint refraction and reflection seismic tomography. *Journal of Geophysical Research*, **105**, 21 591–21 614, https://doi.org/10.1029/2000JB900188

Larsen, B. 1980. *A Marine Geophysical Survey of the Continental Shelf of East Greenland 60°N–71°N: Project 'Dana 79'*. GEUS Report 24285. Geological Survey of Denmark and Greenland, Copenhagen, Denmark.

LARSEN, H.C. 1980. Geological perspectives of the East Greenland continental margin. *Bulletin of the Geological Society of Denmark*, **29**, 77–101.

LARSEN, H.C. & SAUNDERS, A.D. 1998. Tectonism and volcanism at the southeast Greenland rifted margin: a record of plume impact and later continental rupture. *In*: SAUNDERS, A.D., LARSEN, H.C. & WISE, S.W., JR. (eds) *Proceedings of the Ocean Drilling Program, Scientific Results, Volume 152*. Ocean Drilling Program, College Station, TX, 503–533, https://doi.org/10.2973/odp.proc.sr.152.240.1998

LARSEN, M., HAMBERG, L., OLAUSSEN, S., NØRGAARD-PEDERSEN, N. & STEMMERIK, L. 1999. Basin evolution in southern East Greenland: an outcrop analogue for Cretaceous–Paleogene basins in the North Atlantic volcanic margins. *American Association of Petroleum Geologists Bulletin*, **83**, 1236–1261.

LARSEN, M., BJERAGER, M., NEDKVITNE, T., OLAUSSEN, S. & PREUSS, T. 2001. Pre-basaltic (Aptian–Paleocene) of the Kangerlussuaq Basin, southern East Greenland. *Geology of Greenland Survey Bulletin*, **189**, 99–106.

LARSEN, M., KNUDSEN, C., FREI, D., FREI, M., RASMUSSEN, T. & WHITHAM, A.G. 2006. East Greenland and Faroe–Shetland sediment provenance and Palaeogene sand dispersal systems. *Geological Survey of Denmark and Greenland Bulletin*, **10**, 29–32.

LENOIR, X., FÉRAUD, G. & GEOFFROY, L. 2003. High-rate flexure of the East Greenland volcanic margin: constraints from ^{40}Ar/^{39}Ar dating of basaltic dykes. *Earth and Planetary Science Letters*, **214**, 515–528, https://dx.doi.org/10.1016/S0012-821X(03)00392-3

MCINROY, D. & HITCHEN, K. 2008. Geological evolution and hydrocarbon potential of the Hatton Basin (UK sector), north-east Atlantic Ocean (extended abstract). *In*: *Proceedings of the Central Atlantic Conjugate Margins Conference*, Dalhousie University, Halifax, Nova Scotia, Canada, 13–15 August 2008, 132–143.

MILLER, H.G. & SINGH, V. 1994. Potential field tilt – a new concept for location of potential field sources. *Journal of Applied Geophysics*, **32**, 213–217, https://doi.org/10.1016/0926-9851(94)90022-1

MYERS, J.S., GILL, R.C.O., REX, D.C. & CHARNLEY, N.R. 1993. The Kap Gustav Holm Tertiary Plutonic Centre, East Greenland. *Journal of the Geological Society, London*, **150**, 259–276, https://doi.org/10.1144/gsjgs.150.2.0259

NASUTI, A. & OLESEN, O. 2014. Magnetic data. *In*: HOPPER, J.R., FUNCK, T., STOKER, M., ARTING, U., PERON-PINVIDIC, G., DOORNENBAL, H. & GAINA, C. (eds) *Tectonostratigraphic Atlas of the North-East Atlantic Region*. The Geological Survey of Denmark and Greenland (GEUS), Copenhagen, Denmark, 43–53.

NIELSEN, T.F.D., SOPER, N.J., BROOKS, C.K., FALLER, A.M., HIGGINS, A.C. & MATTHEWS, D.W. 1981. *The Pre-Basaltic Sediments and the Lower Basalts at Kangerdlugssuaq, East Greenland: Their Stratigraphy, Lithology, Palaeomagnetism and Petrology*. Meddelelser om Grønland Geoscience, **6**.

NØHR-HANSEN, H. 1993. *Dinoflagellate Cyst Stratigraphy of the Barremian to Albian, Lower Cretaceous, North-East Greenland*. Bulletin Grønlands Geologiske Undersøgelse, **166**.

RIISAGER, P. & RASMUSSEN, T.M. 2014. Aeromagnetic survey in south-eastern Greenland: project Aeromag 2013. *Geological Survey of Denmark and Greenland Bulletin*, **31**, 63–67, open access: https://www.geus.dk/publications/bull

SINTON, C.W. & DUNCAN, R.A. 1998. ^{40}Ar–^{39}Ar ages of lavas from the southeast Greenland margin, ODP Leg 152, and the Rockall Plateau, DSDP Leg 81. *In*: SAUNDERS, A.D., LARSEN, H.C. & WISE, S.W.J. (eds) *Proceedings of the Ocean Drilling Program, Scientific Results, Volume 152*. Ocean Drilling Program, College Station, TX, 387–402, https://doi.org/10.2973/odp.proc.sr.152.234.1998

SMALLWOOD, J.R. & WHITE, R.S. 2002. Ridge–plume interaction in the North Atlantic and its influence on continental breakup and seafloor spreading. *In*: JOLLEY, D.W. & BELL, B.R. (eds) *The North Atlantic Igneous Province: Stratigraphy, Tectonic, Volcanic and Magmatic Processes*. Geological Society, London, Special Publications, **197**, 15–37, https://doi.org/10.1144/GSL.SP.2002.197.01.02

SOPER, N.J., HIGGINS, A.C., MATTHEWS, D.W. & BROWN, P.E. 1976. Late Cretaceous–early Tertiary stratigraphy of the Kangerdlugssuaq area, east Greenland, and the age of the opening of the north-east Atlantic. *Journal of the Geological Society, London*, **132**, 85–104, https://doi.org/10.1144/gsjgs.132.1.0085

SRIVASTAVA, S.P. & ROEST, W.R. 1999. Extent of oceanic crust in the Labrador Sea. *Marine and Petroleum Geology*, **16**, 65–84, https://doi.org/10.1016/S0264-8172(98)00041-5

STOKER, M.S., DOORNENBAL, H., HOPPER, J.R. & GAINA, C. 2014. Tectonostratigraphy: Introduction. *In*: HOPPER, J.R., FUNCK, T., STOKER, M.S., ÁRTING, U., PÉRON-PINVIDIC, G., DOORNENBAL, H. & GAINA, C. (eds) *Tectonostratigraphic Atlas of the North-East Atlantic Region*. Geological Survey of Denmark and Greenland, Copenhagen, Denmark, 129–155.

STOKER, M.S., STEWART, M.A. *ET AL*. 2016. An overview of the Upper Palaeozoic–Mesozoic stratigraphy of the NE Atlantic region. *In*: PÉRON-PINVIDIC, G., HOPPER, J.R., STOKER, M.S., GAINA, C., DOORNENBAL, J.C., FUNCK, T. & ÁRTING, U.E. (eds) *The NE Atlantic Region: A Reappraisal of Crustal Structure, Tectonostratigraphy and Magmatic Evolution*. Geological Society, London, Special Publications, **447**. First published on August 11, 2016, updated August 12, 2016, https://doi.org/10.1144/SP447.2

STOREY, M., DUNCAN, R.A. & TEGNER, C. 2007. Timing and duration of volcanism in the North Atlantic Igneous Province: implications for geodynamics and links to the Iceland hotspot. *Chemical Geology*, **241**, 264–281, https://doi.org/10.1016/j.chemgeo.2007.01.016

TEGNER, C., DUNCAN, R.A., BERNSTEIN, S., BROOKS, C.K., BIRD, D.K. & STOREY, M. 1998. Ar-40-Ar-39 geochronology of Tertiary mafic intrusions along the East Greenland rifted margin: relation to flood basalts and the Iceland hotspot track. *Earth and Planetary Science Letters*, **156**, 75–88, https://doi.org/10.1016/S0012-821X(97)00206-9

THY, P., LESHER, C.E. & LARSEN, H.C. (eds). 2007. *Proceedings of the Ocean Drilling Program, Initial Reports Volume 163X: Transect EG65 Results*. Ocean

Drilling Program, College Station, TX, https://doi.org/10.2973/odp.proc.ir.163x.106.2007

Vallier, T., Calk, L., Stax, R. & Demant, A. 1998. Metamorphosed sedimentary (volcaniclastic?) rocks beneath Paleocene basalt in Hole 917a, East Greenland margin. *In*: Saunders, A.D., Larsen, H.C. & Wise, S.W., Jr. (eds) *Proceedings of the Ocean Drilling Program, Scientific Results, Volume 152*. Ocean Drilling Program, College Station, TX, 129–144, https://doi.org/10.2973/odp.proc.sr.152.206.1998

Wager, L.R. 1934. *Geological Investigations in East Greenland. Part I: General Geology from Angmagssalik to Kap Dalton*. Meddelelser om Grønland, **105**.

Wei, W. 1998. Calcareous nannofossils from the southeast Greenland Margin: biostratigraphy and paleoceanography. *In*: Saunders, A.D., Larsen, H.C. & Wise, S.W., Jr. (eds) *Proceedings of the Ocean Drilling Program, Scientific Results, Volume 152*. Ocean Drilling Program, College Station, TX, 129–144, https://doi.org/10.2973/odp.proc.sr.152.215.1998

White, R.S. 1997. Rift–fplume interaction in the North Atlantic. *Philosophical Transactions of the Royal Society of London, Series A*, **355**, 319–339.

Ziegler, P.A. 1988. *Evolution of the Arctic-North Atlantic and the Western Tethys*. American Association of Petroleum Geologists, Tulsa, OK.

Break-up and seafloor spreading domains in the NE Atlantic

CARMEN GAINA[1]*, AZIZ NASUTI[2], GEOFFREY S. KIMBELL[3] & ANETT BLISCHKE[4]

[1]*Centre for Earth Evolution and Dynamics (CEED), University of Oslo, Sem Sælands vei 24, PO Box 1048, Blindern, NO-0316 Oslo, Norway*

[2]*Geological Survey of Norway (NGU), NO-7491 Trondheim, Norway*

[3]*British Geological Survey, Keyworth, Nottingham NG12 5GG, UK*

[4]*Iceland GeoSurvey, Branch at Akureyri, Rangárvöllum, Akureyri 602, Iceland*

**Correspondence: carmen.gaina@geo.uio.no*

Abstract: An updated magnetic anomaly grid of the NE Atlantic and an improved database of magnetic anomaly and fracture zone identifications allow the kinematic history of this region to be revisited. At break-up time, continental rupture occurred parallel to the Mesozoic rift axes in the south, but obliquely to the previous rifting trend in the north, probably due to the proximity of the Iceland plume at 57–54 Ma.

The new oceanic lithosphere age grid is based on 30 isochrons (C) from C24n old (53.93 Ma) to C1n old (0.78 Ma), and documents ridge reorganizations in the SE Lofoten Basin, the Jan Mayen Fracture Zone region, in Iceland and offshore Faroe Islands. Updated continent–ocean boundaries, including the Jan Mayen microcontinent, and detailed kinematics of the Eocene–Present Greenland–Eurasia relative motions are included in this model.

Variations in the subduction regime in the NE Pacific could have caused the sudden northwards motion of Greenland and subsequent Eurekan deformation. These events caused seafloor spreading changes in the neighbouring Labrador Sea and a decrease in spreading rates in the NE Atlantic. Boundaries between major oceanic crustal domains were formed when the European Plate changed its absolute motion direction, probably caused by successive adjustments along its southern boundary.

Supplementary material: Figures showing the long wavelength of the NAG-TEC magnetic anomaly grid, detailed magnetic anomalies and isochrons, and a Table documenting aeromagnetic surveys for NAG-TEC magnetic compilation are available at https://doi.org/10.6084/m9.figshare.c.3661925

The North Atlantic encompasses the area between Newfoundland–Iberia and the Eurasian Basin in the Arctic Ocean (Fig. 1). It includes active and extinct spreading systems, magma-rich and magma-poor margins, and it witnessed microcontinent formation (Jan Mayen, East Greenland and Hovgaard ridges) and ridge–hotspot interactions linked to the presence of the Iceland hotspot (Fig. 1). The North Atlantic realm underwent episodic continental extension in the Permo-Triassic, Late Jurassic, and Early and mid-Cretaceous, with reactivation and basin formation influenced by pre-existing structures formed during the closure of the Iapetus Ocean (and part of the Rheic Ocean: e.g. Domeier 2015), and subsequent Baltica–Laurentia collision (e.g. Doré *et al.* 1997; Skogseid *et al.* 2000). Seafloor spreading propagated from south to north starting in Cretaceous times in distinct phases that involved the following regions: Newfoundland–Iberia, North America–Porcupine, North America–Greenland (Labrador Sea and Baffin Bay), Greenland–Eurasia and Lomonosov Ridge–Eurasia (Eurasian Basin, Arctic Ocean) (Fig. 1) (see Seton *et al.* 2012 for a review).

The formation of oceanic crust between Greenland and Eurasia was identified and described more than four decades ago (e.g. Pitman & Talwani 1972; Vogt & Avery 1974; Srivastava & Tapscott 1986) as a late Paleocene (*c.* 55 Ma) northwards propagation of seafloor spreading. The mid-ocean ridge from the Central North Atlantic (between North America and Eurasia) was first linked with the plate boundary between the North American and Greenland plates in the Labrador Sea–Baffin Bay (Kristoffersen & Talwani 1974; Roest & Srivastava 1989; Chalmers & Laursen 1995), and later formed a triple junction SW of Greenland with the developing Reykjanes mid-ocean ridge.

In the NE Atlantic, oceanic crust was created continuously from the Early Eocene (*c.* 54–55 Ma)

From: Péron-Pinvidic, G., Hopper, J. R., Stoker, M. S., Gaina, C., Doornenbal, J. C., Funck, T. & Árting, U. E. (eds) 2017. *The NE Atlantic Region: A Reappraisal of Crustal Structure, Tectonostratigraphy and Magmatic Evolution.* Geological Society, London, Special Publications, **447**, 393–417.
First published online February 3, 2017, https://doi.org/10.1144/SP447.12

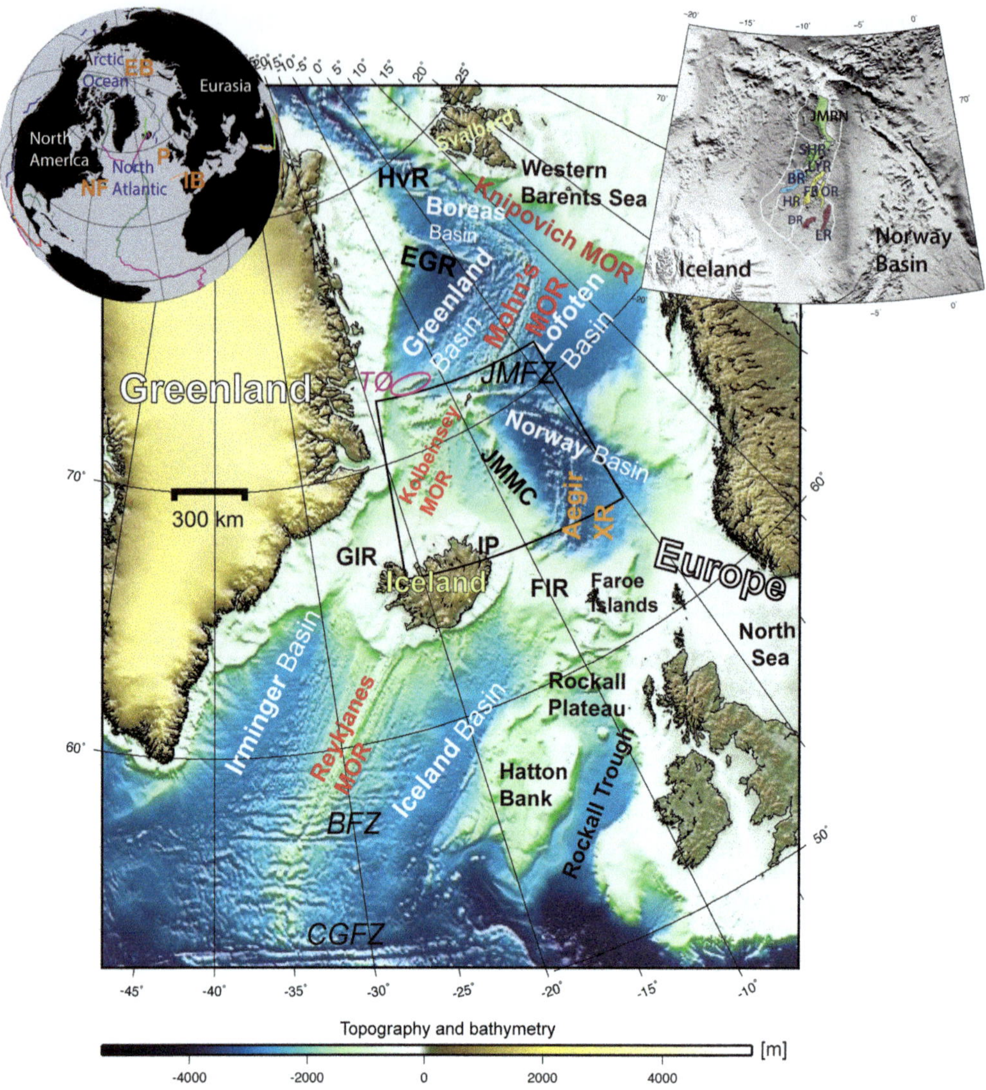

Fig. 1. NAG-TEC bathymetry built on a regional compilation that includes ETOPO1 (Amante & Eakins 2009), IBCAO3 (Jakobsson *et al.* 2012), and a number of local and regional dense bathymetric grids (Hopper & Gaina 2014). The upper-left inset figure shows the present-day plate boundaries in the North Atlantic; and the upper-right inset figure shows a close-up map of the JMMC region, indicated by the black frame on the main map. White lines indicate the tectonic domains within JMMC, with coloured polygons showing basement highs. *Abbreviations*: BFZ, Bight Fracture Zone; BR, Buðli Ridge; CGFZ, Charile Gibbs Fracture Zone; DR, Dreki Ridge; EB, Eurasia Basin; FIR, Faroe–Iceland Ridge; FR, Fáfnir Ridge; GIR, Greenland–Iceland Ridge; EB, Eurasia Basin; EGR, East Greenland Ridge; HR, Högni Ridge; HvR, Hovgard Ridge; IB, Iberia; JMFZ, Jan Mayen Fracture Zone; JMMC, Jan Mayen microcontinent; JMRN, Jan Mayen Ridge North; LR, Langabrún Ridge Ridge; LYR, Lyngvi Ridge; MOR, mid-ocean ridge; NF, Newfoundland; OR, Otur Ridge; P, Porcupine; SHR, Sörlahryggur Ridge; TØ, Trail Ø; XR, extinct mid-ocean ridge.

to the present, forming the Irminger and Iceland basins SW and SE of Iceland, respectively, and the Greenland and Lofoten basins NW and NE of the Jan Mayen Fracture Zone (JMFZ) Complex (Fig. 1). In addition, two basins flank the submerged Jan Mayen Microcontinent (JMMC): the eastern one, called the Norway Basin, was formed from Early Eocene to Early Miocene. The active Reykjanes mid-ocean ridge continues west of the JMMC with the Kolbeinsey Ridge, which in turn connects to the Mohn's Ridge through an approximately 200 km-long segment of the western JMFZ (Fig. 1).

Based on a growing database of geophysical data, many authors have proposed several kinematic models explaining the formation of the NE Atlantic oceanic basins (e.g. Vogt & Avery 1974; Talwani & Eldholm 1977; Srivastava & Tapscott 1986; Vogt 1986; Skogseid & Eldholm 1987; Gaina *et al.* 2002, 2009). The more recent models, which took advantage of new aeromagnetic data collected in the Norway Basin and neighbouring regions, show a more complex kinematic model for the evolution of the NE Atlantic (e.g. Gernigon *et al.* 2008, 2012, 2015; Gaina *et al.* 2009). Although isolated changes in the plate boundary have been previously proposed, the Gaina *et al.* (2009) model presented a comprehensive and integrated regional view of the NE Atlantic Ocean that attempted to explain and quantify the complexities in the evolution of plate boundaries in this region since the Paleocene. They proposed that a series of plate boundary readjustments were expressed by short-lived triple junctions and/or ridge propagations, particularly in the area north and south of the JMMC, and suggested that an Early Eocene tectonic event that has been observed in different regions of the NE Atlantic might be the consequence of regional changes in plate motions.

New studies have documented the detailed configuration of oceanic floor in the Norway Basin and the architecture of the JMMC (e.g. Peron-Pinvidic *et al.* 2012; Gernigon *et al.* 2015; Blischke *et al.* 2016). The seafloor spreading history for the last 20 myr has been modelled in detail by using high-resolution magnetic data along the Reykjanes, Mohn's and Kolbeinsey ridges (Ehlers & Jokat 2009; Merkur'ev *et al.* 2009; Hey *et al.* 2010; Benediktsdóttir *et al.* 2012; Merkouriev & DeMets 2014).

Our present contribution is meant to revisit the detailed evolution of oceanic basins in the NE Atlantic region by using recent magnetic and gravity anomaly maps, and a compilation of magnetic anomaly and fracture zone identifications from studies mentioned above. Based on these data and interpretations, we construct a new regional kinematic model. Finally, we discuss the distinct seafloor spreading domains identified on potential field data and described by our new kinematic model in the context of regional plate motions and associated tectonic events.

New magnetic anomaly grid of the NE Atlantic

Pioneering work that describes regional magnetic anomaly variations in the North Atlantic was made possible by the availability of relatively dense geophysical surveys, and the thorough evaluation and processing of these data by the Geological Survey of Canada in 1995 (Verhoef *et al.* 1996). Part of this digital grid has been renewed by adding data from new aeromagnetic surveys by Olesen *et al.* (2010) and Gaina *et al.* (2011). The NAG-TEC project (Hopper *et al.* 2014) offered the opportunity to revisit the NE Atlantic magnetic anomaly data compilation, and publicly available aeromagnetic data from 1951 to 2012 have been inspected and included in a new magnetic anomaly grid of this region (see the Supplementary material) (Fig. 2).

Data processing

Raw aeromagnetic data (see the Supplementary material) have been processed with algorithms embedded in the commercial software Oasis montaj (https://www.geosoft.com). Firstly, the data for each survey have been interpolated to a regular grid with cell size equal to one-quarter of the flight line spacing. Spikes due to minor noise and artefacts were smoothed with a low-pass filter (cut-off wavelength 30–50 fiducials) in order to keep the signal intact. Outliers and spikes identified in the offshore aeromagnetic measurements were removed manually.

To compute magnetic anomalies from the raw magnetic data, field values calculated using the International Geomagnetic Reference Field (IGRF), or Definitive Geomagnetic Reference Field (DGRF) models have been subtracted from the raw measurements. Several additional corrections, including diurnal corrections, statistical corrections based on profile cross-over analysis and micro-levelling, have also been applied. Individual grids were subsequently merged into the regional magnetic anomaly grid. The long wavelength (larger than 300 km) of the resulting grid has been replaced by the CHAMP satellite magnetic anomaly model MF7 (Maus *et al.* 2009) (see the Supplementary material).

Magnetic anomaly and fracture zone identifications (picks)

For deciphering the architecture and history of oceanic crust formation in individual basins of the NE Atlantic, the two most important pieces of information are the identification of magnetic anomalies (Fig. 3) and mapping of fracture zones within the oceanic crust. Magnetic anomalies are used for dating the oceanic blocks, which are magnetized in the alternating polarities of the Earth's magnetic field as spreading occurs. A database containing a collection of magnetic anomaly picks that indicate the age

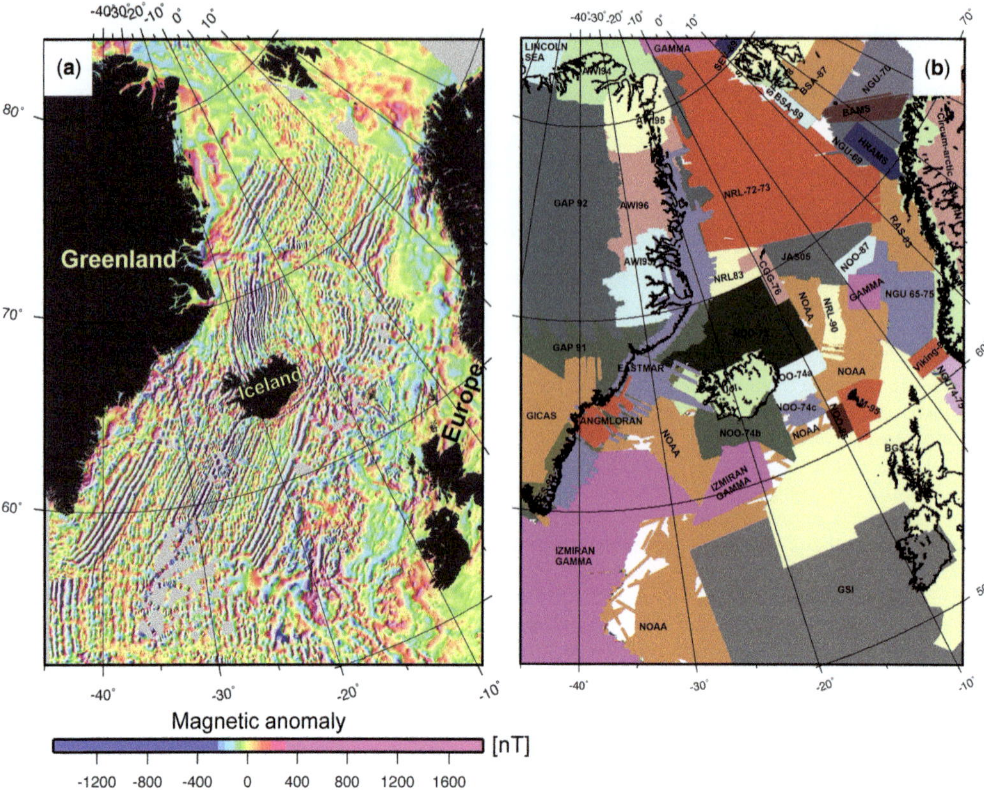

Fig. 2. NAG-TEC magnetic anomaly map (this study and Nasuti & Olesen 2014) (**a**) and location of various local gridded data used in this compilation (**b**). For a complete list of data sources see the Supplementary material.

of oceanic crust at the beginning (y) or end (o) of magnetized blocks of selected magnetic polarity reversals has been assembled and quality checked from various sources (Table 1). For dating the magnetic anomaly picks, the Ogg (2012) geomagnetic polarity scale (part of Gradstein *et al.* 2012) has been adopted.

The gravity anomaly and its derivatives (like the second vertical derivatives, vertical gravity gradient (VGG): see Sandwell & Smith 2009) are used to manually identify the central trough, or the centre of the steepest slope that define the bathymetric and gravity expression of a fracture zone. Using this approach, Matthews *et al.* (2011) interpreted fracture zones in all major oceanic basins. From that study, we have selected the fracture zone segments from the southern area of the NE Atlantic. They were supplemented with more fracture zone segments in the rest of the NE Atlantic region, including the JMFZ Complex, by using the gravity anomaly data (DTU10: Andersen 2010) and various derivatives (Haase & Ebbing 2014). The final database of magnetic anomaly and fracture zone identifications is presented in Figure 3.

A new kinematic model for the NE Atlantic Ocean

The kinematic model is built up by finding rotation parameters that bring the interpreted magnetic anomaly (and fracture zone, when possible) identifications into alignment at a particular time, which essentially defines the active oceanic spreading centre at that time. A constant motion of a crustal block or object on a sphere for a given time interval follows a great circle path and can be described through a rotation around a fixed pivot point, an Euler pole (see Cox & Hart 1986 for a basic introduction). In the case of seafloor spreading and mid-ocean ridge formation, the magnetized bodies parallel to the ridge constitute the meridians that intersect at the Euler pole, and the transform/fracture zones parallel to the direction of motion and perpendicular to the mid-ocean ridges align along small circles of the Euler pole. Rotations derived from these features should therefore bring the anomaly picks on either side of the present-day spreading axis into alignment along the palaeo-spreading centre. An isochron is built by

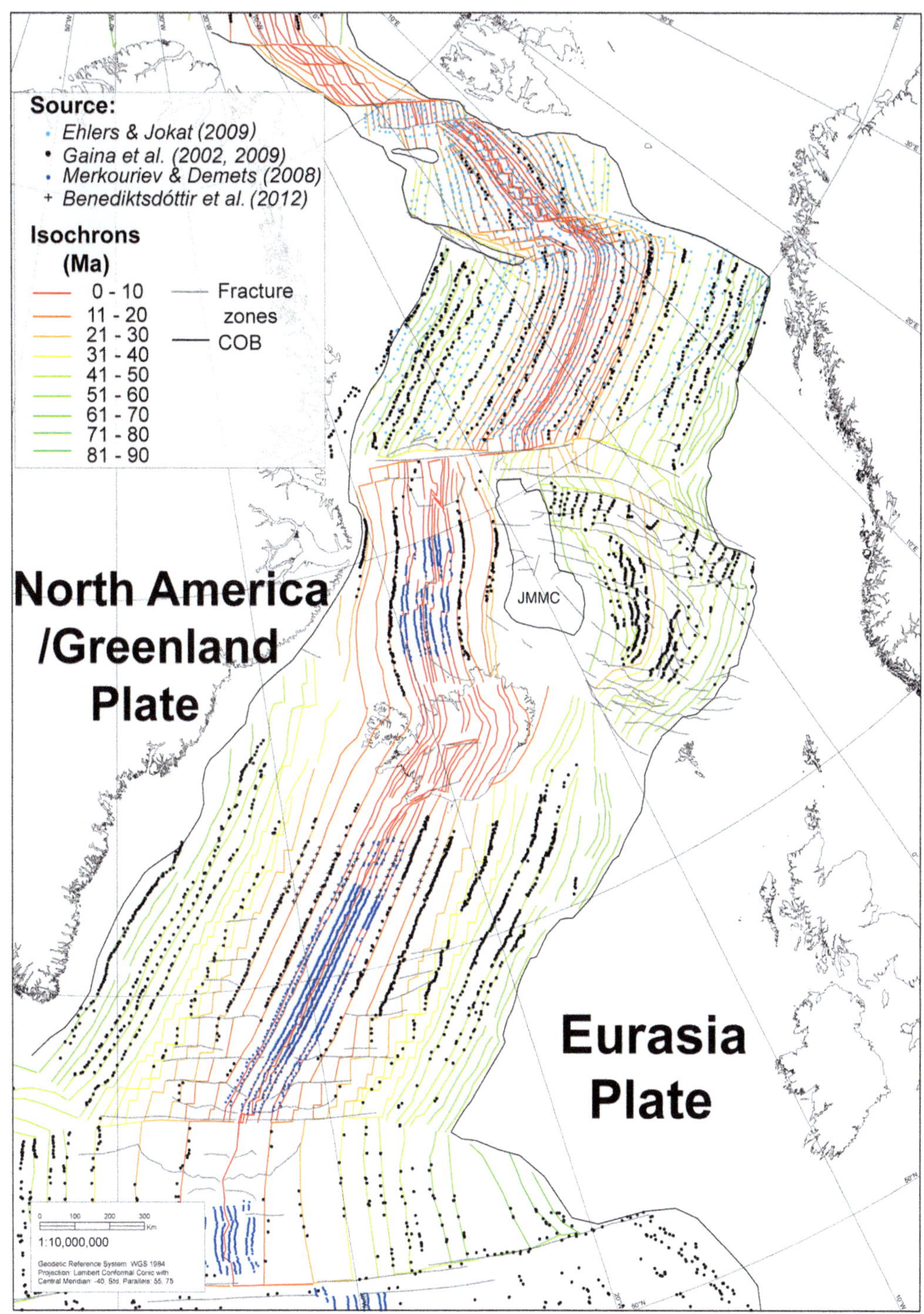

Fig. 3. Magnetic anomaly and fracture zone identifications and interpreted isochrons.

Table 1. *Magnetic anomaly identification according to various geomagnetic timescales*

Chron*	Age [Ma]				Magnetic pick identification (Reference)
	Cande & Kent 1995	Lourens *et al.* 2004	Gee & Kent 2007	Ogg 2012	
1no		0.781	0.780	0.781	Merkur'ev *et al.* (2009)
2An.1ny		2.581	2.581	2.581	Merkur'ev *et al.* (2009)
2An.3no	3.580	3.596	3.580	3.596	Ehlers & Jokat (2009)
3n.4no		5.235	5.230	5.235	Ehlers & Jokat (2009)
3An.1ny = 3ro		6.033	5.894	6.033	Benediktsdóttir *et al.* (2012); Merkur'ev *et al.* (2009)
4n.1ny		7.528	7.432	7.528	Merkur'ev *et al.* (2009)
4Any	8.699	8.769		8.771	Ehlers & Jokat (2009)
5n.2no		11.040	10.949	11.056	Benediktsdóttir *et al.* (2012); Gaina *et al.* (2009); Merkur'ev *et al.* (2009)
5r.2no	11.531			11.657	Ehlers & Jokat (2009)
5ACy		13.734	13.703	13.739	Ehlers & Jokat (2009)
5Cn.1ny = 5Bro		15.974	16.014	15.974	Benediktsdóttir *et al.* (2012); Merkur'ev *et al.* (2009)
5Cn.1no	16.293			16.268	Ehlers & Jokat (2009)
6ny		18.748	19.048	18.748	Ehlers & Jokat (2009)
6no		19.722	20.131	19.722	Gaina *et al.* (2002)
6AAno	21.859			21.159	Gaina *et al.* (2009)
7n.1no	24.730			23.962	Ehlers & Jokat (2009)
9no	27.972			27.439	Ehlers & Jokat (2009)
13ny	33.058		33.058	33.157	Gaina *et al.* (2002)
17n.1no	37.473			37.956	This study
18n.1ny	38.426			38.834	Ehlers & Jokat (2009)
18n.2no	40.130		40.130	40.321	Gaina *et al.* (2009); Gaina *et al.* (2002)
20ny	42.536		42.536	42.301	Ehlers & Jokat (2009)
20no	43.789		43.789	43.432	Gaina *et al.* (2009); Gaina *et al.* (2002)
21ny			46.264	45.683	Ehlers & Jokat (2009)
21no	47.906		47.906	47.329	Gaina *et al.* (2009); Gaina *et al.* (2002)
22no	49.714		49.714	49.335	Ehlers & Jokat (2009); Gaina *et al.* (2009); Gaina *et al.* (2002)
23n.1ny	50.778		50.778	50.613	Ehlers & Jokat (2009)
23n.2no	51.743		51.743	51.826	Ehlers & Jokat (2009)
24n.1ny	52.364		52.364	52.629	Ehlers & Jokat (2009)
24n.3no	53.347		53.347	53.933	Ehlers & Jokat (2009)
25ny	55.904		56.904	57.101	Gaina *et al.* (2002)
31no	68.737			69.269	Gaina *et al.* (2002)
33no			79.08	79.900	Gaina *et al.* (2002)
34ny			83.0		Gaina *et al.* (2002)

* "o" and "y" stand for "old" and "young" sides of normal (n) reverse (r) magnetised oceanic crust.

calculating the best-fitting segments (that are great and small circles using the inferred Euler pole) using the reconstructed magnetic anomaly and fracture zone data, respectively. The locations and geometry of mid-ocean ridges through time are therefore represented by the seafloor isochrons (C) derived above from the magnetic anomaly and fracture zone identifications.

In this study, we have built a new set of relative plate motions between Eurasia and Greenland, and for the various blocks that make up the JMMC (Table 2). The 'relative plate motion' quantitatively describes, through Euler rotations, the position of one tectonic plate relative to another plate that is considered fixed. We use the term 'finite rotation' to quantify the motion between present day and

Table 2. *Finite rotations of main tectonic blocks relative to a fixed Eurasia Plate (a positive sign indicates the northern hemisphere for latitude and eastern hemisphere for longitude)*

Age (Ma)	Chron	Rotation		
		Latitude (+°N)	Longitude (+°E)	Angle (°)
Greenland–Eurasia				
11.100*	C5n.2o	67.5	133.1	2.62
19.722*	C6no	72.2	126.1	5.29
21.16†	C6AAno	72.5	126.75	5.72
23.96†	C7n.1ny	72.23	127.35	6.46
27.44†	C9no	72.01	128.05	7.59
33.160	C13ny	68.3	132.3	7.66
40.320	C18n.2no	61.5	127.8	8.30
43.430	C20no	57.4	127.9	8.59
47.329	C21no	53.7	129.0	9.27
49.335	C22no	55.4	123.5	10.29
53.930	C24n.1no	50.9	123.65	11.09
57.100 (pre-break-up)	C25ny	52.5	123.8	12.32
North Jan Mayen Ridge Complex‡ – Eurasia				
33.160	C13ny	−65.4	167.8	10.51
40.320	C18n.2no	−52.0	150.6	5.73
43.430	C20no	−40.1	143.0	4.78
47.329	C21no	−59.6	159.1	17.51
49.335	C22no	−58.6	157.8	18.02
53.930	C24n.1no	−52.2	151.1	15.32
57.100 (pre-break-up)	C25ny	−52.2	151.1	15.32
Central-west Jan Mayen Ridge Complex§ – Eurasia				
33.160	C13ny	−64.6	167.2	10.52
40.320	C18n.2no	−50.5	150.6	5.78
43.430	C20no	−38.4	143.4	4.86
47.329	C21no	−59.1	158.9	17.53
49.335	C22no	−58.1	157.6	18.05
53.930	C24n.1no	−51.6	151.0	15.37
57.100 (pre-break-up)	C25ny	31.4	−175.8	4.65
Central-east Jan Mayen Ridge Complex‖ – Eurasia				
33.160	C13ny	−65.4	167.8	10.51
40.320	C18n.2no	−52.0	150.6	5.73
43.430	C20no	−40.1	143.0	4.78
47.329	C21no	−59.6	159.1	17.51
49.335	C22no	−58.6	157.8	18.02
53.930	C24n.1no	−52.2	151.1	15.32
57.100 (pre-break-up)	C25ny	32.0	−172.5	4.33
South Jan Mayen Ridge Complex¶ – Eurasia				
33.160	C13ny	−65.6	167.9	10.71
40.320	C18n.2no	−65.2	160.3	14.39
43.430	C20no	−62.3	156.0	13.11
47.329	C21no	−64.2	163.2	26.25
49.335	C22no	−63.4	162.2	26.74
53.930	C24n.1no	−59.9	157.2	23.81
57.100 (pre-break-up)	C25ny	−45.9	−176.5	8.78
33.160	C13ny	−65.6	167.9	10.71
Central-south Norway Basin – Eurasia				
33.160	C13ny	−65.4	167.8	10.51
40.320	C18n.2no	−64.5	169.0	21.16
43.430	C20no	−64.5	168.9	32.10
47.329	C21no	−64.5	168.9	46.28
49.335	C22no	−64.1	168.6	48.45
53.930	C24n.1no	−63.0	166.5	50.29
57.100 (pre-break-up)	C25ny	−63.0	166.5	50.29

*From Merkur'ev *et al.* (2009).
†From Ehlers & Jokat (2009). For times younger than 20 Ma, we have used the North America–Eurasia rotations of Merkur'ev *et al.* (2009). For times between 20 and 33 Ma, we have used the North America–Eurasia rotations of Ehlers & Jokat (2009).
‡The following ridges identified in the JMMC are part of this complex: the Jan Mayen Ridge North (JMRN), the SHR Sörlahryggur Ridge (SHR) and the Lyngvi Ridge (LYR) – for their present-day positions see Figure 1.
§The following ridge identified in the JMMC is part of this complex: Buðli Ridge (BR) – for its present-day position see Figure 1.
‖The following ridges identified in the JMMC are part of this complex: the Högni Ridge (HR), the Fáfnir Ridge (FR) and the Otur Ridge (OR) – for their present-day positions see Figure 1.
¶The following ridges identified in the JMMC are part of this complex: the Dreki Ridge (DR) and the Langabrún Ridge Ridge (LR) – for their present-day positions see Figure 1.

a certain time in the geological past, and 'stage rotation' for the motion between two plates for a selected time interval in the geological past. For pre-break-up times, the position of the tectonic blocks is inferred from rotations based on magnetic anomaly and fracture zones for older oceanic crust in surrounding regions, in particular to the south and to the west, to constrain the motion between Greenland, North America and Eurasia from 55 and 83 Ma (Gaina *et al.* 2002).

Based on the previous interpretation (i.e. Gaina *et al.* 2009), the southern part of the JMMC was deformed as oceanic floor formed at its southern proximity, and the western relocation of the mid-ocean ridge in the Late Eocene–Oligocene gradually detached several microcontinent blocks from Greenland. The JMMC blocks' kinematic parameters were computed by carrying out visual fits (in GPlates: https://www.gplates.org) for four groups of basement ridges mapped by Blischke *et al.* (2016). Note that a separate rotation set was calculated for the oceanic part of the Norway Basin based on magnetic data. Compression described in the SE part of the JMMC demonstrates that relative motion between the oceanic and stretched continental domains took place probably after the seafloor spreading reorganization in the Eocene (see Gernigon *et al.* 2012, 2015; Blischke *et al.* 2016). Although the eastern part of the Norway Basin now has complete aeromagnetic data coverage (described and analysed by Gernigon *et al.* 2015), we did not have access to the new magnetic anomaly data and our interpretation is based on the magnetic grid shown in Figure 2

The magnetic anomaly identification sets (Table 1), combined with the fracture zone segments, were used for constructing densely spaced isochrons for 30 geological times (compared to only six in previous models: e.g. Müller *et al.* 2008) which date the oceanic crust following the timescale of Ogg (2012) (Fig. 3; see also the Supplementary material). The oceanic lithospheric age grid model is constructed using the newly interpreted isochrons (Fig. 3) and the rotation parameters describing the opening of the NE Atlantic (Table 2) following the interpolation technique outlined by Müller *et al.* (2008) and employing a gridding resolution of 0.05°. The age grid has been used to compute seafloor spreading rates and directions, and deviations from symmetrical oceanic crust formation at various intervals as constrained by the kinematic model (Table 2). The asymmetry in oceanic crust accretion is expressed as percentages from 0 to 100%, where 50% indicates symmetrical seafloor spreading (Müller *et al.* 2008). Note that poor age control on various regions of the Greenland–Iceland and Faroe–Iceland ridges (Fig. 1), and between the JMMC and the Iceland Plateau, led to less reliable models of spreading rate and asymmetry, and these areas are masked on our maps (Fig. 4).

Break-up and early seafloor spreading

Following two extensive volcanic episodes, at approximately 62 and 55 Ma, which affected the NE Atlantic margins and formed the North Atlantic Igneous Province (NAIP: e.g. Saunders *et al.* 2007), continental break-up occurred between Greenland and Eurasia before C24 time (*c.* 55 Ma). To show the pre-break-up configuration of the western Eurasian margin and its conjugate margin, we reconstruct the structural elements (major tectonic boundaries, faults and structural highs: Stoker *et al.* 2016) and simplified inferred sedimentary basin ages (Funck *et al.* 2014) at Paleocene–Eocene transition time (Fig. 5a).

The NE Atlantic rift spans a region of more than 3000 km in a north–south direction from the southernmost tip of Greenland to the Western Barents Sea. Devonian–Paleocene multiple rifting events led to the formation of a wide extended area of successive basins and highs confined within the Greenland and Western European Caledonian deformation zone (Fig. 5a) (see also Stoker *et al.* 2016). Four main rifting periods can be identified from sedimentary basins along the NE Atlantic margin: (1) Devonian–Carboniferous; (2) Permian–Triassic; (3) Jurassic–Early Cretaceous; and (4) Late Cretaceous–Paleocene (Fig. 5a). According to Skogseid *et al.* (2000), the Late Palaeozoic rifting is poorly constrained, but the Late Jurassic–Cretaceous rifting caused approximately 50–70 km of crustal extension and subsequent Cretaceous basin subsidence from the Rockall Trough-North Sea areas to the SW Barents Sea. A Late Cretaceous–Paleocene renewed rifting episode caused approximately 140 km of extension (Skogseid *et al.* 2000).

Plate reconstructions of gridded data (present-day magnetic anomaly, isostatic gravity and crustal thickness) at 54 Ma, the time of early seafloor spreading in the NE Atlantic, are shown in Figure 5b, c. Present-day crustal thickness estimated from gravity inversion (Haase *et al.* 2016) gives a first-order approximation of the crustal architecture and amount of margin extension due to rifting. A reconstruction at 54 Ma (the time of early seafloor spreading in the NE Atlantic) shows regions of thin and thick crust (Fig. 5c), and, most importantly, the fact that break-up did not occur where the crust was thinnest.

As it has been postulated that the impingement of the Iceland plume at the base of the lithosphere has created massive volcanism and led to continental break-up (e.g. Morgan 1972), we show the reconstructed position of the Iceland plume (Doubrovine

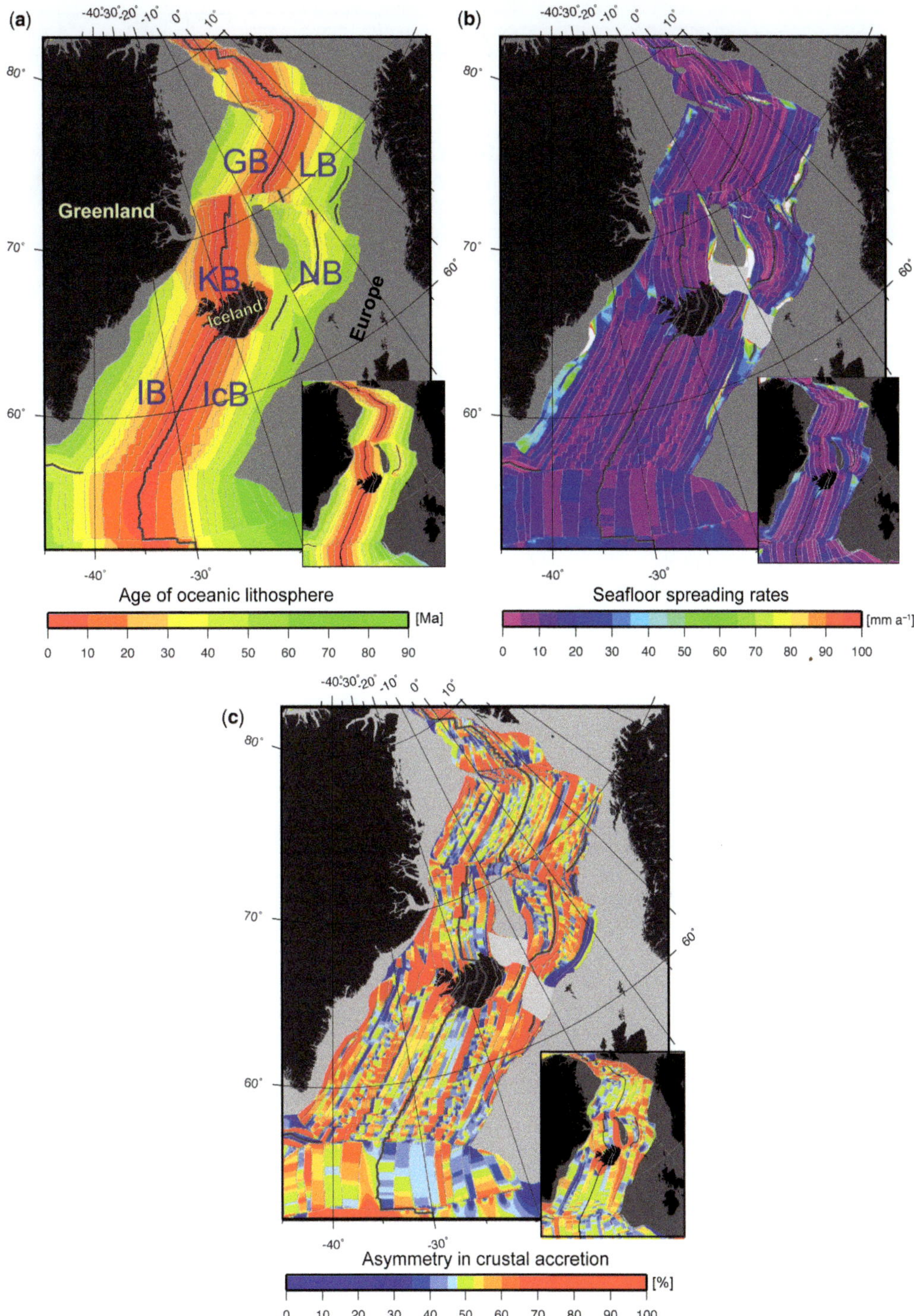

Fig. 4. Models of age of oceanic lithosphere (**a**), half spreading rate (**b**) and asymmetry in crustal accretion (**c**) in the NE Atlantic. Inset figures show the global grids published by Müller *et al.* (2008). Abbreviations: GB, Greenland Basin; IB, Irminger Basin; IcB, Iceland Basin; KB, Kolbeinsey Basin; LB, Lofoten Basin; NB, Norway Basin.

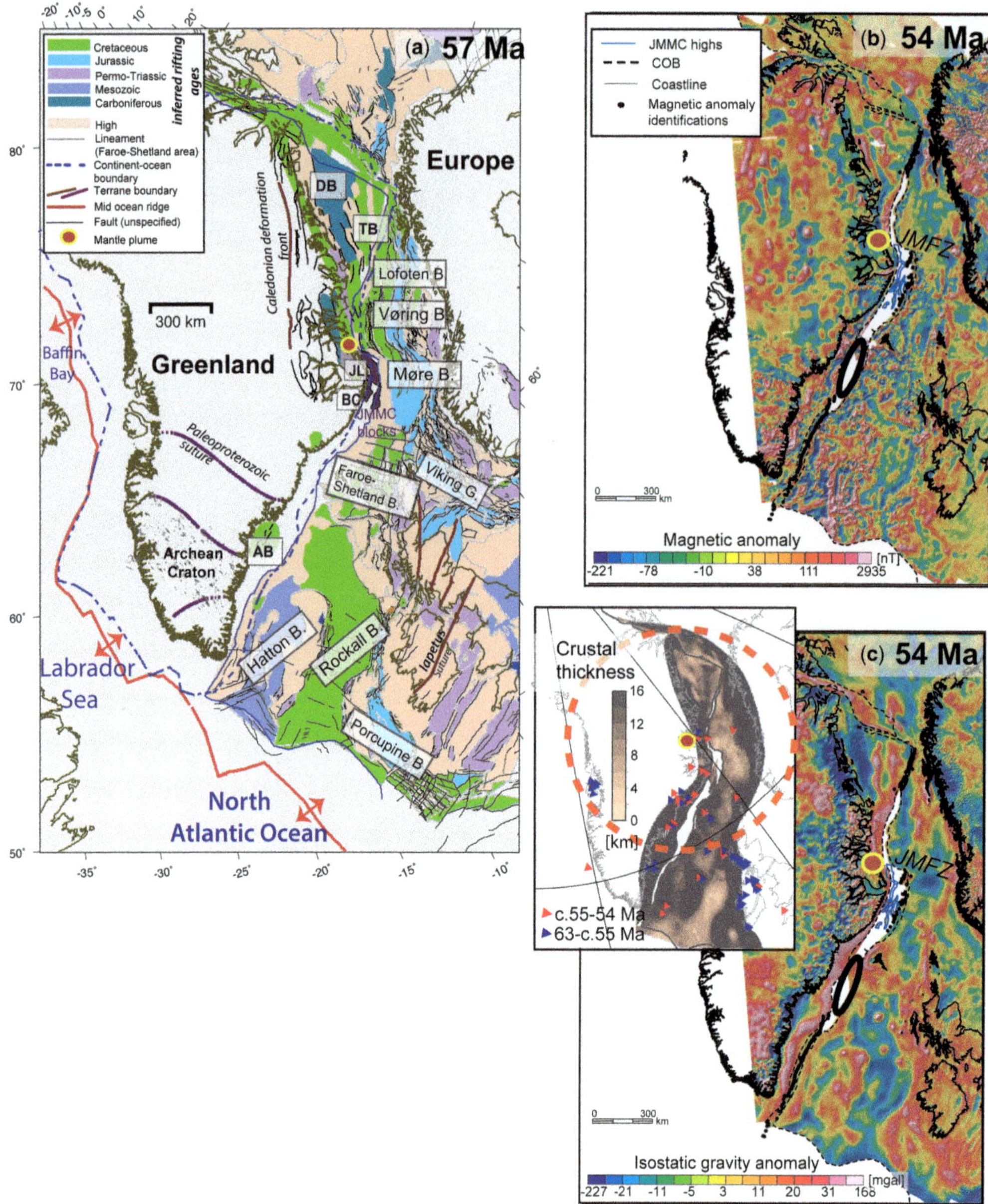

Fig. 5. (**a**) Pre-break-up (57 Ma) reconstruction of Greenland and Western Europe showing major sedimentary basins with inferred rifting ages and tectonic lineaments (Funck *et al.* 2014; Stoker *et al.* 2016), reconstructed plate boundaries, and modelled position of the Iceland hotspot (red and yellow circles) (Doubrovine *et al.* 2012). (**b**) Reconstructed magnetic anomaly grid at C24 (*c.* 54 Ma). Black dots indicate the location of reconstructed magnetic anomaly picks. The JMMC tectonic blocks are also shown with blue outlines. The location of the postulated JMMC extension under present-day Iceland (Torsvik *et al.* 2015) is indicated by the black ellipse; (**c**) Reconstructed isostatic gravity anomaly grid at C24 (*c.* 54 Ma). Inset figure shows the reconstructed crustal thickness (based on gravity inversion: see Haase *et al.* 2016) and reconstructed locations of the dated NAIP rock samples (database of Torsvik *et al.* 2015) indicating the extent of volcanism before (between 63 and 62 Ma and *c.* 55 Ma), and at the break-up and incipient seafloor spreading time (*c.* 55–54 Ma). Orange dashed circle on the inset figure indicates a region of a mantle plume head (1000 km radius) at the base of the lithosphere (e.g. Ernst & Buchan 2002). Abbreviations: AB, Amassalik Basin; B, Basin; BC, Blosseville Coast; COB, continent–ocean boundary; DB, Danmarkshavn Basin; G, Graben; JL, Jameson Land Basin; TB, Thetis Basin.

et al. 2012), which is, indeed, situated in a very close proximity to the future continent–ocean boundary COB) of the Greenland Plate (Fig. 5a). Reconstructed locations of dated Paleocene–Eocene basalts (Fig. 5c) show the areal extent of the Iceland plume volcanism based on rock samples of Early–Late Paleocene (63–55 Ma, in blue) and break-up (55–54 Ma, in red) times, and as a circular area of approximately 1000 km radius as postulated by conceptual models of a mantle plume head extent (Jones *et al.* 2002).

Continental break-up and seafloor spreading occurred parallel to the Jurassic–Cretaceous sedimentary basin axes in the southern NE Atlantic and at a approximately 30° angle (clockwise) in the northern part, which was closer to the Iceland plume at that time (Fig. 5a). Oceanic crust of C24 age (oldest part at 53.93 Ma) is identified in all NE Atlantic sub-basins, but the inception of seafloor spreading may have been first registered in the NE Norway Basin, as also suggested by Gaina *et al.* (2009) and Gernigon *et al.* (2015). According to the reconstructed locations of dated Paleocene–Eocene basalts, the trend of magmatic activity closer to break-up time seems to have been more along the future margin orientation, possibly showing a change in the stress regime (Fig. 5c). The modelled Iceland plume location from 57 to 54 Ma is north of Jameson Land (Fig. 5). Very few dated NAIP samples have been described in the region north of Vøring and the conjugate NE Greenland margins (Fig. 5c), but a large volume of magmatic material has been identified along the margins in the form of seawards-dipping reflectors, inner and outer lava flows, sills intruded in the basement, and lower crustal bodies (e.g. Geissler *et al.* 2016; Horni *et al.*, this volume, in review). The presence of various NAIP volcanic structures is reflected in the high gravity and magnetic data values (Fig. 5b, c).

Gaina *et al.* (2009) suggested that break-up and seafloor spreading between Greenland and Eurasia were different in basins north and south of the Iceland. The new model presented here confirms these results. Part of the tectonic motion resulting from different opening histories of oceanic basins north and south of the Iceland region has been accommodated by extension within the JMMC, which sits at the junction between these two domains, but there are probably other, less well-documented, changes in the centre of the NE Atlantic and associated margins.

Our reconstructions (Fig. 5b, c) show that narrow oceanic basins (25–30 km) opened north of the JMMFZ. Just south of the JMMC reconstructed tectonic blocks is observed a much wider space (*c.* 100–130 km) between the reconstructed COBs. The position of interpreted COBs is not sufficient to infer the presence of an additional continental tectonic block, as COB interpretations could be subjective. However, the gap in the reconstruction, plus evidence of buried continental crust under present-day SE Iceland, led Torsvik *et al.* (2015) to suggest that the JMMC is much larger and its southern fragment, which is possibly buried under present-day SE Iceland, was rifted from the Greenland continental margin situated south of the Blosseville Coast (Fig. 5).

Seafloor spreading domains in the NE Atlantic

In this section we will discuss the relative motion between Greenland and Eurasia since the Paleocene (Fig. 6), and how inferred variations in the rate and direction of spreading between the two plates are reflected in the potential field-data pattern (Fig. 7). Based on these observations, we define several seafloor spreading domains in NE Atlantic sub-basins. A 'seafloor spreading domain' is defined as a region where the oceanic crust displays a certain pattern, or fabric, that can be identified as short wavelength variations mainly in the magnetic anomaly data, but also in the gravity (and sometimes bathymetric) data, and which reflects characteristics of a seafloor spreading regime. Lastly, we inspect changes in absolute plate motion of major plates in the NE Atlantic (Fig. 8), and speculate about possible connections between main kinematic adjustments in the North Atlantic and far-field stresses associated with distant tectonic events (Fig. 9).

Eurasia–Greenland stage rotations calculated from finite rotations listed in Table 2, and recalculated at equal interval of 5 myr, are shown in Fig. 6. At break-up time, the stage pole moved from a position held before 55 Ma near the equator to approximately 46° N (Fig. 6). The NE Atlantic oceanic crust was formed at a rate of 35–40 mm a^{-1} for the first 4–5 myr in all basins, except the Norway Basin, where the spreading rate was lower (*c.* 25 mm a^{-1}) (Fig. 7e–g). This first stage of seafloor spreading resulted in a regular pattern of parallel-magnetized oceanic crust blocks visible on the magnetic gridded data (Fig. 7a, oceanic crust domains D1N, D1C and D1S). A sudden decrease in seafloor spreading rates occurred between 50 and 48 Ma, and coincided with a 30°–40° counter-clockwise change in spreading direction. Gaina *et al.* (2009) suggested that the southern part of the NE Atlantic was influenced more strongly by these changes as they may have been linked to a contemporaneous modification in seafloor spreading direction in the Labrador Sea. That event left its imprint on the orientation of pre- and post-C22 Bight fracture zone segments (see the red segments and arrows near the lower edge of Fig. 7a). Note that this time interval resulted in asymmetrical seafloor spreading in

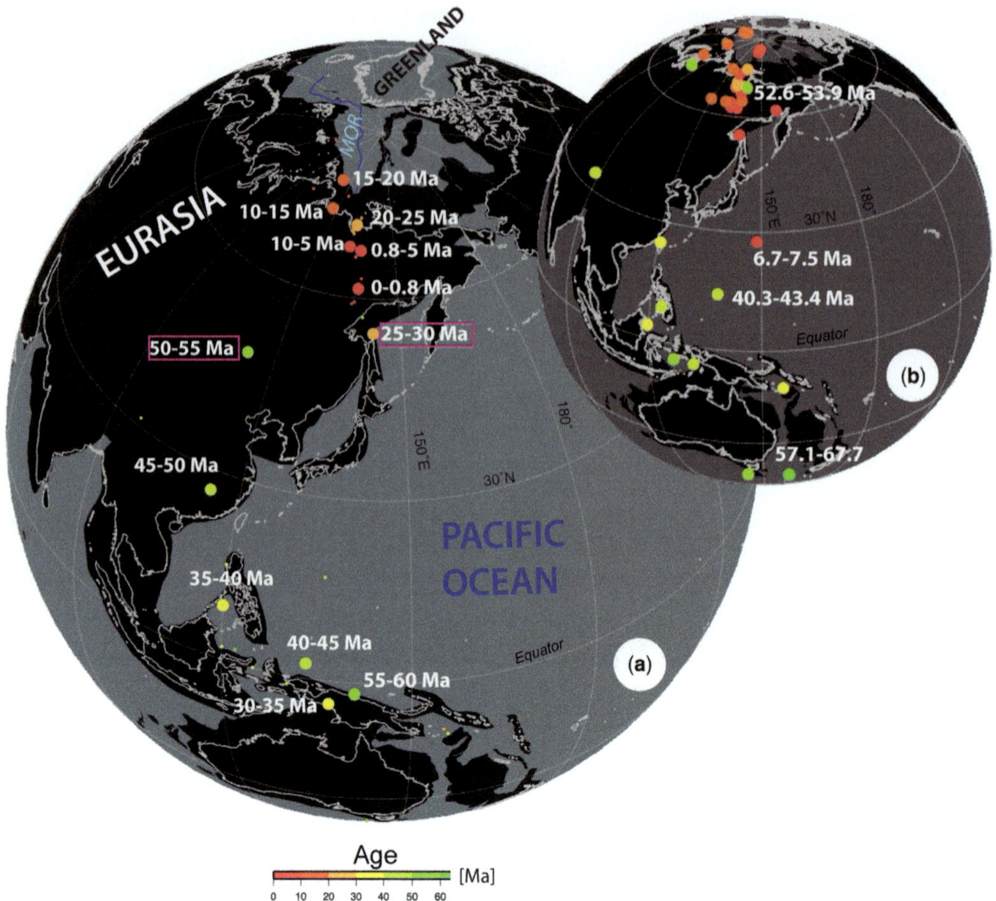

Fig. 6. Present-day location of the Greenland–Eurasia stage poles: (**a**) stage poles at equal interval of 5 myr; and (**b**) original stage poles derived from finite rotations listed in Table 2. MOR, mid-ocean ridge.

the Irminger and Iceland basins, with more crust accreted on the Greenland Plate (Fig. 4). Besides the reorientation of the Bight fracture zone and possible eastwards ridge jumps at C22 time, the seafloor fabric in the oceanic domains D1N and D1S (Fig. 7a) remained virtually the same from early seafloor spreading (*c.* 54 Ma) to C18 (*c.* 40 Ma). In the Norway Basin, the seafloor spreading pattern changed at C21 from parallel to the passive margins to fan-shaped, oblique spreading (Fig. 7a). This was likely to be the result of plate boundary re-locations north and south of the JMMC (e.g. Gaina *et al.* 2009; Gernigon *et al.* 2012, 2015; Blischke *et al.* 2016).

From approximately 50 to 30 Ma, the Greenland–Eurasia stage poles migrated southwards again, further away from the North Atlantic. The seafloor spreading rates continued to decrease in all NE Atlantic basins until approximately 35–33 Ma, when it reached rates of less than 20 mm a^{-1}, with ultra-slow spreading rates of around 10 mm a^{-1} in the Lofoten and Greenland basins (Figs 4 & 7c). During this time interval, a clockwise rotation of the spreading direction was registered at around 45 Ma, and a counter-clockwise rotation at approximately 40 Ma (C18) (Fig. 7b–d). In the Iceland and Irminger basins, the seafloor fabric changed at C18 time from long, continuous parallel magnetic lineations, to shorter magnetic lineations offset by small fracture zone fragments ('staircase'-like domain D2S in Fig. 7a). In the Lofoten and Greenland basins, the seafloor fabric is orientated oblique to the spreading direction, as shown by the magnetic data patterns in domains D2N (Fig. 7a). The gravity data show the development of small offset fracture zones, perpendicular to the direction of spreading, in the southern part of NE Atlantic. These fracture zones are continuous from the Bight Fracture Zone to approximately 60.5° N, where they start to interfere with the V-shaped ridges, as described by numerous earlier studies (e.g. see

Vogt 1971; White *et al.* 1995) and more recently by Hey *et al.* (2010).

A global model that computed the locations of mantle plume at the surface by taking into account global plate motions for the last 130 Ma, and mantle plume conduit deviation due to advection in the mantle (Doubrovine *et al.* 2012), predicts that the Iceland mantle plume head centre crossed the Greenland COB between 40 and 35 Ma, and was located under oceanic crust for times younger than 35 Ma (Fig. 7a). The seafloor spreading domains that formed between C18 (*c.* 40 Ma) and C6 (*c.* 20 Ma) illustrate the complex geodynamics of the NE Atlantic where changes in kinematics and the influence of the Iceland plume left their imprint on the oceanic fabric. Apart from irregular oceanic crust architecture, plume–ridge interactions led to plate boundary relocations (as outlined by Gaina *et al.* 2009), the formation of the JMMC and asymmetrical seafloor spreading (Figs 4 & 7), and may explain the formation of other features observed within the oceanic domain, such as the elongated volcanic Trail Ø complex (Fig. 1) (Geissler *et al.* 2016) and seamount-like oceanic igneous features (SOIFs: see Gaina *et al.* 2016).

After the previously mentioned period of very low spreading rates (less than 20 mm a^{-1}), the JMMC formation and the westwards ridge relocation along the Kolbeinsey Ridge, the NE Atlantic oceanic crust formed again at higher rates (20–28 mm a^{-1}) and, after two changes in spreading directions (at *c.* 14 and 7–8 Ma), it stabilized at approximately 20 mm a^{-1} in a NW–SE direction (Fig. 7e–g). In the Irminger and Iceland basins, oceanic spreading domain D3S gradually formed from north to south between C17 and C4 (38–7.5 Ma), showing a lateral transition from 'staircase' magnetic pattern to linear trends, most probably influenced by the Iceland plume flow. The youngest domain, D4S, shows a steady seafloor spreading regime achieved along the mid-ocean ridges south and north of Iceland (Fig. 7a).

Subduction in the Pacific and Mediterranean realms, and the opening of the NE Atlantic

Until the Silurian, Greenland and its Archaean and Proterozoic crust was part of Laurentia – an amalgamation of cratonic cores surrounded by terranes and deformed tectonic blocks resulting from several Precambrian orogenies (e.g. St-Onge *et al.* 2009). Following the Early Scandian (Caledonian) Orogeny in the Silurian (*c.* 425 Ma) (Torsvik *et al.* 1996), Greenland was confined between North America/Laurentia and Eurasia/Baltica. The Greenland Plate was formed in the Early Eocene, as a result of the NE Atlantic opening, but the dynamics of this tectonic block had to accommodate tectonic changes imposed by its large neighbouring plates.

We endeavour to explore the hypothesis that some changes in the NE Atlantic evolution may have been triggered by tectonic events at the boundaries of either the Eurasia or North America plates. To place the opening of the NE Atlantic in a larger context, we combine our new kinematic model with the global model by Seton *et al.* (2012). To link the relative plate motions to a mantle reference frame, we use two alternatives: Torsvik *et al.* (2008) and Doubrovine *et al.* (2012) models (Fig. 8).

According to the two combined relative-absolute motion models (abbreviated NEATL-T2008 and NEATL-D2012), the direction of absolute motion (relative to the underlying mantle) of Eurasia and North America changed several times in the Cenozoic (Fig. 8). Although the two models differ, we note common features and trends. Both models show latitudinal motion of North America and Eurasia from 60 to 50 Ma. A major change in plate motions is shown by both models at 50 Ma, and another one at 40 Ma, most pronounced for the Eurasian Plate. A smoother change in the absolute motion of North America/Greenland is visible at 20 Ma in the NEATL-D2012 model, but also for the Eurasian Plate which started to move northwards in model NEATL-T2008 or become almost stationary in NEATL-D2012.

Major changes in tectonic plate motion could be triggered by continental collisions (e.g. Patriat & Achache 1984), changes in subduction geometry and subducted slab dynamics (e.g. Austermann *et al.* 2011), and possibly by mantle plume head impacts at the base of the lithosphere (e.g. Cande & Stegman 2011). A recent study proposes abrupt plate accelerations before continental rupture due to a rapid decrease in rift strength (Brune *et al.* 2016).

In the following, we explore whether major tectonic events at the boundaries of Eurasia, North America and Greenland, or changes in these plate motions relative to the mantle, coincide with the time of changes in seafloor spreading regimes. Variations in seafloor spreading direction and rates since the inception of the NE Atlantic are compared with the timing of major tectonic events reported in the literature (Fig. 9a). Changes in absolute plate motion at time intervals that correspond to dated boundaries between various oceanic crust fabric domains (Fig. 7a) are shown in Figure 9b.

Significant changes in seafloor spreading direction in the NE Atlantic were at approximately 50–49, 40, 33–29 and 15 Ma (Fig. 9a). The boundaries between different seafloor spreading fabrics in the NE Atlantic are dated at C21 (*c.* 47 Ma), C17–C18 (*c.* 40 Ma), C6 (*c.* 20 Ma) and C4 (*c.* 8–7 Ma) (Fig. 7a). Greenland's large neighbouring

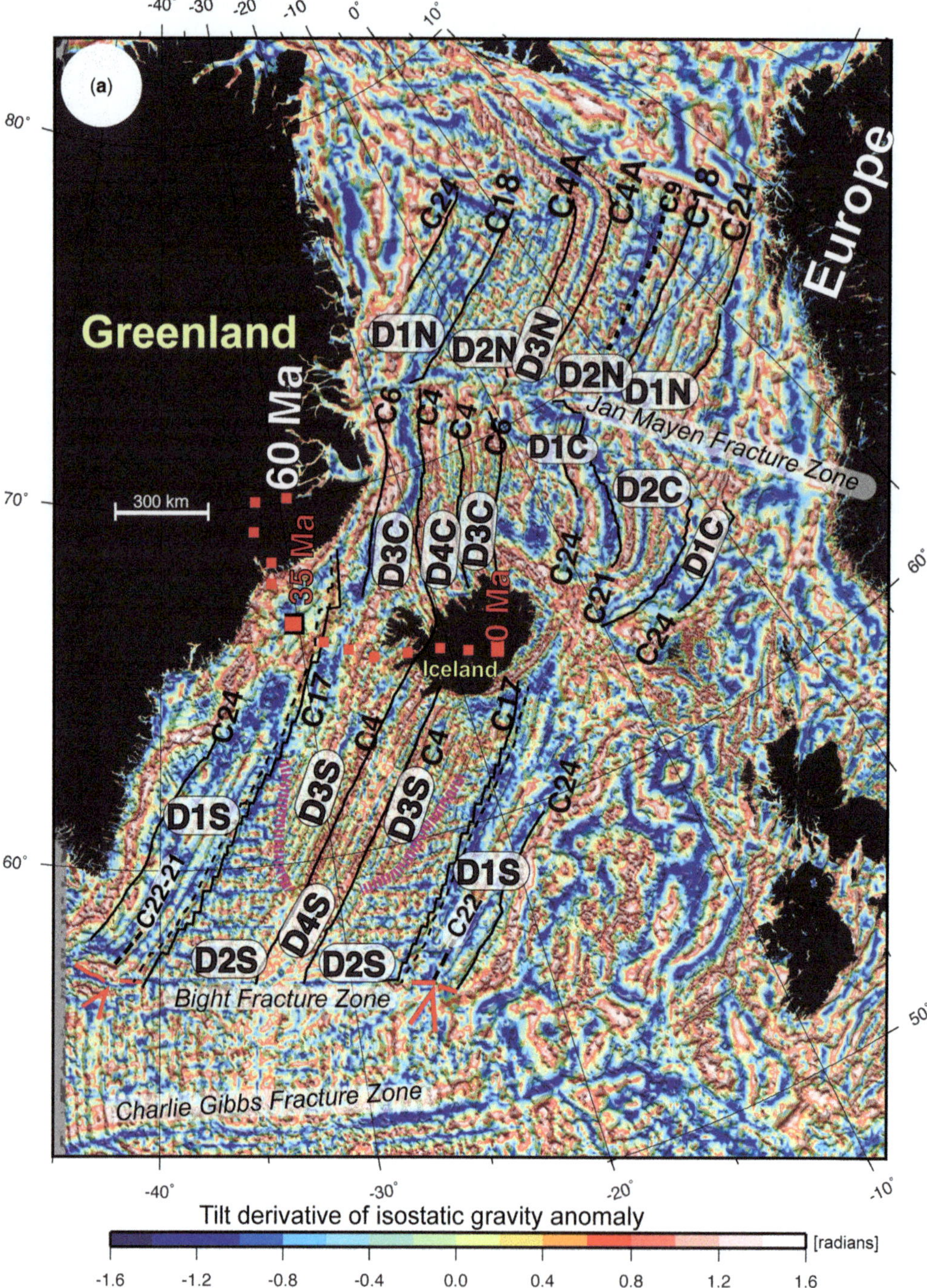

Fig. 7. Identification of seafloor spreading domains in the NE Atlantic Ocean. (**a**) Tilt derivative of isostatic gravity shaded by the tilt derivative of magnetic anomaly (illuminated from 120° N), which highlights the direction of magnetized bodies. Distinctive domains of oceanic crust (D1–D4, N – north, C – central and S – south) morphology are delineated by selected isochrons (age shown by chron (C) number). Red squares show the modelled position of the Iceland plume at 5 myr intervals (Doubrovine *et al.* 2012). The red segments and arrow show a change in spreading direction at C22 visible in the seafloor fabric. Greenland–Eurasia seafloor spreading directions (b)–(d).

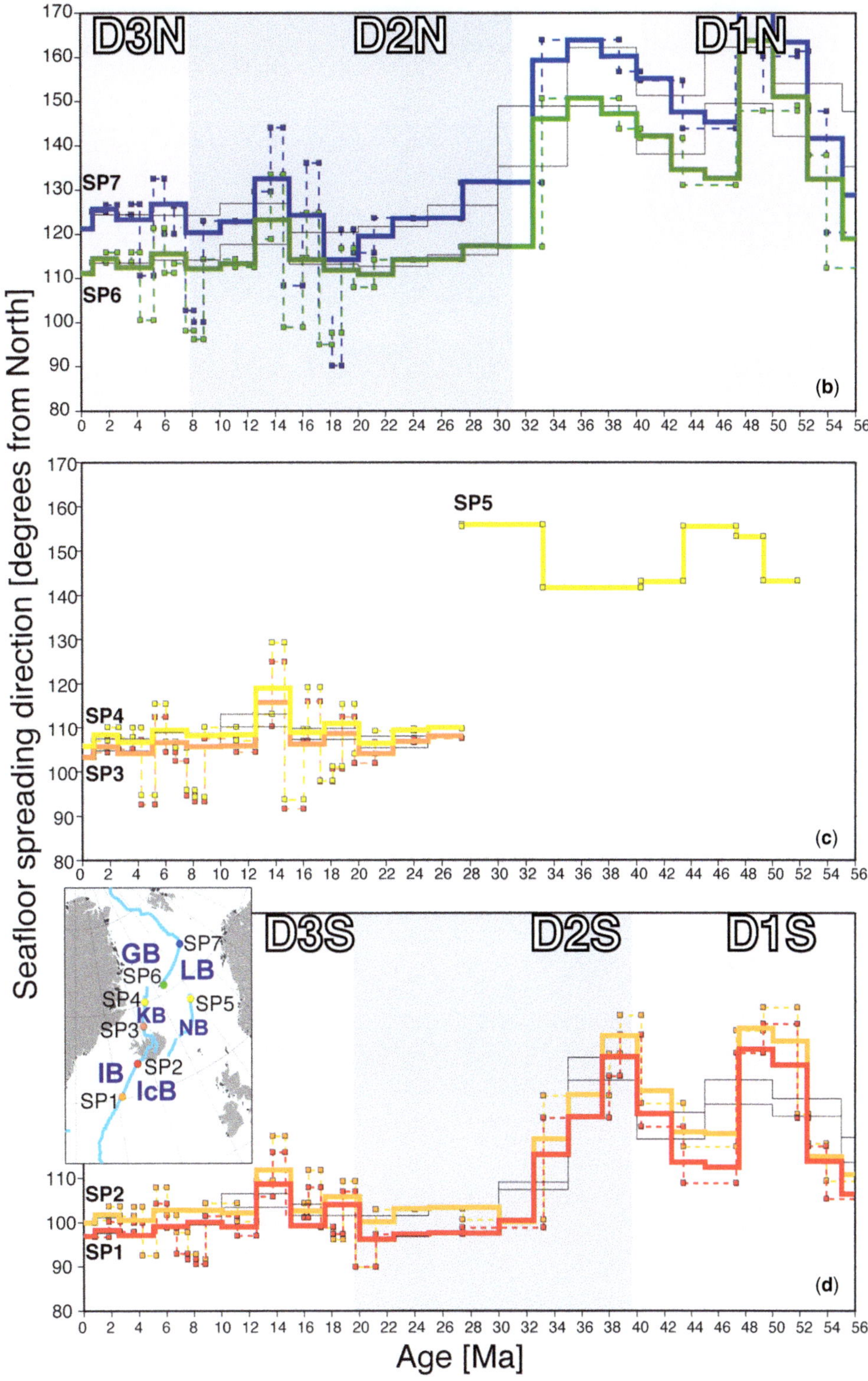

Fig. 7. *(Continued)*

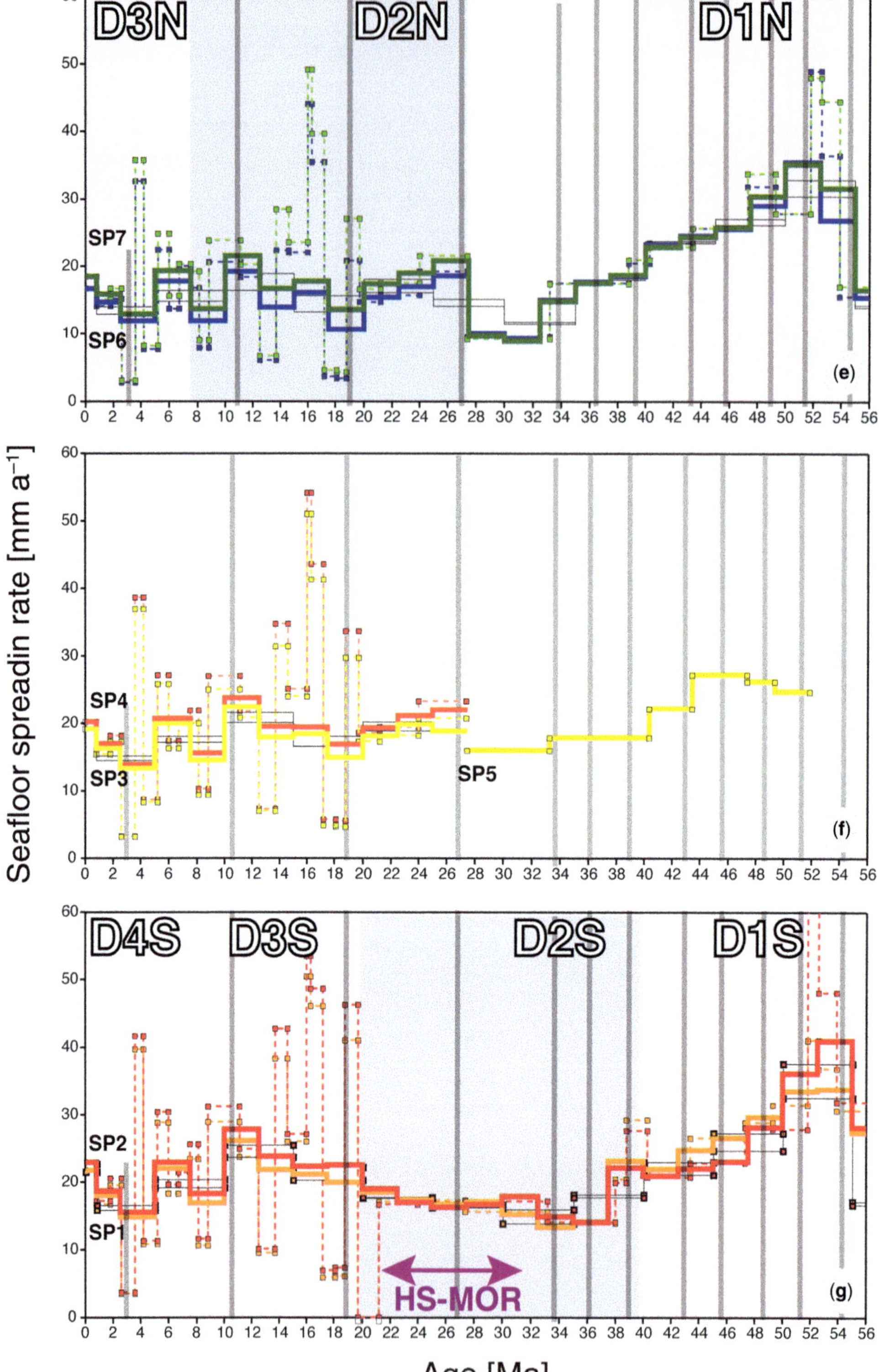
D3N
D2N
D1N
SP7
SP6
(e)
SP4
SP3
SP5
(f)
D4S
D3S
D2S
D1S
SP2
SP1
HS-MOR
(g)
Seafloor spreadin rate [mm a⁻¹]
Age [Ma]

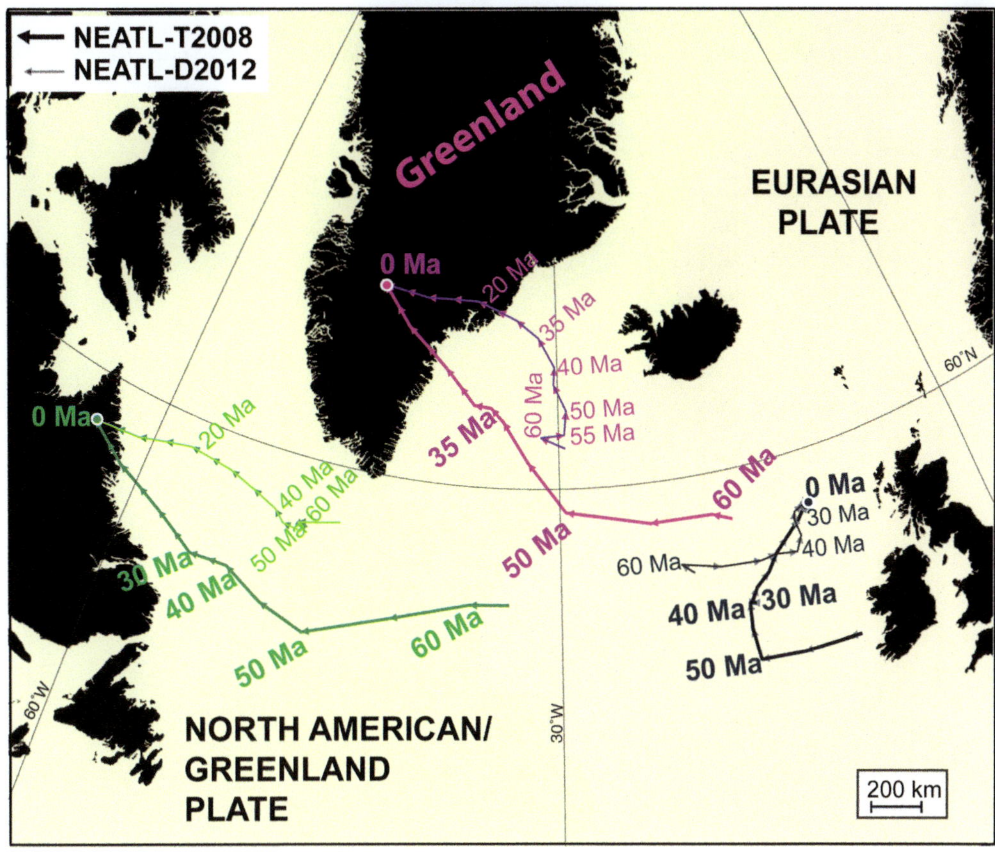

Fig. 8. Motion paths showing the direction of absolute motion (relative to the mantle) for North America (in green), Greenland (in magenta) and Eurasia (in black and grey) calculated at 5 myr intervals from 65 Ma to the present for two global models (thick lines – Torsvik *et al.* 2008; thin lines – Doubrovine *et al.* 2012) where relative plate motions between Eurasia, Greenland and North America have been replaced with the rotations presented in this study.

plates, Eurasia and North America, were bordered throughout the Cenozoic by active plate boundaries where the Pacific and smaller oceanic plates were continuously subducted. For example, the Farallon Plate had a long history of subduction under the western North America. We compare the vectors of absolute plate motion of this plate at 56, 55 and 52 Ma (Fig. 9b), and observe a considerable increase in its absolute plate velocity post-56 Ma. A detailed scrutiny of the slab graveyard under North America revealed a gap in the subducted material identified in tomographical models. This gap, named 'the Big Break', was dated as Paleocene–Eocene (60–40 Ma) (Sigloch 2011). Large oceanic plateau subduction or obduction, slab break-off and the Laramide Orogeny in the western North America were also linked by several authors (e.g. Livaccari *et al.* 1981; Sigloch *et al.* 2008; Liu *et al.* 2010). Although the causal relationship between changes in the subduction regime at the western

Fig. 7. (*Continued*) Greenland-Eurasia full spreading rates (**e**)–(**g**). Seafloor spreading rates and directions are calculated for eight seed points along the active and extinct mid-ocean ridges in NE Atlantic oceanic sub-basins (locations and corresponding seed-point with matching colours are in the (**d**) inset map). The values are calculated based on stage poles at the exact age of identified isochrons (dashed thin lines), averaged at 2.5 myr intervals (thick lines), and at 5 myr intervals (thin horizontal grey lines). Pink and light blue boxes indicate the oceanic domain limits, as in Figure 7a. Grey thick vertical lines show postulated pulses of Iceland plume activity (Parnell-Turner *et al.* 2014). HS-MOR indicates hotspot-ridge interaction interval.

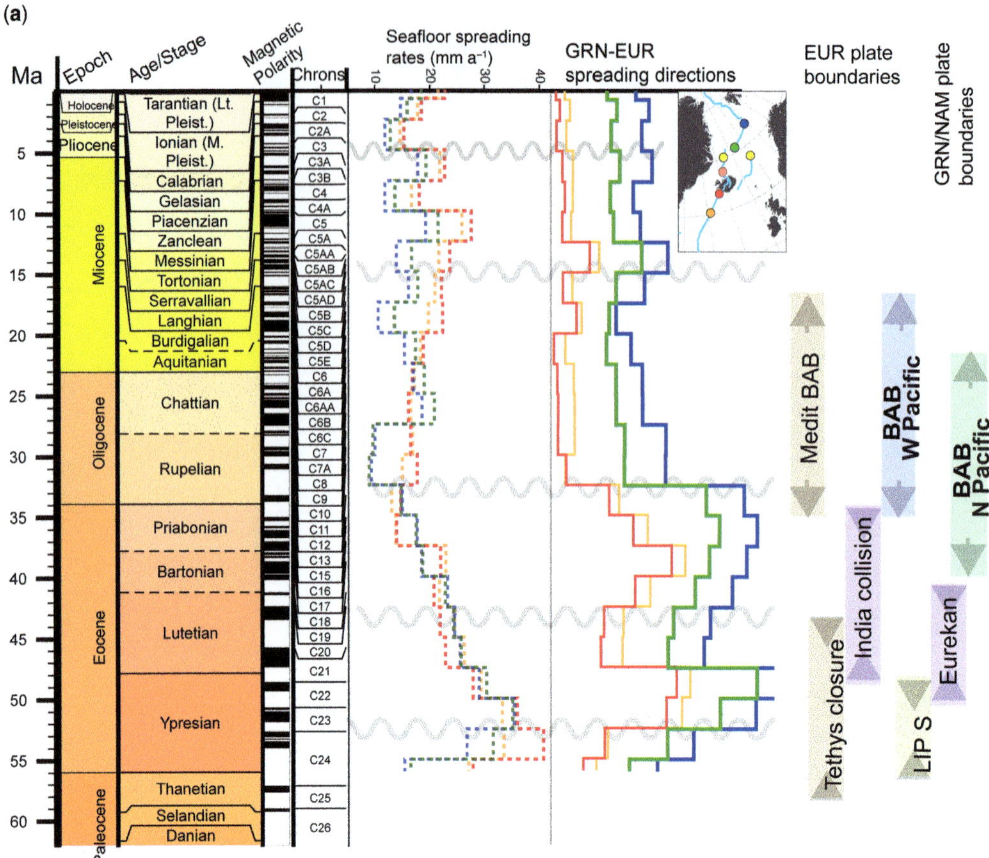

Fig. 9. Chart showing possible correlations between plate motion changes in the NE Atlantic and Cenozoic regional tectonic events. Abbreviations: AFR, Africa Plate; BAB, back-arc basins; EUR, Eurasia Plate; FAR, Farallon Plate; GRN, Greenland Plate; IND, India Plate; KUL, Kula Plate; LIP S, large igneous province subduction; Medit, Mediteranean region; NAM, North America Plate; PAC, Pacific Plate; Sboff, slab break-off; VAN, Vancouver Plate. Seafloor spreading rates and directions (**a**) are calculated for seven locations (seed points) on present-day and extinct mid-ocean ridges in the NE Atlantic (see inset map for seed point locations and Figure 7b–d for more details). Grey undulating lines indicate changes in spreading rates and/or directions that coincide with postulated compressional dome formation along the NE Atlantic continental margins.

North American plate boundary and its subsequent dynamics is not clear, we flag this connection as a future topic to be explored.

The Eurasian Plate, the third largest tectonic plate on Earth, is bordered on the west side by the Atlantic mid-ocean ridge, and to the south and east by a composite plate boundary formed by transform faults, subduction trenches and collisional segments, which in turn continue within NE Asia into an extensional intra-continental boundary that links with the Gakkel Ridge in the Arctic Ocean. In the geological past, the southern part of Eurasia was the locus of several trenches where the Neo-Tethys Ocean was consumed (for a review see Dilek 2006), and where massive continental collisions formed the largest Cenozoic mountain belt: the Alpine-Himalaya (e.g. Suess 1893). The collision of various microblocks in the Mediterranean realm, the India–Asia collision and the Arabia–Asia collision occurred in mid-late Cenozoic, and dramatically affected the southern boundary of the Eurasian Plate (e.g. Yin 2010). In the western and northern Pacific, NE and SW of the junction with the Eurasia and North America plate boundary, the continental margins of North America and Eurasia underwent massive changes in the late Eocene–Oligocene. Numerous marginal back-arc basins were formed during that time period and this was linked to changes in subducted slab dynamics (e.g. Yin 2010).

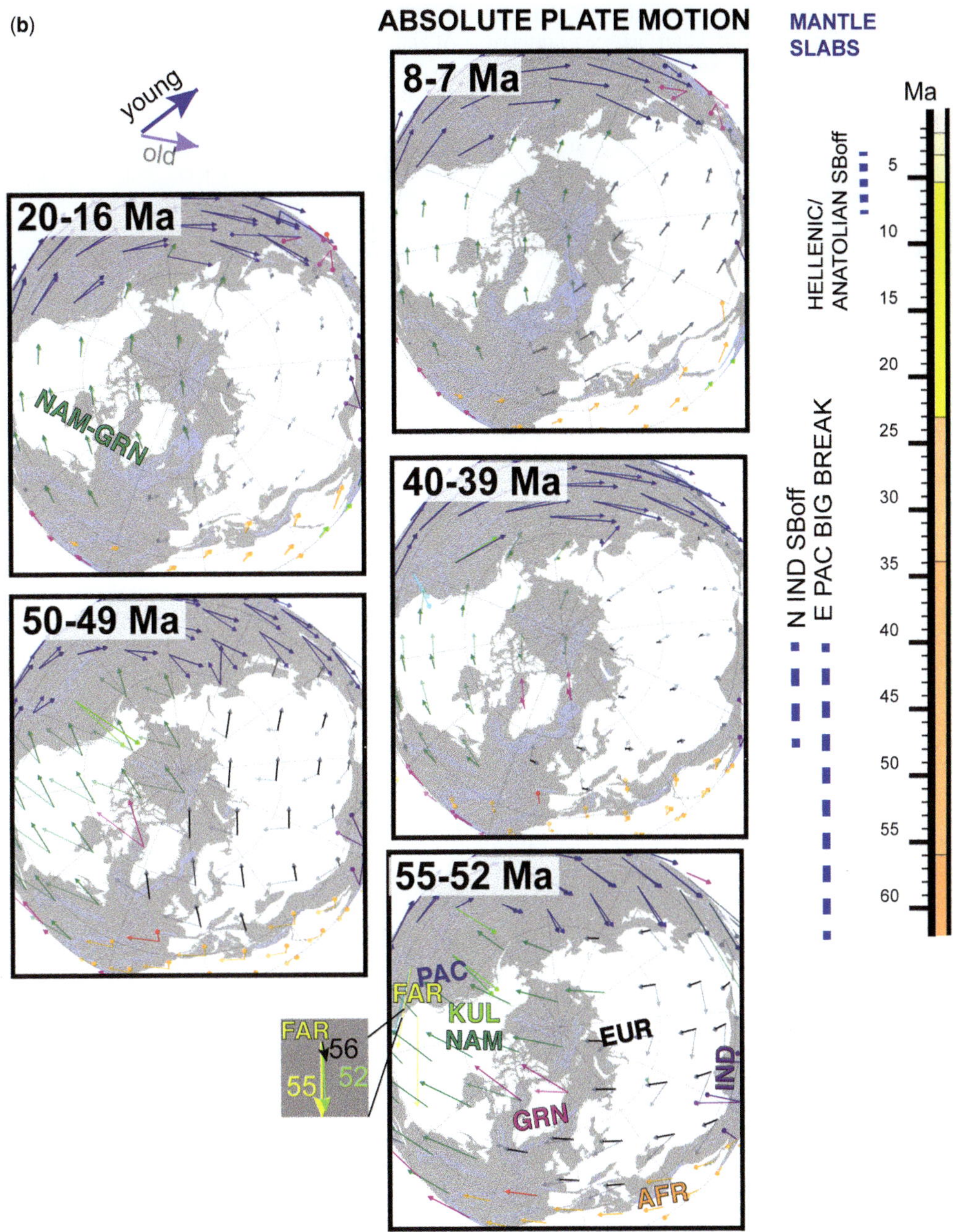

Fig. 9. (*Continued*) (**b**) shows a series of Cenozoic reconstructions with absolute velocity vectors for major tectonic plates in the northern hemisphere. The absolute plate motion is calculated using the Seton *et al.* (2012) global model, where relative plate motions between Eurasia, Greenland and North America have been replaced with the rotations presented in this study.

Around 50 Ma, North America, Greenland and Eurasia changed their absolute motion in a clockwise direction towards north. This may have been triggered by the soft India collision at the southern Eurasian margin and possible readjustment of the North American Plate after oceanic plateau subduction and/or slab break-off. The northwards motion of Greenland due to seafloor spreading in both the Labrador Sea/Baffin Bay and the NE Atlantic led to the Eurekan deformation between northern Greenland and the High Arctic domain (e.g. Piepjohn *et al.* 2013). A change in spreading direction and a decrease in seafloor spreading rate in the NE Atlantic are contemporaneous with compression and deformation on Ellesmere Island (Fig. 9).

The absolute motion vectors of all three main plates in the North Atlantic decreased at approximately 40 Ma (Fig. 9b). The timing coincides with subduction reinitiation along the western margin of North America (Sigloch *et al.* 2008) and initiation of subduction along the Aleutian Trench (Jicha *et al.* 2006), approximately 1000 km south of the previous plate boundary. A change in spreading directions at this time is observed more clearly in the southern part of the NE Atlantic, but the oceanic crust fabric changed along the entire ocean (see the boundaries between oceanic domains 1 and 2 in the northern and southern part of the NE Atlantic – D1N, D1S and D2N and D2S in Fig. 7a).

The first massive volcanism in the NE Atlantic region was recorded around 62 Ma (e.g. Storey *et al.* 2007), and continued intermittently until 55 Ma, when the second large-scale magmatic output led to continental break-up. During the opening of the NE Atlantic Ocean, pulses of magmatic activity affected the NE Atlantic at regular intervals, in the first 20 myr after break-up at every 3 myr and after that only every 8 myr (shown as grey vertical lines in Fig. 7e–g) (Parnell-Turner *et al.* 2014). One of these pulses of magmatic activity occurred at around 40 Ma, and a slight increase in spreading rate can be observed before a relatively sharp reduction in magmatic productivity at 39–37 Ma in the south NE Atlantic (Fig. 7e–g).

For the following 20 myr (from 40 to 20 Ma), oceanic crust in the NE Atlantic seems to go through a series of continuous readjustments of plate boundary directions in an ultra-slow spreading regime. During this time a 'staircase'-like plate boundary developed in the Iceland and Irminger basins (Fig. 7a), and multiple ridge relocations and intracontinental rifts occurred south of and within the JMMC (see Blischke *et al.* 2016). By 22–20 Ma, the Aegir Ridge was extinct and a fully developed Kolbeinsey mid-ocean ridge was building a new oceanic basin west of the JMMC. Interestingly, at around 35–32 Ma, the Euler stage pole describing the opening of NE Atlantic moved northwards by approximately 4500 km (from SE Asia to Japan Sea) (Fig. 6). This is the time when marginal back-arc basins formed in the Japan, Okhotsk and South China seas, apparently as a result of India–Eurasia collision and subsequent lateral extrusion of SE Asia (Yin 2010). Changes at the distal Eurasian plate boundaries may have affected the reorganization of the North Atlantic, as in Early Oligocene (*c.* 35–30 Ma) seafloor spreading in the North Atlantic was gradually focusing on its NE branch (in the NE Atlantic), and eventually the Labrador Sea/Baffin Bay Basin was abandoned.

From 30 Ma to present day, the Eurasia–Greenland/North America stage poles were clustered in NE Asia, along the boundary between the two major plates (Fig. 6). A change in relative plate motions around 7 Ma has been reported by Merkouriev & DeMets (2008) and was accompanied by a sudden decrease in seafloor spreading rate (Fig. 7e–g). The Eurasian absolute plate motion shows a change in a clockwise direction around that time (Fig. 9b), with larger angles in its SW part. The third largest orogenic plateau in the world, the Anatolian Plateau, was also formed in latest Miocene times. Recently, Schildgen *et al.* (2014) proposed a model where the inception of this plateau was directly linked with oceanic slab break-off and tearing, and rapid Hellenic Trench retreat. These circumstantial overlaps in time between various tectonic events that occur at boundaries of a large tectonic plate, such as collision, large topographical feature formation, slab break-off and subsequent change in plate motions, are noted here and may be used as hypotheses that can be tested by future studies and geodynamic modelling.

The opening of the NE Atlantic and Cenozoic compressional events on NE Atlantic passive margins

For the last two decades an on-going discussion has attempted to elucidate the nature and processes involved in the Early Paleocene–Present reactivation of NE Atlantic passive margins (e.g. Doré & Lundin 1996; Japsen & Chalmers 2000; Lundin & Doré 2002; Ritchie *et al.* 2003; Gomez & Verges 2005; Doré *et al.* 2008; Stoker *et al.* 2010; Tuitt *et al.* 2010; Japsen *et al.* 2012; Yamato *et al.* 2013; Døssing *et al.* 2016). Cenozoic compressional domes have been described on the Vøring, Faroe–Shetland and Hatton margins, in the Rockall Basin (for a short summary see Kimbell *et al.* 2016), and in the NE Greenland margin (Price & Whitham 1997; Hamann *et al.* 2005). Few distinct phases of compressional dome formation along the west European margin are identified for post-break-up times in the NE Atlantic: Early and Mid-Eocene, Mid-Eocene–Oligocene and Mid-Miocene (e.g. Doré *et al.* 2008; and see the summary in Kimbell *et al.* 2016). Previous studies have postulated a range of possible causes for post-break-up compression along the NE Atlantic margins, and they range from, for example, ridge-push and gravity potential forces, mantle convection (including small-scale convection along passive margins), far-field stresses, and differential compaction (for a comprehensive review see Doré *et al.* 2008).

The time intervals with relatively sharp changes in seafloor spreading rates and/or directions are highlighted in Figure 9. Notable variation in spreading rates and spreading directions between Eurasia and Greenland derived from our study are observed for the Early Eocene (*c.* 52 Ma), Mid-Eocene (ca 42 Ma), Early Oligocene (*c.* 33 Ma), Mid-Miocene (*c.* 15 Ma) and Late Miocene–Pliocene (*c.* 7–5 Ma). We observe that they coincide with postulated formation of compressional domes along the NE Atlantic continental margins. Other studies have pointed to this correlation, emphasizing the role of different spreading rates along the NE Atlantic spreading axes: for example, on the formation of Miocene domes in the Vøring and Faroe–Shetland basins (e.g. Mosar *et al.* 2002).

We do not attempt to discuss in detail the correlation between the occurrences of compressional domes along the NE Atlantic margins and the timing of regional tectonic events, but we would like to point out that changes at the distant plate boundaries of the Eurasian and North American plates discussed in the previous section, and schematically shown in Figure 9, have a temporal connection with NE Atlantic continental margin reactivation. In particular, periods of slab break-off described along the southern European plate margin and western Pacific plate shortly predate changes in plate motion that may result in seafloor spreading rate readjustments and therefore a reorientation in intra-plate stresses.

Conclusions

An updated magnetic anomaly grid of the NE Atlantic, together with a new interpretation of the oceanic crust age in the Irminger, Iceland, Lofoten, Greenland and Norway basins, and west of the Jan Mayen microcontinent (JMMC), have prompted us to revisit the kinematic history of this region. Continental break-up occurred parallel to the Mesozoic rift axes in the southern NE Atlantic, but obliquely to the previous rifting trend in the northern NE Atlantic. The modelled location of the Iceland plume at 57–54 Ma is below the East Greenland continental margin, just north of the Jameson Land Basin, and very close to the break-up line, which can probably explain the rupture of continental lithosphere at an angle with previous rifted basin axes. Oceanic lithospheric ages, spreading rates and asymmetries in crustal accretion were calculated based on a dense set of isochrons. The new model includes a detailed kinematic history, including ridge jumps which led to the asymmetrical crustal accretion in the following areas: SE Lofoten Basin, the Jan Mayen Fracture Zone (JMFZ) region, Iceland and offshore Faroe Islands.

We describe various spreading regimes and associated oceanic crust fabrics, and we attempt to link changes in spreading directions and rates with large-scale tectonic events. Boundaries between major oceanic crust domains were mainly formed at the time of changes in the European absolute plate motion, which, in turn, may have been caused by successive changes in the subduction or collisional regime along its eastern and southern boundaries. Variations in the subduction regime in the NE Pacific, followed by a change in the absolute motion of the North American Plate could have caused the Eurekan deformation between NW Greenland and Ellesmere Island. These events caused seafloor spreading changes in the neighbouring Labrador Sea/Baffin Bay and a decrease in spreading rates in the NE Atlantic. The collision between India and Eurasia, crustal extrusion, and the formation of marginal back-arc basins along the eastern boundary of Eurasia coincide with a northwards jump of the rotation pole between Eurasia and Greenland, and the gradual demise of the triple junction between North America, Greenland and Eurasia.

Distal changes in plate motions triggered by collision and changes in subduction dynamics may have determined the rates and spreading directions in the NE Atlantic, but magmatic productivity reflected in oceanic crust fabric was also influenced by the local mantle dynamics. The new kinematic model, as well as geophysical data, confirm that several oceanic crust domains in the NE Atlantic bear the imprint of the Iceland hotspot pulsations, as shown by other previous studies.

We acknowledge the support of the NAG-TEC industry sponsors (in alphabetical order): Bayerngas Norge AS; BP Exploration Operating Company Ltd; Bundesanstalt für Geowissenschaften und Rohstoffe (BGR); Chevron East Greenland Exploration A/S; ConocoPhillips Skandinavia AS; DEA Norge AS; Det norske oljeselskap ASA; DONG E&P A/S; E.ON Norge AS; ExxonMobil Exploration and Production Norway AS; Japan Oil, Gas and Metals National Corporation (JOGMEC); Maersk Oil; Nalcor Energy – Oil and Gas Inc.; Nexen Energy ULC, Norwegian Energy Company ASA (Noreco); Repsol Exploration Norge AS; Statoil (UK) Ltd; and Wintershall Holding GmBH. C.G. acknowledges support from the Research Council of Norway through its Centres of Excellence funding scheme, project number 223272. GSK publishes with permission of the Executive Director of the British Geological Survey (Natural Environment Research Council).

Part of this manuscript has been published in Chapter 4 (Nasuti and Olesen) and Chapter 5 (Gaina) of the NAG-TEC Atlas (Hopper *et al.* 2014). Dr Gwenn Péron-Pinvidic (NGU, Norway), NAG-TEC Work Package 2 coordinator, is thanked for comments on NAG-TEC Atlas Chapter 5. C. DeMets and A. Benediktsdóttir are thanked for sharing their magnetic anomaly interpretations at an early stage of the NAG-TEC project. We are very grateful to the Editor, Thomas Funck, for his detailed comments. The authors thank to Tony Doré and an anonymous reviewer for excellent suggestions.

References

AMANTE, C. & EAKINS, B.W. 2009. *ETOPO1 1 Arc-Minute Global Relief Model: Procedures, Data Sources and Analysis*. NOAA Technical Memorandum NESDIS NGDC-24. National Geophysical Data Center, NOAA, Boulder, CO, https://doi.org/10.7289/V5C8276M

ANDERSEN, O.B. 2010. The DTU10 gravity field and mean sea surface. Presented at the *Second International Symposium of the Gravity Field of the Earth (IGFS2)*, 20–22 September 2010, Fairbanks, Alaska.

AUSTERMANN, J., BEN-AVRAHAM, Z., BIRD, P., HEIDBACH, O., SCHUBERT, G. & STOCK, J.M. 2011. Quantifying the forces needed for the rapid change of Pacific plate motion at 6 Ma. *Earth and Planetary Science Letters*, **307**, 289–297, https://doi.org/10.1016/j.epsl.2011.04.043

BENEDIKTSDÓTTIR, A., HEY, R., MARTINEZ, F. & HOSKULDSSON, A. 2012. Detailed tectonic evolution of the Reykjanes Ridge during the past 15 Ma. *Geochemistry, Geophysics, Geosystems*, **13**, Q02008, https://doi.org/10.1029/2011GC003948

BLISCHKE, A., GAINA, C. ET AL. 2016. The Jan Mayen microcontinent: an update of its architecture, structural development, and role during the transition from the Ægir Ridge to the mid-oceanic Kolbeinsey Ridge. *In*: PÉRON-PINVIDIC, G., HOPPER, J.R., STOKER, T., GAINA, C., DOORNEBAL, H., FUNCK, T. & ÁRTING, U. (eds) *The NE Atlantic Region: A Reappraisal of Crustal Structure, Tectonostratigraphy and Magmatic Evolution*. Geological Society, London, Special Publications, **447**. First published online September 8, 2016, https://doi.org/10.1144/SP447.5

BRUNE, S., WILLIAMS, S.E., BUTTERWORTH, N.P. & MÜLLER, R.D. 2016. Abrupt plate accelerations shape rifted continental margins. *Nature*, **536**, 201–204, https://doi.org/10.1038/nature18319

CANDE, S.C. & KENT, D.V. 1995. Revised calibration of the geomagnetic polarity timescale for the Late Cretaceous and Cenozoic. *Journal of Geophysical Research*, **100**, 6093–6095. https://doi.org/10.1029/94JB03098

CANDE, S.C. & STEGMAN, D.R. 2011. Indian and African plate motions driven by the push force of the Reunion plume head. *Nature*, **475**, 47–52, https://doi.org/10.1038/nature10174

CHALMERS, J.A. & LAURSEN, K.H. 1995. Labrador Sea: the extent of continental and oceanic crust and the timing of the onset of seafloor spreading. *Marine and Petroleum Geology*, **12**, 205–217.

COX, A. & HART, R.B. 1986. *Plate Tectonics: How It Works*. Wiley, Chichester, UK.

DILEK, Y. 2006. Collision tectonics of the Mediterranean region: causes and consequences. *In*: DILEK, Y. & PAVLIDES, S. (eds) *Postcollisional Tectonics and Magmatism in the Mediterranean Region and Asia*. Geological Society of America, Special Papers, **409**, 1–13, https://doi.org/10.1130/2006.2409(01)

DOMEIER, M. 2015. A plate tectonic scenario for the Iapetus and Rheic oceans. *Gondwana Research*, **36**, 275–295, https://doi.org/10.1016/j.gr.2015.08.003

DORÉ, A.G. & LUNDIN, E.R. 1996. Cenozoic compressional structures on the NE Atlantic margin: Nature, origin and potential significance for hydrocarbon exploration. *Petroleum Geoscience*, **2**, 299–311, https://doi.org/10.1144/Petgeo.2.4.299

DORÉ, A.G., LUNDIN, E.R., BIRKELAND, O., ELIASSEN, P.E. & JENSEN, L.N. 1997. The NE Atlantic Margin: implications of late Mesozoic and Cenozoic events for hydrocarbon prospectivity. *Petroleum Geoscience*, **3**, 117–131, https://doi.org/10.1144/Petgeo.3.2.117

DORÉ, A.G., LUNDIN, E.R., KUSZNIR, N.J. & PASCAL, C. 2008. Potential mechanisms for the genesis of Cenozoic domal structures on the NE Atlantic margin: pros, cons and some new ideas. *In*: JOHNSON, H., DORÉ, A.G., GATLIFF, R.W., HOLDSWORTH, R., LUNDIN, E.R. & RITCHIE, J.D. (eds) *The Nature and Origin of Compression in Passive Margins*. Geological Society, London, Special Publications, **306**, 1–26, https://doi.org/10.1144/SP306.1

DØSSING, A., JAPSEN, P., WATTS, A.B., NIELSEN, T., JOKAT, W., THYBO, H. & DAHL-JENSEN, T. 2016. Miocene uplift of the NE Greenland margin linked to plate tectonics: seismic evidence from the Greenland Fracture Zone, NE Atlantic. *Tectonics*, **35**, 257–282, https://doi.org/10.1002/2015TC004079

DOUBROVINE, P.V., STEINBERGER, B. & TORSVIK, T.H. 2012. Absolute plate motions in a reference frame defined by moving hot spots in the Pacific, Atlantic, and Indian oceans. *Journal of Geophysical Research: Solid Earth*, **117**, B09101, https://doi.org/10.1029/2011JB009072

FUNCK, T., HOPPER, J.R. ET AL. 2014. Chapter 6: Crustal structure. *In*: HOPPER, J.R., FUNCK, T., STOKER, T., ARTING, U., PERON-PINVIDIC, G., DOORNEBAL, H. & GAINA, C. (eds) *Tectonostratigraphic Atlas of the North-East Atlantic Region*. Geological Survey of Denmark and Greenland (GEUS), Copenhagen, Denmark, 69–126.

EHLERS, B. & JOKAT, W. 2009. Subsidence and crustal roughness of ultra-slow spreading ridges in the northern North Atlantic and the Arctic Ocean. *Geophysical Journal International*, **177**, 451–462, https://doi.org/10.1111/j.1365-246X.2009.04078.x

ERNST, R.E. & BUCHAN, K.L. 2002. Maximum size and distribution in time and space of mantle plumes: evidence from large igneous provinces. *Journal of Geodynamics*, **34**, 309–342, https://doi.org/10.1016/S0264-3707(02)00025-X

GAINA, C., ROEST, W.R. & MÜLLER, R.D. 2002. Late Cretaceous–Cenozoic deformation of northeast Asia. *Earth and Planetary Science Letters*, **197**, 273–286, https://doi.org/10.1016/S0012-821X(02)00499-5

GAINA, C., GERNIGON, L. & BALL, P. 2009. Paleocene–Recent plate boundaries in the NE Atlantic and the formation of Jan Mayen microcontinent. *Journal of the Geological Society, London*, **166**, 601–616, https://doi.org/10.1144/0016-76492008-112

GAINA, C., WERNER, S.C., SALTUS, R. & MAUS, S. & CAMP-GM GROUP 2011. Chapter 3: Circum-Arctic mapping project: new magnetic and gravity anomaly maps of the Arctic. *In*: Spencer, A.M., Embry, A.F., Gautier, D.L., Stoupakova, A.V. & Sørensen, K. (eds) *Arctic Petroleum Geology*. Geological Society, London, Memoirs, **35**, 39–48, https://doi.org/10.1144/M35.3

Gaina, C., Blischke, A., Geissler, W.H., Kimbell, G.S. & Erlendsson, Ö. 2016. Seamounts and oceanic igneous features in the NE Atlantic: a link between plate motions and mantle dynamics, *In*: Péron-Pinvidic, G., Hopper, J.R., Stoker, T., Gaina, C., Doornebal, H., Funck, T. & Árting, U. (eds) *The NE Atlantic Region: A Reappraisal of Crustal Structure, Tectonostratigraphy and Magmatic Evolution*. Geological Society, London, Special Publications, **447**. First published online September 8, 2016, https://doi.org/10.1144/SP447.6

Gee, J.S. & Kent, D.V. 2007. Source of oceanic magnetic anomalies and the geomagnetic polarity timescale. *In*: Schubert, G. (ed.) *Treatise on Geophysics*. Elsevier, Amsterdam, 455–507, https://doi.org/10.1016/B978-044452748-6.00097-3

Geissler, W.H., Gaina, C. *et al.* 2016. Seismic volcanostratigraphy of the NE Greenland continental margin. *In*: Péron-Pinvidic, G., Hopper, J.R., Stoker, T., Gaina, C., Doornebal, H., Funck, T. & Árting, U. (eds) *The NE Atlantic Region: A Reappraisal of Crustal Structure, Tectonostratigraphy and Magmatic Evolution*. Geological Society, London, Special Publications, **447**. First published online December 14, 2016, https://doi.org/10.1144/SP447.11

Gernigon, L., Olesen, O. *et al.* 2008. Geophysical insights and early spreading history in the vicinity of the Jan Mayen Fracture Zone, Norwegian–Greenland Sea. *Tectonophysics*, **468**, 185–205, https://doi.org/10.1016/j.tecto.2008.04.025

Gernigon, L., Gaina, C., Olesen, O., Ball, P.J., Peron-Pinvidic, G. & Yamasaki, T. 2012. The Norway Basin revisited: from continental breakup to spreading ridge extinction. *Marine and Petroleum Geology*, **35**, 1–19, https://doi.org/10.1016/j.marpetgeo.2012.02.015

Gernigon, L., Blischke, A., Nasuti, A. & Sand, M. 2015. Conjugate volcanic rifted margins, seafloor spreading, and microcontinent: insights from new high-resolution aeromagnetic surveys in the Norway Basin. *Tectonics*, **34**, 907–933, https://doi.org/10.1002/2014TC003717

Gomez, M. & Verges, J. 2005. Quantifying the contribution of tectonics v. differential compaction in the development of domes along the Mid-Norwegian Atlantic margin. *Basin Research*, **17**, 289–310, https://doi.org/10.1111/j.1365-2117.2005.00264.x

Gradstein, F.M., Ogg, J.G., Schmitz, M. & Ogg, G. 2012. *The Geologic Time Scale 2012*. Elsevier, Amsterdam.

Haase, C. & Ebbing, J. 2014. Gravity data. *In*: Hopper, J.R., Funck, T., Stoker, M., Árting, U., Peron-Pinvidic, G., Doornenbal, H. & Gaina, C. (eds) *Tectonostratigraphic Atlas of the North-East Atlantic Region*. Geological Survey of Denmark and Greenland (GEUS), Copenhagen, Denmark, 29–39.

Haase, C., Ebbing, J. & Funck, T. 2016. A 3D regional crustal model of the NE Atlantic based on seismic and gravity data. *In*: Péron-Pinvidic, G., Hopper, J.R., Stoker, T., Gaina, C., Doornebal, H., Funck, T. & Árting, U. (eds) *The NE Atlantic Region: A Reappraisal of Crustal Structure, Tectonostratigraphy and Magmatic Evolution*. Geological Society, London, Special Publications, **447**. First published online October 12, 2016, https://doi.org/10.1144/SP447.8

Hamann, N.E., Whittaker, R.C. & Stemmerik, L. 2005. Geological development of the Northeast Greenland Shelf. *In*: Doré, A.G. & Vining, B.A. (eds) *Petroleum Geology: North-West Europe and Global Perspectives – Proceedings of the 6th Petroleum Geology Conference*. Geological Society, London, 887–902, https://doi.org/10.1144/0060887

Hey, R., Martinez, F., Hoskuldsson, A. & Benediktsdóttir, A. 2010. Propagating rift model for the V-shaped ridges south of Iceland. *Geochemistry, Geophysics, Geosystems*, **11**, Q03011, https://doi.org/10.1029/2009GC002865

Hopper, J. & Gaina, C. 2014. Chapter 2: Bathymetry and elevation. *In*: Hopper, J.R., Funck, T., Stoker, T., Arting, U., Peron-Pinvidic, G., Doornebal, H. & Gaina, C. (eds) *Tectonostratigraphic Atlas of the North-East Atlantic Region*. Geological Survey of Denmark and Greenland (GEUS), Copenhagen, Denmark, 23–28.

Hopper, J.R., Funck, T., Stoker, M.S., Árting, U., Peron-Pinvidic, G., Doornenbal, H. & Gaina, C. (eds). 2014. *Tectonostratigraphic Atlas of the North-East Atlantic Region*. Geological Survey of Denmark and Greenland (GEUS), Copenhagen, Denmark.

Horni, J., Hopper, J.R. *et al.* In review. Regional distribution of volcanism within the North Atlantic Igneous Province. *In*: Péron-Pinvidic, G., Hopper, J.R., Stoker, T., Gaina, C., Doornebal, H., Funck, T. & Árting, U. (eds) *The NE Atlantic Region: A Reappraisal of Crustal Structure, Tectonostratigraphy and Magmatic Evolution*. Geological Society, London, **447**.

Jakobsson, M., Mayer, L.A. *et al.* 2012. The International Bathymetric Chart of the Arctic Ocean (IBCAO) Version 3.0. *Geophysical Research Letters*, **39**, L12609, https://doi.org/10.1029/2012GL052219

Japsen, P. & Chalmers, J.A. 2000. Neogene uplift and tectonics around the North Atlantic: overview. *Global and Planetary Change*, **24**, 165–173, https://doi.org/10.1016/S0921-8181(00)00006-0

Japsen, P., Chalmers, J.A., Green, P.F. & Bonow, J.M. 2012. Elevated, passive continental margins: not rift shoulders, but expressions of episodic, post-rift burial and exhumation. *Global and Planetary Change*, **90–91**, 73–86, https://doi.org/10.1016/j.gloplacha.2011.05.004

Jicha, B.R., Scholl, D.W., Singer, B.S., Yogodzinski, G.M. & Kay, S.M. 2006. Revised age of Aleutian Island Arc formation implies high rate of magma production. *Geology*, **34**, 661–664.

Jones, S.M., White, N. & Maclennan, J. 2002. V-shaped ridges around Iceland: Implications for spatial and temporal patterns of mantle convection. *Geochemistry, Geophysics, Geosystems*, **3**, 1059, https://doi.org/10.1029/2002GC000361

Kimbell,G.S., Stewart, M.A. *et al.* 2016. Controls on the location of compressional deformation on the

NW European margin. *In*: Péron-Pinvidic, G., Hopper, J.R., Stoker, T., Gaina, C., Doornebal, H., Funck, T. & Árting, U. (eds) *The NE Atlantic Region: A Reappraisal of Crustal Structure, Tectonostratigraphy and Magmatic Evolution*. Geological Society, London, **447**. First published online August 12, 2016, https://doi.org/10.1144/SP447.3

Kristoffersen, Y. & Talwani, M. 1974. Extinct triple-junction south of Greenland. *Transactions of the American Geophysical Union*, **55**, 295–295.

Liu, L.J., Gurnis, M., Seton, M., Saleeby, J., Müller, R.D. & Jackson, J.M. 2010. The role of oceanic plateau subduction in the Laramide orogeny. *Nature Geoscience*, **3**, 353–357, https://doi.org/10.1038/NGEO829

Livaccari, R.F., Burke, K. & Sengor, A.M.C. 1981. Was the Laramide orogeny related to subduction of an oceanic plateau? *Nature*, **289**, 276–278.

Lourens, L., Hilgen, F., Laskar, J., Shackleton, N. & Wilson, D. 2004. The neogene period. *In*: Gradstein, F., Ogg, J. & Smith, A. (eds) *A Geologic Time Scale 2004*. Cambridge University Press, Cambridge, UK, 409–440.

Lundin, E. & Doré, A.G. 2002. Mid-Cenozoic post-breakup deformation in the 'passive' margins bordering the Norwegian–Greenland Sea. *Marine and Petroleum Geology*, **19**, 79–93, https://doi.org/10.1016/S0264-8172(01)00046-0

Matthews, K.J., Müller, R.D., Wessel, P. & Whittaker, J.M. 2011. The tectonic fabric of the ocean basins. *Journal of Geophysical Research: Solid Earth*, **116**, B12109, https://doi.org/10.1029/2011JB008413

Maus, S., Barckhausen, U. *et al.* 2009. EMAG2: a 2-arc min resolution Earth Magnetic Anomaly Grid compiled from satellite, airborne, and marine magnetic measurements. *Geochemistry Geophysics Geosystems*, **10**, Q08005, https://doi.org/10.1029/2009GC002471

Merkouriev, S. & DeMets, C. 2008. A high-resolution model for Eurasia-North America plate kinematics since 20 Ma. *Geophysical Journal International*, **173**, 1064–1083, https://doi.org/10.1111/j.1365-246X.2008.03761.x

Merkouriev, S. & DeMets, C. 2014. High-resolution Quaternary and Neogene reconstructions of Eurasia-North America plate motion. *Geophysical Journal International*, **198**, 366–384, https://doi.org/10.1093/gji/ggu142

Merkur'ev, S.A., Demets, C. & Gurevich, N.I. 2009. Geodynamic evolution of crust accretion at the axis of the Reykjanes Ridge, Atlantic Ocean. *Geotectonics*, **43**, 194–207.

Morgan, W.J. 1972. Deep mantle convection plumes and plate motions. *American Association of Petroleum Geologists Bulletin*, **56**, 203–213.

Mosar, J., Lewis, G. & Torsvik, T.H. 2002. North Atlantic sea-floor spreading rates: implications for the Tertiary development of inversion structures of the Norwegian–Greenland Sea. *Journal of the Geological Society, London*, **159**, 503–515, https://doi.org/10.1144/0016-764901-135

Müller, R.D., Sdrolias, M., Gaina, C. & Roest, W.R. 2008. Age, spreading rates, and spreading asymmetry of the world's ocean crust. *Geochemistry, Geophysics, Geosystems*, **9**, Q04006, https://doi.org/10.1029/2007GC001743

Nasuti, A. & Olesen, O. 2014. Chapter 4: Magnetic data. *In*: Hopper, J.R., Funck, T., Stoker, T., Arting, U., Peron-Pinvidic, G., Doornebal, H. & Gaina, C. (eds) *Tectonostratigraphic Atlas of the North-East Atlantic Region*. Geological Survey of Denmark and Greenland (GEUS), Copenhagen, Denmark, 41–51.

Ogg, J.G. 2012. The Geomagnetic Polarity Timescale. *In*: Gradstein, F.M., Ogg, J.G., Schmitz, M. & Ogg, G. (eds) *The Geologic Time Scale 2012*. Elsevier, Amsterdam, 85–115.

Olesen, O., Bronner, M. *et al.* 2010. New aeromagnetic and gravity compilations from Norway and adjacent areas: methods and applications. *In*: Vining, B. & Pickering, S.C. (eds) *From Mature Basins to New Frontiers: Proceedings of the 7th Petroleum Geology Conference*. Geological Society, London, 559–586, https://doi.org/10.1144/0070559

Parnell-Turner, R., White, N., Henstock, T., Murton, B., Maclennan, J. & Jones, S.M. 2014. A continuous 55-million-year record of transient mantle plume activity beneath Iceland. *Nature Geoscience*, **7**, 914–919.

Patriat, P. & Achache, J. 1984. India-Eurasia collision chronology has implications for crustal shortening and driving mechanisms of plates. *Nature*, **311**, 615–621.

Peron-Pinvidic, G., Gernigon, L., Gaina, C. & Ball, P. 2012. Insights from the Jan Mayen system in the Norwegian–Greenland sea – I. Mapping of a microcontinent. *Geophysical Journal International*, **191**, 385–412, https://doi.org/10.1111/j.1365-246X.2012.05639.x

Piepjohn, K., von Gosen, W., Laufer, A., McClelland, W.C. & Estrada, S. 2013. Ellesmerian and Eurekan fault tectonics at the northern margin of Ellesmere Island (Canadian High Arctic). *Zeitschrift der Deutschen Gesellschaft für Geowissenschaften*, **164**, 81–105, https://doi.org/10.1127/1860-1804/2013/0007

Pitman, W.C., III & Talwani, M. 1972. Seafloor spreading in the North Atlantic. *Geological Society of America Bulletin*, **83**, 619–646, https://doi.org/10.1130/0016-7606(1972)83[619:SSITNA]2.0.CO;2

Price, S.P. & Whitham, A.G. 1997. Exhumed hydrocarbon traps in East Greenland: analogs for the Lower–Middle Jurassic play of northwest Europe. *American Association of Petroleum Geologists Bulletin*, **81**, 196–221.

Ritchie, J.D., Johnson, H. & Kimbell, G.S. 2003. The nature and age of Cenozoic contractional deformation within the NE Faroe–Shetland Basin. *Marine and Petroleum Geology*, **20**, 399–409, https://doi.org/10.1016/S0264-8172(03)00075-8

Roest, W.R. & Srivastava, S.P. 1989. Seafloor spreading in the Labrador Sea: a new reconstruction. *Geology*, **17**, 1000–1004.

Sandwell, D.T. & Smith, W.H.F. 2009. Global marine gravity from retracked Geosat and ERS-1 altimetry: ridge segmentation v. spreading rate. *Journal of Geophysical Research: Solid Earth*, **114**, B01411, https://doi.org/10.1029/2008JB006008

Saunders, A.D., Jones, S.M., Morgan, L.A., Pierce, K.L., Widdowson, M. & Xu, Y.G. 2007. Regional uplift associated with continental large igneous provinces: the roles of mantle plumes and the lithosphere. *Chemical Geology*, **241**, 282–318, https://doi.org/10.1016/j.chemgeo.2007.01.017

Schildgen, T.F., Yildirim, C., Cosentino, D. & Strecker, M.R. 2014. Linking slab break-off, Hellenic trench retreat, and uplift of the Central and Eastern Anatolian plateaus. *Earth-Science Reviews*, **128**, 147–168, https://doi.org/10.1016/j.earscirev.2013.11.006

Seton, M., Müller, R.D. *et al.* 2012. Global continental and ocean basin reconstructions since 200 Ma. *Earth-Science Reviews*, **113**, 212–270, https://doi.org/10.1016/j.earscirev.2012.03.002

Sigloch, K. 2011. Mantle provinces under North America from multifrequency P wave tomography. *Geochemistry, Geophysics, Geosystems*, **12**, Q02W08, https://doi.org/10.1029/2010gc003421

Sigloch, K., McQuarrie, N. & Nolet, G. 2008. Two-stage subduction history under North America inferred from multiple-frequency tomography. *Nature Geoscience*, **1**, 458–462, https://doi.org/10.1038/ngeo231

Skogseid, J. & Eldholm, O. 1987. Early Cenozoic crust at the Norwegian continental-margin and the conjugate Jan-Mayen Ridge. *Journal of Geophysical Research: Solid Earth and Planets*, **92**, 11,471–11,491, https://doi.org/10.1029/JB092iB11p11471

Skogseid, J., Planke, S., Faleide, J.I., Pedersen, T., Eldholm, O. & Neverdal, F. 2000. NE Atlantic continental rifting and volcanic margin formation. *In*: Nøttvedt, A. (ed.) *Dynamics of the Norwegian Margin*. Geological Society, London, Special Publications, **167**, 295–326, https://doi.org/10.1144/GSL.SP.2000.167.01.12

Srivastava, S.P. & Tapscott, C.R. 1986. Plate kinematics of the North Atlantic. *In*: Vogt, P.R. & Tucholke, B.E. (eds) *The Western North Atlantic Region*. Geological Society of America, Boulder, CO, 379–405.

Stoker,M., Stewart, M. *et al.* 2016. An overview of the Upper Paleozoic–Mesozoic stratigraphy of the NE Atlantic region. *In*: Péron-Pinvidic, G., Hopper, J.R., Stoker, T., Gaina, C., Doornebal, H., Funck, T. & Árting, U. (eds) *The NE Atlantic Region: A Reappraisal of Crustal Structure, Tectonostratigraphy and Magmatic Evolution*. Geological Society, London, **447**. First published online August 11, 2016, https://doi.org/10.1144/SP447.2

Stoker, M.S., Holford, S.P., Hillis, R.R., Green, P.F. & Duddy, I.R. 2010. Cenozoic post-rift sedimentation off northwest Britain: recording the detritus of episodic uplift on a passive continental margin. *Geology*, **38**, 595–598, https://doi.org/10.1130/G30881.1

St-Onge, M.R., van Gool, J.A.M., Garde, A.A. & Scott, D.J. 2009. Correlation of Archaean and Palaeoproterozoic units between northeastern Canada and western Greenland: constraining the pre-collosional upper plate accretionary history of the Trans- Hudson orogen. *In*: Cawood, P.A. & Kroner, A. (eds) *Earth Accretionary Systems in Space and Time*. Geological Society, London, Special Publications, **318**, 193–235.

Storey, M., Duncan, R.A. & Tegner, C. 2007. Timing and duration of volcanism in the North Atlantic Igneous Province: implications for geodynamics and links to the Iceland hotspot. *Chemical Geology*, **241**, 264–281, https://doi.org/10.1016/j.chemgeo.2007.01.016

Suess, E. 1893. Are ocean depths permanent? *Natural Science: A Monthly Review of Scientific Progress*, **2**, 180–187.

Talwani, M. & Eldholm, O. 1977. Evolution of the Norwegian-Greenland Sea. *Geological Society American Bulletin*, **88**, 969–999, https://doi.org/10.1130/0016-7606

Torsvik, T.H., Smethurst, M.A. *et al.* 1996. Continental break-up and collision in the Neoproterozoic and Palaeozoic – a tale of Baltica and Laurentia. *Earth-Science Reviews*, **40**, 229–258.

Torsvik, T.H., Müller, R.D., Van der Voo, R., Steinberger, B. & Gaina, C. 2008. Global plate motion frames: toward a unified model. *Reviews of Geophysics*, **46**, 1–44, https://doi.org/10.1029/2007RG000227

Torsvik, T.H., Amundsen, H.E.F. *et al.* 2015. Continental crust beneath southeast Iceland. *Proceedings of the National Academy of Sciences of the United States of America*, **112**, E1818–E1827, https://doi.org/10.1073/pnas.1423099112

Tuitt, A., Underhill, J.R., Ritchie, J.D., Johnson, H. & Hitchen, K. 2010. Timing, controls and consequences of compression in the Rockall–Faroe area of the NE Atlantic Margin. *In*: Vining, B. & Pickering, S.C. (eds) *From Mature Basins to New Frontiers: Proceedings of the 7th Petroleum Geology Conference*. Geological Society, London, 963–977, https://doi.org/10.1144/0070963

Verhoef, J., Roest, W.R., Macnab, R. & Arkani, H.J. 1996. *Magnetic Anomalies of the Arctic and North Atlantic Oceans and Adjacent Areas*. Geological Survey of Canada, Open File, **3125** (CD compilation).

Vogt, D. 1971. Asthenosphere motion recorded by the ocean floor south of Iceland. *Earth and Planetary Science Letters*, **13**, 153–160.

Vogt, P.R. 1986. Magnetic anomalies of the North Atlantic Ocean. *In*: Vogt, P.R. & Tucholke, B.E. (eds) *The Western North Atlantic Region*. Geological Society of America, Boulder, CO, 229–256.

Vogt, P.R. & Avery, O.E. 1974. Detailed magnetic surveys in the northeast Atlantic and Labrador Sea. *Journal of Geophysical Research*, **79**, 363–389, https://doi.org/10.1029/JB079i002p00363

White, R.S., Bown, J.W. & Smallwood, J.R. 1995. The temperature of the Iceland plume and origin of outward-propagating V-shaped ridges. *Journal of the Geological Society, London*, **152**, 1039–1045, https://doi.org/10.1144/GSL.JGS.1995.152.01.26

Yamato, P., Husson, L., Becker, T.W. & Pedoja, K. 2013. Passive margins getting squeezed in the mantle convection vice. *Tectonics*, **32**, 1559–1570, https://doi.org/10.1002/2013TC003375

Yin, A. 2010. Cenozoic tectonic evolution of Asia: a preliminary synthesis. *Tectonophysics*, **488**, 293–325, https://doi.org/10.1016/j.tecto.2009.06.002

Seamounts and oceanic igneous features in the NE Atlantic: a link between plate motions and mantle dynamics

CARMEN GAINA[1]*, ANETT BLISCHKE[2], WOLFRAM H. GEISSLER[3], GEOFFREY S. KIMBELL[4] & ÖGMUNDUR ERLENDSSON[5]

[1]*Centre for Earth Evolution and Dynamics (CEED), University of Oslo, Sem Sælands vei 24, PO Box 1048, Blindern, NO-0316 Oslo, Norway*

[2]*Iceland GeoSurvey, Branch at Akureyri, Rangárvöllum, 602 Akureyri, Iceland*

[3]*Alfred Wegener Institute, Helmholz Centre for Polar and Marine Research, Am Alten Hafen 26, 27568 Bremerhaven, Germany*

[4]*British Geological Survey, Keyworth, Nottingham, NG12 5GG, UK*

[5]*Iceland GeoSurvey, Grensásvegi 9, 108 Reykjavík, Iceland*

**Correspondence: carmen.gaina@geo.uio.no*

Abstract: A new regional compilation of seamount-like oceanic igneous features (SOIFs) in the NE Atlantic points to three distinct oceanic areas of abundant seamount clusters. Seamounts on oceanic crust dated 54–50 Ma are formed on smooth oceanic basement, which resulted from high spreading rates and magmatic productivity enhanced by higher than usual mantle plume activity. Late Eocene–Early Miocene SOIF clusters are located close to newly formed tectonic features on rough oceanic crust in the Irminger, Iceland and Norway basins, reflecting an unstable tectonic regime prone to local readjustments of mid-ocean ridge and fracture zone segments accompanied by extra igneous activity. A SOIF population observed on Mid-Miocene–Present rough oceanic basement in the Greenland and Lofoten basins, and on conjugate Kolbeinsey Ridge flanks, coincides with an increase in spreading rate and magmatic productivity. We suggest that both tectonic/kinematic and magmatic triggers produced Mid-Miocene–Present SOIFs, but the Early Miocene westwards ridge relocation may have played a role in delaying SOIF formation south of the Jan Mayen Fracture Zone. We conclude that Iceland plume episodic activity combined with regional changes in relative plate motion led to local mid-ocean ridge readjustments, which enhanced the likelihood of seamount formation.

Supplementary material: Figures detailing NE Atlantic seamounts and SOIF distribution, and the location of earthquake epicentres are available at https://doi.org/10.6084/m9.figshare.c.3459729

The NE Atlantic oceanic basins have been formed since Early Eocene times following the break-up between Eurasia and Greenland. Various types of volcanic edifices (including seamounts) were emplaced on stretched continental crust before final break-up and seafloor spreading (Jones *et al.* 1994; Marty *et al.* 1998; O'Connor *et al.* 2000). The formation of oceanic crust was preceded by high magmatic activity, which resulted in additional igneous material being emplaced at the base and on top of stretched continental margins (Storey *et al.* 2007). Seamount volcanism and the emplacement of igneous centres on oceanic crust continued after seafloor spreading was established in various basins between Greenland and Eurasia. A considerable number of volcanic edifices have been identified in the NE Atlantic, mainly on remote sensing data including bathymetry and gravity data derived from satellite altimetry (e.g. Hillier & Watts 2007; Kim & Wessel 2011; Yesson *et al.* 2011) (Fig. 1).

Seamount volcanism is attributed to magmatic processes connected to the formation of new ocean floor/oceanic crust (seafloor spreading), or to the modification of this crust by subsequent intra-plate volcanism. A classic example of intra-plate volcanism is the plume-related creation of linear chains of age-progressing volcanic edifices on oceanic or continental crust (e.g. Morgan 1971). Intra-plate volcanism may also be the result of local processes such as lithosphere cracking or melt extraction from heterogeneous mantle (e.g. Forsyth *et al.* 2006), small-scale sublithospheric convection (e.g. Ballmer

From: Péron-Pinvidic, G., Hopper, J. R., Stoker, M. S., Gaina, C., Doornenbal, J. C., Funck, T. & Árting, U. E. (eds) 2017. *The NE Atlantic Region: A Reappraisal of Crustal Structure, Tectonostratigraphy and Magmatic Evolution*. Geological Society, London, Special Publications, **447**, 419–442.
First published online September 8, 2016, https://doi.org/10.1144/SP447.6

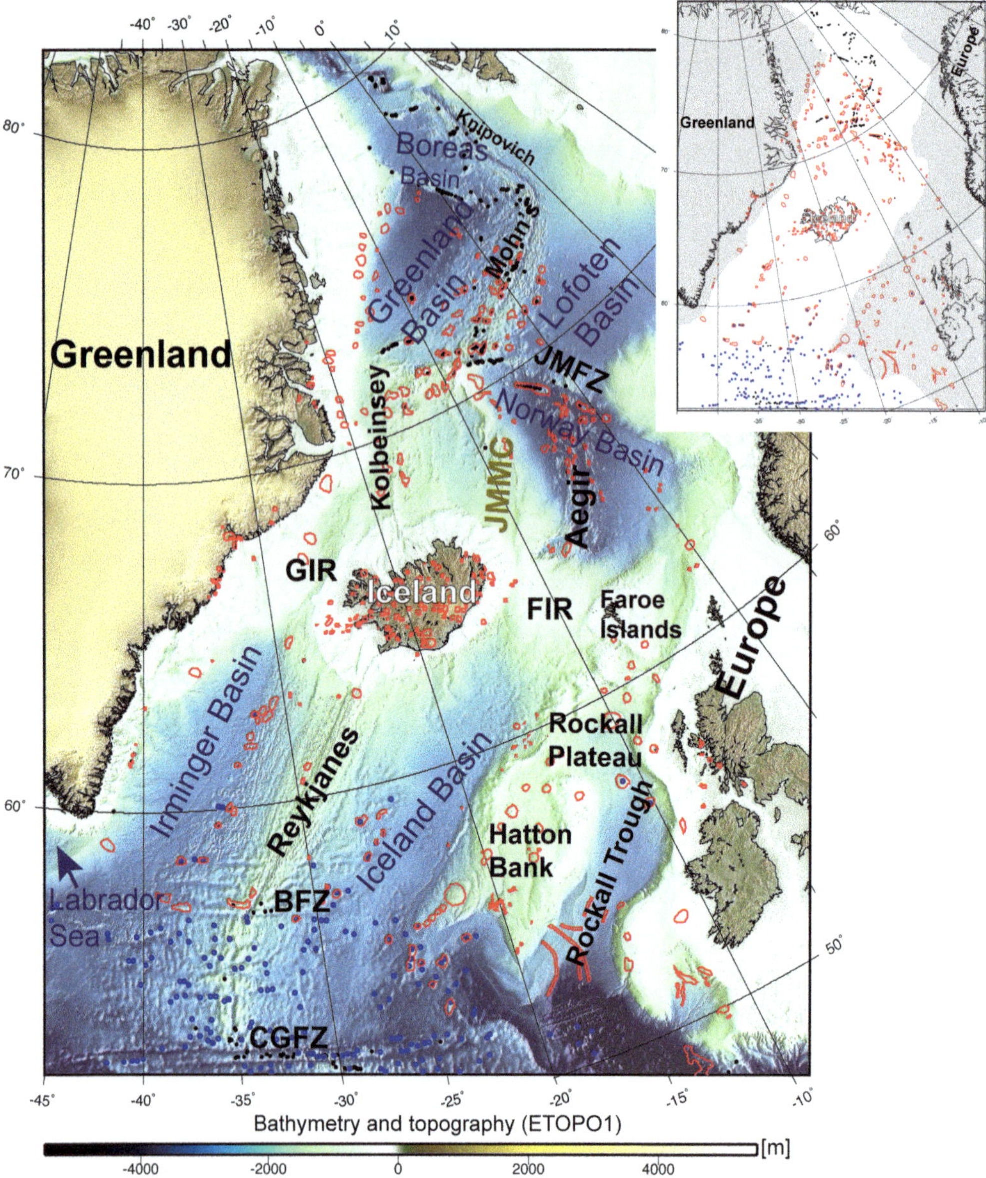

Fig. 1. Distribution of volcanic edifices (red, this study; blue, Kim & Wessel 2011; black, Yesson *et al.* 2011) in the NE Atlantic region superimposed on bathymetry (ETOPO1: Amante & Eakins 2009). Abbreviations are: BFZ, Bight Fracture Zone; CGFZ, Charlie Gibbs Fracture Zone; FIR, Faroe Iceland Ridge; GIR, Greenland Iceland Ridge; JMFZ, Jan Mayen Fracture Zone. Inset to the figure shows the distribution of seamounts on oceanic (white) and continental and extended continental crust (grey).

et al. 2009), or shear-induced melting of low-viscosity pockets of asthenospheric mantle located along the base of the lithosphere (Conrad *et al.* 2010).

This study aims to evaluate the correlations between a new database of oceanic volcanic features (seamounts and other small igneous edifices), and the oceanic crust morphology and evolution as established within the international NAG-TEC project (Hopper *et al.* 2014). We will first present the current knowledge of seamount and volcanic feature distribution in the NE Atlantic. The occurrence of these volcanic features on oceanic crust of various ages and structure is subsequently described. Possible links between NE Atlantic variations in seafloor spreading, mantle dynamics and seamount formation since the Eocene is also discussed. Our results may help in understanding the spatial and temporal interplay between volcanism and tectonics in a region that has also been heavily influenced by a pulsating mantle plume since the inception of oceanic crust formation.

Regional distribution of seamounts and volcanic edifices in NE Atlantic oceanic basins

According to the International Hydrographic Organization (IHO 1994, pages 211 and 121), a seamount is 'an isolated or comparatively isolated elevation rising 1000 m or more from the seafloor and of limited extent across the summit', whereas a knoll is 'a relatively small isolated elevation of a rounded shape rising less than 1000 m from the seafloor and of limited extent across the summit'. The first regional count of seamounts in the North Atlantic was carried out by Epp & Smoot (1989), who used multibeam data to identify approximately 800 seamounts between the equator and Iceland. More recently, seamount-like features interpreted on ship-track bathymetry data (Hillier & Watts 2007), gridded bathymetric data (Yesson *et al.* 2011), and satellite-derived gravity anomaly data and its vertical gradients (Wessel 2001; Kim & Wessel 2011) were catalogued in regional and global databases.

The igneous centres from the NE Atlantic identified in the NAG-TEC study (Hopper *et al.* 2014) is a collection of 429 features that has been divided into six subunits: offshore seamounts; igneous complexes; inactive calderas; active calderas; inactive central volcanoes; and active central volcanoes (Horni *et al.*, this volume, in review). They occur both onshore and offshore, on oceanic and on continental crust (Figs 1 & 2). An overview of the complete NE Atlantic igneous centre compilation is presented in Hopper *et al.* (2014).

In this contribution, we will focus on seamounts and igneous edifices situated on NE Atlantic oceanic crust (Fig. 1). These features were identified on published multichannel and single-channel seismic reflection profiles as mounded or bank features with dipping flanks and commonly erosional features on top (Fig. 3). In areas with no seismic control, bathymetry (SRTM30_PLUS: Becker *et al.* 2009), gravity (Andersen 2010) and magnetic gridded data (Gaina *et al.*, this volume, in review) were used for locating seamounts and other volcanic-like edifices, which are usually characterized by circular or elliptical anomalies in potential field data (Fig. 2). The seamount-like features were first manually identified on bathymetry and gravity data, and the interpretation was cross-checked with the magnetic anomaly maps. It has been assumed that a magnetic source will result in a distinct magnetic anomaly, and therefore only features with clear signatures on bathymetry, gravity and magnetic data have been considered in this database. Only features that rise more than 500 m above their surroundings and have a subcircular or well-defined base were included in the NAG-TEC database. Their structure varies and some are flat-topped, while others are more peaked. In addition, the volcanic features described in the EarthRef database (Earthref.org/SC) were also included (Fig. 3). Altogether, 175 identified features have an elevation of more than 500 m (therefore they fall into the 'knoll' category), but only 12 of them are over 1000 m in height (and can be called 'seamounts'). We suggest labelling the volcanic edifices discussed in this paper as 'seamount-like oceanic igneous features' (SOIFs), an acronym that is used in the rest of the paper.

For studying the geodynamic context of SOIF formation in various sub-basins of the NE Atlantic, we have scrutinized four main regions that display various volcanic activity patterns (Fig. 4). The structure and evolution of these sub-basins differ depending on their geographical location relative to the Iceland plume and on the proximity to additional plate boundaries – including the ones created by the formation of the Jan Mayen microcontinent (JMMC) (Fig. 1) (e.g. Gaina *et al.* 2009). In this study, we do not discuss in detail the magmatic history of Iceland or volcanism formed on extended continental crust.

Region I: south of Iceland and north of the Bight Fracture Zone

The identification of SOIFs in region I (Fig. 5a) is based on bathymetry (in most cases, satellite-derived altimetry and multibeam data for a few cases: see EarthRef.org/SC) only. The sediment thickness (Funck *et al.* 2014) in these basins is less than 2 km (Fig. 2d). The majority of SOIFs are located in three distinct areas. Few edifices (seven out of 42) are on Early Eocene crust (*c.* 52–54 Ma), close to the identified continent–ocean boundary (COB). Most of the volcanic features (28 out of 42) are located on Late Eocene–Oligocene crust (38–20 Ma), and are distributed almost symmetrically on both conjugate oceanic ridge flanks. SOIFs are not identified on the conjugate European flank in the northern Iceland Basin, at approximately 63° N, in a region with higher sediment thickness than on the Greenland flank.

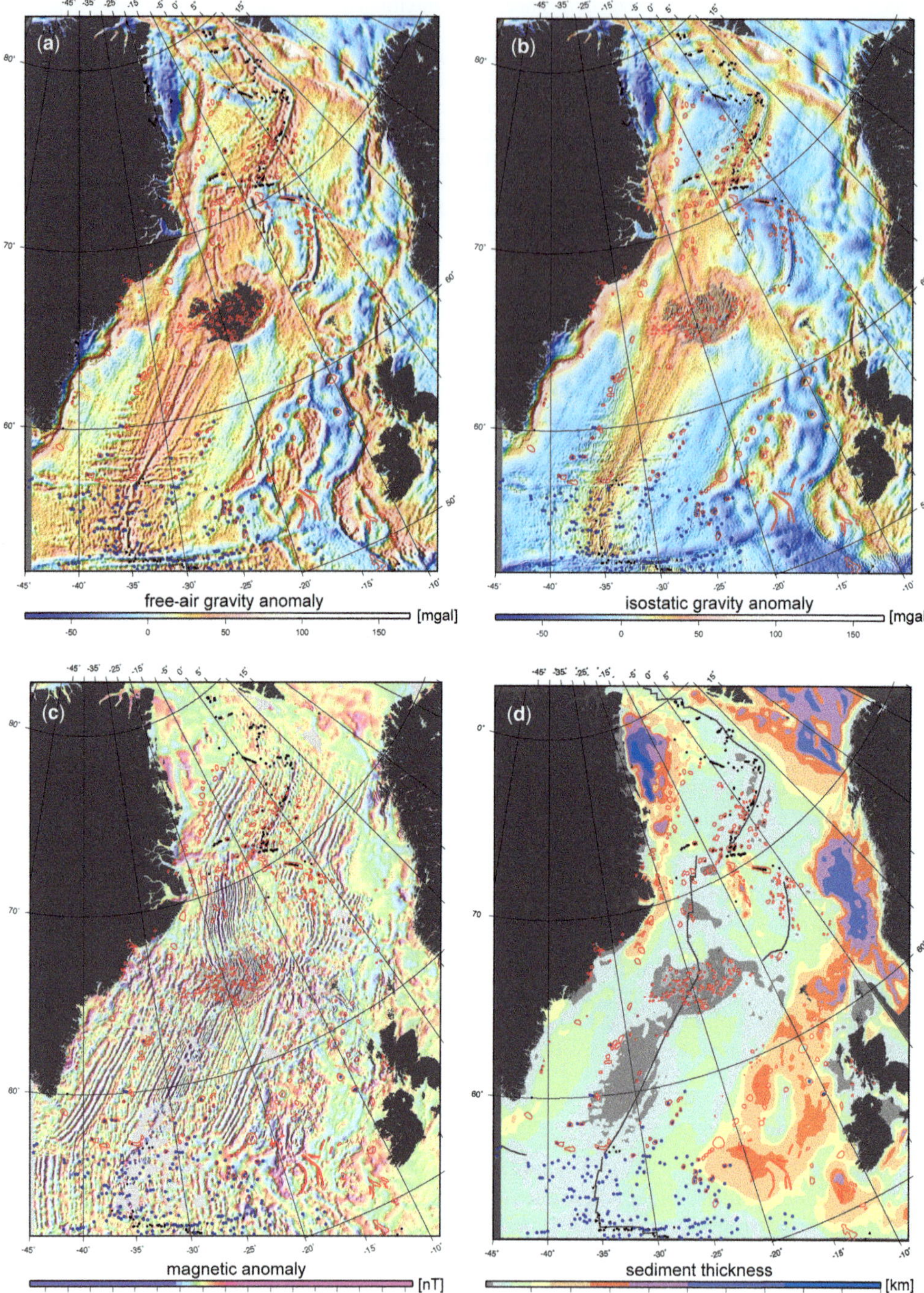

Fig. 2. (**a**) Free-air gravity (DTU10: Andersen 2010); (**b**) isostatic gravity anomaly (this was computed using the Airy–Heiskanen model, where the compensation is accomplished by variations in thickness of the constant density layers: the root is calculated using the ETOPO1 topography and bathymetry: Haase *et al.*, this volume, in press); (**c**) magnetic anomaly (Nasuti & Olesen 2014; Gaina *et al.*, this volume, in review); and (**d**) sediment thickness (Funck *et al.* 2014). Distribution of volcanic edifices as in Figure 1. Dark grey lines indicate the active and extinct plate boundaries.

Note that part of that thick sediment succession constitutes Late Miocene and younger drift deposits resulting from the onset of deep-water circulation in the NE Atlantic. Two recent seismic reflection lines crossing the northern Irminger and Iceland basins from west to east (fig. 1 in Parnell-Turner *et al.* 2015) imaged prominent high bathymetric features, including V-shaped ridges, under the sediment pile on both flanks of the Reykjavik Ridge (Parnell-Turner *et al.* 2015).

In region I, less than 20% of seamounts/volcanic edifices (seven out of 42) are on Mid-Miocene–Recent oceanic crust. Here, we do not discuss the volcanic edifices observed in the Rockall region on continental or extended continental crust: they have been described in detail in previous studies (e.g. Jones *et al.* 1994; O'Connor *et al.* 2000).

In the Irminger and Iceland basins, a clear change in the seafloor spreading regime occurred at C17 time (*c.* 38 Ma). The seafloor spreading direction changed by 20°–25° counterclockwise and the spreading rate decreased by about 30% (see Gaina *et al.*, this volume, in review). As a result, the oceanic crust was transformed from a linear, fracture-zone-free fabric to a 'stair-case'-like fabric due to the appearance of small offset fracture zones, especially in the area south of 60° N and north of the Bight Fracture Zone (Fig. 1).

A closer look at the second group of SOIFs described above reveals that the volcanic edifices are mostly elliptical in shape, and some of them coincide with the intersection between fracture zones and palaeo-mid-ocean ridges (identified as magnetic isochrons) that formed between C13 and C6. Further observations related to the oceanic crust characteristics in the regions linked to SOIF occurrences are summarized in Table 1. The crustal thickness obtained with two different methods (Fig. 5), and the seafloor spreading rates and asymmetry (Fig. 4), are described in detail in Funck *et al.* (2016) and Gaina *et al.* (this volume, in review).

Region II: the Norway Basin

The NAG-TEC SOIF database contains 34 seamounts in the Norway Basin (Fig. 5b), which were identified on the gravity, magnetics and bathymetry gridded data, with five of them cross-checked on 2D seismic reflection data. In addition, 13 SOIFs were identified as igneous centres, four of them on the eastern JMMC, in the vicinity of the COB (Peron-Pinvidic *et al.* 2012; Blischke *et al.*, this volume, in press), and therefore linked to break-up volcanism. The emplacement of these igneous centres occurred during and immediately after the initial formation of seawards-dipping reflectors (SDRs). The igneous centres cut through the SDR section and are located close to fracture/fault zones. Peron-Pinvidic *et al.* (2012) and Blischke *et al.* (this volume, in press) suggest that the igneous centres located in the vicinity of the JMMC eastern margin are related to break-up and volcanic margin formation. Six igneous centres were identified on old oceanic crust (*c.* C24) close to the JMMC and four along the Norwegian margin (Table 1).

The majority of SOIFs are on Late Eocene–Early Oligocene crust (C20–C18 to C13: i.e. 40–33 Ma), flanking the Aegir extinct spreading ridge. A small number of large, elongated seamount chains or isolated rounded seamounts are also visible along the Jan Mayen Fracture Zone (JMFZ) situated in the northern Norway Basin (Table 1). One large feature was identified at the southernmost tip of the Aegir Ridge as a possible central volcano, formed by ridge propagation just prior to its extinction (Vogt & Jung 2009).

The SOIF production in the Norway Basin may have started in post-C21 (*c.* 47 Ma) time and continued until the Early Oligocene (C13, *c.* 33 Ma). This is illustrated by the fact that only a few isolated seamounts were identified on Early Eocene crust (C24–C22). Seamounts in the Norway Basin can also be seen along the oblique SSE–NNW pseudofaults, features visible on gravity anomaly maps (e.g. Fig. 2), and described by Breivik *et al.* (2006) and Gernigon *et al.* (2012). Note that in Region II, the appearance of SOIFs in Late Eocene time coincides with a change in the spreading regime, when a drop in the spreading rate and a change in the spreading direction resulted in a fan-shaped basin geometry.

Region III: Kolbeinsey Ridge and associated oceanic basin

A continuous mid-ocean ridge (MOR) was established west of the JMMC about 20 myr ago (Nunns 1983; Kuvaas & Kodaira 1997; Gaina *et al.*, this volume, in review), and oceanic crust continued to form until today along the Kolbeinsey Ridge (Fig. 5c). A few igneous centres (five out of 22) were identified on seismic reflection, bathymetry, gravity and magnetic anomaly gridded data along the Greenland margin (and may be linked to the Oligocene break-up processes). One of these igneous centres is located on continental crust. No igneous features are visible on the conjugate western JMMC margin.

Two prominent SOIF populations are distinguished on Late Miocene–Pliocene crust (6 Ma and younger) (Fig. 5c), both of which are located in the vicinity of fracture zones (four SOIFs near the Spar Fracture Zone) or a ridge propagator tip (four SOIFs next to the southern propagator). The third distinct population (seven out of 22) is grouped SW and west of Jan Mayen Island, and includes the Eggvin Bank – a plateau with young (<1 Ma), scattered volcanic peaks (Mertz *et al.* 2004).

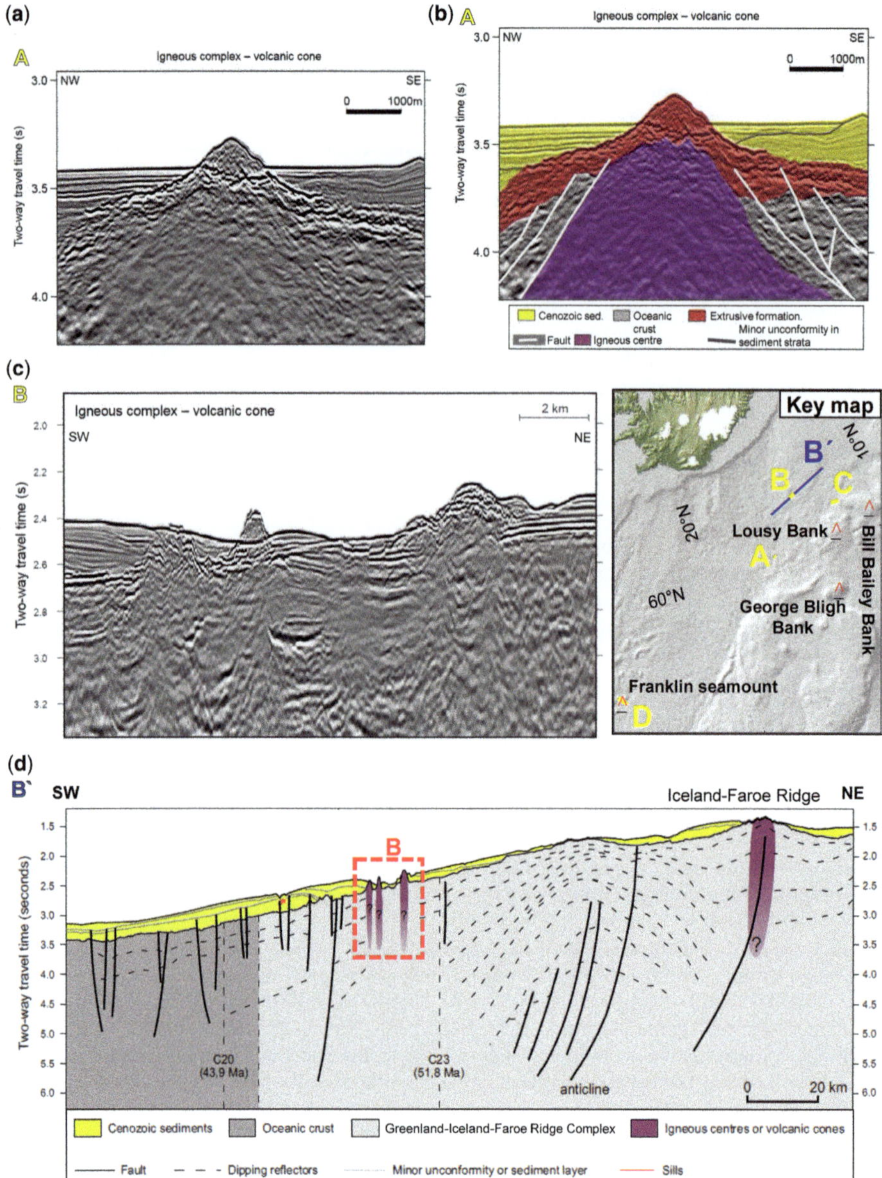

Fig. 3. Examples of seamount-like oceanic igneous features (SOIFs). These features were identified on 2D multichannel seismic reflection data (2D MCS) (National Energy Authority of Iceland; Elliott & Parson 2008) as mound or bank features with dipping flanks, sometimes with evidence of erosion at the top (**a–g**). SOIFs were first localized using gravity (Andersen 2010) and bathymetry data (SRTM30_PLUS: Becker *et al.* 2009). The key map shows the position of four seamounts registered in the EarthRef.org database (red open triangles), and the location of the 2D MCS profiles (A–D) as yellow and blue lines superimposed on a bathymetry map (SRTM30_PLUS: Becker *et al.* 2009). Seismic profiles A (in a & b) and C (in e & f) show examples of SOIFs situated on Early Eocene oceanic crust or very close to the COB. The Bill Bailey Bank (SMNT-606N-0103W from EarthRef.org) intrusive complex (line C in e & f) is an eroded seamount covered by Cenozoic sediments. The igneous centre imaged by profile B (in c) is also shown on a longer SW–NE-orientated profile (B' shown in d) that is crossing the transition from normal oceanic crust in the Iceland Basin to the thicker Iceland–Faroe Ridge. Profile D (in g) shows the Franklin Seamount (SMNT-578N-0266W from EarthRef.org). (**h**) shows the free-air gravity anomaly (Andersen 2010) in the background, the location of the Franklin Seamount and other identified SOIFs (thin black contours), and the location of profile D (thick black line).

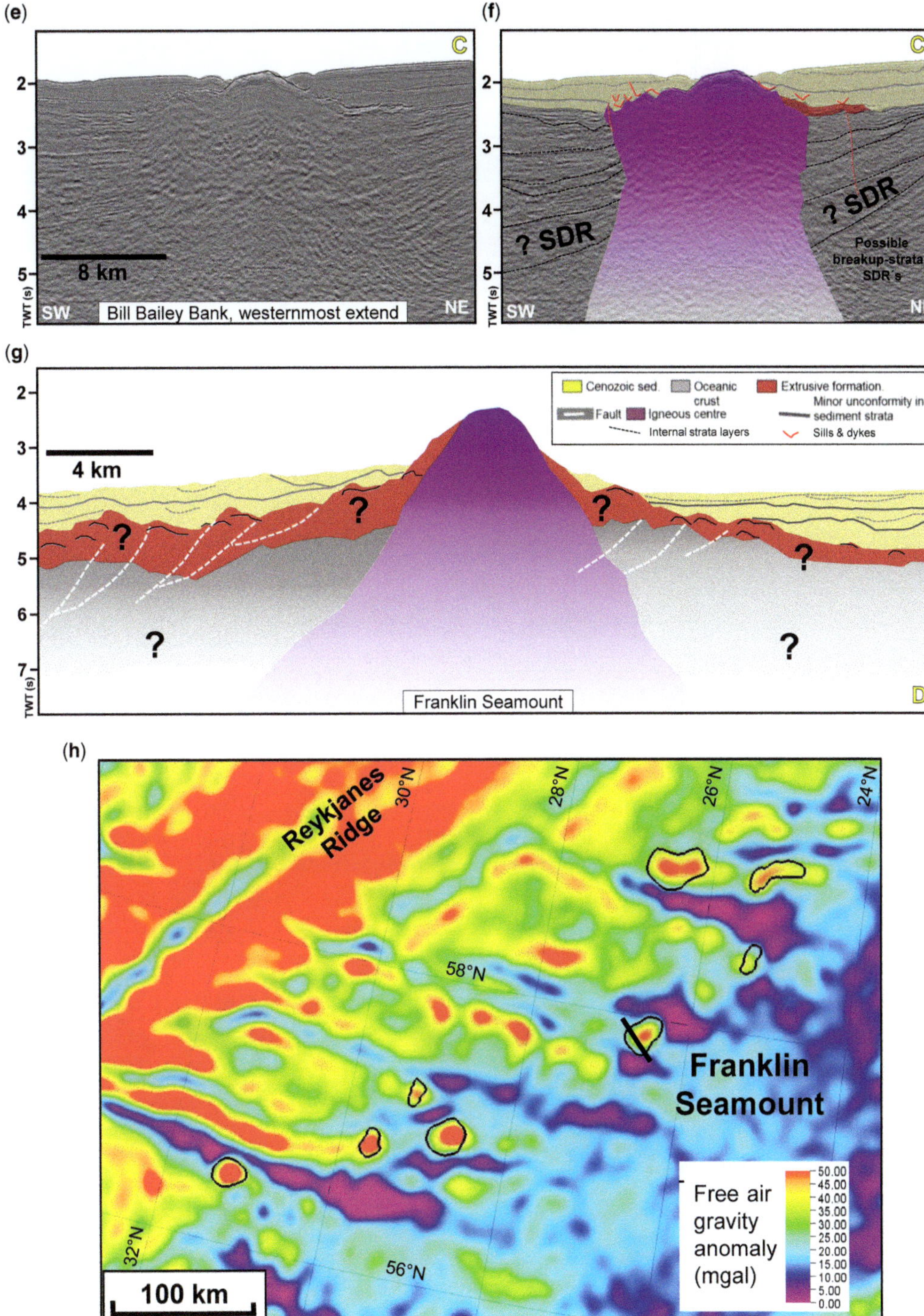

Fig. 3. *Continued.*

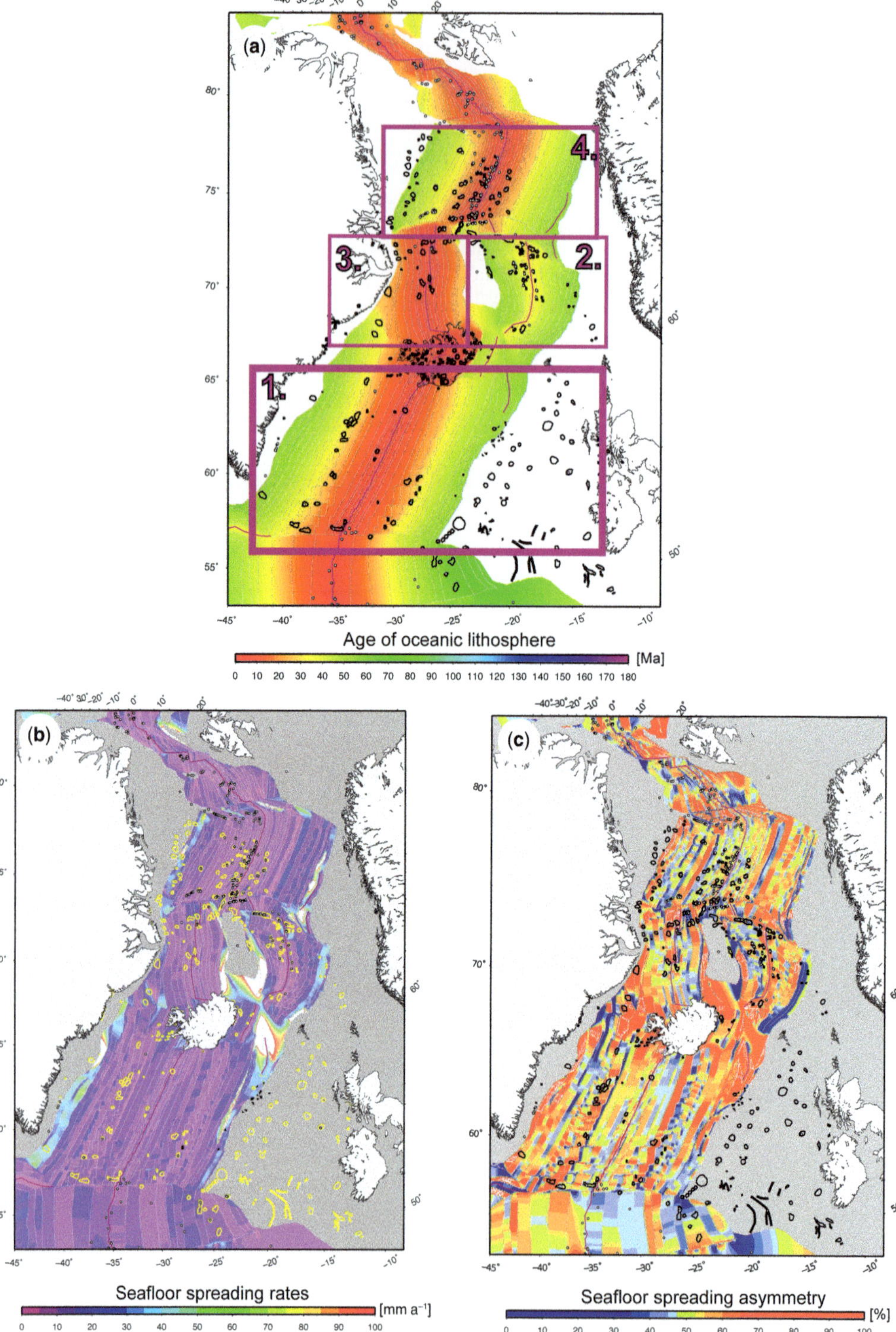

Fig. 4. Age grid (**a**), half spreading rates (**b**) and asymmetry in seafloor spreading (**c**) of the NE Atlantic oceanic crust (Gaina *et al.*, this volume. in review). The distribution of volcanic edifices is as in Figure 1. Rectangles indicate the location of the four regions discussed in the text.

Region IV: Mohn's Ridge and associated oceanic basin

The SOIFs are divided in two distinct groups within the Greenland and Lofoten basins, which have formed along the Mohn's Ridge since C24 (*c.* 54 Ma) onwards (Fig. 5d). The first group (22 out of 70) is distributed along the Greenland margin on both extended continental crust and Early Eocene oceanic crust. The second group comprises 43 out of 70 SOIFs, and is scattered on oceanic crust younger than 28 Ma on both conjugate flanks of Mohn's Ridge. Only one large seamount is located on the present-day MOR in the southern Mohn's Ridge, referred to as the Troll Wall–Soria Moria (Pedersen *et al.* 2010), and four SOIF are located on the JMFZ.

Four seamounts are outside these two SOIF groups and are located on 44–33 Ma crust on the Greenland side. One of these seamounts is the Vesteris Seamount (Cherkis *et al.* 1994; Haase & Devey 1994), a young, large intra-plate volcano of non-plume origin.

North of Region IV, the compilation by Yesson *et al.* (2011) shows a few seamounts along the Knipovich Ridge, mostly in the Boreas Basin on the Greenland Plate. The NAG-TEC study has not included this area into its SOIF database and we will not discuss them further, as there is sparse information about volcanic centres in that region.

Discussion

SOIFs and oceanic crust formation

The distribution of SOIFs in the NE Atlantic oceanic basins, and links to the age of oceanic crust, seafloor spreading rates, asymmetry of oceanic crustal accretion and oceanic crustal thickness, are summarized in Table 1. A new grid for the oceanic lithospheric age has been constructed based on updated magnetic anomaly identification in the NE Atlantic (Gaina *et al.*, this volume, in review). The oceanic lithospheric age grid model, together with rotation parameters describing the opening of the NE Atlantic, have been used to compute seafloor spreading rates, directions and deviations from symmetrical oceanic crust formation at various intervals, as constrained by the kinematic model (see Gaina *et al.*, this volume, in review) (Fig. 4). Seafloor spreading asymmetry can be described as the percentage of crustal accretion (values from 0 to 100%) on conjugate flanks along a MOR (Müller *et al.* 2008). Symmetrical seafloor spreading is expressed as 50% asymmetry, values smaller than 50% indicate less oceanic crust on one flank, which is compensated for on the conjugate flank with a crustal accretion percentage greater than 50%. Besides the age of oceanic crust, crustal thickness and seafloor spreading parameters, we also inspected the oceanic basement seismic reflection characteristics, as described by Horni *et al.* (this volume, in review). The NE Atlantic oceanic basement has been divided into six categories: smooth, transitional, rough, very rough, rubbly and igneous provinces (Funck *et al.* 2014) (Fig. 6). We observe that most SOIFs (75%) are associated with rough basement, and only a few with smooth basement. The rough basement type, as described by Horni *et al.* (this volume, in review), is present in areas with significant basement relief of the order of 1 s two-way travel time (TWT) on seismic reflection sections. They suggest that rough basement may result from tectonic processes (e.g. faulting) and volcanic processes (e.g. intrusions, seamounts or locally robust volcanism). The smooth basement type is distinguished by long continuous, high-amplitude, seismic reflections and often appears as a single reflection, although sometimes there may be packages of strong subplanar continuous reflections. The two types of basement morphology are also reflected in bathymetry (Fig. 1) and gravity data (Fig. 2).

Note that the transition from smooth to rough oceanic crust has been mainly associated with the boundary between areas affected by higher magma supply from the Iceland plume (situated on V-shaped regions south and north of Iceland) and 'colder' areas, which were less affected, or unaffected, by the hotter mantle (e.g. Poore *et al.* 2009). Hey *et al.* (2010) postulated that a seafloor spreading asymmetry-producing mechanism is rift propagation. Rift propagation produces V-shaped ridges, a 'ridge and trough' geometry within the V-shaped elevated area, and crustal thickness variations. According to Hey *et al.* (2010), this mechanism can be triggered independently of the presence of a mantle plume, and better explains the relief of V-shaped regions in the NE Atlantic. More recently, Jones *et al.* (2014) used geophysical and geochemical data and modelling to confirm the original idea of Vogt (1971) that the V-shaped ridges are generated by the plume stem pulses that spread radially from the central plume location. The model of Jones *et al.* (2014) explains both the seafloor spreading roughness and the geochemical signatures revealed by oceanic crust samples.

SOIFs or oceanic core complexes (OCC)? We indicated that some SOIFs are located at the intersection between fracture zones and former MORs (as depicted by isochrons). The occurrence of some of these features also coincides with a decrease in spreading rate (Fig. 7a–c). These conditions would suggest that the identified bathymetric features might be oceanic core complexes (OCC). Oceanic core complexes are bathymetric features composed

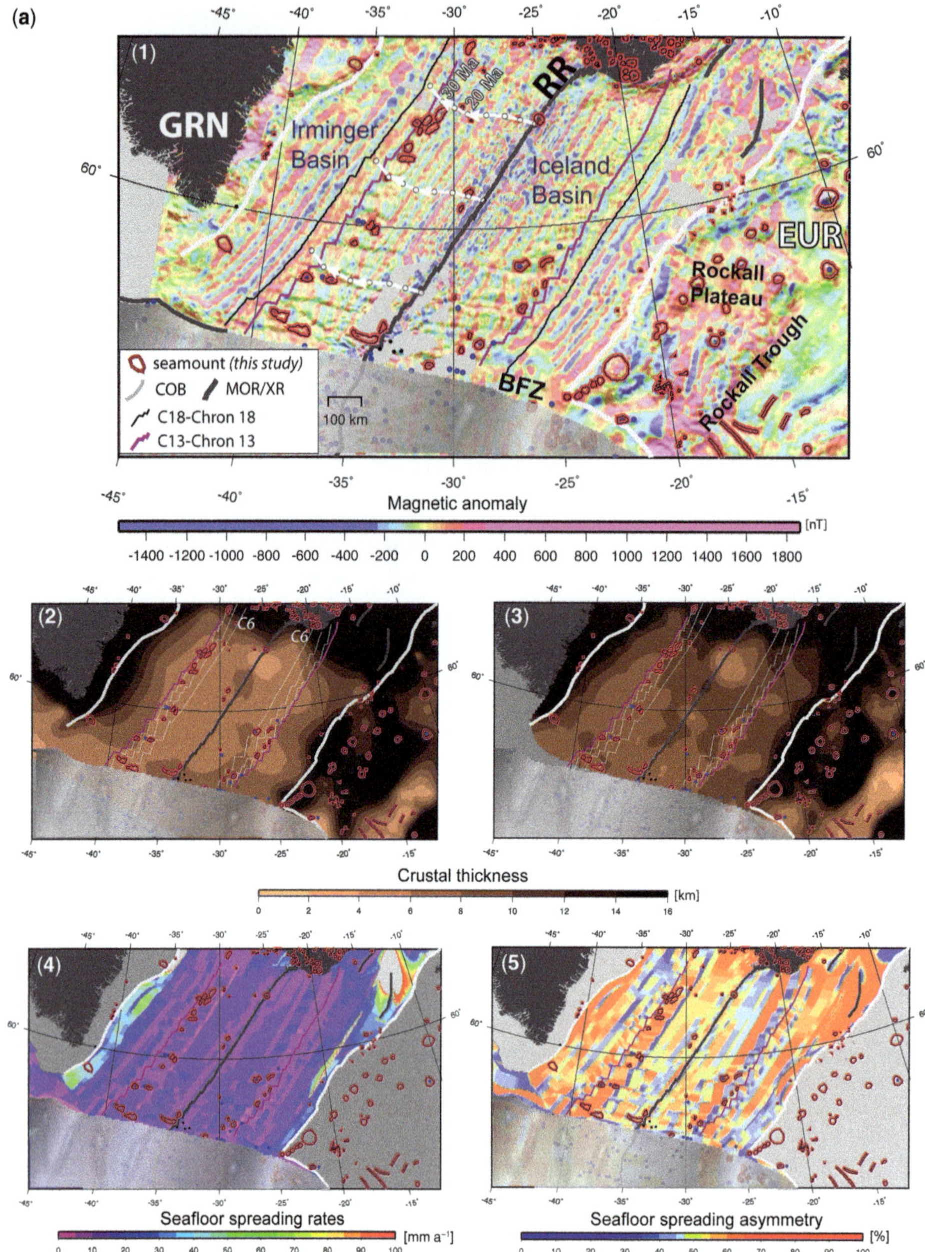

Fig. 5. Distribution of volcanic edifices (see Fig. 1) in the four NE Atlantic main regions superimposed on various geophysical data and models. (**a**) Iceland and Irminger basins (region I), (**b**) Norway Basin (region II), (**c**) east of Jan Mayen microcontinent (region III); (**d**) Lofoten and Greenland basins (region IV). Background images are: (**1**) the magnetic anomaly grid (Gaina *et al.*, this volume, in review); (**2**) the crustal thickness derived from seismic refraction data (Funck *et al.* 2016); (**3**) the crustal thickness from gravity inversion (Haase *et al.*, this volume, in press); (**4**) half seafloor spreading rates; and (**5**) seafloor spreading asymmetry (Gaina *et al.*, this volume, in review). Dark grey lines show the location of active and extinct plate boundaries; light grey is the interpreted COB. Isochron C13young (33.2 Ma) is shown in magenta, and other selected isochrons are shown as thin blue (or white) lines. White thick arrows indicate the motion of Greenland relative to the underlying mantle for the last 40 myr. Abbreviations: AR, Aegir Ridge; EUR, Europe; GRN, Greenland; JMFZ, Jan Mayen Fracture Zone; JMMC, Jan Mayen microcontinent; KnR, Knipovich Ridge; KR, Kolbeinsey Ridge; MR, Mohn's Ridge; RR, Reykjanes Ridge.

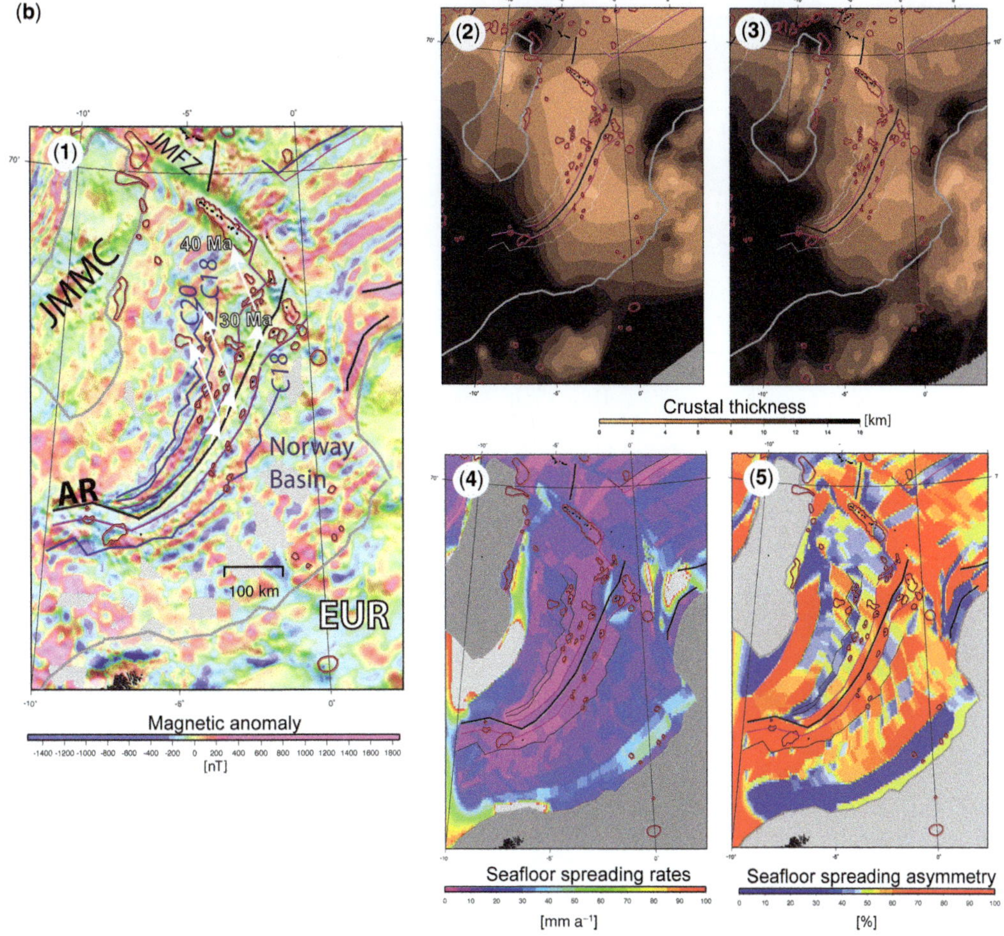

Fig. 5. *Continued.*

of mantle rocks exposed on the seafloor by large detachment faulting. These tectonic features can have lengths up to 150 km and widths up to 15 km, with a height of between 500 and 1500 km (MacLeod *et al.* 2009). It has been postulated that OCCs can be associated with serpentinized peridotites and, therefore, have very weak magnetization (e.g. Sato *et al.* 2009). The general view about OCC suggests that they form in a tectonic regime with very low magma supply (e.g. MacLeod *et al.* 2009). Detailed studies of OCCs at the Mid-Atlantic Ridge (e.g. Ildefonse *et al.* 2007; Mallows & Searle 2012) show that OCC formation is discontinued when the magma supply increases, although models (e.g. Olive *et al.* 2010) and observations (e.g. in Cayman Trough: Hayman *et al.* 2011) postulate that they can form under a spectrum of magma injection rates.

Our criteria to identify SOIFs included height above 500 m and relatively high total magnetic field values (>100 nT on the NAG-TEC magnetic map, which shows the total magnetic field 2 km upwards continued from the original measurement position). Therefore, if any of the identified SOIFs in this study happen to be an OCC, then that feature was formed in a tectonic regime able to generate rounded to elliptical bathymetric structures that are surrounded, covered or intruded by basaltic rocks that have remanent magnetization.

Previous detailed studies on seamount volcanism in the NE Atlantic focused on the enigmatic Vesteris Seamount (Mertz & Renne 1995) located in the Greenland Basin (Fig. 5d), and the igneous centres situated in the Rockall region (Fig. 5a). These seamounts have been dredged and a few petrological studies have been published (e.g. Cherkis *et al.* 1994; Haase & Devey 1994; O'Connor *et al.* 2000). The young episodic alkaline volcanism of the Vesteris Seamount has been attributed to intra-plate

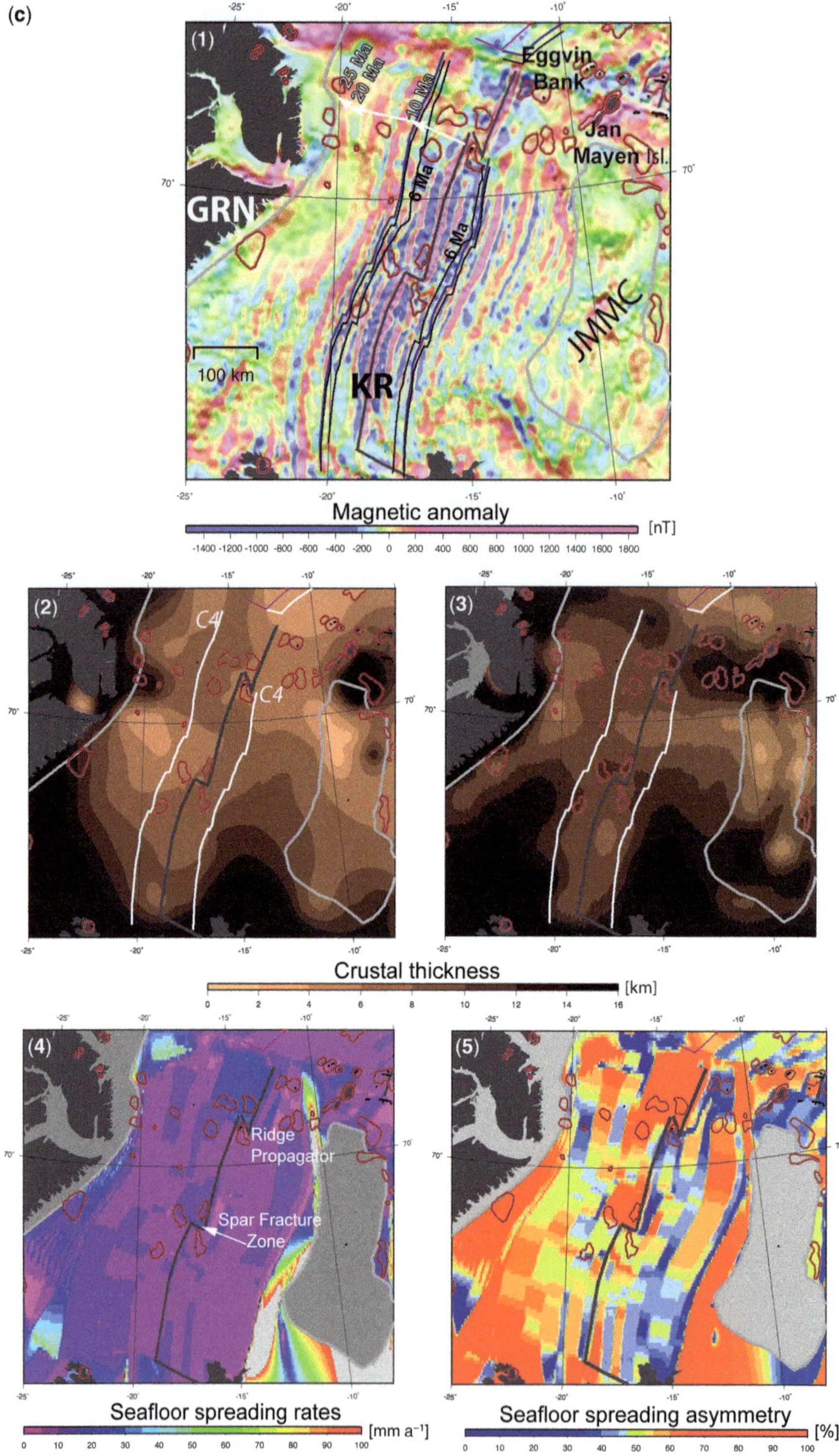

Fig. 5. *Continued.*

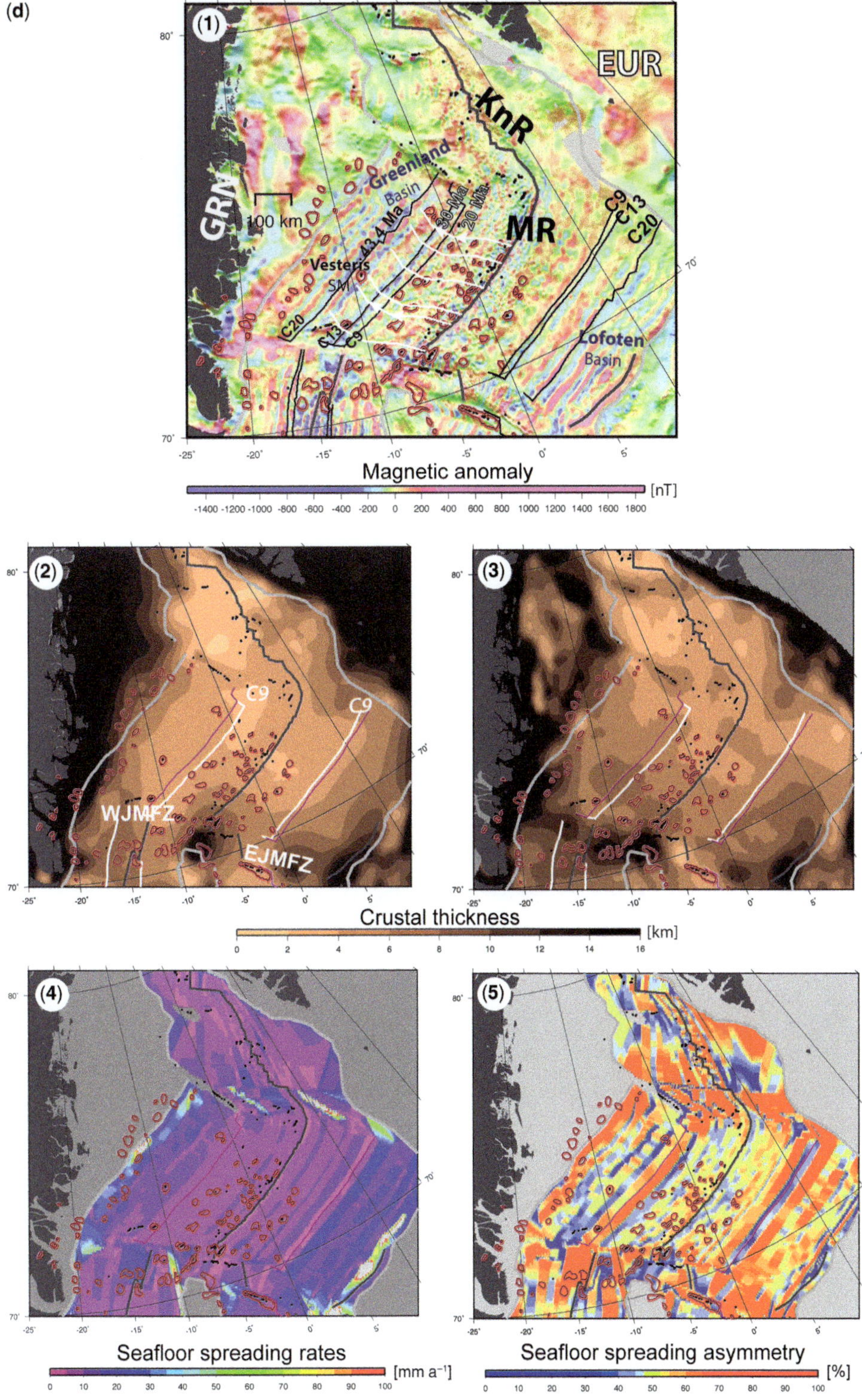

Fig. 5. *Continued.*

Table 1. *Summary of SOIF distribution in regions I–IV and underlying oceanic crust characteristics*

Region/subregion	Oceanic crust age (Ma)	Number of SOIFs	Seafloor spreading rate (mm a^{-1})	Seafloor spreading asymmetry (%)	Oceanic basement type (Funck *et al.* 2014)	Oceanic crustal thickness (Funck *et al.* 2016) (km)	Oceanic crustal thickness (Haase *et al.*, this volume, in press) (km)	Tectonic features
I-1	54–52	7	45–40	Eastern flank <50%	Smooth	>10 km	>10 km	
I-2	34–20	28	15–20	Eastern flank <50%	Rough	8–10	8–11	Intersection with fracture zones (FZ)
I-3	8–0	7	15–23	Eastern flank >50% (SOIFs associated with boundaries between excess/deficit)	Rough	6–9	8–11	
Region I total		42						
II-1	COB-54	5						
II-2	54–52	6	50–40	Eastern flank <50%	Rough Norwegian margin; smooth JMMC	8–10	10–12	Some associated with oblique FZ
II-3	47–30	33	28–18	Eastern flank >50% (SOIFs associated with excess crustal production or boundaries between excess/deficit)	rough	4–6 (seems to be confined to thinner crust)	6–8	Some associated with oblique FZ
Region II total		44						
III-1	COB-25	5	22–20	Eastern flank <50%	Smooth	10–14	10–12	
III-2	19–14	7	15–20		Rough	6–8	8–12	
III-3	6–2.6	10	21–15	Eastern flank <50% (SOIFs associated with excess crustal production or boundaries between excess/deficit)	Rough	8–10	12–14	Fracture zones, rift propagator
Region III total		22						
IV-1	COB-49	22	15–33	Eastern flank <50% (SOIF associated with excess crustal production or boundaries between excess/deficit)	Smooth	6	8–10	
IV-2	27.4–0	43	22–10	27–21 Ma eastern flank <50%; 21–0 Ma eastern flank >50% or symmetrical spreading (SOIFs associated with excess crustal production or boundaries between excess/deficit)	Rough	2–6	6–10	
Region IV total		65						

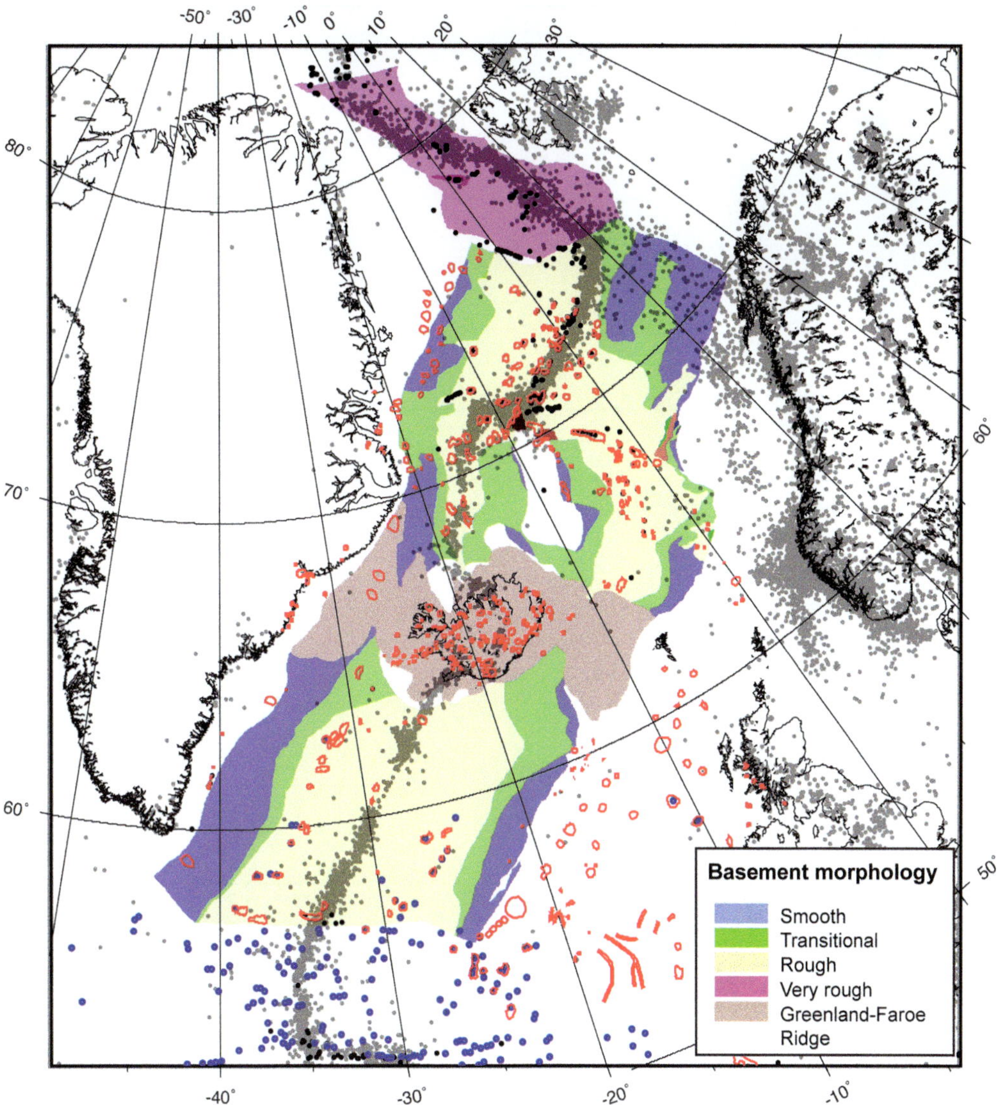

Fig. 6. Oceanic basement morphology based on seismic characteristics (Funck *et al.* 2014). The distribution of volcanic edifices is as in Figure 1. Grey dots indicate the location of earthquake epicentres (0.1–6.5 magnitude) from January 1978 to October 2012 according to the International Seismological Centre database (http://www.isc.ac.uk/).

stresses, as the volcanic edifice is situated on older oceanic crust, at the intersection of major structural features (Haase & Devey 1994). In contrast, the episodic volcanism (from the Late Cretaceous to the Mid-Eocene) that formed the igneous edifices situated on extended continental crust in the Rockall region was linked to the Iceland plume pulsations which may have occurred at 5–10 myr intervals (O'Connor *et al.* 2000). The third group of NE Atlantic seamounts/submarine volcanic edifices documented by petrological studies is located in the Eggvin Bank region, situated between Jan Mayen Island and the Kolbeinsey Ridge (Fig. 5c). The off-axis, transitional to alkaline lavas are similar to the Jan Mayen Island basalts, but the near-axis tholeiites resemble the SE Iceland lavas, indicating a change in the mantle composition (Trønnes *et al.* 1999; Mertz *et al.* 2004), or different magma sources along the Kolbeinsey Ridge.

Most of the NE Atlantic SOIFs located on oceanic crust lack absolute ages, as only few seamounts or volcanic edifices have been dredged and dated.

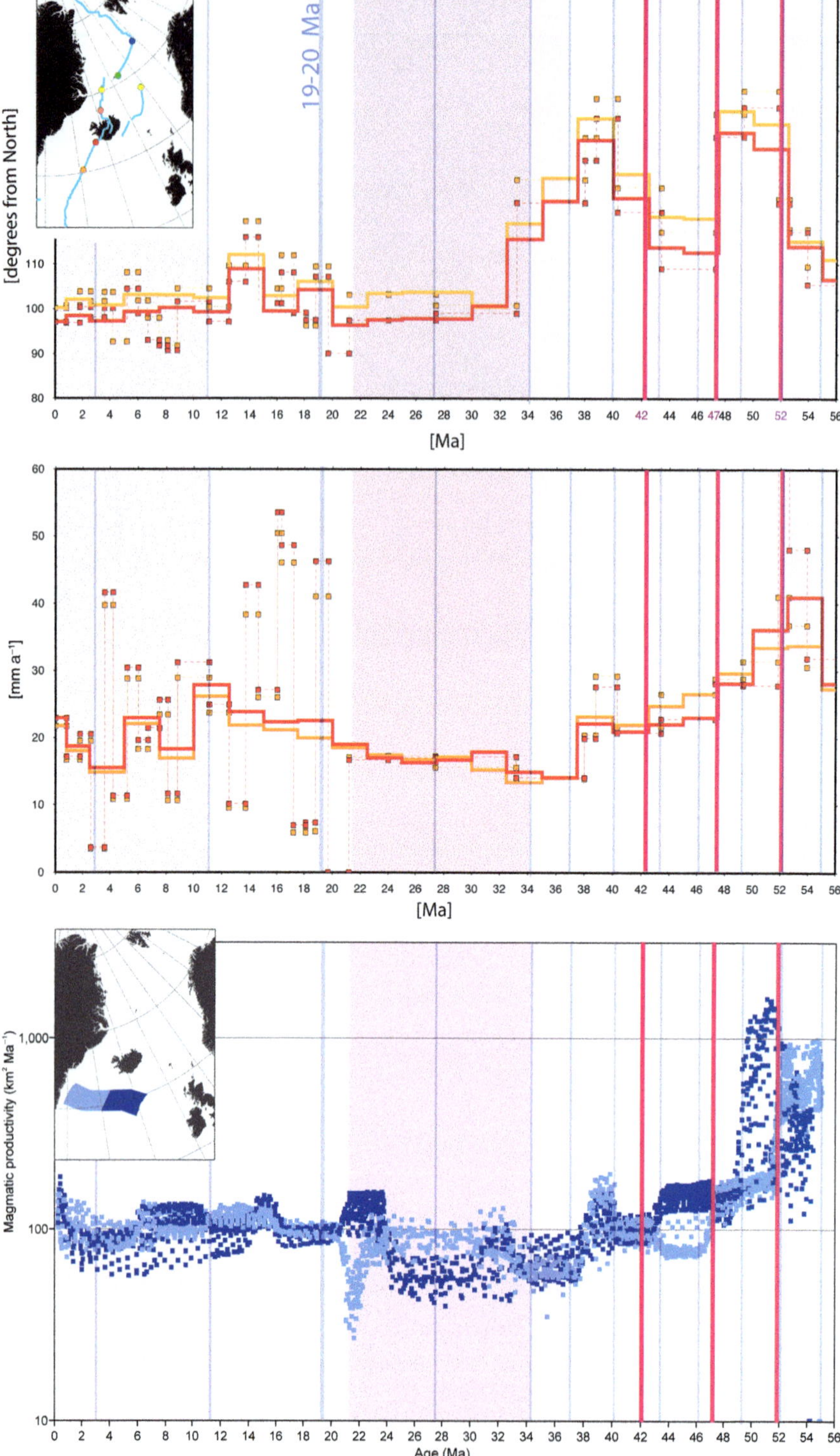

Fig. 7. Spreading direction (upper panel), spreading rates (middle panel, modified from Gaina *et al.*, this volume, in review) and magmatic crustal production (lower panel) calculated values in (**a**). The Iceland and Irminger basins.

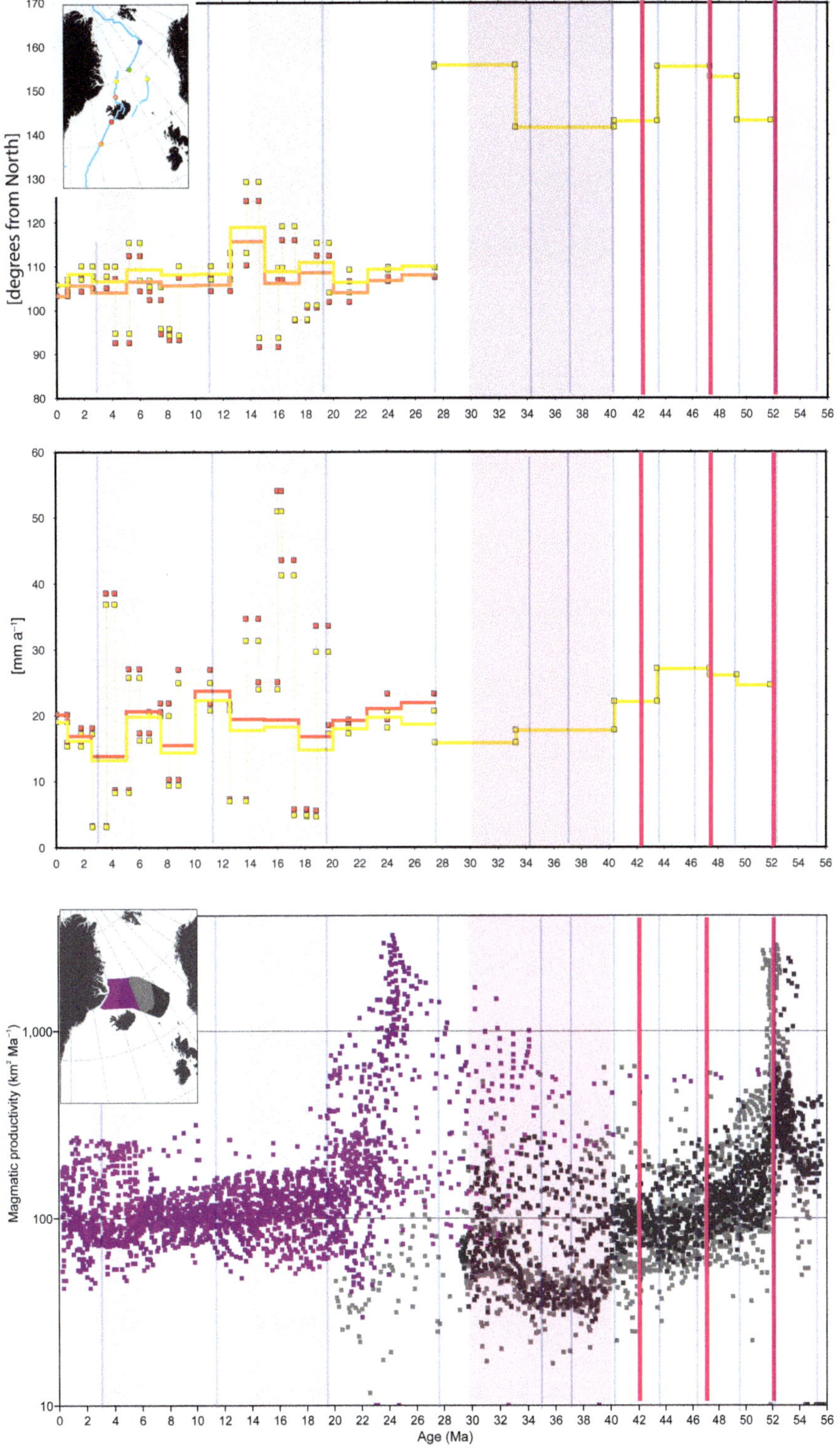

Fig. 7. (**b**) the Norway Basin and Kolbeinsey region.

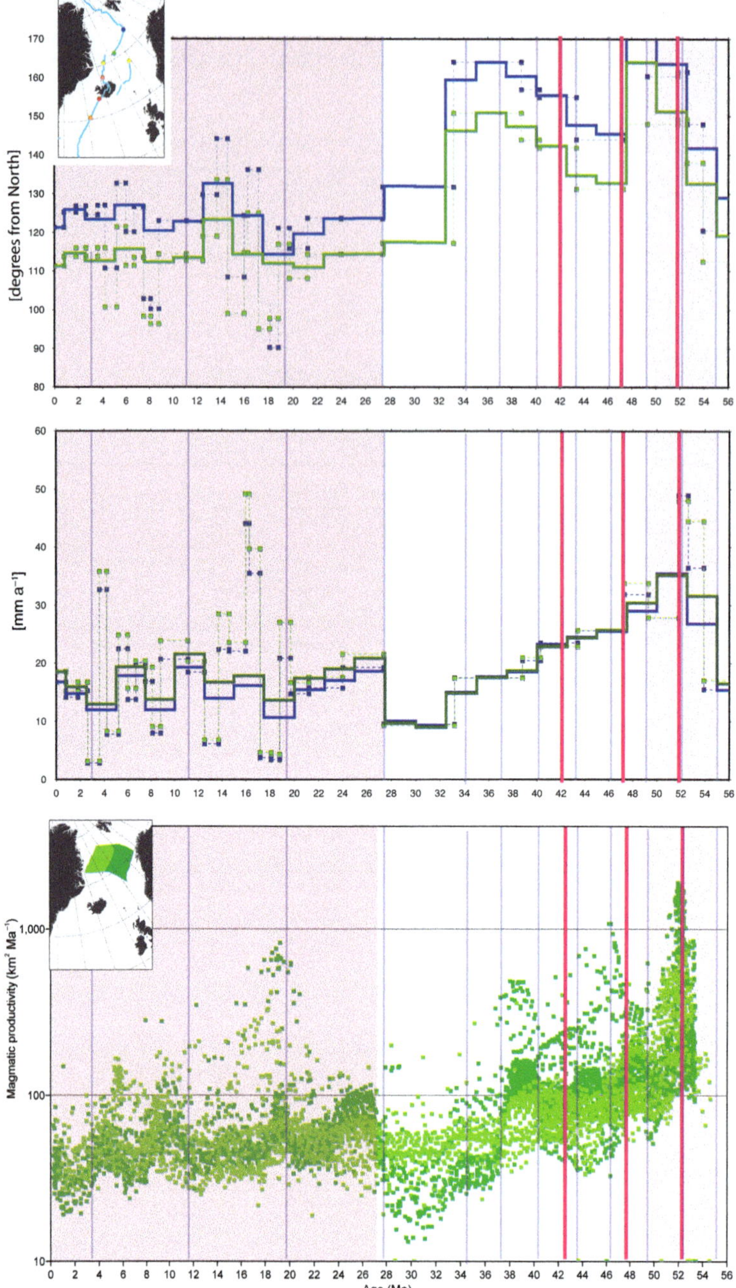

Fig. 7. (**c**) the Greenland and Lofoten basins. Inset map in the upper panels show the location of 'seed points' along the plate boundaries used as starting points to compute seafloor spreading rates and directions (look for matching colours). The magmatic crustal production is calculated by multiplying gridded data of oceanic spreading rates (Gaina *et al.*, this volume, in review) and crustal thickness (Haase *et al.*, this volume, in press). Inset maps on the lower panels show the selected corridors selected for this calculation. The results from different corridors illustrate similarities and contrasts in the crustal productivity variations. Different colours are used for conjugate sides in each corridor to illustrate asymmetries. Time intervals of SOIF formation are marked in various shades of pink (darker for higher SOIF production). Seamount ages from the Rockall region (O'Connor *et al.* 2000) are shown by magenta vertical lines. Times of Iceland plume magmatic pulses (Parnell-Turner *et al.* 2014) are indicated by thin light blue lines.

The amount of lithospheric flexure due to the formation of a volcanic edifice can reveal the age of loading and, indirectly, the approximate age of the volcanic feature built on oceanic crust. We have not attempted to establish accurate ages for the oceanic volcanic edifices in the oceanic basins of the NE Atlantic, but consider that they were formed not long after the underlying oceanic crust. This hypothesis is probably more accurate in the case of SOIFs developed as a result of tectonic changes accommodated at ridge–fracture zone junctions or as ridge propagation.

Based on our data compilations (Table 1) and SOIF description (Figs 5 & 6), we infer that: (1) 75% of SOIFs are associated with rough basement formed at low spreading rates, and with seafloor spreading asymmetries; (2) 35% of NE Atlantic SOIFs cluster on Late Eocene–Early Oligocene crust and 38% on Miocene–Present oceanic crust; and (3) SOIFs are associated with tectonic changes accommodated at ridge–fracture zone junctions or at ridge propagators. These observations lead to the conclusion that the SOIF magmatic activity is mostly linked to the normal oceanic crust production at times of spreading direction readjustments. We note that SOIFs occur on crust produced at intermediate spreading rates and on smooth to transitional basement, but most of them are associated with low spreading rates, asymmetry in crustal accretion and rough basement (Figs 5 & 6; Table 1). Batiza (2001) postulated that off-ridge volcanism is mainly encountered in regions with intermediate to high spreading rates and abundant melt supply. However, Standish & Sims (2010) described young off-axis volcanism along the ultraslow spreading SW Indian Ridge and suggest that this volcanism is the result of magma rising along faults, which contributes to off-axis accretion of oceanic crust. Some SOIFs can be OCCs formed in a slow seafloor spreading regime that was affected by episodes of higher than normal magma injection supply.

SOIFs and the Iceland plume. The NE Atlantic region has been strongly influenced by the Iceland mantle plume before, during and after continental break-up and subsequent seafloor spreading. Massive volcanism that predated and assisted the continental break-up formed the North Atlantic Igneous province (NAIP) in two pulses, at approximately 62 and 55 Ma, and was spread along the Greenland and NW European margin (e.g. Storey *et al.* 2007). Subsequently, the Iceland plume-related magmatic activity formed the Greenland–Faroe province (GIR and FIR in Fig. 1) from the Eocene onwards (Soager & Holm 2009). However, age-progressing seamount chains are absent from the NE Atlantic region.

About 100 igneous centres have been identified in Iceland and the surrounding regions (Fig. 1), and most of them show the evolution of volcanism connected to mid-ocean ridge–plume interactions. All other seamount and igneous feature clusters described in this study do not show distinct linear trends, and we infer that they cannot fall into the hotspot seamount chain category. Some of the NW–SE-trending seamounts in the Greenland Basin may resemble linear seamounts formed in the direction of plate motion relative to the mantle (so-called 'hot lines' or melting anomalies elongated in the direction of plate motion). The motion of Greenland relative to the mantle for the last 40 myr (shown by the white arrow in Fig. 5d) appears to have been sub-parallel to these seamount trends in the Greenland Basin.

Plume-related volcanism has been invoked to explain the formation of seamounts in the Rockall region (O'Connor *et al.* 2000), and the elevated bathymetry and V-shaped ridges south of Iceland in the Iceland and Irminger basins (White *et al.* 1995; Jones *et al.* 2002). In order to study any possible correlations between episodic pulsations of the Iceland plume (Jones *et al.* 2002; Parkin *et al.* 2007; Parnell-Turner *et al.* 2014) and the formation of seamounts in the basins situated south and north of Iceland, we examined the evolution of oceanic crust production. We computed the amount of oceanic crustal accretion in our selected regions (Figs 4 & 5) by taking into account the crustal thickness derived from gravity anomaly inversion (Haase *et al.*, this volume, in press) and the spreading rates (Gaina *et al.*, this volume, in review), and compared the magmatic production and spreading rates fluctuations with suggested episodes of high plume activity (Fig. 7). According to recent studies based on high-quality seismic reflection data, the Iceland plume activity had peaks every 3 myr from 55 to 35 Ma, and every 8 myr from 35 Ma to the present day (Parnell-Turner *et al.* 2014). We observed that the identified periods of SOIF production south and north of Iceland and the Greenland–Faroe Ridge (GIR and FIR in Fig. 1) fall within pulses of higher activity of the Iceland plume, but not every pulse resulted in SOIF cluster production. The three periods of seamount formation in the Rockall region, dated at 52, 47 and 42 Ma by O'Connor *et al.* (2000), coincide with pulses in the magmatic crustal production in all oceanic basins (mostly asymmetrical on conjugate flanks), decreases in spreading rates and changes in spreading directions, but have no obvious connection with SOIF production (Fig. 7).

We notice that SOIF formation in the basins north of the JMFZ coincides with an increase in spreading rates and magmatic production, reflected in higher crustal thickness. South of the JMFZ, Late Eocene–Early Miocene SOIF formation occurred when the spreading rates and magmatic crustal production first decreased and became highly asymmetrical (Fig. 7). Parnell-Turner *et al.* (2014) observed that the Iceland plume fluctuations are superimposed on a rapidly cooling temperature structure manifested by a northwards shift from smooth to rough crust in the Iceland and Irminger basins. The NE Atlantic oceanic basement morphology model (Funck *et al.* 2014) and kinematic history (Gaina *et al.*, this volume, in review) also show changes at the times of Iceland plume pulsations, as postulated by Parnell-Turner *et al.* (2014). We infer that SOIF formation south of Iceland occurred as plume activity decreased, but coincided with a considerable change in plate motion, which resulted in transtensional motion that allowed localized additional melting along the newly formed fracture zones and ridge propagators. Interestingly, recent seismic activity (from ISC catalogue: http://www.isc.ac.uk); appears to cluster in some of the SOIF locations (Fig. 6). Apart from the seismicity associated with active mid-ocean ridges, we observed seismic activity in Region I in the proximity of the SOIF cluster located on Oligocene oceanic crust, in Region II in the Norway Basin and in Region IV close to SOIFs situated on Oligocene–Miocene oceanic crust. A more conservative seismic event catalogue, the EHB Bulletin (http://www.isc.ac.uk/ehbbulletin), shows much less intraplate seismic activity in the southern part of the NE Atlantic Ocean. Note that the seismic events extracted from the EHB Bulletin indicate only earthquakes with magnitude greater than 3, which occurred before 2008. We suggest that a link may exist between seismic events and SOIF distribution, and indicates that lithosphere weakening due to tectonic activity combined with subsequent volcanic loading may leave a long-lasting imprint and facilitate subsequent crustal deformation.

'Paired' SOIFs

Although SOIF formation in all NE Atlantic subbasins occurs on conjugate flanks of the same age,

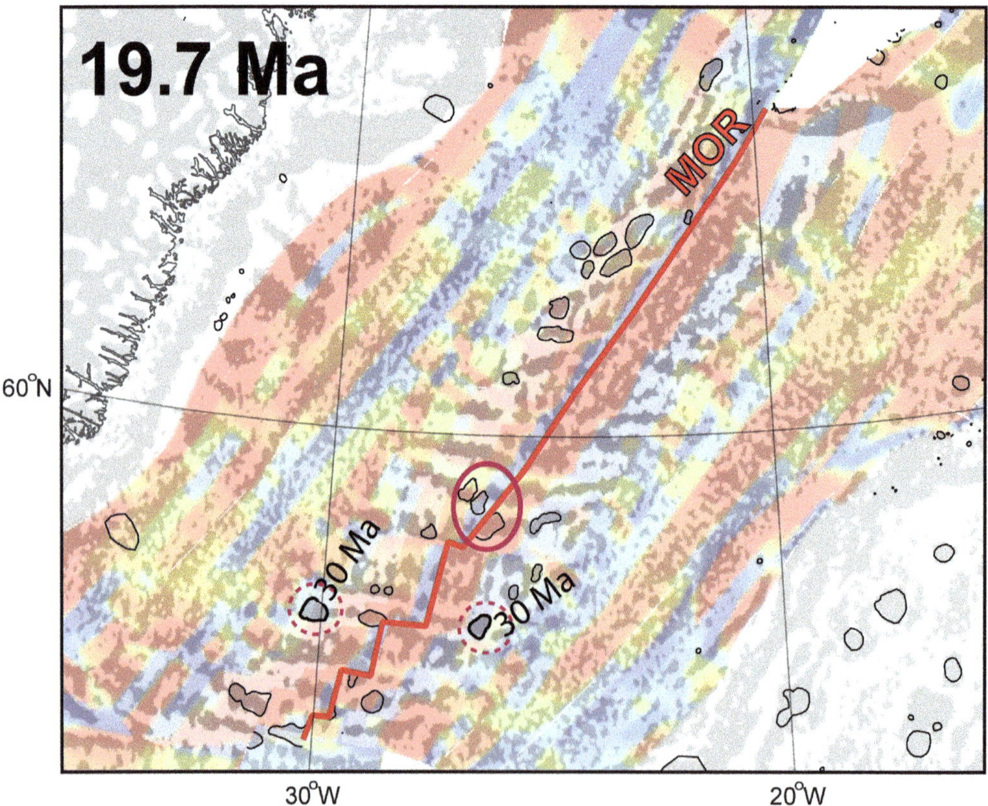

Fig. 8. Plate reconstruction at C6 (19.7 Ma). Background gridded data are the free-air gravity curvature (Sandwell & Smith 2009) in the grey palette and the spreading asymmetry (colour palette as in Fig. 4). SOIF outlines are in black; the pink ellipse shows the position of possible paired SOIFs along Reykjanes Ridge at C6 time.

there are very few 'paired basement ridges', which are volcanic edifices built at mid-ocean ridges (MORs) and equally split on both flanks by subsequent MOR evolution. Vogt & Jung (2005) described a series of paired basement ridges in the North Atlantic realm, including V-shaped ridges of Reykjanes Ridge, and suggested that small-scale conjugate ridge pairs are generated by axial magmatic centres that are not fixed to a mantle frame. They also postulate that the magma centres pulsate at 0.2–1.0 myr intervals and may be active for at least 1 myr, creating several off-axis ridge pairs. Among our identified SOIF, we observed two cases of small-scale conjugate volcanic pairs. Figure 8 shows the reconstructed locations of one pair of such conjugate volcanic features at the intersection between the Reykjanes MOR and a non-offset fracture zone at approximately 20 Ma. The underlying oceanic crust shows asymmetry on conjugate flanks (Fig. 8), and was formed at the time when the seafloor spreading direction changed to clockwise and the spreading rate increased (Fig. 7a). According to the Parnell-Turner *et al.* (2014), the model of the Iceland plume pulsation episodes, the 19–20 Ma formation of the paired SOIF along the Reykjanes axis, coincides with the increase in mantle plume activity. However, this may only be a coincidence, as the size and number of paired SOIF cannot justify a clear link between the two processes. Changes in plate motion and transtension at the MOR–fracture zone intersection seem to be a more realistic explanation for the formation of these small-scale paired SOIFs along the Reykjanes Ridge.

Conclusions

We have inspected the new NE Atlantic database of oceanic volcanic edifices including seamounts and igneous centres (abbreviated as SOIFs in this study) in four different regions situated south and north of the Greenland–Faroe Ridge. SOIF occurrences are distributed differently in the four regions, but we have identified three distinct 'pulses' of abundant seamount cluster formation. SOIFs on older oceanic crust (54–50 Ma) are situated on smooth oceanic basement. If the seamounts were formed at the time of, or shortly after, seafloor spreading, then their emplacement coincides with an increase in spreading rates and higher magmatic productivity. Large seamounts were formed on the Rockall plateau and in the Rockall Trough around 52 Ma, and the Iceland plume activity increased at 55 and 52 Ma, which could indicate that the Early Eocene SOIF formation in the NE Atlantic may have resulted from higher than usual mantle plume activity.

The second SOIF group is located on Late Eocene–Early Miocene oceanic crust of the Irminger, Iceland and Norway basins. This group is located on rough oceanic basement, in the proximity of newly formed tectonic features, such as fracture zones, ridge propagators and V-shaped ridges. In the Late Eocene, seafloor spreading rates dropped and the crustal production became highly asymmetrical. We suggest that the formation of these volcanic edifices is mostly related to kinematic changes, which led to local readjustments of MOR segments and fracture zones. The third SOIF population is observed on Mid-Miocene–Present oceanic crust, mostly located in the Greenland and Lofoten basins, but also on conjugate flanks of the Kolbeinsey Ridge. In addition, these SOIF clusters are associated with rough oceanic basement and fracture zones, V-shaped ridges, and ridge propagators. Early–Mid-Miocene (*c.* 27 Ma) oceanic crust registered an increase in spreading rate and magmatic productivity, which appears to coincide with a burst of seamount formation north of the JMFZ. South of the JMFZ, this activity was delayed until about 11 Ma, as shown by the higher number of seamounts emplaced on crust of this age and younger. It is not clear which processes initiated and dominated this third period of SOIF formation, as it seems that both tectonic/kinematic and magmatic triggers were present. However, the Early Miocene relocation of the mid-ocean ridge from the Ægir Ridge to the Kolbeinsey Ridge, west of the Jan Mayen microcontinent, may have played a role in delaying SOIF formation south of the JMFZ.

We note that the identified periods of seamount production south and north of Iceland and the Greenland–Faroe Ridge fall within Iceland plume pulses of higher activity, but not every pulse resulted in SOIF cluster production. We conclude that the episodic activity of the Iceland plume, combined with regional changes in relative plate motion, led to local MOR readjustments that enhanced the likelihood of seamount and other igneous feature formation. Further studies aimed at dating the identified SOIFs will help in understanding the connection between tectonic and magmatic activity in the NE Atlantic.

We acknowledge the support of the NAG-TEC industry sponsors (in alphabetical order): Bayerngas Norge AS; BP Exploration Operating Company Limited, Bundesanstalt für Geowissenschaften und Rohstoffe (BGR); Chevron East Greenland Exploration A/S; ConocoPhillips Skandinavia AS; DEA Norge AS; Det norske oljeselskap ASA; DONG E&P A/S; E.ON Norge AS; ExxonMobil Exploration and Production Norway AS; Japan Oil, Gas and Metals National Corporation (JOGMEC); Maersk Oil; Nalcor Energy – Oil and Gas Inc.; Nexen Energy ULC, Norwegian Energy Company ASA (Noreco); Repsol Exploration Norge AS; Statoil (U.K.) Limited, and

Wintershall Holding GmBH. C.G. acknowledges support from the Research Council of Norway through its Centres of Excellence funding scheme, project number 223272. GSK publishes with permission of the Executive Director of the British Geological Survey (Natural Environment Research Council). The authors thank Thomas Funck, Christian Berndt and an anonymous reviewer for their suggestions that greatly improved the manuscript.

References

Amante, C. & Eakins, B.W. 2009. *ETOPO1 1 Arc-Minute Global Relief Model: Procedures, Data Sources and Analysis*. NOAA Technical Memorandum NESDIS NGDC-24 National Geophysical Data Center, NOAA, https://doi.org/10.7289/V5C8276M, https://www.ngdc.noaa.gov/mgg/global/global.html

Andersen, O.B. 2010. The DTU10 gravity field and Mean sea surface. Paper presented at the *Second International Symposium of the Gravity Field of the Earth (IGFS2)*, 20–22 September 2010, Fairbanks, Alaska, USA, http://www.space.dtu.dk/english/Research/Scientific_data_and_models/downloaddata

Ballmer, M.D., Van Hunen, J., Ito, G., Bianco, T.A. & Tackley, P.J. 2009. Intraplate volcanism with complex age-distance patterns: a case for small-scale sublithospheric convection. *Geochemistry Geophysics Geosystems*, **10**, Q06015, https://doi.org/10.1029/2009gc002386

Batiza, R. 2001. Seamounts and off-ridge volcanism. *In*: Steele, J., Thorpe, S. & Turekian, K. (eds) *Encyclopedia of Ocean Sciences*. Academic Press, San Diego, CA, 2696–2708.

Becker, J.J., Sandwell, D.T. *et al.* 2009. Global bathymetry and elevation data at 30 Arc seconds resolution: SRTM30_PLUS. *Marine Geodesy*, **32**, 355–371, https://doi.org/10.1080/01490410903297766

Blischke, A., Gaina, C. *et al.* In press. The Jan Mayen microcontinent: an update of its architecture, structural development, and role during the transition from the Ægir Ridge to the mid-oceanic Kolbeinsey Ridge. *In*: Péron-Pinvidic, G., Hopper, J.R., Stoker, M.S., Gaina, C., Doornenbal, J.C., Funck, T. & Árting, U.E. (eds) *The NE Atlantic Region: A Reappraisal of Crustal Structure, Tectonostratigraphy and Magmatic Evolution*. Geological Society, London, Special Publications, **447**, https://doi.org/10.1144/SP447.5

Breivik, A.J., Mjelde, R., Faleide, J.I. & Murai, Y. 2006. Rates of continental breakup magmatism and seafloor spreading in the Norway Basin–Iceland plume interaction. *Journal of Geophysical Research: Solid Earth*, **111**, B07102, https://doi.org/10.1029/2005JB004004

Cherkis, N.Z.P., Steinmetz, S., Schreiber, R., Thiede, J. & Theiner, J. 1994. Vesteris Seamount – an Enigma in the Greenland Basin. *Marine Geophysical Researches*, **16**, 287–301, https://doi.org/10.1007/Bf01224746

Conrad, C.P., Wu, B.J., Smith, E.I., Bianco, T.A. & Tibbetts, A. 2010. Shear-driven upwrelling induced by lateral viscosity variations and asthenospheric shear: a mechanism for intraplate volcanism. *Physics of the Earth and Planetary Interiors*, **178**, 162–175, https://doi.org/10.1016/j.pepi.2009.10.001

Elliott, G.M. & Parson, L.M. 2008. Influence of margin segmentation upon the break-up of the Hatton Bank rifted margin, NE Atlantic. *Tectonophysics*, **457**, 161–176, https://doi.org/10.1016/j.tecto.2008.06.008

Epp, D. & Smoot, N.C. 1989. Distribution of seamounts in the North Atlantic. *Nature*, **337**, 254–257, https://doi.org/10.1038/337254a0

Forsyth, D.W., Harmon, N., Scheirer, D.S. & Duncan, R.A. 2006. Distribution of recent volcanism and the morphology of seamounts and ridges in the GLIMPSE study area: implications for the lithospheric cracking hypothesis for the origin of intraplate, non-hot spot volcanic chains. *Journal of Geophysical Research: Solid Earth*, **111**, B11407, https://doi.org/10.1029/2005jb004075

Funck, T., Geissler, W.H., Kimbell, G.S., Gradmann, S., Erlendsson, Ö., McDermott, K. & Petersen, U.K. 2016. Moho and basement depth in the NE Atlantic Ocean based on seismic refraction data and receiver functions. *In*: Péron-Pinvidic, G., Hopper, J.R., Stoker, M.S., Gaina, C., Doornenbal, J.C., Funck, T. & Árting, U.E. (eds) *The NE Atlantic Region: A Reappraisal of Crustal Structure, Tectonostratigraphy and Magmatic Evolution*. Geological Society, London, Special Publications, **447**. First published online July 13, 2016, https://doi.org/10.1144/SP447.1

Funck, T., Hopper, J.R. *et al.* 2014. Crustal structure. *In*: Hopper, J.R., Funck, T., Stoker, M., Árting, U., Peron-Pinvidic, G., Doornenbal, H. & Gaina, C. (eds) *Tectonostratigraphic Atlas of the North-East Atlantic region*. Geological Survey of Denmark and Greenland (GEUS), Copenhagen, Denmark, 69–126.

Gaina, C., Gernigon, L. & Ball, P. 2009. Paleocene–Recent plate boundaries in the NE Atlantic and the formation of Jan Mayen microcontinent. *Journal of Geological Society, London*, **166**, 601–616, https://doi.org/10.1144/0016-76492008-112

Gaina, C., Nasuti, A., Kimbell, G.S. & Blischke, A. In review. Break-up and seafloor spreading domains in the NE Atlantic. *In*: Péron-Pinvidic, G., Hopper, J.R., Stoker, M.S., Gaina, C., Doornenbal, J.C., Funck, T. & Árting, U.E. (eds) *The NE Atlantic Region: A Reappraisal of Crustal Structure, Tectonostratigraphy and Magmatic Evolution*. Geological Society, London, Special Publications, **447**.

Gernigon, L., Gaina, C., Olesen, O., Ball, P.J., Peron-Pinvidic, G. & Yamasaki, T. 2012. The Norway Basin revisited: from continental breakup to spreading ridge extinction. *Marine and Petroleum Geology*, **35**, 1–19, https://doi.org/10.1016/j.marpetgeo.2012.02.015

Haase, C., Ebbing, J. & Funck, T. In press. A 3D crustal model of the NE Atlantic based on seismic and gravity data. *In*: Péron-Pinvidic, G., Hopper, J.R., Stoker, M.S., Gaina, C., Doornenbal, J.C., Funck, T. & Árting, U.E. (eds) *The NE Atlantic Region: A Reappraisal of Crustal Structure, Tectonostratigraphy and Magmatic Evolution*. Geological Society, London, Special Publications, **447**, https://doi.org/10.1144/SP447.6

Haase, K.M. & Devey, C.W. 1994. The petrology and geochemistry of Vesteris Seamount, Greenland

Basin – an intraplate alkaline volcano of non-plume origin. *Journal of Petrology*, **35**, 295–328, https://doi.org/10.1093/petrology/35.2.295

Hayman, N.W., Grindlay, N.R., Perfit, M.R., Mann, P., Leroy, S. & De Lépinay, B.M. 2011. Oceanic core complex development at the ultraslow spreading Mid-Cayman Spreading Center. *Geochemistry, Geophysics, Geosystems*, **12**, Q0AG02, https://doi.org/10.1029/2010GC003240

Hey, R., Martinez, F., Hoskuldsson, A. & Benediktsdottir, A. 2010. Propagating rift model for the V-shaped ridges south of Iceland. *Geochemistry, Geophysics, Geosystems*, **11**, https://doi.org/10.1029/2009GC002865

Hillier, J.K. & Watts, A.B. 2007. Global distribution of seamounts from ship-track bathymetry data. *Geophysical Research Letters*, **34**, L13304, https://doi.org/10.1029/2007GL029874

Hopper, J.R., Funck, T., Stoker, T., Arting, U., Peron-Pinvidic, G., Doornebal, H. & Gaina, C. (eds) 2014. *Tectonostratigraphic Atlas of the North-East Atlantic Region.* Geological Survey of Denmark and Greenland (GEUS), Copenhagen, Denmark.

Horni, J., Blischke, A. et al. In review. Seismic volcanostratigraphy of the North Atlantic Igneous Province (NAIP): the volcanic facies units. *In*: Péron-Pinvidic, G., Hopper, J.R., Stoker, M.S., Gaina, C., Doornenbal, J.C., Funck, T. & Árting, U.E. (eds) *The NE Atlantic Region: A Reappraisal of Crustal Structure, Tectonostratigraphy and Magmatic Evolution.* Geological Society, London, Special Publications, **447**.

Ildefonse, B., Blackman, D.K., John, B.E., Ohara, Y., Miller, D.J., Macleod, C.J. & Integrated Ocean Drilling Program Expeditions 304/305 Science, P. 2007. Oceanic core complexes and crustal accretion at slow-spreading ridges. *Geology*, **35**, 623–626, https://doi.org/10.1130/G23531A.1

IHO 1994. *Hydrographic Dictionary, Volume 1.* International Hydrographic Bureau Special Publications, **32**. International Hydrographic Bureau, Monaco.

Jones, E.J.W., Siddall, R., Thirlwall, M.F., Chroston, P.N. & Lloyd, A.J. 1994. Seamount Anton Dohrn and the evolution of the Rockall Trough. *Oceanologica Acta*, **17**, 237–247.

Jones, S.M., White, N. & Maclennan, J. 2002. V-shaped ridges around Iceland: implications for spatial and temporal patterns of mantle convection. *Geochemistry, Geophysics, Geosystems*, **3**, 1059, https://doi.org/10.1029/2002GC000361

Jones, S.M., Murton, B.J., Fitton, J.G., White, N.J., Maclennan, J. & Walters, R.L. 2014. A joint geochemical-geophysical record of time-dependent mantle convection south of Iceland. *Earth and Planetary Science Letters*, **386**, 86–97, https://doi.org/10.1016/j.epsl.2013.09.029

Kim, S.S. & Wessel, P. 2011. New global seamount census from altimetry-derived gravity data. *Geophysical Journal International*, **186**, 615–631, https://doi.org/10.1111/J.1365-246x.2011.05076.X

Kuvaas, B. & Kodaira, S. 1997. The formation of the Jan Mayen microcontinent: the missing piece in the continental puzzle between the Møre–Vøring basins and East Greenland. *First Break*, **15**, 239–247, https://doi.org/10.3997/1365-2397.1997008

MacLeod, C.J., Searle, R.C. et al. 2009. Life cycle of oceanic core complexes. *Earth and Planetary Science Letters*, **287**, 333–344, https://doi.org/10.1016/j.epsl.2009.08.016

Mallows, C. & Searle, R.C. 2012. A geophysical study of oceanic core complexes and surrounding terrain, Mid-Atlantic Ridge 13°N–14°N. *Geochemistry, Geophysics, Geosystems*, **13**, Q0AG08, https://doi.org/10.1029/2012GC004075

Marty, B., Upton, B.G.J. & Ellam, R.M. 1998. Helium isotopes in early tertiary basalts, northeast Greenland: evidence for 58 Ma plume activity in the north Atlantic Iceland volcanic province. *Geology*, **26**, 407–410, https://doi.org/10.1130/0091-7613(1998)026<0407:Hiietb>2.3.Co;2

Mertz, D.F. & Renne, P. 1995. Quaternary multi-stage alkaline volcanism at Vesteris seamount (Norwegian–Greenland Sea): evidence from laser step heating ^{40}Ar/^{39}Ar experiments. *Journal of Geodynamics*, **19**, 79–95.

Mertz, D.F., Sharp, W.D. & Haase, K.M. 2004. Volcanism on the Eggvin Bank (Central Norwegian–Greenland Sea, latitude similar to ~71°N: age, source, and relationship to the Iceland and putative Jan Mayen plumes. *Journal of Geodynamics*, **38**, 57–83, https://doi.org/10.1016/j.jog.2004.03.003

Morgan, W.J. 1971. Convection plumes in the lower mantle. *Nature*, **230**, 42–43, https://doi.org/10.1038/230042a0

Müller, R.D., Sdrolias, M., Gaina, C. & Roest, W.R. 2008. Age, spreading rates, and spreading asymmetry of the world's ocean crust. *Geochemistry, Geophysics, Geosystems*, **9**, Q04006, https://doi.org/10.1029/2007GC001743

Nasuti, A. & Olesen, O. 2014. Magnetic data. *In*: Hopper, J.R., Funck, T., Stoker, M., Árting, U., Peron-Pinvidic, G., Doornenbal, H. & Gaina, C. (eds) *Tectonostratigraphic Atlas of the North-East Atlantic Region.* The Geological Survey of Denmark and Greenland (GEUS), Copenhagen, Denmark, 43–53, 340 pp.

Nunns, A. 1983. The structure and evolution of the Jan Mayen Ridge and surroundings regions. *In*: Watkins, J.S. & Drake, C.L. (eds) *Studies in Continental Margin Geology.* American Association of Petroleum Geologists, Memoirs, **34**, 193–208.

O'Connor, J.M., Stoffers, P., Wijbrans, J.R., Shannon, P.M. & Morrissey, T. 2000. Evidence from episodic seamount volcanism for pulsing of the Iceland plume in the past 70 Myr. *Nature*, **408**, 954–958, https://doi.org/10.1038/35050066

Olive, J.-A., Behn, M.D. & Tucholke, B.E. 2010. The structure of oceanic core complexes controlled by the depth distribution of magma emplacement. *Nature Geoscience*, **3**, 491–495, https://doi.org/10.1038/ngeo888

Parkin, C.J., Lunnon, Z.C., White, R.S. & Christie, P.A.F. 2007. Imaging the pulsing Iceland mantle plume through the Eocene. *Geology*, **35**, 93–96, https://doi.org/10.1130/G23273A.1

Parnell-Turner, R., White, N., Henstock, T., Murton, B., Maclennan, J. & Jones, S.M. 2014. A continuous 55-million-year record of transient mantle plume activity beneath Iceland. *Nature Geoscience*, **7**, 914–919, https://doi.org/10.1038/ngeo2281

Parnell-Turner, R., White, N., Mccave, N., Henstock, T., Murton, B. & Jones, S.M. 2015. Architecture of North Atlantic contourite drifts modified by transient circulation of the Icelandic mantle plume. *Geochemistry, Geophysics, Geosystems*, **16**, 3414–3435, https://doi.org/10.1002/2015GC005947

Pedersen, R.B., Thorseth, I.H., Nygård, T.E., Lilley, M.D. & Kelley, D.S. 2010. Hydrothermal activity at the Arctic Mid-Ocean ridges. *In*: Rona, P.A., Devey, C.W., Dyment, J. & Murton, B.J. (eds) *Diversity of Hydrothermal Systems on Slow Spreading Ocean Ridges*. American Geophysical Union, Washington, D.C., https://doi.org/10.1029/2008GM000783

Peron-Pinvidic, G., Gernigon, L., Gaina, C. & Ball, P. 2012. Insights from the Jan Mayen system in the Norwegian-Greenland sea, I. Mapping of a microcontinent. *Geophysical Journal International*, **191**, 385–412, https://doi.org/10.1111/j.1365-246X.2012.05639.x

Poore, H.R., White, N. & Jones, S. 2009. A Neogene chronology of Iceland plume activity from V-shaped ridges. *Earth and Planetary Science Letters*, **283**, 1–13, https://doi.org/10.1016/j.epsl.2009.02.028

Sandwell & Smith. 2009. Global marine gravity from retracked Geosat and ERS-1 altimetry: ridge segmentation v. spreading rate. *Journal of Geophysical Research: Solid Earth*, **114**, B01411, https://doi.org/10.1029/2008JB006008

Sato, T., Okino, K. & Kumagai, H. 2009. Magnetic structure of an oceanic core complex at the southernmost Central Indian Ridge: analysis of shipboard and deep-sea three-component magnetometer data. *Geochemistry, Geophysics, Geosystems*, **10**, Q06003, https://doi.org/10.1029/2008gc002267

Soager, N. & Holm, P.M. 2009. Extended correlation of the Paleogene Faroe Islands and East Greenland plateau basalts. *Lithos*, **107**, 205–215, https://doi.org/10.1016/j.lithos.2008.10.002

Standish, J.J. & Sims, K.W.W. 2010. Young off-axis volcanism along the ultraslow-spreading Southwest Indian Ridge. *Nature Geoscience*, **3**, 286–292, https://doi.org/10.1038/NGEO824

Storey, M., Duncan, R.A. & Tegner, C. 2007. Timing and duration of volcanism in the North Atlantic Igneous Province: implications for geodynamics and links to the Iceland hotspot. *Chemical Geology*, **241**, 264–281, https://doi.org/10.1016/j.chemgeo.2007.01.016

Trønnes, T., Planke, S., Sundvoll, B. & Imsland, P. 1999. Recent volcanic rocks from Jan Mayen: low degree melt fractions of enriched north-east Atlantic mantle. *Journal of Geophysical Research: Solid Earth*, **104**, 7153–7167, https://doi.org/10.1029/1999JB900007

Vogt, D. 1971. Astenosphere motion recorded by the ocean floor south of Iceland. *Earth and Planetary Science Letters*, **13**, 153–160, https://doi.org/10.1016/0012-821X(71)90118-X

Vogt, P.R. & Jung, W.-Y. 2005. Paired basement ridges: spreading axis migration across mantle heterogeneities? *In*: Foulger, G.R., Natland, J.H., Presnall, D.C. & Anderson, D.L. (eds) *Plates, Plumes, and Paradigms*. Geological Society of America, Special Papers, **388**, 555–579.

Vogt, P.R. & Jung, W.Y. 2009. Treitel ridge: a unique inside corner hogback on the west flank of extinct Aegir spreading ridge, Norway basin. *Marine Geology*, **267**, 86–100, https://doi.org/10.1016/j.margeo.2009.09.006

Wessel, P. 2001. Global distribution of seamounts inferred from gridded Geosat/ERS-1 altimetry. *Journal of Geophysical Research: Solid Earth*, **106**, 19,431–19,441, https://doi.org/10.1029/2000jb000083

White, R.S., Bown, J.W. & Smallwood, J.R. 1995. The temperature of the Iceland Plume and origin of outward-propagating V-shaped ridges. *Journal of the Geological Society, London*, **152**, 1039–1045, https://doi.org/10.1144/GSL.JGS.1995.152.01.26

Yesson, C., Clark, M.R., Taylor, M.L. & Rogers, A.D. 2011. The global distribution of seamounts based on 30 arc seconds bathymetry data. *Deep-Sea Research Part I: Oceanographic Research Papers*, **58**, 442–453, https://doi.org/10.1016/J.Dsr.2011.02.004

Geology and seepage in the NE Atlantic region

GEERT-JAN VIS

TNO, Geological Survey of The Netherlands, PO Box 80015, 3508 TA, Utrecht, The Netherlands, geert-jan.vis@tno.nl

Abstract: Exploration for hydrocarbons in the NE Atlantic mainly focuses on the central eastern margin. The western margin has remained virtually unexplored, with no exploration wells drilled so far. A cost-efficient way to infer the presence of natural hydrocarbons in the poorly explored regions of the NE Atlantic is the application of synthetic aperture radar (SAR). This study presents four areas, the Western Barents Sea Margin, the Irish Atlantic Margin, East Greenland and Jan Mayen, where clustered oil-slick data indicate possible active oil seepage. The eastern margin of the NE Atlantic contains numerous oil-slick observations, but along the western margin the number of observations is limited, partly due to a persistent sea-ice coverage. Based on the tectonostratigraphic setting, it is suggested that Triassic and Jurassic source rocks are the most likely candidates for the generation of seeps in the areas studied. Near Jan Mayen and East Greenland, Cenozoic source rocks could also be present. SAR data are a useful tool in an early stage of exploration, but further work is needed to improve the understanding of the subsurface below the observed oil slicks in the NE Atlantic to determine the origin of the seepage.

The North Atlantic has a long history of major basin-forming events. Therefore, it is an area ideally suited for hydrocarbon exploration. By the 1950s, significant oil and gas production had started in many places, including in the enormous Groningen gas field in The Netherlands. This field triggered exploration for stratigraphic equivalents offshore in the North Sea towards England (Breunese & Rispens 1996). By the early 1960s, these were confirmed and the era of major exploration in the North Sea was underway. Coincidently, the theory of plate tectonics was developed, providing a new paradigm that would prove to be fundamental for unravelling the geological history and development of these economically important areas.

Today, the North Sea is considered a mature area for exploration and for the last decade production has been in decline. In response, exploration has turned towards frontier areas along the eastern margin of the NE Atlantic – less explored and, possibly, higher risk areas such as the mid-Norway margin, further offshore around Rockall, the Faroes and the Barents Sea (see Brennand *et al.* 1998).

Approximately 260 oil and gas fields have been discovered within the NE Atlantic area. Most of them are concentrated along the eastern margin of the NE Atlantic Ocean in the Tromsø-Hammerfest Basin, Trøndelag Platform, Viking Graben and Shetland Basin (Fig. 1). Along the western margin in East Greenland, five exploration and exploitation licences are currently active, despite tough circumstances such as the challenging (sub-) Arctic environmental conditions (Fig. 1). To date, 14 wells have been drilled offshore Greenland, all of them scientific wells, of which the deepest reached 1310 m below seabed (Ocean Drilling Program (ODP), Leg 152, Site 918). No hydrocarbon exploration wells have been drilled and no 3D seismic surveys have been shot.

Both margins of the NE Atlantic share a similar pre-break-up geological history due to their conjugate nature. The collision of Baltica and Laurentia during the Late Silurian caused the Caledonian Orogeny. Subsequent collapse of the Caledonian belt was followed by a long history of extension linked to the Mesozoic break-up of Pangaea and, ultimately, to the early Cenozoic opening of the NE Atlantic Ocean at approximately 56 Ma. A detailed overview of the Upper Palaeozoic–Mesozoic stratigraphy of this region is provided by Stoker *et al.* (2016).

One cost-efficient way to infer the presence of natural hydrocarbons in the poorly explored regions of the NE Atlantic is the application of remote-sensing images. Synthetic aperture radar (SAR) is a tool which has proven to be reliable for the detection of oil slicks at the ocean surface (e.g. MacDonald *et al.* 2002; Williams & Lawrence 2002; Garcia-Pineda *et al.* 2010; Crooke *et al.* 2015). The advantages of SAR images are their large areal coverage and the repeatability of observations during multiple consecutive time intervals. Combining

From: Péron-Pinvidic, G., Hopper, J. R., Stoker, M. S., Gaina, C., Doornenbal, J. C., Funck, T. & Árting, U. E. (eds) 2017. *The NE Atlantic Region: A Reappraisal of Crustal Structure, Tectonostratigraphy and Magmatic Evolution*. Geological Society, London, Special Publications, **447**, 443–455.
First published online April 7, 2017, https://doi.org/10.1144/SP447.16

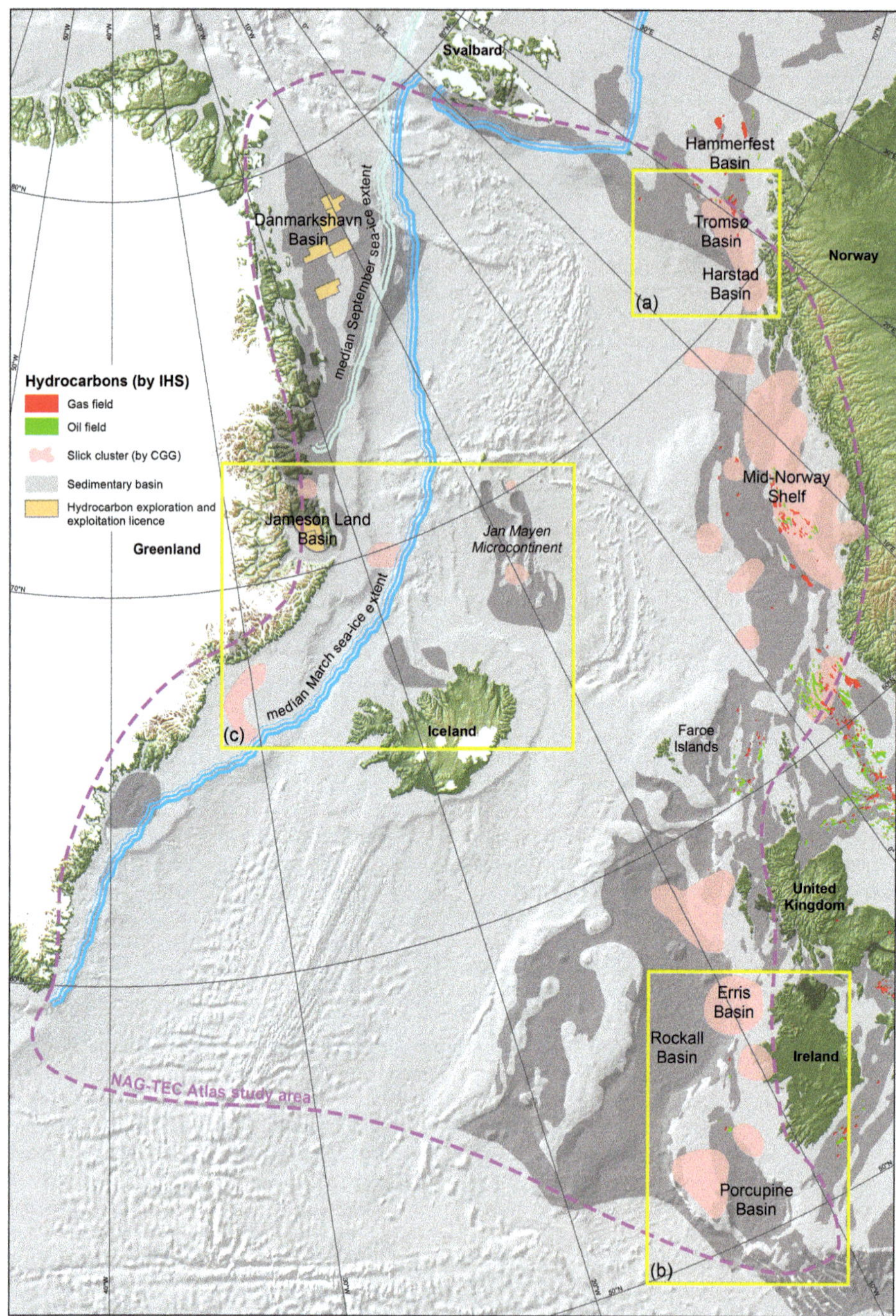

Fig. 1. The NAG-TEC Atlas study area. Yellow boxes mark the locations of the study areas: (**a**) Western Barents Sea Margin; (**b**) Irish Atlantic Margin; and (**c**) East Greenland and Jan Mayen. Slick clusters represent areas of enhanced slick density and may include natural oil seeps. See the text for data explanation (oil-slick data provided by CGG: GOSD 2014). The absence of slick clusters NE of Greenland is mainly due to the year-round sea-ice cover in that area, hampering satellite observations (sea-ice extent from Fetterer *et al.* 2002). Hydrocarbon field polygons were kindly provided by IHS Energy in 2014 (http://www.ihs.com) and Greenland licences come from the Government of Greenland (Minerals and Petroleum Licence Map, http://licence-map.bmp.gl). Bathymetry and elevation data were compiled in the NAG-TEC project.

oil-slick observations with publically available knowledge of the geology of the underexplored regions in the NE Atlantic provides insight into their hydrocarbon potential. The aim of this work is to identify areas of active oil seepage and thus the existence of possible active petroleum systems using clustered oil-slick data from the Global Offshore Seepage Database of CGG's NPA Satellite Mapping Group (GOSD 2014).

The study focuses on four underexplored regions: (1) the Western Barents Sea Margin; (2) the Irish Atlantic Margin; (3) East Greenland; and (4) Jan Mayen (Fig. 1).

Methods

Oil-slick data

Oil slicks on the ocean may be the surface expression of oil seeps rising from the seafloor and can therefore be used for source de-risking in new ventures exploration. Seepage data in this study derive from the Global Offshore Seepage Database (GOSD 2014), which contains over 20 000 records of satellite-sourced SAR images. Conventional SAR satellites are single-wavelength, side-looking radars with polar orbits and return visit times of about 1 month, although newer SAR platforms have both higher spatial resolution and shorter revisit times. The ocean surface is therefore imaged multiple times, which, under optimal weather conditions, allows repeat slicks to be located. In the vast majority of cases, these slicks derive from leaking hydrocarbon traps. Important exceptions include leaking wrecks or pipelines on the seabed. NPA's seep-detection methodology identifies three types of slicks: pollution, natural film (e.g. from plankton, algal blooms) and natural oil seeps. The data presented here show only generalized clusters of natural oil seeps with multiple repeats within the NAG-TEC Atlas study area (Fig. 1). For proprietary reasons, individual seep locations are not shown. Details on this methodology are given in Williams & Lawrence (2002).

The SAR methodology is much less effective in regions with temporal or permanent sea-ice cover or in areas with strong currents. The absence of slicks along the NE Greenland Margin (e.g. in the region of the Danmarkshavn Basin) is probably due to the nearly year-round sea-ice cover (Fig. 1). The onshore geology, nonetheless, has important similarities to the mid-Norway margins and the Barents Sea, suggesting that highly prospective areas may be present offshore NE Greenland.

It should be emphasized that SAR oil-seep detection has certain limitations, and geochemical sampling of either the surface slicks or of the seabed vents generating the slicks (by shallow coring) will be required to prove a hydrocarbon source. Repeating slicks only imply the presence of a working petroleum system of some kind; further validation will be required to confidently de-risk the presence of a mature source rock and oil-filled reservoirs in the unexplored basins of the study area.

Geological data

An up-to-date assessment of the present understanding of the Devonian and younger stratigraphy of the NE Atlantic conjugate continental margins was made for the NAG-TEC Atlas using published and publicly available material (Hopper *et al.* 2014). In addition, knowledge and expertise of contributors, unpublished information from contributing geological surveys, and released commercial wells were used for the atlas. Lithological columns summarizing key lithologies of the three areas were produced using the more detailed stratigraphic columns presented in Stoker *et al.* (2016). In addition, country-specific petroleum-geology-related data were taken from the Norwegian Petroleum Directorate (NPD: http://factpages.npd.no/factpages/), the Irish Petroleum Affairs Division (Department of Communications, Climate Action and Environment, Petroleum Affairs Division (PAD): DCCAE: http://www.dccae.gov.ie/natural-resources/en-ie/Oil-Gas-Exploration-Production/Pages/home.aspx#) and the Government of Greenland (Minerals and Petroleum Licence Map, http://licence-map.bmp.gl).

The underexplored status of the study areas implies that there is an inherent large degree of uncertainty with respect to stratigraphy and geology. Many locations have few or no wells. In such cases, the understanding of the subsurface is mostly based on interpretation of seismic lines or outcrops.

Distribution of oil slicks

In general, oil-slick observations are most numerous along the eastern margin of the NE Atlantic and limited along the western margin. Below, the locations of clustered oil slicks in four underexplored regions of the NE Atlantic are presented.

The Western Barents Sea Margin contains many observations of oil slicks. The cluster region covers an area of roughly 25 000 km^2. The density of 2D seismic reflection surveys in the region is relatively high, reaching up to 7.5–10 line km km^{-2} (fig. 6.1 in Funck *et al.* 2014), and there are about 100 wells. The cluster of oil slicks measures roughly 340 km in length and up to 90 km in width (Fig. 2). Because the oil-slick dataset used only covers the NAG-TEC study area, no oil slicks are shown in waters above the Hammerfest Basin, where, for example, the Goliat oil field is situated. Within the

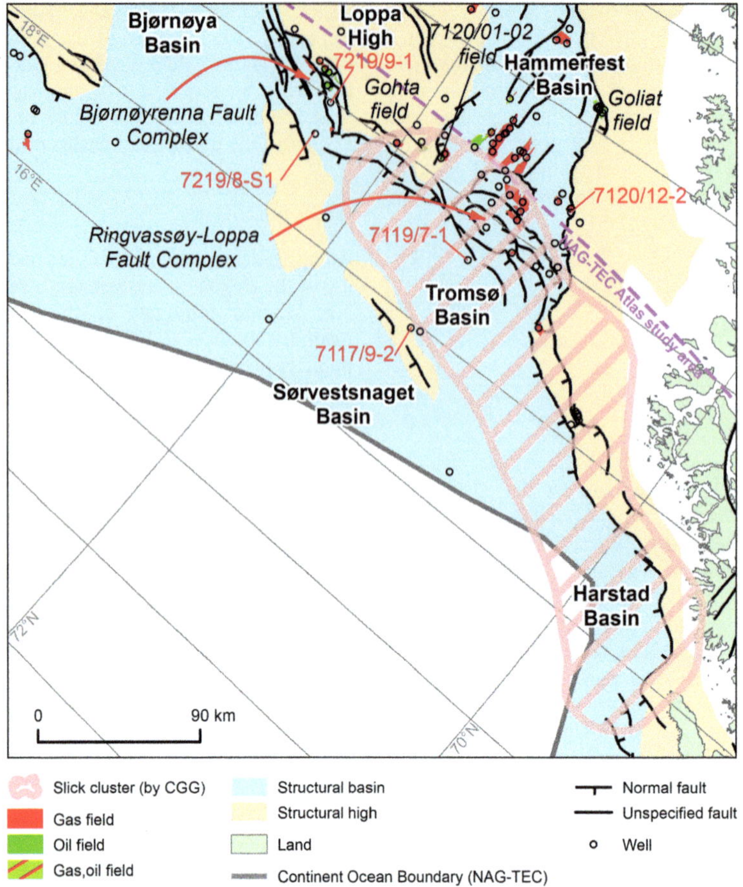

Fig. 2. Detailed map of the Western Barents Sea Margin, showing regional structural basins and highs (Funck *et al.* 2014), and the area within which oil slicks have been observed (oil-slick data provided by CGG: GOSD 2014). The presence of oil and gas fields demonstrates that an active petroleum system is in place, mainly in the Hammerfest Basin.

oil-slick cluster, two oil fields are present: Gohta and 7120/01-02, both on the Loppa High. The slicks occur in waters above the Tromsø and Harstad basins, especially above the Ringvassøy–Loppa Fault Complex (Fig. 2).

The oil slicks identified offshore Ireland occur in four clusters (Fig. 3). A cluster offshore NW Ireland measures 170 km in diameter, and is located above the Erris Basin and the Eastern Rockall Basin. Further south, a cluster is located close to the coastline and is centred around Achill Island. It measures roughly 100 km in diameter and lies east of the Slyne Basin, relatively close to the shoreline. Southeast of that lies the smallest cluster, which measures roughly 100 km in length and 70 km in width. It is located along the eastern margin of the Porcupine Basin and the bordering basins. Finally, the largest cluster is present above the Porcupine High, and measures 200 km in length and up to 160 km in width. Three oil fields have been discovered in this region: Bandon, Connemara and Burren. They are all located outside of the four oil-slick clusters.

Along the western margin of the NE Atlantic no oil fields have been discovered yet (Fig. 4). In this region, the detection of oil slicks is hampered by sea-ice (Fig. 4). This has its largest extent in March and its minimum extent in September (Fetterer *et al.* 2002). Repeated satellite radar observations of this region are limited to the ice-free summer period between minimum and maximum ice-extent. Three oil-slick clusters have been observed offshore East Greenland. Two of these occur nearly completely seawards of the continent–ocean boundary (COB) as defined by Funck *et al.* (2016). The smallest East Greenland cluster is located in the approximately 20 km-wide King

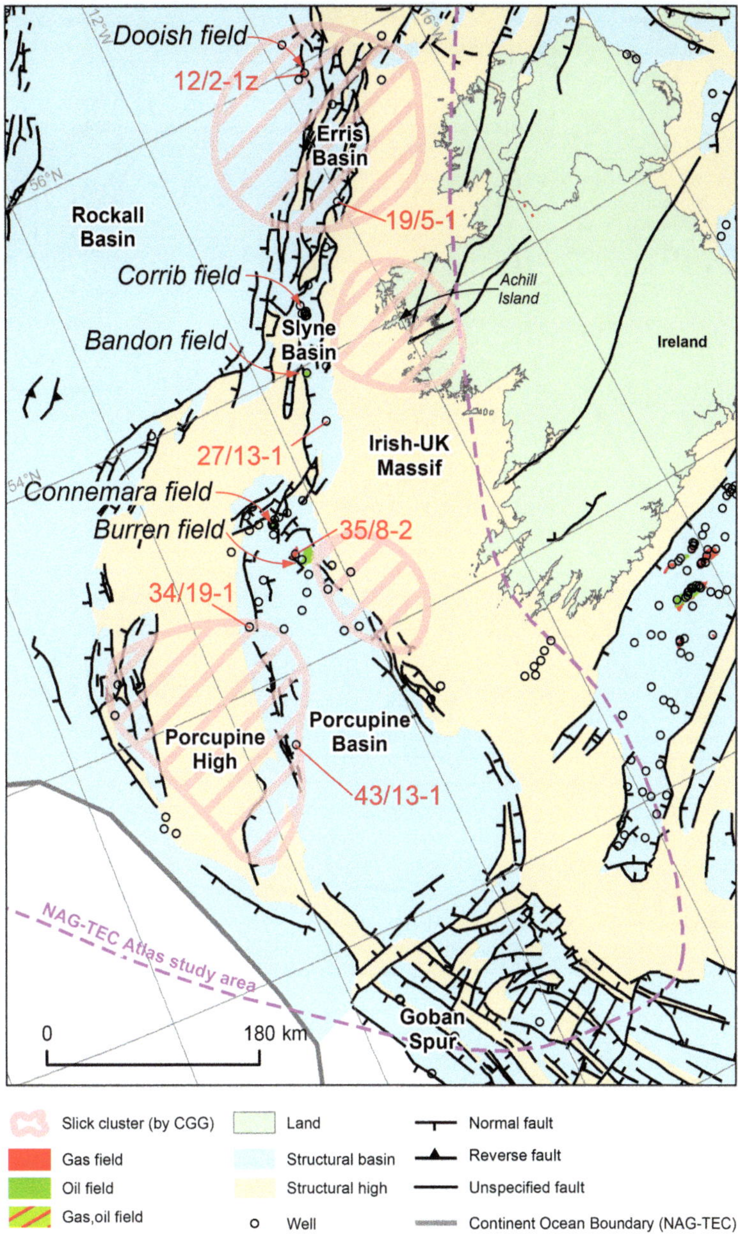

Fig. 3. Detailed map of the Irish Atlantic Margin, showing regional structural basins and highs (Funck *et al.* 2014), and the areas where oil slicks have been observed (oil-slick data provided by CGG: GOSD 2014). The presence of some oil and gas fields demonstrates that an active petroleum system is locally in place in the Slyne and Porcupine basins.

Oscar Fjord. This area lies to the north of the Jameson Land Basin in which two exploration and exploitation licences were granted by the Greenland Government in June 2015 (Fig. 4). The next oil-slick cluster lies approximately 100 km off Scoresby Sund and measures roughly 90 km in diameter. It is not located above a known basin, but lies 65 km seawards from the COB. The southernmost cluster lies south of the onshore Kangerlussuaq Basin (Fig. 4). The L-shaped cluster is located around

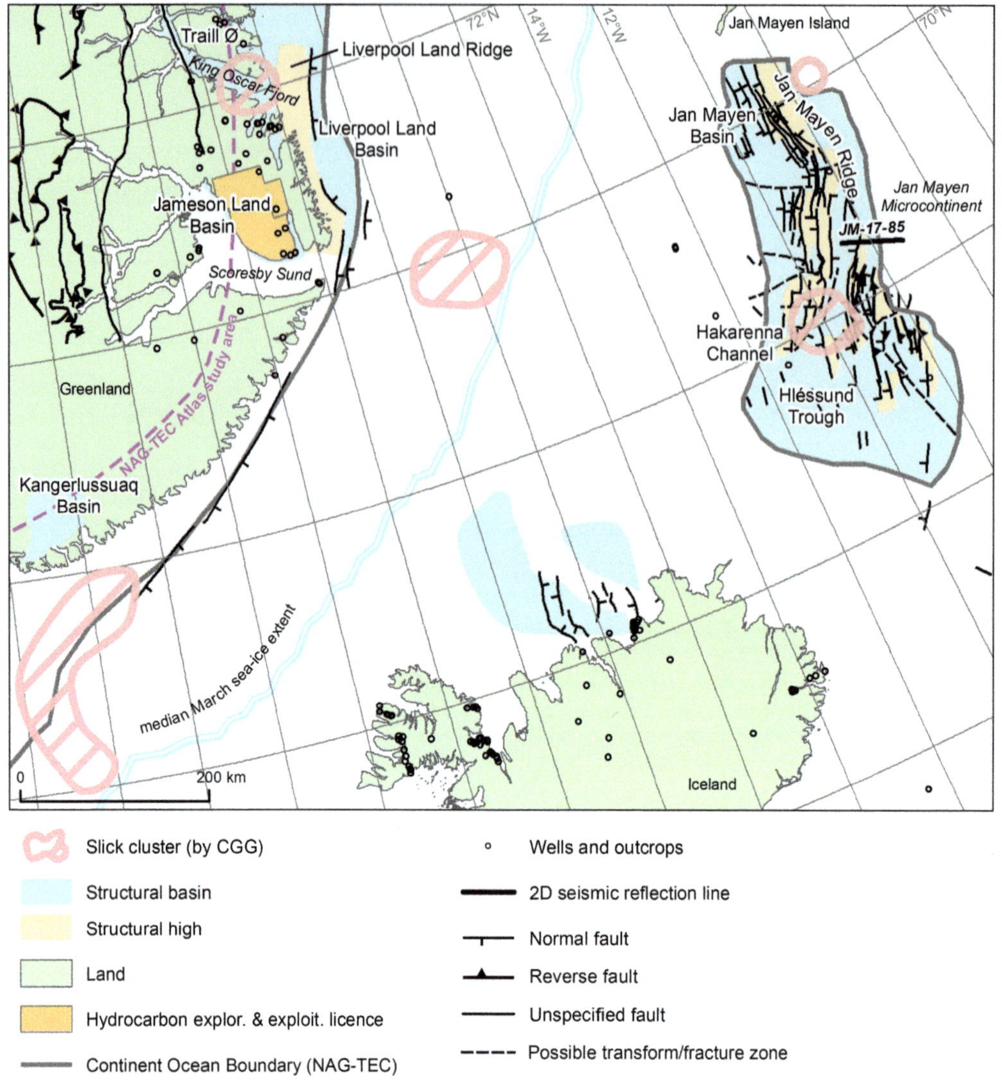

Fig. 4. Detailed map of East Greenland and Jan Mayen, showing regional structural basins and highs (Funck *et al.* 2014), and the areas where oil slicks have been observed (oil-slick data provided by CGG: GOSD 2014). Although no hydrocarbons have been discovered so far, the presence of oil slicks and seafloor oil-seepage observations suggest that petroleum is being generated and leaking towards the surface locally.

the COB and measures 275 km in length along its longest axis.

In the area of the Jan Mayen microcontinent, two small oil-slick clusters were identified. The northernmost cluster is located NE of the Jan Mayen Ridge, and lies seawards of the COB (Fig. 4). The cluster in the south is located above the structural elements of the Hléssund Trough and the Hakarenna Channel. Within the Jan Mayen microcontinent, only five wells have been drilled, which are all scientific Deep Sea Drilling Project (DSDP) wells.

Discussion

Western Barents Sea Margin

The oil slicks along the western edge of the Hammerfest Basin – outside of the NAG-TEC study area – coincide with several oil fields in this active petroleum system (Fig. 2). The source rocks of the Goliat Field are considered to be located in the Lower–Middle Triassic Sassendalen Group and in the Upper Jurassic Hekkingen Formation (Fig. 5)

(Bjorøy *et al.* 2009). Marine shales of the Sassendalen Group were deposited under anoxic conditions and have a high organic content, making them a potentially important hydrocarbon source rock (Worsley 2008; Bjorøy *et al.* 2009). The Upper Triassic Fruholmen Formation has been identified in various wells in the Hammerfest Basin (Fig. 5). A potential Triassic oil source rock is present in the Kapp Toscana Group. The Hekkingen Formation also has a marine origin with organic shales deposited under anoxic conditions (Bjorøy *et al.* 2009). Analyses of the source rocks from the Hekkingen Formation in the central Barents Sea (well 7430/10-U-01: 74° 12′ 47.79″ N, 30° 14′ 44.22″ E) show total organic carbon (TOC) values ranging between 3 and 36 wt% (Langrock *et al.* 2003). This unit is up to 350 m thick in the Hammerfest Basin, but thins into the centre of the basin: a consequence of active doming along the axis of the basin (Dallmann 1999). The deposits from the Hekkingen Formation are oil mature. The overlying Lower Cretaceous Kolje Formation also shows an initial oil potential (Pedersen 2014).

Oil slicks have been observed in waters above the Tromsø Basin. However, their origin is difficult to assess due to the poorly explored status of the basin, where no well penetrates its centre (Fig. 2). Seismic data show a series of tilted fault blocks, of which the faults do not appear to reach the surface (fig. 13a in Stoker *et al.* 2016). East of the basin centre, well 7119/7-1 (total depth (TD) = 3134 m total vertical depth (TVD) in Permian deposits) did not encounter Lower Cretaceous or Upper Jurassic oil-source prone sediments (NPD: http://factpages.npd.no/factpages/). Along the western edge of the Tromsø Basin, the Cretaceous succession is very thick but no hydrocarbons were found (3625 m in vertical well 7117/9-2). Jurassic sediments are inferred in the Tromsø Basin but are not proven. Because the Hammerfest Basin and the intermediate Ringvassøy–Loppa Fault Complex contain both Jurassic and Cretaceous rocks, they are also expected to be present in the Tromsø Basin, where they probably generate the hydrocarbons related to the observed seepage.

Wells further away from the Tromsø Basin show various Mesozoic deposits. In exploration well 7219/9-1 (TD = 4286 m TVD in Late Triassic deposits) in the Bjørnøyrenna Fault Complex (Fig. 2), a relatively complete succession of at least 2050 m of the Upper Triassic Fruholmen Formation was found. It consists of interbedded mudstone, shale, sandstone and thin coal (Hjelstuen 2014). In particular, the basal grey to dark grey shales may be candidates for oil generation. Black shale samples from this well have TOC values of 1.3–3.6% (Hoare & Bathurst 2010). The well also contains the Lower Cretaceous Kolje Formation containing residual hydrocarbons (NPD: http://factpages.npd.no/factpages/). To the east, well 7219/8-S1 (TD = 4404 m TVD in Early Jurassic deposits) found Upper Jurassic and Lower Cretaceous sandstones and mudstones (Fig. 2). The Hekkingen Formation (Fig. 5) only contained mudstones without potential oil-source rocks, while the silty parts of the overlying source-rock-barren Knurr Formation contained some hydrocarbon shows (NPD: http://factpages.npd.no/factpages/).

East of the Tromsø Basin in the Hammerfest Basin, sediments belonging to the Lower–Middle Triassic Sassendalen Group were found in well 7120/12-2 (TD = 4667 m TVD in pre-Devonian rocks). They consist of sand and siltstones but not shales with organic-rich intervals. This well also contains oil-prone source rocks belonging to the Hekkingen and Kolje formations (Fig. 5). However, all sediments in this well down to 2500–3000 m are immature and unlikely to generate hydrocarbons (NPD: http://factpages.npd.no/factpages/).

The other oil slicks in the region are located in the Harstad Basin, where no public well data are available (Fig. 2). Based on seismic data and 3D modelling studies, the total thickness of the sedimentary succession in the Harstad Basin is expected to reach more than 10 km (Ebbing & Olesen 2010; Funck *et al.* 2016). According to Halland *et al.* (2014), the basin is part of a structurally defined deep Cretaceous basin that stretches northwards to the Bjørnøya Basin. This large sediment thickness and the presence of oil slicks suggest that source rocks and leaky seals may be present and that petroleum may be generated from the Harstad Basin.

In general, exploration in the SW Barents Sea region has yielded abundant gas and relatively little oil. This may be explained by Cenozoic exhumation and depressurization of hydrocarbon-bearing reservoirs, which possibly results from glacial erosion during Pleistocene glaciations (Corcoran & Doré 2002; Cavanagh *et al.* 2006). However, a recent study based on thermochronology challenges this idea by presenting results which suggest that the last important phase of exhumation occurred during the late Miocene–early Pliocene (Zattin *et al.* 2016).

Summarizing, upwards-migrating hydrocarbons are likely to be found on the flanks of basins in the Barents Sea where partly leaking seals occur (Ohm *et al.* 2009). The presence of oil slicks, faults which may act as conduits for oil migration from leaky reservoirs or source rocks, and the occurrence of formations with source rocks in the area suggest that in this region petroleum is being generated and migrates towards the surface.

Irish Atlantic Margin

The Irish Atlantic Margin contains two large basins bordered by several smaller basins. The Rockall

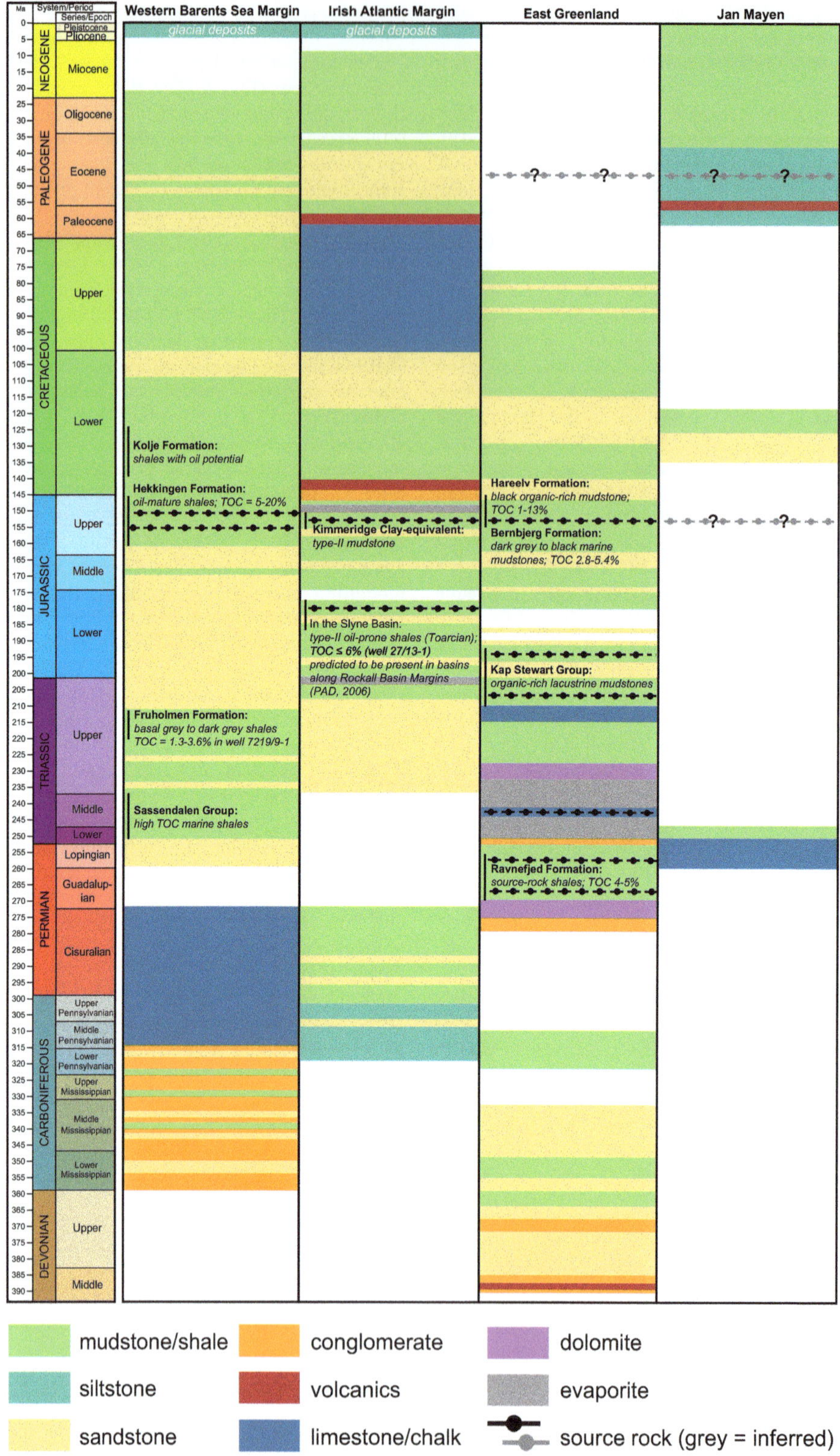
Ma
System/Period
Series/Epoch
Western Barents Sea Margin
Irish Atlantic Margin
East Greenland
Jan Mayen
NEOGENE
PALEOGENE
CRETACEOUS
JURASSIC
TRIASSIC
PERMIAN
CARBONIFEROUS
DEVONIAN
Pleistocene
Pliocene
Miocene
Oligocene
Eocene
Paleocene
Upper
Lower
Upper
Middle
Lower
Upper
Middle
Lower
Lopingian
Guadalupian
Cisuralian
Upper Pennsylvanian
Middle Pennsylvanian
Lower Pennsylvanian
Upper Mississippian
Middle Mississippian
Lower Mississippian
Upper
Middle
glacial deposits
glacial deposits
Kolje Formation:
shales with oil potential
Hekkingen Formation:
oil-mature shales; TOC = 5-20%
Kimmeridge Clay-equivalent:
type-II mudstone
In the Slyne Basin:
type-II oil-prone shales (Toarcian);
TOC ≤ 6% (well 27/13-1)
predicted to be present in basins
along Rockall Basin Margins
(PAD, 2006)
Hareelv Formation:
black organic-rich mudstone;
TOC 1-13%
Bernbjerg Formation:
dark grey to black marine
mudstones; TOC 2.8-5.4%
Kap Stewart Group:
organic-rich lacustrine mudstones
Fruholmen Formation:
basal grey to dark grey shales
TOC = 1.3-3.6% in well 7219/9-1
Sassendalen Group:
high TOC marine shales
Ravnefjed Formation:
source-rock shales; TOC 4-5%
?
?
?
?
?
?
mudstone/shale
siltstone
sandstone
conglomerate
volcanics
limestone/chalk
dolomite
evaporite
source rock (grey = inferred)

Basin is the largest basin in this area (Figs 1 & 3) and is filled with a thick sedimentary succession of Cenozoic age. The smaller Porcupine Basin is located further south (Fig. 3), and contains a thick succession of Cretaceous and Cenozoic deposits (fig. 15 in Stoker *et al.* 2016). In both basins, Jurassic and Permo-Triassic rocks have been encountered along the basin margins. The two basins experienced rapid subsidence and relatively starved sedimentation since the mid-Cenozoic, following the regional sagging event described by Praeg *et al.* (2005). According to current knowledge, only the Jurassic sedimentary succession along the Irish Atlantic Margin (Fig. 5) contains oil-prone source rocks (PAD 2006).

Oil and gas fields in Atlantic Ireland are rare, but the presence of an active hydrocarbon system is demonstrated by numerous shows and several discoveries (PAD 2006). Important discoveries include the Dooish gas condensate on the NE Rockall margin, the Corrib gas field in the Slyne Basin, the Connemara oil field and Jurassic discoveries in the Porcupine Basin (Fig. 3). Indications of hydrocarbons are seen based on the presence of gas chimneys and seeps in areas not yet explored (PAD 2006). The source rock of the Dooish gas condensate discovery is thought to be a Jurassic oil-prone mudstone.

The northernmost cluster of oil slicks along this margin is located near the Erris Basin and the Eastern Rockall Basin (Fig. 3). Two wells penetrating the sedimentary succession in this area (19/5-1 and 12/2-1z) encountered Jurassic sediments. In well 19/5-1, a succession of limestones and mudstones of 255 m thickness has been interpreted and dated to the Hettangian–Sinemurian based on biostratigraphic data (Odell & Thomas 1978). Well 12/2-1z encountered 136 m of sandstones and gravels, which are inferred to be of Middle Jurassic age based on regional data (Mecklenburgh 2004; MELLENIA 2004). Although no source rocks were found in the Jurassic succession of this well, it does have a hydrocarbon column with a net thickness of 187 m in Upper Permian and overlying Middle Jurassic sandstones right below the base-Cretaceous unconformity (3951 m along hole). The well was plugged and abandoned as a gas condensate discovery in 2003 (PAD 2006). The seeping hydrocarbons are probably generated from Jurassic source rocks in this region.

The cluster of oil slicks east of the Slyne Basin is difficult to assess since no wells penetrate the underlying succession and no seismic data were available to the NAG-TEC project. Modelling studies indicate that the total thickness of the sedimentary succession in this area may be of the order of 1–2 km (fig. 6.22 in Funck *et al.* 2014). This succession may contain oil-prone source rocks which are as yet unknown. West of this oil-slick cluster in the Slyne Basin, Toarcian Type-II oil-prone shales were encountered in well 27/13-1A (Figs 3 & 5). Analyses showed TOC values of up to 6% and a hydrocarbon index (HI) of 205–539 (PAD 2006). To the south, deposition of these shales occurred in a more open-marine environment and TOC values drop to approximately 1%. In the Rockall Basin, thermogenic oil seeps (from gravity core geochemistry) and gas chimneys on seismic data have been observed, providing evidence for an active petroleum system (PAD 2006). Despite careful screening of the oil-slick observations, it is also possible that the slicks east of the Slyne Basin have been transported there from the Slyne Basin area by strong winds and currents. Alternatively, they may originate from other sources than seepage. However, shipping is of moderate intensity in this region and not many shipwrecks are known according to the Irish programme INFOMAR (2016). Therefore further work is needed to determine the origin of the slicks here.

Oil slicks were identified along the eastern margin of the Porcupine Basin and on both sides of the Porcupine High (Fig. 3). On both sides of that high, deposits of Kimmeridgian age have been encountered: for example, in wells 43/13-1, 34/19-1, 35/19-1 and 35/8-2 (PAD 2006). Seismic and well data suggest the presence of tilted fault blocks on both sides of the Rockall and Porcupine basins that potentially host Jurassic source rocks and stratigraphic traps. Hydrocarbons may use the faults for migration towards the surface to generate seepage slicks. Some of these faults reach up into the Neogene succession, although they do not appear to reach the surface (fig. 15c in Stoker *et al.* 2016). Regional 3D basin modelling demonstrates the potential for generation and expulsion of substantial volumes of hydrocarbons from these source horizons, with generation still continuing at the present time (PAD 2006). Shows in numerous wells across the Porcupine Basin and throughout the stratigraphic column have been reported. Although well data are lacking in the southern Porcupine Basin, thermogenic oil seeps provide positive evidence for a petroleum system in that area. Gas chimneys

Fig. 5. Schematic lithostratigraphic correlation diagram showing the general lithology and (potential) source rocks of the study areas. Vertical black lines indicate the presence of a lithostratigraphic unit containing source rock. The diagram was mainly drawn using regional stratigraphic correlation panels in Stoker *et al.* (2016) and data from PAD (2006).

are also found in the Porcupine Basin (PAD 2006). Oil and gas fields are present to the north of the Porcupine Basin, but no oil slicks were found in that area. This suggests that the seal there is working. The Connemara oil field, the Spanish Point gas and oil field, and well 35/8-2 (oil) are known to have their source in Upper Jurassic mudstones, which are age-equivalent to the Kimmeridge Clay Formation of the North Sea (Fig. 5) (PAD 2006). In summary, the observed oil slicks add to the available data which imply that petroleum is being generated in these basins.

East Greenland

The basins along the eastern, NE and northern margin of Greenland resulted from Palaeozoic to Mesozoic rifting (Fig. 1). Borehole control for the offshore basins is absent in the areas where oil slicks have been identified. For details on the Greenland margin tectonostratigraphy, the reader is referred to Stoker *et al.* (2016).

The oil-slick cluster in King Oscar Fjord is structurally located in the Jameson Land Basin. The stratigraphy of this basin is relatively well known as a result of outcrop studies and onshore–offshore correlation. Because of a lack of well data, it is unknown which parts of the Jameson Land Basin stratigraphy are present underneath the King Oscar Fjord. A geoseismic section in the basin shows the presence of Palaeozoic and Mesozoic rocks (fig. 12d in Stoker *et al.* 2016). Outcrop studies in onshore Greenland indicate the presence of source-rock shales of the Upper Permian Ravnefjeld Formation belonging to the Foldvik Creek Group (Stemmerik *et al.* 1998). Two of the five units identified in the shales are organic-rich and laminated. They were deposited under anoxic conditions and their hydrocarbon quality is considered good to excellent, with TOC values of between 4 and 5%, and a HI of 300–400 (Christiansen *et al.* 1993).

Besides the Permian source rocks, there are the dark grey to black marine mudstones of the Bernbjerg Formation which are Late Jurassic in age (Fig. 5). They were deposited in a proximal marine setting, based on plant debris and coal fragments in the sediment (Alsgaard *et al.* 2003). Measured TOC values range between 2.8 and 5.4%, and HI varies between 32 and 143 (Alsgaard *et al.* 2003). The slightly younger Hareelv Formation consists of black organic-rich mudstones (Fig. 5). The sediments were deposited in a slope and base-of-slope setting, and form the youngest preserved sediments in much of southern Jameson Land (Surlyk 1987). Besides massive sandstones, the formation contains mudstones that were deposited in poorly oxygenated environments. The TOC varies from 1 to 13% and the HI is <350 due to the dominantly Type-II kerogen (Stemmerik *et al.* 1998). Because the oil-slick cluster in the King Oscar Fjord lies at the northern edge of the Jameson Land Basin, one may speculate that oil possibly leaks to the surface along basin-bounding faults that appear to reach the surface in the geoseismic section (fig. 12d in Stoker *et al.* 2016).

The oil-slick cluster located east of the Scoresby Sund is seawards of the COB where only oceanic crust is present (fig. 6.19 in Funck *et al.* 2014). A Cenozoic sediment package in which the thickness decreases in an offshore direction from 3000 to 200 m is estimated to overlie the oceanic crust here (fig. 6.22 in Funck *et al.* 2014). Owing to a lack of well or seismic data, the uncertainty remains large, as indicated by the estimates by Voss *et al.* (2009) which present sediment-thickness values of up to 5000 m for this area. The oil slicks may originate from a yet unknown source rock in the thick Cenozoic succession that is visible on a geoseismic section just north of this location (fig. 12d in Stoker *et al.* 2016). The thinning of the total sediment thickness may point to a similar Cenozoic fan-type deposit below the observed oil slicks. No oil-source rocks are known from the Cenozoic in this region.

The southernmost oil-slick cluster south of the Kangerlussuaq Basin is also located seawards of the COB (fig. 6.19 in Funck *et al.* 2014). The cluster is underlain by oceanic crust, and modelling studies estimate a total sediment thickness of between 0 and 700 m (fig. 6.22 in Funck *et al.* 2014). However, there is a large range of estimated thickness values, with the highest ones reaching up to approximately 1500 m (Voss *et al.* 2009), implying a great degree of uncertainty due to a lack of well and seismic data. For both East Greenland clusters seawards of the COB, no structurally defined sedimentary basins are expected at those locations (Stoker *et al.* 2016). This implies that either the mapped location of the COB needs revision or the seepage origin for these slicks is debatable. Further work is needed to verify the origin of these slicks.

Jan Mayen

The Jan Mayen microcontinent (Fig. 4) is a distinct structural entity located between the volcanic complex of Jan Mayen Island in the north and the NE coastal shelf area of Iceland in the south (Svellingen & Pedersen 2003; Blischke *et al.* 2016). The stratigraphic succession of the Cenozoic displays a marked east–west asymmetry (Blischke *et al.* 2014). On the eastern margin of the microcontinent, Palaeogene rocks dip steeply towards the Norway Basin as a result of normal faulting (fig. 16 in Stoker *et al.* 2016). On the western margin, a west-facing listric normal fault system is present, with rotated crustal blocks that are downfaulted along major detachment faults towards the Jan Mayen Basin (Blischke *et al.* 2014).

The oil-slick cluster in the north of the Jan Mayen microcontinent lies seawards of the COB where no basin is thought to be present (Fig. 4). However, the total sediment thickness is modelled to range between 2500 and 5500 m (fig. 6.22 in Funck *et al.* 2014). This suggests the presence of some kind of basin or accommodating structure that may contain source-rock-bearing deposits. Unfortunately, no well or seismic data can confirm the presence and internal structure of the thick sediment package. Potential oil-source rocks are most likely to be no older than the Eocene, since east of the Jan Mayen Ridge older successions are covered by a thick layer of Palaeogene volcanic rocks (fig. 16 in Stoker *et al.* 2016), which probably hamper upwards oil migration from underneath. No information on Cenozoic source rocks in this region is currently available.

To the south of the Jan Mayen microcontinent, an oil-slick cluster has been identified above the Hléssund Trough and Hakarenna Channel where a sediment package of 300–3000 m is inferred (Funck *et al.* 2014). The structural setting of the microcontinent suggests that its pre-Palaeogene stratigraphic succession might be comparable to the Jameson Land Basin in East Greenland and the Møre Basin in SW Norway (Blischke *et al.* 2014). Consequently, the organic-rich Upper Jurassic Hareelv Formation may be present here and sourcing the seepage (Fig. 5).

A seafloor sampling campaign along the Jan Mayen Ridge in 2011 found indications for active oil seepage (without the observation of oil slicks at the ocean surface) and the presence of petroleum basins (Polteau *et al.* 2012). Sampling was carried out at a location along a 2D seismic line, JM-17-85, which is located approximately 100 km NE of the area where oil slicks were observed (Fig. 4). This supports the suggestion that petroleum is being generated to the south of the Jan Mayen microcontinent.

Conclusions

This study presents the distribution of oil-slick clusters in the NE Atlantic Ocean. Oil slicks were detected using synthetic aperture radar (SAR), which provides a cost-effective way to infer the presence of natural hydrocarbons in a frontier area, such as the NE Atlantic. Most clusters occur along the eastern margin of the NE Atlantic – the western margin shows fewer oil slicks. This unbalanced distribution can be explained by the presence of sea-ice near Greenland, which hampers satellite radar observations of seepage.

Four areas were studied in more detail. Despite the underexplored status of these areas, some preliminary conclusions may be drawn:

- Western Barents Sea Margin: seepage in the Tromsø Basin is most likely to originate from Triassic and/or Jurassic source rocks embedded in tilted fault blocks. The origin of seepage in the Harstad Basin could not be determined due to a lack of subsurface data.
- Irish Atlantic Margin: the only known source rocks offshore Ireland are of Jurassic age. The observed oil slicks are probably seeping from them along bounding faults of tilted fault blocks.
- East Greenland: seepage in the King Oscar Fjord is inferred to originate from source rocks as found in outcrops, since no wells penetrate the sediments in the Jameson Land Basin. These source rocks are of Late Permian, Late Triassic–Early Jurassic and Late Jurassic age.
- Jan Mayen: seepage to the south of the Jan Mayen microcontinent may be from a source rock of Late Jurassic age, equivalent to source rocks in the Møre and Jameson Land basins in Norway and Greenland, respectively;
- Oil slicks observed seawards of the COB in East Greenland and the north of the Jan Mayen microcontinent may indicate the presence of yet unknown basins or Cenozoic source rocks.

The current knowledge of the petroleum systems over a large part of the NE Atlantic where seeps are observed is still too limited, so further work is necessary. A first step could be the localization and sampling of the observed oil slicks to come to a first-order ranking of the most promising prospective areas. In addition, the collection of subsurface data from the underexplored regions is crucial for a better understanding of their petroleum geology.

The support of the industry sponsors is hereby acknowledged (in alphabetical order): Bayerngas Norge AS; BP Exploration Operating Company Ltd; Bundesanstalt für Geowissenschaften und Rohstoffe (BGR); Chevron East Greenland Exploration A/S; ConocoPhillips Skandinavia AS; DEA Norge AS; Det norske oljeselskap ASA; DONG E&P A/S; E.ON Norge AS; ExxonMobil Exploration and Production Norway AS; Japan Oil, Gas and Metals National Corporation (JOGMEC); Maersk Oil; Nalcor Energy – Oil and Gas Inc.; Nexen Energy ULC, Norwegian Energy Company ASA (Noreco); Repsol Exploration Norge AS; Statoil (UK) Ltd; and Wintershall Holding GmbH. The author would like to thank two anonymous reviewers, and John Hopper and Martyn Stoker for their constructive comments on earlier drafts of this paper. João Araújo is thanked for providing inspiring accommodation during the writing of this paper. Lastly, I would like to thank Alan Williams and Michael King of CGG's NPA Satellite Mapping group for kindly supplying a version of their NE Atlantic seep data for inclusion into this study. Please contact them (alan.williams@cgg.com;

michael.king@cgg.com) if you would like more specific seep information from any of these basins.

References

ALSGAARD, P.C., FELT, V.L., VOSGERAU, H. & SURLYK, F. 2003. The Jurassic of Kuhn Ø, North-East Greenland. *Geological Survey of Denmark and Greenland Bulletin*, **1**, 865–892.

BJORØY, M., HALL, P.B., FERRIDAY, I.L. & MØRK, A. 2009. Triassic source rocks of the Barents Sea and Svalbard. Poster presented at the *AAPG Convention*, 7–10 June 2009, Denver, Colorado, USA.

BLISCHKE, A., PÉRON-PINVIDIC, G., ERLENDSSON, E. & ÁRNADÓTTIR, S. 2014. The Jan Mayen Microcontinent. *In*: HOPPER, J.R., FUNCK, T., STOKER, M., ÁRTING, U., PÉRON-PINVIDIC, G., DOORNENBAL, J.C. & GAINA, C. (eds) *Tectonostratigraphic Atlas of the North-East Atlantic Region*. Geological Survey of Denmark and Greenland (GEUS), Copenhagen, Denmark, 186–188.

BLISCHKE, A., GAINA, C. *ET AL*. 2016. The Jan Mayen microcontinent: an update of its architecture, structural development, and role during the transition from the Ægir Ridge to the mid-oceanic Kolbeinsey Ridge. *In*: PÉRON-PINVIDIC, G., HOPPER, J.R., STOKER, T., GAINA, C., DOORNEBAL, H., FUNCK, T. & ÁRTING, U. (eds) *The NE Atlantic Region: A Reappraisal of Crustal Structure, Tectonostratigraphy and Magmatic Evolution*. Geological Society, London, Special Publications, **447**. First published online 8 September, 2016, https://doi.org/10.1144/SP447.5

BRENNAND, T.P., VAN HOORN, B., JAMES, K.H. & GLENNIE, K.W. 1998. Historical review of North Sea exploration. *In*: GLENNIE, K.W. (ed.) *Petroleum Geology of the North Sea: Basic Concepts and Recent Advances*. 4th edn. Blackwell Science, Oxford, 1–41, https://doi.org/10.1002/9781444313413.ch1

BREUNESE, J.N. & RISPENS, F.B. 1996. Natural gas in the Netherlands: exploration and development in historic and future perspective. *In*: RONDEEL, H.E., BATJES, D.A.J. & NIEUWENHUIJS, W.H. (eds) *Geology of Gas and Oil under the Netherlands*. Kluwer Academic, Dordrecht, The Netherlands, 19–30.

CAVANAGH, A.J., DI PRIMIO, R., SCHECK-WENDEROTH, M. & HORSFIELD, B. 2006. Severity and timing of Cenozoic exhumation in the southwestern Barents Sea. *Journal of the Geological Society, London*, **163**, 761–774, https://doi.org/10.1144/0016-76492005-146

CHRISTIANSEN, F.G., PIASECKI, S., STEMMERIK, L. & TELNÆS, N. 1993. Depositional environment and organic geochemistry of the Upper Permian Ravnefjeld Formation source rock in East Greenland. *American Association of Petroleum Geologists Bulletin*, **77**, 1519–1537.

CORCORAN, D.V. & DORÉ, A.G. 2002. Depressurization of hydrocarbon-bearing reservoirs in exhumed basin settings: evidence from Atlantic margin and borderland basins. *In*: DORÉ, A.G., CARTWRIGHT, J.A., STOKER, M.S., TURNER, J.P. & WHITE, N.J. (eds) *Exhumation of the North Atlantic Margin: Timing, Mechanisms and Implications for Hydrocarbon Exploration*. Geological Society, London, Special Publications, **196**, 457–483, https://doi.org/10.1144/GSL.SP.2002.196.01.25

CROOKE, E., TALUKDER, A. *ET AL*. 2015. Determination of sea-floor seepage locations in the Mississippi Canyon. *Marine and Petroleum Geology*, **59**, 129–135.

DALLMANN, W.K. 1999. *Lithostratigraphic Lexicon of Svalbard: Review and Recommendations for Nomenclature Use: Upper Palaeozoic to Quaternary Bedrock*. Norwegian Polar Institute, Tromsø, Norway,

EBBING, J. & OLESEN, O. 2010. New compilation of top basement and basement thickness for the Norwegian continental shelf reveals the segmentation of the passive margin system. *In*: VINING, B. & PICKERING, S.C. (eds) *From Mature Basins to New Frontiers: Proceedings of the 7th Petroleum Geology Conference*. Geological Society, London, Petroleum Geology Conference Series, **7**, 885–897, https://doi.org/10.1144/0070885

FETTERER, M., KNOWLES, K., MEIER, W. & SAVOIE, M. 2002. *Daily Updated Sea Ice Index*, http://nsidc.org/data/G02135; https://doi.org/10.7265/N5QJ7F7W

FUNCK, T., HOPPER, J.R. *ET AL*. 2014. Crustal structure. *In*: HOPPER, J.R., FUNCK, T., STOKER, M., ÁRTING, U., PÉRON-PINVIDIC, G., DOORNENBAL, J.C. & GAINA, C. (eds) *Tectonostratigraphic Atlas of the North-East Atlantic Region*. Geological Survey of Denmark and Greenland (GEUS), Copenhagen, Denmark, 69–126.

FUNCK, T., GEISSLER, W.H., KIMBELL, G.S., GRADMANN, S., ERLENDSSON, Ö., MCDERMOTT, K. & PETERSEN, U.K. 2016. Moho and basement depth in the NE Atlantic Ocean based on seismic refraction data and receiver functions. *In*: PÉRON-PINVIDIC, G., HOPPER, J.R., STOKER, M.S., GAINA, C., DOORNENBAL, J.C., FUNCK, T. & ÁRTING, U.E. (eds) *The NE Atlantic Region: A Reappraisal of Crustal Structure, Tectonostratigraphy and Magmatic Evolution*. Geological Society, London, Special Publications, **447**. First published online 13 July, 2016, https://doi.org/10.1144/SP447.1

GARCIA-PINEDA, O., MACDONALD, I., ZIMMER, B., SHEDD, B. & ROBERTS, H. 2010. Remote-sensing evaluation of geophysical anomaly sites in the outer continental slope, northern Gulf of Mexico. *Deep Sea Research Part II: Topical Studies in Oceanography*, **57**, 1859–1869.

GOSD 2014. *Global Offshore Seepage Database*. CGG, Paris, http://www.cgg.com/en/What-We-Do/Multi-Client-Data/Geological/Seep-Explorer

HALLAND, E.K., BJØRNESTAD, A. *ET AL*. 2014. Chapter 6: The Barents Sea. *In*: *Compiled CO_2 Atlas for the Norwegian Continental Shelf*. Norwegian Petroleum Directorate (NPD), Stavanger, Norway.

HJELSTUEN, B. 2014. The Møre–mid-Norway–western Barents Sea–Svalbard Margin. *In*: HOPPER, J.R., FUNCK, T., STOKER, M., ÁRTING, U., PÉRON-PINVIDIC, G., DOORNENBAL, J.C. & GAINA, C. (eds) *Tectonostratigraphic Atlas of the North-East Atlantic Region*. Geological Survey of Denmark and Greenland (GEUS), Copenhagen, Denmark, 168–177.

HOARE, R. & BATHURST, P. 2010. A hot spot in the Barents Sea. 3D seismic in the Barents Sea reveal promising traps and convincing flat spots. *GeoExpro*, **7**, 62–63.

HOPPER, J.R., FUNCK, T., STOKER, M., ÁRTING, U., PÉRON-PINVIDIC, G., DOORNENBAL, J.C. & GAINA, C. (eds). 2014. *Tectonostratigraphic Atlas of the North-East Atlantic Region*. Geological Survey of Denmark and Greenland (GEUS), Copenhagen, Denmark.

INFOMAR 2016. INtegrated Mapping FOr the Sustainable Development of Ireland's MArine Resource (INFOMAR). Geological Survey of Ireland, Dublin, Ireland, http://www.infomar.ie

Langrock, U., Stein, R., Lipinski, M. & Brumsack, H.-J. 2003. Paleoenvironment and sea level change in the early Cretaceous Barents Sea – implications from near-shore marine sapropels. *Geo-Marine Letters*, **23**, 34–42, https://doi.org/10.1007/s00367-003-0122-5

MacDonald, I.R., Leifer, I., Sassen, R., Stine, P., Mitchell, R. & Guinasso, N. 2002. Transfer of hydrocarbons from natural seeps to the water column and atmosphere. *Geofluids*, **2**, 95–107.

Mecklenburgh, R. 2004. *Ireland Operations, Well IRE 12/2-1z 'Dooish Deepwater Exploration'*. Final Report Shell, The Hague, The Netherlands.

MELLENIA 2004. Biostratigraphic analysis of the Enterprise Wildcat Exploration Well 12/2-1z (interval 3841 m–4471 m), Rockall Trough, Licence 2/94, West of Ireland.

Odell, R.T. & Thomas, I.W. 1978. *Rockall Basin 19/5-1*. Completion Report, Amoco.

Ohm, S.E., Karlsen, D.A. & Austin, T.J. 2009. Geochemically driven exploration models in uplifted areas: Examples from the Norwegian Barents Sea. AAPG Search and Discovery Article 40470, presented at the *AAPG Annual Convention and Exhibition*, 7–10 June 2009, Denver, Colorado, USA.

PAD 2006. *Petroleum Systems Analysis of the Rockall and Porcupine Basins Offshore Ireland – Digital Atlas*. Petroleum Affairs Division (PAD), Special Publications, **3/06**.

Pedersen, J.H. 2014. Source rocks of the northern Hammerfest Basin, Barents Sea: new insights from well 7120/6-3 S. Paper presented at the *Seminar on Hydrocarbon Habitats*, March 2014, Geological Society of Norway. http://hydrocarbonhabitats.no/previous-seminars/source-rocks-in-the-barents-sea/

Polteau, S., Mazzini, A., Trulsvik, M. & Planke, S. 2012. *JMRS11 – Jan Mayen Ridge Sampling Survey 2011*. Volcanic Basin Petroleum Research TGS Report JMRS11

Praeg, D., Stoker, M.S., Shannon, P.M., Ceramicola, S., Hjelstuen, B., Laberg, J.S. & Mathiesen, A. 2005. Episodic Cenozoic tectonism and the development of the NW European 'passive' continental margin. *Marine and Petroleum Geology*, **22**, 1007–1030, https://doi.org/10.1016/j.marpetgeo.2005.03.014

Stemmerik, L., Dam, G., Noe-Nygaard, N., Piasecki, S. & Surlyk, F. 1998. Sequence stratigraphy of source and reservoir rocks in the Upper Permian and Jurassic of Jameson Land, East Greenland. *Geology of Greenland Survey Bulletin*, **180**, 43–54.

Stoker, M.S., Stewart, M.A. *et al.* 2016. An overview of the Upper Paleozoic–Mesozoic stratigraphy of the NE Atlantic region. *In*: Péron-Pinvidic, G., Hopper, J.R., Stoker, T., Gaina, C., Doornenbal, H., Funck, T. & Árting, U. (eds) *The NE Atlantic Region: A Reappraisal of Crustal Structure, Tectonostratigraphy and Magmatic Evolution*. Geological Society, London, **447**. First published online 11 August, 2016, updated August 12, 2016, https://doi.org/10.1144/SP447.2

Surlyk, F. 1987. Slope and deep shelf gully sandstones, Upper Jurassic, East Greenland. *American Association of Petroleum Geologists Bulletin*, **71**, 464–475.

Svellingen, W. & Pedersen, R. 2003. Jan Mayen: a result of ridge–transform–micro-continent interaction. *Geophysical Research Abstracts*, **5**, 12993.

Voss, M., Schmidt-Aursch, M.C. & Jokat, W. 2009. Variations in magmatic processes along the East Greenland volcanic margin. *Geophysical Journal International*, **177**, 755–782, https://doi.org/10.1111/j.1365-246X.2009.04077.x

Williams, A. & Lawrence, G. 2002. The role of satellite seep detection in exploring the South Atlantic's ultradeep water. *In*: Schumacher, D. & LeSchack, L.A. (eds) *Surface Exploration Case Histories: Applications of Geochemistry, Magnetics, and Remote Sensing*. American Association of Petroleum Geologists, Studies in Geology, **11**, 327–344.

Worsley, D. 2008. The post-Caledonian development of Svalbard and the western Barents Sea. *Polar Research*, **27**, 298–317, https://doi.org/10.1111/j.1751-8369.2008.00085.x

Zattin, M., Andreucci, B., de Toffoli, B., Grigo, D. & Tsikalas, F. 2016. Thermochronological constraints to late Cenozoic exhumation of the Barents Sea Shelf. *Marine and Petroleum Geology*, **73**, 97–104, https://doi.org/10.1016/j.marpetgeo.2016.03.004

Index

Page numbers in *italics* refer to Figures. Page numbers in **bold** refer to Tables.